International Association of Geodesy Symposia

Klaus-Peter Schwarz, Series Editor

Springer
Berlin
Heidelberg
New York
Barcelona
Hong Kong
London
Milan
Paris
Singapore
Tokyo

International Association of Geodesy Symposia

Klaus-Peter Schwarz, Series Editor

Symposium 101: Global and Regional Geodynamics
Symposium 102: Global Positioning System: An Overview
Symposium 103: Gravity, Gradiometry, and Gravimetry
Symposium 104: Sea Surface Topography and the Geoid
Symposium 105: Earth Rotation and Coordinate Reference Frames
Symposium 106: Determination of the Geoid: Present and Future
Symposium 107: Kinematic Systems in Geodesy, Surveying, and Remote Sensing
Symposium 108: Application of Geodesy to Engineering
Symposium 109: Permanent Satellite Tracking Networks for Geodesy and Geodynamics
Symposium 110: From Mars to Greenland: Charting Gravity with Space and Airborne Instruments
Symposium 111: Recent Geodetic and Gravimetric Research in Latin America
Symposium 112: Geodesy and Physics of the Earth: Geodetic Contributions to Geodynamics
Symposium 113: Gravity and Geoid
Symposium 114: Geodetic Theory Today
Symposium 115: GPS Trends in Precise Terrestrial, Airborne, and Spaceborne Applications
Symposium 116: Global Gravity Field and Its Temporal Variations
Symposium 117: Gravity, Geoid and Marine Geodesy
Symposium 118: Advances in Positioning and Reference Frames
Symposium 119: Geodesy on the Move
Symposium 120: Towards an Integrated Global Geodetic Observation System (IGGOS)
Symposium 121: Geodesy Beyond 2000: The Challenges of the First Decade

Geodesy Beyond 2000

The Challenges of the First Decade

IAG General Assembly
Birmingham, July 19-30, 1999

Edited by
Prof. Dr. Klaus-Peter Schwarz

Springer

Volume Editor

Prof. Dr. Klaus-Peter Schwarz
University of Calgary
Department of Geomatics Engineering
2500 University Drive N. W.
Calgary, Aberta T2N 1N4
Canada

Series Editor

Prof. Dr. Klaus-Peter Schwarz
University of Calgary
Department of Geomatics Engineering
2500 University Drive N. W.
Calgary, Aberta T2N 1N4
Canada

ISSN 0939-9585
ISBN-13: 978-3-642-64105-3 Springer-Verlag Berlin Heidelberg New York

CIP data applied for
Die Deutsche Bibliothek - CIP-Einheitsaufnahme
Geodesy beyond 2000: the challenges of the first decade; IAG general assembly, Birmingham, July 19-30, 1999/convened and ed. by Klaus-Peter Schwarz. - Berlin; Heidelberg; New York; Barcelona; Hong Kong; London; Milan; Paris; Singapore; Tokyo: Springer, 2000
(International Association of Geodesy symposia; Symposium 121)

ISBN-13: 978-3-642-64105-3 e-ISBN-13: 978-3-642-59742-8
DOI: 10.1007/978-3-642-59742-8

Springer-Verlag is a company in the specialist publishing group BertelsmannSpringer

Softcover reprint of the hardcover 1st edition 2000

Typesetting: Camera ready by editor/author
Cover layout: design & production GmbH, Heidelberg

SPIN: 10757269 32/3136xz - 5 4 3 2 1 0 - Printed on acid-free paper

PREFACE

The 35th General Assembly of the International Association of Geodesy (IAG) took place in Birmingham (UK) from July 19-30, 1999, in the framework of the 22nd General Assembly of the International Union for Geodesy and Geophysics (IUGG). The scientific program in which the IAG was involved consisted of two distinct parts, Inter-Association symposia and IAG symposia. The IAG participated in 15 of the Inter-Association symposia and took the lead in three of them. These symposia cover areas of research, which require co-operation with other associations, and they clearly are a growing trend at General Assemblies (a total of 49 this time). Because of the strong involvement in Inter-Association symposia, the IAG-specific symposia were restricted to six which took place during four days in the first week and three days in the second. The first five of them (G1 to G5) were devoted to research and scientific organisation of the Sections. In the last one (G6), highlights of IAG research were presented and possible changes to the current IAG structure were discussed with view to a clearer organisational profile.

At its meeting in March 1999, the IAG Executive decided that a representative selection of papers from the six IAG symposia should be published in the IAG series. Since a total of about 450 papers had been submitted to the six symposia, this was not an easy task. It was agreed that ten papers from each section symposium and all papers of the G6-symposium would be eligible for inclusion in this volume. Section II, which had organised a major symposium in December 1998, decided to not use its quota, because the complete proceedings of that meeting had been accepted for publication in the same series. The presidents of the other sections made the selection of papers in their area of expertise and the editor attempted to merge these papers with the ones of the G6-symposium, under the following six headings:

- Modelling the Earth at the Part-per-billion Level
- Seamless Gravity
- Advances in Theory and Techniques
- Geodynamics
- Kinematic Systems and Precision Engineering
- An IAG Structure to Meet Future Challenges

Since the six parts do not follow the original session structure, the symposium convenors and the session chairpersons have been listed in alphabetical order under the heading where most of 'their' papers ended up. This is the reason why the number of convenors varies for each part.

The papers of the last part deal with organisational rather than scientific questions. They were invited to stimulate discussion and have therefore not been reviewed. They have been included in this volume as background material to document an important decision that has been taken at this General Assembly and that may influence the future of the Association considerably. This decision is to set up an IAG Review Committee that will examine the IAG structure and submit its recommendations to a special Council meeting at the next Scientific Assembly in two years time.

The paper review process and the first editing of the manuscripts for the scientific sessions were done by the Section presidents and the session chairpersons. They were:

Fritz Brunner (Section 1); Rene Forsberg (Section 3); Peter Holota (Section 4) Martine Feissel (Section 5);

and for the scientific sessions of the G6-symposium:

Gerhard Beutler; Paul de Jonge; Bernhard Heck; Heinz Ilk; Heribert Kahmen; Chris Rizos; Michael Sideris.

They deserve a special vote of thanks for organising the reviews under a very tight time schedule. The co-operation of both authors and reviewers in trying to meet this schedule is also very much appreciated. It will make a timely publication of this volume possible.

The 71 papers contained in this volume give a representative overview of problems relevant to geodesy today. They show both the state of the art in the more established areas and the challenges of the future in those areas where new instrumentation or advances in modelling promise further progress. One area that will certainly blossom during the next decade is gravity field determination. The dedicated satellite missions, which have recently been approved, will change our knowledge in this field dramatically. Exciting new results can also be expected in many other areas. It is with view to these opportunities that the title of this volume has been chosen: 'Geodesy Beyond 2000 – The Challenges of the First Decade'.

A final word of thanks goes to Mr. Pavel Novák who organised the copy editing in Calgary. He had to deal with the many problems that electronic manuscript transfer still has and spent many extra hours to avoid the return of a paper to the author. He was ably assisted by Alex Bruton, Brad Groat, and Michael Kern. To all of them, my thanks.

November 1999 Klaus-Peter Schwarz

CONTENTS

PART 1

Modelling the Earth at the Part per Billion Level

Gerhard Beutler

Earth scale below a part per billion from Satellite Laser Ranging

D. E. Smith and R. Kolenkiewicz
Laboratory for Terrestrial Physics, NASA GSFC, Greenbelt, MD 20771, USA

P. J. Dunn and M. H. Torrence
Raytheon Corp., 7701 Greenbelt Rd., Greenbelt, MD 20770, USA

Abstract. Since the LAGEOS I satellite was launched in 1976, the systematic instrument error of the best satellite laser ranging observatories has been steadily reduced to the current level of only a few millimeters. Advances in overall system accuracy, in conjunction with improved satellite, Earth, orbit perturbation and relativity modeling, now allows us to determine the value of the geocentric gravitational coefficient (GM) to less than a part per billion (ppb). This precision has been confirmed by observations of the LAGEOS II satellite, and is supported by results from Starlette, albeit at a lower level of precision. When we consider observations from other geodetic satellites orbiting at a variety of altitudes and carrying somewhat more complex retro-reflector arrays, we obtain consistent measures of scale, which however must be based upon empirically determined, satellite-dependent detector characteristics. We arrive at an estimate of GM of 398600.44187 +/- .00020 km3/sec2, which lies within the 2 ppb uncertainty of the current standard, but differs from it by more than the error of the new estimate. Both the current standard and our recommended value fall comfortably within the ten ppb uncertainty of that determined from the most accurate alternative from lunar laser ranging observations. The precision of the estimate of GM from satellite laser ranging has improved by an order of magnitude in each of the last two decades, and we will discuss projected advances which will result in further refinements of this measure of Earth scale.

Keywords. Satellite Laser Ranging, Earth's scale.

1 Introduction

Satellite geodesy depends on the integrity of the geocentric gravitational coefficient for its definition of scale. An accurate value of the universal gravitational constant times the mass of the Earth (GM) enables us to monitor the behavior of the Earth, as sensed by satellite observations, in an absolute reference frame. This work will help to define the long-term stability of the SLR reference system and improve the positioning capability of the network. The rate of sea level rise caused by global warming is currently measured over decade time scales with tide gauges which provide observations relative to the Earth's surface. The scale inherent in laser ranging observations to stable high-altitude satellites enables us to determine accurate geocentric height at the observatories. Improved orbit scale definition will therefore contribute to the refinement of the positioning of altimeter satellites, which carry stringent radial accuracy requirements in order to define sea level relative to the Earth's center, and. will provide an important component to the definition of the geopotential model. The precision of the estimate of GM has improved by an order of magnitude in each of the last two decades, and refinements by another order of magnitude will allow us to test contemporary limits on Gdot/G. Furthermore, we anticipate that routine monitoring of Earth processes to which ranging scale is sensitive, such as seasonal and secular variability in Earth albedo, can also provide important indicators of the effects of global change.

2 Background

The technical applications of SLR data cover a variety of scientific areas. The accurate satellite position defined by a network of SLR stations enables us to improve the gravity model of the Earth and to investigate other force model effects on the orbit. The instruments can also be employed to define high resolution Earth orientation parameters from observations of geodetic satellites in stable orbits, such as the LAGEOS and ETALON constellations, and the scale of the measurements allows very accurate definition of the center of mass of the Earth, as well as the dimensions of the planet and its gravitational constant. The International Laser Tracking Service (ILRS) has collected measurements from retro-reflector carrying satellites for over three decades, and tracking of the first passive laser

International Association of Geodesy Symposia, Vol. 121
Schwarz (ed.), Geodesy Beyond 2000 – The Challenges of the First Decade

geodetic satellite, Starlette, which orbits at 950 km altitude, was initiated in 1975. The LAGEOS I satellite, launched in 1976, provided a more stable platform at the higher altitude of about one Earth radius, and was a more sensitive monitor of the scale of the measurements. Several other passive, retroreflector-carrying satellites have since been launched, in particular, LAGEOS II, which occupies a stable orbit at the same one-Earth-radius altitude as LAGEOS I, but at a lower inclination. ETALON I and II were put in higher, twelve-hour orbits with even better intrinsic stability and scale sensitivity than LAGEOS. Ajisai occupies a near-Earth orbit of 1400 km, a little higher than Starlette at 950 km., and Stella, the most recent addition to the constellation, is in the lowest orbit at 800 km. altitude.

3 SLR scale measurement

The ability of the network to define Earth scale is directly proportional to the ranging accuracy, and by the time of LAGEOS I launch, this had reached the decimeter level in most of the systems deployed in 1976. During the 1980s, advances in technology further improved instrument accuracy to centimeters and recent developments indicate millimeter capability (Degnan, 1993) in the most advanced systems. The distribution of the network also has an effect on its ability to determine global parameters such as the geopotential, the motion of the geocenter and GM. The MERIT campaign of 1983 initiated an era of improved data coverage from a worldwide distribution of stations. Regular up-grades in the instruments have taken place during the 1980s (see, for example, Bosworth et al., 1993). Significant campaign-related changes in the accuracy of several important stations also occurred in the early 1990s, in preparation for the support for the TOPEX/Poseidon mission, for which SLR was chosen as the prime tracking system. Advancements in SLR instrument technology now allow a precise enough determination of the geocentric gravitational coefficient that a different value should be applied in analyses which expect to exploit the full accuracy of these modern SLR systems.

4 The evolution of scale definition

During the 1970s, the determination of the Earth's geocentric gravitational coefficient was independently determined from observations of several interplanetary spacecraft, including the Ranger, Surveyor, Lunar Orbiter, Pioneer, Mariner and Viking flights. Lerch et al (1978) give a summary of these results, as well as those from laser ranging to lunar retroreflectors, in the introduction to their analysis of the first six months of LAGEOS I data. Perhaps it should be noted that the vertical scale which was used for the plot of prevailing estimates had a full range of 1 km3/sec2, corresponding to the sixth decimal digit of GM, and equivalent to about one part per million. They described the influence of first six months of early decimeter accuracy LAGEOS observations in the development of the GEM-L2 gravity model, most of whose data was collected in the 1970's, when the prevailing scale knowledge was based on the speed of light 299729.5 km/sec. An uncertainty of .02 km3/sec2 (50 parts per billion) was assigned to the new estimate for GM with a value of 398600.44 km3/sec after appropriate scaling for the speed of light currently adopted (299729.458 km./sec). Lerch et al. noted a variation of .05 km3/sec2 in estimates for individual satellites, which nonetheless established a significant improvement on earlier GM determinations during that decade and moved the certainty another decimal place to the right. Smith et al. (1985) list a variety of error sources which affect LAGEOS I determinations of GM at the level of a few parts per billion, which include the relativity model, instrument bias, refraction, and several Earth model parameters. Their published annually determined estimates of GM varied within a range of .005 km3/sec2, and this variability provided a measure of the effect of unmodelled error on the prevailing resolution of orbit scale. Tapley et al. (1985) conducted an analysis of a full 7.7-year span of LAGEOS data to determine a value of 398600.440 km3/sec2, which was subsequently assigned an uncertainty of .002 km3/sec2. Tapley et al. note the important influence of the adopted value of GM on the scale of laser station coordinates, particularly when comparing positioning results from other techniques, in that case from Very Long Baseline Interferometry measurements.

Table 1 summarizes some of the historical values of GM based on analyses of observations from LAGEOS I. The plot of Figure 1 shows annual GM estimates determined from this satellite with early Earth and orbit models, together with the overlapping series of recent results which we describe in detail below. The evolution of the network in extent and accuracy manifests itself in the year-by-year variations. Our analysis suggests that the largest variations are due to the changing tracking distribution, with some influence from our evolving knowledge of the Earth and satellite perturbation model. The values listed in Table 1 and Figure 1 between 1976 and 1990 were determined with the satellite, Earth and force models prevailing in 1990.

These early results have been compensated with a correction for the satellite center-of-mass offset which was too small in the original work of Lerch et al., Smith et al, and Tapley et al. At the level of data and modeling accuracy of 1990, this correction is equivalent to only about 15 % of the quoted .002 km3/sec2 uncertainty of these later estimates. The value determined by Tapley et al. included measurements from the improved instruments of the MERIT campaign and therefore was considered the most accurate estimate in 1985. It agreed with the lunar laser ranging estimate (Ferrari, 1980) well within the LLR uncertainty of .004 km3/sec2. At this SLR accuracy level, the shorter satellite ranges now shared the sensitivity of the lunar ranging results to the relativity model.

5 The effect of general relativity

The analysis of the SLR observations was conducted in a geocentric reference frame, which is more suitable for the treatment of Earth-orbiting satellites than the solar system barycentric coordinate system in which the relativistic equations of motion are formulated (see, for example, Ries et al, 1988). General relativistic corrections to the satellite accelerations, the light time measurement, station clock times and station coordinates were made according to Martin et al., 1985. The measurement computation of the analysis software (GEODYN, Putney, 1977) allows us to operate in a geocentric frame for which the secular component of Earth's rotation is defined by Very Long Baseline Interferometric quasar observations. General relativistic corrections to the satellite accelerations, the light time measurement, station clock times and station coordinates were made according to the treatment of Martin et al., 1985. Ries et al., 1989, showed that the determination of GM at the level of accuracy of .001 km3/sec2 requires careful consideration of all relativistic effects. Their comparison of the barycentric and geocentric formulations for the effect of general relativity on near-Earth satellites established that relativity could be fully modeled in the geocentric reference frame if the light-time correction is applied to the ranging data.

The analysis of Ries et al, (1992) included all significant effects of general relativity, as well as improved knowledge of the LAGEOS satellite center-of-mass correction, which had been indicated in pre-launch tests of the LAGEOS-II satellite (Minott et al., 1993). The GM thus determined, employing data collected between November 1986 and November 1991, was 398600.44150 +/- .00080 km3/sec2. This value was well within the uncertainty assigned to the lunar laser ranging (LLR) estimate of Newhall et al. (1987), after the required scale difference of 1.4808*10-8 between barycentric dynamical time and terrestrial dynamic time had been applied for a comparison in compatible reference frames. Dickey et al. (1994) combined an accurate value of the mass ratio of the sun/(Earth+moon) from LLR with the solar GM and the lunar GM from lunar-orbiting spacecraft (Ferrari et al, JGR, 85,1980). They arrive at a value of GM in an Earth centered reference frame of 398600.44300 +/- .00400, an accuracy of ten parts per billion. The estimate of Ries et al., 1992 was the basis of the IERS standard (IERS92), but they employed measurements from systems which underwent regular upgrade after 1986, during the period of their analysis. It is apparent from Figure 1 that annual estimates starting in 1990, after improvements to many systems had been made in anticipation of the TOPEX/Poseidon mission, show more stable behavior, and it is in this period which we shall concentrate. In particular, we will focus on GM determinations which can take advantage of the observations of LAGEOS II, which was launched in 1992, for the confirmation of any revised GM estimate from LAGEOS I.

6 Model of Earth and orbit

The goal of this analysis is to use the scale properties of modern SLR observations tracking LAGEOS I and II to determine GM and to test the consensus between different satellite estimates to improve our determination and to give a realistic measure of its uncertainty. All modern SLR observations from LAGEOS I, LAGEOS II, ETALON I, ETALON II, Starlette and Stella satellites were analyzed with a comprehensive solution parameter model. A ten-day orbital arc length was adopted, except for the ETALON satellites, for which a thirty-day span is needed to obtain a strong enough solution with their limited tracking coverage. The solution parameters are based on a speed of light of 299792.458 km/sec., and the force model includes geopotential perturbations to degree and order 20 from a recently determined gravity model (Lemoine et al., 1997) which had been determined from observations from many different satellites. Full Earth and ocean tidal models are employed, and third body perturbations from the Sun, Moon, and Mercury through Neptune are included. The effects of general and special relativity were modeled, together with Earth albedo, and ocean loading and solid Earth tides are modeled at each station. Tidally coherent diurnal and semi-diurnal geocenter and Earth Orientation parameters are applied for fourteen tidal frequencies.

Non-conservative pertubations contribute a significant component to the orbit error budget of the higher satellites, LAGEOS and ETALON. Solar radiation pressure and Earth albedo are applied to all satellites using solar reflectivity coefficients which remain constant for each satellite. Thermal thrusting effects due to solar and Earth-reflected radiation (Rubincam, 1988, Ries et al., 1991, Scharroo, 1991) are accommodated by empirical force model parameterization which is satellite-dependent. In the cases of LAGEOS I and II, along-track and once-per revolution along-track accelerations are estimated at 5-day intervals and the solar radiation pressure coefficient held at 1.13. For ETALON I and II, along-track and once-per revolution accelerations are estimated at 15 day intervals, but the solar radiation coefficient were held at 1.21 for ETALON I and 1.25 for ETALON II. Starlette and Stella have no generalized accelerations adjusted, the solar radiation pressure adjusted once per arc, and a coefficient of atmospheric drag adjusted daily. In the main analysis, annual estimates of GM are made simultaneously with orbit and force model parameters for each satellite with fixed station positions and Earth orientation parameters.

7 Error analysis

The determination of parameters using the Bayesian least-squares process (Putney et al., 1990) which we adopt in this analysis depends on a priori assumptions for the Earth and satellite model, the ability of the adjusted parameters to accommodate errors which are not correlated to the sought–for variables, and the quality of the data. In order to test the sensitivity of the analysis to errors in our assumed Earth, force model and instrument models a series of experiments were conducted. The simultaneous estimation of station position and velocity vectors, as well as the adjustment of both components of polar motion and length-of day at daily intervals were investigated. The influence of the improved gravity model and the thorough parameterization of non-conservative perturbations on each satellite must also be considered. The model currently employed gives a total root mean square range residual fit to the estimated orbits amounting to about 3 cm. for the LAGEOS satellites, 5 cm. for ETALON I and II, 6 or 7 cm. for Ajisai and Starlette, and about 15 cm. for Stella. Improvements to these values are anticipated as improved orbit parameterization is applied, in particular to the lower satellites, for which the effects of atmospheric drag dominate the residuals, but are unrelated to Earth scale.

The strong dependence of the GM estimates on instrument accuracy is shown in Table 2, demonstrating their sensitivity to instrument range bias and atmospheric refraction error. The correction of a ranging measurement for atmospheric refraction is approximately 2 meters at zenith, and as large as 7 meters at 20 degrees elevation, which is the usually observed minimum for most systems. The average effect on the measurement is about 4 meters and so a .25% refraction error is equivalent on average to a one-centimeter range difference. The similar influence of refraction error and range bias is seen in Table 2, which shows an empirically determined effect on the estimated value of GM in the data reduction process using observations collected between 1992 and 1996. The global estimate of GM from each satellite changes by amounts which depend on the satellite altitude. The sensitivity is compared in Table 2 to a theoretical estimate (labeled in the Table as a one centimeter height change) based on the assumption that the satellite's period is independently determined by the epoch timing properties in the data and GM is directly related to a change in satellite height (semi-major axis for these near-circular orbits) according to Kepler's third law. Any error in the adopted center-of-mass offset for the Ajisai and Etalon satellites, which have been seen to exhibit system-dependent variations of several centimeters, will significantly affect their estimate of GM, and we have therefore computed the effective offset correction which would correspond to the well-determined GM estimates from LAGEOS I and II, as these targets show minimal station-dependent effects. The resulting values of center-of-mass offset for each satellite are shown in Table 3. The possibility of errors in the adopted station position and velocity tracking complement must also considered although it is expected that the estimates of GM are insensitive to these parameters. Improvement of the Earth and force model (gravity, tidal and non-conservative) will, however probably improve the formal error estimate for GM. There is one perturbation to the satellite which has a direct effect on the scale of the solution: the effect of Earth-reflected radiation. The albedo will exhibit seasonal and long-term variations, which provide a measure of global 'health', as it monitors the cloud cover which results from global temperature changes. Martin et al. (1988) have investigated the albedo radiation and find that the effect is significant as it affects orbital evolution through interaction with spacecraft properties and that albedo should be considered if the full accuracy of the observations is to be utilized. The total effect of the Earth albedo amounts to about one-tenth ppb.

8 Satellite signature

The range measured by an SLR instrument to a spherical satellite is always longer than the distance to its exterior surface. Retroreflector cubes are recessed into the body of the target, and the refractive properties of the visible array cause the laser light to effectively penetrate the target. Reflections from LAGEOS, which has an outside radius of 298 mm., would theoretically return from points between 200 and 258 mm. from the satellite center (Neubert, 1990). The signal which arrives at the station's receiving telescope will be a convolution of the optical transfer function with the laser pulse shape. The finally measured return distribution will be skewed towards longer ranges by an amount depending on the laser pulse width and the response characteristics of the photo-detector.

Systems operating at the multi-photon return level generally use a detector, such as a micro channel plate, which is sensitive to return pulse shape. The measured time-of-flight is typically defined by the triggering of a discriminator at the pulse's leading edge. A system, which is calibrated with a similarly detected measurement from a terrestrial target at a known distance, will produce accurate observations with a noise level of a few millimeters (Degnan, 1985). The evolution of future systems such as SLR2000 (Degnan et al., 1997) will take a different emphasis and will rely on low light-level, eye-safe instruments, which must detect a much weaker return signal.

Single photon systems detect returns with a probability proportional to the density profile of the reflected pulse, and so individual range observations will be influenced by the skewness of the satellite signature. The noise level of the resulting measurements is higher than those from the high-energy instruments, but consistent performance can be maintained by calibrating with terrestrial measurements collected at the same energy level as the satellite returns. Accuracy can be achieved in these systems if any difference in the satellite and ground target data distribution is accommodated in the computation of the final, 'normal' measurements. The formation of normal points from the full-rate observations is prescribed by a process which effectively takes the mean of the data distribution as the normal range. When avalanche photo-diode detectors are used for increased light sensitivity, another skew tail is added to the satellite signature which will increase the noise level of the data and impose a further requirement for accurate calibration. The instruments of the ILRS undergo continuous improvements in performance. Early transmitters with longer pulses and early detection systems with lower return rates measured longer than the improved systems which now provide measurements corresponding to the expected satellite signature model. In particular, extensive up-grades of NASA-developed SLR systems were implemented in the early 1990's in preparation for the TOPEX/Poseidon altimeter mission. An elaborate TOPEX laser retroreflector array model allowed the improved systems to successfully produced TOPEX orbit height definition of centimeter accuracy and helped to focus attention on satellite signatures which can affect results at the millimeter level.

9 Improvement in scale definition

In Figure 2 we show a plot of independent annual determinations of GM from LAGEOS I and II data over ten years, as reported in Dunn et al. (1999). The data between 1986 and 1992 were those used to determine the IERS92 standard, at which the center-line is set, and the annual values are consistent with the results of Ries et al. (1992). The scatter of values during the first five years is also compatible with the quoted uncertainty of .0008 km3/sec2. The data collected since 1990 suggest higher estimates of GM with generally better consistency from each LAGEOS satellite. The average value for LAGEOS I is 398600.44187 km3/sec2., which we choose as the most appropriate value from this analysis. Figure 3 shows the scatter of annual determinations of GM from several satellites for a four-year period. The values for LAGEOS I, LAGEOS II and Starlette were independently estimated from the ranging observations. The Etalon I and II, Ajisai and Stella estimates were derived using satellite center-of-mass values based on the best GM determination from LAGEOS I. The larger and more complex satellites thus contribute no absolute scale information, but their scatter provides an indication of overall network consistency for the tracking of these satellites, and is considered in the error estimate which we assign to the recommended GM value. Table 3 gives a summary of the range of possible center-of-mass corrections for each satellite, and the estimated values from this analysis.

The results from Stella are difficult to explain, as they suggest a different center-of-mass offset from that of Starlette, its identical twin in all senses but age. The aging of the satellite might contribute to the difference in the results from these two targets, but the older twin appears to give the stronger and more consistent results. After considering this supporting information, we arrive at a conservative estimate of .00020 km3/sec2 for the uncertainty in the estimate of GM from these observations. The center-of-mass values listed in Table 3 are generally lower than would be expected from analytical satellite signature models, and should be refined for the most precise application of global SLR observations of these satellites in gravity and tidal analysis.

Table 4 provides a summary of the GM estimates from a four year span (1993-1996), together with two measures of uncertainty. The scatter (standard deviations) about the means of ten-day (monthly for ETALON I and II) values gives a realistic error measure, and the formal errors of single estimates from seven years of data give optimistic assessments. The formal errors would hold if the ranging observations were randomly distributed about the orbits at the level of the final residual fit for each satellite. As noted above, the orbit fits were about 3 cm for LAGEOS I and II, 5 cm for ETALON I and II, 6 cm for Ajisai, 8 cm for Starlette, and 13 cm for Stella, and the orbital residuals were, of course, far from random. The high range residual level is caused by unmodeled Earth, satellite and orbit errors, and as these models improve in the future, the associated uncertainty in the scale parameters which we seek will be reduced. If orbit fits were to reach the millimeter intrinsic data accuracy, formal errors of multi-year estimates would become an order of magnitude smaller than those listed in Table 4, that is, a few parts per trillion (ppt). At this accuracy level, improved bounds on any change in the gravitational constant (G) in time could be determined. Lunar range data has provided such a test through the lunar orbit sensitivity to solar longitude (Dickey et al., 1994), who cite the suggestion of La and Steinhardt (1989), that the early history of the universe has seen quite large changes in G. Recent estimates of limits on Gdot/G range from 20 ppt per year from binary pulsar data (Damour et al., 1988) to 4 ppt per year from solar system data (Heilings et al, 1983). An important part of the advances which would allow us to test these limits will require full consideration of the properties of individual receiver characteristics within the network.

10 Summary

The value of GM which can be determined with recent LAGEOS observations (398600.44187 +/- .00020 km3/sec2) is almost one part per billion (i.e. .00037 km3/sec2) higher than the currently adopted standard, and spacecraft trajectories computed with that standard will orbit about 3 millimeters lower than those computed with our recommended value. The adopted value of GM affects the application of the observations to studies of Earth orientation, regional and tectonic motions, particularly in the vertical, as well as gravity field analysis. It is directly dependent on our assumptions of the effective size of the satellite, and advances in our knowledge of satellite and photo-detector characteristics will allow us to refine the determination of scale from improving SLR instruments. The consistency of the network's determination of GM is a strong measure of overall data quality, and can thus provide a monitor of SLR system health, as well as to provide indications of subtle changes in the behavior of the Earth.

Table 1: Estimates of GM from LAGEOS I

Data period	GM-398600 km3/sec2	Error in km3/sec2	Source
1976	0.44	0.02	Lerch et al.(1978)
1976-1984	0.441	0.002	Tapley et al. 1985(1)
1976-1982			Smith et al. 1985(1)(2)
1986-1992	0.4415	0.0008	Ries et al. 1992
1992-1996	0.44187	0.00020	Current estimate

(1) + .001 km3/sec2 (for a center-of-mass offset of 251 mm)
(2) + .006 km3/sec2 (for coordinate time scale factor in relativity model)

Table 2: Effect of orbit or data errors on GM determinations from several satellites

Satellite	Altitude in km.	Effect one cm height	in km3/sec2 one cm range	of .25 % refraction
Etalon I	19,000	.00047	.00047	.00040
Etalon II	19,000	.00047	.00047	.00040
LAGEOS I	7,000	.00092	.00104	.00092
LAGEOS II	7,000	.00092	.00105	.00087
Ajisai	1,400	.00146	.00190	.00175
Starlette	950	.00155	.00215	.00201
Stella	800	.00157	.00230	.00221

Table 3: Center of Mass offset estimates

Satellite	Ajisai	Etalon I/II	Lageos I/II	Starlette	Stella
Standard com offset	1010	558	251	75	75
Single photon (Neubert, 1995)	978	590	248	77	77
Detector range (Appleby, 1995)		581-610	236-252		
Detector range (Otsubo,1999)	982-1010				
Current estimate	994	584	251	75	72

Table 4: Recovered and adopted GM values from several satellites with data between 1993 and 1996

Satellite	GM	scatter km3/sec2	(ppb)	formal km3/sec2	(ppb)
LAGEOS I	.44187	.00011	(0.3)	.00001	(.03)
LAGEOS II	.44190	.00017	(0.4)	.00001	(.04)
ETALON I	(.44187)	.00058	(1.5)	.00006	(.15)
ETALON II	(.44187)	.00065	(1.6)	.00005	(.17)
Ajisai	(.44187)	.00040	(1.0)	.00003	(.07)
Starlette	.44207	.00087	(2.1)	.00005	(.12)
Stella	(.44187)	.00064	(1.6)	.00013	(.32)

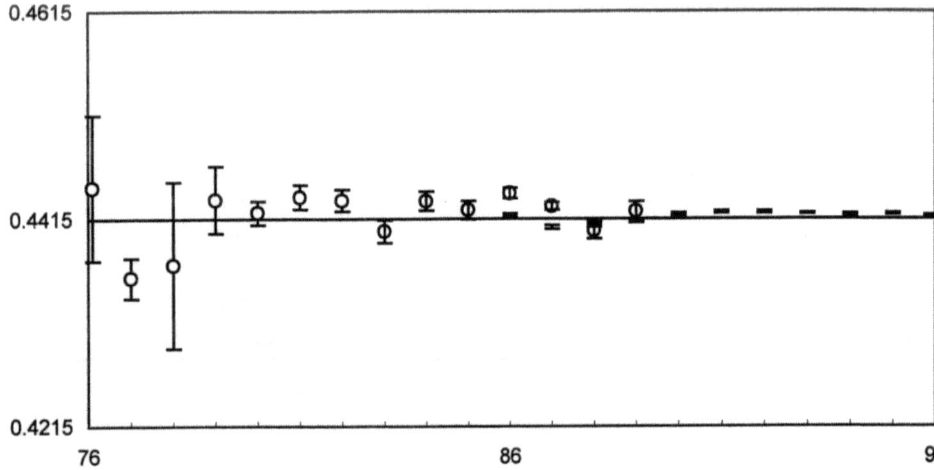

Fig. 1: Annual determinations of (GM-398600 km3/sec2) from LAGEOS I data over 20 years. Open circles show estimates based on models available in 1990; later estimates used current models. The full vertical scale is +/- .02 km3/sec2, the magnitude of the error quoted by Lerch et al. (1978) for the first GM estimate from LAGEOS data. Instrument improvements, changes in network distribution and modelling advances all serve to reduce the inter-annual variation, and by 1985 the precision of GM was +/- .002 km3/sec2.

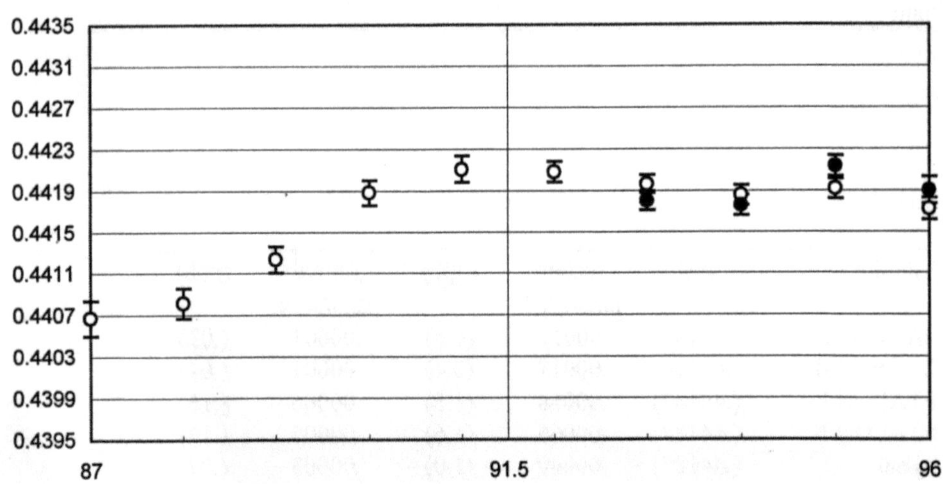

Fig. 2: Annual determinations of (GM-398600 km3/sec2) from LAGEOS I (open circles) and LAGEOS II data over ten years, shown on a full vertical scale (+/- .002 km3/sec2) corresponding to the error estimate for GM determined from LAGEOS data in the early 1980's. The data between 1987 and 1992 was the basis of the IERS92 standard (398600.4415 +/- .0008 km3/sec2) at which the centerline is set. The data collected from each LAGEOS satellite since 1990 indicates a higher value of GM, with a precision estimate of .0002 km3/sec2, an order of magnitude improvement in precision since 1985.

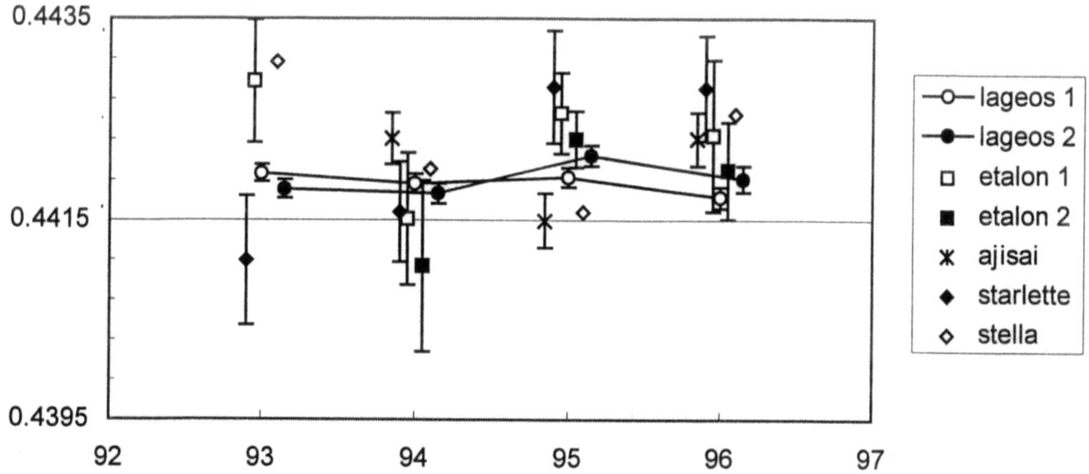

Fig. 3: The scatter of annual determinations of (GM-398600 km3/sec2) from several satellites for the four years since LAGEOS II was launched. LAGEOS I, LAGEOS II and Starlette values were independently estimated from the ranging observations. Estimates for Etalon I and II, Ajisai and for Stella (shown without its very large error bars) used satellite characteristics derived from the 1991-1996 LAGEOS I GM determination, and their scatter provides an indication of overall network consistency for the tracking of these satellites.

References

Appleby, G, Center of Mass Corrections for LAGEOS and Etalon for Single-Photon Ranging Systems, Proc. Eurolas Meeting, Munich, 1995

Bosworth, J. M., R. J. Coates & T. L. Fischetti, The Development of NASA's Crustal Dynamics Project, Contributions of Space Geodesy in Geodynamics: Crustal Dynamics, AGU Geodynamics Series, 25, 1-20, 1993

Damour, T., G.W. Gibbons, J.H. Taylor, Phys. Rev. Lett., 61,1151, 1988

Degnan, J. J., Satellite Laser Ranging: Current Status and Future Prospects, IEEE Transactions on Geoscience and Remote Sensing, GE- 23, 4, 398-413, 1985.

Degnan, J. J., Millimeter Accuracy Satellite Laser Ranging, Contributions of Space Geodesy to Geodynamics: Technology, AGU Geodynamics Series, 25, 133-162, 1993.

Degnan, J. J.and J. McGarry, SLR2000: Eyesafe and autonomous satellite laser ranging at kilohertz rates, Proc. Conf. on Laser Radar Techniques, European Symposium on Aerospace Remote Sensing, 1997

Dickey, J. O., P.L. Bender, J.E. Faller, X X Newhall, R.L. Ricklefs, J. G. Ries, P.J. Shelus, C. Veillet, A.L. Whipple, J.R. Wiant, J. G. Williams and C.F. Yoder, Lunar Laser Ranging: A Continuing Legacy of the Apollo Program, Science, 265, 22 July, 1994.

Dunn,P., M.Torrence, R.Kolenkiewicz and D.Smith, Earth Scale defined by Modern Satellite Ranging Observations, Geophys. Res. Lett., 26, 10,1489-1492, 1999

Heilings, R.W. et al., Phys. Rev. Lett., 51, 1609, 1983.

La, D and P. J. Steinhardt, Phys. Rev. Lett , 62,376, 1989.

Lemoine, F.G., D. E. Smith, L. Kunz, R. Smith, E. C. Pavlis, N. K. Pavlis, S. M. Klosko, D. S. Chinn, M. H. Torrence, R. G. Williamson, C. M. Cox, K. E. Rachlin, Y. M Wang, S. C. Kenyon, R. Salman, R. Trimmer, R. H. Rapp, and R. S. Nerem, "The Development of the NASA GSFC and NIMA Joint Geopotential Model", Gravity, Geoid and Marine Geodesy, Vol. 117, International Association of Geodesy Symposia, J. Segawa, H. Fujimoto, and S. Okubo (editors), pp 461-469, 1997.

Lerch, J.F., R.E Laubscher, S.M. Klosko, D.E. Smith, R. Kolenkiewicz, B.H. Putney, J.G. Marsh and J.E. Brownd, Determination of the geocentric gravitational constant from laser ranging on near-earth satellites, Geophys. Res. Let., 5(12), December, 1978.

Martin, C. F., M. H. Torrence & C. W. Meisner, Relativistic Effects on an Earth-Orbiting Satellite in the Barycenter Coordinate System, JGR 90, 9403-9410, 1985.

Martin, C.F. and D. P. Rubincam, Earth albedo affects on LAGEOS I satellite based on Earth radiation Budget (ERBE) satellite measurements, EOS, paper G32A-3, Spring Meeting of the AGU, Baltimore, MD, 1993.

Minott , P. O, T. W. Zagwodzki, T. Varghese, and M. Selden, " Prelaunch Optical Characterization of the Laser Geodynamics Satellite (LAGEOS 2)", NASA TP-3400, Sept, 1993.

Neubert, R., Satellite signature model: Application to Lageos and Topex, Proc. Eurolas Meeting, Munich, 1995.

Otsubo, T., J. Amagi and H. Kunimoori, The Center-of-mass Correction of the Geodetic Satellite AJISAI for Single-photon Laser Ranging, JGR 1999, in press.

Putney, B. H., R. Kolenkiewicz, D. E. Smith, P. J. Dunn and M. H. Torrence, "Precision Orbit Determination at the NASA Goddard Space Flight Center", Adv. Space Res., V.10, No.3, pp.197-203, 1990

Ries, J. C., R. J. Eanes, C. K. Shum, and M. M. Watkins, Progress in the determination of the gravitational coefficient of the Earth, Geophys. Res. Lett., 19(6), 529-531, 1992.

Ries, J.C., R.J. Eanes, C. Huang, B.E. Schutz, C.K. Shum,B.D. Tapley, M.M. Watkins and D. N. Yuan, Determination of the gravitational coefficient of the Earth from near-Earth satellites. Geophys. Res. Lett.,16(4),271-274, April 1989.

Ries, J.C., C. Huang and M.M. Watkins, Effect of General Relativity on a Near-earth Satellite in the Geocentric and Barycentric Reference Frames, Phys. Rev. Ltrs., 61,8, 1988

Smith, D. E., D. C. Christodoulidis, R. Kolenkiewicz, P.J.Dunn, S.M. Klosko, M.H. Torrence, S. Fricke and S. Blackwell, A Global Geodetic Reference Frame from LAGEOS Ranging (SL5.1AP), JGR, 90 (B11) 9221-9234,1985.

Tapley, B. D., B. E. Schutz & R. J. Eanes, Station Coordinates, Baselines, and Earth Rotation from LAGEOS Laser Ranging: 1976-1984,1985 JGR - V.90, pp.9235-9248.

Long-term stability of altimetric data with applications to mean sea-level change

P. Moore, S. Carnochan, M.D. Reynolds, P.E. Sterlini
Department of Geomatics
University of Newcastle-upon-Tyne, NE1 7RU, UK

Abstract. The oceanographic community has access to near 14 years of altimetric data ranging from GEOSAT in 1985-1989 to the current TOPEX/Poseidon and ERS-2 missions. For sea-level change studies careful validation is required to avoid contamination of the results by spurious non-oceanographic effects. In this study we first utilise *in situ* tide gauge data to monitor stability of the TOPEX/Poseidon data base. This series is linked to both ERS satellites by dual satellite crossovers. The analysis is then extended to incorporate GEOSAT ERM data through tide gauge enhanced dual satellite crossovers with TOPEX. Results of the stability analysis are used to consider global mean sea-level change over the period 1986-1998.

Keywords. Altimetry, mean sea-level change.

1 Introduction

In the past, sea-level change over extended time-periods has relied on the global network of tide gauges with corrections applied for post-glacial rebound. However, short-term studies from tide gauge data, as required for evidence of sea-level rise acceleration due to global warming, is susceptible to inter-annual and decadal basin wide fluctuations that can mask or accentuate any rise in data spanning forty years or less (Sturges, (1987)). Alternatively, satellite altimetry can monitor sea-level change on a truly global scale but is prone to systematic errors due to instrumental and other effects. The global nature of the data renders analyses less susceptible to localised effects but long-term studies are still required to separate the dominant inter-annual variability from the underlying trend.

To date, the altimetric community has access to continuous monitoring of the ocean surface from ERS-1 (Aug. 1991 - Jun. 1996); TOPEX/Poseidon (Sep. 1992 - present) and ERS-2 (Apr. 1995 - present) with near continuous coverage back to 1985 from GEOSAT (Apr. 1985 - Dec. 1989). With planned launches of ENVISAT and JASON-1 around the turn of the century the data sets will give 20 years of data by 2005. The combined data set will yield an unprecedented history of global ocean variability of which sea-level change studies place the greatest demand on the integrity of the data. Long-term study of sea-level change from multi-satellite altimetric missions requires precise inter- and intra-calibrations if systematic errors within a particular mission are not to be aliased into the derived signal. This study addresses the issue of inter-calibrations of TOPEX/Poseidon (T/P), GEOSAT and the ERS missions and seeks to extract mean sea-level change between 1986 and 1998.

2 Review of inter- and intra-calibrations

Let h_{cor} be the raw altimeter measurement corrected for all effects except sea-surface topography and the inverse barometric effect. Assume this measurement has a bias h_{bias} then the misclosure equation defining the bias is given by

$$h_{cor} + h_{ssh} - h_{bias} - h^{T}_{orb} \approx 0 , \qquad (1)$$

where h_{ssh} and h^{T}_{orb} are the instantaneous sea-surface height and true orbital height above the mean reference ellipsoid respectively. At a repeat pass location, with epochs t_1 and t_2, equation (1) gives

$$\{[h_{cor}(t_1) - h^{T}_{orb}(t_1)] - [h_{cor}(t_2) - h^{T}_{orb}(t_2)]\} +$$

$$[h_{ssh}(t_1) - h_{ssh}(t_2)] - [h_{bias}(t_1) - h_{bias}(t_2)] \approx 0. \qquad (2)$$

International Association of Geodesy Symposia, Vol. 121
Schwarz (ed.), Geodesy Beyond 2000 – The Challenges of the First Decade

Equation (2) can be estimated from the altimetric and tide gauge data and thus exhibits signatures due to any systematic errors in that data. In particular, for the ERS satellites at an altitude of 780 km, orbital error still remains a problem despite the advances in gravity field tuning (e.g. Scharroo and Visser (1998)). Several authors (e.g. Rosborough, (1986)) have investigated the effect of gravity field mis-modelling on satellite positioning and have shown that the radial height error, Δr, can be expressed as

$$h^T_{orb} = h_{orb} + \Delta r = h_{orb} + \Delta f + \delta_{A/D}\Delta v \qquad (3)$$

where h_{orb} is the estimated orbit height, Δf the geographically correlated error, Δv the anti-correlated error and $\delta_{A/D}=\pm 1$ with sign depending on whether the first epoch is from an ascending or a descending arc. Δv represents the variation about Δf between ascending and descending arcs.

On rearranging equations (2) and (3) we have

$$\Delta_{RP}(t_1,t_2) + \Delta_{TG}(t_1,t_2) \approx [h_{bias}(t_1) - h_{bias}(t_2)] + \delta_{RP} \quad (4)$$

where $\Delta_{RP}(t_1,t_2)$ is the repeat pass residual, $\Delta_{TG}(t_1,t_2)$ the tide gauge correction and δ measurement noise etc. An important consequence of repeat pass data is the effective cancellation of orbit error. A similar expression holds at a dual crossover point (DXO), i.e. the intersection of ground tracks of two altimetric satellites, namely

$$\Delta_{DXO}(t_1,t_2)+\Delta_{TG}(t_1,t_2) = [h^{(1)}_{bias}(t_1) - h^{(2)}_{bias}(t_2)] +$$

$$(\Delta f^{(1)} - \Delta f^{(2)})+ (\delta_{A/D}{}^{(1)}\Delta v^{(1)} - \delta_{A/D}{}^{(2)}\Delta v^{(2)}) + \delta_{DXO} \quad (5)$$

where, for example, the epoch t_1 corresponds to satellite 1. Equation (5) illustrates the importance of orbital modelling for inter-calibrations using dual crossovers as the geographically correlated error, Δf, is indistinguishable from the bias at a given geographical location. The anti-correlated error, Δv, tends to cancel given the equal likelihood of t_1 lying on an ascending/descending arc. Further, on identifying satellite 2 with T/P one can assume that $\Delta f^{(2)}= \Delta v^{(2)}=0$ as T/P has been established to have accuracy of 3-4 cm rms or better (Tapley et al. (1994)). Thus, orbital error in equation (5) can be attributed to the lower satellite (ERS or GEOSAT).

Equations (4) and (5) form the basis for monitoring stability of altimetric range biases at points close to tide gauges. In particular, Nerem et al. (1997) recovered global mean sea-level variations from T/P repeat pass data with the trend adjusted for an observed drift of -2.3±1.2 mm/yr in the altimetric signal. Most of this drift is attributed to a drift in the TOPEX microwave radiometer (TMR). A TMR drift rate of -1.9±1.2 mm/yr (Kubitschek et al, (1997)) was deduced by comparison against an upward-looking water radiometer and is in close agreement with the value of -1.2±0.4 mm/yr deduced by Haines and Bar-Sever (1998) from terrestrial GPS receivers. In early 1999 an anomaly was identified in the side A altimeter due to slow erosion of the point-target response. After careful validation of side B, TOPEX altimetry was switched in Feb. 1999.

For ERS, Moore et al. (1999a, 1999b) investigated the stability of ERS-1 and ERS-2 altimetry. ERS-1 is the most problematic mission to date as the various phases, with repeat periods of 3, 35 and 168 days, precluded the use of repeat pass data for the entire mission. The ERS-1 study thus utilised both tide gauge enhanced single satellite crossovers (SXO) and dual crossovers between ERS-1 and T/P to examine long-term variability in the geographically correlated error. Failure to incorporate this long-term variability lead to a spurious drift of 2-3 mm/yr for the period Apr. 1992 to Aug. 1995. On utilising corrections for the variability the observed drift rate of -0.43±1.40 mm/yr is not statistically different from zero.

From the launch of ERS-2 in 1995 it become apparent that the altimetric data exhibited a spurious trend. Repeat pass differences between ERS-1 and ERS-2 over the tandem period (May 1995 - Jun. 1996) revealed a systematic difference equivalent to a slope of 11 mm/yr (Moore et al., 1999b). The anomalous behaviour in ERS-2, and to a lesser extent ERS-1, was attributed to bias jumps associated with the so-called single point target response (SPTR) corrections. These jumps occur when the altimeter is placed in its safe-mode and then reactivated at a later date. The temperature differential over the stand-by period leads to a discontinuity in the range measurement as the clock stabilises to a different temperature regime.

Apart from important considerations of stability, time series from multi-satellite missions are effectively offset relative to each other due to electronic delays in the altimetric hardware. This so-called electronic bias is usually determined absolutely by dedicated campaigns involving precise orbit determinations, tide gauge data, radiosondes etc. to quantify all parameters within the altimeter mis-closure equation. Thus for ERS-1 a dedicated campaign using the Venice Tower (Francis, (1993)) estimated a bias of -41.50±2.0 cm.

Similar studies for T/P (Christensen et al, (1994)) used data over the Harvest Platform and Lampedusa. No dedicated absolute calibration was performed for ERS-2 although relative calibration exercises were undertaken against ERS-1 and TOPEX/Poseidon (Benveniste, (1997)).

Comparatively little attention has been devoted to GEOSAT over the past few years given the relatively imprecise nature of the altimetry due to orbit determinations based on TRANET Doppler measurements; the gravity gradient stabilisation and the absence of a radiometer for water vapour data. However, GEOSAT altimetry links the late 1980's with data from the 1990's and has the potential to quantify sea-level change from 1985 onwards.

In the following sections we utilise T/P as the basis of inter-calibrations of both ERS satellites and GEOSAT. The stability of T/P is first revisited (Mitchum (1998)) through comparisons against tide gauge data. Once validated T/P can be used within dual satellite crossovers to monitor the stability of ERS-1 and ERS-2. The methodology is based on the contemporaneous nature of the missions enabling dual crossovers with epochs differing by 5 days or less to be effectively independent of sea-level variability. Finally, an attempt is made to incorporate GEOSAT through the use of dual crossovers with T/P. Here, the sea-surface variability in the vicinity of tide gauges is removed by reference to the tide gauge time series.

3 Intra-calibration of TOPEX

For this study TOPEX/Poseidon version C data was used throughout with the Gaspar et al. (1994) 4 parameter model for sea-state bias, the CSR 3.0 (Eanes and Bettapur, (1995)) ocean tide model and the internal calibration of Hayne et al. (1994) applied as default. The NASA JGM-3 orbit was selected.

Tide gauge data is available for 109 gauges in the FASTWOCE data set. Hourly data, filtered for tidal signatures, reflect variability in the ocean surface from currents, meteorological forcing etc. Tide gauges thus provide a simple mechanism for both monitoring and replicating the temporal signatures in satellite altimetry.

To examine the validity of each gauge, repeat pass TOPEX altimetry was extracted at the latitude of the gauge for the two ascending and two descending passes straddling the gauge. Cycles 1-219, Sep. 1992 to Sep. 1998, were used. Time series of ocean variability at each of the 330 repeat pass locations were compared directly against the tide gauge time series interpolated to the required epochs. A correlation analysis and linear regression of the differences supplied a measure of the reliability of a particular gauge. Repeat pass locations were accepted if

- the number of cycles contributing to the time series exceeded 140 out of the maximum 198. Note that 21 of the total 219 cycles were entirely Poseidon
- the correlation coefficient exceeded 0.3
- the rms difference between altimetry and tide gauge time series was 10.0 cm or less
- linear regression of the differences resulted in a slope of 15 mm/yr or less in absolute terms.

204 of the original 330 repeat pass time series passed these acceptance criteria but 30 tide gauges were eliminated. The geographical location of the 79 accepted tide gauges is given in Figure 1.

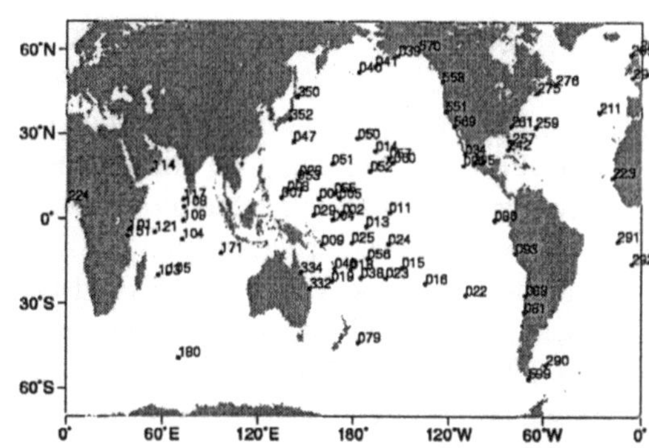

Fig. 1: WOCE tide gauges used to monitor stability of TOPEX.

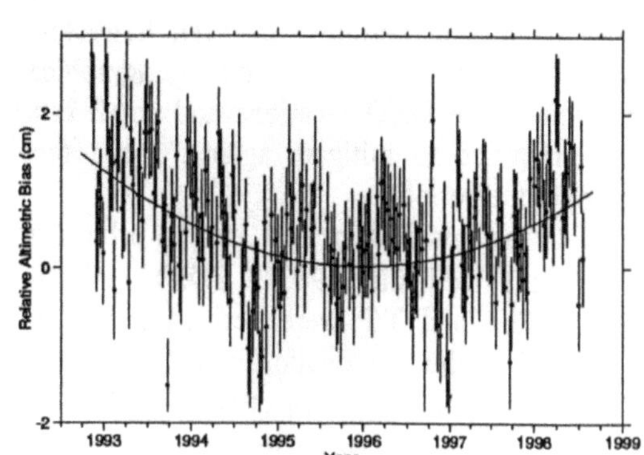

Fig. 2: TOPEX bias drift determined as a step function constant over each cycle with quadratic fit.

The 204 altimetric repeat pass time series were modified for the observed ocean variability by reference to the tide gauge time series. To combine these time series into a single data set, the mean sea-level height at each location was derived as the mean of the second and third quartiles. The process eliminated outliers but still left over 70 measurements to establish the mean offset. Once corrected for the appropriate mean sea-level height the residuals were analysed for the TOPEX altimetric bias. The bias was recovered as a step function with values constant over each near 10 day cycle. In practice, the single value per cycle was recovered from over 170 tide gauge enhanced altimetric measurements. A plot of the TOPEX bias drift and error bars per cycle is presented as Figure 2. TOPEX altimetry is characterised by a quasi-secular decrease from launch until early 1997 after which the range bias tends to increase as a consequence of the deterioration of Side A of the altimeter. The quadratic fit of Figure 2 will be used later to correct ERS relative to TOPEX.

4 Inter-calibration of ERS and TOPEX

Equation (5) forms the basis for inter-calibration of ERS and TOPEX. For this study we utilised ERS orbits derived at Aston with respect to the AGM98 gravity field. AGM98 was recovered from satellite laser ranging to ERS-1 and T/P; ERS-1 SXO data; DORIS tracking to T/P and dual crossovers between ERS-1 and T/P. DXOs provided a measure of the crucial geographically correlated ERS error with AGM98 designed to minimise this error component.

DXO residuals were derived for epochs differing by 5 days or less over which sea-surface variability was assumed negligible; i.e. $\Delta_{TG}(t_1,t_2)\approx 0$ in equation (5). The choice of 5 days is somewhat arbitrary being a compromise between large time intervals required for a good geographical distribution of the DXO locations and small time intervals required to mitigate against sea-surface change over the period.

ERS-1 was launched in 1991 into an orbit near 780 km and inclination 98.6°. DXO residuals between ERS-1 and T/P were derived from Oct. 1992 through to the deactivation of ERS-1 in Jun. 1996. The period thus encompassed ERS-1 cycles 5-18 of the first multidisciplinary phase C (35 day repeat); the second ice phase D (3 day repeat) ; the two 168 day repeat periods for the geodetic phases E and F and the 12 cycles of the second multidisciplinary phase G. In total the data set comprised of 1339010 DXO residuals. It is important to note that altimetry for phases C through F were released as version 3 with phase G as version 6. The latter included upgrades of certain correction fields and a major data processing change and is inconsistent with version 3. The change to version 6 data coincided with the launch of ERS-2 in April 1995. ERS-2 was launched into an identical (35 day repeat) orbit to ERS-1. Both satellites orbited in tandem and followed the same ground track with the ERS-2 sub-satellite points one day behind ERS-1. 1062575 version 6 DXO residuals between ERS-2 and T/P were formed for the first 33 cycles of ERS-2 data (May 1995 - Jul. 1998).

In addition to temporal variations in the altimeter bias, DXO residuals of equation (5) may exhibit other trends, δ_{DXO}. In particular, the altimetric centre of figure for ERS and T/P may not coincide exactly due to the geographical distribution of the tracking data available for precise orbit determinations. Also, T/P benefits from a dual frequency altimeter for recovery of the ionospheric correction whereas ERS is reliant on a model. Furthermore, the predominance of high latitude DXO data emphasises the importance of sea-state bias models given the high wave heights and wind speeds in the Southern Ocean. With empirical sea-state bias models inferred globally from SXO it is not inconceivable that the models favour the dominant regimes with a possible zonal signature to the error. It was thus considered advisable to solve for additional correction terms, namely

$$\delta_{DXO} = D_1\cos\phi\cos\lambda + D_2\cos\phi\sin\lambda + D_3\sin\phi + D_4 r_\tau + D_5(\cos^2\phi - 0.5) \qquad (6)$$

where D_i, i=1, 2, 3 are the coefficients of first order spherical harmonics to correct for variation in the centre of figure; r_τ the radial height correction associated with the timing bias, D_4, and D_5 the coefficient of an empirical second order zonal term with zero mean. DXO residuals were also corrected for the TOPEX mean drift taken to be the quadratic of Figure 2.

Recognising the inadequacy of some SPTR corrections (Moore (1999b)) equations (5) and (6) have been applied to intervals between consecutive SPTR events with all parameters recovered as a constant value over the period. The derived ERS altimetric range bias is plotted in Figure 3. The figure illustrates the inconsistency of range bias between ERS-1 versions 3 and 6 and the small offset between the ERS altimeters.

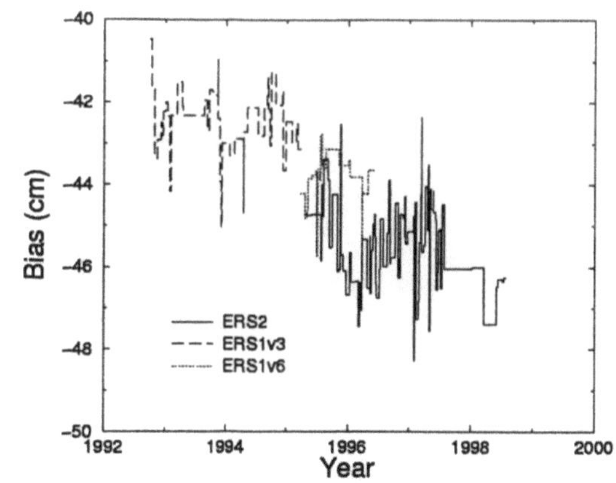

Fig. 3: ERS range bias determined as a step function constant over the period between SPTR events.

5 Inter-calibration of GEOSAT and TOPEX

GEOSAT was launched in Mar. 1985 into an orbit of inclination 108°. Data collection commenced 30 Mar. 1985 and continued for 18 months as the orbit drifted under natural forces. Due to its military significance this geodetic phase was originally classified but became freely available in 1995. Manoeuvres in Oct./Nov. 1986 placed GEOSAT into a 17 day repeat orbit and the so called ERM (Exact Repeat Mission) commenced 8 Nov. 1986. GEOSAT was a precursor to the more advanced missions of the 1990's with nadir pointing of the altimeter antennae maintained by gravity gradient stabilisation. This resulted in loss of lock on numerous occasions particularly as solar activity approached its maximum. The satellite suffered electronic failure on Oct. 4 1989. Upgraded GEOSAT data was released in 1997 with JGM-3 orbits and improved geophysical corrections. The JGM-3 orbits yield crossover residuals near 13 cm rms for the first 40 ERM cycles but accuracy deteriorated thereafter as air drag increased. Only the ERM data are considered in this study.

Wagner et al. (1997) have considered the connection between GEOSAT and T/P altimetry. The authors concluded that the relatively poor determination of the centre of mass from the Tranet Doppler tracking network gave rise to large height differences as characterised by coefficients D_1-D_3 of equation 6. We have thus followed their example and recovered the offset by comparing global sea heights from GEOSAT (1987-8) with TOPEX (1993-8). The comparison only used complete years of data to reduce aliasing by the dominant annual variation. The derived corrections (D_1= -8.1 cm; D_2 = -6.0 cm; D_3 = 1.8 cm) were subsequently removed from GEOSAT sea-surface heights. The process may also remove long-wavelength oceanographic signal, but, until GEOSAT orbits are improved, some empirical correction scheme will always be necessary.

The methodology of the preceding section is applicable to contemporaneous altimetric missions. With a two year gap between GEOSAT and T/P (and ERS-1) the sea-surface change must be considered. An option is to utilise equation (5) in full and to consider dual crossovers in the vicinity of tide gauges. Figure 4 plots the location of the 53 gauges used for T/P that were also operational during the lifetime of GEOSAT. On utilising repeat pass data from both satellites and the tide gauge enhanced DXOs, the GEOSAT altimetric range bias was recovered. Within this procedure TOPEX data was given a high weight to preserve the solution of Figure 2 with the GEOSAT bias estimated from the repeat pass data. DXOs were assigned a low weight but still supplied the connection between the two missions. A plot of the GEOSAT range bias is given in Figure 5. The figure illustrates that GEOSAT altimetry is offset from TOPEX by near 13 cm with obvious deficiencies from mid 1988 onwards. There is also some evidence of an annual trend throughout the 4 years.

6 Sea-level change

The stability analyses can be used to connect the various altimeter missions for sea-level change studies. In the first instance we restrict ourselves to TOPEX and then attempt to complement the study by utilising GEOSAT. For TOPEX we use the first 225 cycles (Sep. 1992 - Oct. 1998) with the data adjusted for the quadratic fit of Figure 2. Data were binned into 2° lat. by 2.835° long. pixels; the longitude span being the equatorial spacing between two ascending TOPEX arcs. The mean sea-level was removed by reference to the EGM96 geoid and sea surface topography. For each bin a linear and annual trend were recovered with the gradient providing the measure of sea-level change. An equal area weighting strategy gave a global *increase* of 1.37 mm/yr for 1992-1998 with the global distribution of sea-level change presented as Fig. 6.

A similar study was performed after TOPEX altimetry was complemented with GEOSAT data. Again TOPEX altimetry was adjusted for the mean trend of Figure 2 with GEOSAT adjusted for a constant offset of 12.79 cm. Sea-level change was then estimated as described for TOPEX data. The equal area weighting gave a global *decrease* of 0.62 mm/yr for the period 1986 - 1998. Figure 7 plots the global distribution of sea-level change for this period.

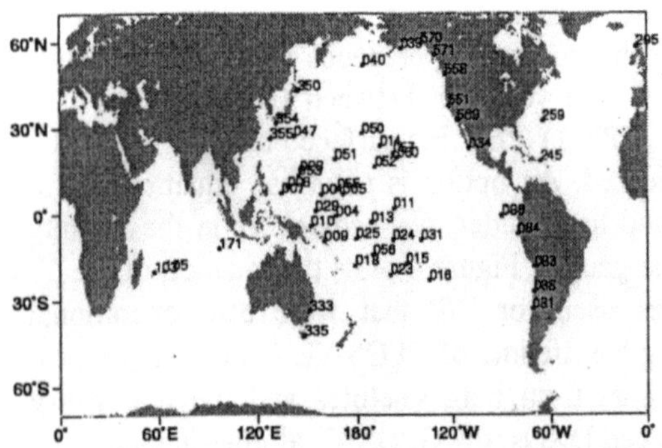

Fig. 4: Tide gauges used to remove sea-level change in GEOSAT and TOPEX dual crossover data.

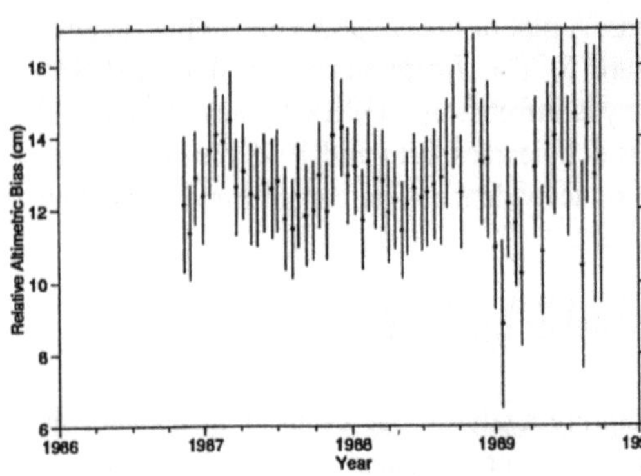

Fig. 5: GEOSAT bias drift determined as a step function constant over each ERM.

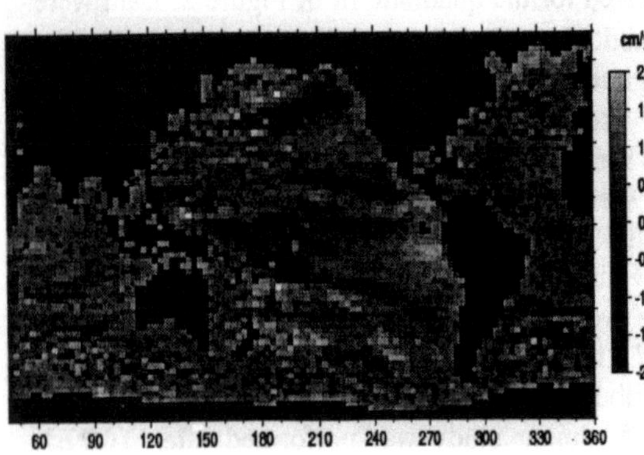

Fig. 6: Global distribution of sea-level change for 1992-1998 from TOPEX data adjusted for bias drift.

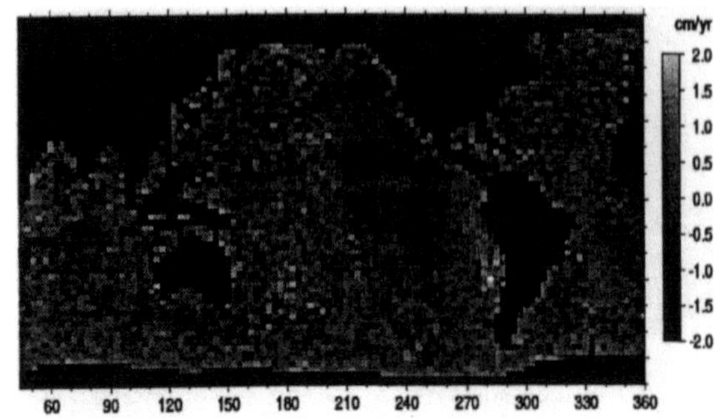

Fig. 7: Global distribution of sea-level change for 1986-1998 from GEOSAT and TOPEX data adjusted for bias drift.

7 Conclusions

The study shows that ERS and TOPEX altimetry can be unified through additional corrections for the ERS range bias jumps, geocentre offset, time tag bias and an empirical second order zonal term. The cause of the latter needs further investigation with sea-state bias and the wet tropospheric correction as possible candidates. GEOSAT orbits, similarly, require correction for the centre of mass. Further, empirical terms such as a 1 cy/rev correction may be required to reduce GEOSAT orbital errors and allow separation of orbital and oceanographic signatures. Global sea-level rise deduced from TOPEX gave a value of 1.37 mm/yr for 1992-1998 on allowing for the observed range drift. On utilising data for 1986-1998 the combined GEOSAT and TOPEX altimetry indicated a small sea-level fall of 0.62 mm/yr.

References

Benveniste, J (1998). Calibration/validation of ERS-2 altimetry. *ESA publication*, in press.

Christensen, E. J. et al. (1994). Calibration of TOPEX/Poseidon at Platform Harvest. *J. Geophys. Res.*, 99, pp 24465-24486.

Eanes R J and S V Bettadpur (1995). The CSR 3.0 global ocean tide model. *CSR-TM-95-06*, Center for Space Research, University of Texas at Austin.

Francis, C R (1993). The height calibration of the ERS-1 radar altimeter. *Proceedings of the First ERS-1 Symposium*, ESA SP-359, pp 381-394.

Gaspar P, F Ogar, P-Y Le Traon and O Z Zanife (1994). Estimating the sea state bias of the TOPEX and Poseidon altimeters from crossover differences. *J. Geophys. Res.*, 99, pp 24981-24994.

Haines B J and Y E Bar-Sever (1998). Monitoring the TOPEX microwave radiometer with GPS; Stability of columnar water vapour measurements. *Geophys. Res. Lett.*, *25*, No. 19, pp 3563-3566.

Hayne G, S Hancock III and C Purdy (1994). TOPEX altimeter range stability estimates from calibration mode data. *TOPEX/Poseidon Research News*, 3, pp 18-20.

Kubitschek D G et al. (1997). Calibration methods for the TOPEX and Poseidon altimetry systems (abstract). *Annales geophys.*, 15, Suppl. 1, p. C196.

Mitchum G T (1998). Monitoring the stability of satellite altimeters with tide gauges. *J. Atmos. Oceanic Tech.*, 15, pp 721-730.

Moore P et al. (1999a). Investigation of the stability of the ERS-1 range bias through tide gauge augmented altimetry. *J. Geophys. Res.*, in press.

Moore P, S Carnochan and R J Walmsley (1999b). Stability of the ERS altimetry during the tandem mission. *Geophys. Res. Lett.*, 26, No 3, pp373-376.

Nerem R S et al. (1997). Improved determination of global mean sea-level variations using TOPEX/Poseidon data. *Geophys. Res. Lett.*, 24 , No. 11, pp 1331-1334.

Rosborough G W (1986). Satellite orbit perturbations due to the geopotential. *CSR-86-1*, Center for Space Research, University of Texas at Austin.

Scharroo R and P Visser (1998). Precise orbit determination and gravity field improvement for the ERS satellites. *J. Geophys. Res.*, 103, pp 8113-8127.

Sturges, W. (1987) Large scale coherence of sea-level at very low frequencies. *J. Phys. Ocean.*, 17, pp 2084-2094.

Tapley B D et al. (1994). Precision orbit determination for TOPEX/Poseidon. *J. Geophys. Res.* 99, pp 24383-24404.

Wagner C A, J Klokocnik and R Cheney (1997). Making the connection between GEOSAT and TOPEX/Poseidon. *J. Geodesy*, 71, pp 272-281.

CORE: Continuous, high accuracy Earth orientation measurements for the new millennium

N.R. Vandenberg, C.C. Thomas
NVI, Inc./Goddard Space Flight Center, Code 920.1, Greenbelt, MD 20771

J.M. Bosworth, B. Chao, T.A. Clark, C. Ma
Laboratory for Terrestrial Physics, Code 926, Goddard Space Flight Center, Greenbelt, MD 20771

Abstract. The international geodetic VLBI community will inaugurate during 1999 a new phase of the program called CORE: Continuous Observations of the Rotation of the Earth. The capabilities of the new Mark IV correlators, available in late 1999, will enable greater sensitivity, more frequent observing sessions and improved system throughput.

Keywords. Earth orientation, VLBI, Earth system science.

1 Introduction

The CORE program (Continuous Observations of the Rotation of the Earth) is designed to provide continuous, high accuracy measurements of Earth orientation using VLBI (very long baseline interferometry). This paper describes the CORE program and how it will evolve into the next millennium.

2 What is CORE?

CORE is a VLBI program designed to provide, ultimately, continuous Earth Orientation Parameter (EOP) results. It is a cooperative scientific observing program, let by the VLBI group at NASA Goddard Space Flight Center. The goal of the program is to provide daily and sub-daily values for polar motion and UT1 with accuracy goals of sub-microsecond UT1 and 25 μas pole position.

The design of the observing program uses different VLBI networks on different days of the week. This approach is designed to spread the usage of VLBI resources among participating countries and agencies so that no one bears an undue burden.

The CORE program has been evolving since 1997 and the full program is anticipated to be in place by 2003.

3 How is CORE related to IVS?

CORE is an internationally sponsored measurement program, whereas IVS (International VLBI Service) is an international organization. The CORE program was proposed in response to the NASA Announcement of Opportunity in 1996. Tom Clark of Goddard Space Flight Center was the Principal Investigator with co-investigators from more than 20 international VLBI groups. CORE received official approval from NASA and we began observations in 1997.

With the start of the CORE program the geodetic VLBI community shifted its focus from measuring contemporary plate tectonics, which we pioneered in the early 1980s, to measuring Earth rotation. Earth rotation is one area of Earth system science where VLBI makes a unique contribution because VLBI is the only technique that can measure long-term UT1.

IVS is an organization recently formed by the international VLBI community. The many VLBI groups that participate in IVS contribute their resources to many different observing programs, of which CORE is one.

4 Evolution of CORE

Observing for the CORE program began in January 1997 with proof-of-concept sessions. We needed to answer first whether the different VLBI networks that would observe on different weekdays would actually give the same EOP result. That is, could different networks act as interchangeable measuring tools. The second goal of the initial sessions was to demonstrate the ability and performance of the stations that would participate in the full CORE program.

The CORE-A sessions were run simultaneously with NEOS (a program of the U. S. National Earth Orientation Service) sessions to try to answer the first question. We still have an enigma in that we do not understand the differences between the two measurement series. For 43 sessions in 1997 and 1998, there are significant differences between the EOP measurements from NEOS and from CORE-A. The average difference between the two

International Association of Geodesy Symposia, Vol. 121
Schwarz (ed.), Geodesy Beyond 2000 – The Challenges of the First Decade

measurements is non-zero in all three components of EOP. As an example, the X-pole average bias is -150 μas and the 1-σ error estimate on this value is ~20 μas. For both Y-pole and UT1 the bias value is ~4σ. For the CORE-B sessions compared to NEOS, the X- and Y-pole differences are not significantly different from zero, but the UT1 bias is 10σ.

The next step for CORE is to expand the number of observing days per week. Until the Mark IV correlator is available, we are strictly limited in the number of observing days that can be processed. Table 1 shows the planned growth of CORE in terms of the number of observing days per week.

Table 1: Future evolution of CORE

Date	Start	Avg. days/wk
2000 Jan 1	CORE-3 weekly (NEOS is on day 2)	2.5
2000 Jul 1	CORE-4 bi-weekly	3.0
2001 Jan 1	CORE-1 bi-weekly	3.5
2001 Jul 1	CORE-4 weekly	4.0
2002	CORE continuous	7.0

Table 2 shows the evolution of the accuracy of the CORE measurements.

Table 2: Evolution of CORE accuracy goals

	1997-1999	2000-2001	2002 ...
UT1	3.5 μs	1.5 μs	1.0 μs
X,Y pole	100 μas	50 μas	25 μas
SNR/point	55	45	
Obs. length	95 sec	25 sec	
# sources	45	130	
Obs/day	1500	3500	
Coverage	1.5 d/wk	3.5 d/wk	7 d/wk

During the past three years we have been achieving 3.5 μs for UT1 and 100 μas for pole position, using 1500 observations in a 24-hour period. Each observation has an average length of 95 seconds and achieves an average SNR of 55. We use approximately 45 celestial sources. During the first years of the millennium, as the Mark IV correlator begins operation, we will be able to use weaker sources because of the increased sensitivity of the Mark IV system. Since weaker sources are more numerous, our observing schedules can be expanded to achieve uniform sky coverage without sacrificing observing time to antenna slewing.

The next step toward increased accuracy will require improvements in the observing strategy so that we can take advantage of the increased sensitivity.

5 Conclusions

During the initial years of the new millennium the internationally-based CORE program will evolve toward providing daily and sub-daily VLBI measurements of UT1 and pole position. More continuous VLBI measurements will enable detailed comparisons of GPS and VLBI polar motion results, hopefully leading to better understanding of systematic effects in both techniques.

Challenges for CORE in the new millennium are 1) to increase temporal coverage and 2) improve measurement accuracy. Increased temporal coverage means that we have to increase the number of station observing days, particularly on weekends. Since VLBI is largely a manned operation, such operations mean increased resource usage. Improved measurement accuracy will come through improved sensitivity and better understanding of error sources.

The IGEX-98 campaign: Highlights and perspective

P. Willis
Institut Géographique National, ENSG/LAREG, 6-8, Avenue Blaise Pascal, 77455 Marne-la-Vallée, France

J. Slater
National Imagery and Mapping Agency, 4600 Sangamore Rd., Bethesda, MD 20816, USA

G. Beutler, W. Gurtner
Astronomical Institute of Bern, Sidlerstrasse 5, CH-3012 Bern, Switzerland

C. Noll
NASA, Goddard Space Fligth Center, Code 920-1, Greenbelt, MD 20771, USA

R. Weber
Technische Universität Vienna, Institut für Theoretische Geodäsie und Geophysik,
Gusshausstrasse 27-29 /1281, A-1040 Wien, Austria

R.E. Neilan
Jet Propulsion Laboratory, IGS Central Bureau, MS 238-540, 4800 Oak Grove Drive, 91109 Pasadena, USA

G. Hein
Universitat der Bundeswehr München, Institute of Geodesy and Navigation, Werner-Heisenberg-Weg 39,
D-85579 Neubiberg, Federal Republic of Germany

Abstract. Starting in October 1998, a large international campaign of GLONASS observations, called IGEX-98, has been conducted during a 6 month period. This campaign has involved a large international cooperation worldwide, including GLONASS, GPS and Laser observations and data processing. The goal of this paper is to present the scientific goals of the IGEX-98 campaign, to describe briefly its organization and also to present the preliminary results obtained for precise orbit determination, terrestrial reference frame and timing issues.

Keywords. Satellite geodesy, orbit determination, satellite positioning, time synchronization, GLONASS, GPS.

1 Introduction

The GLONASS system is a Russian navigation system using a constellation of medium Earth orbiting (MEO) satellites. Its constellation and signals are very similar in their principles to the American GPS system making it possible for manufacturers to provide combined GPS/GLONASS equipment for navigation, time metrology or geodetic applications [Langley, 1997; Zarraoa, 1998]. However, the GLONASS is far from being fully operational as only a limited number of satellites are available (14 in early 1999). The number of operational GLONASS has even been decreasing in the recent years.

Asides from the system's limited availability, GLONASS has several aspects that makes it particularly attractive for scientific and professional navigation (integrity applications: the precise P-code is fully available to the civil community and the dual utilization of GPS and GLONASS, by augmenting the number of available satellites, has great advantages for monitoring) and for precise rapid positioning (rapid ambiguity fixing).

In order to really validate these aspects, a global international GLONASS campaign, called IGEX-98 (International GLONASS Experiment 1998), was organized for a six-month period starting in October 1998.

The purpose of this paper is to present the major scientific goals of this campaign, to review its current status, describe a few organizational aspects and to highlight a few preliminary results that will be presented in the near-to-come IGEX-98 workshop that will be held in Nashville, USA, September 13-14, 1999.

International Association of Geodesy Symposia, Vol. 121
Schwarz (ed.), Geodesy Beyond 2000 – The Challenges of the First Decade

2 Scientific goals of the IGEX-98 campaign

The purposes of the IGEX-98 GLONASS campaign are numerous [P. Willis et al., 1999], due to the broad range of possible applications of combined GLONASS/GPS receivers for navigation, precise positioning and time synchronization.

First of all, by providing a important worldwide data sets of GLONASS/GPS data, it helps all these communities to enhance and test their software on actual data.

The first scientific goal of this campaign is to provide precise GLONASS orbits (at least at a sub-meter level) after some post-processing. This was really never done as all the other campaigns were usually local, or regional (over Europe) but never global.

The second goal is to provide for real-time users the best estimate of the terrestrial reference frames transformations between PZ-90 (as derived from broadcast GLONASS orbits) and WGS-84 (as derived from the GPS broadcast orbit). It is then important to collocate all the GPS data available at each GLONASS receiver site (worldwide).

On the other hand, as precise GLONASS orbits would be available, it would then become important to test the capabilities of the GLONASS system to provide precise geodetic results (long baseline determinations estimated daily).

Finally, using the precise GLONASS orbits should enhance the timing capabilities, allowing time transfer with better accuracy for longer distances.

3 Organizational aspects

The IGEX-98 has been sponsored by the IAG Commission VIII (CSTG), the International GPS Service (IGS), the Institute of Navigation (ION) and the International Earth Rotation Service (IERS).

A Steering Committee was formed at the early stage of the preparation of the campaign in fall 1997, after the IAG Scientific Assembly in Rio de Janeiro, Brazil. It is presently composed of the following individuals: G. Beutler/Switzerland, W. Gurtner/Switzerland (Network Coordinator and liaison with the laser ranging community), G. Hein/Germany, C. Noll/USA (Data Flow Coordinator), R.E. Neilan/USA (liaison with the IGS), J.A. Slater/USA (liaison with the ION and the navigation community in general [Misra et al., 1998]), R. Weber/Austria (Analysis Coordinator), P. Willis/France (Chair).

Several technical documents were drafted in fall 1998 (still available at the following Web site: http://lareg.ensg.ign.fr), leading to an International Call for Participation [P. Willis et al., 1998].

The campaign itself has been organized in close cooperation with the IGS, trying to mimic the IGS efficient organization: tracking stations, data centers, analysis centers [Beutler et al. 1995; Beutler et al., 1999].

4 Status of the IGEX-98 campaign

The response from the international community far exceeded our most optimistic expectations. After a few technical problems, due to a delay in the availability of GLONASS receivers, the campaign was postponed by one month and officially started on October 19, 1998.

During the data span of the campaign (October 19, 1998 to April 19, 1999), 74 receivers were deployed at 62 sites, involving 26 countries. Figure 1 presents the geographic distribution of the IGEX-98 network, distinguishing dual-frequency GLONASS receivers from single-frequency receivers. It must be noted that all receivers were either combined GPS/GLONASS equipment or were closely (within few meters) collocated with a precise dual-frequency GPS receiver.

IGEX-98 NETWORK

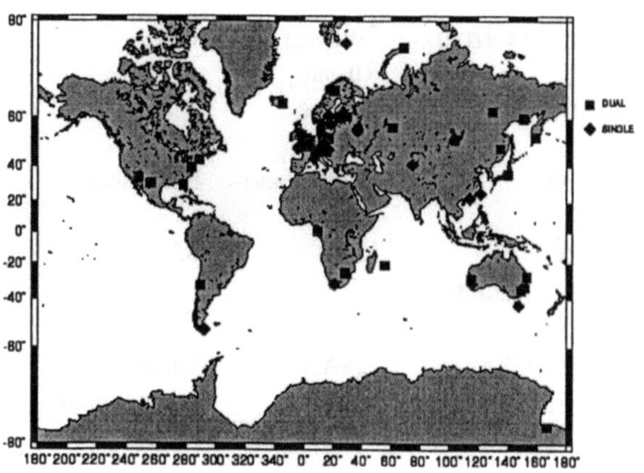

Fig. 1: IGEX-98 tracking network

All the data (GLONASS and GPS) were collected on an almost daily basis and were distributed to two Global Data centers (NASA/CDDIS and IGN) where these data are still freely available to the scientific community for further investigation. Table 1 presents the list of all 8 data centers that were used during this campaign.

Table 1: IGEX-98 Data Centers

Organization	Country	Type
AUSLIG	Australia	Regional
IGN	France	Global
DLR	Germany	Operational
ESA	Germany	Operational
BKG	Germany	Regional
GFZ	Germany	Operational
NCKU	Taiwan	Operational
NASA/CDDIS	USA	Global

Table 2: IGEX-98 Analysis Centers

Organization	Country	Software
AUSLIG	Australia	Microcosm
UNSW	Australia	
TU Vienna	Austria	
IGN	France	
ACRI/CERGA	France	
DLR	Germany	
ESA/ESOC	Germany	Bahn
GFZ	Germany	EPOS P
BKG	Germany	Bernese
NPL	India	
Univ. Of Olztyn	Poland	TOP
MCC	Russia	
MIIGAIK	Russia	
OSO	Sweden	
CODE	Switzerland	Bernese
NCKU	Taiwan	
NERC	UK	SATAN
Univ. E. London	UK	
U. Leeds	UK	
Univ. Of Texas	USA	Gipsy/Oasis
JPL	USA	Gipsy/Oasis

It should also be noted that thanks to the newly created ILRS (International Laser Ranging Service), satellite laser ranging observations from 30 observatories worldwide were obtained from 9 GLONASS satellites. These data are available at the ILRS data centers

Table 2 gives the list of groups who answered the IGEX-98 call for participation as Analysis Centers. From all these groups, six were able during the time of the campaign to upgrade their software and to produce precise GLONASS orbits on a regular basis (BKG/Federal Bureau for Cartography and Geodesy in Germany, CODE/Center for Orbit Determination in Europe et the University of Bern in Switzerland, ESOC/European Space Operations Center in Germany, GFZ/Geoforschungzentrum Potsdam in Germany, JPL/Jet Propulsion Laboratory in USA, MCC/Mission Control Moscow in Russia). The group TU Vienna in Austria produced combined precise orbits from these individual solutions for the whole campaign data span.

From these tables, it is quite obvious that there was a large positive response from the community covering all aspects from geodesy, to navigation and time metrology.

5 Preliminary results

The first goal of the campaign was to produce precise post-processed GLONASS orbits. Several groups were able to obtain orbits with precision in the 10-30 cm range. These orbits were compared to one another (internal precision) and also were validated using the available SLR measurements (external precision). Furthermore, the Analysis Coordinator (R. Weber) was able to produce a combined IGEX-98 orbit for the all duration of the campaign [Ineichen et al, 1999].

Several groups provided precise geodetic coordinates of the stations (using the IGEX-98 precise orbits), showing repeatability of 1 cm to 5 cm for long baseline over 1000 km using only the GLONASS data.

More detailed investigations concerning the terrestrial reference frame aspects were done using the large amount of data provided by the IGEX-98 campaign, and some are still under investigation. The main purpose is to provide real-time users a simple algorithm (Helmert 7-parameter transformation) in order to convert PZ-90 coordinates to WGS-84 (or the reverse).

Several methods have been tried using (1) point positioning derived with broadcast GLONASS orbits compared to WGS-84 (or ITRF-96 [Boucher et al., 1996]) point positions derived from GPS precise orbits and (2) direct comparison of PZ-90 broadcast orbits with the precise WGS-84 (or ITRF) orbits. The results of these investigations seem to agree on a large rotation around the Z-axis.

Precise orbits will also benefit the time transfer applications. Results should be forthcoming after a worldwide receiver calibration campaign is completed and also some problems concerning ionosphere are taken into account [Lewandowski, 1999].

6 Conclusions and plans for the future

In conclusion, the IGEX-98 has been conducted successfully from October 1998 to April 1999. In fact, the observations and data analysis have continued since April on a best effort basis, until a decision can be made at the upcoming IGEX-98 workshop (Nashville, USA, September 1999).

This campaign was only possible thanks to the large international cooperation between geodesists, navigation users and time metrologists. It benefited greatly from the experience gained with the establishment of the IGS as a permanent Service since 1994.

The worldwide GPS/GLONASS/SLR network deployed during this campaign led to precise orbits (20 cm precision) that are available to scientific users on a post-processing basis. It also demonstrated the possible use of GLONASS (in combination with GPS) for precise geodetic applications. The issue of GLONASS and GPS terrestrial reference frames incompatibilities and time differences should now be resolved leading to a real inter-operability of the GPS and GLONASS systems (and possibly Galileo in the future).

It is also probable that there will be a continuation of this experiment after the Nashville workshop as long as the GLONASS constellation is at least partially maintained in the future.

References

Beutler, G., R.E. Neilan, I.I. Mueller (1995), The International GPS Service (IGS): the Stroy. In *IAG Proc.*, Springer-Verlag, pp.3-13.

Beutler, G, M. Rothacher, T. Springer, J. Kouba, R.E. Neilan (1999). The International GPS Service (IGS): An interdisciplinary service in support of Earth sciences. In *Advances in Space Research*, Pergamon, 23, 4, pp 631-653.

Boucher C., Z. Altamimi (1996). International Terrestrial Reference Frame, In *GPS World*, 7, 9, pp 71-74.

Ineichen D., M. Rothacher, T. Springer, G. Beutler (1999). Computation of Precise GLONASS Orbits for IGEX-98' .In *Proc. IAG General Assembly*, Birmingham, July 1999, Springer-Verlag, in press. for IUGG Birmingham.

Langley R. (1997). GLONASS: Review and update. In *GPS World*, 8, 7, pp 46-51.

Lewandowski (1999). Private communication.

Misra, M., J.A. Slater (1998). A report of the third meeting of the GLONASS/GPS interoperability working group. In *Proc of ION GPS-98*, Nashville, USA, Institute of Navigation, pp 2103-2106.

Willis P., G. Beutler, W. Gurtner, G. Hein, C. Noll, R.E. Neilan, J. Slater (1998). International call for participation to the IGEX-98 campaign. In *Journal of Geodesy*, Springer-Verlag, 72, 5, p 313.

Willis, P., G. Beutler, W. Gurtner, G. Hein, R.E. Neilan, C. Noll, J. Slater (1999). IGEX: International GLONASS Experiment: Scientific objectives and preparation. In *Advances in Space Research*, Pergamon, 23, 4, pp 659-663.

Zarraoa, N., W. Mai, E. Sardòn, A. Jungstand, Preliminary evaluation of the Russian GLONASS system as a potential geodetic tool. In *Journal of Geodesy*, 72, 6, pp 356-363.

Computation of precise GLONASS orbits for IGEX-98

D. Ineichen, M. Rothacher, T. Springer, G. Beutler
Astronomical Institute, University of Berne, Sidlerstrasse 5, CH-3012 Bern, Switzerland

Abstract. On October 19, 1998, at the beginning of the International GLONASS Experiment (IGEX-98), the Center for Orbit Determination in Europe (CODE) has started to compute precise orbits for all active GLONASS satellites. The campaign was initially scheduled for three months, but the activities still continue in September, 1999. One of the main reasons for this extension was the launch of three new GLONASS satellites at the end of the year 1998.

The processing of the IGEX network is done on a routine basis at CODE and precise ephemerides are made available through the global IGEX Data Centers. The improved GLONASS orbits are referred to the International Terrestrial Reference System (ITRF 96) and to GPS system time. They are therefore fully compatible with GPS orbits and allow a combined processing of both satellite systems.

All GLONASS satellites are equipped with a laser reflector array and the SLR ground network is tracking most of the GLONASS satellites. Comparisons of the GLONASS orbits computed by CODE with the SLR measurements show that the orbit accuracy is better than 20 cm.

Keywords. GLONASS, IGEX-98, orbit determination, system time difference.

1 The IGEX-98 campaign

The main purpose of the IGEX-98 campaign was to conduct the first global GLONASS observation campaign for geodetic and geodynamic applications. The experiment took place under the auspices of the International Association of Geodesy (IAG), the International GPS Service (IGS), the Institute of Navigation (ION), and the International Earth Rotation Service (IERS). The campaign started on October 19, 1998 (GPS week 980), with a planned duration of 3 months.

The IGEX-98 steering committee decided in a first step to extend the campaign for an additional three months, and in a second step to continue on a best effort basis the campaign activities till the IGEX-98 workshop in Nashville, USA (September 13–14, 1999).

The main objectives of the campaign (Willis et al. (1998)) are to

- test and develop GLONASS post-processing software,
- determine GLONASS orbits with an accuracy of 1 meter or below, realized in a well defined Earth-fixed reference frame,
- determine transformation parameters between the GLONASS reference frame (PZ-90) and the GPS reference frame (ITRF 96),
- investigate the system time difference between GLONASS and GPS, and
- collaborate with the SLR community to evaluate the accuracy of the computed GLONASS orbits.

A map of the IGEX-98 network as used by CODE for orbit determination processing may be found in Figure 1. Only the sites providing dual-frequency GLONASS data are shown on the map (and only those sites were used for the processing). Most of the sites are located in Europe. The connection to the receivers located in other parts of the World is quite weak. The map shows all used sites during the campaign (about 35 sites). For some weeks, however, the number of available sites decreased to 20.

The measurement data of these sites are collected and made available at five Regional and two Global Data Centers (Noll (1998)). Until now, six Analysis Centers were or are making use of the data for computing improved GLONASS orbits. The precise GLONASS orbits may be found at the CDDIS Global Data Center (CDDIS, 1999).

International Association of Geodesy Symposia, Vol. 121
Schwarz (ed.), Geodesy Beyond 2000 – The Challenges of the First Decade

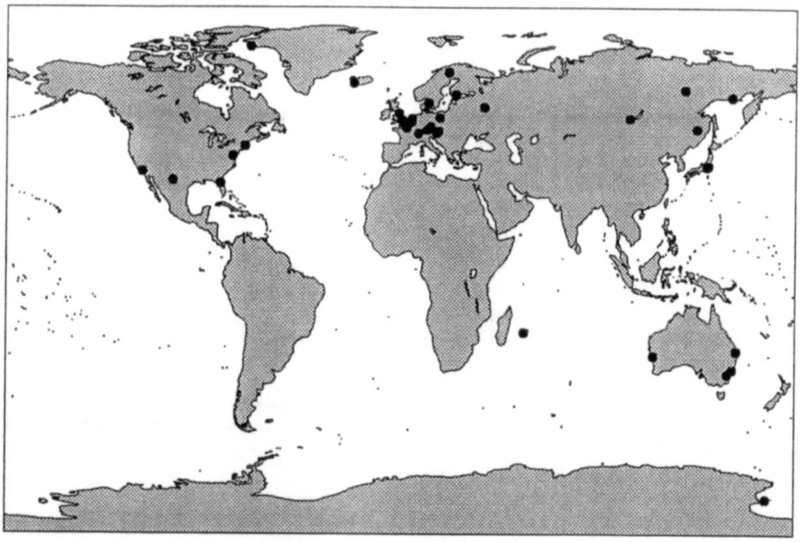

Fig. 1: The IGEX observation network as used by the CODE analysis center

2 Determination of precise GLONASS orbits

2.1 Processing strategies

For the combined processing of GLONASS and GPS data the enhanced Version 4.1 of the Bernese GPS Software is used, see Rothacher and Mervart (1996), Habrich (1999). The analysis is done by fixing both, the GPS orbits and Earth rotation parameters to CODE's final IGS solutions. The number of available sites within the IGEX network would not allow to estimate these parameters with an accuracy comparable to the IGS solutions. The orbital parameters of the GLONASS satellites are estimated using double difference phase observations (including double differences between GLONASS and GPS satellites). The processing of the IGEX network is done without fixing the ambiguities to their integer values.

Six initial conditions and nine radiation pressure parameters are determined for each satellite and arc (Springer (1999)). Pseudo-stochastic pulses have been set up every 12 hours for test purposes, but have been constrained to zero for the official CODE solution. Only receivers providing dual-frequency GPS and GLONASS data or dual frequency GLONASS data are included in the processing procedure. The final precise orbits stem from the middle day of a 5-day arc. The satellite clock values included in the precise orbit files are broadcast clock values for the GLONASS satellites, because no satellite clock estimation is performed so far. In order to align the IGEX network to the terrestial reference frame the coordinates of five sites (Greenbelt, Kiruna, Metsahovi, Onsala, Yaragadee, and Zimmerwald), are constrained to their ITRF 96 coordinates.

2.2 Quality assessment

Long arc fits

In order to check the internal consistency of our precise GLONASS orbits, we perform a long-arc fit for each processed week. For each satellite, one orbital arc is fitted through the seven consecutive daily solutions of the week. As an example, the result of such a long arc fit is given in Table 1 for GPS week 1002. The Table shows the rms of this fit for each satellite and day. In the last line ("Week") the rms of the whole 7-day fit is included for each satellite.

The values of this overall rms are in general between 5 and 20 cm. On the one hand it must be stated that these values might be too optimistic because we fit the middle days of 5-day arcs with a 7-day arc. On the other hand the small values indicate that the adopted orbit model is well suited to describe the motion of the GLONASS satellites over a time period of several days.

Table 1: Orbit repeatability from a 7-day fit through daily orbit solutions (days 80–86, 1999, RMS values in [cm])

Slot No.	1	3	4	6	7	8	9	10
DOY 80	5	6	10	11	8	8	7	18
DOY 81	5	4	5	6	7	7	7	9
DOY 82	6	5	3	8	5	6	7	9
DOY 83	5	3	6	7	6	6	4	6
DOY 84	6	6	5	6	5	6	6	6
DOY 85	6	4	6	6	6	4	7	9
DOY 86	6	4	5	9	6	5	10	18
Week	6	5	6	8	6	6	7	12

Slot No.	11	13	16	17	20	22
DOY 80	11	8	15	8	10	8
DOY 81	6	7	9	5	5	6
DOY 82	9	8	7	7	8	6
DOY 83	7	11	10	5	5	6
DOY 84	12	9	7	5	6	4
DOY 85	7	7	5	4	4	5
DOY 86	16	11	14	7	5	6
Week	10	9	10	6	7	6

Comparison with the precise orbits of other IGEX analysis centers

The IGEX Analysis Center Coordinator R. Weber (Technical University of Vienna) is in charge of comparing the precise GLONASS orbits stemming from the six IGEX-98 Analysis Centers providing precise GLONASS orbits. In addition, he combines the Analysis Centers' precise GLONASS orbits into one official IGEX orbit product. The results of this combination procedure are distributed via IGEX mail and can be found on the following Web page:

`http://lareg.ensg.ign.fr/IGEX/IGEXMAIL/`

At present, combined orbits of ten weeks (0980–0989) are made available at the Global IGEX Data Center at CDDIS. The results of the first ten weeks of orbit comparison confirm that the reached orbit quality is of the order of 20 *cm*.

Comparison with SLR measurements

The comparison of CODE's precise GLONASS orbits with SLR measurements is a fully independent quality check and therefore very valuable for checking the quality of GLONASS orbits determined by means of microwave signals. This method of quality assessment also shows that the precise GLONASS orbits of CODE are on a 10 – 20 *cm* accuracy level. More details are given in Section 5 of this report.

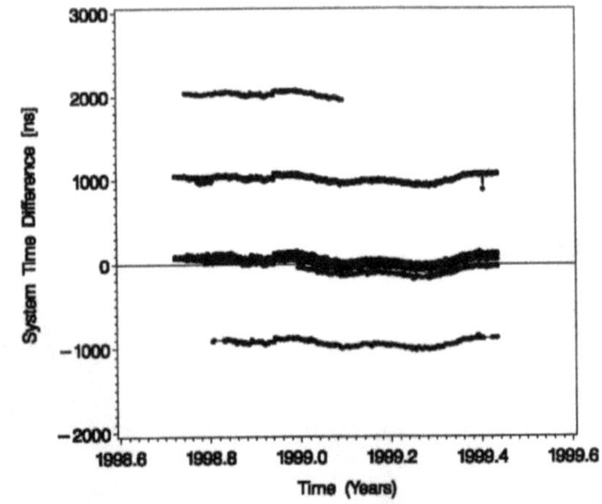

Fig. 2: System time difference estimated with different receiver types

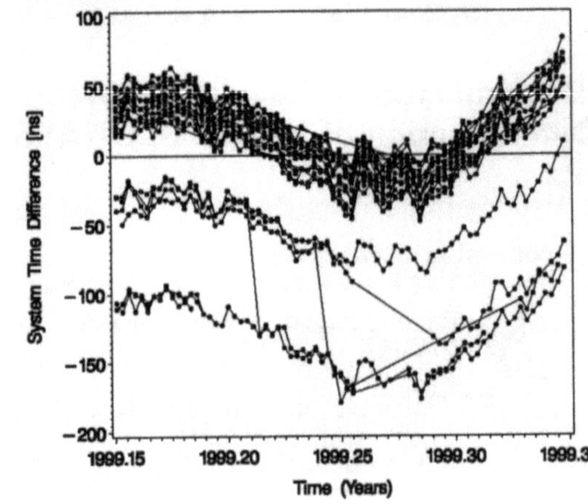

Fig. 3: Detail of Fig. 2 – system time differences of the Z18 and JPS receivers

3 System time difference between GLONASS and GPS

When processing GLONASS and GPS data we are setting up one additional parameter for each station and session in our pseudorange pre-processing step: the difference between GLONASS and GPS system time. The estimation is done in the following way: we use broadcast orbits for both systems, estimate the site coordinates and the time offset between GPS and GLONASS once per session and, as usual, one receiver clock correction for each epoch. In order to account for the different reference systems, the GLONASS broadcast orbits are rotated around the z-axis by −330 *mas* (Habrich (1999)).

What kind of components are contributing to

the estimated system time difference? On the one hand we have the difference between the national realizations of UTC (Universal Time Coordinated) on which the GLONASS and GPS system times are based: UTC (USNO, Washington DC) and UTC (SU, Moscow). Values for the difference between these national time references and UTC are published in the Circular T of the Bureau International des Poids et Mesures (BIPM (1999)). In July 1999, the difference between UTC (USNO) and UTC is below 10 nanoseconds and the difference between UTC (SU) and UTC below 100 nanoseconds. On the other hand we have to take into account the differences between GPS system time and UTC (USNO) and GLONASS system time and UTC (SU).

When comparing the time offsets resulting from the IGEX network processing, it becomes clear that we do not have direct access to the pure difference between GPS and GLONASS system time, but that receiver type specific offsets have to be taken into consideration as well. Figure 2 shows the estimated system time differences for different receiver types covering a time span from September 20, 1998 to June 6, 1999 (260 days). Each line represents a different receiver type. Starting from bottom to top:

- ESA/ISN GNSS (about -900 *ns*)
- JPS receivers (about -50 *ns*)
- Ashtech Z18 receivers (about 50 *ns*)
- 3S Navigation receivers (about 1000 *ns*)
- Ashtech GG24 receiver (about 2100 *ns*)

The differences between different receivers types are of the order of one microsecond. Figure 3 shows a detail of Figure 2: the time differences of the Z18 and JPS receivers during a time period of 73 days. The lower two bands represent the JPS receivers, the upper one the Z18 receivers. The time series of the individual stations are highly correlated. It is interesting to note the jumps of three JPS receivers from the medium level to the lower level. These jumps are correlated with software upgrades. The firmware of the JPS receiver at Zimmerwald, Switzerland was, e.g., upgraded from Version 1.4 to Version 1.5 and the RINEX converter software from Version 1.01 to Version 1.02. At that time the estimated system time differences show a jump of -40 *ns* (between doy 093 and doy 106, 1999).

Table 2: Mean values and RMS of the daily Helmert transformation parameters for the transition from PZ-90 to ITRF 96

Parameter		Mean	RMS
X–Translation	[m]	-0.03	0.23
Y–Translation	[m]	-0.02	0.27
Z–Translation	[m]	-0.45	0.47
X–Rotation	[mas]	37	6
Y–Rotation	[mas]	-10	8
Z–Rotation	[mas]	-350	21
Scale Value	[ppb]	13	3

4 Transformation parameters between the reference systems

In principle, there are two possibilities for the determination of transformation parameters between the GLONASS and the GPS reference system: One is based on coordinate sets which are determined in both systems, and the other is based on the comparison of satellite orbits available in both systems. Here, we present results stemming from the orbit comparison method.

Seven Helmert transformation parameters were determined using precise GLONASS orbits in the ITRF 96 reference frame and the broadcast GLONASS orbits in the PZ-90 reference frame. For each day one set of parameters (three translations, three rotations, and one scale factor) was established. Figure 4 shows the time series of the rms values, the translation parameters, and the rotation parameters. Using the described method, the accuracy of the transformation parameters is limited by the quality of the GLONASS broadcast orbits. The rms of the daily Helmert transformations (between 3 *m* and 6 *m*) may be interpreted as indicators of the broadcast orbit quality.

Mean values and standard deviations for each of the seven Helmert parameters and for the entire time series are summarized in Table 2.

The rotation around the z-axis definitely has to be taken into account when processing combined GLONASS and GPS data using broadcast orbits. A rotation of -350 *mas* around the z-axis is determined highly significant and corresponds to a satellite position offset of up to 45 *m* (if the satellite is close to the equatorial plane). The 3 translations and the rotation around the y-axis are not significantly different from zero. The influence of the x-rotation and the scale value on the GLONASS broadcast orbits is of the order of the broadcast orbit accuracy itself and may therefore be neglected.

5 Comparison of precise GLONASS orbits with SLR measurements

All GLONASS satellites are equipped with a laser retroreflector array. It is an interesting and important aspect of the IGEX-98 campaign that the SLR community was and still is very active in observing the GLONASS satellites: measurements to nine GLONASS satellites were performed during the first six months of the IGEX-98 campaign. At present, during the extended phase of the test campaign, three GLONASS satellites are still tracked by the SLR community. The SLR measurements are completely independent on the orbit determination process based on microwave signals. Comparisons between SLR measurements and improved orbits are therefore an important measure for the achieved quality of the precise GLONASS orbit determination using the microwave observations.

For the comparison of GLONASS orbits with SLR measurements, the residuals between SLR measurements and computed distances (derived from our GLONASS precise orbits and the SLR site coordinates) are analyzed. In addition, one constant offset is estimated for all SLR distances and removed from the residuals.

Figure 5 shows the residuals of the SLR measurements with respect to the GLONASS broadcast orbits and with respect to the CODE final IGEX orbits (middle day of a 5-day arc) over a time span of 230 days (October 10, 1998 to May 29, 1999). The rms decreases from 1,67 m (broadcast orbits) to 0.16 m (precise orbits), which proves that we are not only changing the orbits, but actually improve them. An offset of 39 mm between improved orbits and SLR measurements was determined (SLR distances are shorter than the distances derived from the CODE orbits). It is interesting to note that this offset agrees well with the offset found for SLR measurements with respect to the GPS orbits (55 mm). The reason for this offset is not yet understood.

A comparison of SLR measurements was also done with respect to daily orbits stemming from the mid day of a 3-day arc. The smaller rms of our 5-day solution compared to the 3-day solution is the reason for submitting the mid day of a 5-day arc as CODE official IGEX orbit product.

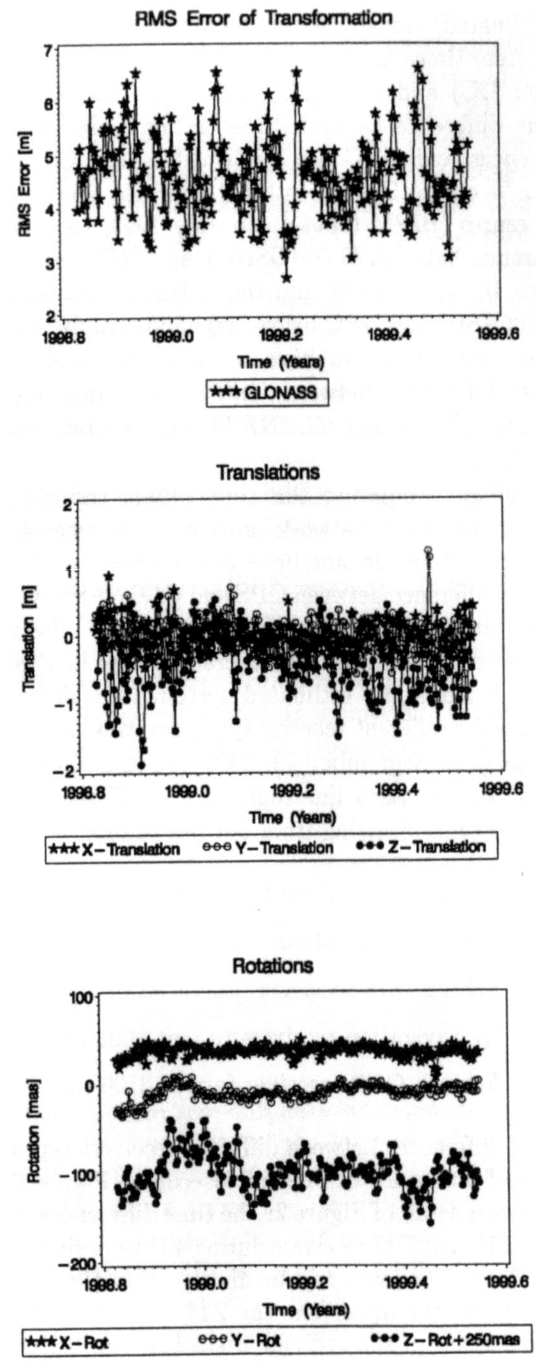

Fig. 4: RMS, translation parameters, and rotation parameters of the Helmert transformation between broadcast orbits in the PZ-90 system and CODE's precise orbits in the ITRF 96 system

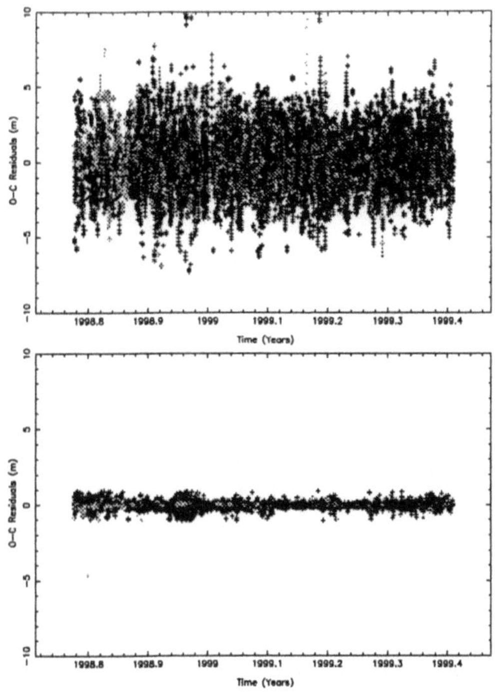

Fig. 5: Comparison of broadcast GLONASS orbits (top) and CODE's precise orbits (bottom) with SLR measurements

6 Outlook

In September 1999 an IGEX Workshop will be held at Nashville, USA, where the IGEX Analysis Centers have the opportunity to present their results using IGEX network data. In addition, at this workshop the decision will be made whether or not this global GLONASS experiment will continue and what will be the organizational form for such a project.

On the technical side, there are several issues waiting for investigation, such as:

- Tests concerning the parameterization of GLONASS orbits (reduction of the number of estimated radiation pressure parameters, estimation of stochastic pulses).
- Introduction of ambiguity fixing for GLONASS phase measurements within the IGEX network.
- Processing IGEX and IGS data in one step. Study of the impact of GLONASS data on estimated global parameters like, e.g., Earth rotation parameters (different orbit inclinations of GPS and GLONASS, no 2:1 resonance of GLONASS revolution period with Earth rotation).

A densification of the global dual-frequency receiver network would certainly significantly contribute to the improvement of the accuracy of the estimated GLONASS orbit.

References

BIPM, (1999). Bureau International des Poids et Mesures, Sevres, France. *http://www.bipm.fr/enus/5_Scientific* .

CDDIS, (1999). Crustal Dynamics Data Information System, Greenbelt, USA. *http://cddisa.gsfc.nasa.gov/glonass_datasum.html.*

Habrich, H. (1999). Geodetic Applications of the Global Navigation Satellite System (GLONASS). *Ph. D. thesis, Astronomical Institute, University of Berne.* In preparation.

Noll, C. (1998). IGEX98 Data Center Information. *http://lareg.ensg.ign.fr/IGEX/ix_datac.html.*

Rothacher, M., and L. Mervart (1996). Bernese GPS Software, Version 4.0. *Printing Office of the University of Berne.*

Springer, T., G. Beutler, M. Rothacher (1999). Improving the Orbit Estimates of GPS Satellites. *Journal of Geodesy*, Vol. 73, pp. 147-157.

Willis, P., G. Beutler, W. Gurtner, G. Hein, R. Neilan, J. Slater (1998). The International Glonass Experiment IGEX-98. *http://lareg.ensg.ign.fr/IGEX/goals.html*

Ocean loading tides in GPS and rapid variations of the frame origin

H.-G. Scherneck, J.M. Johansson
Onsala Space Observatory, Chalmers University of Technology, SE-439 92 ONSALA, Sweden

F.H. Webb
Jet Propulsion Laboratory, 4800 Oak Grove Drive, Pasadena, CA 91109, USA

Abstract. Ocean loading tides introduce earth surface displacements at the centimetre level. We show that these effects can be recovered from GPS data using long time-series from Precise Point Positioning solutions. The formal uncertainty of the observed ocean loading parameters can be as low as half a millimetre. Observations from 11 stations are used.

We investigate whether the GPS time-series show any impact of geocentre tides. We find that this is not the case and attribute the negative finding to the way the orbit data are prepared.

Large perturbations exist at sidereal periods. We suppose that they are related to once and twice per revolution orbit errors. Other perturbations might arise due to multipathing near a receiver antenna.

Keywords. GPS, ocean loading tides, geocenter motion.

1 Introduction

This article addresses the problem of small, rapid perturbation of site position in GPS data analysis. We ask the question whether ocean tide loading effects can be detected in GPS data. As ocean tides also cause a translational mode of motion of the solid earth we also try to resolve whether such perturbations can be detected.

As an effect of elastic deformation of the earth, ocean tides cause displacements of the surface, both in the vertical and the horizontal directions. A typical order of magnitude is one centimetre (vertical) or 3 millimetre (horizontal), and coastal extremes are found up to five centimetres (vertical). However, the deformation field does not disappear at the millimetre level, not even in the interior of a large continent like Asia.

If ocean loading effects turn out significant in GPS we can ask further questions as to alternative routes of computation in site motion models, differences of computed loading effects due to using different ocean tide models, local refinements etc.

Apart from the deformation effects there exists a translational mode that involves the planet as a whole. Because of the ocean-land geometry ocean tides do possess a mass dipole moment. In order to preserve the joint mass centre, the non-oceanic parts of earth, in particular crust and mantle, must perform a counter-movement to the motion of the ocean. This effect is expected at the order of one centimetre and has the typical tidal frequency spectrum. Since the geodetic stations are attached to the solid earth, this movement should become visible in range measurements to satellites as the actions of the mass dipoles on the orbit cancel each other. Here we inherently assume that accurate satellite orbit datum relates positions to the joint mass centre.

In reality, the orbit data references the frame origin. Therefore, a clear statement (recommendation) is required to what extent the origin of the frame is to match the position of the physical mass centre of the earth. Alternatively we might allow small rapidly fluctuating offsets around a stable mean position. The crucial point of the matter is only that assumptions and methods applied by users of the system need to be consistent with the orbit data that is provided by international services.

The orbit data will depend on the models that have been applied at the stage of data processing, including eventual systematic perturbations at the tracking stations. If many tracking stations move in a coherent way while the effect is unaccounted for in the orbit data preparation, the orbits will be slightly biased. Users of sta-

International Association of Geodesy Symposia, Vol. 121
Schwarz (ed.), Geodesy Beyond 2000 – The Challenges of the First Decade

tions in the region of the tracking stations will find that their data residuals are lower if they use a similarly incomplete site motion model as compared to a more full-fledged one. This could occur for instance in situations where large air pressure systems load and deform the crust in a wide area.

Obviously, a global, rigid translation like in frame origin tides is an extreme case of coherent site motion. On the other hand, the translational motion in GPS is significant at most in single point positioning measurements; differencing methods as well as network processing will cancel the effect of translations efficiently if not identically. Different techniques might implement frame origin definitions and constraints in different ways to the extent that offsets might occur in measurements involving GPS and another techniques (SLR, c.f. Watkins and Eanes, 1997), and orbit integration techniques might be adhering to the centre of gravity constraint with varying efficiency.

Ocean loading deformations have mainly short spatial correlation, on the order of hundreds of kilometres. Hence, for avoiding systematic biases it appears that models for site displacement should be applied at the stage of orbit computation, and that users are encouraged to apply a compatible model in the analysis of their own stations.

1.1 Ocean tide models

There are a large number of ocean tide models available. For application to loading deformations in global geodetic networks we demand that the models provide geographically homogeneous coverage, especially on the continental shelves, for a representative number of frequency constituents. Under these conditions the number of useful tide models is quickly reduced to just a few. Coverage of long-period tides is provided at present only by (Schwiderski, 1980) and (Seiler, 1991). However, as these effects are small, we will not consider them in this article.

The models that are employed in this study comprise the following partial tide setup: (Schwiderski, 1980) M2 S2 N2 K2 K1 O1 P1 Q1; (Le Provost et al., 1994) M2 S2 N2 K2 2N2 K1 O1 Q1 and several more interpolated by admittance; (Eanes and Bettadpur, 1996) diurnal and semidiurnal orthotides.

We use the 1994 version of (Le Provost et al., 1994) rather than 1995 because of problems in the latter in the areas of the Sunda Islands and Australia.

The (Eanes and Bettadpur, 1996) model (CSR) employs the frequency domain representation by orthotides (degree-two complex polynomials) according to (Groves and Reynolds, 1975). We used two successive generations of CSR, versions 3.0 and 4.0.

The global ocean models differ quite much in their centre-of-mass moments. We show results for one diurnal and one semidiurnal lunar tide in Table (1).

Table 1: Coefficients for geocentre tides, cos and sin factors, respectively for motion along the three Cartesian components Z, X and Y. Unit is millimetre. The data are based on ocean tide models due to L - (Le Provost et al., 1994), C - CSR3.0 (Eanes and Bettadpur, 1996), S - (Schwiderski, 1980). Model CW lists the geocentre results of (Watkins and Eanes, 1997). C4 denotes CSR4.0, the recent update of CSR3.0

Model	Z		X		Y	
	cos	sin	cos	sin	cos	sin
Tide M2						
S	-0.60	-0.84	-4.29	-0.68	2.98	-1.71
L	-1.68	-2.23	-0.49	-0.67	0.04	-0.77
C	-1.44	-1.53	-1.46	1.30	1.34	0.05
CW	-2.72	-1.50	-2.38	0.34	1.67	0.62
C4	-1.43	-1.57	-1.31	0.86	1.05	-0.17
Tide O1						
S	-0.52	2.47	-1.00	-0.03	-1.02	-1.09
L	-1.04	3.87	-0.88	-0.61	-0.93	-0.99
C	-0.40	2.93	-1.11	-0.19	-0.97	-0.85
CW	-0.46	3.27	-2.85	-0.64	-1.65	0.45
C4	-0.25	2.8	-1.18	-0.21	-0.92	-0.77

1.2 Ocean loading parameters

Deformation of the crust is concentrated around the point where the earth is loaded. The process of elastic deformation under loading is largely linear, thus the field of loading deformations can be conceived as an integral over all simultaneous loads producing a field of loading displacements. Also, the methods of harmonic decomposition in both space (spherical harmonics) and time (tidal frequency spectrum) apply.

The basic method to compute ocean loading parameters from the global ocean tide models and their use in space geodesy is described in (Scherneck, 1991) and (McCarthy, 1996). In short, the loading effects can be computed using the point load integration method (Farrell, 1972) and using a load representation on a regular grid. The ocean tide is approximated by a sum over a number of frequency constituents (partial waves).

The displacement field is particularly complicated at the coasts since there the distribution of tidal pressure is intercepted by an oftentimes complicated coastline. Also, tide loads tend to be large on the continental shelves. The point-load integration method can quite easily be adapted to complicated coastlines, as opposed to methods that work with fixed grid intervals like sphercial harmonic development and resynthesis (Ray and Sanches, 1989).

Accidentally, many geodetic stations of international interest are near the coast. For this reason, the standard method to parameterize the ocean loading tides is on a per-site basis using a catalogue of coefficients (McCarthy, 1996). The parameters represent radial, west and south displacement amplitudes and phase at a station. The phase coefficient states the lag of the local loading tide with respect to the solid earth tide at the zero meridian. The parameters are accurate at the millimetre level only within a narrow range around the station, on the order of 25 km for a conservative limit.

The ocean loading parameters that are derived with the original definitions of the load Love numbers of (Farrell, 1972) carry only the deformation mode. The transformation for the conservation of the mass centre of the entire earth consists simply of adding the corresponding degree-one vector spherical harmonic representation of the inverse of the oceanic mass dipoles to the ocean loading parameters at each station. We will use the abbreviation CMC (Centre of Mass Correction) to signify this process and the associated tide loading parameters. In the case of the CSR model version 3.0 we can alternatively add a degree-one tide elevation field in order to match the results of (Watkins and Eanes, 1997); this model version will be called CSR3.1_CMC.

GIPSY/OASIS (Webb and Zumberge, 1993) and other GPS processing software, e.g., BERNESE, see (Rothacher, 1998) is capable of computing predicted ocean loading tide displacements.

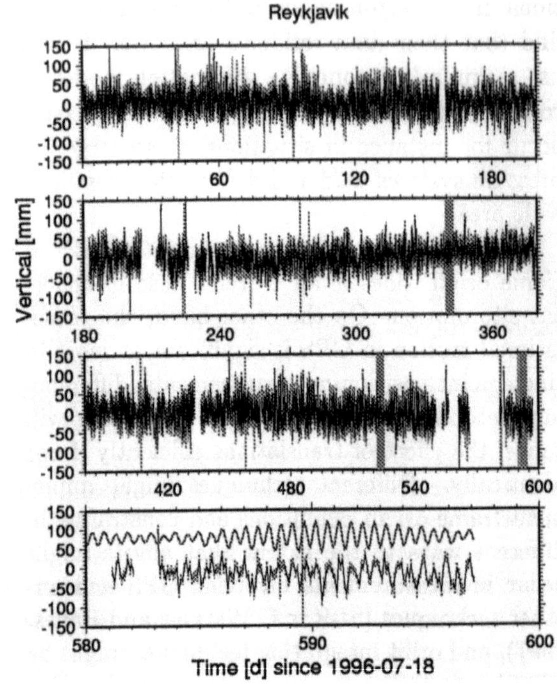

Fig. 1: Time series at 2 hour sampling rate, vertical component, from GPS Precise Point Positioning at Reykjavik. Gray background shows one-sigma formal error. The last frame zooms in on the last 17 days and also shows the predicted loading tides

2 Precise Point Positioning solutions

The GIPSY/OASIS analysis program for GPS is able to determine non-fiducial solutions for a number of parameters, among them geodetic position, using only observations at a single site. This method, termed Precise Point Positioning (PPP, see Zumberge et al., 1997), requires that accurate corrections for the satellite clocks are provided together with the satellite positions. To obtain a time series of positions with GIPSY/OASIS, the Kalman filter state vector is read out at regular intervals (e.g., one or two hours). A seven parameter Helmert transform is determined once for the whole time-series and applied to obtain geodetic positions within a terrestrial reference frame. The JPL orbit data that is used in our analysis was generated with the ocean loading parameters as contained in the IERS Standards 1992 (McCarthy, 1992).

Among other state vector elements worth mentioning is the neutral, wet atmosphere. The wet delay is set up as a random walk with a process noise parameter of 1.7×10^{-4} m/$\sqrt{s}$ and state vector updates at five minute intervals. A 7 deg elevation mask was applied. The

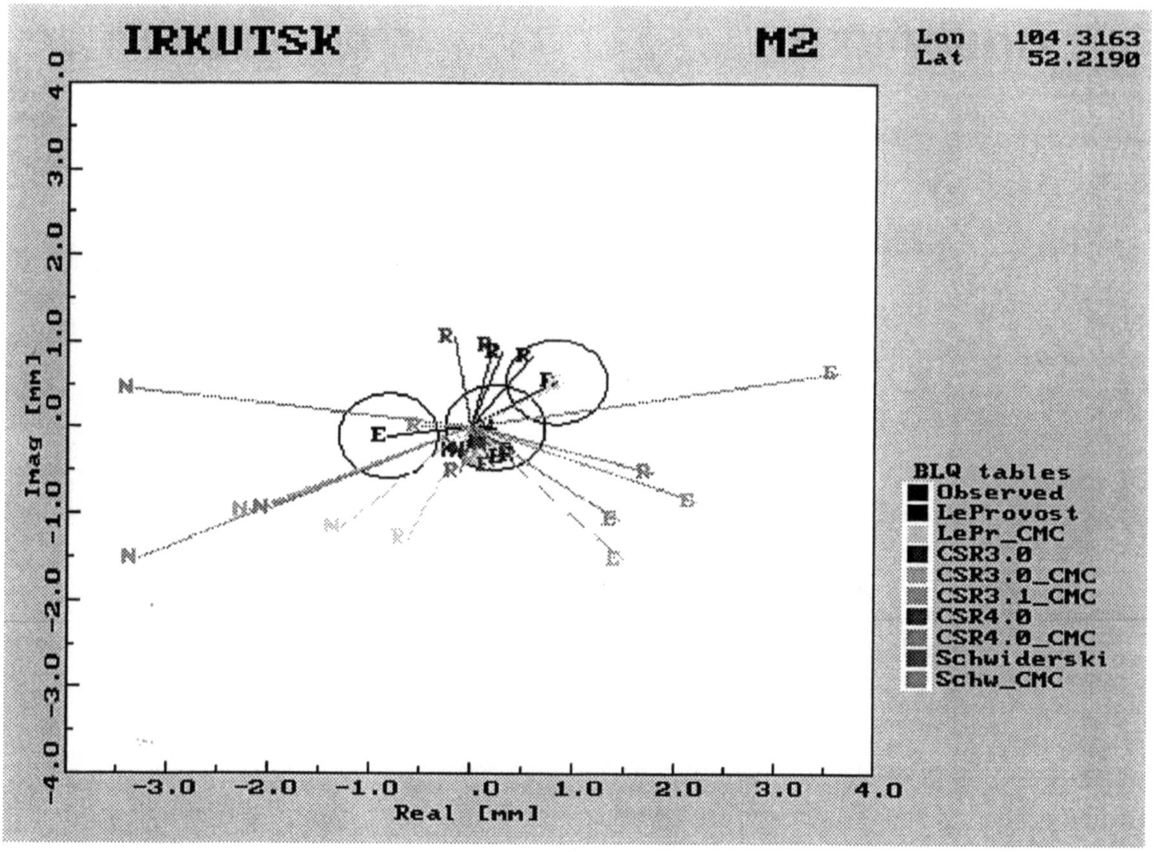

Fig. 2: Ocean loading phasors of the lunar tide M2 at Irkutsk, Siberia, Russia. Observations are given with a 95% confidence circle. Units are millimetres. All three spatial components are shown, signified by a character R - radial (vertical), E - east and N - north. We compare a number of ocean models. The Centre-of-Mass-Corrected parameters (CMC) are represented with light grays; they all appear to be significantly contradicted by the observations

site motion model includes the solid earth tides but not ocean loading tides.

Figure (1) shows as an example the data from Reykjavik, Iceland. Data from the following IGS stations have been used for this study: Reykjavik, Iceland (REYK), Mauna Kea, Hawaii (MKEA); Irkutsk, Russia (IRKT); Mendeleevo, Russia (MDVO); Hartebeesthook, South Africa (HARK); Yaragadee, Australia (YAR1); Pie Town, New Mexico (PIE1); Ascension Island (ASC1); Westford, Massachusetts (WEST). We also included two regional stations in Sweden, Onsala (AONS) and Sundsvall (ASUN).

3 Tide analysis

In order to detect tidal effects in the data we fit predicted tides to the observations using least squares. The predicted tides consist of in-phase and quadrature-phase signals computed from a theoretical, harmonic development of the tide potential (HDTP). We estimate amplitude coefficients representative for narrow ranges of frequencies (wave-groups) in the two fundamental bands, diurnal and semidiurnal. The spacing of the wave-groups is designed to resolve two beat periods per year in the data sets, which all have at least 6 months duration. Thus we can discern the P1 from the K1 tide. In the cases when the duration of the data series is longer than one year an oscillation of solar day period can be discriminated between the K1 and P1 tides. Effects with this period could arise mostly due to solar-synchronous environmental disturbances at the station.

The low signal to noise ratio suggests that a rather short HDTP like (Tamura, 1987) will suffice with a wide margin. For the same reason

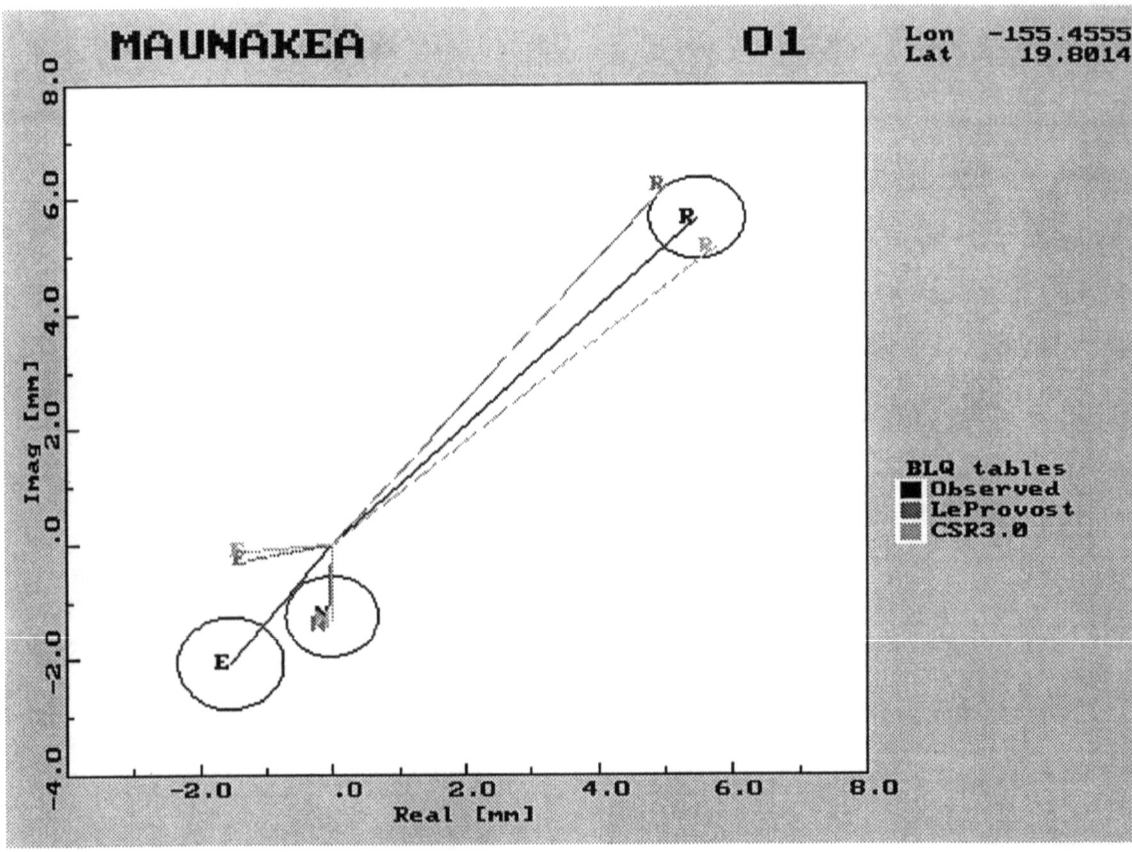

Fig. 3: Ocean loading phasors of the lunar tide O1 at Mauna Kea, Hawaii. Comparison with the loading models based (Le Provost et al., 1994) (LePr) and (Eanes and Bettadpur, 1996) (CSR3.0)

we have restricted analysis to the amplitude-wise most important wave groups. This means that the wave-groups that pertain to the edges of the fundamental bands have been widened, i.e. Q1, J1, 2N2 and K2 also comprise all partial tides up respectively down to the edge.

Eventually we fit predicted air pressure loading effects simultaneously in the least-squares solution. In some cases the post-fit χ^2 is slightly reduced. We do not explore this matter further in this article.

3.1 Error sources

What kinds of discrepancies between models and GPS observations might be expected ? Due to the long duration and tight sampling of the data, the major perturbations that come of interest are cotidal (oscillatory with frequencies coinciding or near tidal frequencies). Candidates are

- **Solar cycles in the environment,** e.g. thermal expansion of pillars. The stationarity of such processes can be quite low so that spill-over into neighbouring frequency bands might be considerable.

- **Signal scattering and multipath** on obstacles near the antenna. If anomalous signal scattering occurs, the nearly-repetitive satellite geometry from day to day tends to introduce spikes in the position estimates that repeat from day to day during certain sections of the data. The periodicity is almost sidereal, but the spectral quality of the process is very low, and large spill-over in the diurnal band must be expected.

- **Satellite orbit errors.** If the orbits contain once- and twice-per-revolution errors etc., the range is perturbed with nearly sidereal period and the associated upper harmonics.

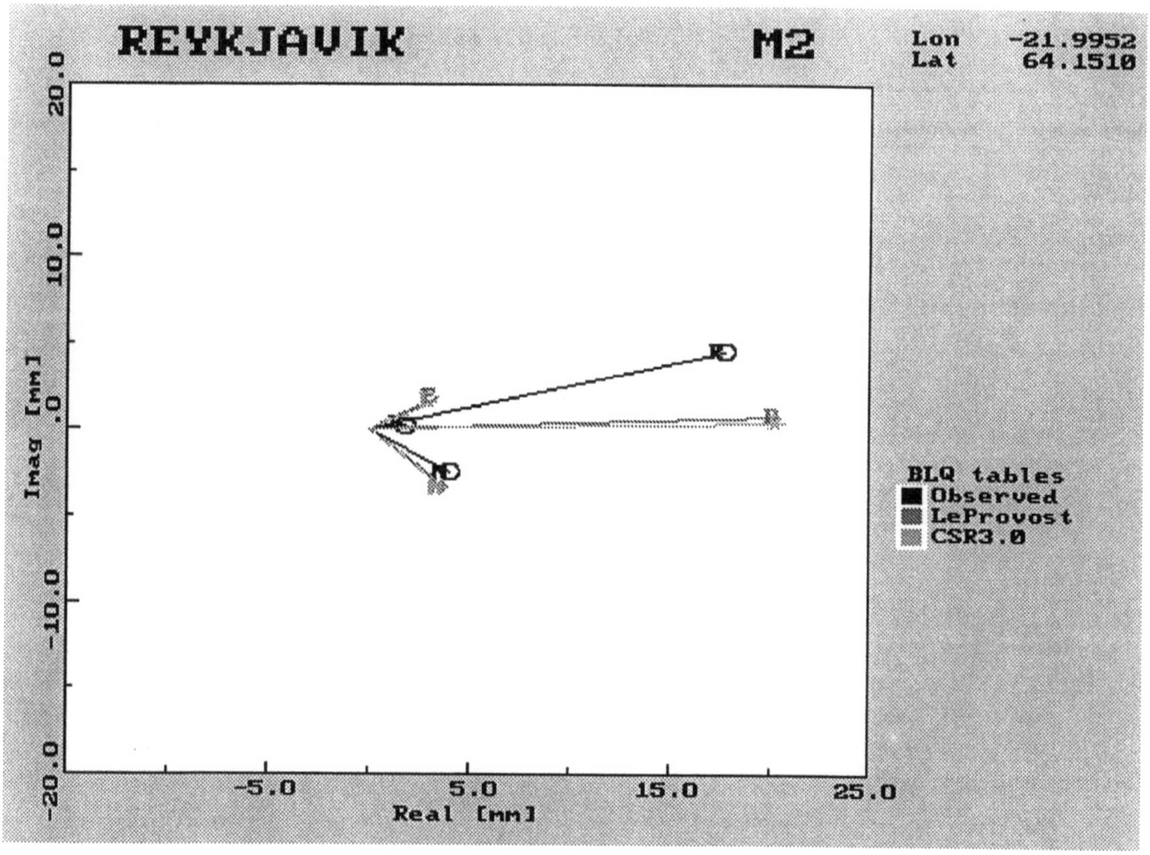

Fig. 4: Ocean loading phasors of the lunar tide M2 at Reykjavik, Iceland. Otherwise, c.f. Fig. (3)

- **Loading effects** in ocean basins might not be accurate. A clear example is the Gulf of Maine, which is not represented in some of the global models, or not resolved well enough in others. The loading effect on a station like Westford, on the order of magnitude 100 km away from the basin, might be significantly underestimated.

The complications introduced by the satellite motion suggest that one should focus on detecting tides of lunar origin as the associated periods are sufficiently separated from the sidereal periods (we hope).

4 Results from tidal analysis

The response coefficients from the least-squares analysis are converted into complex coefficients for the amplitude and the phase of the displacements using the same conventions as in the ocean loading coefficients. We compare the results with model predicitions in a few illustrative cases. The phasor plots display the complex-valued ocean loading coefficients of a specific partial tide; values on the positive real axis represent an exact in-phase relation with the corresponding solid earth tide, negative imaginary values indicate a phase lag.

Figure (2) shows the situation for Irkutsk. Here, in the middle of Siberia, ocean loading deformation is predicted to be very small. The observed parameters are within 1.5 mm amplitude. In the vertical component the 95 percent confidence limit in the figure should be increased by a factor of two as the post-fit, normalized χ^2 was found near four; the horizontal parameters were solved with a normalized χ^2 near unity. In place like Irkutsk it is suitable to test whether frame origin tides are present in the GPS ranges and whence in the geodetic positions. We find that the sets of ocean loading coefficients that are corrected for the center-of-mass conservation (CMC) all show significant discrepancies with re-

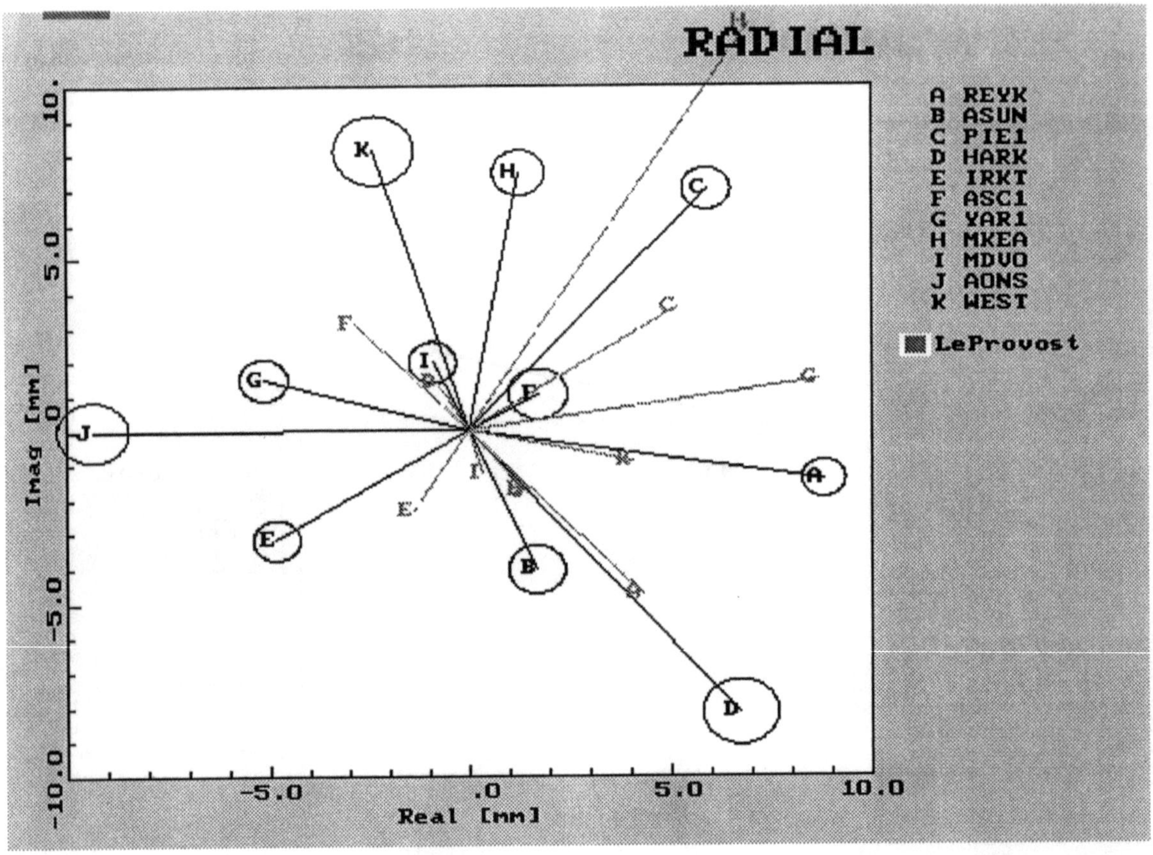

Fig. 5: Ocean loading phasors of the sidereal tide K1 at all sites included in this study. Comparison with the loading model based on (Le Provost et al., 1994)

spect to the observations. The CMC-coefficients also show the large internal owing to the mass moments being largely different as shown in Table (1). Very similar results are found the another station in the interior of the Eurasian land mass, Mendeleevo.

We show two other cases where ocean loading is large, and where the motion can be recovered sucessfully. We demonstrate this in Fig. (3) for the O1 loading tide at Mauna Kea, Hawaii, and in Fig. (4) for the M2 loading tide at Reykjavik, Iceland. In each figure we restrict comparison with predictions from the models of (Le Provost et al., 1994) and (Eanes and Bettadpur, 1996), CMC not applied, for better legibility.

As a last example with phasor plots we show the situation for the K1 vertical loading tide (period is one sidereal day) at all stations of this study and compare with predictions based on the model of (Le Provost et al., 1994), cf. Fig. (5). We find that there are large discrepancies at all stations.

This finding suggests that perturbations of once per orbital revolution cause slightly larger offsets, order of 10 mm, in site position estimates than the magnitude of ocean loading effects. Similar results are found at the K2 tide (two cycles per sidereal day).

To summarize our results we compare observations with predictions in three variants. For each station we compute

$$\chi^2(\mathcal{S},\mathcal{M}) = \sum_{i\in\mathcal{S},j=1,3} \frac{\left|\zeta_{ij}^{(o)} - \zeta_{ij}^{(\mathcal{M})}\right|^2}{\sigma_{ij}^{(o)2}} \qquad (1)$$

where superscript (o) denotes observations, $\mathcal{S}$ a set of tide wave-groups, and $(\mathcal{M})$ a loading model. To the four models mentioned in section 1.1 we add as a null case, illustrative of the signal content in the measurements, by setting $\zeta_i^{(\mathcal{M})} \equiv 0$. Note that j runs over the three spa-

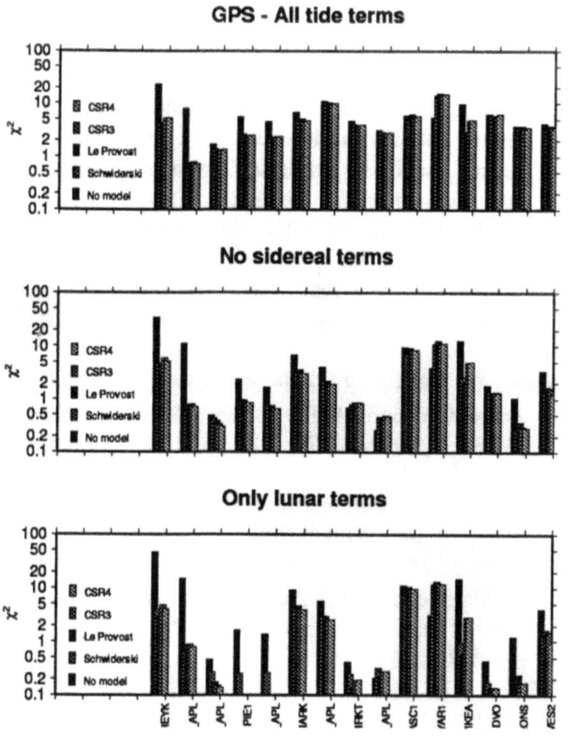

Fig. 6: Comparison of observed tide parameters and ocean loading models. Notice the reduction of a χ^2 measure (see Eq. 1) due to the ocean loading models as compared to a null model. Station codes are shown along the abscissa, subscript _APL shows whether air pressure loading was included in the data analysis

tial components. First we include in the set S the tide wave groups (M2 S2 N2 K2 K1 O1 P1 Q1), then we remove the sidereal waves K1 and K2, and finally we include only the lunar species (M2, N2, O1, Q1). The logarithm of the χ^2 is shown for the 3×5 cases in Figure (6).

We note the general confirmation of lunar loading tides being observed in all three spatial components whereas systematic errors at sidereal periods are overwhelming. Only the results for Yaragadee, Australia, (YAR1) contradict the proposition of tidal loading according to the model (Scherneck, 1991).

5 Conclusions

We are able to show that site displacement due to ocean loading can be recovered from GPS data. However, orbit perturbations appear to shroud these effects at sidereal periods. We also found sites where the level of perturbations is high probably due to signal scattering. We do not find frame origin tides which are expected to be excited by the global ocean tide mass moments. We believe that this negative finding is attributed to the fact that frame origin tides are neglected (or suppressed) at the stage of orbit computation.

It appears clear that more complete models for site motion should be implemented consistently at the stage of orbit computation by international services and at the stage of geodetic parameter estimation by a user. Particularly in the cases where perturbations are not strongly correlated over distance we can expect sources of systematic errors to be removed through the modelling efforts. This would both mean less orbit bias and less errors in geodetic parameters. Thus one could conceive of an orbit product in addition to standard IGS and rapid orbits, namely high-accuracy (perhaps slowly disseminated) orbits the generation of which could involve extensive tide modelling, pressure loading effects, and other processes of (rapid or slow) crustal deformation.

In principle one could formally recommend that also frame origin tides should be treated in this way. However, the following considerations argue against this scheme:

- The centre of mass terms of different ocean tide models show comparatively large differences.
- The motion is a rigid translation, therefore strictly coherent across the whole planet.
- GPS orbit adjustment seems to be relatively insensitive to a physical constraint of preservation of the joint mass centre location of the earth.

Thus, geocentre tides can be avoided in the data processing else—in the best case—a user would exactly remove the very frame translation that an orbit data provider has introduced. (In the worst case, any additional model could cause confusion, problems with implementation, in short: errors.) Since all satellites are affected in an equivalent way, the effect of the translation would anyway be eliminated in double-difference operations or in network processing.

The recommendation to ignore the frame origin tides is along the recommendations of the IERS Analysis Campaign to Investigate Motions of the Geocenter (Ray, 1999), (Scherneck, 1999).

Future users are cautioned, however, in situations beyond the standard GPS techniques, e.g.

using alternative routes to integrate GPS orbits, or comparing laser ranging and GPS results.

Acknowledgements. Figures in this paper were made with GMT software (Wessel, and Smith, 1995). This project is supported by grants from the Natural Science Research Council of Sweden under account G-AA/GU 03590.

References

Eanes, R. and S. Bettadpur (1996). *The CSR 3.0 global ocean tide model,* Center for Space Research, Technical Memorandum, CSR-TM-96-05.

Farrell, W. E. (1972). Deformation of the Earth by Surface Loads, *Rev. Geophys. Space Phys.*, 10, pp. 761–797.

Groves, G.W. and R.W. Reynolds (1975). An orthogonalized convolution method of tide prediction. *J. Geophys. Res.*, 80, pp. 4131–4138.

Le Provost, C., M. L. Genco, F. Lyard, P. Vincent, and P. Canceil (1994). Spectroscopy of the world ocean tides from a finite element hydrodynamical model, *J. Geophys. Res.*, 99, pp. 24,777–24,798.

McCarthy, D.D. (1992). *IERS Standards* IERS Technical Note 13, Observatoire de Paris, France.

McCarthy, D.D. (1996). *IERS Conventions* IERS Technical Note 21, Observatoire de Paris, France.

Ray, J.R. (1999). Foreword. In: *IERS Analysis Campaign to Investigate Motions of the Geocentre.* Ray, J.R. (ed.), IERS Technical Note 25, Observatoire de Paris, France, pp. 1–2.

Ray, R.D. and B.V. Sanchez (1989). *Radial Deformation of the Earth by Oceanic Tidal Loading,* NASA Technical Memorandum 100743, NASA, Greenbelt, MD, USA.

Rothacher, M. (1998). *Recent Contributions of GPS to Earth Rotation and Reference Frames,* Habilitation (2nd Ph.D.), University Press, University of Berne, Switzerland.

Scherneck, H.-G. (1991). A Parameterized Solid Earth Tide Model and Ocean Tide Loading Effects for Global Geodetic Baseline Measurements, *Geophys. J. Int.*, 106, pp 677–694.

Scherneck, H.-G., and F.H. Webb (1999). Ocean tide loading and diurnal tidal motion of the solid Earth centre. In: *IERS Analysis Campaign to Investigate Motions of the Geocentre.* Ray, J.R. (ed.), IERS Technical Note 25, Observatoire de Paris, France, pp. 83–89.

Schwiderski, E. W. (1980). On charting global ocean tides, *Rev. Geophys. Space Phys.*, 18, pp. 243–268.

Seiler, U. (1991). Periodic changes of the angular momentum budget due to the tides of the world ocean, *J. Geophys. Res.*, 96, pp. 10,287-10,300.

Tamura, Y. (1987). A harmonic development of the tide-generating potential, *Bull. d'Inform. Marées Terr.*, 99, pp. 6813–6855.

Watkins, M.M. and R.J. Eanes (1997). Observations of tidally coherent diurnal and semidiurnal variations in the geocenter, *Geophys. Res. Letters*, 24, pp. 2231–2234.

Webb, F.H. and J.F. (1993). An Introduction to GIPSY/OASIS-II Precision Software for the Analysis of Data from the Global Positioning System, *JPL Publ. No. D-11088,* Jet Propulsion Laboratory, Pasadena, Cal.

Wessel, P., and W.H.F. Smith (1995). New version of the Generic Mapping Tools released, *EOS Trans. AGU,* 76, p.329.

Zumberge, J.F., M.B. Heflin, D.C. Jefferson, M.M. Watkins, F.H. Webb (1997). Precise point positioning for the efficient and robust analysis of GPS data from large networks. *J. Geophys. Res.*, 102, pp. 5005–5017

Use of GPS carrier phase for high precision frequency (time) comparison

Z. Jiang, G. Petit
Bureau International des Poids et Mesures, Pavillon de Breteuil, F92312, Sèvres Cedex, France

P. Uhrich, F. Taris
Laboratoire Primaire du Temps et des Fréquences, F75014 Paris, France

Abstract. Quality in time metrology depends on the performance of the atomic clocks and the means for time-frequency comparison. New generation of caesium standards have accuracy below 2×10^{-15} and the classical common-view method is no longer satisfactory. Use of GPS carrier phase combined with code measurements is a promising technique for the accurate comparison of remote clocks. This paper discusses some error sources disturbing the uncertainty of the comparison using the code-phase method, mainly on the equipment capability and the data processing strategy. In fact, the receiver system, comprised with the main unit, the antenna and cables, is very sensitive to the environmental temperature and this may seriously influence the comparison result. Due to problems with the raw data or some defeats in the data processing, there often exist phase discontinuities in comparison results, ranging from several hundred ps up to several ns, which may be mistaken as drifts produced by the compared clocks and increase the uncertainty of the carrier-phase method so as to limit its application for highly accurate frequency transfer. A number of equipment setting up conditions and a multi-day data processing strategy are proposed in which the temperature effect and the discontinuities are significantly reduced and the frequency stability is improved. An uncertainty of 1 $\times10^{-15}$ for an averaging time of 1 day is expected.

Keywords. Atomic clock, time and frequency comparison, GPS, carrier phase, frequency stability.

1 Introduction

One of the principal disciplines in time metrology is time-frequency transfer between remote clocks, that is, clock comparison. Since the appearance of GPS, it becomes the dominating technique for clock comparison (Allan and Weiss 1980). The principal idea is quite similar as the geodetic baseline measurement: two distant GPS receivers, between e.g. Paris and Birmingham driven respectively by two clocks (T_{Paris} and T_{Bham}) to be compared, observe a satellite in *common view* (Fig. 1). Using a simple difference of an observation, the pseudo-range (*PsR*) for example, we can establish the following equation between the two clocks via the GPS time (T_{GPS}):

$$\begin{aligned} PsR\ Paris/c &= T_{Paris} - T_{GPS} + d_P/c \\ PsR\ Bham/c &= T_{Bham} - T_{GPS} + d_B/c \\ \hline PsR\ (Paris\text{-}Bham)/c &= T_{Paris} - T_{Bham} + (d_P - d_B)/c \end{aligned}$$

here d_P and d_B are distances between the satellite and the antennas and c the light speed. Obviously, the GPS time and hence its errors are cancelled this way. This is the famous common view method, actually using only the GPS CA code as its observable and its uncertainty is limited to about 1 ~ 2×10^{-14} for an averaging time of 1 day.

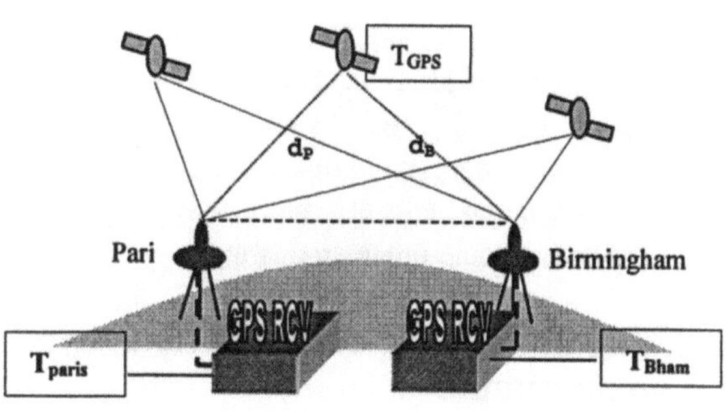

Fig. 1: Time comparison via a satellite in common view

Recently developed caesium standard (fountain) reach the frequency stability of $6\times10^{-14}\times\tau^{-1/2}$ and the accuracy of 1.4×10^{-15}. The classical common-view method is unfortunately no longer satisfactory for such a high accuracy. On the other hand the GPS phase measurements have been widely used in the

International Association of Geodesy Symposia, Vol. 121
Schwarz (ed.), Geodesy Beyond 2000 – The Challenges of the First Decade

geodetic field. Dunn (1993) and Baeriwyl et al. (1995) first proposed to use the GPS carrier-phase measurement collected from geodetic receivers as an additional observation to the code for time metrology application at continental scale. No doubt, time metrology has its own characteristics, the use of GPS carrier phase implies a series of new subjects to study, for example, the instrument electronic delay including its variation and real-time continuity in the comparison result. Some special requirements other than the geodetic applications are then needed for data collection and data treatment.

In December 1997, a joint IGS (International GPS Service) and BIPM (International Bureau of Weight and Measurement) Pilot Project was authorised. Its central goal is to investigate and develop operational strategies to exploit GPS phase and code measurements for improved availability of accurate time and frequency comparisons world-wide, especially for maintaining the international UTC time-scale as a new generation of frequency standards emerges with accuracy of 1×10^{-15} or even better.

The IGS contributions are:

- global tracking network with high quality data,
- efficient data delivery system and,
- state-of-art data analysis groups, methods and products.

The BIPM contributions are:

- high accuracy metrology standards and measurements,
- time scale and calibration methods and,
- formation and dissemination of UTC.

Many authors are being involved in this project (e.g. Bruyninx et al. 1999, Taris et al. 1999, Dudle et al. 1999; Petit et al. 1998, Larson 1998, Petit et al. 1999, Jiang and Petit 1999 etc.). This paper presents some works carried out at the BIPM. Main attention is focused on the limit of the equipment precision and the data processing strategy.

2 Equipment precision and sensitivity to temperature

Unlike most geodetic positioning, clock comparison is characterised by continuity with time. Equipment, consisted of the main unit, the antenna and cables etc., caused electronic delay and its unexpected variation should be carefully taken into account. Among the disturbing factors, the dominating one is the temperature. Petit et al. (1998) reported a coefficient of 1 ps/m/C° for the usual cables. This is not negligible in accurate comparison.

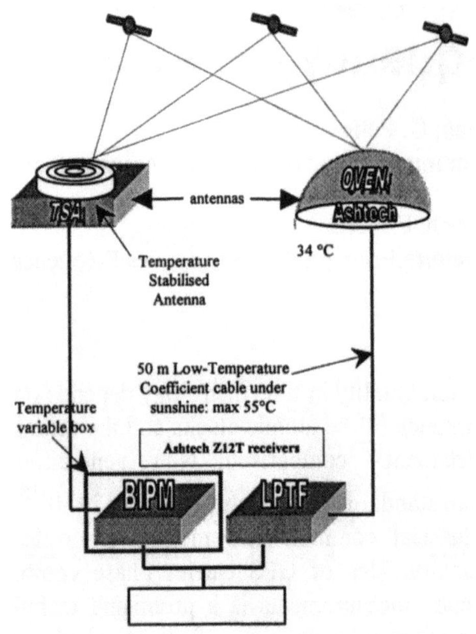

Fig. 2: A short baseline setting up

In order to estimate the temperature effect, we used two Ashtech Z12T geodetic-like receivers (modified Z12 for time comparison, Petit et al. 1998) and zero and short (several meters) baselines to investigate equipment behaviours. On a zero baseline, there is only the two main units that are independent each other and all the other conditions (antennas, cables and an external clock) are identical so that the single difference of an observable is nothing but the white noise. This setting up allows us to study main unit's precision limit. On a short baseline (cf. Fig. 2), several configurations are possible by using different antennas (Ashtech Standard antennas, a temperature controlled oven protected Ashtech antenna and a 3S Navigation Temperature Stabilised Antenna, TSA-100), different cables (Ashtech standard cable and Low-Temperature Coefficient LTC cable) and the main units put in a temperature variable box. In this way, we observed the behaviours of different parts of the equipment under constant-variable temperature fields. The two Z12T belong respectively to the BIPM and the LPTF (Lab. Primaire du Temps et des Fréquences, France).

The zero baseline tests result in that the main unit noise is about 3 ps (1 ps = 0.001 ns = 1×10^{-12} second) for the carrier phase observations. This is perhaps the limit of this observable. Fig. 3 ~ 5 show the correlation between different observable and the

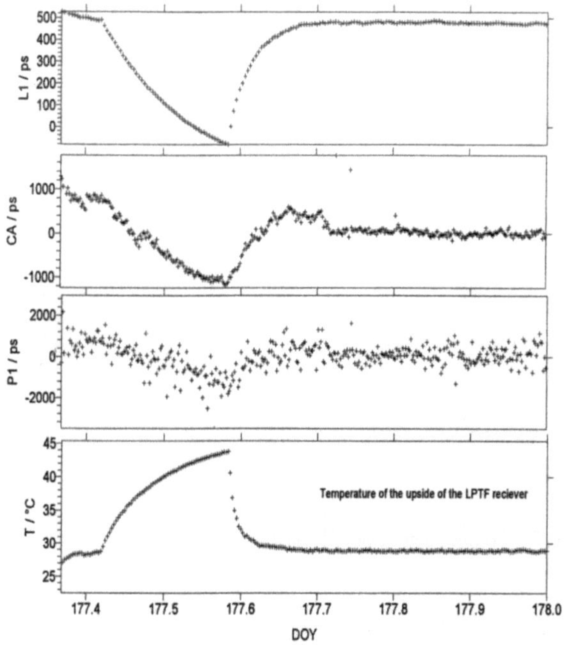

Fig. 3: Temperature effect on main unit

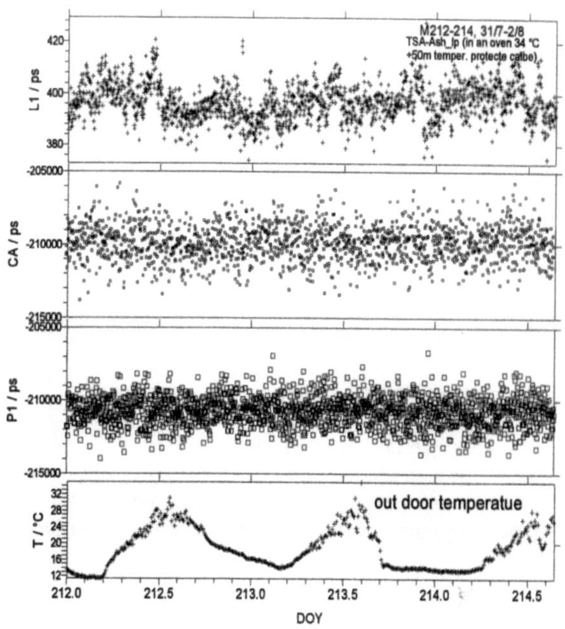

Fig. 4: Temperature effect on the LTC cables

temperature variation via different parts of the equipment system. These figures are only part of many other tests and similar results were obtained. Temperature gives a strong influence on the main units (Fig. 3). The averaged coefficients for the phase, CA and P codes (L1) are respectively: 20 ~ 40 ps/C°, 120 ~ 140 ps/C° and 130 ~ 160 ps/C°; An earlier study (Petit et al. 1998) showed an important coefficient on the Ashtech standard cable: 1 ps/m/C° for phase measurements. If there are 20 m cable exposed to out-door temperature variation (10 ~ 20 C° daily), this effect is considerable. However, if the LTC cable is used, as shown in Fig. 4, this effect, against 20 °C temperature changes, becomes negligible for all the three observable. The results shown in Fig. 5 are a little surprising. With an extreme temperature change up to ± 35 °C within some minutes, the standard Ashtech antenna against TSA has only a slight effect on the phase, some ±10 ps and no visible effect on the codes compared with their noise levels. This may imply that compared with the Ashtech antenna the expensive and complex TSA or the temperature- controlled oven might not be absolutely necessary.

Under idea conditions, that is, the main units located in air conditioned laboratories with the LTC cables (TSA is a plus), the frequency stability of the phase method, given by modified Allan Deviation, goes down to several part in 10^{-17} for an averaging time of 18 hours, see Fig. 6.

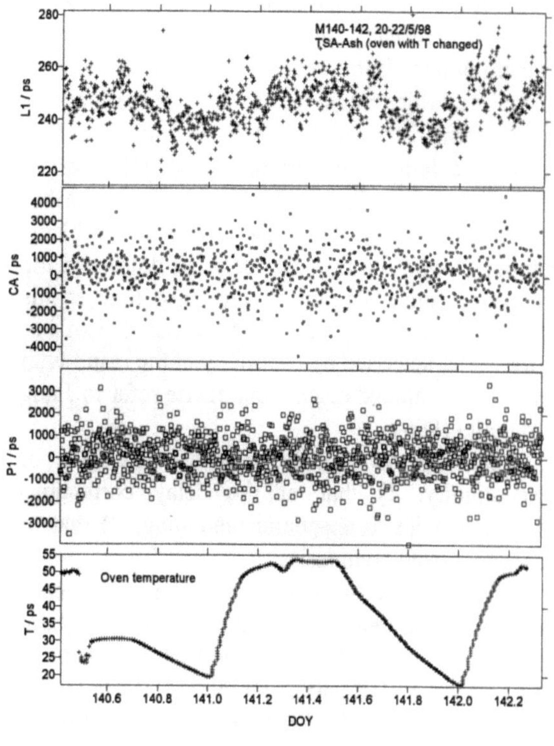

Fig. 5: Temperature effect on Ashtech standard antenna

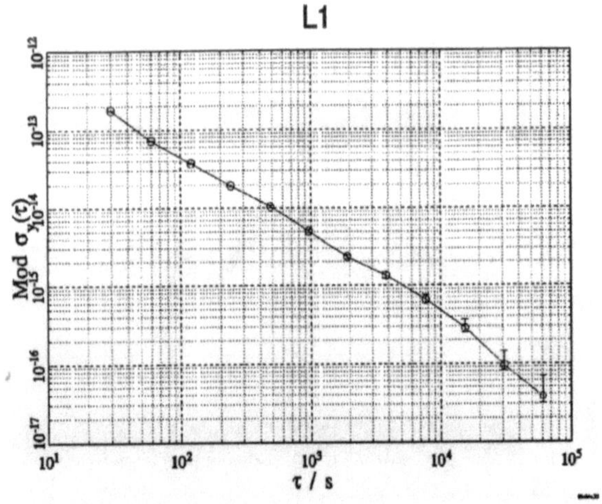

Fig. 6: Frequency stability of the phase method under idea condition

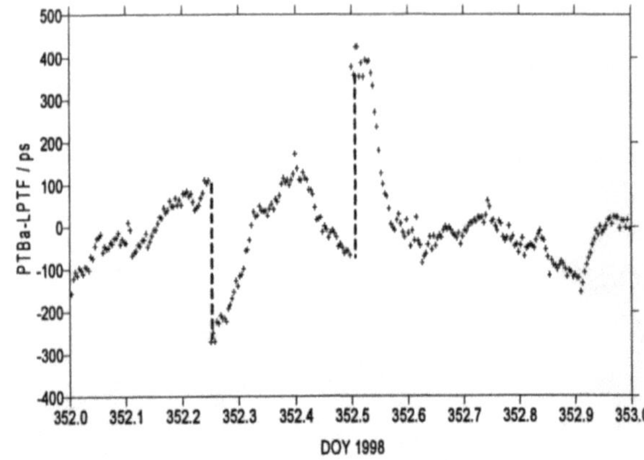

Fig. 7: Comparison of two H-masers by standard processing

3 Discontinuities in phase comparison and data processing strategy

Although a frequency stability of 1×10^{-15} over an averaging time of 1 day may be expected for distant primary frequency standard comparison (Petit and Thomas 1996), the results of such comparisons (Petit et al. 1999; Douglas and Popelar 1994), however, reveal the presence of phase discontinuities from several hundred ps up to several ns, which might be mistaken as the long- or middle-term drifts of the clocks in question. This becomes a limit to the application of GPS carrier phase for time-frequency comparison. Further investigation found that this is due to problems associated with the raw data or some defeats in the data processing. We here give a quick look at this problem, keeping in mind that the purpose of our work is characterised by continuity of the result with time. For a detailed discussion, please refer to Jiang and Petit (1999).

Generally we can distinguish two kinds of discontinuity: 1) that of the delay corrections associated with tropospheric modelling; 2) that of data processing boundaries.

Fig. 7 shows an one-day result of a frequency comparison between two distant H-masers (692km) located at the PTB (Physikalisch Technische Bundesanstalt, Germany), and the LPTF. Up to 300 ps jumps appear (a linear drift was removed). Without proving, we point out that these jumps are introduced by the 6 hours interval tropospheric zenith delay corrections as done by usual geodetic processing (Rothacher and Mervart 1996).

On the other side, we recognise that discontinuities must happen on the boundaries of data storing and data processing. As an example, the IGS RINEX observation and orbit information, etc. are daily stored. So using these data directly, a day-separated processing may produce a day-boundary discontinuity. An example of a 'hidden' boundary is a slip which may produce, automatically by the software used, a resetting of the ambiguity for the satellite in question. A problem caused by a satellite in a bad status or a receiver failure might also produce a jump in the data processing boundaries. Most jumps appear on the setting boundaries of the unknown parameters, such as the tropo-corrections, the ambiguities, the phase-code smoothing segments etc. Our experiences prove that the discontinuities can be significantly reduced by using a careful processing procedure. Let us discuss this with the help of the following example, an European-transatlantic frequency comparison campaign carried out using a three day data set collected from 6 time laboratories with the lengths of the baselines vary from 180 m up to 6275 km (PTB-USNO). Ashtech Z12T GPS receivers were installed at the LPTF, the LHA (Lab. de l'Horologe Atomique, France), the PTBa, the NPLb (National Physical Laboratory, UK), and the USNb (US Naval Obs.). Rogue SNR-12 receivers were used at Brus (Royal Obs. of Belgium, Brussels) and the USNo. All the receivers were driven respectively by a Hydrogen maser. The Bernese GPS software (Rothacher and Mervart 1996) was used for data treatment.

Firstly, we use a usual geodetic-like processing for the 3-day frequency comparison and found jumps of several hundred ps to several ns on the poorer frequency stabilises for each pair of compared clocks. Then we used the following continuous 3-day procedure:

a). Pre-processing:

- generate an 3-day RINEX observation file
- generate an 3-day smoothed observation file by the phase-code smoothing
- generate an 3-day orbit file

b). Pre-adjustment:

- first adjustment using the standard processing with 6-hour intervals
- smooth and interpolate the pre-tropo corrections into the one-hour intervals
- second adjustment using the above tropo-corrections as the *a priori* observation with σ=1 cm
- analyse the results and reject the 'bad' observations

c). Final adjustment.

When necessary, reset the unknown *a priori* parameters and redo an adjustment. Sometimes it is necessary to redo the code-phase smoothing.

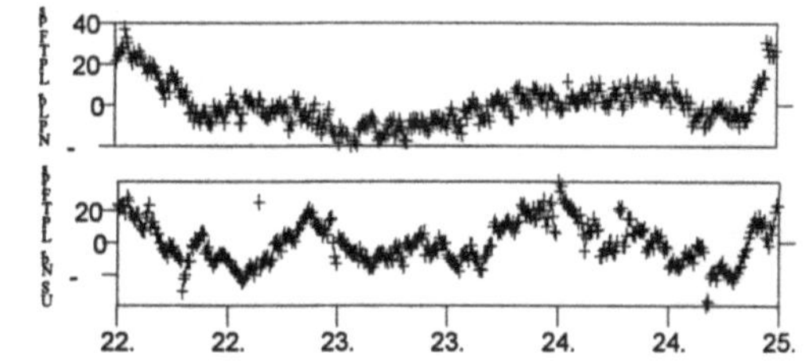

Fig. 8: Clock comparison results by the 3-day processing

The discontinuities were considerable reduced and the uncertainty were therefore improved. Figs. 8 and 9 give the comparison results and the corresponding frequency stability between NPL-LPTF (345km) and the transatlantic USNb-LPTF (5943km). The obtained frequency stability for the two baseline is of 1.4 ~ 1.6×10^{-15} for an averaging time of 18 hours. This is of the order that can be expected from the comparison of the two H-masers. In fact, many other numerical tests have been carried out and in most cases the above processing gives an improved comparison result.

4 Conclusion

The new generation of frequency standards requires a comparison accuracy of 1x10^{-15} for an averaging time of 1 day. GPS carrier phase has proven its high accuracy potential. However, special attentions should be paid at the data collecting and data treatment strategies, especially to reduce the temperature effect on the equipment and the data processing boundary discontinuities in the comparison results. They may decrease significantly the frequency stability of the comparison result and limit the application of the carrier phase method.

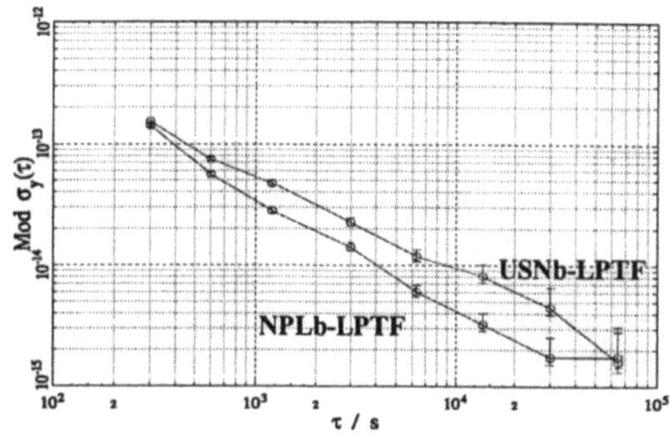

Fig. 9: Modified Allan deviations

The precision limit of the Ashtech Z12T GPS receiver is about 3 ps for the L1/L2 frequency phase measurements. Under idea conditions, that is, air conditioned laboratory to put the main unit with LTC cables, a frequency stability of several parts in 10^{-17} was reached on short baseline. The European-transatlantic test shows that a frequency stability of 1.5 ~ 2 x10^{-15} over an averaging time of 18 hours is possible for H-maser comparison over long distance up 6000 km. An uncertainty of 1x10^{-15} for an averaging time of one day is expected to be operational for the comparison of primary frequency standards with each other or with TAI (International Atomic Time) at continental scale.

Further studies on the time-frequency transfer such as absolute calibration of the GPS geodetic receivers, comparison between the geodetic receivers and the time transfer receivers, comparison with other independent methods like Glonass P-code or phase and two-way time-frequency transfer etc. are necessary and being addressed at the BIPM and other laboratories.

Acknowledgements. The authors are grateful to the colleagues from the NPL, PTB, AIUB, USNO and IGS for making available their high-quality data and for constructive discussions with many of them.

References

Allan D., Weiss M. (1980). Accurate time and frequency transfer during common-view of a GPS satellite. In: *Proc. IEEE Freq. Contr. Symp.*, Philadelphia, USA, pp 334-356.

Baeriswyl P., Schildknecht T., Utzinger J., Beulter G. (1995). Frequency and time transfer with geodetic GPS receivers: first result. In: *Proc. 9th EFTF*, pp.46-51.

Douglas R.J., Popelar J. (1994). PTTI applications at the limit of GPS, *Proc. 26th PTTI*, pp 141-152.

Dudle G., Overney F., Prost L., Schildknecht T., Springer T. (1999). Transatlantic time and frequency transfer by GPS carrier phase. In: *Proc. EFTF, IEEE IFC Symposium.*

Dunn C., Jefferson D., Lichten S., Thomas J.B., Vigue Y. (1993). Time and positioning accuracy using codeless GPS. In: *Proc. 25th Precise Time and Time Interval Application*, Planning Mtg., Marina Del Rey, Ca, pp 169-182.

Jiang Z., Petit G. (1999). Accurate frequency comparison using GPS carrier phase, submitted to *Special Issue of the IEEE Transactions on Ultrasonics, Ferroelectrics and Frequency Control on Frequency Control and Precision Timing.*

Larson K.M. (1998). Time transfer using the phase of the GPS carrier. In: *Proc. IEEE FCS*, pp292-297.

Petit G., Thomas C. (1996). GPS frequency transfer using carrier phase measurements. In: *Proc. 50th IEEE FCS*, pp 1151-1158.

Petit G., Thomas C., Jiang Z., Uhrich P., Taris F. (1998). Use of GPS Ashtech Z12T receivers for accurate time and frequency comparison. In: *Proc. IEEE FCS*, pp306-314

Petit G., Jiang Z., Taris F., Uhrich P., Barillet R., Hamouda F. (1999). Processing strategies for accurate frequency comparison using GPS carrier phase. In: *Proc. EFTF, IEEE IFC Symposium*

Rothacher M., L. Mervart (1996). Bernese GPS software version 4.0 (Published by Astronomical Institute University of Bern)

The European Reference System coming of age

Josef Adam, Wolfgang Augath, Claude Boucher, Carine Bruyninx, Paul Dunkley, Erich Gubler, Werner Gurtner, Helmut Hornik, Hans v.d. Marel, Wolfgang Schlüter, Hermann Seeger, Martin Vermeer, Janusz B. Zieliński
Members of the EUREF Technical Working Group

EUREF Subcommission Secretary: DGK, Marstallplatz 8, D-80539 München

Abstract. More than ten years ago, the advantages of the GPS technology were recognized and a first GPS campaign covering the western part of Europe was organized in order to establish a uniform European Reference Frame (EUREF). Through successive GPS campaigns, the network has been extended towards eastern parts of Europe and various countries have undertaken densification campaigns. The international co-operation within Europe has resulted in the establishment of a high accuracy, three-dimensional geodetic network with links to global and national reference systems.

Strategies and guidelines have been developed for network densification, observation procedures, data flow and data analysis. This has resulted in today's permanent GPS network comprising in excess of more than 80 stations, a data handling service and supported by 12 analysis centers. The results show an accurate and consistent network (+/-3mm in the horizontal component, +/-6mm in the height component).

Since 1995, emphasis has been placed on the height component, resulting in an extended and improved adjustment of the United European Leveling Network (UELN) and the establishment of the European Vertical GPS Reference Network (EUVN). Today, the EUREF Network contributes towards multi-disciplinary activities such as the estimation of meteorological parameters and links to tide gauges.

Keywords. EUREF, reference system, reference frame, GPS, permanent GPS networks, leveling network.

1 General remarks

At the end of the eighties, the requirement for the provision of geoinformation data on a uniform geodetic reference system grew tremendously due to the availability of GPS and its versatile application in many areas of surveying, navigation, transportation and logistics amongst others. Demand for uniform maps covering Europe, for example, from the car industry for navigation purposes or from EUROCONTROL for precise positions at airports and navigation aids, the survey agencies in Europe were forced to establish a uniform reference frame. At that time, the uniform network over Europe was the European Datum ED50 resp. ED87, derived by the IAG Commission RETrig, as a result of the combination and readjustment of the national triangulation networks, which never fulfilled the new requirements. Also the World Geodetic System 1984 (WGS 84) with its realization via GPS of only a few meters did not fulfill the expectations.

Regarding the future needs of precise basic reference networks for both practical and scientific applications and for the investigation of geokinematic and geodynamic aspects the IAG at its General Assembly in August 1987 formed the new subcommission EUREF, which should continue the work of RETrig, employing new space techniques for the implementation of a European Reference Frame. One month later the Comité Européen des Responsables de la Cartographie Officielle (CERCO), which was faced with similar problems - more from the view of digital maps and practical applications - established the CERCO Working group VIII (WG VIII), which should focus on the application of GPS in the national land survey agencies. (H. Seeger became the president of the CERCO WG VIII). In order to avoid the duplication of work a joint meeting of both groups was held to analyze the requirements and to set up the steps necessary to realize the European Reference Frame (EUREF). A steering committee was established. Members were Augath/Germany, Bordley/UK, Boucher/F, Engen/N, Gurtner/CH, Seeger/Germany and Sigl/Germany.

Some investigations (e.g. EUNAV) using GPS, which at that time was in the test phase with only seven satellites, were conducted to study its application for the realization of EUREF. A timeslot

International Association of Geodesy Symposia, Vol. 121
Schwarz (ed.), Geodesy Beyond 2000 – The Challenges of the First Decade

of only a few hours per day with more than 4 satellites was available. The receivers were still in a development phase (dual frequency, code- and phase measurements on a few channels only etc.). GPS observations were carried out in collocation with SLR (Satellite Laser Ranging) and VLBI stations, in order to compare and estimate the accuracy. The maximum deviation of only 3cm between GPS and SLR/VLBI solutions encouraged the start of the EUREF Project.

It has to be stated that in the beginning, not all 24 GPS-satellites were in orbit, observation windows of only several hours per day with 4 or more satellites in view were available. The full GPS constellation improved the observation strategies and the accuracy in the positioning. The early EUREF GPS campaigns achieved less accuracy than the campaigns observed since 1992.

Initial EUREF activities were based on GPS campaigns of several days of observation, whilst nowadays a permanent GPS network has been set up covering the European area with more than 80 sites. Data links and analysis procedures have been established for the daily determination of the positions, in order to monitor changes with highest accuracy. Guidelines for the establishment and operation of the observing stations along with strategies for the data transmissions and reductions have been set up, which are in accordance with IGS (International GPS Service).

Dependant on the kind of observation (obtained during campaigns before 1992, campaigns after 1992 or as a permanent sites) three classes of accuracy were defined:

- **Class A**: 1cm accuracy for each component of the three dimensional position in ETRS (1 sigma level) independent of the epoch, guaranteed by permanent GPS observations.
- **Class B**: 1cm accuracy, but guaranteed only at a specific epoch (case in active zones where space geodetic estimates of the velocity are not sufficiently accurate), obtained by GPS campaigns since 1993.
- **Class C**: 5cm accuracy, obtained in the first GPS campaigns from 1989 to 1992.

Today EUREF bodies are

- the IAG Subcommission for Europe (EUREF) (chaired by Erich Gubler), which has organized the EUREF Symposium every year since 1992.
- the EUREF Technical Working Group (Chaired by Claude Boucher) which meet twice to three times a year to review the campaigns and released the analysis results under the flag of EUREF.

The EUREF-TWG has set up working groups focusing on special topics such as height related problems

- the EUVN Working Group (chaired by Wolfgang Schlüter) for the establishment of the European Vertical GPS Network, and
- the UELN95 Working Group (chaired by Wolfgang Augath) for the refinement of the leveling networks aiming in a dynamical height system.

All the EUREF activities are summarized at the yearly Symposia. Proceedings were published EUREF 1991-1999. Moreover most of the results are available through the WWW or ftp:

- ftp:/ftpserver.oma.be/pub/astro/euref
- http://www.oma.be/KSB-ORB/EUREF/
- http://lareg.ensg.ign.fr
- http://gibs.leipzig.ifag.de
- http://hpiers.obspm.fr
- ftp://igs.ifag.de/pub/IFG/EURO
- http://igscb.jpl.nasa.gov

2 Concept, objectives, and relation to ETRS and ITRS

In 1987 the IAG at its General Assembly in Vancouver and the CERCO at its Plenary Assembly in Athens decided independently to develop a GPS based new European Geodetic Reference Frame which fulfills the following requirements:

1. representing a geocentric reference frame for any precise geodetic-geodynamic projects on the European plate.
2. being a precise reference very near to the WGS84 to be used for geodesy as well as for all sorts of navigation in the area of Europe
3. being a continent wide modern reference for multinational Digital Cartographic Data sets, no longer derived from multiple national datums across Europe.

ED50 rep. ED 87 did not fulfill the requirements especially concerning the overall accuracy and the three-dimensional global position and orientation. The WGS 84 on the other hand could not guarantee the required very high precision, being mainly derived from Doppler observations at that time.

As in the late eighties, the IERS combined with SLR/VLBI solutions (ITRF) provided the best global realization of a geodetic reference system. The IAG Sub-commission EUREF and the CERCO WG VIII agreed to base the European Reference System on the ITRF and to select about 35 European SLR and

VLBI-sites to form part of the ITRF-solution computed for the epoch 1989.0. This would establish the basic set of geocentric coordinates defining the ETRF89 (European Terrestrial Reference Frame) as the first realization of the European Reference System ETRS89. Selecting epoch 1989.0, ETRF89 is a subset of the global solution ITRF89.

Due to plate tectonics the coordinates of the European subset of stations slowly change in the order of about 2.5cm/year. Therefore it was decided that ETRS89 should rotate with the stable part of Europe, so that the station to station relations are kept fixed. Of course, from such a decision it results that the relationship to positions defined in another reference system may slightly change. Transformation parameters for the conversion of ITRF to ITRF89 resp. ETRF89 are derived on a regular basis. As a consequence, the transformation parameters determined between ETRF and WGS84 will vary slightly and will have to be regularly modified at 10 year intervals, accounting for the influence of the rotation of Europe which cannot be neglected.

Investigations have shown, that at the beginning ITRF89 resp. ETRF89 agreed with WGS84 within 1-2m. Today WGS84 realization has been improved, so that the agreement is within a few centimeters. The reference ellipsoid for EUREF is the GRS 80 ellipsoid, which differs only slightly from the WGS84-ellipsoid.

3 EUREF realization through campaigns

3.1 Observation campaigns

The transformation of national coordinates to other systems requires at least three but better 6-8 identical points with coordinates in both reference systems (national reference system and ETRS). In October 1988 it was decided to perform a European GPS densification campaign (***EUREF89***), with the goal to establish EUREF-stations at distances of about 300 to a maximum of 500 km (as soon as possible). All campaigns are summarized in figure 1. The final network consisted of 92 sites which were observed by 2-frequency GPS-receivers in 2 different campaigns:

- phase A: lasting from May 16 to May 21, 1989 with 62 sites and
- phase B: covering the period from May 23 to May 28, 1989 with 55 sites.

Amongst them were 21 SLR- and four VLBI-sites. The subdivision into two phases turned out to be necessary as in spring 1989 only 69 2-frequency GPS-receivers were available in Western Europe. Whilst the precision of the SLR/VLBI-ETRF-coordinates was at that time 13-23 mm per component, the precision of the additional GPS-sites turned out to be around 30-40 mm in the horizontal components and about 50-60 mm in the vertical (the formal errors were with 5 exceptions smaller than 1 cm in all components).

The EUREF Subcommission, recognizing that the coordinates of the stations would be subject to improvement, that the existing network would be extended and that improvements and extensions could affect the homogeneity of EUREF 89, decided in 1992 that new campaigns to fulfill EUREF standards must include observations of a sufficient number of primary stations (SLR, VLBI) and other neighbouring EUREF sites. The Technical Working Group (EUREF-TWG) was established to define standards, to monitor the results and to advise the EUREF PLENARY whether the results fulfill EUREF standards and can then be endorsed.

In July 1990 the network was extended to the Northwest of Europe (***EUREF-NW***) by including 15 additional stations on different islands including Greenland and Island.

The densification and extension campaign in 1991 covered the *Czech Republic, Slovakia and Hungary* (***EUREF-East***). It consisted of 11 new points. In 1992 three campaigns, one for *Poland* (***EUREF-Pol***), one for the Baltic States (*Lithuania, Latvia and Estonia*) (***EUREF-Bal***) and one for *Bulgaria* (***EUREF-Bul***) were performed. In Poland 11 new stations were established while in Baltic countries 13 new sites were measured. In Bulgaria 15 stations formed the EUREF network. In 1994 *Croatia and Slovenia* entered to EUREF with 18 stations and *Romania* with 7 stations. In all of these campaigns the involvement and support of IfAG (today: BKG) was substantial, except for the Baltic stations campaign, where the engagement of Scandinavian countries was very helpful.

As the results of the EUREF-NW-Campaign due to ionospheric activities the observations in Iceland from 1991 turned out to be less accurate than for the other parts of Europe. From August 03 to August 13, 1993 a post-campaign (***Iceland 93***) was performed which also included the positioning of additional 115 stations all over Iceland.

From May 30 to June 3, 1994 (four days of 24 hours of observations) *Croatia and Slovenia* were connected to EUREF. The network was connected to the IGS-stations in Wettzell (D), Graz (A) and Matera (I). From 25.09. to 02.10.1995 the area of Slovenia was re-measured as part of a geodynamics campaign including all first order stations. As in 1994 only parts of the Croatian Area could be measured, a new campaign was observed from 29.08. to 12.09.1996 including also all the 63 stations of the Croatian first order network. During this campaign 3 additional EUREF-stations were determined.

From June 19 to June 23, 1995 EUREF was extended into the area of the *Ukraine* (***Ukraina-95***), where 15 stations were observed by Ukrainian/German-teams with Trimble SSE-receivers. The campaign could not be completed, as security forces intervened. The data were confiscated at the border, when the IfAG teams left the Ukraine. Meanwhile our Ukrainian colleagues are trying to perform their own computations but at present there is no chance to publish the results.

From 12.08. to 17.08.1996 *FYROM (Former Yugoslavia Republic of Macedonia)* (***EUREF FYROM***) was connected to EUREF (6 stations). On the same days Macedonian colleagues measured at additional 18 first order stations of this country and at sites at the airports in Skopje and Ohrid using IfAG-equipment. As further control points further the EUREF-stations Dionysos (Greece), Sofia (Bulgaria) and Ilin Vrh (Croatia) were re-observed. The campaign included 5 sessions of 24 hours each.

Malta was connected to EUREF ***(EUREF-Malta-96)*** from 29.10. to 03.11.1996 (6 stations); control observations were performed at Lampedusa (SLR-site). The campaign included 5 sessions of 24 hours each.

Within the EUREF 98 campaigns GPS measurements were performed in *Albania, Bosnia and Herzegovina and Yugoslavia* between September 4 and 9, 1998 at 29 stations. The selection to become EUREF points is currently under decision.

Covering Europe with the uniform reference network, only a few areas are still missing:

- Belorussia
- Russia (west of the Ural).

3.2 EUREF post-campaigns

After it turned out that the accuracy of the EUREF-1989-Campaign was limited to about 3-4 cm in position, several countries urgently required an improvement as far as the accuracy is concerned. Such activities were performed at least in

- Germany (1993)
- the Netherlands (1993)
- Denmark (1994)
- Belgium (1994)
- Iceland (1993 and 1995)
- UK 1992
- Ireland (1995)
- France (1993)
- Switzerland (1992) and
- Austria
- Iberia
- Norway
- Finland and
- Estonia.

From 1992 until May 1995 the EUREF Technical Working Group urged every group performing EUREF-Campaigns to refer such networks to the surrounding SLR/VLBI-stations and to neighbouring EUREF-sites. It eventually turned out that this may not be the optimal technique in all the situations where the reference coordinates or their velocities are less accurate than the precision of modern GPS-Campaigns. It was therefore agreed by the EUREF-TWG- and Plenary-Meeting in Helsinki (May 1995), that future EUREF-campaigns should be referred to the surrounding IGS-stations and should also include other neighbouring EUREF-sites to be handled as control points.

3.3 Processing strategies and transformation into the ETRS-89

The computation of the coordinates nowadays is performed following the procedures specified by the EUREF Technical Working Group. The following principles are applied:

- use the orbits from the IGS Final Orbit Combination including the associated earth rotation parameters,
- use coordinates for the fixed stations which refer to the current reference frame of the orbits (ITRF yy). The coordinates of the fixed stations are then rotated to the observation epoch (t(obs)) by using velocities given for the ITRF yy (published in the respective IERS Annual Report 19yy),
- Computation of the coordinates at the epoch of the observation (t(obs)),
- Transformation and back-rotation of these coordinates into ETRS89 with official parameters.

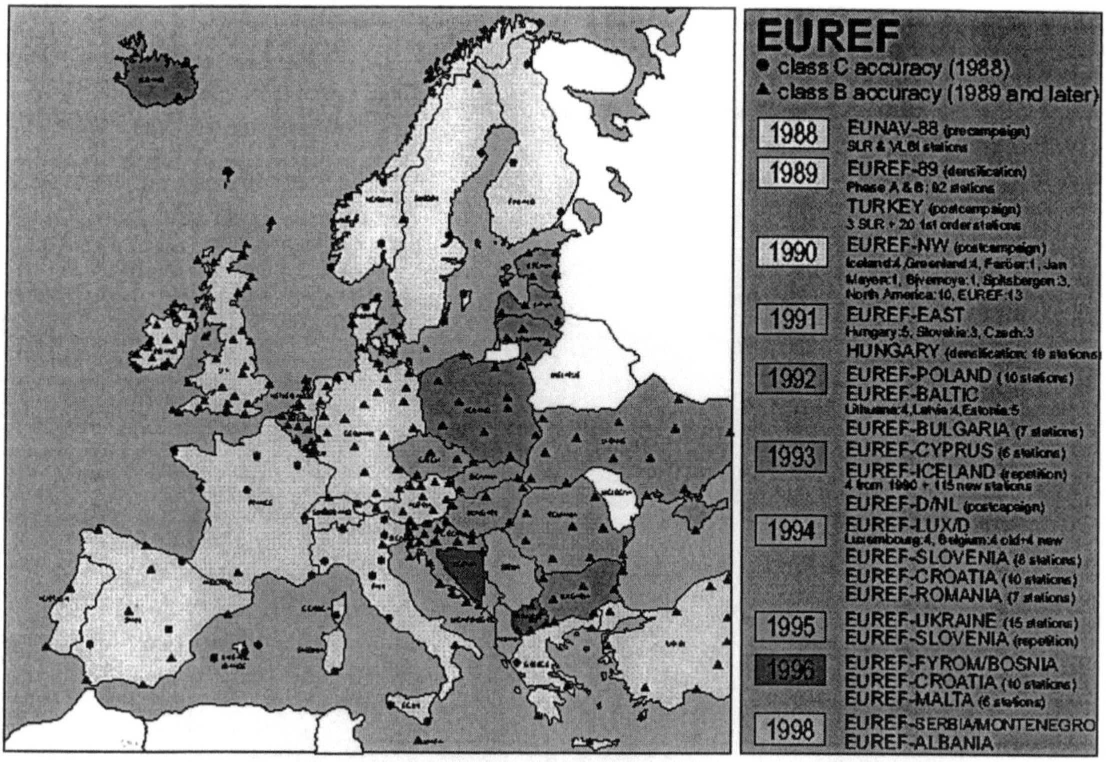

Fig. 1: EUREF network from 1989 to 1999

The conversion into ETRS89 at epoch 1989.0 is performed as follows:

x(SO)=x(S1)+T(S1)+R x(S1) dt

with

x(SO): coordinates in ETRS89 epoch 1989.0
x(S1): coordinates in ITRFyy at the epoch of observation t(obs)
T(S1): shifts T1, T2, T3, based on a global transformation from ITRFyy to ITRF89
R: rotations (no network rotation) back to epoch 1989.0 due to the motion of the European plate with the motion model NNR-NUVEL (IERS-Technical Note 13) or with individual velocities (IERS Annual Report 19yy)
dt: time difference t(obs) minus 1989.0.

4 EUREF permanent GPS network

4.1 Relation to IGS

Recognizing the growing number of permanent installation of GPS receivers in Europe, which were collecting continuously GPS tracking data, following the IGS (International GPS Service) regulations, the EUREF Subcommission made use of the situation for the maintenance of the EUREF. Werner Gurtner (University of Bern) proposed in 1995 in accordance to IGS, the organization of the EUREF permanent GPS network consisting of the following components:

- Permanent GPS Stations
- Operational Centers (OC)
- Local Data Centers (LDC)
- Regional Data Center (RDC)
- Local Analysis Centers (LAC)
- Regional Analysis Center (RAC)
- Network Coordinator

As most of the components already existed, the realization was more or less a question of the coordination. In October 1995 Carine Bruyninx (Royal Observatory of Belgium) has been appointed as network coordinator. Following the IGS rules, the EUREF permanent GPS network could be regarded as a densification of the global IGS network in the European area. In January 1996 IGS released a "Call

Fig. 2: EUREF permanent GPS network

for participation as IGS regional networks associate analysis center (RNAAC) for regional station position analyses", to which the EUREF Subcommission responded and expressed the willingness of CODE (Center for Orbit Determination Europe -- a joint cooperation of the University of Bern/CH, Bundesamt für Landestopographie, Wabern/CH, the Institut Géographique National, Paris/F and the Bundesamt für Kartographie und Geodäsie, Frankfurt/D) to act as IGS Regional Associated Analysis Center. CODE delivers weekly free-network solutions for the European region to the IGS global network associate analysis centers. The free-network solutions delivered from CODE are obtained by combining weekly solutions from the Local Analysis Centers (LAC). The IGS has officially accepted the EUREF proposal in May 1996.

The EUREF products are -next to the data of the tracking stations- weekly estimates of the coordinates of the EUREF permanent stations and their covariance information as a combined solution of sub-network solutions, submitted by EUREF Local Analysis Centers (LAC). The LACs process their sub-network following specific strategies and exchange the results employing the Software Independent Exchange Format (SINEX). To align the EUREF weekly solution with the International Terrestrial Reference Frame (ITRF) a selected set of "reference stations" is fixed to their successive realizations of their ITRS coordinates. In addition, the coordinates are updated monthly using the corresponding ITRF velocity field. Applying guidelines for reference frame fixing, see chapter 3.3, the weekly EUREF solutions, available in ITRFyy, at the epoch of observation can be linked to the ETRS89.

4.2 Network stations and operation centers

The Network is shown in figure 2. More than 80 stations, permanently operating geodetic GPS receivers with antennae mounted on suitable geodetic markers. The stations have to fulfill the EUREF specifications before they obtain the label as a permanent EUREF stations. The criteria are strong in order to ensure the data quality, the timeliness and the reliability of the provision of data, the stability of monumentation and the availability of documentation. Guidelines and data file conventions have been set up which strictly have to be fulfilled. Data provision is required on daily basis via local data centers to the regional data center. Some stations today are able to provide data-files every hour. Operational Centers, mainly identical with the agencies responsible for the stations, perform data validation, conversion of raw data into RINEX (Receiver Independent Exchange Format), data compression, and data upload to a data center through the Internet.

4.3 Data centers

Local data centers (LDC) are collecting data from all local network stations, then distribute the data or provide access to the data. Not all of the local stations need to be EUREF stations. The LDC forwards the data or a selection of data, to a Regional Data Center (RDC), which collects the data from all EUREF stations. While the LDC's in general are identical with the operational centers, only one RDC exists within Europe (Table 1).

Table 1: Local (L), Regional (R), and Global (G) data centers

ASI,	Centro di Geodesia Spaziale/I	L	
BKG,	Bundesamt für Kartographie und Geodäsie/D	L R	
DUT,	University Delft/NL	L	
FGI,	Geodetic Institut Finland/SF	L	
GRAZ	Austrian Space Agency/A		
IGN,	Institut Géographique National/F	L	G
NLS;	Nation. Landsurvey Sweden/S	L	
OSO;	Onsala Space Observatory/S	L	
ROB;	Royal Observatory Belgium/B	L	

4.4 Analysis centers

Local Analysis Centers process subnetworks out of the EUREF permanent network following the rules and guidelines as set up by the IGS and supplemented by the EUREF TWG. They submit weekly solutions, which CODE is combining to the EUREF solution. Currently a transition is ongoing concerning the combinations of the solutions. BKG will take over the routine combination process from CODE, after parallel analysis of the data demonstrated identical results and guarantees the same quality and continuation. Today 12 local Analysis Centers are involved in the data reduction procedure:

1. **ASI,** Centro di Geodesia Spaziale - Matera/I
2. **BEK,** Bayerische Kommission für die Internationale Erdmessung - München/D
3. **BKG,** Bundesamt für Kartographie und Geodäsie - Frankfurt/D
4. **COE,** Center for Orbit Determination in Europe - Bern /CH (CODE)
5. **GOP,** Geodetic Observatory Pecny -Pecny/CR
6. **IGN,** Institut Géographique National - Paris/F
7. **LPT,** Bundesamt für Landestopographie - Wabern/CH
8. **NKG,** Nordic Geodetic Commission GPS data Analysis Center - Onsala/S
9. **OLG,** Institute for Space Research - Graz/A
10. **ROB,** Royal Observatory of Belgium - Brussels/B
11. **UPA,** University of Padova - Padova/I
12. **WUT,** Warsaw University of Technology - Warsaw/Pl

The subnetworks are organized in such a way, that the data of a permanent EUREF station will be analyzed on the average by three LOC's in order to avoid unrealistic weighting in the analysis procedure.

5 Contribution of EUREF to height systems

5.1 The European Height System UELN

After a break of ten years, the work on the United European Leveling Network (UELN) resumed in 1994 under the name UELN-95. The objectives of the UELN-95 project have been to establish a unified vertical datum for Europe at the one decimeter level with the simultaneous enlargement of UELN as far as possible to include Central and Eastern European countries.

More than 3000 nodal points were adjusted, constraint-free, in geopotential numbers linked to the reference point of UELN-73 (gauge Amsterdam). The new heights in the system UELN-95/98 are available for more than 20 participating countries. More information in the detail you can find in (J. Adam et. al. 1999).

5.2 The EUVN European Vertical GPS Network

Within the frame of the IAG-Subcommission for Europe (EUROPE, former EUREF) the activities for the European Vertical GPS-Reference Network (EUVN) have been carried out on the basis of

- Resolution No. 2 of the EUREF Symposium in Helsinki, Mai 1995,
- Resolution No. 3 of the EUREF Symposium in Ankara, Mai 1996.

The final objective of the EUVN is to provide a set of coordinates for all EUVN sites consisting of three dimensional coordinates X, Y, Z resp. latitude, longitude, ellipsoidal height and the "physical" height derived through leveling and gravity measurements with respect to the UELN and/or UPLN and/or to the national height system.

6 Prospects

The EUREF activities have been a strong driving force for the implementation of the GPS-technique since the very beginning in Europe. A lot had to be learned, investigated and developed on the technique itself (observations, analysis, data handling, and communications). The objective to establish a uniform reference frame, at first on the basis of campaigns, today on the basis of a permanent network forced all the European countries, involved in EUREF to cooperate, to coordinate their activities and to follow the standards which have been set up by the EUREF-TWG. EUREF today is the best organized regional network world wide and fulfill the strong geodetic requirements for a reference network on the most accurate level. EUREF is the backbone for national activities e. g. for the establishment of network reference frames and provide the basis for new research such as atmospheric or ionospheric investigations as well as for the monitoring of geodynamical changes (crustal movements, sea level changes).

References

EUREF 1991, Report on the Symposium of the IAG Sub-commission for the European Reference Frame (EUREF) held in Florence, May 28 - 31, 1990; Veröff. d. Bay. Komm. f. d . intern. Erdm.; Astron.-Geod. Arbeiten, Heft Nr. 52, 1992

EUREF 1992, Report on the Symposium of the IAG Sub-commission for the European Reference Frame (EUREF) held in Berne, March 4-6, 1992; Veröff. d. Bay. Komm. f. d . intern. Erdm.; Astron.-Geod. Arbeiten, Heft Nr. 52, 1992

EUREF 1993, Report on the Symposium of the IAG Subcommission for the European Reference Frame (EUREF) held in Budapest, May 17-19, 1993; Veröff. d. Bay. Komm. f. d . intern. Erdm.; Astron.-Geod., Heft Nr. 53, 1993

EUREF 1994, Report on the Symposium of the IAG Sub-commission for the European Reference Frame (EUREF) held in Warsaw, June 8-11, 1994; Veröff. d. Bay. Komm. f. d . intern. Erdm.; Astron.-Geod. Arbeiten, Heft Nr. 54, 1994

EUREF 1995, Report on the Symposium of the IAG Sub-commission for the European Reference Frame (EUREF) held in Helsinki, May 3-6, 1995; Veröff. d. Bay. Komm. f. d . intern. Erdm.; Astron.-Geod. Arbeiten, Heft Nr. 56, 1995

EUREF 1996, Report on the Symposium of the IAG Sub-commission for the European Reference Frame (EUREF) held in Ankara, May 22-25, 1996; Veröff. d. Bay. Komm. f. d . intern. Erdm.; Astron.-Geod. Arbeiten, Heft Nr. 57, 1996

EUREF 1997, Report on the Symposium of the IAG Sub-commission for the European Reference Frame (EUREF) held in Sofia, June, 1996; Veröff. d. Bay. Komm. f. d . intern. Erdm.; Astron.-Geod. Arbeiten, Heft Nr. 58, 1997

EUREFF 1999, Report on the Symposium of the IAG Sub-commission for the European Reference Frame (EUREF) held in Ahrweiler, June 10-13, 1996; Reports of the EUREF Technical Working Group; (EUREF Publication Nr.7/1); Mitteilungen des Bundesamtes für Kartographie und Geodäsie, Band 6 und 7; Frankfurt/M 1999

Adam, J.; Augath, W.; Brouwer, F.; Engelhardt, G.; Gurtner, W.; Harsson, B.G.; IHDE, J.; Ineichen, D.; Lang, H.; Ludthardt, J.; Sacher, M.; Schlüter, W.; Springer, T.; Wöppelmann, G.: Status and Development of the European Height System; presented at the IUGG General Assembly, Birmingham 1999

Status and development of the European height systems

J. Adam, W. Augath, F. Brouwer, G. Engelhardt, W. Gurtner, B. G. Harsson, J. Ihde, D. Ineichen, H. Lang, J. Luthardt, M. Sacher, W. Schlüter, T. Springer, G. Wöppelmann
Members and associated members of the EUREF-Technical Working Group

Contact address: Secretary of the Deutsche Geodätische Kommission, Marstallplatz 8, D-80539 München, Germany, Phone: +49 89 23 031 113, Fax: +49 89 23 031 100/-240, E-mail: hornik@dgfi.badw-muenchen.de

Abstract. After a break of ten years, the work on the United European Levelling Network (UELN) resumed in 1994 under the name UELN-95. The objectives of the UELN-95 project are to establish an unified vertical datum for Europe at the one decimeter level with the simultaneous enlargement of UELN as far as possible to include Central and Eastern European countries. More than 3000 nodal points were adjusted linked to the reference point of UELN-73 (gauge Amsterdam). The new heights in the system UELN-95/98 are available for more than 20 participating countries.

The European Vertical Reference Network (EUVN) is designed to contribute to the UELN project along with the connection of European tide gauge benchmarks as contribution to monitoring absolute sea level variations, the establishment of fiducial points for the European geoid determination. The EUVN includes 196 points all over Europe. At every EUVN point, three-dimensional coordinates in ETRS89 and levelling heights primarily in the system of the UELN-95 have to be derived. The GPS computations are finalised, some levelling connections still have to be realised. At the tide gauge stations of EUVN additional sea level observations have to be included.

The height systems will be developed as a combination of GPS permanent observations, levelling, and geoid information under consideration of well known vertical movements towards an European kinematic height reference system.

Keywords. European height system, levelling, GPS.

1 Introduction

The IAG Subcommission for Europe (EUREF) started in 1994 with its activities for development and establishment of European height systems.

Since 1994 the work at the UELN has been continued after a break of 10 years under the name of UELN-95. In accordance with the Resolution No. 3 of the EUREF Symposium 1994 in Warsaw, the objective of the UELN project is to establish an unified vertical datum for Europe at the one-decimeter level with simultaneous enlargement of UELN as far as possible to the Eastern European countries. The results of the adjustment with status of end 1998 were handed over to each participating country under the name UELN-95/98.

The European Vertical Reference Network (EUVN) was prepared in parallel to the UELN. It is an integrated network of GPS, levelling and tide gauge observations.

In May 1997 the EUVN GPS campaign with more than 200 stations was realized. Most of the countries were able to support EUVN in their national area with own receivers, with own staff and at their own cost.

The final results of the successful EUVN GPS campaign as a set of coordinates in ITRF96 (Epoch 1997.4) and ETRS89 (ETRF96, Epoch 1997.4) for all sites are available. It was the result of the excellent cooperation of the observing agencies, of the preprocessing and analysis centres. The EUVN activity will be successfully finished if levelling heights in the system of UELN-95/98 (as result of levelling connections to the UELN or to the national height system) for all sites and sea level in tide gauges in the system of UELN-95/98, Epoch 1997.4 are available. The next step to an European kinematic height network (EVS 2000) is in preparation.

2 United European Levelling Network (UELN)

Starting point of UELN-95 project were the data of UELN-73 with which in a first step the adjustment 1986 was repeated. The weights were derived from

International Association of Geodesy Symposia, Vol. 121
Schwarz (ed.), Geodesy Beyond 2000 – The Challenges of the First Decade

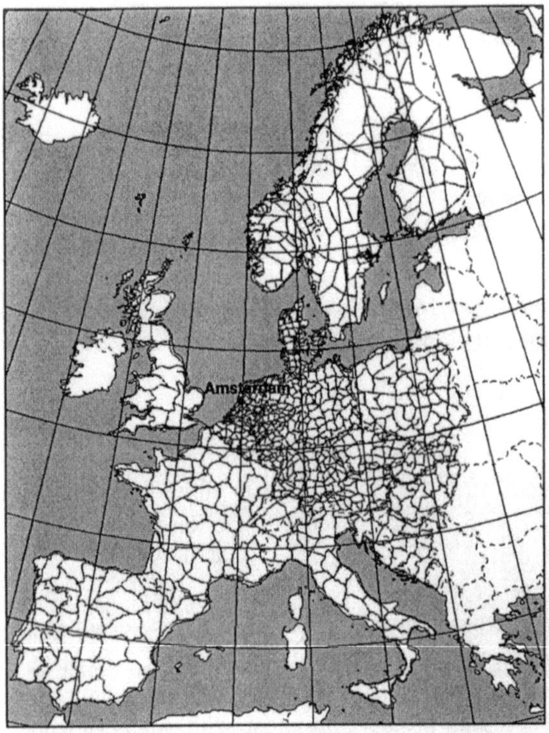

Fig. 1: UELN with status of December 1998

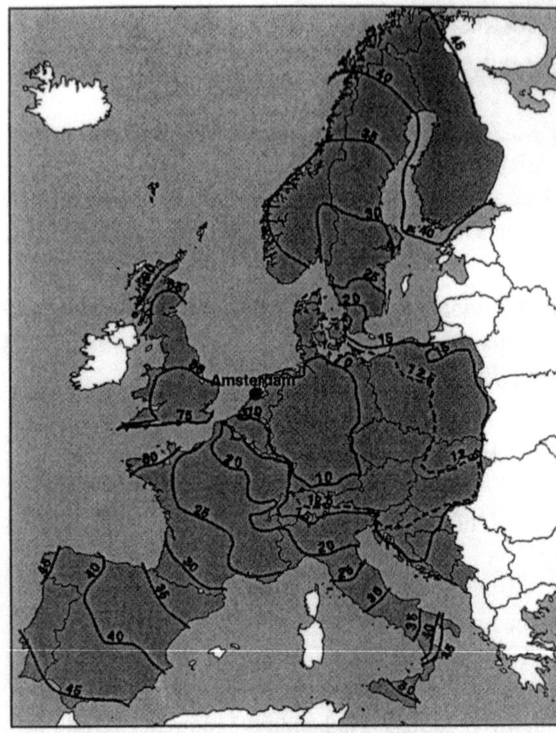

Fig. 2: Error propagation of the UELN (in mm)

a variance component estimation of the observation material which was actually introduced into the adjustment and where each national network was regarded as one group.

The enlargement to UELN-95 is performed in two qualitatively different steps:

- Substitution of data material of such network blocks which were already part of UELN-73 but show current new measurements with improved (mostly more dense) network configuration (intensive enlargement)
- Adding new national network blocks of Central and Eastern Europe which were not part of UELN-73 (extensive enlargement)

At the UELN data and computing centre at the Bundesamt für Kartographie und Geodäsie (BKG) in Leipzig the data handling and adjustment are carried out.

The adjustment in geopotential numbers is performed as constraint-free adjustment linked to the reference point of UELN-73 (gauge Amsterdam). For the UELN adjustment the program system HOENA developed at BKG is used.

The parameters of the last adjustment version. UELN-95 are the following:

- number of fixed points: 1
- number of unknowns: 3063
- number of measurements: 4263
- degrees of freedom: 1200
- average redundancy: 0.281
- a posteriori standard deviation referred to the levelling distance of 1 km:1.10 kgal · mm.

Figure 2 shows the error propagation.

In January 1999 the last adjustment version of UELN-95 was handed over to the participating countries as the UELN-95/98 solution (Figure 1).

Looking at the current network configuration of the UELN-95/98 some problem areas are found and should be solved in the next steps:

- Some of the oldest data of the UELN are found in the parts of Western Europe which were already included in the UELN-73. Several countries are repeating the observation of their levelling networks or have established new networks with better configuration. The concerning network blocks in the UELN should be replaced in order to increase the precision of UELN. On the other hand the repetition of the observation of a levelling network presents the chance to take a first step on the way to a geokinematic height network. The data bank of UELN is prepared to store more than one epoch.
- A request in the Resolution No. 4 of the EUREF-Symposium 1998 is the extension of the UELN to the Black Sea.

- The closing of the network around the Baltic Sea would be an important step to improve the reliability of the Scandinavian part of the UELN. The next condition for that is the inclusion of the Baltic States into the UELN.
- Inclusion of a second height difference between France and Great Britain by using the EURO Tunnel measurement.

In 1999 representatives from Estonia and Latvia carried out the adjustment of their national levelling networks at the UELN data and computing center at the BKG. The results were delivered to the UELN database.

3 The European Vertical Reference Network (EUVN)

The initial practical objective of the EUVN project is to unify different European height datums within few centimeters. In addition this project is to prepare a geokinematic height reference system for Europe and to connect levelling heights with GPS heights.

At all EUVN points P three-dimensional coordinates in the ETRS89 $(X_p, Y_p, Z_p)_{ETRS}$ and geopotential numbers $c_p = W_{o\ UELN} - W_p$ will be derived. Finally the EUVN is representing a geometrical-physical reference frame. In addition to the geopotential numbers c_p normal heights $H_n = c_p / \bar{\gamma}$ will be provided ($\bar{\gamma}$ is the mean normal gravity value between the ellipsoid and the telluroid.).

The application of the GPS technique for practical levelling would dramatically extend if the geoid would be known precisely enough in relation to the concerned GPS reference system and the levelling reference system. To derive such a geoid, an European reference geoid is required in the reference system ETRS89 and the reference system of UELN. Up to now there is no precise geoid available for Europe with an accuracy of a few centimeters which fulfils the requirement for the practical applications. This proposal points out a possibility to derive a geoid tailored for the GPS-levelling methods by combining the existing reference network EUREF/ETRS89 with the UELN95.

The EUVN project contributes to the realization of an European vertical datum and to connect different sea levels of European oceans with respect to the work PSMSL (Permanent Service of Mean Sea Level) and of anticipated accelerated sea level rise due to global warming. The project provides a contribution to the determination of an absolute world height system.

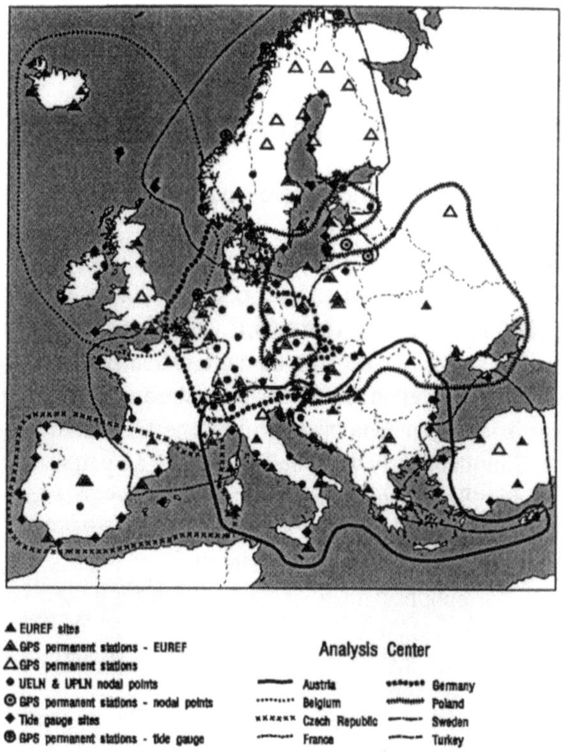

Fig. 3: EUVN distribution of EUVN points and analysis center areas

Three kinds of observation groups are necessary:

- GPS measurements for the determination of the ellipsoidal heights of all defined EUVN points,
- levellings between the EUVN sites and the UELN nodal points for the determination of the physical height of all defined EUVN points,
- observations of sea level at tide gauge stations.

In total the EUVN consists of about 196 sites: 66 EUREF and 13 national permanent sites, 54 UELN and UPLN (United Precise Levelling Network of Central and Eastern Europe) stations and 63 tide gauges (Figure 3).

The GPS observations for the EUVN were carried out in the period from May 21 to May 29, 1997. Three types of receiver were used: 35 Turbo Rogue Receivers, 134 Trimble SSI or SSE and 51 Ashtech Z12. The time interval was set to 30 s, the elevation mask was 5°. The campaign was running very smoothly and everybody who participated in the campaign supported the action successfully.

The data preprocessing after the EUVN campaign performed by 9 EUVN Preprocessing Centers (PPC) was mainly a check concerning completeness and consistency of the data and auxiliary information. The PPCs were requested to prepare complete access information and/or data flow guidelines for the observing agencies before the start of the campaign (Luthardt et al., 1998). The task of the EUVN GPS Analysis Center (AC) was to process the data of a special subnetwork. A subdivision of the whole EUVN Network was done under the aspect of receiver type and regions.

10 European institutions were ready to contribute as Analysis Centers. On the Analysis Center Workshop in September 1997 in Leipzig the subdivision of EUVN was discussed and decided (see Figure 3) (Ineichen et al., 1999). The AC of Croatia was responsible for the analysis of the collocation points and the investigation of the biases introduced by using different antenna types within one GPS network. Simultaneously with the EUVN 97 Campaign the Baltic Sea Level (BSL) GPS campaign was performed. The BSL 97 GPS campaign was processed by the Finnish Geodetic Institute.

The Astronomical Institute of the University of Bern (AIUB) and the BKG were responsible for the computation of the final GPS solution of EUVN (Ineichen et al., 1998).

The analysis centers produced different solution types in order to investigate the influence of the processing strategy on the results. Mainly the following three types of solutions were looked at:

- 15 degrees without weighting: The 'standard' solution with the highest priority. Data down to an elevation cut-off angle of 15 degrees were used for generating this solution type. All observations were introduced with the same weight. This solution type corresponds to the processing strategy used for the permanent EUREF network at the time of the EUVN campaign.
- 5 degrees with elevation-dependent weighting: Measurements down to an elevation cut-off angle of 5 degrees were used for this solution type. In addition, the observations were weighted with $w = \cos^2(z)$, where z is the zenith angle of the observed satellite.
- Satellite-specific weighting: The IGS precise orbit files (in SP3 format) contain accuracy codes for each satellite. These accuracy codes can be used by the Bernese GPS Software to weight the corresponding observations. Not all analysis centers delivered solutions of this type, and therefore no combined solution was generated.

The question which solution to choose as the official EUVN97 solution (the unweighted 15-degree solution or the weighted 5-degree solution), was discussed during the Analysis Center Workshop at Wettzell (April 2-3, 1998): The unweighted 15 degree solution was selected as the official one. The following aspects had to be taken into account:

- The comparison of the height component of redundant points in both solution types showed a slightly better repeatability for the unweighted 15-degree solution.
- Not all sites within EUVN97 were tracking satellites below 15 degrees with the same quality and quantity. For some sites the number of observations is hardly increasing when changing to the lower cut-off angle, whereas for others the number of observations increased by up to 20 %. Therefore the site coordinates within the EUVN97 GPS network could be more inhomogeneous in the 5-degree solution.
- The elevation-dependent antenna phase center variations are not yet well known below 10 degrees. Introduction of poorly defined corrections could lead to additional systematic errors.
- We do not have enough experience yet with the performance of the tropospheric mapping functions at very low elevations.

The final solution was constrained to ITRF96 coordinates (epoch 1997.4) of 37 stations with an a-priori standard deviation of 0.01 mm for each coordinate component. As a consequence of these tight constraints the resulting coordinates of the reference points are virtually identical with the ITRF96 values.

A comparison for the combined solutions of BKG and AIUB showed that these two solutions were identical.

For many practical purposes it is useful to have the ETRS89 coordinates available. To get conformity with other projects, the general relations between ITRS and ETRS were used. The coordinate transform formula from ITRF96 to ETRF96 and the final coordinates are given in Ineichen et al 1999.

In order to reach the goal it is necessary to connect the EUVN stations by levellings with nodal points of relevant levelling networks. So it is possible to use levelling observations to update the gravity related EUVN heights in context with the new adjustment of UELN. At present for about 80% of the EUVN points levelling heights are available.

As the EUVN is a static height network it is necessary to know the value of the mean sea level in relation to the tide gauge bench mark at the epoch of EUVN GPS campaign 1997.5. However for future tasks it is useful to have available the monthly mean values over a period of some years.

The Permanent Service for Mean Sea Level (PSMSL), as member of the Federation of the Astronomical and Geophysical Data Analysis Service (FAGS), is in principle in charge of the data collection. The information which are sent to the PSMSL databank in general should also be made available for the EUVN project.

EUVN is a step to establish a fundamental network for a further geokinematic height reference system such as European Vertical System (EVS 2000) under the special consideration of the Fennoscandian uplift and the uplift in the Carpathian-Balkan region.

4 Transformation relations between national European height systems and the UELN

In Europe three different kinds of heights (normal heights, orthometric heights and normal-orthometric heights) are used. Examples for the use of orthometric heights are Belgium, Denmark, Finland, Italy and Switzerland. Today normal heights are used in France, Germany, Sweden and in the most countries of Eastern Europe. In Norway, Austria and in the countries of the former Yugoslavia normal-orthometric heights are used.

The vertical datum is determined by the mean sea level, which is estimated at one or more tide gauge stations. The tide gauge stations of the national European height systems in Europe are located at various oceans and inland seas: Baltic Sea, North Sea, Mediterranean Sea, Black Sea, Atlantic Ocean. The differences between these sea levels can come up to several decimeters. They are caused by the various separations between the ocean surface and the geoid.

Figure 4 shows the distribution of the mean transformation parameters between the national height systems and the UELN.

If the differences in one country are not sufficiently constant then parameters for a 3-parameter-transformation are determined. For more information see Sacher et al. (1999b).

5 The European Vertical System (EVS2000) - Outlook

The European Vertical System is planned as geokinematic height network as combination of the European GPS permanent station network, the UELN with repeated levellings, the European gravimetric geoid and tide gauge measurements along European coast lines as well as repeated gravity measurements. In May 1999 a special working group was formed to determine the direction of future work. At the first working group meeting 3

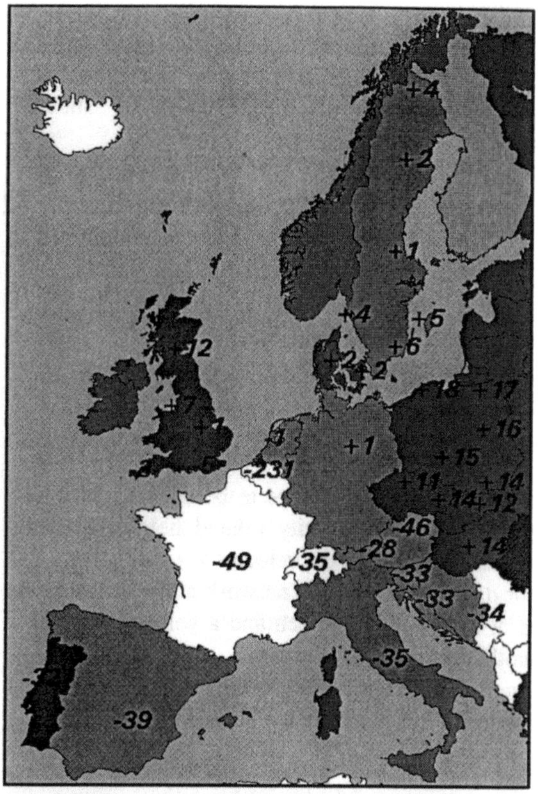

Fig. 4: Preliminary transformation parameters from national height systems to UELN

first tasks were established:

- analysis of available repeated levelling measurements and store the data base in the UELN data
- development of software as base for test computation
- testing of the principles in a test area (Netherlands, Denmark, northern part of Germany).

The GPS observations of about 80 European permanent stations are available. The analysis of 10 European GPS permanent stations shows daily repeatabilities between 7 to 9 mm in the height component. This is in good agreement with the special GPS height campaigns in Germany for deriving GPS levelling geoidal heights ($m_h = \pm 7$ mm).

Furthermore the linear height regression analysis gives for a three year period an accuracy of a GPS height difference of about

$$m_{V_h} = m_h \sqrt{2} / \sqrt{365} \,/\, \text{year} = \pm 0.5 \,\text{mm/year},$$

that means from a statistical point of view that a vertical movement of V_h = 1.0 mm/year can be significantly determined after a three years GPS observation period ($m_{V_h} = \pm 0.3$ mm/year).

Repeated precise levellings (1 mm · km$^{-1/2}$) with an epoch difference of 20 years give velocities for height differences with an accuracy of about ± 0.07 mm · km$^{-1/2}$ / year.

From this follows, that GPS permanent stations in a distance of about 300 km can significantly support repeated levellings with above mentioned suppositions. This combination of GPS and levelling is promising for a stable kinematic height reference system (Ihde, 1999).

The observation equation for levelling observations $\Delta h_{ij,k}$ between points i and j at the epoch k is:

$$\Delta h_{ij,k} = H_j - H_i + V_j(t_k - t_0) - V_i(t_k - t_0). \quad (1)$$

Two unknowns per point are to be determined: the levelling height H (gravity related height) at the reference epoch t_0 and the velocity V.

For datum fixing of the network a height for one point at a determined epoch and a velocity for this or another point shall be given.

The relation between levelling heights H and GPS heights h is given by the geoid height N

$$h = H + N. \quad (2)$$

Since the accuracy of the geoid heights resp. geoid height differences is not in the same order like the levelling observations, GPS heights cannot be used as observations. But under the condition of no significant geoid height changes, velocities v derived from GPS permanent station observations can be used as additional observation type in levelling points i

$$v_i = V_i. \quad (3)$$

The unknown velocities V are to be determined in combination with the repeated levellings. It is necessary, that the variance-covariance matrix of the observed GPS velocities is given.

The EVS project will start in late 1999 with a circular letter of the IAG Subcommission for Europe to all European countries with a call for participation.

References

Augath, W. (1996). UELN-2000 - Possibilities, Strategy, Concepts - or: How should we realize a European Vertical System? *Presented at the EUREF Symposium.* Ankara, May 22-25, 1996. In: *Veröffentlichung der Bayerischen Kommission für die Internationale Erdmessung.* Heft Nr. 57, S. 170-174.

Augath, W. et al. (1999). European Vertical System (EVS) 2000 - Status and Proposals. *Presented at the Symposium of the IAG Subcommission for Europe (EUREF).* Prague, June 2-4, 1999.

Ihde, J., Adam, J., Gurtner, W., Harsson, B.G., Schlüter, W., Wöppelmann, G. (1999a). The Concept of the European Vertical GPS Reference Network (EUVN). In: *Mitteilungen des Bundesamtes für Kartographie und Geodäsie, EUREF Publication No. 7/II,* Frankfurt am Main, 1999, Bd. 7, S. 11 - 22.

Ihde, J. et al. (1999). Combination of repeated Levellings and GPS Permanent Observations for the Realization of a Kinematic Height Reference System. *Presented at the First Meeting of the EVS Working Group.* Dresden, May 20-21, 1999.

Ihde, J., Adam, J., Gurtner, W., Harsson, B. G., Schlüter, W., Wöppelmann, G. (1999b). Status report of the EUVN project. *Presented at the Symposium of the IAG Subcommission for Europe (EUREF).* Prague, June 2-4, 1999.

Ineichen, D., Gurtner, W., Springer, T., Engelhardt, G., Luthardt, J., Ihde, J. (1999). EUVN97 - Combined GPS Solution. In: *Mitteilungen des Bundesamtes für Kartographie und Geodäsie, EUREF Publication No. 7/II,* Frankfurt am Main, 1999, Bd. 7, S. 23 - 46.

Luthardt, J., Harsson, B., Jaworski, L., Jivall, L., Kahveci, M., Molendijk, R. E., Pesec, P., Simek, J., Vermeer, M. Report on the GPS Processing and EUVN Data Center. In: *Mitteilungen des Bundesamtes für Kartographie und Geodäsie, EUREF Publication No. 7/II,* Frankfurt am Main, 1999, Bd. 7, S. 47-51.

Sacher, M., Lang, H., Ihde, J. (1999a). Status and Results of the adjustment and enlargement of the United European Levelling Network 1995 (UELN-95). In: *Mitteilungen des Bundesamtes für Kartographie und Geodäsie,* Frankfurt am Main, 1999, Bd.6, S. 131 - 141.

Sacher, M., Ihde, J., Seeger, H. (1999b). Preliminary transformation relations between national European height systems and the United European Levelling Network (UELN). *Prepared for the CERCO Plenary, Oslo, September 1999. Presented at the Symposium of the IAG Subcommission for Europe (EUREF).* Prague, June 2-4, 1999.

Sacher, M., Ihde, J., Celms, A., Ellmann, A. (1999c). The first UELN stage is achieved, further steps are planned. *Presented at the Symposium of the IAG Subcommission for Europe (EUREF).* Prague, June 2-4, 1999.

PART 2

Seamless Gravity

Rene Forsberg
Karl-Heinz Ilk
Michael G. Sideris

Recovering the global gravitational field from satellite measurements of the full gravity gradient

M.S. Petrovskaya
Main (Pulkovo) Astronomical Observatory of Russian Academy of Sciences
Pulkovskoe Shosse 65, St. Petersburg, 196140, Russia; e-mail: petrovsk@gao.spb.ru

J.B. Zieliński
Space Research Center of Polish Academy of Sciences, Barticka 18A 00-716, Warsaw, Poland; e-mail: jbz@cbk.waw.pl

Abstract. The problem of recovering the earth's potential coefficients $\bar{C}_{n,m}$ from satellite gradiometry missions is studied by the space-wise approach. The conventional procedure of solving this problem is based on constructing from the start several boundary value (*BV*) relations, each of them corresponding to a separate second order potential derivative. In the present paper an optimal approach is elaborated for solving the spaceborne scalar boundary value problem in which the external gravitational field is reproduced from the total magnitude Γ of the gravity gradient measured in a spacecraft. Correspondingly, instead of constructing a set of *BV* equations, a unique one is derived combining the components of the gravity gradient tensor and the first order potential derivatives entering the expression for Γ. This equation is solved analytically in form of a quadrature formula for $\bar{C}_{n,m}$, depending on the observed anomaly $\Delta\Gamma$ of the gravity gradient magnitude Γ.

Keywords. Gravitational potential, satellite gradiometry, spaceborne boundary value problem.

1 Introduction

The most promising for the future progress in studying the gravity field structure is satellite gradiometry from which the geopotential approximation will be improved by one order with respect to the accuracy and resolution. There are two main approaches to processing satellite gradiometry data: space-wise and time-wise ones. Fundamental analysis of different aspects of these problems is given in (Rummel and Colombo, 1985), (Rummel, 1986), (Rummel *et al.*, 1993), (Koop, 1993). The first approach is a spatial analogy of formulation and solution of the surface gravimetric boundary value problem. For a long time duration (6–12 months) of a satellite gradiometry mission the measurement data, reduced to a mean orbital sphere, will densely cover it. For each observed second order potential derivative a linear relation is compiled between the observations and the unknown potential coefficients. These relations are combined to a set of observation equations which are solved simultaneously by the least squares (*LS*) technique. The time-wise approach is based on presenting a partial potential derivative in form of a trigonometric series with two time-dependent arguments whose coefficients are functions of satellite orbital elements. The so-called lumped coefficients are linear functions of the unknown potential coefficients. The lumped coefficients are determined from a set of observation equations by the *LS* procedure.

If during a satellite gradiometry mission only one second order potential derivative is measured then an explicit quadrature formula can be derived for $\bar{C}_{n,m}$. Such is the *STEP* project, in which only the across-track potential derivative is supposed to be measured. For this case Albertella *et al.* (1995a,b) developed the theory of bi-orthogonal series. A bi-orthogonal basis was constructed numerically (by applying the minimum norm criterion in L^2), corresponding to a finite set of the unknown potential coefficients $\bar{C}_{n,m}$. Petrovskaya (1997) proposed a more simple approach for solving the *STEP* problem which is close to the one described above. Instead of a bi-orthogonal basis, a modified *STEP* observable is introduced into consideration which allows to retain the standard basis of the solid spherical functions. A simple quadrature formula was derived for recovering the geopotential coefficients.

International Association of Geodesy Symposia, Vol. 121
Schwarz (ed.), Geodesy Beyond 2000 – The Challenges of the First Decade

In order to find an optimal approach to evaluating the coefficients $\bar{C}_{n,m}$ from spatial gradiometry data it is quite natural to take the full magnitude Γ of the gravity gradient as a basic quantity for constructing a unique *BV* equation. This quantity contains most information on the gravity field due to combining the components of the gravity gradient tensor. In (Petrovskaya, 1996) and (Petrovskaya and Zieliński, 1997) some aspects of such *BV* problem were discussed within the frames of the space-wise approach. In the present paper the concept of the generalized observable Γ is further developed and an explicit quadrature formula is derived for evaluating the potential coefficients $\bar{C}_{n,m}$.

2 Spaceborne boundary-value relation

Two local coordinate systems will be used, connected with the gravity center of a satellite (Koop., 1993). One of them is the north-oriented triad $\{x,y,z\}$, which is simply connected with the geocentric spherical coordinates $\{r,\theta,\lambda\}$. The observational data are referred to the local orbital reference set $\{u,v,w\}$.

Designating by G the modulus of the potential gradient, one has

$$G = |grad\, V| = \sqrt{V_x^2 + V_y^2 + V_z^2} \tag{1}$$

where V_x, V_y and V_z are the first order derivatives of the potential V.

The absolute value of the gravitational gradient vector is

$$\Gamma = \left|grad\left(|grad\, V|\right)\right| = |grad\, G| = \sqrt{G_x^2 + G_y^2 + G_z^2}.$$

The explicit form of Γ is presented by the relation:

$$G^2\Gamma^2 = (V_xV_{xx} + V_yV_{xy} + V_zV_{xz})^2 + \\ + (V_xV_{xy} + V_yV_{yy} + V_zV_{yz})^2 + (V_xV_{xz} + V_yV_{yz} + V_zV_{zz})^2. \tag{2}$$

As usual, the potential V is presented in the form:

$$V = U + T \tag{3}$$

where T is the disturbing potential and the normal potential U is

$$U = \frac{\mu}{r}\left[1 - \frac{a^2}{r^2}J_2P_{2,0}(\cos\theta)\right],\; P_{2,0} = \frac{3}{2}\cos^2\theta - \frac{1}{2}. \tag{4}$$

Here μ is the gravitational constant multiplied by the earth's mass, r is the geocentric distance to the measurement point, a is the semi-major axis of the ellipsoid, and θ is the polar angle.
One has

$$J_2 = 0.00108263 = \frac{1}{6}e^2 + O(10^{-5}), \\ e^2 = 0.00669439810568 \tag{5}$$

where e is the ellipsoid first eccentricity.

The external disturbing potential T on and outside the sphere Σ_R of the minimum radius $r = R$, bounding the earth, can be presented as the convergent spherical harmonic series

$$T(r,\theta,\lambda) = \frac{\mu}{R}\sum_{n=2}^{\infty}\sum_{m=-n}^{n}\left(\frac{R}{r}\right)^{n+1} C_{n,m}Y_{n,m}(\theta,\lambda), \\ Y_{n,m}(\theta,\lambda) = P_{n,|m|}(\cos\theta)\,Q_m(\lambda), \\ Q_m(\lambda) = \begin{cases} \cos m\lambda, & m \ge 0 \\ \sin|m|\lambda, & m < 0. \end{cases} \tag{6}$$

Here $Y_{n,m}(\theta,\lambda)$ are the surface spherical functions, $P_{n,|m|}(\cos\theta)$ are the non-normalized Legendre functions, and λ is the longitude. By $C_{n,m}$ the corresponding non-normalized harmonic coefficients are designated.

Proceeding from formulas (1) – (5), after cumbersome transformations, a very simple formula was derived by Petrovskaya (1996) for the gravity gradient anomaly $\Delta\Gamma = \Gamma - \Gamma_0$:

$$\Delta\Gamma = T_{zz} + \frac{3}{4}J_2\sin 2\theta\left(5T_{xz} - \frac{1}{r}T_x\right). \tag{7}$$

Here Γ_0 corresponds to the normal field. Its explicit expression was developed in (ibid.). The second term on the right hand side of (7) has a factor of the order 10^{-3}, according to (5). All the other of 9 second order derivatives, as well as the first order potential derivatives, which initially have presented in the expression for $\Delta\Gamma$, are disregarded in (7) because their coefficients were of higher order of smallness (10^{-5}) than J_2. Let us note that this neglecting higher order terms does not mean at all that such errors will be automatically introduced into the *BV* solution which will be further derived. It is explained by the fact that the BV equation will represent an identity between the observed («known») combination of the potential derivatives with respect to coordinates $\{u,v,w\}$ and the same expression in terms of $\{x,y,z\}$ (which was later expressed in terms of the «unknown» coefficients $\bar{C}_{n,m}$). The small terms will be neglected simultaneously in both sides of this identity.

From (7), by applying the transformations between the potential derivatives with respect to the north-oriented and spherical coordinates (Koop, 1993), it is derived:

$$\Delta\Gamma = T_{rr} + \frac{3}{4r} J_2 \sin 2\theta \left(\frac{6}{r} T_\theta - 5T_{r\theta} \right). \tag{8}$$

The substitution of the series (6) in (8) gives:

$$\Delta\Gamma = \frac{\mu}{r^3} \sum_{n=2}^{\infty} \sum_{m=-n}^{n} \left(\frac{R}{r} \right)^n C_{n,m} [(n+1)(n+2) + \\ + \frac{3}{4}(5n+11) J_2 \sin 2\theta \frac{d}{d\theta}] P_{n,|m|}(\cos\theta) Q_m(\lambda). \tag{9}$$

The right hand side contains the operator of differentiation which can be «removed» by means of the relation

$$\sin\theta \cos\theta \frac{d P_{n,|m|}}{d\theta} = \bar{a}_{n,m} P_{n+2,|m|} - \bar{b}_{n,m} P_{n,|m|} - \\ - \bar{c}_{n,m} P_{n-2,|m|} \tag{10}$$

where $\bar{a}_{n,m}$, $\bar{b}_{n,m}$ and $\bar{c}_{n,m}$ are certain numerical constants.

After substituting (10) in (9) and transformations the following expression is derived:

$$\Delta\Gamma = \Delta\Gamma_0 + \delta\Gamma, \tag{11}$$

$$\Delta\Gamma_0 = \frac{\mu}{r^3} \sum_{n=2}^{\infty} \sum_{m=-n}^{n} \left(\frac{R}{r} \right)^n (n+1)(n+2) \bar{C}_{n,m} \bar{Y}_{n,m}(\theta,\lambda), \tag{12}$$

$$\delta\Gamma = \frac{3}{2} \frac{\mu}{r^3} J_2 \sum_{n=0}^{\infty} \sum_{m=-n}^{n} \left(\frac{R}{r} \right)^n \times \\ \times [(n-2)(5n+1) \alpha_{n,m} \bar{C}_{n-2,m} - (5n+11) \beta_{n,m} \bar{C}_{n,m} - \\ - (n+3)(5n+21) \alpha_{n+2,m} \bar{C}_{n+2,m}] \bar{Y}_{n,m}(\theta,\lambda). \tag{13}$$

Here $\bar{C}_{n,m}$ and $\bar{Y}_{n,m}(\theta,\lambda)$ are the fully normalized quantities and $\alpha_{n,m}$, $\beta_{n,m}$ are certain numerical constants.

Let us designate by $\Delta\bar{\Gamma}$ the observed magnitude of the gravity gradient anomaly $\Delta\Gamma$. This generalized observable will be expressed below in terms of the orbital coordinates $\{u,v,w\}$.

The optimum BV relation for the determination of the earth's potential from satellite gradiometry measurements will be

$$\Delta\bar{\Gamma} = \Delta\Gamma \tag{14}$$

where the right hand side is represented by the series (11) – (13). Relation (14) can be written in form

$$\Delta\bar{\Gamma} = \Delta\Gamma^N \left(\bar{Y}_{n,m}(\theta,\lambda), \bar{C}_{n,m} \right) \tag{15}$$

where $\Delta\Gamma^N$ is the truncation (up to degree $n = N$) of the above series for $\Delta\Gamma$. This equation can be solved by the LS technique.

Below from the BV equation (14) a quadrature formula for $\bar{C}_{n,m}$ will be derived, depending on the observable $\Delta\bar{\Gamma}$.

3 Quadrature formula for the potential coefficients

Let us suppose that the observational data are reduced to a mean orbital sphere whose radius is designated by the same symbol r, as was used for the geocentric distance.

After substituting the series (11) – (13) in the right hand side of (14) and equating the harmonics of the same degree and order from both sides of it, the following formula is derived:

$$\bar{C}_{n,m} = K_{n,m} - \frac{3}{2} J_2 \left[\tilde{a}_{n,m} \bar{C}_{n-2,m} - \tilde{b} \bar{C}_{n,m} - \tilde{c} \bar{C}_{n+2,m} \right]. \tag{16}$$

Here

$$K_{n,m} = \frac{r^3}{4\pi\mu} (n+1)^{-1} (n+2)^{-1} \left(\frac{r}{R} \right)^n \times \\ \times \int_\sigma \Delta\bar{\Gamma} \bar{Y}_{n,m}(\theta,\lambda) d\sigma, \quad n = 2,3,\ldots, \ |m| \le n. \tag{17}$$

The quantities $\tilde{a}_{n,m}$, $\tilde{b}_{n,m}$ and $\tilde{c}_{n,m}$ are certain numerical constants.

The right hand side of (16) contains correction terms proportional to the small quantity J_2 given in (5). Then as an initial approximation for an iteration calculation one can accept

$$\overline{C}_{n,m} = K_{n,m}, \quad n = 2,3,\ldots, |m| \le n. \tag{18}$$

After substituting (18) in the correction terms in (16) one has

$$\overline{C}_{n,m} = \left(1 + \frac{3}{2} J_2 \tilde{b}_{n,m}\right) K_{n,m} - \frac{3}{2} J_2 \times \\ \times [\tilde{a}_{n,m} K_{n-2,m} - \tilde{c}_{n,m} K_{n+2,m}]. \tag{19}$$

By substituting the integrals $K_{p,m}$ from (17) into (19) the simple final quadrature formula is derived:

$$\overline{C}_{n,m} = (n+1)^{-1}(n+2)^{-1}\left(\frac{r}{R}\right)^n I_{n,m}. \tag{20}$$

Here

$$I_{n,m} = \frac{r^3}{4\pi\mu} \int_\sigma \Delta\overline{\Gamma} F_{n,m}(\theta,\lambda)\, d\sigma,$$
$$F_{n,m} = \left(1 + \frac{3}{2} J_2 a_{n,m}\right) \overline{Y}_{n,m}(\theta,\lambda) - \tag{21}$$
$$- \frac{3}{2} J_2 \left[b_{n,m} \overline{Y}_{n-2,m}(\theta,\lambda) - c_{n,m} \overline{Y}_{n+2,m}(\theta,\lambda)\right]$$

where

$$a_{n,m} = \frac{(5n+11)(n^2 - 3m^2 + n)}{(n+1)(n+2)(2n+3)(2n-1)},$$
$$b_{n,m} = \frac{(n-2)(5n+1)}{n(n-1)} \alpha_{n,m}, \quad n \ne 3; \; b_{3,m} = 0,$$
$$c_{n,m} = \frac{5n+21}{n+4} \alpha_{n+2,m}, \quad n = 2, 3, \ldots, |m| \le n,$$
$$\alpha_{n,m} = \frac{\sqrt{n^2 - m^2}\sqrt{(n-1)^2 - m^2}}{(2n-1)\sqrt{(2n+1)(2n-3)}}.$$

The right hand side of (21) depends on the observable $\Delta\overline{\Gamma}$. In (Petrovskaya, 1996), by applying (7) and using the relations between the coordinates $\{x,y,z\}$ and $\{u,v,w\}$, the observable $\Delta\overline{\Gamma}$ was expressed in terms of the potential derivatives with respect to the orbital coordinates:

$$\Delta\overline{\Gamma} = T_{ww} - \frac{3}{4r} J_2 \sin 2\theta \; (\cos\alpha \; T_u - \sin\alpha \; T_v) + \\ + \frac{15}{4} J_2 \sin 2\theta \; (\cos\alpha \; T_{uw} - \sin\alpha \; T_{vw}). \tag{22}$$

Here α is the satellite track azimuth, i.e. the angle between the tangent to the satellite orbit in the flight direction and the tangent to the local meridian in the north direction.

The derivative T_{ww} in (22) can be replaced by some combinations of the diagonal derivatives (due to Laplace relation). In particular, one of two expressions can be used:

$$T_{ww} = -T_{uu} - T_{vv}, \quad T_{ww} = \frac{1}{3}(2T_{ww} - T_{uu} - T_{vv}).$$

In the latter case all the diagonal derivatives of the disturbing potential will present in the solution (20) – (23).

4 Conclusions

The derived solution of the scalar spaceborne *BV* problem is an optimal one since in it the contribution of the gravity gradient magnitude is reproduced most completely. The generalized observable $\Delta\overline{\Gamma}$ depends on the derivatives with respect to the orbital coordinates: T_{uu}, T_{vv}, T_{ww}, T_{uw}, T_{vw}, T_u and T_v. All of them, except the diagonal derivatives, have small factors proportional to the quantity J_2. The first order derivatives can be simultaneously excluded from both sides of the *BV* relation, given by (7), (14) and (22), because it represents actually an identity. This operation was performed in (ibid.). The corresponding integral formula, similar to the above one, can be developed for $\overline{C}_{n,m}$.

The derived solution for $\overline{C}_{n,m}$ can be applied for performing spherical synthesis to derive the closed integral expressions for different functions of the potential at the earth's surface. The procedures will be similar to those developed by Petrovskaya and Zieliński (1998, 1999), when solving airborne gradiometry problems.

Acknowledgments. This research was supported by the Russian Foundation of the Fundamental Research (Project № 99-01-007-39) and the Polish Committee for Scientific Research (Project № SM1-94-164).

References

Albertella, A., F. Migliaccio, F. Sansó (1995a). Global gravity field recovery by use of STEP observations. In: *Proc. of IAG Symposia,* Vol. 113, *«Gravity and Geoid».* Sünkel, H. and Marson, I. (eds), Springer, Berlin, Heidelberg, New York, pp. 111-115.

Albertella, A., F. Migliaccio, F. Sansó (1995b). Application of the concept of biorthogonal series to a simulation of a gradiometric mission. In: *Proc. of IAG Symposia,* Vol. 114, *«Geodetic Theory Today».* Sansó, F. (ed), Springer, Berlin, Heidelberg, New York, pp. 350-361.

Koop, R. (1993). Global gravity field modeling using satellite gravity gradiometry. *Netherlands Geodetic Commission, New Series,* No 38.

Petrovskaya, M.S. (1996). Optimal approach to the investigation of the Earth's gravitational field by means of satellite gradiometry. *Artificial Satellites,* Vol. 31, No 1, Warsawa, pp. 1-23.

Petrovskaya, M.S. (1997). Global gravity field determination from *STEP* mission gradiometry observations. In: *Proc. of IAG Symposia «Gravity, Geoid and Marine Geodesy».* Segawa, J., Fujimoto, H., Okubo, S. (eds) Vol. 117, Springer, Berlin, Heidelberg, New York, pp. 188-195.

Petrovskaya, M. S. and J. B. Zieliński (1997). Determination of the global and regional gravitational fields from satellite and balloon gradiometry observations. *Advances in Space Research,* Vol. 19, No 11, pp. 1723-1728.

Petrovskaya, M. S. and J. B. Zieliński (1998). Evaluation of the regional gravity anomaly from balloon gradiometry. *Bollettino di Geodesia e Scienze Affini,* No 2, Firenze, Italy, pp. 141-164.

Petrovskaya, M. S. and J. B. Zieliński (1999). Solution of the airborne gradiometry boundary value problem for the height anomaly and gravity anomaly. *Bollettino di Geodesia e Scienze Affini,* No 1, Firenze, Italy, pp. 33-51.

Rummel, R. (1986). Satellite Gradiometry. In: *Lecture Notes in Earth Science «Mathematical and Numerical Techniques in Physical Geodesy»,* Sünkel, H. (ed), Vol. 7, Springer-Verlag, Berlin.

Rummel, R. and O. L. Colombo (1985). Gravity field determination from satellite gradiometry. *Bulletin Geodesique,* Vol. 59, pp. 233-246.

Rummel, R., F. Sansó, M. Van Gelderen, P. Koop, E. Schrama, M. Brovelli, F. Miggliaccio, F. Sacerdote (1993). Spherical harmonic analysis of satellite gradiometry. Netherlands Geodetic Commission, New Series, No 39.

Data analysis for the GOCE mission

R. Klees, R. Koop, P. Visser, J. van den IJssel
Delft Institute for Earth-Oriented Space Research (DEOS),
Delft University of Technology, Thijsseweg 11, NL-2629 JA Delft, The Netherlands

R. Rummel
Institut für Astronomische und Physikalische Geodäsie,
Munich University of Technology, Arcisstrasse 21, D-80290 Munich, Germany

Abstract. We investigate the time-wise approach to the data anlaysis for the GOCE mission. The number of observations collected during the mission, the number of potential coefficients to be estimated, and the complexity of the mathematical model for the time-wise approach require a special strategy, which has to reduce the CPU-time and storage requirements considerably. Our approach is based on (1) the iterative solution of the normal equations using a Richardson-iteration scheme and (2) the approximation of the design matrix in order to assemble the right-hand side in each iteration step efficiently. We demonstrate the performance of the approach for white noise and coloured noise observations along a simulated GOCE orbit up to degree and order 180. We provide error estimates and show that the solution is unbiased. We also prove that the method does not converge to the solution of the normal equations. However, the approximation error can be neglected in our simulations.

Keywords. Gravity field determination, satellite gravity gradiometry, GOCE, data analysis.

1 Introduction

The Gravity Field and Steady State Ocean Circulation Explorer (GOCE) is one of four candidate missions under assessment by the European Space Agency (ESA) for the implementation in ESA's Earth Explorer Programme. The main objective of the mission is to provide a high-accuracy high-resolution global model of the Earth's static gravity field from a combination of spaceborne gravity gradiometry (SGG) and high-low satellite-to-satellite tracking (SST). A homogeneous accuracy of about 1 cm in terms of geoid heights and better than $2 \cdot 10^{-5}$ m/s^2 in terms of gravity anomalies will be feasible for wavelengths down to 200 km.

In the past numerous studies were done, which addressed the fundamental problems regarding the gravity field recovery from SGG and SST observations. Among them are the well documented studies by the Consortium for the Investigation of Gravity Anomaly Recovery (CIGAR) (CIGAR I, 1989; CIGAR II, 1990; CIGAR III/1, 1993; CIGAR III/2, 1995; CIGAR IV, 1996). In the framework of these studies, consistent mathematical models were developed for the gravity field recovery from SGG and SST observations, potential data analysis strategies were proposed and investigated, and numerous simulations and covariance propagation studies were carried out for various mission and system designs. Recently, the focus of attention was directed towards sophisticated end-to-end closed-loop simulation studies aiming at a detailed modelling of the satellite and its instruments, and the contribution of GOCE end products to oceanography and geodynamics. Part of the work was done in the framework of the Phase A study, which focussed on the definition of an optimal mission architecture for GOCE, and the demonstration of its feasibility, performance, and cost-effectiveness (cf. Alenia, 1999). Currently, the so-called Eötvös to Milligal study concentrates on all steps needed to provide the mission products for realistic mission scenarios. Among them are the data processing approach, the data processing algorithms, the data production quality assessment, and a number of special issues such as polar gap, regularization, and temporal variations of the Earth's gravity field. These activities have to be seen as important steps towards a scientific data centre for the GOCE mission, cf. Fig. (1). In this figure, the activities described under "sensor unit", which cover e.g. instrument readouts, calibration, corrections, and stochastic models of the observations, are likely to be the responsibility of ESA. The data processor unit, the end product unit, and the quality

International Association of Geodesy Symposia, Vol. 121
Schwarz (ed.), Geodesy Beyond 2000 – The Challenges of the First Decade

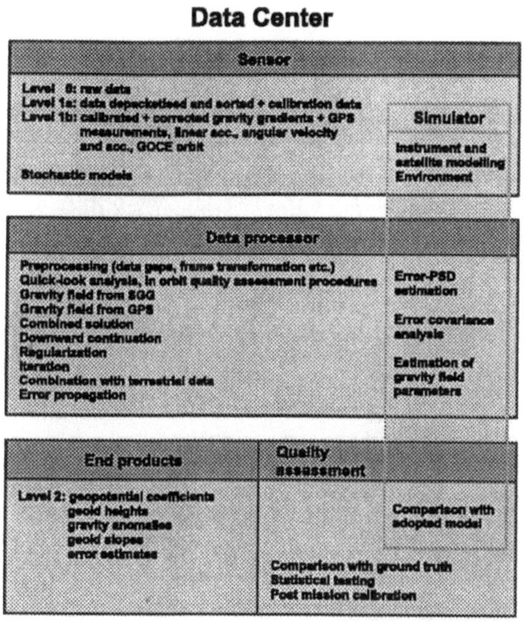

Fig. 1: Sketch of a scientific data centre for the GOCE mission

assessment unit will likely be the responsibility of an international scientific consortium. All tasks listed in Fig. (1) are key issues and the subject of numerous scientific studies. Practical solutions to them, i.e. solutions which can handle real mission conditions, are currently being developed or have still to be developed in the first years of the next millenium. Among them are for instance (1) the assessment of different approaches for gravity field recovery from real data, (2) the development of efficient algorithms for the different approaches, (3) the development and analysis of approaches for data quality assessment on the fly and a posteriori, and (4) special tasks such as regularization techniques and the combination with other space, airborne or ground data.

The objective of this paper is to study data analysis methods for the estimation of potential coefficients from SGG observations for realistic mission conditions. In particular we want to investigate the so-called time-wise in the time domain approach and propose a fast iterative solution strategy for the gravity field recovery from GOCE SGG data. In Section 2 we briefly discuss the two main approaches of GOCE data analysis, i.e. the space-wise and the time-wise approach. Section 3 contains the mathematical model of the time-wise in the time domain approach and the stochastic model of the SGG observations. Moreover, we formulate the parameter estimation problem as a standard Gauss-Markov model and discuss the numerical problems of this straightforward approach. In Section 4 we propose an iterative solution strategy, which overcomes the numerical problems and may be used under real mission conditions. We also summarize the main results concerning error estimates, convergence, and bias of the approximate solution. Finally, in Section 5 we present results of some numerical experiments done in order to assess the performance and accuracy of the iterative solution method.

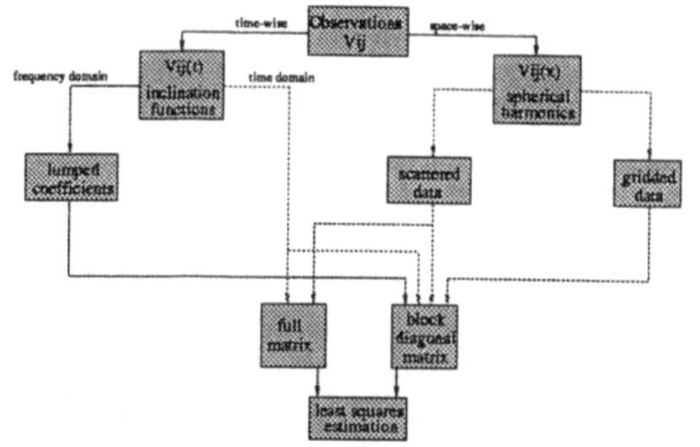

Fig. 2: The space-wise and time-wise approach to the gravity field recovery from SGG observations

2 Data analysis methods

There are two main approaches that were proposed so far for the estimation of potential coefficients from SGG (and SST) observations: the space-wise approach, and the time-wise approach, cf. Fig. (2).

The space-wise approach considers the measurements as a function of position. The solution techniques fully coincide with the techniques well-established in the field of physical geodesy for the solution of geodetic boundary value problems. If the data are reduced to a spherical surface and interpolated on a regular grid, fast numerical techniques for harmonic analysis can be applied in order to estimate the potential coefficients. Alternatively, we may first consider the data as a continuum and apply quadrature formulas in order to estimate the potential coefficients.

The time-wise approach considers the measurements as a discrete time series collected along the GOCE orbit. It requires that the spherical harmonic representation of the earth's gravity field is transformed from the earth-fixed frame to the orbit frame. The so-called time domain variant directly operates on the time series of measurements; the frequency domain variant first applies a discrete Fourier transform, and then considers the Fourier coefficients, the so-called *lumped coefficients*, as pseudo-observables from which the potential coefficients are determined by linear parameter estimation techniques. For a detailed comparison of both approaches, see (Rummel et al., 1993).

3 The time-wise in the time domain approach

The time-wise approach represents the gravitational potential as function of the Kepler orbit elements. For nearly circular orbits this reads (Koop, 1993)

$$V(P) = \sum_{l=0}^{\infty} \sum_{m=0}^{l} \sum_{k=-l[2]}^{l} H_{lmk}(r, I) \tag{1}$$

$$[\alpha_{lm} \cos \psi_{km} + \beta_{lm} \sin \psi_{km}] \ ,$$

with

$$H_{lmk}(r, I) = \frac{GM}{R} \left(\frac{R}{r}\right)^{l+1} \bar{F}_{lm}^{k}(I) \ , \tag{2}$$

$$\alpha_{lm} = \begin{cases} \bar{C}_{lm} & \text{if } l-m \text{ is even} \ , \\ -\bar{S}_{lm} & \text{if } l-m \text{ is odd} \ , \end{cases}$$

$$\beta_{lm} = \begin{cases} \bar{S}_{lm} & \text{if } l-m \text{ is even} \ , \\ \bar{C}_{lm} & \text{if } l-m \text{ is odd} \ . \end{cases}$$

$\bar{F}_{lm}^{k}(I)$ are the normalized inclination functions, $\bar{C}_{lm}, \bar{S}_{lm}$ the 4π-normalized potential coefficients, $\psi_{km} = \psi_{km}(t) = k(\omega + M) + m(\Omega - \Theta_G) =: k\omega_o + m\omega_e$ the inertial argument, I, ω, M, Ω Kepler orbit elements, r the distance from the coordinate origin, and Θ_G the earth's argument of longitude. t is the time. In the gravity field recovery from SGG observations the GOCE orbit is assumed to be known. This is justified because the orbit can be determined in a dynamic-kinematic procedure at the cm-level. The remaining gradient differences are of the order 0.1 mE, about a factor 10 smaller than the expected performance of the measured gravity gradients. Moreover, in order to avoid the same ψ for different (k, m), the maximum degree of the spherical harmonic expansion must be smaller than half the number of orbital revolutions in one repeat cycle, which is easily fulfilled for realistic mission scenarios. The measured gradients can now be written as linear functions of the potential coefficients. For instance, we obtain for the second radial derivative, V_{rr}:

$$V_{rr}(P) = \sum_{l=0}^{\infty} \sum_{m=0}^{l} \sum_{k=-l[2]}^{l} H_{lmk}^{rr}(r, I) \tag{3}$$

$$[\alpha_{lm} \cos \psi_{km} + \beta_{lm} \sin \psi_{km}] \ ,$$

with

$$H_{lmk}^{rr}(r, I) = (l+1)(l+2)/r^2 \, H_{lmk}(r, I) \ . \tag{4}$$

Assuming (i) certain components of the gradient tensor are measured along the orbit and they are collected in the observation vector Y, (ii) the potential coefficients form a vector X, and (iii) the harmonic expansion is truncated at some maximum degree L, we may write the mathematical model as $E\{Y\} = AX$, with $E\{\cdot\}$ the expectation operator. That is, the mathematical model is linear in the unknown potential coefficients. If N observations are collected, the design matrix A has dimension $N \times u$, where $u = (L+1)^2 - 3$ is the number of unknown potential coefficients.

A quite realistic stochastic model of SGG observations can be obtained from a detailed analysis of possible error sources, their characteristics and interactions. This is possible for instance with the end-to-end closed-loop simulator developed under contract of ESA by the SID consortium, a cooperation between the Delft Institute for Earth-Oriented Space Research (DEOS), the Netherlands Space Research Organisation (SRON), and the Institute for Astronomical and Physical Geodesy at the Technical University of Munich (IAPG). It allows to study, among others, sensor errors, control unit errors, and environmental effects, taking the interaction between the sensors, control loops, and other subsystems into account. A typical result of this type of simulations in terms of the power spectral density of the observation errors is shown in Fig. (3). It confirms that the error has not the character of white noise over the entire spectrum. Only in the band between about 0.1 Hz and $5 \cdot 10^{-3}$ Hz, the so-called measurement bandwidth, the spectrum is almost flat at a level of about $1-2$ mE.

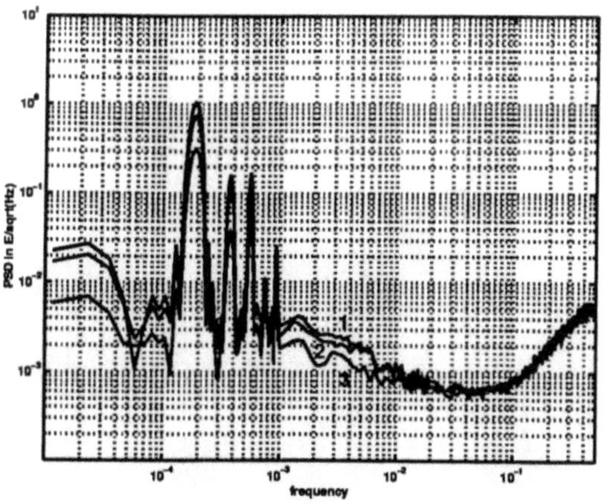

Fig. 3: The square root of the error power spectral density function of SGG observations in $E/\sqrt{Hz}$ from an end-to-end closed-loop simulation. 1: V_{zz}, 2: V_{yy}, 3: V_{xx}

This means that the variance-covariance matrix of the measured gravity gradients is a full matrix. The error power spectral density shown in Fig. (3) show the same characteristic features as the adopted error model for the GOCE Phase A study (cf. Alenia, 1999).

Now, as functional model of the gravity field recovery from SGG observations we may use the standard Gauss-Markov model (Grafarend and Schaffrin, 1993):

$$E\{Y\} = AX, \qquad D\{Y\} = \sigma^2 Q \ . \tag{5}$$

If X_0 contains the potential coefficients of the adopted normal geopotential model, we may also write

$$E\{y_0\} = Ax_0, \qquad D\{y_0\} = \sigma^2 Q \ , \tag{6}$$

with the reduced observation vector $y_0 = Y - AX_0$, and the new vector of unknowns $x_0 = X - X_0$, which contains the potential coefficients of the disturbing potential. Note that X_0 is assumed to be a real non-stochastic vector. Now, the Best Linear Uniformly Unbiased Estimation w.r.t. the Q-norm of x_0 is

$$\xi_0 = N^{-1} z_0, \qquad D\{\xi_0\} = \sigma^2 N^{-1} \ , \tag{7}$$

with the normal equation matrix $N = A^T PA$, the right-hand side $z_0 = A^T P y_0$, and $P = Q^{-1}$. When using a sophistiacted mathematical model A, which allows for instance for real perturbed orbits, satellite maneuvers and data gaps, the normal equation matrix N is full. Moreover, the design matrix A is almost impossible to compute explicitly due to the huge number of observations during the mission and the costly computation of the matrix elements. For instance, for an observation V_{rr} acquired at time epoch t_j at position $\omega_o(t_j), \omega_e(t_j), r(t_j)$, the coefficient of A assigned to a potential coefficient of degree l and order m requires the computation of an expression like

$$\sum_{k=-l[2]}^{l} H^{rr}_{lmk}(r(t_j), I) arg(\psi_{km}(t_j)) \ , \tag{8}$$

where arg is either cos or $\pm$sin, depending on whether $l-m$ is even or odd and whether we are dealing with a coefficient $\bar{C}_{lm}$ or $\bar{S}_{lm}$. Thus, the corresponding element of the vector $A^T y$ requires the computation of an expression like

$$\sum_{j+1}^{N} \sum_{k=-l[2]}^{l} H^{rr}_{lmk}(r(t_j), I) arg(\psi_{km}(t_j)) \ , \tag{9}$$

and the computation of an element of for instance $A^T A$

$$\sum_{j+1}^{N} \left(\sum_{k_1=-l_1[2]}^{l_1} H^{rr}_{l_1 m_1 k_1}(r(t_j), I) arg_1(\psi_{k_1 m_1}(t_j)) \right)$$

$$\left(\sum_{k_2=-l_2[2]}^{l_2} H^{rr}_{l_2 m_2 k_2}(r(t_j), I) arg_2(\psi_{k_2 m_2}(t_j)) \right) .$$

The complexity of setting up the normal equations makes the estimation problem for the time-wise in the time domain approach very demanding in terms of computer power and storage requirements. Compared to the space-wise approach the fundamental difference is the extra summation over the index k, which is needed in the time-wise approach to link the gravitational frequencies to the orbital frequencies. Any fast solution strategy has to address the numerical complexity and storage requirements.

4 Iterative solution strategy

Let us assume that we have suitable approximations M to N and A_0 to A at our disposal. Suitable means that M and A_0 allow (1) to set up and solve the normal equations very quickly, (2) require only moderate memory, and (3) introduce sufficiently small approximation errors. Then, we may replace the normal equations $N\xi_0 = z_0$ by the approximate normal equations

$$M\tilde{\xi}_0 = \tilde{z}_0, \qquad \tilde{z}_0 = A_0^T P y_0 \,, \tag{10}$$

and try to reduce the approximation error by iteration: From $\tilde{\xi}_0$ we obtain the improved set of potential coefficients $\tilde{X}_0 = X_0 + \tilde{\xi}_0$. We take this as the new normal geopotential model $X_1 = \tilde{X}_0$. Then, X_1 is considered as a real non-stochastic vector, and the reduced observation vector becomes $y_1 = Y - AX_1$. Finally, we compute the new right-hand side $\tilde{z}_1 = A_0^T P y_1$ and solve the system $M\tilde{\xi}_1 = \tilde{z}_1$. So, the iteration scheme is:

Given Y, P, M, X_0. For $i = 0, 1, \ldots$, compute

$$y_i = Y - AX_i$$

$$\tilde{z}_i = A_0^T P y_i$$

$$\tilde{\xi}_i = M^{-1}\tilde{z}_i$$

$$\tilde{X}_i = X_i + \tilde{\xi}_i$$

$$X_{i+1} = \tilde{X}_i \tag{11}$$

The iteration stops until there is no significant improvement anymore. Note that both M and A_0 are kept fixed in the iteration, and the reduced observation vector $y_i = Y - AX_i = y_{i-1} - A\tilde{\xi}_{i-1}$ is updated correctly. Moreover, the vector X_i is considered as a real non-stochastic vector!

(Klees et al., 1999) have investigated the convergence of the iteration scheme. They showed that the approximation error after iteration step i, $\varepsilon_i = \tilde{X}_i - X$, is

$$\varepsilon_i = N^{-1}(z_i - \tilde{z}_i) + N^{-1}(I - NM^{-1})\tilde{z}_i \,, \tag{12}$$

with $z_i = A^T P y_i$. Moreover, the iteration scheme converges in the limit $i \to \infty$ to $\tilde{X}_\infty = X_0 + M_0^{-1}\tilde{z}_0$ iff the spectral radius of $I - M^{-1}M_0$, with $M_0 = A_0^T P A$, is smaller than 1. Finally, the limit is unbiased, i.e. $E\{\tilde{X}_\infty\} = X$.

Up to now we have left open how appropriate approximations M to N and A_0 to A can be defined. In our approach, the matrix M is the normal equation matrix we would obtain if the following conditions are fulfilled:

1. the orbit is a circular repeat orbit,

2. we have an uninterrupted time series of measurements,

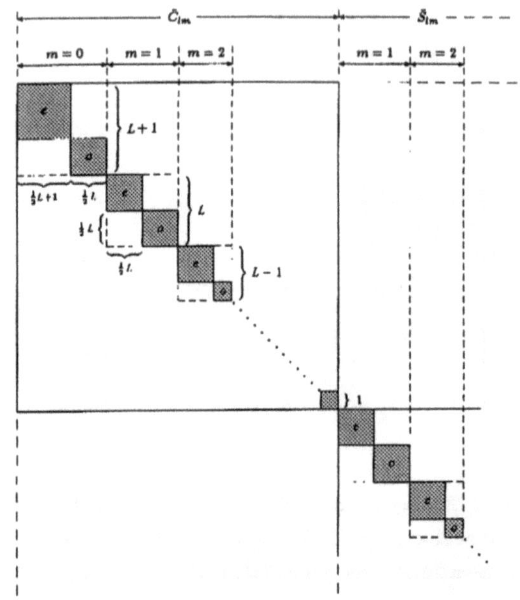

Fig. 4: Structure of the matrix M used in the iterative solution. L denotes the maximum degree of the spherical harmonic expansion to be solved for

3. the number of nodal days and the number of orbital revolutions in one repeat cycle have no common divisor,

4. the maximum degree and order L of the spherical harmonic expansion of the potential we want to solve for is at most half the number of orbital revolutions in a repeat cycle.

Then, the normal equation matrix would be a block-diagonal matrix with the maximum sub-block of size $(L/2+1) \times (L/2+1)$, cf. Fig. 4. Moreover, each individual non-zero entry of the normal equation matrix can be computed analytically. For instance, the computation of one entry (observation type V_{rr}, coefficients (l_1, m) and (l_2, m), $l - m$ even, $m \neq 0$) requires the evaluation of an expression like:

$$\sum_{k=-\min(l_1,l_2)[2]}^{\min(l_1,l_2)} \bar{F}_{l_1 m}^k \bar{F}_{l_2 m}^k / \sigma_{km}^2 \,, \tag{13}$$

where σ_{km}^2 is the variance element of the band-limited and coloured noise model. The block-diagonal structure allows to solve the approximate normal equations in each iteration step very fastly order by order; each block can be inverted separately.

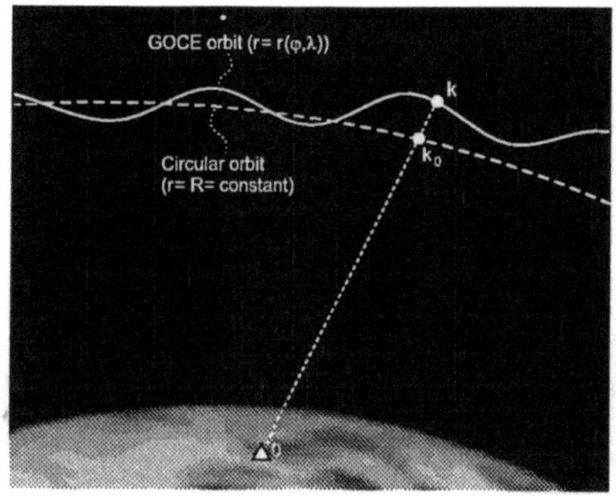

Fig. 5: Choice of the approximation A_0 to A. The entries of A_0 are computed along a circular orbit, which speeds up the computation significantly

The matrix A_0 is computed along a *circular* orbit, which approximates the GOCE orbit (cf. Fig. 5), i.e. instead of computing the entries at points k, we compute them at points k_0. Then, the terms $\left(\frac{R}{r}\right)^{l+1}$ are constant, which speeds up the assembly of A_0 considerably.

5 Results and discussion

In order to investigate the performance and the accuracy of the iterative solution method we did a number of test calculations. We used the geopotential model *OSU91A* (cf. Rapp et al., 1991) as reference model and generated observations along various orbits. Then, we applied the iterative scheme and computed after each iteration step the (averaged) degree-order error, $e_l^{(i)}$, defined by

$$\left(e_l^{(i)}\right)^2 = \frac{1}{2l+1} \sum_{m=0}^{l} \left[\left(\bar{C}_{lm} - \bar{c}_{lm}^{(i)}\right)^2 + \left(\bar{S}_{lm} - \bar{s}_{lm}^{(i)}\right)^2\right], \qquad (14)$$

$$l = 2, \dots L, \quad i = 0, 1, \dots ,$$

where $\bar{C}_{lm}, \bar{S}_{lm}$ are the "true" OSU91A geopotential coefficients (i.e. the elements of the vector X) and $\bar{c}_{lm}^{(i)}, \bar{s}_{lm}^{(i)}$ are the estimated geopotential coefficients for the iteration step i, i.e. the coefficients of the vector $\tilde{X}_i$.

Fig. (6) shows the results for a 29-day circular orbit without observation noise. All coefficients up to degree and order $L = 180$ were estimated. Since the orbit is circular, we have in fact done the iteration $M\tilde{\xi}_i = z_i, \quad i = 0, 1, \dots$, i.e. $A_0 = A$, which means that the right-hand side was not approximated. Then, the iteration converges to the solution of the normal equations $N\xi_0 = z_0$ iff the spectral radius of $I - M^{-1}N$ is smaller than (Klees et al., 1999). The speed of convergence is quite fast. After 7 iteration steps no significant improvement was obtained anymore. Compared to the solution $\tilde{X}_0$ the error could be improved by about 4 orders of magnitude. More accurate results cannot be expected for a 29-day orbit.

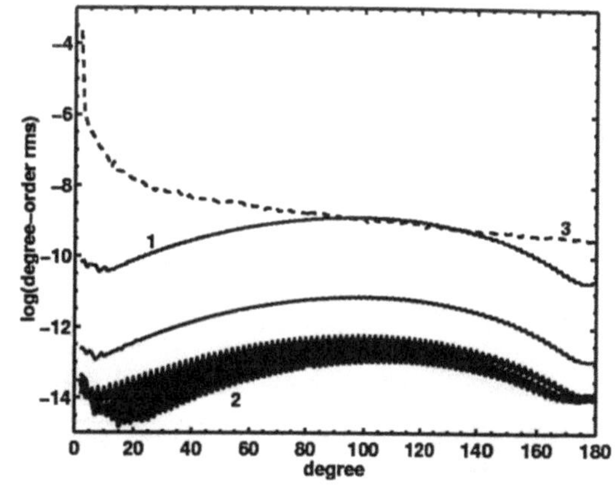

Fig. 6: Averaged degree-order error for a 29-day circular orbit ($i = 96.5^o$, repeat period = 29 days); no observation noise, no regularization applied. 1: solution after step 0 ($\tilde{X}_0$), 2: solution after step 7 ($\tilde{X}_7$), 3: averaged degree-order signal of OSU91A. Solution up to degree and order $L = 180$

Fig. (7) shows the results for a 29-day GOCE orbit. It took 8 iterations until no further improvements were obtained. Again, the error is reduced by about 4 orders of magnitude. Obviously, the contribution to the total error ε_8 of the approximation A_0 by A is insignificant compared to the error due to the limitation to a 29-day orbit.

Fig. (8) shows the effect of the realistic coloured and band-limited noise model shown in Fig. (3). Coloured noise seems to have no significant influence on the convergence of the iteration scheme. Of course, the relative accuracy of the solution per degree is low (about 10^{-3}), which is due to the choice of a 10-day orbit and due to the coloured noise model. The latter clearly shows up when comparing the degree-order error with measurement noise with the degree-order error without measurement noise.

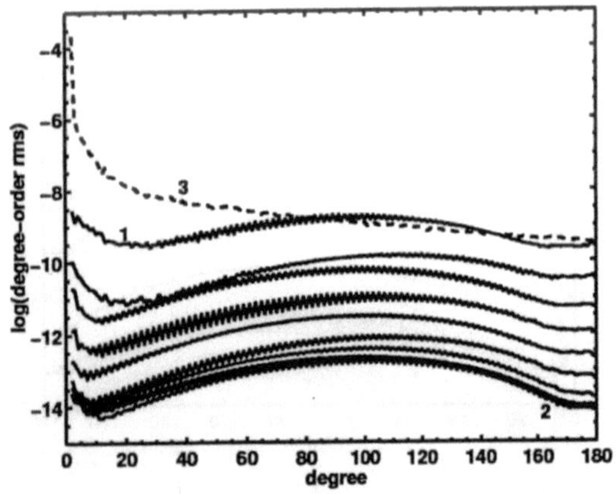

Fig. 7: Averaged degree-order error for a 29-day GOCE orbit ($i = 96.5°$, repeat period = 29 days); no observation noise, no regularization. 1: solution after step 0 ($\bar{X}_0$), 2: solution after step 8 ($\bar{X}_8$), 3: averaged degree-order signal of OSU91A. Solution up to degree and order $L = 180$

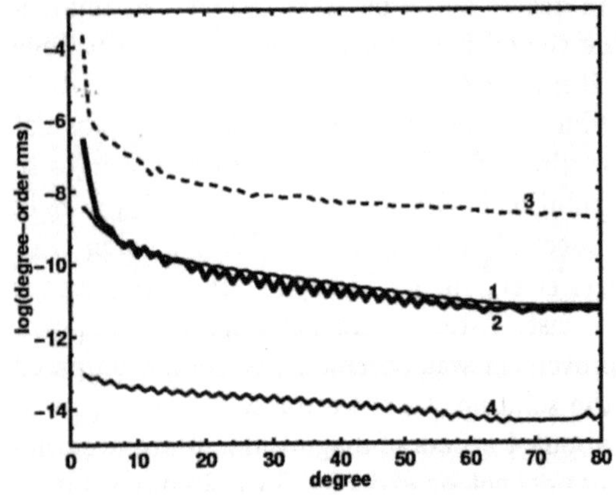

Fig. 8: Averaged degree-order error for a 10-day GOCE orbit ($i = 96.5°$, repeat orbit = 10 days); realistic coloured noise model, no regularization applied. 1: propagated measurement error (covariance propagation), 2: final solution after iteration (coloured noise model), 3: averaged degree-order signal of OSU91A, 4: final solution after iteration (no noise). Solution up to degree and order $L = 80$

Acknowledgements. A. Selig from SRON and J. Müller from IAPG provide us with the error PSD shown in Fig. (3). Computing resources were provided by the Centre for High Performance Applied Computing (HPαC) of Delft University of Technology. Their support is gratefully acknowledged.

References

Alenia (1999). Gravity field and ocean circulation explorer. Phase A Study, Executive Summary, DOC: GOC-RP-Al-0006, Alenia Aerospazio.

CIGAR I (1989). CIGAR I: Study on precice gravity field determination methods and mission requirements (phase 1). Final Report. ESA contract No. 7521/87/F/F/FLI, European Space Agency.

CIGAR II (1990). CIGAR II: Study on precice gravity field determination methods and mission requirements (phase 2). Final Report. ESA contract No. 8153/88/F/FL, European Space Agency.

CIGAR III/1 (1993). CIGAR III: Study on precice gravity field determination methods using gradiometry and GPS (phase 1). Final Report. ESA contract No. 9877/92/F/FL, European Space Agency.

CIGAR III/2 (1995). CIGAR III: Study on precice gravity field determination methods and mission requirements (phase 2). Final Report. ESA contract No. 10713/87/F/FL, European Space Agency.

CIGAR IV (1996). CIGAR IV: Study of advanced reduction methods for spaceborne gravimetry data, and of data combination with geophysical parameters. Final Report. ESA contract No. 152163 - ESA study ESTEC/JP/95-4-137/MS/nr, European Space Agency.

Grafarend, E. and Schaffrin, B. (1993). *Ausgleichungsrechnung in linearen Modellen.* BI Wissenschaftsverlag, Mannheim.

Klees, R., Koop, R., and van der IJssel, J. (1999). Fast gravity field recovery from GOCE gravity gradient observations. (In preparation).

Koop, R. (1993). *Global gravity field modelling using satellite gravity gradiometry.* Publications on Geodesy, New Series, Number 38, Netherlands Geodetic Commission.

Rapp, R., Wang, Y., and Pavlis, N. (1991). *The Ohio State 1991 geopotential and sea surface topography harmonic coefficient model.* Report 410, Dept. of Geodetic Science and Surveying, The Ohio State University.

Rummel, R., van Gelderen, M., Koop, R., E, E.S., Sansó, F., Brovelli, M., Migliaccio, F., and Sacerdote, F. (1993). *Spherical harmonic analysis of satellite gradiometry.* Publications on Geodesy, New Series, Number 39, Netherlands Geodetic Commission.

Direct and local comparison between different satellite missions for the gravity field on-the-fly

A. Albertella, F. Migliaccio, F. Sansò
DIIAR, Sez. Rilevamento, Politecnico di Milano, Piazza Leonardo da Vinci 32, I-20133 Milano
e-mail: gradio@ipmtf4.topo.polimi.it

Abstract. Many different satellite missions are in project at the moment to estimate the global gravity field of the earth and its time variations. The principles of measurement are basically three, namely: the measurement of accurate orbit anomalies and of non-gravitational forces acting on the satellite; the measurement of differential orbit anomalies and of their time variations along the line of sight of twin satellites; the measurement of the gradiometric tensor. In all three cases it is interesting to discuss the possibility of being able to compare the observations performed during a mission at different times and at points close in space, as well as to compare different kinds of observations from different missions approximately in the same area. This could produce on-the-fly calibration procedures and cross-calibration procedures. To reach this goal one can try to "localize" the observations so that the problem is reduced to the comparison of different functionals of the anomalous potential T in the same zone of space.

Keywords. Calibration procedures, gravity field estimation, satellite accelerometry, satellite gradiometry.

1 Introduction

In these days we can perceive a new international effort directed to the determination of a reliable model of the gravity field of the earth by spatial means, based on a set of homogeneous and fairly even distributed observations.

The only drawback is that to measure from space one has to carry an instrument into an orbital altitude which has to be maintained for a suitably long time (e.g. 6 months) at least above the limb of the atmosphere (e.g. about 250 km). Of course this reduces the power of the signal to be measured, namely the gravity field, due to the natural attenuation of harmonic functions inside their domain.

In an effort to capture as much information as possible from measurements, different strategies combining the lowermost orbits with various types of observations have been proposed and are under development, like CHAMP, GRACE and GOCE (ESA, 1998). With such a wealth of measurements spread in space, we will be able to compute several different models of the gravity field, which therefore will have to be compared to one another, and the question arises whether we could or not directly compare and cross-check observations in space by taking advantage of the marked spatial correlation of the gravity field.

According to the present study, it seems that observations made at significantly different altitudes cannot be directly compared with an accuracy acceptable with respect to measurement noise, nevertheless observations at the same altitude can be used to "self calibrate" the instruments when the satellite passes over the same area at different times.

This is most easily done when the direct observations are first preprocessed in order to produce intermediate observables which are expressed by "local" functionals of the gravity field, which can then be treated in the manner of a boundary value problem (globally) or of the collocation concept (locally).

2 Space observations and their localization

When reasoning only on the feasibility of a direct comparison of observations we can in principle start with a very simplified model. Working with mission "M", we can write the equation of the reference orbit as

$$\ddot{\tilde{x}}_M = \nabla u_0(\tilde{x}_M) \tag{1}$$

where u_0 represents the known model potential.

International Association of Geodesy Symposia, Vol. 121
Schwarz (ed.), Geodesy Beyond 2000 – The Challenges of the First Decade

Hill's perturbation equations of motion are written as

$$\ddot{\underline{\xi}}_M + 3n_M^2(I - 3P_r)\underline{\xi}_M = \nabla T(\tilde{\underline{x}}_M) + \underline{f}_p \quad (2)$$

n being the mean motion of the satellite, $\underline{f}_p$ the perturbation forces acting on the satellite (both non-gravitational and gravitational) and P_r a radial projection operator.

The tracking or in situ observations Q_t (performed by mission "M") are non linear functions of the position $\underline{x}_M$ and velocity $\dot{\underline{x}}_M$ of the satellite and of the anomalous potential T. Following the procedure described in (Betti and Sansò, 1989), we represent Q_t in its linearized form

$$Q_t = F_t(\underline{x}_M; \dot{\underline{x}}_M; T) =$$

$$\tilde{Q}_t + K_{Mt}^T \underline{\xi}_M(t) + H_{Mt}^T \dot{\underline{\xi}}_M(t) + L_{Mt} \cdot T \quad (3)$$

where

$\tilde{Q}_t$ = approximate value of the observation;

$\underline{\xi}(t)$ and $\dot{\underline{\xi}}(t)$ = orbit anomaly and its time derivative;

K, H, L = matrices of the (first) derivatives with respect to $\underline{\xi}_M$, $\dot{\underline{\xi}}_M$ and T.

The following relation holds

$$\left| \begin{array}{c} \underline{\xi}_M(t) \\ \dot{\underline{\xi}}_M(t) \end{array} \right| = \left| \begin{array}{c} F_t \\ \dot{F}_t \end{array} \right| \cdot \left| \begin{array}{c} \underline{\xi}_M(0) \\ \dot{\underline{\xi}}_M(0) \end{array} \right| +$$

$$\int_{(\tilde{\underline{x}}_M)}^t \left| \begin{array}{c} G(t,\tau) \\ \dot{G}(t,\tau) \end{array} \right| \cdot \nabla T(\tilde{\underline{x}}_M(\tau)) d\tau \quad (4)$$

where

$G(t,\tau)$ and $\dot{G}(t,\tau)$ = kernels for the particular solutions of the dynamic differential equations.

Localization can be achieved by inspecting the observation equations and suitably exploiting Hill's equations in such a way as to eliminate the need of the orbital parameters $\underline{\xi}_0, \dot{\underline{\xi}}_0$ corresponding to initial data.

In the following we will specify Eq. (3) for the different cases of observation strategies under development.

- *High-Low SST (CHAMP)*

The observations are represented by

$$\underline{Q}_t = \underline{x}_t = \tilde{\underline{x}}_t + \underline{\xi}_t \quad (5)$$

The stochastic order of magnitude (SO), which can be defined through the root mean square of random variables in analogy to the O of Landau, for the observed values of $\underline{\xi}_t$ is given by (Balmino et al., 1998)

$$SO(\underline{\xi}_t) \sim 1\ cm\ . \quad (6)$$

Moreover, bases joining two observation points can be observed with higher precision, for instance $SO \sim 0.5\ cm$.

If we now make the hypothesis that the time interval between observations is $\Delta t = 10\ s$, which gives a distance along the orbit arc of about 100 km, having

$$\frac{\underline{\xi}_{t+2\Delta t} + \underline{\xi}_t - 2\,\underline{\xi}_{t+\Delta t}}{(\Delta t)^2} \sim \ddot{\underline{\xi}} \quad (7)$$

we obtain

$$SO(\ddot{\underline{\xi}}) \sim 0.5 \cdot 10^{-2}\ cm\ s^{-2} = 5\ mGal \quad (8)$$

$$SO(n^2\ \underline{\xi}) \sim 10^{-6}\ cm\ s^{-2} = 1\ \mu Gal \quad (9)$$

- *Low-Low SST (GRACE)*

For this mission, three kinds of observations must be taken into account, and their corresponding SO (Balmino et al., 1998):

a) by GPS observations, considering that the satellite passes from position 1 to position 2 in $\Delta t = 10\ s$ we have

$$\underline{Q}_t = \underline{x}_2 - \underline{x}_1 = \tilde{\underline{x}}_2 - \tilde{\underline{x}}_1 + \underline{\xi}_2 - \underline{\xi}_1 \quad (10)$$

$$SO(\underline{\xi}_2 - \underline{\xi}_1) \sim 5\ mm \quad (11)$$

which is practically the same kind of observable discussed in the previous part for CHAMP;

b) by distance observations we have

$$Q_t = |\underline{x}_2 - \underline{x}_1| = \tilde{Q}_t + \underline{\tau} \cdot (\underline{\xi}_2 - \underline{\xi}_1) \quad (12)$$

where $\underline{\tau}$ represents the unit vector in the along track (tangential) direction to the orbit, hence the index τ will denote this direction;

$$SO(\xi_{2\tau} - \xi_{1\tau}) \sim 5 \cdot 10^{-2}\ mm \quad (13)$$

$$SO(\ddot{\xi}_{2\tau} - \ddot{\xi}_{1\tau}) \sim \frac{\sqrt{2}\cdot 5\cdot 10^{-2}}{\Delta t^2} =$$

$$10^{-4}\ cm\ s^{-2} = 0.1\ mGal \tag{14}$$

c) by Doppler observations we have

$$Q_t = \frac{\partial}{\partial t}|\underline{x}_2 - \underline{x}_1| = \tilde{Q}_t + \dot{\xi}_{2\tau} - \dot{\xi}_{1\tau} \tag{15}$$

which combined with the difference of the radial components $(\xi_{2r} - \xi_{1r})$ in Hill's equations gives

$$\ddot{\xi}_\tau + 2n\dot{\xi}_r = \underline{\tau}\cdot\nabla T \tag{16}$$

The following stochastic orders of magnitude are expected:

a) for $\ddot{\xi}_\tau$

$$SO(\dot{\xi}_{2\tau} - \dot{\xi}_{1\tau}) \sim 5\ 10^{-3}\ mm\ s^{-1}$$

$$SO\left(\frac{\dot{\xi}_{2\tau} - \dot{\xi}_{1\tau}}{\Delta t}\right) \sim 5\ 10^{-4}\ mm\ s^{-2} \tag{17}$$

b) for $2\ n\ \dot{\xi}_r$

$$SO\left(2n\ \frac{\xi_{2r} - \xi_{1r}}{\Delta t}\right) \sim 10^{-5}\ mm\ s^{-2} \tag{18}$$

As a consequence we obtain

$$SO(\underline{\tau}\cdot\nabla T) \sim 5\ mGal \tag{19}$$

- *Gravity gradiometry (GOCE)*

The observations are represented by (ESA, 1999)

$$Q = M \equiv \left[\partial^2_{ik} T\right] \tag{20}$$

In this case it must be noticed that Q has the great advantage of being always localized.

The stochastic order of magnitude of M is given by

$$SO(M) \sim 10^{-12}\ s^{-2} \sim$$

$$1mE/\sqrt{Hz}\ in\ MBW\ . \tag{21}$$

3 Comparison between spatially distributed observations coming from different missions

In this section we will consider two cases: first we will show how to compare gravity models derived from different satellite missions (with a testing procedure); afterwards we will discuss the problem of predicting one observable from another of a different kind and we will produce an elementary, yet meaningful, example.

3.1 Comparison of derived gravity field models

In this case we want to compare the estimated coefficients $\hat{T}^{(1)}_{nm}$ of model (1) with the estimated coefficients $\hat{T}^{(2)}_{nm}$ of model (2), which in general have been estimated using data coming from different missions and have different resolutions:

$$\hat{T}^{(1)}_{nm} \Rightarrow \hat{\underline{T}}^{(1)}_m = \left\{T^{(1)}_{nm}\ ,\ n = m, \ldots, N^{(1)}\right\} \tag{22}$$

$$\hat{T}^{(2)}_{nm} \Rightarrow \hat{\underline{T}}^{(2)}_m = \left\{T^{(2)}_{nm}\ ,\ n = m, \ldots, N^{(2)}\right\} \tag{23}$$

We will use the symbols $C^{(1)}_m, C^{(2)}_m$ to represent the covariance matrices of the coefficients, order by order. In fact it must be noticed that typically correlations exist among estimated coefficients with different degree but same order. A testing procedure can be established as follows. Under the hypothesis that $N^{(2)} > N^{(1)}$, a projection must be performed of $\hat{\underline{T}}^{(2)}_m$ on $\hat{\underline{T}}^{(1)}_m$, using the projector Π_{21}

$$\Pi_{21}\ \hat{\underline{T}}^{(2)}_m = \hat{\underline{T}}^{(21)}_m \quad ,\ n = m, \ldots, N^{(2)} \tag{24}$$

Introducing the covariance of the $(N^{(1)} - m)$ components of $\hat{\underline{T}}^{(2)}_m$, $C^{(21)}_m$, we can write

$$\Pi_{21}\ C^{(2)}_m\ \Pi^T_{21} = C^{(21)}_m \tag{25}$$

and finally set the test

$$\left[\hat{\underline{T}}^{(21)}_m - \hat{\underline{T}}^{(1)}_m\right]^T \left[C^{(1)}_m + C^{(21)}_m\right]^{-1}\cdot$$

$$\left[\hat{\underline{T}}^{(21)}_m - \hat{\underline{T}}^{(1)}_m\right] = \chi^2_{(N^{(1)}-m)} \tag{26}$$

which holds if

$$H_0\ :\ E\{\hat{\underline{T}}^{(21)}_m\} = E\{\hat{\underline{T}}^{(1)}_m\} \tag{27}$$

$$SO(\ddot{\xi}_{2\tau} - \ddot{\xi}_{1\tau}) \sim \frac{\sqrt{2} \cdot 5 \cdot 10^{-2}}{\Delta t^2} =$$

$$10^{-4}\ cm\ s^{-2} = 0.1\ mGal \qquad (14)$$

c) by Doppler observations we have

$$Q_t = \frac{\partial}{\partial t}|\underline{x}_2 - \underline{x}_1| = \tilde{Q}_t + \dot{\xi}_{2\tau} - \dot{\xi}_{1\tau} \qquad (15)$$

which combined with the difference of the radial components $(\xi_{2r} - \xi_{1r})$ in Hill's equations gives

$$\ddot{\xi}_\tau + 2n\dot{\xi}_r = \underline{\tau} \cdot \nabla T \qquad (16)$$

The following stochastic orders of magnitude are expected:

a) for $\ddot{\xi}_\tau$

$$SO(\dot{\xi}_{2\tau} - \dot{\xi}_{1\tau}) \sim 5\ 10^{-3}\ mm\ s^{-1}$$

$$SO\left(\frac{\dot{\xi}_{2\tau} - \dot{\xi}_{1\tau}}{\Delta t}\right) \sim 5\ 10^{-4}\ mm\ s^{-2} \qquad (17)$$

b) for $2\ n\ \dot{\xi}_r$

$$SO\left(2n\ \frac{\xi_{2r} - \xi_{1r}}{\Delta t}\right) \sim 10^{-5}\ mm\ s^{-2} \qquad (18)$$

As a consequence we obtain

$$SO(\underline{\tau} \cdot \nabla T) \sim 5\ mGal \qquad (19)$$

- *Gravity gradiometry (GOCE)*

The observations are represented by (ESA, 1999)

$$Q = M \equiv \left[\partial^2_{ik} T\right] \qquad (20)$$

In this case it must be noticed that Q has the great advantage of being always localized.

The stochastic order of magnitude of M is given by

$$SO(M) \sim 10^{-12}\ s^{-2} \sim$$

$$1mE/\sqrt{Hz}\ in\ MBW\ . \qquad (21)$$

3 Comparison between spatially distributed observations coming from different missions

In this section we will consider two cases: first we will show how to compare gravity models derived from different satellite missions (with a testing procedure); afterwards we will discuss the problem of predicting one observable from another of a different kind and we will produce an elementary, yet meaningful, example.

3.1 Comparison of derived gravity field models

In this case we want to compare the estimated coefficients $\hat{T}^{(1)}_{nm}$ of model (1) with the estimated coefficients $\hat{T}^{(2)}_{nm}$ of model (2), which in general have been estimated using data coming from different missions and have different resolutions:

$$\hat{T}^{(1)}_{nm} \Rightarrow \hat{\underline{T}}^{(1)}_m = \left\{T^{(1)}_{nm}\ ,\ n = m, \ldots, N^{(1)}\right\} \quad (22)$$

$$\hat{T}^{(2)}_{nm} \Rightarrow \hat{\underline{T}}^{(2)}_m = \left\{T^{(2)}_{nm}\ ,\ n = m, \ldots, N^{(2)}\right\} \quad (23)$$

We will use the symbols $C^{(1)}_m, C^{(2)}_m$ to represent the covariance matrices of the coefficients, order by order. In fact it must be noticed that typically correlations exist among estimated coefficients with different degree but same order. A testing procedure can be established as follows. Under the hypothesis that $N^{(2)} > N^{(1)}$, a projection must be performed of $\hat{\underline{T}}^{(2)}_m$ on $\hat{\underline{T}}^{(1)}_m$, using the projector Π_{21}

$$\Pi_{21}\ \hat{\underline{T}}^{(2)}_m = \hat{\underline{T}}^{(21)}_m \quad , n = m, \ldots, N^{(2)} \qquad (24)$$

Introducing the covariance of the $(N^{(1)} - m)$ components of $\hat{\underline{T}}^{(2)}_m$, $C^{(21)}_m$, we can write

$$\Pi_{21}\ C^{(2)}_m\ \Pi^T_{21} = C^{(21)}_m \qquad (25)$$

and finally set the test

$$\left[\hat{\underline{T}}^{(21)}_m - \hat{\underline{T}}^{(1)}_m\right]^T \left[C^{(1)}_m + C^{(21)}_m\right]^{-1} \cdot$$

$$\left[\hat{\underline{T}}^{(21)}_m - \hat{\underline{T}}^{(1)}_m\right] = \chi^2_{(N^{(1)} - m)} \qquad (26)$$

which holds if

$$H_0\ :\ E\{\hat{\underline{T}}^{(21)}_m\} = E\{\hat{\underline{T}}^{(1)}_m\} \qquad (27)$$

is satisfied.

3.2 Prediction of one observable from another

The direct prediction by collocation of one spatial observable from another between very different altitudes not only when going downward, but even when going upward by some 100 km, seems to be unable to recover the level of a very accurate local measurement.

Here we will discuss an example. Let us take one observation of the second radial derivative T_{zz} at a point P at an altitude of 250 km and try to predict the first radial derivative T_z at a point Q at an altitude of 400 km along the same radius.

Setting $U_0 = \frac{\mu}{R}$ we can write the usual formula for the gravity potential

$$T = U_0 \sum_{N_{min}}^{N_{max}} T_{nm} \left(\frac{R}{r}\right)^{n+1} Y_{nm} \qquad (28)$$

$$O(T_{nm}) = \frac{10^{-5}}{n^2} \quad \text{(Kaula's rule)} \qquad (29)$$

The covariance function of the gravity potential can be written as

$$C_{TT}(P,Q) =$$

$$U_0^2 \sum_{N_{min}}^{N_{max}} \left(\frac{R^2}{r_P r_Q}\right)^{n+1} \sigma_n^2 (2n+1) P_n(\cos\psi) \quad (30)$$

where σ_n^2 is the variance of the individual coefficient, hence $(2n+1)\ \sigma_n^2 = C_n$ represents the degree variance (full power),

$$C_n = \frac{2n+1}{n^4}\, 10^{-10} \sim \frac{2\ 10^{-10}}{n^3}\ . \qquad (31)$$

The first derivative of the potential and its covariance function are (setting $\Gamma_0 = \frac{\mu}{R^2}$)

$$\delta g = T_z = -\Gamma_0 \sum_{n,m} T_{nm}(n+1) \left(\frac{R}{r}\right)^{n+2} Y_{nm} \quad (32)$$

$$C_{\delta g \delta g}(P,Q) =$$

$$\Gamma_0^2 \sum_n C_n (n+1)^2 \left(\frac{R^2}{r_P r_Q}\right)^{n+2} P_n(\cos\psi) \quad (33)$$

The second derivative of the potential, the cross-covariance between δg and T_{zz} and the covariance of T_{zz} are (setting $G_0 = \frac{\mu}{R^3}$)

$$T_{zz} = G_0 \sum_{n,m} (n+1)(n+2) T_{nm} \left(\frac{R}{r}\right)^{n+3} Y_{nm} \quad (34)$$

$$C_{\delta g T_{zz}}(P,Q) =$$

$$-\Gamma_0 G_0 \sum_n C_n (n+1)^2 (n+2) \cdot \left(\frac{R}{r_P}\right)^{n+2} \cdot \quad (35)$$

$$\left(\frac{R}{r_Q}\right)^{n+3} P_n(\cos\psi) \qquad (36)$$

$$C_{T_{zz}T_{zz}}(P,Q) =$$

$$G_0^2 \sum_n C_n (n+1)^2 (n+2)^2 \cdot \qquad (37)$$

$$\left(\frac{R^2}{r_P r_Q}\right)^{n+3} P_n(\cos\psi) \qquad (38)$$

The prediction of T_z at point Q starting from the observed value of T_{zz} at point P can then be achieved by the formula

$$\hat{T}_z(Q) = C_{\delta g T_{zz}}(Q,P)\ C^{-1}_{T_{zz}T_{zz}}(P,P)\ T_{zz}(P) \ (39)$$

and the mean square error of the prediction is represented by

$$\mathcal{E}^2 = C_{\delta g \delta g}(Q,Q) +$$

$$C_{\delta g T_{zz}}(Q,P)\ C^{-1}_{T_{zz}T_{zz}}(P,P)\ C_{T_{zz}\delta g}(P,Q) \quad (40)$$

An easy computation leads to the following formula for $\mathcal{E}^2$

$$\mathcal{E}^2 = 2\ 10^{-10}\ \Gamma_0^2 \cdot$$

$$\cdot \left\{ \sum_n \frac{1}{n} (q^2)^{n+2} - \frac{\left[\sum p^{n+3}\ q^{n+2}\right]^2}{\sum n (p^2)^{n+3}} \right\} \qquad (41)$$

where $q = \frac{R}{r_Q}$, $p = \frac{R}{r_P}$. The numerical evaluation of (39) for $N_{min} = 10$, $N_{max} = 200$ gives

$$\mathcal{E} = 2.206\ mGal\ . \qquad (42)$$

A remark has to be made: roughly speaking, based on this elementary example we can say that this kind of prediction cannot for instance be used to check the fine Doppler GRACE observations by GOCE observations. Nevertheless it could be used to check the GPS tracking (CHAMP and GOCE).

Of course one could argue that this figure does not justify that the direct cross-control can be applied to the two missions, because the computation has been performed for one measurement only. However, even considering 100 measurements we would reduce the error by a factor 10 if the prediction error had been white noise, what is not, because the collocation error is very correlated. Then even a figure like 0.2 *mGal* would not be enough to perform the sought check.

4 Self calibration of a mission

In this section we will compare two observables of the same kind under two different hypotheses, namely: the observables are taken on parallel orbits laying at the same altitude, and the observables are taken on crossing orbits almost at the same altitude.

4.1 Comparison of two observables of the same kind on parallel orbits at the same altitude

Again, we want to discuss an example consisting in making the simple prediction at one point starting from a single observation, as discussed before. Let us take one observation of the second radial derivative T_{zz} at a point P along an orbit arc at 250 *km* and let us predict the value of T_{zz} at a point Q along an orbit arc laying at the same altitude, at an (angular) distance from the first arc equal to ψ. In this case we have $\frac{R}{r_P} = \frac{R}{r_Q} = q$ and the prediction error is represented by

$$\mathcal{E}^2 = C_{T_{zz}T_{zz}}(Q,Q) - \frac{C^2_{T_{zz}T_{zz}}(P,Q)}{C_{T_{zz}T_{zz}}(P,P)} =$$

$$2\ 10^{-10} G_0^2 \cdot$$

$$\frac{\left[\sum_n n(q^2)^{n+3}\right]^2 - \left[\sum_n n(q^2)^{n+3} P_n(t)\right]^2}{\sum_n n(q^2)^{n+3}} \quad (43)$$

The numerical evaluation of (41) for $N_{min} = 10$, $N_{max} = 200$ gives

$$\begin{aligned} \mathcal{E} &= 0.024\ E \qquad \text{for } \psi = 0.25^\circ \\ \mathcal{E} &= 0.012\ E \qquad \text{for } \psi = 0.125^\circ \end{aligned} \quad (44)$$

The angular distance $\psi = 0.25^\circ$ between two tracks will be certainly attained after 6 months of observations.

It must be remarked that the *same* computation holds for the prediction from one observation to the other along the orbit, although in this case *real* observations are even much closer to one another.

4.2 Comparison of two observables of the same kind on orbits crossing at almost the same altitude

It often happens that ascending and descending orbits cross at almost the same altitude. In this case radial derivatives can be directly compared for both GRACE and GOCE missions. We will produce two examples corresponding to the two cases.

a) GRACE

We consider the T_z observations derived from GPS and accelerometers. We want to determine whether we can simply write

$$T_z(P) \cong T_z(Q) \quad (45)$$

when the two points P, Q are on the same radius at distance

$$\delta r \leq 10\ km\ , \quad (46)$$

which is not a very stringent condition. The relation (45) can be accepted if

$$SO(T_{zz}) \cdot 10\ km < \sigma_{noise}\ . \quad (47)$$

We have

$$SO(T_{zz})^2 = 2\ 10^{-10}\ G_0^2 \sum_n n(q^2)^{n+3} =$$

$$2\ 10^{-10}\ G_0^2\ F(q^2) \quad (48)$$

For $N_{min} = 10$, $N_{max} = 200$, it is $F(q^2) \cong 32$, therefore

$$SO(T_{zz})^2 \sim 64\ 10^{-10}\ G_0^2\ . \quad (49)$$

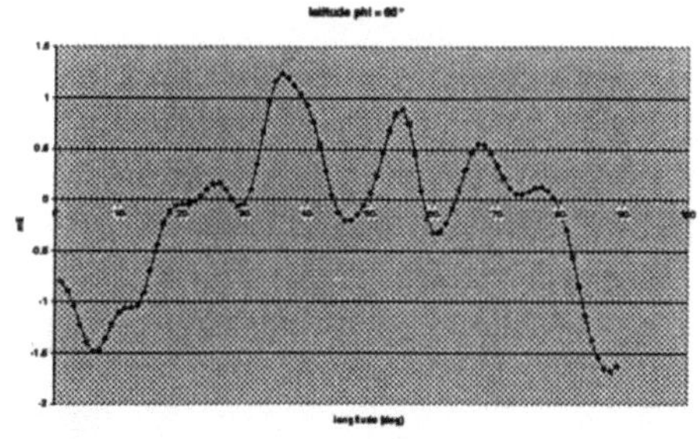

Fig. 1: Differences $T_{zz}(P) - T_{zz}(Q)$ at points along the parallel $\varphi = 60°$; values on the y axis range from $-2\ mE$ to $1.5\ mE$

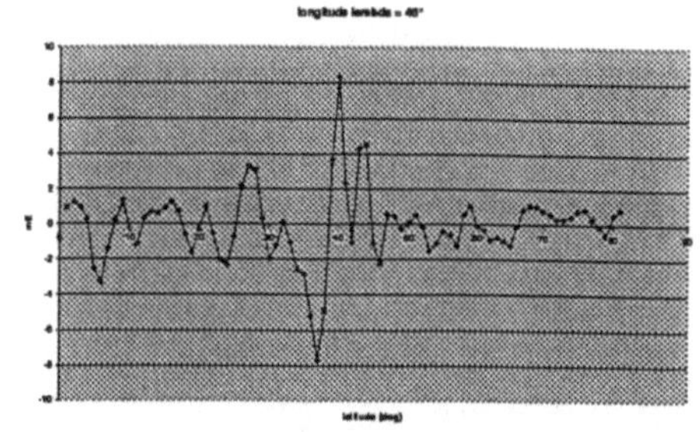

Fig. 2: Differences $T_{zz}(P) - T_{zz}(Q)$ at points along the meridian $\lambda = 40°$; values on the y axis range from $-10\ mE$ to $10\ mE$

In the end we obtain

$$SO(T_{zz}) \cdot 10\ km \sim 0.12\ mGal \tag{50}$$

which shows that we can at least control the radial derivative $\delta g \sim T_z$ (cf. 14).

b) GOCE

In this case we want to verify the applicability of the control test

$$T_{zz}(P) \cong T_{zz}(Q) \tag{51}$$

when P, Q are on the same radius at distance

$$\delta r \leq 1\ km\ . \tag{52}$$

Orbit simulations of the GOCE mission show that this condition is met at least twice per cycle.

In this case it is possible to demonstrate both analitically and numerically that

$$SO(T_{zzz}) \sim 1.24\ 10^{-15}\ s^{-2}\ m^{-1} \tag{53}$$

therefore for a radial distance between two crossing orbits smaller than 1 km one has

$$SO(T_{zzz}) \cdot 1km \simeq 1.24\ 10^{-12}\ s^{-2}\ =$$

$$1.24\ mE\ , \tag{54}$$

showing that it is possible to test in this way the functionality of the (zz) channel of the gradiometer, since $SO(T_{zzz}) \cdot 1\ km$ is smaller than the measurement noise.

To further prove this fact, we numerically computed the difference $T_{zz}(P) - T_{zz}(Q)$ at a number of points, obtaining values smaller than the noise standard deviation. This is illustrated in Fig. 1 and Fig. 2 where the difference is computed, without any measurement noise, along one quarter of the parallel $\varphi = 60°$ and along the meridian $\lambda = 40°$.

5 Conclusions

The conclusions that we can draw can be summarized as follows.

Regarding the cross checking of the observations taken from two different missions, it seems evident that in general the prediction of one observable from another of a different kind is not feasible, even when going upward. The only possible use could be the control of GPS - accelerometric observations from gradiometric measurements. Of course, it is always possible to compare two gravity potential models obtained from observations of different satellite missions in terms of their coefficients, using a χ^2 test, introducing suitable covariance matrices for the coefficients of the two models.

Regarding the so called "self calibration", it means that observables of the same kind coming from the same mission are compared by predicting the value at one point starting from the value observed at a different point. It is possible to check observations taken at cross-over points, i.e. at points having the same geographic coordinates but laying on crossing orbits at different altitudes. In particular, for GOCE it is possible to predict values of T_{zz} at cross over points with a maximum distance of 1 km in altitude, thus establishing a control test. The feasibility of this test has also been numerically controlled for series of points providing results in the expected range.

Table 1: Comparison between direct computation of $T_{zz}(P)$ and the results of the linear interpolation: the values are in mE

φ_P, λ_P	$T_{zz}(P)$ (from interp.)	$T_{zz}(P)$ (direct)	Difference
$2°, 2°$	-188.11	-189.46	1.35
$10°, 10°$	235.89	235.96	-0.07
$20°, 20°$	-28.47	-26.31	-2.16
$30°, 30°$	-76.41	-76.97	0.56
$40°, 40°$	533.63	530.39	3.24
$60°, 60°$	245.37	245.22	0.15
$70°, 70°$	-71.74	-72.25	0.61
$80°, 80°$	-69.40	-68.63	-0.77

Appendix

Regarding the comparison of two observables of the same kind performed at cross-over points, a fact to be highlighted is that in general the instrument on board the satellite will not exactly measure at cross-over points P, Q. The nearest measurements available will be made at points A, B and A_0, B_0 immediately following and preceding point P and point Q along the orbit.

In the case of GOCE, we have made the hypothesis that the observation points A, B and A_0, B_0 along the orbit are so close to each other that it is possible to perform a linear interpolation to obtain the values of $T_{zz}(P)$ and $T_{zz}(Q)$,

$$T_{zz}(P) = \lambda T_{zz}(A) + (1-\lambda) T_{zz}(B)$$

$$T_{zz}(Q) = \lambda_0 T_{zz}(A_0) + (1-\lambda_0) T_{zz}(B_0)$$

where

$$\lambda = \frac{AP}{AB} \quad ; \quad \lambda_0 = \frac{A_0Q}{A_0B_0}$$

To check the systematic error in the linear interpolation, given that for GOCE $AB \cong A_0B_0 \cong d = 8\ km$ ($\Delta t = 1\ s$), we decided to make a direct computation of $T_{zz}(P)$ and compare its the results with the results of the interpolation at a number of points geographically spread on the sphere. The values obtained are reported in Table 1.

From our results we derive the reasonable guess that the systematic error is here well below the noise level.

References

ASI (1998). SAGE Phase A Final Report (Albertella and Migliaccio eds.), Agenzia Spaziale Italiana.

Balmino, G., Perosanz F., Rummel R., Sneeuw N., Suenkel H. (1998). CHAMP, GRACE and GOCE: Mission Concepts and Simulations, II Joint Meeting of the International Gravity Commission and the International Geoid Commission, Trieste, Italy, in print.

Betti, B. and Sansò F. (1989). The integrated approach to satellite geodesy, Lecture notes in earth sciences n. 25, "Theory of satellite geodesy and gravity field determination" (Sansò and Rummel eds.), Springer-Verlag.

ESA (1998). European Views on Dedicated Gravity Field Missions: GRACE and GOCE, ESD-MAG-REP-CON-001, ESA, May 1998.

ESA (1999). GOCE Phase A Preliminary Requirements Review Presentation, ESA, June 1999.

Kaula, W.M. (1966). Theory of Satellite Geodesy, Blaisdell Publishing Company Waltham, Massachusetts-Toronto-London.

Rummel, R., Van Gelderen M., Koop R., Schrama E., Sansò F., Brovelli M., Migliaccio F., Sacerdote F. (1993). Spherical harmonic analysis of satellite gradiometry. *Netherlands Geodetic Commission, Publications on Geodesy, New Series*, no. 39, Delft.

Calibration/validation methods for GRACE

Christopher Jekeli
Department of Civil and Environmental Engineering and Geodetic Science
Ohio State University, 2070 Neil Ave., Columbus, OH 43210, USA

Abstract. The upcoming GRACE (Gravity Recovery and Climate Experiment) satellite mission is expected to map Earth's global gravitational field with unprecedented accuracy, especially at the long wavelengths. The measurements are line-of-sight range rates between two co-orbiting, low-altitude satellites that can be transformed into differences of geopotential and horizontal gravitation. One method to attempt the calibration and validation of GRACE data is to compare these in situ measurements using very accurate local terrestrial gravity fields that are upward continued to satellite altitude. This paper presents a rudimentary analysis of the requirements in gravity anomaly data spacing, accuracy, and extent to compute differences in disturbing potential and horizontal gravity disturbance at GRACE altitude. These studies show that to achieve the anticipated in situ accuracy of 0.01 m^2/s^2 in potential and 0.01 mgal in gravitation the terrestrial spacing and accuracy requirements are rather loose, e.g., 0.5° spacing and 0.25 mgal accuracy. Data extent, on the other hand, is much more important, where caps with radius larger than 10° are required, and then only to validate wavelengths smaller than 1200 km (30-deg reference field). Furthermore, it is shown that terrestrial data requirements are significantly less stringent in the case of computing horizontal gravitation differences at altitude rather than potential differences.

Keywords. GRACE calibration and validation, upward continuation, geopotential differences, gravitation differences.

1 Introduction

The upcoming GRACE (Gravity Recovery and Climate Experiment; Tapley and Reigber, 1998) satellite mission is expected to map Earth's global gravitational field with unprecedented accuracy, especially at long wavelengths, and with time resolution of one to three months. With these temporally-spaced, estimated gravity fields, GRACE will allow the study of Earth's dynamics due to mass motions within the atmosphere, cryosphere, ocean, lithosphere, mantle, and the core, as well as their dynamic interactions (MDR, 1998). These products should contribute to the detection and study of hydrologic, oceanic, solid Earth, and cryospheric mass changes over scales of a thousand kilometers, or even shorter resolution. The GRACE mission is scheduled to follow the CHAMP (Challenging Mini-satellite Payload for geophysical research applications; Reigber et al., 1998) satellite mission, also dedicated, in part, to static gravity field mapping.

Because of their anticipated high accuracy, GRACE products must undergo careful calibration and validation to maximize their potential for scientific investigations. *Calibration* may be defined as the determination of the systematic errors in a measuring device by comparing its data with measurements of a device that is considered correct, or at least more accurate (NGS, 1986); and *validation* as the verification that the measurements of a device lead to results with the resolution and accuracy consistent with expected scientific objectives.

Several proposals to achieve these essential first steps have been already discussed by the GRACE Science Team. In general, any such procedure is problematic if the measuring device to be calibrated and validated is supposed to be more accurate than any existing comparable device. To some extent, this is the case with GRACE and our existing knowledge of the *global* gravitational field. Figure 1 shows the predicted accuracy of the GRACE-derived harmonic model compared to our current best model, EGM96. Furthermore, since GRACE measurements are made on orbit and calibration and validation methods must utilize terrestrial measurements, one deals with analytic continuations of the data and the problems associated with such extrapolations. It is likely that different methods of calibration and/or validation will be optimal for different spectral bands of the gravity field; and both downward and upward continuation methods have their uses.

For example, some putative methods (using downward continuation) utilize very accurate

International Association of Geodesy Symposia, Vol. 121
Schwarz (ed.), Geodesy Beyond 2000 – The Challenges of the First Decade

absolute gravity and ocean bottom pressure measurements, both being sensitive to mass variations and possessing an internal accuracy presumed commensurate with the predicted GRACE accuracy (Bettadpur, 1999).

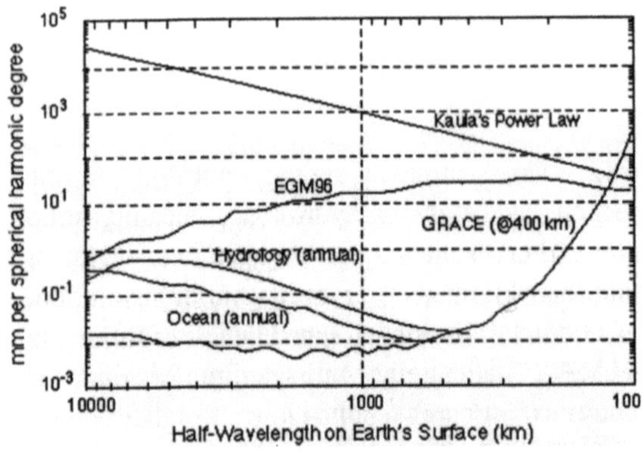

Fig. 1: Geoid uncertainty in EGM96 vs. predicted GRACE mission, vis-à -vis total gravitational signal (Kaula's rule) and hydrology and ocean signals (after UT-CSR, 1998).

The calibration and validation of data products (such as geopotential spectrum, and between-satellite range, range-rate, line-of-sight (LOS) acceleration) can be achieved at several levels. Direct comparisons to like quantities derived from terrestrial data fall under the calibration umbrella. Ultimately, validation should show that GRACE data yield the time-varying components of the geopotential to an accuracy that will allow inference of corresponding hydrological and ocean mass fluxes at time scales of several months or longer (Wahr et al., 1998).

Calibration and validation based on terrestrial gravity data, with both static and time-varying components, involves three aspects. A comparison between GRACE data and terrestrial data requires a functional continuation in altitude based on potential theory. It is well known that downward continuation is less stable numerically than upward continuation, at least at the short wavelengths. Second, according to its definition, calibration should be directed as close as possible to the measured quantity. Therefore, the calibration methods should relate terrestrial gravity data to in situ (on-orbit) measured range-rate and LOS acceleration. Finally, the extremely high accuracy predicted for the recovered low-degree gravity spectrum is possible because the presumed white noise in the GRACE measurement errors is filtered through global harmonic analysis. Calibration and validation of the low-degree spectrum could be done either by verifying the white noise nature of the measurement errors or by comparison with truth models of similar bandwidth.

In this paper some of these aspects are considered on the premise of an upward continuation of regional, accurate gravity fields to estimate the potential and gravitation differences between the two GRACE satellites, thus providing a direct comparison with in situ measurements. It is noted that upward continuation also underlay a simulation study conducted by Arabelos and Tscherning (1996) on the calibration of a satellite gravity gradiometry mission.

2 GRACE mission parameters

The fundamental GRACE measurement is the near-instantaneous range between two satellites following each other in almost identical orbits. The ranging is accomplished with a microwave (K-band) link between the two satellites that provides two one-way ranges, each obtained by comparing an on-board generated phase to the received phase. Both the transmitted and on-board phases are generated by the same ultra-stable oscillator (USO). In addition, accelerometers measure non-gravitational accelerations of the center of mass of each satellite, and star cameras ensure accuracy in the orientation of the transmitting and receiving K-band horns. The range values are to be obtained at a sampling rate of 10 Hz. These will be decimated and filtered to produce range-rates and range-rate-rates (LOS accelerations) at a sampling rate of 0.1 Hz (1 sample every 10 seconds). The predicted accuracy in the decimated, filtered range-rate is 10^{-6} m/s . A second band (Ka-band) will be used to calibrate first-order ionospheric delays in the signal. Other laboratory procedures and possibly on-orbit maneuvers will be used to model calibration functions for these instruments.

In addition, each satellite is equipped with a GPS antenna and receiver, utilizing the same USO. The GPS data will be used for orbit determination and atmospheric sounding on the basis of GPS satellite occultations. The GPS data rate is nominally 1 Hz, but greater for the occultation measurements. The K-band ranges can be used with orbit determination algorithms to infer the spherical harmonic coefficients (spectral components) of the Earth's gravitational field. Figure 1 shows the predicted accuracy in the spectrum in terms of geoid height.

3 The in situ measurements

The object is to determine the systematic errors in the data products that represent the gravitational field. These products include the gravitational component determined from LOS accelerations combined with accelerometer measurements, the geopotential difference determined from range-rate and accelerometer measurements, and a truncated spherical harmonic model inferred from range measurements and orbit determination. The formula relating gravitational differences, $\mathbf{g}_{12}$, to the LOS acceleration, $\ddot{\rho}_{12}$, is well known (e.g., Rummel, 1979) and is given by

$$\ddot{\rho}_{12} = \mathbf{e}_{12}^{T}\left(\mathbf{g}_{12} + \mathbf{F}_{12}\right) + \frac{1}{\rho_{12}}\left(\left|\dot{\mathbf{x}}_{12}\right|^{2} - \dot{\rho}_{12}^{2}\right), \qquad (1)$$

where subscripts "12" indicate the difference between quantities associated with the two satellites. In (1), $\mathbf{e}_{12}$ is the unit vector from one to the other satellite; $\mathbf{F}_{12}$ is the specific force vector difference; ρ_{12} is the range between the satellites; $\dot{\rho}_{12}$ is the range-rate; and $\dot{\mathbf{x}}_{12}$ is the velocity vector difference.

The range-rates together with the accelerometer measurements provide an in situ measurement of the geopotential difference. A model for this was developed by Jekeli (1998) and is given by

$$V_{12} = V_{12}^{0} + \left|\dot{\mathbf{x}}_{1}^{0}\right| \Delta\dot{\rho}_{12} - v$$

$$- \sum_{k=1}^{3} \int_{t_0}^{t} \left(F_{2k}\dot{x}_{2k} - F_{1k}\dot{x}_{1k}\right) dt - \Delta VR_{12} + E, \qquad (2)$$

where superscript "0" indicates reference quantities; $\Delta\dot{\rho}_{12}$ is the residual range rate with respect to the reference; v accounts for potential terms depending on velocity components orthogonal to the line of sight; t is time; ΔVR_{12} accounts for the time-variation of Earth's gravitational potential in inertial space; and E is a constant.

A similar model exists for the measurement of in situ geopotential differences as determined from GPS baseline *vector* velocities (low-low, satellite-to-satellite tracking (SST); Jekeli, 1998), also applicable to high-low GPS SST, and for gravitational *vector* determination from GPS high-low SST (Jekeli and Upadhyay, 1990). The latter model would apply to the CHAMP mission.

Finally, the truncated spherical harmonic expansion of the gravitational field (a defined science product for GRACE) might also serve as the quantity to be calibrated. While EGM96 is not sufficiently accurate for calibration purposes (see Figure 1), gravitational signals inferred from very precise determinations of global-scale geodynamic processes, such as ocean and hydrological cycles, may be suitable, as discussed by Wahr et al. (1998). These will not be discussed further in this paper.

Figure 2 (Jekeli, 1998), derived on the basis of equation (2), shows (among others) the accuracy relationship between geopotential differences and range-rate measurements. The geopotential accuracy of $10^{-2}\,m^2/s^2$ is consistent with a range-rate accuracy of $10^{-6}\,m/s$. This is roughly equivalent to an accuracy of 1 mm in the at-altitude geoid difference. Rummel (1980) estimated the corresponding accuracy in the LOS acceleration to be better than $10^{-7}\,m/s^2$.

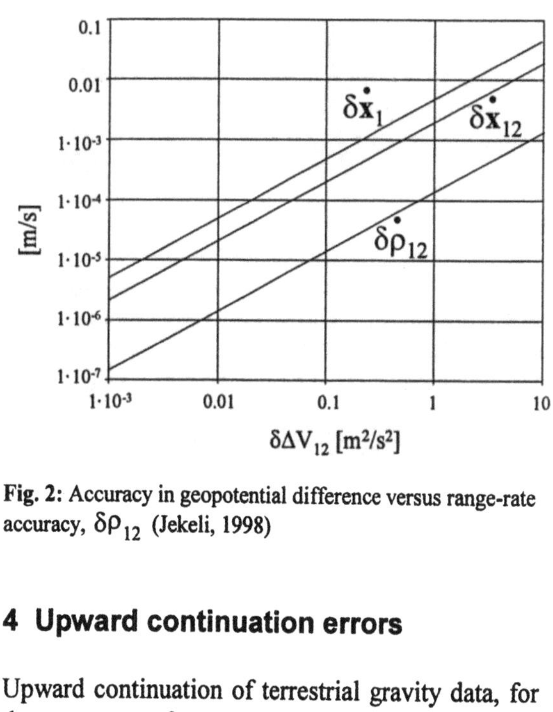

Fig. 2: Accuracy in geopotential difference versus range-rate accuracy, $\delta\dot{\rho}_{12}$ (Jekeli, 1998)

4 Upward continuation errors

Upward continuation of terrestrial gravity data, for the purpose of comparison with on-orbit data, follows according to the principles of potential theory, where the potential in free space is the solution to Laplace's differential equation for the potential subject to boundary values on the Earth's surface (or geoid), for example, Poisson's and Stokes' integral formulas. Given a solution for the potential, its derivative is also easily found. In terms of gravity anomalies, Δg, on the geoid, approximated by a sphere of radius, R, we have the well known solutions for the disturbing potential, T, and its derivative:

$$T(r,\theta,\lambda) = \frac{R}{4\pi} \iint_{\sigma} \Delta g(R,\theta',\lambda')\, S(R,r,\psi)\, d\sigma \qquad (3)$$

$$\frac{\partial T}{\partial \rho}(r,\theta,\lambda) = \frac{R}{4\pi} \iint_{\sigma} \Delta g(R,\theta',\lambda')\, \frac{\partial S}{\partial \rho}(R,r,\psi)\, d\sigma \qquad (4)$$

where ρ is a horizontal coordinate, ψ is the central angle subtended by the points (r,θ,λ) and (R,θ',λ'), that are separated by the distance, l; S is Stokes' function ($S \cong 1/l$, $\partial S/\partial\rho \cong 1/l^3$); and $d\sigma$ is an area element on the unit sphere. In principle, the inverse to these operations exists, which then allows downward continuation, but it belongs to the class of solutions to ill-posed problems, and is associated with numerical difficulties.

In either case, the continued function (T or $\partial T/\partial\rho$) theoretically requires that the integration be done over the entire globe. However, because of the attenuation of the Green's functions, $1/l$ or $1/l^3$, (3) and (4) may be implemented in practice using a limited local region of data (and a high-degree reference field), depending on the altitude of continuation and the required accuracy. The other requirement, that of continuous boundary values, is also strictly not met since only discrete data are available.

Both of these practical limitations incur model errors that may be analyzed using the appropriate transfer functions, as follows. From (Jekeli and Rapp, 1980; see also (Dickey et al., 1997)), we have for the effects of discretization and data noise:

$$\varepsilon_d(\Delta T_{nm}) \approx \sqrt{\frac{\Delta\sigma}{4\pi}}\, \frac{R\,\varepsilon(\Delta g)}{n-1}\, d_n \left(\frac{R}{r}\right)^{n+1}, \qquad (5)$$

$$\varepsilon_d(\Delta(\partial T/\partial\rho)_{nm}) \approx \sqrt{\frac{\Delta\sigma}{4\pi}}\, \frac{\sqrt{n(n+1)}}{n-1}\, \varepsilon(\Delta g)\, d_n \left(\frac{R}{r}\right)^{n+2}; \qquad (6)$$

where

$$d_n = \sqrt{2(1 - P_n(\cos\psi_{12}))}\,; \qquad (7)$$

and where $\varepsilon()$ stands for error; ΔT_{nm} and $\Delta(\partial T/\partial\rho)_{nm}$ are harmonic coefficients (degree n, order m) of the differences of, respectively, the disturbing potential and its horizontal derivative, between satellite points 1 and 2; and $\Delta\sigma$ is the data grid size. In (7), P_n is the Legendre polynomial and ψ_{12} is the central angle between the two satellites. In the absence of global data coverage, the error spectra (5) or (6) are optimistic since they assume white noise over the entire spectrum of the measurements.

The effect due to limited integration area (the truncation error), likewise, can be formulated in terms of harmonic constituents (Jekeli, 1980):

$$\varepsilon_t(()_{nm}) \approx \sqrt{\frac{\sigma_n^2(\Delta g)}{2n+1}}\, \tilde{k}_n(), \qquad (8)$$

where $()$ is either (ΔT) or $\Delta(\partial T/\partial\rho)$, $\sigma_n^2(\Delta g)$ is the degree variance of the gravity anomaly, and

$$\tilde{k}_n() = \frac{1}{2} \sum_{r=2}^{\infty} (2r+1)\, k_r()\, e_{r,n}, \qquad (9)$$

with

$$e_{r,n} = \int_{\psi_0}^{\pi} P_r(\cos\psi)\, P_n(\cos\psi)\, \sin\psi\, d\psi\,. \qquad (10)$$

For the difference in disturbing potential at altitude we have

$$k_n(\Delta T) \approx \frac{R}{n-1}\, d_n \left(\frac{R}{r}\right)^{n+1}; \qquad (11)$$

and for the difference in its horizontal derivative,

$$k_n(\Delta\partial T/\partial\rho) \approx \frac{\sqrt{n(n+1)}}{n-1}\, d_n \left(\frac{R}{r}\right)^{n+2}. \qquad (12)$$

The approximations indicated in (8) through (12) are due primarily to identifying a difference with its root-mean-square (rms) over all azimuths, as constructed in (Jekeli and Rapp, 1980).

On the basis of equation (5), Figure 3 shows the expected accuracies in the upward continued disturbing potential differences at an altitude of 400 km from terrestrial gravity anomalies. Similarly, using equation (6), Figure 4 shows accuracies for LOS accelerations at altitude from terrestrial gravity anomalies. These figures show that only modest degradation in estimation error occurs with a decrease in ground data resolution.

The more significant error comes from the truncation of the integrals (3) and (4), that is, the neglect of the remote zone. Using the EGM96 model (Lemoine et al., 1998) for the degree variances, including its analytic continuation for degrees greater than 360 (Jekeli, 1998), Figure 5 and 6 show the corresponding rms errors for different a priori reference fields.

5 Time-varying gravity calibration

The time-varying gravitational signal of interest concerns only the long-wavelength spectral band, down to half-wavelengths of about 1000 km, or

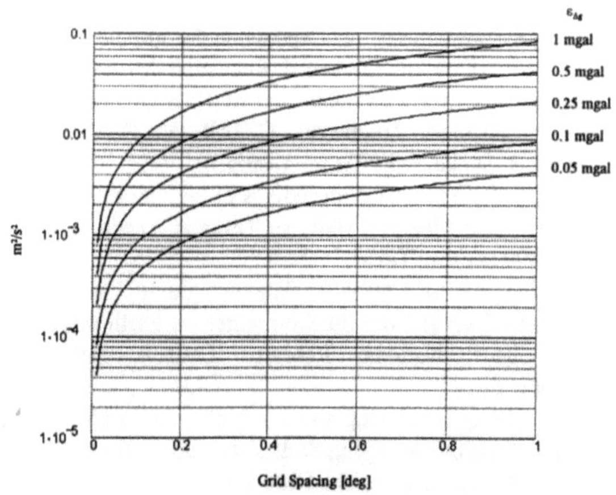

Fig. 3: Predicted accuracy in geopotential differences computed from upward continued, globally gridded gravity anomalies with accuracy, $\varepsilon_{\Delta g}$.

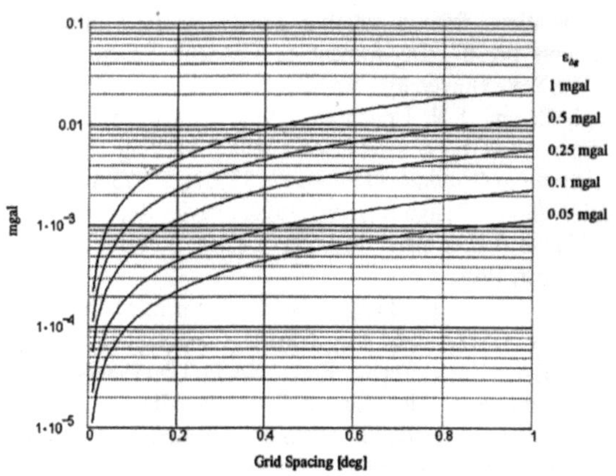

Fig. 4: Predicted accuracy in horizontal gravitational differences computed from upward continued, globally gridded gravity anomalies with accuracy, $\varepsilon_{\Delta g}$.

harmonic degrees $n = 2, \ldots, 20$ (Figure 1). To calibrate and validate the measured signal (satellite range-rate) it should be compared in the corresponding spectral band of a terrestrial signal of better accuracy. Analogous to (5), the relationship between the standard error of gravity anomaly and the degree-and-order spectrum of the geoid undulation is given by

$$\varepsilon(N_{nm}) = \sqrt{\frac{\Delta\sigma}{4\pi}}\,\frac{R\,\varepsilon(\Delta g)}{n-1}\,, \tag{13}$$

where $\Delta\sigma$ is the grid size of the gravity anomaly points evenly distributed on the sphere. Assuming a global grid of absolute gravity measurements accurate to $10^{-8}\,\mathrm{m/s^2}$, a grid size of $\Delta\sigma \le 6°\times 6°$ is

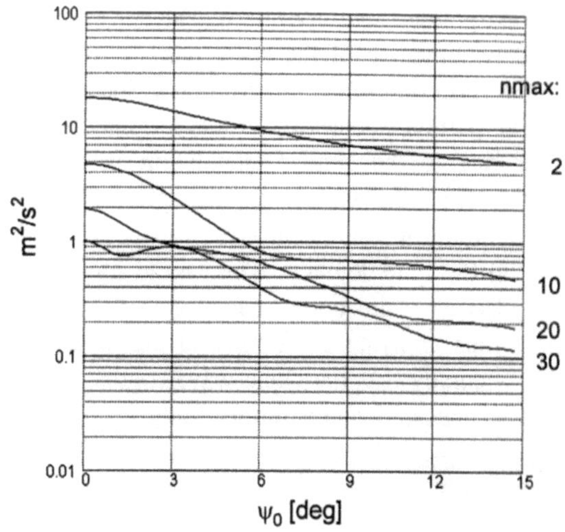

Fig. 5: Data extent error in disturbing potential differences with respect to reference field up to degree n_{max} and computed from upward continued gravity anomalies within a cap of radius, ψ_0 .

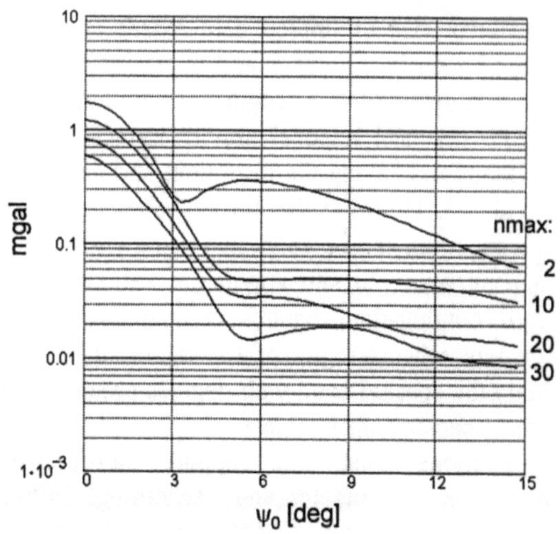

Fig. 6: Data extent error in horizontal gravitational differences with respect to reference field up to degree n_{max} and computed from upward continued gravity anomalies within a cap of radius, ψ_0 .

required to achieve $\varepsilon(N_{nm}) \approx 10^{-2}$ mm, $n \le 20$, as predicted in Figure 1. This is not practical, and other types of global measurements must be analyzed similarly to determine their utility for validation.

An alternative validation concept makes use of the time series generated by the in situ measurements as the satellites repeatedly return to a particular region that is well surveyed in terms of the gravity field, including its temporal variations. This time series can be compared to the upward continued gravity field. While the model error in upward continuation of the static part, as seen in Figure 6, is on the order of 0.1 mgal, that of the temporal part is about two orders of magnitude smaller due to the similar attenuation in signal strength. Thus, the white noise character of the in situ gravitation measurement error can be verified locally in the time domain. This would indirectly verify also its white noise character in the space domain thus implying conformity to the predicted geoid accuracy of the long-wavelengths. This validation concept requires further study.

5 Summary

To achieve $10^{-2}\,m^2/s^2$ error in the calculation of geopotential differences at altitude and $2\times10^{-8}\,m/s^2$ error in LOS accelerations at altitude requires a discretization of 20 km in a terrestrial gravity data grid and about 0.5 mgal data noise, as seen from Figures 3 and 4.

The error due to limited data extent is much larger, as seen in Figures 5 and 6, where calibration/validation at all wavelengths appears not feasible from practical regional gravity models up to 30° -extent and having the resolution and accuracy stated above. However, the shorter-wavelength part of the spectrum can be calibrated using reasonable ground surveys. In situ gravitation validation by upward continuation requires comparatively less data extent, where a region of 15° radius suffices to validate the spectral band with degrees greater than 30 at an accuracy of $10^{-7}\,m/s^2$.

Simulations must still support these error analyses, as in (Arabelos and Tscherning, 1996). Preliminary tests using EGM96 to simulate ground and at-altitude geopotential signals and applying the method of least-squares collocation have confirmed the general trends depicted here.

Also note that registration errors of $10^{-2}\,m^2/s^2$ and $2\times10^{-9}\,m/s^2$ per millimeter of orbit error are associated, respectively, with the geopotential and gravitation. Thus, orbit error is more significant in calibrating the in situ geopotential than the LOS gravitation measurements.

Acknowledgments. The author is indebted to Mr. Ramon Garcia for conducting the simulation computations and to the referees for careful reviews. This work was supported by the Center for Space Research, University of Texas, Austin; Contract No. UTA98-0223.

References

Arabelos, D. and C.C. Tscherning (1996): Support of spaceborne gravimetry data reduction by ground based data. In: Study of Advanced Reduction Methods for Spaceborne Gravimetry Data and of Data Combination with Geophysical Parameters, CIGAR IV, ESA Final Report, H. Sünkel (ed.), Graz, Austria.

Bettadpur, S. (ed.) (1999): Proceedings of the Third GRACE Science Team Meeting, Univ. of Texas at Austin, 23-24 June 1999, Austin, Texas.

Dickey, J.O., et al. (1997): Satellite gravity and the geosphere. Report from the Committee on Earth Gravity from Space, National Research Council, National Academy Press.

Jekeli, C. (1980): Reducing the error of geoid undulation computations by modifying Stokes' function. Report 301, Dept. of Geod. Science, The Ohio State University.

Jekeli, C. (1998): The determination of gravitational potential differences from satellite-to-satellite tracking." submitted to *Celestial Mechanics and Dynamical Astronomy*, 10/1998.

Jekeli, C. and R.H. Rapp (1980): Accuracy of the determination of mean anomalies and mean geoid undulations from a satellite gravity mapping mission. Report no.307, Department of Geodetic Science, The Ohio State University.

Jekeli, C. and T.N. Upadhyay (1990): Gravity estimation from STAGE, a satellite-to-satellite tracking mission. *Journal of Geophysical Research*, **95**(B7), 10973-10985.

MDR (Mission Design Review), 1998: Gravity Recovery and Climate Experiment Mission Design Review, 15-16 October, 1998.

Lemoine, F.G. et al. (1998): The development of the joint NASA GSFC and the NIMA Geopotential Model EGM96. Nasa Tech. Rep. NASA/TP-1998-206861, Goddard Space Flight Center, Greenbelt, MD.

NGS (1986): Geodetic Glossary. National Geodetic Information Center, NOAA, Rockville, MD.

Reigber, C., H. Luehr, and P. Schwintzer (1998): The CHAMP geopotential mission. Presented at 2nd Joint Meeting of the International Gravity Commission and the International Geoid Commission, 7-12 September 1998, Trieste, Italy.

Rummel, R. (1979): Determination of short-wavelength components of the gravity field from satellite-to-satellite tracking or satellite gradiometry - an attempt to an identification of problem areas. *Manuscr Geod* 4: 107-148.

Rummel, R. (1980): Geoid heights, geoid height differences, and mean gravity anomalies from "low-low" satellite-to-satellite tracking - an error analysis. Report 306, Dept. of Geodetic Science., Ohio State University.

Tapley, B.D. and C. Reigber (1998): A satellite-to-satellite tracking geopotential mapping mission. Presented at 2nd Joint Meeting of the International Gravity Commission and the International Geoid Commission, 7-12 September 1998, Trieste, Italy.

UT-CSR (1998): GRACE Preliminary Design Review. University of Texas, Center for Space Research, Austin, 19-20 August, 1998.

Wahr, J., M. Molenaar, F. Bryan (1998): Time variability of the Earth's gravity field: hydrological and oceanic effects and their possible detection using GRACE. *J. Geophys. Res.*, **103**(B12), 30205-30229.

The 1999 GFZ pre-CHAMP high resolution gravity model

T. Gruber, C. Reigber, P. Schwintzer
GeoForschungsZentrum Potsdam (GFZ), Division 1, Telegrafenberg A17, D-14473 Potsdam, Germany
e-mail: gruber@gfz-potsdam.de

Abstract. In preparation of the CHAMP gravity and magnetic field satellite mission, a completely new satellite-only gravity field solution named GRIM5 is currently under development by GFZ Potsdam and GRGS in Toulouse. Based on the full variance-covariance matrix of this solution and terrestrial and altimetry derived surface gravity data, a new high resolution global gravity field model up to degree and order 359 will be computed. The solution will be calculated by a rigorous combination of full variance-covariance matrices from the satellite-only model and the land and ocean surface gravity data, with block-diagonal structured variance-covariance matrices from the high resolution terrestrial and altimetric gravity information provided by the U.S. National Imagery and Mapping Agency (NIMA) and, in case of altimetry derived information, by GFZ Potsdam and CLS in Toulouse. For this new solution the spherical harmonic degree and order for the full variance-covariance matrix will be extended to 140, compared to degree 100 for previous GFZ solutions and degree 70 for the EGM96 model. It can be shown by preliminary investigations, based on former satellite solutions, that, due to the extension of the full normal equation system, the quality of the final model will be increased significantly for this frequency range. Quality parameters for such investigations are derived from the internal error parameters from the least squares approach and from comparisons with external independent gravity and geoid information. The new high resolution combination model will represent a major improvement towards a series of new gravity field models with unprecedented accuracy and resolution from the CHAMP, GRACE and GOCE gravity field missions. With these missions long, medium and even parts of the short wavelengths will be dominated by satellite data, while the rest of the spectrum will be determined solely from surface data. Therefore the presented combination technique can be extended also to the computation of high resolution CHAMP, GRACE and GOCE based gravity field models.

Keywords. Global gravity model, gradiometry.

1 Introduction

In view of the CHAMP magnetic and gravity field mission, which is due for launch in early 2000 (Reigber et al, 1996), and with the approved GRACE mission (Tapley, 1997) and the proposed GOCE gradiometry mission (Schuyer, 1997) a new era of gravity field determination from space will begin. The error predictions for all three missions show, that each of them will drastically improve the geoid accuracy in different spectral domains, compared to our current knowledge, which is represented by the EGM96 gravity field model (Lemoine et al, 1998) (figure 1). But it also can be seen by these error curves, that for CHAMP and GRACE, the inclusion of terrestrial and altimetry derived gravity field information can improve the higher frequencies (>degree 70 for CHAMP,

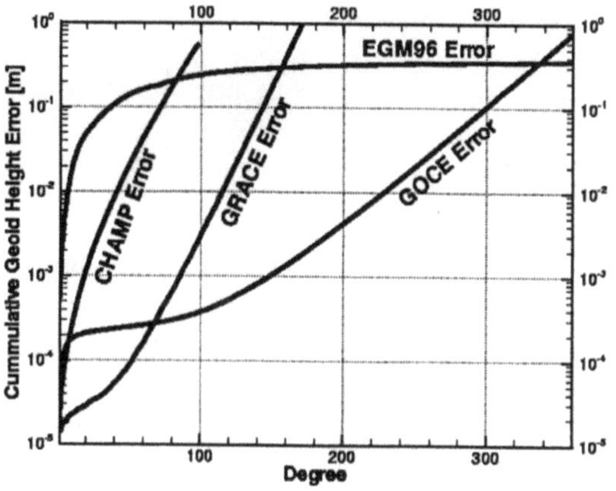

Fig. 1: Comulative geoid error for EGM96 gravity field and predictions for satellite missions

International Association of Geodesy Symposia, Vol. 121
Schwarz (ed.), Geodesy Beyond 2000 – The Challenges of the First Decade

>degree 140 for GRACE). If GOCE will be approved this situation could change. Surface gravity field observations can only contribute to the highest frequencies (>degree 250) of the regarded spectrum, assuming that global data are available not denser than with 30′ resolution. If GRACE and GOCE are combined (what means a combined approach of orbit perturbation theory and satellite gradiometry), surface data probably could not improve the global geoid accuracy for the complete spectrum up to degree 360.

Consequently for the near future surface gravity and altimeter data still will be valuable for the determination of global high resolution combined gravity field models. For the combination of satellite-only models with surface data the main task is the optimal combination of data, which are given in different domains (e.g. satellite tracking data with gravity anomalies and geoid heights) and which are complementary in frequency and space domain. To illustrate this the partial redundancies of the GRIM4-S4 satellite-only model (Schwintzer et al, 1997) are shown (figure 2). A value of 1 indicates, that this coefficient is completely determined by satellite tracking information, while a value of 0 means, that the coefficient is determined solely from a-priori information (Kaula rule of thumb) (Schwintzer, 1990). There, clearly the resonance orders for some of the satellites included in the model are visible. The optimal combination approach fills in all fields, where the a-priori information dominates. Chapter 3 describes more detailed this approach as it is used for the GFZ combined models (and the GFZ contribution to the GRIM5 project). Table 1 shows the complementarily of both data types by specifying advantages and drawbacks of both data sources.

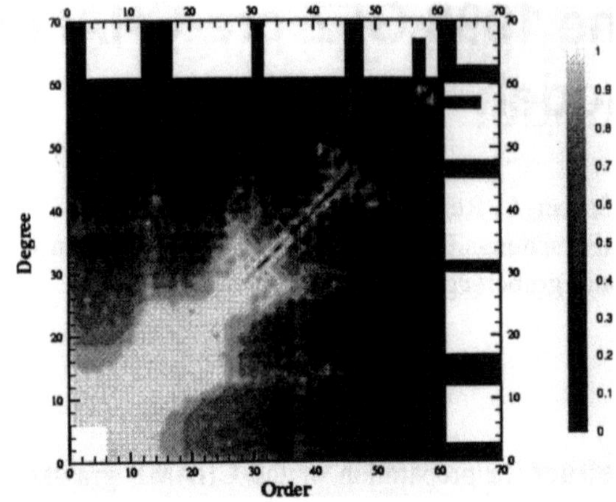

Fig. 2: Partial redundancies of GRIM4-S4 satellite-only gravity model (mirrored at diagonal)

2 Data processing

As data sources the following data sets are used:

a. NIMA 30'x30' mean terrestrial and airborne mean gravity anomalies including standard deviations and topographic heights (Lemoine et al, 1998)
b. NIMA 1x1 degree mean terrestrial and ship gravity anomalies including standard deviations and topographic heights (Lemoine et al, 1998)
c. NIMA 30'x30' mean altimetric ocean/sea ice gravity anomalies including standard deviations (Lemoine et al, 1998)
d. GFZ 15'15' point altimetric ocean gravity anomalies (Rentsch 1999, personal communication)
e. CLS 30'x30' mean sea surface heights including standard deviations (Biancale 1999, personal communication)

The determination of the high resolution gravity field model is a two-step procedure. In a first step combined gravity models up to degree and order 120 based on the full variance-covariance matrix are computed. For this each data set has to be preprocessed individually, such that residual gravity field quantities from which consistent normal equations can be generated are available. The preprocessing steps are (letters in brackets specify the data sets for which the processing step is applied):

- From 15'x15' point gravity anomalies, 30'x30' means are determined by simple mean value computation of all 9 points (at least 75% must be defined) *(d)*

Table 1: Advantages and drawbacks of satellite and surface data for combined global gravity field determination

Satellite Data	Surface Data
• consistent per satellite • quasi global coverage dependent on orbit • resonances for specific orders	• full gravity sensitivity on earth • consistency over oceans from altimetry
• limited gravity sensitivity in satellite altitude • limited number of tracking stations; no global observations	• inconsistencies over land for different data sources with different height systems • polar data gaps

- Remove stationary sea surface topography (POCM) and apply permanent tidal correction to remove influence of sun and moon on geoid heights *(e)*
- Fit data coverage of ocean data sets to NIMA ocean data set to avoid gaps or double occupation of blocks with respect to land data set *(d,e)*
- Apply ellipsoidal correction to gravity anomalies due to spherical differential equation for gravity anomalies definition *(a,b,c,d)*
- Apply 5 degree coast mask to ship gravity anomalies to be used together with altimetry derived quantities for stabilizing the solutions at coastlines *(b)*
- Compute 1x1 degree mean values from 30'x30' means to reduce computational effort during normals generation *(a,b,c,d,e)*
- Transform normal field to satellite-only standards and subtract reference field used for satellite-only model *(a,b,c,d,e)*
- Apply lowpass filter to reduce frequencies above degree and order 120 in the data set (see further comments below) *(a,b,c,d,e)*

In addition also the available standard deviations and topographic heights have to be preprocessed, such that they can be used as weights for the generation of the full normal equation systems.

- Compute 1x1 degree block mean values from 30'x30' means of topographic heights and set all negative heights to 0. Topographic heights are used as height information when generating surface normals *(a)*
- Transform standard deviations to weights by applying minimum, maximum operators *(a = 1.6-5.6 mgal, b = 5.0-15.0 mgal, c = 1.0-3.8 mgal, d=c, e = 25.0-35.0 cm)*

For each data set a normal equation system up to degree and order 120 is generated and used for combination.

In the second step, based on the combined model, a high resolution model up to degree and order 359 is computed. To reduce the computational effort only block-diagonal structured normal equation systems are set up. Such systems are generated if global uniform data sets on the geoid with longitude independent weights are available (Colombo, 1981). Consequently only 3 of the 5 data sets can be further used. The preprocessing steps for the high resolution models are (the first four steps are the same as above):

- Apply terrain correction to surface gravity anomalies using Helmert's condensation method to reduce them to the geoid *(a)*
- Eliminate overlapping blocks. Give priority to altimeter derived anomalies *(a)*
- Fill gaps with gravity anomalies computed from the 120 combined gravity model *(a,c,d)*
- Transform normal field to satellite-only standards and subtract reference field used for satellite-only model *(a,c,d)*
- Apply lowpass filter to reduce frequencies above degree and order 359 in the data set (see further comments below) *(a,c,d)*

For the combinations of data set a+c and a+d block-diagonal normal equations to degree 359 were generated. To avoid underdetermination in the least squares procedure the maximum degree is limited to 359, because for a global 30'x30' block-means grid only 360 latitude bands are available to determine the 360 zonal coefficients.

2.1 Lowpass filtering

To avoid the folding of high frequency parts in the data sets to be analyzed into the lower spectrum of the spherical harmonic series, which is caused by aliasing effects and/or the cut-off of the spherical harmonic series, a lowpass filter was applied to the input data sets. For testing this filter a closed loop scenario was developed to visualize the improvement, which can be addressed to the filter.

We started with the GRIM4-C4 combined gravity model complete to degree and order 72 (Schwintzer et al, 1997). From the coefficients a global set of 1x1 degree mean gravity anomalies was computed. Then the field was recomputed using the block-diagonal technique for the least squares approach up to degree and order 59 with and without applying the lowpass filter. This means that a cut-off degree with respect to the frequency content in the data set was introduced for the test. For reference also the case without cut-off and aliasing was computed (solve series up to degree and order 59 and recompute the coefficients). Finally coefficient differences between the original and the resulting spherical harmonic series were regarded in terms of degree variances (see figure 3). The filter for this test case was designed in the following way:

- The input data set was analyzed by numerical quadrature up to degree and order 100
- From all coefficients above degree 59 residual gravity anomalies were computed (degree 60-100).
- The resulting residual gravity anomalies were subtracted from the input data set before it was analyzed by the block-diagonal approach.

Figure 3 clearly shows the influence of the filter on the final result. While, with respect to the reference case, quasi no differences are visible up to degree 57 when the lowpass filter is applied, clear differences can be found if the filter is not applied. Starting around degree 35, the differences, which can be interpreted as estimation errors, are increasing significantly. The drastic increase of differences for the last two degrees (58 and 59) can not be explained by this. Even if they can be reduced by applying the filter they still show large differences compared to the reference case. This phenomenon must be related to the cut-off of the spherical harmonic series and the frequency range in the gridded data set. Consequence of this could be, that the last two degrees of a gravity field spherical harmonic series should be regarded more uncertain than all the others. As final result it can be concluded, that the filter is adequate to reduce the coefficient estimation error significantly and, that it should be applied to the gravity field input data sets prior to the spherical harmonic analysis, because their spectral decomposition generally is unknown.

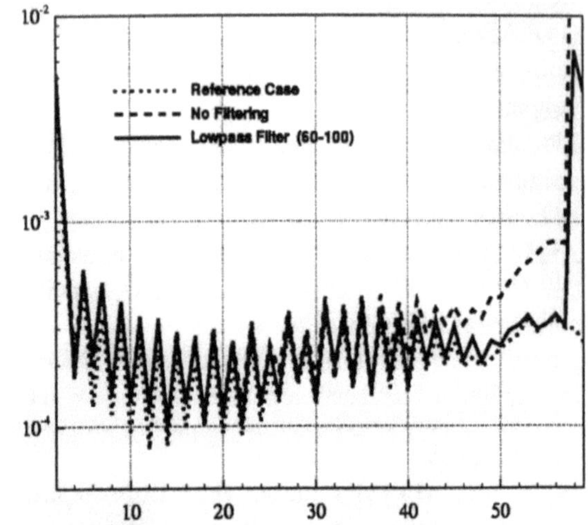

Fig. 3: Square root of degree variances of coefficient differences in geoid height [m]

3 Models and quality control

As mentioned above the high resolution gravity field is estimated in a two step procedure with a full variance-covariance matrix and a reduced block-diagonal structured variance-covariance matrix for the degrees above 120.

3.1 Combined base solutions with full variance-covariance matrix (120x120)

By combination of different sets of normal equations, which have been generated from the preprocessed input data sets, and by adding information from the satellite-only gravity field three different combined gravity field solutions were computed The solutions differ only by the ocean data set used (small letters are related to the preprocessed data sets in chapter 2).

- A = a + b(coast) + c + grim4-s4(gaps) + grim4-s4
- B = a + b(coast) + e + grim4-s4(gaps) + grim4-s4
- C = a + b(coast) + d + grim4-s4(gaps) + grim4-s4

As visible, the satellite-only model GRIM4-S4 plays different roles in the final combination. First it is used to fill remaining gaps in the surface and ocean data sets (about 10%). Second the normal equation system is fully added to the surface and altimetry normal equation systems. And third, coefficients, which are determined quasi exclusively from satellite observations and not from the a-priori information in the satellite model are added as additional observation with high weight to the overall normal equation system. This ensures, that good long wavelength information in the satellite-only model is not superimposed with bad long wavelength information from the surface data. To determine which coefficients are determined quasi solely from satellite observations the partial redundancies, shown in figure 2, are used. All coefficients for which the partial redundancy is larger than 0.9 are assumed to be well determined from the satellite data and are added as additional information to the overall combined normal equation system. This approach helps a lot to hold consistency for the long wavelengths.

For testing the quality of a gravity field model internal quality parameters, based on the results of the least squares solution (variance-covariance matrix), and external quality parameters from comparisons with independent data sets can be used.

Table 2: Orbital fits for different combined gravity field solutions to degree and order 120 (full variance-covariance matrix) [cm]

	ERS-1		GFZ-1	Ajisai	Starl.
	Laser	Cross.	Laser	Laser	Laser
A	7.3	9.83	42.3	8.1	10.9
B	7.3	9.70	42.9	7.8	10.5
C	7.3	9.82	39.9	8.1	10.5
EGM96	8.3	10.35	47.2	7.6	7.9

Internal quality parameters for example are:

- Quotient of surface and satellite normals diagonal elements to check contribution of each source. For long wavelengths result should reflect the partial redundancies of the satellite-only model.
- Coefficient standard deviations and significance of coefficients from quotient of standard deviation and coefficient value.
- Correlations of coefficients.
- Degree and error degree variances.
- Error propagation of variance-covariance matrix to geoid heights or gravity anomalies.

As a sample figure 4 shows the correlation of one of the base models. Clearly visible are the side-band correlations, which are caused by the surface data. The block structure along the diagonal is mainly caused by the satellite normal equation system.

External quality parameters are:

- Orbital fits to a variety of satellites using independent arcs not included in satellite-only gravity field solution.
- Comparisons with independent point geoid heights from GPS and spirit levelling.
- Comparisons with independent geoid heights and
- geoid height gradients derived from altimetry after subtracting a sea surface topography model.

As a sample table 2 shows orbital fits for a number of independent arcs for 4 different satellites.

From the orbital fits it can be concluded, that for the long wavelengths solution B (with the CLS sea surface) is the best and that for the full spectrum, for which GFZ-1 is sensitive, solution C (with the GFZ altimetric gravity anomalies) is better. For all satellites except Starlette solutions A,B and C show better fits than EGM96. The problem with Starlette is caused by an error, which happened during the GRIM4-S4 satellite-only processing. The problem can not be reduced by including information from surface and altimeter data. From the geoid comparisons on GPS stations (not shown here) solution C shows slightly better results than solution A (with NIMA altimetric gravity anomalies) and solution C. Summarizing it can be concluded that all three 120x120 combined models perform good, but that more independent data sets for comparisons are necessary.

3.2 High resolution models (359x359)

As mentioned above in the second step the high resolution gravity field components are estimated with the block-diagonal approach. Because a mixture of gravity anomalies and geoid heights would destroy this structure the CLS sea surface data set (e) is not further used for the high resolution model. So the following two models were computed (letters are related to the previous definitions):

- HA = a + c + solution A (gaps and normals)
- HC = a + d + solution C (gaps and normals).

The combination of the full normal equation system up to degree 120 and the block-diagonal system up to degree 359 is done according to the strategy described in (Bosch, 1993). This method uses a block matrix technique after the reordering of the normal equation system. The computational amount for the additional high resolution components is small with respect to the inversion of the full matrix, because they can be solved independently order by order.

As for the combined base solutions, again internal and external quality parameters can be derived to assess the overall quality of the final models. Figure 5 shows the correlation coefficients for the combination of the full and block-diagonal normals. Now the side-band correlations are not any more visible. Only small correlations outside the block-diagonals along the main diagonal can be identified. Reason for this is the strong decorrelation of the coefficients by introducing the assumptions to be made for the generation of block-diagonal normals (global coverage with longitude independent weights). As a ample for external quality control we want to compare the model geoid heights with the largest available GPS/levelling derived geoid height data set over the U.S. (Milbert, 1998). RMS values for the complete data set and after elimination of data below +30 degree latitude are shown in table 3. The reason for eliminating data below +30

Table 3: RMS of geoid height differences with U.S. benchmark data set [m]

Model	RMS full set (5168 points)	RMS > +30° (4693 points)
HA	0.485	0.395
HC	0.516	0.402
EGM96	0.458	0.425

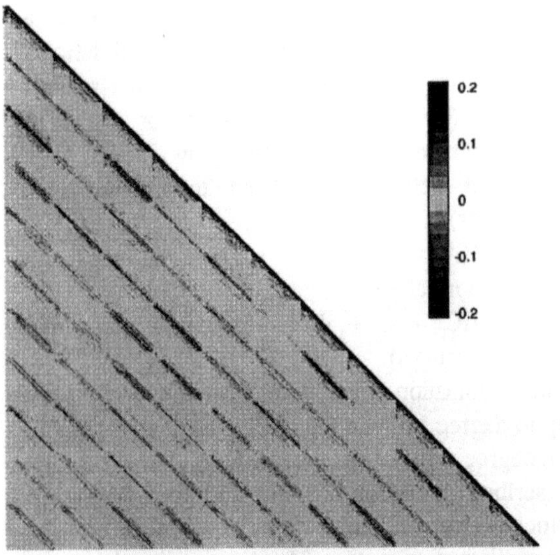

Fig. 4: Correlation coefficients for combined model (sample up to degree 5; coefficients ordered primary wrt. increasing degree, secondary wrt. increasing order)

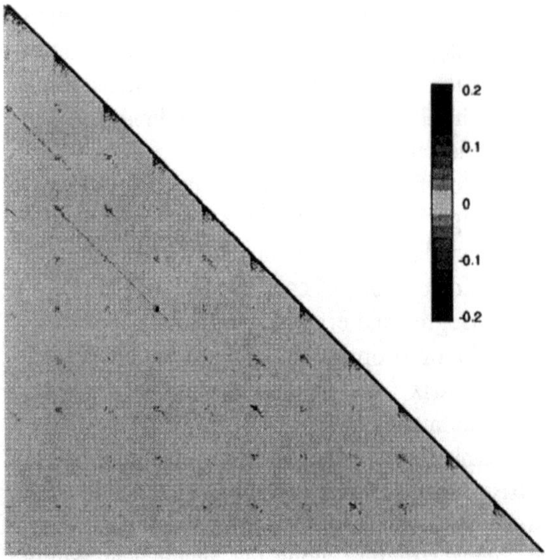

Fig. 5: Correlation coefficients for high resolution model (sample up to degree 5)

degree latitude was the partially bad data quality of GPS heights in Florida (Milbert, 1998).

Results show clearly that solution HA is superior to solution HC. Looking to the degree variances of both solutions (not shown here) it can be found that the signal above degree 120 for HC is too small. This is caused by too much lowpass filtering during processing of the GFZ altimetric anomalies. Comparing results for HA and EGM96 contradicting results are visible. Such comparisons make clear that a simple RMS value of differences can not be taken as a final quality estimate. One always has to look to a wide range of test procedures. Also the knowledge of quality of test data sets is essential for the interpretation of gravity field tests.

4 Conclusions

The following conclusions from the different investigations can be drawn:

- Surface and altimetric gravity information is still valuable for high resolution modelling, even when CHAMP and GRACE are in orbit. But satellite-only models will provide much more and better information than now. With GOCE nearly the complete spectrum of current high resolution models will be covered. Surface data will be only valuable if higher spatial resolutions are available.
- Lowpass filtering of input data sets to reduce aliasing and cut-off effects improves final gravity field model estimates.
- Several combined and high resolution models have been computed. For the combined base models up to degree 120, improvements with respect to EGM96 can be achieved. Higher terms are too smooth because of the block-diagonal technique and/or the application of a non adequate lowpass filter during estimation of gravity anomalies from altimeter data.
- The block-diagonal technique for high resolution components is only an intermediate step towards the full variance-covariance matrix up to degree 359. With current computers 180x180 solutions with full normal equation systems are possible.

References

Bosch W. (1993); A Rigorous Least Squares Combination of Low and High Degree Spherical Harmonics; *Paper presented at XIX IUGG General Assembly*, Vancouver

Colombo O. (1981); Numerical Methods for Harmonic Analysis on the Sphere; *The Ohio State University, Department of Geodetic Science, Report No. 310*; Columbus

Lemoine F., Kenyon S.C., Factor J.K., Trimmer R.G., Pavlis N.K., Chinn D.S., Cox C.M., Klosko S.M., Luthcke S.B., Torrence M.H., Wang Y.M., Williamson R.G., Pavlis E.C., Rapp R.H., Olson T.R. (1998); The Development of the Joint NASA GSFC and the National Imagery and Mapping Agency (NIMA) Geopotential Model EGM96; *NASA Technical Paper NASA/TP-1998-206861*; Goddard Space Flight Center, Greenbelt.

Milbert D.G. (1998); Documentation of the GPS Benchmark Data Set of 23-July 1998; *U.S. National Geodetic Survey*, NOAA, SSMC3, Silver Spring

Reigber Ch., Bock R., Förste Ch., Grunwaldt L., Jakowski N., Lühr H., Schwintzer P., Tilgner C. (1996); CHAMP Phase B: Executive Summary; *Scientific Technical Report STR96/13*; GeoForschungsZentrum Potsdam.

Schuyer N. (1997); Capabilities and Prospects for a Spaceborne Gravimetric Mission; *in: Sansò, Rummel (eds): Geodetic Boundary Value Problems in View of the One Centimeter Geoid; Lecture Notes in Earth Sciences No. 65, p. 569-589*; Springer Verlag

Schwintzer P. (1990); Sensitivity Analysis in Least Squares Gravity Field Modelling by Means of Redundan Decomposition of Stochastic A-Priori Information; *Internal Report, PS/51/90/DGFI*; Deutsches Geodätisches Forschungsinstitut Abt. 1, München.

Schwintzer P., Reigber Ch., Bode A., Kang Z., Zhu S.Y., Massmann F.H., Raimondo J.C., Biancale R., Balmino G., Lemoine J.M., Moynot B., Marty J.C., Barlier F., Boudon Y. (1997); Long Wavelength Global Gravity Field Models: GRIM4-S4, GRIM4-C4; *Journal of Geodesy 71/4: 189-208.*

Tapley B.D. (1997); The Gravity Recovery and Climate Experiment (GRACE); *Supplement to EOS Transactions of American Geophysical Union, 78 (46)*.

Comparison and evaluation of the new Russian global geopotential model to degree 360

G. Demianov, A. Maiorov
Central Research Institute of Geodesy, Air Surveying and Cartography, Federal Service of Geodesy and Cartography of Russia, 26, Onezhskaya, 125413 Moscow, Russia, e-mail: gleb@space.ru

P. Medvedev
Geophysical Center, Russian Academy of Sciences, Molodezhnaya 3, 117296, Moscow, Russia
e-mail: pmedv@wdcb.ru

Abstract. A new Russian global geopotential model was created with the use of detailed gravity data, satellite tracking and altimetry data. Through the research and operations executed, the global Earth gravity field model GAO-98 with the resolution level corresponding to the geopotential expansion into spherical harmonics up to 360th degree has been developed. The main parameters of the model developed are a set of mean anomalies for 30′ x 30′ - blocks covering the whole surface of Earth and a set of spherical harmonics to 360th degree resulted from the expansion of the geopotential. The model is destined mainly for use in highly accurate quasigeoidal height determination, which will significantly improve the efficiency of implementing GPS/GLONASS satellite equipment unto the routine geodetic operations of the Federal Service of Geodesy and Cartography of Russia. The accuracy of absolute quasigeoid heights by use of the model developed is for most regions of Russia $M_\varsigma = \pm 20\text{-}25$ cm. The test data that we used for model evaluations include geoid undulations (or height anomalies) determined from GPS positioning and level observations, global geopotential models EGM96, global marine gravity field from Russian GEOIK altimetry data and from GEOSAT, ERS-1, TOPEX/POSEIDON data. Future problems geoid determinations discussed.

Keywords. Global geopotential model.

1 Introduction

The wide development of geodetic measurement techniques by the use of the global satellite navigation systems GPS/GLONASS requires an exceptionally high accuracy of determination of the Earth gravity field (EGF) parameter. A higher accuracy of EGF-parameter determination makes it possible to considerably expand the applicability of the satellite systems GPS/GLONASS and to resolve thus a number of important problems of the up-to-date geodesy, foremost such as:

1. Establishment of a single high precision world heights systems;
2. Geocentricity checking of the coordinate systems established by satellite techniques;
3. Determination of dimensions of the general earth ellipsoid with the accuracy of present geodetic measurements (5-10 cm);
4. Study of deviations of the oceanic and sea topography from geoid;
5. Improvement of the satellite leveling technique to be used for improving, checking and in some cases as substitute of costly and manpower consuming traditional leveling techniques.

The above problems may be resolved if accurate quasi-geoid heights determinations from gravity data would be ensured. The gravimetric quasi-geoid heights are calculated by the combined method, in which the effect of gravity data in distant zones is taken into account by global gravity model parameters.

To secure up-to-date requirements to the accuracy of quasi-geoid height determination, EGF global models of high precision and level from all the data about the Earth gravity field should be established.

Great attention all over the world is nowadays given to the problems of establishing precise gravity models of a high level of detail. A number of such models, with degree of detail corresponding to the expansion of the geopotential into spherical harmonics up to 360-th degree, have been derived in the last years in some countries. Among them are such models as OSU-91, GFZ-1 and 2, and, finally, EGM-96 developed from all gravity and satellite altimetry data available in the USA up to 1996.

International Association of Geodesy Symposia, Vol. 121
Schwarz (ed.), Geodesy Beyond 2000 – The Challenges of the First Decade

At first, the global EGF models have been mainly used for precise satellite orbit prediction. This fact stimulated the choice of the method of combined gravity data adjustment, i.e. of the parametric adjustment method. This adjustment method has been traditionally retained in the development of most of the foreign models, though the application sphere of these models have been considerably changed.

In the report presented here the development of the model GAO-98 is described. The detailability level of this model corresponds to the gravity anomaly expansion into spherical harmonics up to 360-th degree.

In TcNIIGAiK (Central Research Institute of Geodesy, Air Surveying and Cartography) the attention has been mainly paid to problems of precise determination of gravity parameters on the Earth's surface. The condition (correlate) adjustment method has been found to be more effective because it allows to reduce detailed gravity and satellite altimetry data to a single level of global parameters, and to considerably diminish the effect of systematic errors of these data on the accuracy of guasi-geoid heights determination.

2 Source data

The bulk of source sea gravity and altimetry data used for this model has considerably increased in comparison with former TcNIIGAiK global models, with a degree of detail of 180-th degree. While for the preceding TcNIIGAiK model (GAO-95), sea gravity database TcNIIGAiK, sea gravity database TcNIIGAiK (ZKMGD) and SEASAT and GEOSAT altimetry data have been mainly used, for the models. GAO-97 and 98, sea gravity database GEODAS (about 1200 survey routes), gravity anomalies averaged for 30′ x 30′ blocks for the territory of foreign countries taken from terrestrial and aerial gravity surveys and used in the model EGM-96 (USA), as well as considerably greater bulk of altimetry data have been additionally used.

In the model GAO-98, gravity anomalies for 15′ x 15′ blocks on the territory of Europe, processed under leadership of R. Forsberg and U. Schafer as part of cooperative work within the project INTAS-93-1779-ext, have additionally been used.

Besides, an analysis of altimetry and sea gravity data in the model GAO-98 has been carried out. About 20% of GEODAS data have been rejected because of inadmissible amounts of random and systematic errors. The main obstacle for further improvement of accuracy of the gravity models remains the lack of gravity data for some non- or insufficiently gravimetrically explored areas of land and sea. To these hard-accessible areas, in which gravity measurements by traditional techniques are difficult, belong, first of all, the high mountain regions of South America and Central Asia, as well as continental and coastal Antarctic areas.

Three types of information have used for development of GAO-98 model:

- gravity data from aerial and route gravity surveys on land in the World ocean;
- altimetry data in form of gravity anomalies for 3′ x 3′ and 30′ x 30′ blocks computed from GEOSAT (geodetic mission) and ERS-1 altimetry;
- satellite orbital data in form of coefficient set up to 60-th degree of purely satellite EGF models derived from satellite orbit perturbation analysis.

The choice of such a high degree of EGF model resolution as 360 made it necessary to use as source data mean gravity anomalies for 30′ x 30′ blocks. Gravity anomalies are calculated with the free-air reduction in respect to the international normal gravity formula 1980 and reduced to the gravimetric system IGSN-71.

Source data preparation for the GAO-98 model is characterized by a great bulk of information, as well as by complicated and labor consuming processing, including analysis, checking, accuracy estimation, and rejection of inadequate data, especially for foreign gravity measurements in the World Ocean.

3 Adjustment technique

Unlike the widely used parametric adjustment technique, which is optimal for determination of only EGF model parameters that approximate in the best way measurement data, the condition adjustment technique traditionally used for model parameter determination turns out to be more effective for resolution of geodetic problems, for which mutual concordance of gravimetric and satellite altimetry data and corrections to measurements due to measuring errors are of importance.

As gravity measurements (especially marine) and altimetry are prone to systematic errors of small amounts, but spread over considerable areas, such a concordance of global parameters allows a significant dismissing of the effect of systematic errors on the determination of gravity field parameters, first of all, quasigeoid height. Nowadays, it is especially important due to wide introduction of GPS/GLONASS measurements into routine (production) geodetic operation.

The combined adjustment of gravity and satellite orbital data results in a set of corrections to the initial gravity values. Presentation of the parameters in form of corrections to the initial gravity anomaly values allows to detect and reject the systematic errors of individual gravity surveys, and to get information on eventual incompatibility of weights in the combined adjustment of heterogeneous information.

The adjustment technique adopted in TcNIIGAiK provides for several consecutive stages:

1. Initial gravity values for 30′ x 30′ - blocks from all the gravity and satellite altimetry data are used for calculation of mean anomalies for 3° x 3° – blocks covering the total surface of the Earth;
2. The anomalies for 3° x 3° – blocks are used for expansion into spherical harmonics up to 60-th degree;
3. The harmonics thus abstained are adjusted with the purely satellite model JGM-2S. The differences between the coefficients a_{nk}^{gr}, b_{nk}^{gr} from the expansion and the harmonics a_{nk}^{s}, b_{nk}^{s} of the satellite model are used as obsolete terms of the condition equations;
4. From the adjustment results, using the condition equation of the type:

$$\sum_{j=1}^{j=n} V\Delta g_{3^\circ} \cos B_j P_{nk}(\cos B_j)\begin{Bmatrix}\cos kL\\ \sin kL\end{Bmatrix} - \begin{Bmatrix}V^s_{a_{nk}}\\ V^s_{b_{nk}}\end{Bmatrix} + L_j = 0$$

the corrections to mean anomalies of the blocks 3°×3° - $\overline{V\Delta g}_{3^\circ}$, are calculated, where $V^s_{a_{nk}, b_{nk}}$ - corrections to the satellite coefficient values.

5. The corrections $\overline{V\Delta g}_{3^\circ}$ calculated for 3°×3° - blocks at the stage 4 are distributed among the initial 30′×30′ - blocks.
6. By expanding the adjusted anomaly set for 30′×30′ -blocks into spherical harmonics up to 360-th degree we get the parameters of the EGF model in form of harmonic coefficients.

We will have thus the main parameters set of the EGF model in form of a full adjusted set of mean gravity anomalies for each 30′×30′ -block.

4 Principal results and perspective

Through the research and operations executed, the global Earth gravity field model GAO-98 with the resolution level corresponding to the geopotential expansion into spherical harmonics up to 360th degree has been developed. The main parameters of the model developed are a set of mean anomalies for 30′×30′ -blocks covering the whole surface of Earth and a set of spherical harmonics up to 360th degree resulted from the expansion of the geopotential. The model is destined mainly for a highly accurate quasigeoid height determination, which will significantly improve the efficiency of implementing GPS/GLONASS satellite equipment into the routine geodetic operations of the Federal Service of Geodesy and Cartography of Russia.

Presentation of the EGF model GAO-98 in the form of mean gravity anomalies for 30′×30′ -blocks allows to make gravity data compatible with the global EGF parameters, which contributes to detection and rejection of systematic errors of detailed gravity surveys.

The accuracy of absolute quasigeoid heights by use of the model developed is for most regions of Russia M_ζ = ±20-25 cm. The a priori accuracy estimation of these data are confirmed by the results of precise GPS measurements along the traverse Moscow-Sankt Petersburg on a high-precision leveling line [6], results of TcNIIGAiK in Project "Level of Baltic Sea"[3], and international project IGEX [1].

The anomaly corrections from the adjustment allowed to estimate the accuracy of gravity data for different areas of the globe, to substantially assign the weights of the initial data, and to more realistically estimate the accuracy of the parameters determined.

A further progress in improvement of EGF models may be achieved through practical realization of satellite gradiometry.

The possibilities of improving the accuracy of terrestrial gravity data are far from being exhausted. The accuracy of the fundamental gravity net being established by absolute (ballistic) gravimeters is ± 6 μGal. If all the detailed gravity surveys would be set on the basic net of such accuracy level, then we can speak of achieving the accuracy of quasigeoid height determinations of 5 to 10 cm as quite realistic in the near future. Such accuracy is comparable with present measurement accuracy on the GPS station for geodynamics.

At present, different height systems are used in different continents or in some groups of countries, even on the territory of Europe two height systems (Amsterdam and Kronshtadt ones) are available. This factor in its turn is an additional cause of errors in high-precision computation quasigeoid heights from gravity data.

The use of one initial water gauge as zero for height reckoning (for example Amsterdam or Kronshtadt) is now the same anachronism as the fundamental station Pulkovo in coordinate system 1942.

In our case leveling height (normal height) will equal zero in that point on Earth in which the real geopotential will equal normal on ellipsoid surface $W_i = U_{00}$ [2],[7].

As a unique datum both horizontal and vertical, a World permanent station net of satellite observation of GPS/GLONASS systems should be taken into consideration. Quasigeoid heights from gravity data should be determined with a high accuracy for each of these stations. The problem will be determination of conformity of reference ellipsoids of the WGS and PZ-90 systems, GPS and GLONASS coordinate systems as well as of the general earth ellipsoid in the gravimetric sense of

$$\int_{\sigma} \zeta \, d\sigma = 0 .$$

This problem may be resolved through a combined adjustment of quasigeoid heights, high-accurate leveling data and satellite observation data from the world permanent station net GPS/GLONASS (for instance IGS).

The accuracy of this data are confirmed by results of comparison between GAO-98 and EGM-96 models also for 7 regions of the world. Results of comparison are presented in table 1. Moreover quasigeoid heights determined by GAO-98 and EGM-96 models was checked by comparison with detail gravity geoid for Russian territory and with altimetry data for sea surface [4], [5].

Results of this estimation are given in the tables 2 and 3. These results confirm once more a priory accuracy of geoid heights determined by use of GAO-98 model: $m_\zeta = \pm 0.2\text{-}0.4$ m for most regions and $m_\zeta = \pm 0.6\text{-}1.0$ m for anomalous regions.

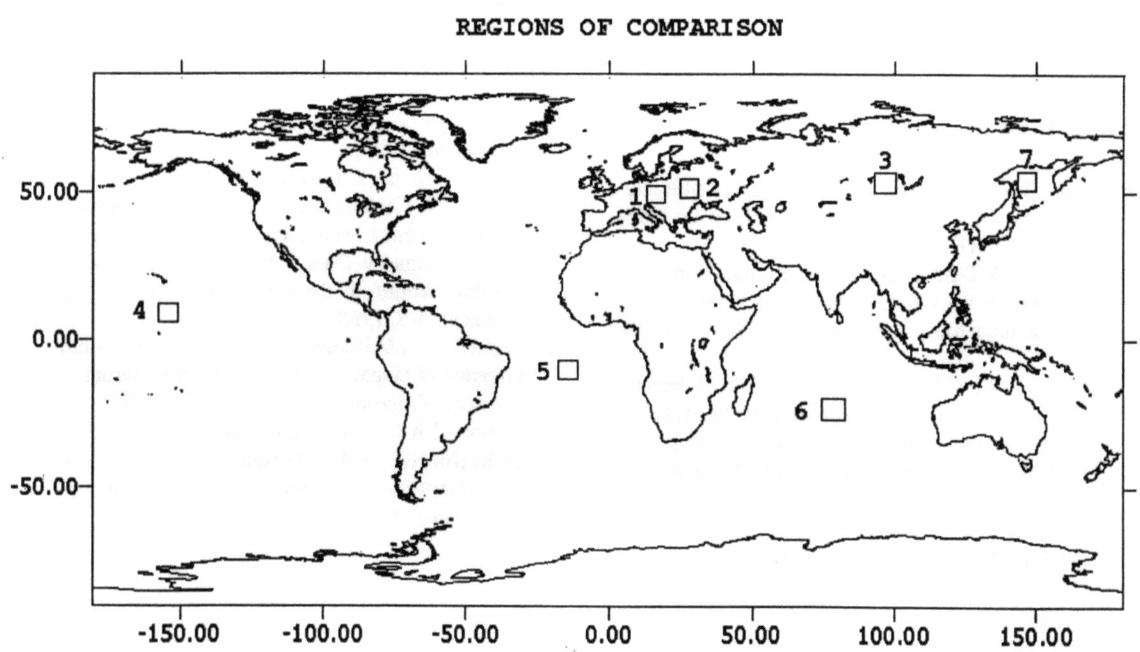

Fig 1: Regions of comparison GAO-98 vs. EGM96

Table 1: Comparison GAO-98 vs. EGM96

Region	1	2	3	4	5	6	7
Max	-0.38	0.57	0.18	0.30	0.90	0.25	0.91
Min	-2.23	-1.61	-3.53	-0.32	0.09	-0.75	-0.57
Mean	-1.06	-0.48	-1.30	0.02	0.43	-0.11	0.14
Standard Dev.	0.42	0.42	0.39	0.12	0.15	0.22	0.32

Table 2: Comparison GAO-98 and EGM-96 gravity models with detail geoid for Russia

Region	2		3	
Model	GAO-98	EGM-96	GAO-98	EGM-96
Mean	0.05	-0.25	0.07	-0.50
RMS	0.18	0.20	0.30	0.45

Table 3: Comparison geoid heights of GAO-98 and EGM-96 models with mean sea surface heights

Region	4		5		6		7	
Model	GAO	EGM	GAO	EGM	GAO	EGM	GAO	EGM
Max	0.98	0.60	0.85	0.60	0.26	0.42	0.90	0.90
Min	-0.52	-0.50	-0.48	-0.47	-0.60	-0.52	-0.70	-0.75
Mean	0.15	0.13	0.38	0.75	-0.13	0.13	0.08	0.13
RMS	0.48	0.37	0.27	0.80	0.24	0.17	0.33	0.30

References

Ashjaee J., G.V. Demyanov et all. "Precise positioning with the use of JPS legacy GPS/GLONASS Receivers". IUGG Birmingham.

Bursa M., G.V.Demyanov, M.I.Yurkina "On the determination of the Earth's model – the mean equipotential surface". Studia geoph. et geod. 42 (1998).

Demianov G.V., V.I. Kaftan, V.I. Zubinsky "Participation of the Central Research Institute of Geodesy, Aerial Surveying and Cartography in the Third Baltic Sea Level GPS Campaign, Paper on the Second workshop "Results of the Third Baltic Sea Level GPS Compaign,1997" of the sub-commission IAG SSG 8.1 Studies of the Baltic Sea.

Medvedev P., Yu S., Tyupkin, S.A. Lebedev "An Integrated Satellite Altimetry, Gravity, Geodesy Data Base: Architecture, Verification Data Processing, Data Base Management System. Presented to XXII General Assembly of IUGG, Birmingham, 1999.

Medvedev P., S.A. Lebedev, Yu.S. Tyupkin, V.F. Galazin, D.I. Pleshakov, A.N. Zueva. An Integrated Satellite Altimety Data Base and Final Results of Russian Altimetry Data Processing. Presented to 2nd Joint Meeting of the IGC/IGeC, Trieste, Italy, September, 5-9, 1998.

Zubinsky V.I., G.V.Demianov et al. "Experimental estimation of heights accuracy by GPS construction big extent".Proceeding of TcNIIGAiK. 1999.

Zhalkovsky E.A., G.A. Demyanov, B.V. Brovar, A.N.Mayorov, V.A. Taranov, M.I. Yurkina "Gravimetry and Modern Satellite Geodesy". Proceeding of TcNIIGAiK. 1999.

Assessing the global land one-km base elevation DEM

D. A. Hastings, P. K. Dunbar and A. M. Hittelman
World Data Center - A, Boulder Centers
National Oceanic and Atmospheric Administration
National Geophysical Data Center, 325 Broadway, Boulder CO 80303, USA
Email: dah@ngdc.noaa.gov & pkd@ngdc.noaa.gov

Abstract. The Global Land One-kilometer Base Elevation (GLOBE) digital elevation model is the most thoroughly designed, reviewed, and documented global digital elevation dataset to date. Developed by an international group of specialists, GLOBE comprises a global 30 arc-second latitude-longitude array, with land areas populated with integer elevation data. GLOBE is available on the World Wide Web, and on CD-ROM in a format convenient for image processing and raster geographic information systems (GIS). Full GLOBE documentation accompanies the CD-ROM and appears on the Website (http://www.ngdc.noaa.gov/seg/topo/globe.html). This paper summarizes some findings made during GLOBE's development.

Keywords. Digital elevation model.

1 Objectives and design of GLOBE

When GLOBE was conceived almost a decade ago, the definitive global digital elevation model was ETOPO5, managed by the National Geophysical Data Center (NGDC). ETOPO5 is a mosaic of the Digital Bathymetric Data Base 5 global bathymetric model (derived by interpolating elevation contours derived from several million point soundings), TERDAT (a 10-minute gridded global DEM) and selected regional 5-minute DEMs. TerrainBase (Row and Hastings, 1994, Row and others, 1995) was being designed at NGDC with more 5-minute data output from a collection of co-registered grids of various resolutions corresponding to those of original data sources.

GLOBE's objective was to open an empty two-dimensional latitude-longitude 30 arc-second computer array, then populate it with the best available data that could be distributed with minimal restrictions. In addition, a completely unrestricted version of GLOBE was envisaged. It would do this by opening several two-dimensional arrays with gridding corresponding to available source materials, prioritizing source data by quality, developing blending methods where appropriate, then mosaicking and documenting the best possible grid as GLOBE Version 1.0. All actions in developing GLOBE were reviewed and approved by the GLOBE Task Team of the Committee on Earth Observation Satellites Working Group on Information Systems and Services (CEOS-WGISS), with participation by the International Geosphere-Biosphere Programme Data and Information System (IGBP-DIS) and International Society for Photogrammetry and Remote Sensing Working Group IV/6 on Global Datasets Supporting Environmental Monitoring.

2 Sources and adaptation methods

GLOBE sources were divided into two categories, in two different ways:

2.1 Raster vs. vector source data

Raster data sources: Though DEMs can include point elevations, digitized vector contours, and raster grids, most current DEMs are in raster format. With the exception of raster DEMs from the USA, the GLOBE project was not optimistic about receiving many raster DEMs due to traditional copyright and security restrictions imposed by most traditional makers of such data.

Vector data sources: The U. S. Defense Mapping Agency (DMA, now the National Imagery and Mapping Agency, NIMA) designed and released the Digital Chart of the World (DCW) in 1992. DCW included digitized contour lines and selected point elevations for much of its global coverage. The U. S. Geological Survey and University College London designed techniques for

International Association of Geodesy Symposia, Vol. 121
Schwarz (ed.), Geodesy Beyond 2000 – The Challenges of the First Decade

converting these data to 30" grids using techniques similar to those discussed by Hutchinson (1996). Originally, these were the main hope for improved data coverage in GLOBE.

2.2 Resolution higher or lower than GLOBE's

Source resolution at 30" or finer: Most raster DEMs fit this category. At GLOBE's inception, available data included USGS and NGDC data for the USA, and data for Italy contributed by the Servizio Geologico Nazionale of Italy to TerrainBase. Other DEMs were considered essential, but difficult to obtain for GLOBE.

Source resolution coarse than 30": DCW was considered slightly coarser than 30", but close enough to serve where better data were lacking. In the absence of finer data, sources for ETOPO5 or TerrainBase could complete global coverage.

2.3 Creative arrangements with data sources

Fortunately, several major producers of raster DEMs were willing to negotiate varying types of contributions to GLOBE:

The National Imagery and Mapping Agency (then the Defense Mapping Agency) and GLOBE jointly designed a prototype 30" extraction from its 3" Digital Terrain Elevation Data, which eventually became NIMA's publicly distributed DTED Level 0. This pioneering data set offers "maximum," "minimum," mean and spot elevation values for each included 30" grid cell. It covers over half of the Earth's land surface. Although GLOBE quality review found problems (see section 3.2 below) with first-release DTED Level 0 "minimum" and mean values, the other two parameters are important for users of GLOBE.

The Australian Surveying and Land Information Group offered to let NGDC produce a 30" grid from AUSLIG's large point elevation database. The resultant grid would still be copyright by AUSLIG, but licensed to NGDC for distribution with GLOBE.

The Geographical Survey Institute of Japan designed a publicly releasable 30" synopsis of its higher-resolution copyright data, and contributed it to the scientific community, including GLOBE.

Manaaki Whenua Landcare Research of New Zealand contributed a DEM to public access via the U. S. Geological Survey, which was also used in GLOBE.

The Scientific Committee on Antarctic Research offered to let its Antarctic Digital Database be used for such purposes. USGS converted these digitized contours to grids.

Elevation grids were contributed for othe areas. Nevertheless, some areas remained without coverage. For these areas, the Geographical Survey Institute of Japan adapted selected maps to digital form. Then the U. S. Geological Survey converted those digital data to grid form for use in USGS' GTOPO30 model. GLOBE then used selected data (including some of the data noted just above) from GTOPO30.

NIMA released DTED Level 0 in 1996. USGS released its own GTOPO30 in 1997; GLOBE continued with its own enhancement of coverage, quality control, and documentation.

2.4 Future enhancements: contributions are still valuable

The GLOBE Secretariat is still working on enhancements to GLOBE data, documentation, and access. If you have DEM coverage for any area, we would like to discuss possible arrangements for access to 30" or better versions of these data for GLOBE.

3 Quality assessment

Available for assembly into GLOBE Version 1.0 were data from 11 original sources, some of which had more than one form of post-source processing. A total of 18 combinations of source/lineage were used in GLOBE.

3.1 Types of quality review techniques

Quality review included (1) viewing source imagery with various color palettes in NGDC's GeoVu public domain browse and visualization utility (http:// www.ngdc.noaa.gov/seg/geovu/ geovu.html) and the GRASS GIS. Among the most useful palettes is GRASS' "random" color palette, which assigns each elevation a color rather different than neighboring elevations. In addition, GIS and image processing techniques were used to produce color-coded shaded-relief images, which were browsed for terrain characteristics. These tools allow one to view non-terrain-like "geometric" features in a DEM, such as ungraceful mosaicking sutures; (2) viewing slope, aspect, and shaded relief images of each source data set; (4) principal components analysis inputting all available sources for a given area; (5) histogram analysis of each source data set, and (6) statistical comparisons between sources, to see if any source had a particular elevation bias.

3.2 Examples of findings

Visualizations showed near-vertical offsets at edges of many 1°x1° cell blocks in the mosaicked DEM. Such offsets appear due to the challenges of rectifying misdocumented or incorrectly adjusted vertical datums, either in source maps or grids, or in various stages of conversion. These offsets were typically on the order of a few meters - within the de-facto apparent accuracies of the data sets. Offsets were usually not vertical, but appeared to be blended with linear blending filters a few grid cells wide.

Visualizations of shaded-relief images also exhibited striping in a west-northwesterly direction, likely perpendicular to orbital tracks of source satellite images. These appear to be artifacts of many products of digital scanner imagery. Similar striping in shaded-relief images can also result from undersampled stereoprofiling techniques - but that striping is usually in a more north-south or east-west orientation.

In some areas, visualizations showed variations in character within 1°x1° cell blocks of DTED. These suggested the need to use more than one source to complete such blocks. These occurrences were relatively rare.

Visualizations also showed varying styles of "layer-cake" effects in areas with cartographic sources. A few areas had relatively coarse layering, suggestive of widely spaced contour lines that may be describing the terrain less well than in areas with finer layering. Relatively coarse layering is less common than originally anticipated, due to higher quality of sources than originally envisaged.

Visualizations showed that the USGS conversion of the Antarctic Digital Database contained dropouts near 180 degrees longitude and near the South Pole. These appear to be edge effects in a contour-to-grid conversion. These dropouts were repaired for GLOBE.

Visualizations also found that some islands do not exist in source data; some others are not consistently located in all source data. For example, Isla de Guadalupe (Mexico, about 118.3° west longitude and 29° north latitude) appears to be offset several minutes of longitude between DTED and Digital Chart of the World computed to grids by USGS. The result is a "siamese-twin" effect in GTOPO30, as well as in even late prototypes of GLOBE. However, a detailed visualization of central North America prominently showed this artifact. World Vector Shoreline agreed with DTED. GLOBE Version 1.0 has thus removed the "siamese twin" at Isla de Guadalupe. Similarly with another nearby island, browse imagery from Space Imaging was used to relocate DTED Level 0 data to the location seen by the Indian Remote Sensing Satellite (IRS).

When a similar color-shaded-relief image was made of Europe to help illustrate our presentation for this IAG meeting, we discovered that a piece of the southern coast of Crete (Greece) was missing. A review of source materials noted that DTED coverage was lacking for this area, and that attempts at global DEMs from other institutions also missed this area. However, a recent revision of DTED Level 0, which now fills gaps in DTED Level 1 coverage with DCW (now called VMAP) conversions, has coverage for this single 1°x1° DTED file block. GLOBE data were updated to fill this gap, just before GLOBE Version 1.0 CD-ROMs were sent for mastering. Thus GLOBE Version 1.0 appears to be the first global compilation other than DTED Level 0 (which keeps the world parsed into thousands of files covering 1°x1° each) to include coverage of this area.

Sample visualizations can be found on the Web at: ftp://ftp.ngdc.noaa.gov/GLOBE_DEM/pictures/ Despite imperfections in the data, most first-time viewers are enthralled with the large-scale details provided by the visualizations.

Statistical comparisons (Hastings and Dunbar (1999) pp. 36-37) confirmed that DTED Level 0 "minima" were frequently *higher* than comparable "mean" values from the same data set. "Mean" values were lower than spot values more than should occur. Independent evaluations of DTED Level 0 data with full 3"-gridded DTED Level 1 data suggested that the spot values were the most representative DTED level 0 values, not the originally intended computed means.

Histogram analysis (Hastings and Dunbar (1999) pp. 19-35,39-40,42,44,45,47,50-51,52, 54,56) was used to estimate vertical accuracies for sources without documented statements of precision. Histogram peaks suggested that implied contour intervals in Japanese and Australian sources were about 20m, giving these countries nominal estimates of 10m (6m Root-Mean-Square-Error) vertical precision. At the other end of the scale, DCW and some other sources in frontier areas have RMSE of about 100m or greater. Although DTED's design criteria aim for 18m RMSE, considerable amounts of DTED were developed under conditions that could not meet its design criteria.

Most DEMs derived from contour maps had histograms that favored contour values in the original maps. This well-known but less-than-ideal characteristic was, however, not evident in the Geographical Survey Institute's contributed DEM (Hastings and Dunbar (1999) pp. 26, 41-42, 58). GSI's techniques for DEM development may be worth wider circulation, as they may contain techniques worth adopting by other agencies.

Shaded relief analysis showed that the Servizio Geologico Nazionale's DEM of Italy was much more terrain-like for low elevations than alternative candidates (Hastings and Dunbar (1999) pp. 27, 42-44, 58-59). Shaded relief analysis helped in the selection of best available candidates for several areas.

Comparisons with reference points. Muller (1997), and Muller (oral communication, 1996, 1997, 1998) compiled point reference values from several sources, including geophysical measurement sites from the National Geophysical Data Center, airport locations and heights from Jeppeson databases, geodetic control, etc. Muller used these data for comparison with three regional/global DEMs. These evaluations helped to guide the selection of sources for GLOBE, but also demonstrated that the "control" data were of varying reliability. Muller once wondered aloud "which data are the control, and which are being evaluated?"

Comparisons with altimetry. Berry (1999) made several comparisons between GLOBE, other DEMs, and elevations she derived from satellite altimetry. Berry found systematic discrepancies between many of her altimetry-based derivations and the DEMs, which she attributed to the DEMs. Some discrepancies (http://www.cse.dmu.ac.uk/geomatics/) suggest errors in navigating the satellite altimeter over the Netherlands and over north-central Siberia. Berry also cites a discrepancy of about 250m extending 25° in longitude and 5°-10° in latitude in Siberia. One can certainly anticipate challenges in determining an accurate vertical datum in certain areas, such as within the influence of the Himalayas - but such challenges also affect those working with satellite altimetry (Thomas Logan, Jet Propulsion Laboratory, oral communication (1999)). It may be premature to claim that one dataset is a "control," as described in the previous paragraph. Despite these apparent questions about the satellite altimetric presentations to date, some discrepancies between the altimetry and GLOBE appear graphically: for example, difficult to quantify discrepancies between sources used in GLOBE. Further work with the altimetry might help guide vertical adjustments of some vertical discrepancies noted in the first paragraph of this section 3.3.

3.3 Summary of the quality assessment

Overall, perhaps half of GLOBE exceeds 20m RMSE, where data for Antarctica may be as much as 300m off.

Some may be disappointed by the worst-case vertical accuracies for some areas in GLOBE. However, discrepancies between different geodetic models can create elevation differences in this order of magnitude. Some maps and digital data sets lack full documentation details, and errors have been detected in the projection and datum information in some maps and digital data. An example of such an apparent misfit was reported for Japan at a GLOBE Task Team meeting (Hiroshi Murakami, 1995, oral communication). These discontinuities appear to be mostly small and rare. With such mislabeling of source materials in mind, GLOBE's overall accuracy appears generally to be a pleasant surprise.

Please remember that specific features may be missed or stylized in such a data set. Critical applications should be designed with caution. Treat all data sets such as GLOBE as you might treat hot coffee. Expect some pain in learning how to handle GLOBE or hot coffee, if you do not already know how to handle them. The pain, and the benefit from learning, are technology's gift to the user.

4 Assembly

All candidate data sets were placed in the GRASS GIS, and assembled using the r.patch program with appropriate masking at each stage to avoid having non-zero values in a lower-quality subsequent patch overwrite higher-quality zero elevations. From previous experience (Row and others, 1995) NGDC had learned that blending replaces abrupt discontinuities with "smeared out" discontinuity zones. It was able to avoid most such blending. When blends were needed to improve data continuity at sutures, these were done in advance of the patching process.

Testing of the final GLOBE mosaic found a surprisingly good fit in most cases. However, it found an apparent vertical misfit between DEMs derived from Digital Chart of the World and other adjacent DEMs from other sources in some areas. This may be caused by inadequate documentation of vertical datum in source materials for DCW. They are under consideration for adjustment in future enhancements of GLOBE.

5 Data access

There are two versions of the GLOBE data set: (1)

Best Available Data (BAD), which allows the inclusion of copyright data if those data may be distributed without restriction by the GLOBE Task Team. (2) Globally Only Open Access (GOOD), which contain data without any copyright or proprietary restrictions.

GOOD data are available from the GLOBE Website (noted above in the abstract), and on CD-ROM. Arrangements are being made for additional distributors, and for Web mirror sites. The CD-ROM version is accompanied by NGDC's GeoVu software, enabling one to browse, visualize, assess, and do limited processing (such as limited reformatting and subsetting) of the data. In addition, the CD-ROM contains a "plastic Website" that can be accessed with popular Web browsers. Extensive documentation is available in hardcopy, as well as in various print-file (.pdf) and Web-based (.html) options.

BAD GLOBE data are currently available only via CD-ROM. Currently the only area with BAD GLOBE data coverage is Australia (there is overlapping GOOD coverage for the same area), though other possible copyright enhancements are being considered for future versions of GLOBE. Web access to GLOBE data and documentation is free. CD-ROMs have a charge based on a cost-recovery formula.

6 Future prospects for global DEMs

While GLOBE was being developed a climate of improved cooperation between several producers of copyright, security-restricted, and/or completely unrestricted DEMs also seemed to evolve. Three promising developments are underway:

1. NIMA and the National Aeronautics and Space Administration (NASA) have agreed to conduct the joint Shuttle Radar Topography Mapper (SRTM) mission. Using the Shuttle Imaging Radar system, this dedicated mission hopes to record complete global coverage of interferometric synthetic aperture radar (SAR) data, from which DEMs will be derived. 3 arc-second global DEMs are anticipated for public release early in the next decade, covering areas covered by SRTM (about 60° North to South latitudes).

2. Radarsat was operated to provide interferometric looks at Antarctica, for the development of a South Polar DEM to complement SRTM coverage in the southern hemisphere.

3. Several private satellite initiatives include stereo-optical coverage in their design. Hopes are high for local DEMs of high resolution, relatively quick production speed, and relatively user-selectable repeat cycles from such initiatives.

In short, in the best of all worlds, GLOBE will be obsolete sometime in the next decade. However, before this happens, SRTM and other missions will have to combine to create complete global coverage, and that coverage will have to pass a quality assessment exercise. GLOBE will probably have an important role to play in that assessment.

Acknowledgements. The CEOS-WGISS GLOBE Task Team was founded by Gunter Schreier of the Deutsches Fernerkundungsdatenzentrum (DFD, the German Remote Sensing Data Center), part of the Deutsches Zentrum Fur Luft- und Raumfahrt (DLR, the German Aerospace Center), in conjunction with University College London and NGDC. Participants also included the U. S. Defense Mapping Agency (later National Imagery and Mapping Agency), Geographical Survey Institute of Japan, Australian Surveying and Land Information Group, the Jet Propulsion Laboratory, the U. S. Geological Survey, IGBP-DIS, and ISPRS Working Group IV/6.

References

AUSLIG (1996 Point elevation data file for Australia. Belconnen, ACT, Australia; Australian Surveying and Land Information Group

AUSLIG and NGDC (1998) 30 arc-second digital elevation model for Australia. IN GLOBE Task Team and others, ed. (1998)

Berry, Philippa A. M. (1999) Global digital elevation models - fact or fiction? Astronomy and Geophysics 40:3.10-3-13

Defense Mapping Agency (1992) Digital Chart of the World. Fairfax, Virginia, Defense Mapping Agency. (four CD-ROMs)

Gittings, Bruce (1997) Digital Elevation Data Catalogue. Department of Geography, Univ. of Edinburgh. (http://www.geo.ed.ac.uk/home/ ded.html)

GLOBE Task Team and others (Hastings, David A., P. K. Dunbar, G. M. Elphingstone, M. Bootz, H. Murakami, Peter Holland, Nevin A. Bryant, Thomas L. Logan, J.-P. Muller, Gunter Schreier and John S. MacDonald), eds. (1998) The Global Land One-kilometer Base Elevation (GLOBE) Digital Elevation Model, Version 1.0. National Oceanic and Atmospheric Administration, National Geophysical Data Center, 325 Broadway, Boulder, Colorado 80303, USA. Digital Database on the WWW (URL: http://www.ngdc.noaa.gov/seg/topo/ globe.html) and CD-ROMs

GSI (1995) 30 arc-second digital elevation model of Japan. Tsukuba, Japan; Geographical Survey Institute. IN GLOBE Task Team and others (1998)

Hastings, David A., and P. K. Dunbar (1997) The development of global digital elevation data. Proceedings, 18th Conference on Remote Sensing, Kuala Lumpur. pp. JS-3-1 to JS-3-6

Hastings, David A., and P. K. Dunbar (1998) Development and assessment of the Global Land One-km Base Elevation digital elevation model (GLOBE). International Archives for Photogrammetry and Remote Sensing, v.32, Part 4: 218-221

Hastings, David A., and Paula K. Dunbar (1999) Global Land One-kilometer Base Elevation (GLOBE). NOAA National Geophysical Data Center, Boulder, Colorado, Publication KGRD-34. 138pp.

Hastings, David A., A. M. Hittelman, and P. K. Dunbar (1995) Digital representations of topography, attempting to characterize the coastal environment. Proceedings, Third ERIM Thematic conference: Remote Sensing for Marine and Coastal Environments, Seattle. I:334-343

Hutchinson, M.F. (1996) A locally adaptive approach to the interpolation of digital elevation models. In: Proceedings, Third International Conference/Workshop on Integrating GIS and Environmental Modeling, Santa Fe, New Mexico, January 21-26, 1996. National Center for Geographic Information and Analysis, Santa Barbara, California

National Imagery and Mapping Agency (1996) Digital Terrain Elevation Data Level 0. Fairfax, Virginia, NIMA Published on the World Wide Web (http://www.nima.mil) and distributed on CD-ROM in a format enhanced for image processing and GIS as GLOBE Prototype ver. 0.5 by the NOAA National Geophysical Data Center, Boulder, Colorado

Row, L.W., and D.A. Hastings (1994) TerrainBase Worldwide Digital Terrain Data on CD-ROM, Release 1.0. NOAA National Geophysical Data Center, Boulder, Colorado

Row, L.W., D.A. Hastings, and P.K. Dunbar (1995) TerrainBase Worldwide Digital Terrain Data - Documentation Manual, CD-ROM Release 1.0. NOAA National Geophysical Data Center, Boulder, Colorado

USGS, ed. (1997) GTOPO30 digital elevation model. Sioux Falls, South Dakota, U. S. Geological Survey EROS Data Center

Recent advances in the acquisition and use of terrain data for geoid modelling over the United States

D.A. Smith, D.R. Roman
NOAA/National Geodetic Survey
1315 East-West Highway, SSMC3, N/NGS5, Silver Spring, MD, 20910, USA

Abstract. For much of the past decade, high resolution geoid models have been produced at the National Geodetic Survey (NGS). The use of digital elevation models (DEMs) has been important to the computation of geoid models, but has also been one of the primary limitations of the geoid accuracy due to both poor data quality and a series of computational approximations. Recent steps have been taken at NGS to replace existing DEM data with data of finer resolution and higher quality. Additionally, new computational tools are being implemented which no longer rely on accuracy-reducing approximations. By moving toward more accurately computed terrain-induced gravity signals, errors in gravity anomalies exceeding tens of mGals have been removed, and subsequently centimeters of geoid accuracy have been gained.

Keywords. Geoid, elevation models, terrain effects.

1 Introduction

The need for accurate terrain data in the computation of an accurate geoid model has long been known. Whether one uses the Stokes-Helmert approach or alternative methods, such as that of Molodensky, there has always been a need for an accurate digital elevation model (DEM). Although DEMs with horizontal resolution as fine as three arcseconds have been available since the 1960's, their general accuracy has been limited due to their data sources (small-scale digitized cartographic maps) and lack of significant photogrammetric coverage. In addition to DEM accuracy, the use of a DEM in computing geoid models is further limited by computational resources (RAM, CPU speed, disk space) and computational shortcuts (planar approximations, flat-topped prisms, constant density, Faye anomalies approximating Helmert anomalies). At NGS, even as late as 1996, the gravitational terrain signal was modelled as planar, 2-D FFT terrain corrections at a horizontal resolution of 30 arcseconds (Smith and Milbert, 1999). Initial tests at NGS indicate that significant (10 cm and greater) errors were being made in the geoid model due to the coarse resolution of the DEM, existing large-area biases in the DEM, the computational approximations in computing terrain corrections, and the use of Faye anomalies instead of Helmert anomalies. Two separate efforts were begun at NGS in 1998 to improve this situation. The first was a search for a more accurate DEM for the United States, and the second was taking some initial steps toward removing classical approximation in terrain corrections. Both were successful, and may be implemented in the GEOID99 model, anticipated for release in September 1999.
(http://www.ngs.noaa.gov/GEOID/GEOID99/geoid99.html)

2 Testing new digital elevation models

In the United States, there are only two significant sources of high resolution (better than 30 arcseconds) DEMs. The first is 7.5 minute quadrangle DEM from the United States Geological Survey (USGS) and the other is Digital Terrain Elevation Data (DTED) Level 1 from the National Imagery and Mapping Agency (NIMA). The comparison between these two available DEMs is made in Table 1.

It was clear from the initial comparison that the USGS data was somewhat superior to the NIMA data, but in order to be sure, two tests were conducted on these data: 1) Agreement with point heights and 2) Terrain correction tests.

International Association of Geodesy Symposia, Vol. 121
Schwarz (ed.), Geodesy Beyond 2000 – The Challenges of the First Decade

Table 1: Comparison of USGS and NIMA high resolution Digital Elevation Models

Category	USGS	NIMA
DEM Name	7.5 minute quadrangle DEM	DTED Level 1
Horizontal Resolution	30 meters	3 arcseconds
Grid	UTM	Geographic (ϕ / λ)
Availability	Free on WWW	Proprietary
Format	SDTS	NIMA DTED
Vertical Datum	Well-defined (often NGVD 29)	Unknown (named "Mean Sea Level")
Horizontal Datum	Well-defined (often NAD 27)	Unknown (WGS72? WGS84?)
Sources	1:24,000 scale maps and photogrammetry	1:250,000 scale maps and 25 other sources
Biggest Problems	a) 1.1% of files have blunders b) Limited quality control c) No data outside the USA d) Partially complete (85%) over USA	a) Does not always contain true 3 arcsecond spectral information b) Poorly identified datum information c) Some cells have biases over 10 meters

2.1 DEM heights versus point heights

The first test on the DEMs was to compare their gridded heights to point heights stored in the NGS database. This is important because both sources of heights are used in geoid modeling, and consistency is therefore desirable. The point heights were predominantly scaled from topographic maps by surveyors during gravity surveys. Although these heights can be erroneous, the variety of surveyors who gathered such heights supports the idea that in a large sample of point heights, errors are expected to be more or less random. Two separate 1 x 1 degree areas of the Pacific Northwest area of the USA were chosen for this test, based on the ruggedness of the topography in that region. The geographic limits of the two areas were -- Area A: 47-48° N and 237-238° E, Area B: 46-47° N and 240-241° E. Area A had 1076 point heights in the NGS database and Area B had 397 point heights. In each area, the gridded DEM heights were biquadratically interpolated to the location of the point heights. A difference between the heights was made, and histograms of these differences made. In order to fairly compare the USGS DEM to the NIMA DTED, the USGS DEM was decimated to 3 arcseconds first, and then height comparisons were made. Figures 1 and 2 show the histograms for Area A, and Figures 3 and 4 show them for Area B.

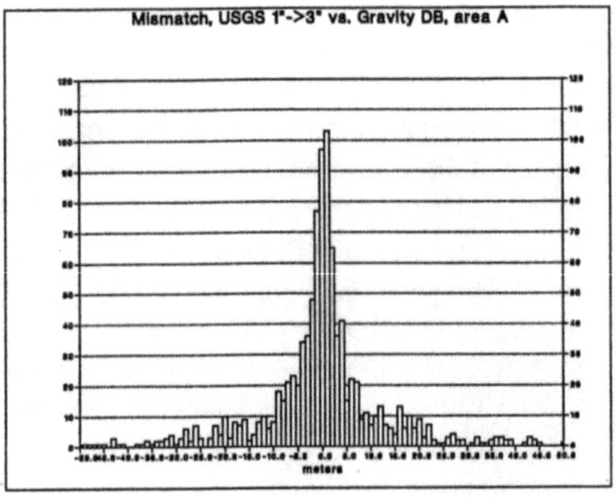

Fig. 1: Mismatch between USGS 30 meter DEM and point heights in the NGS Gravity database, Area A.

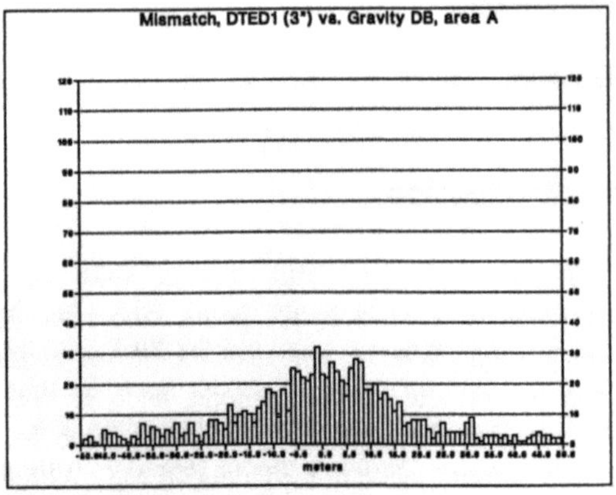

Fig. 2: Mismatch between NIMA Level 1 DTED and point heights in the NGS Gravity database, Area A.

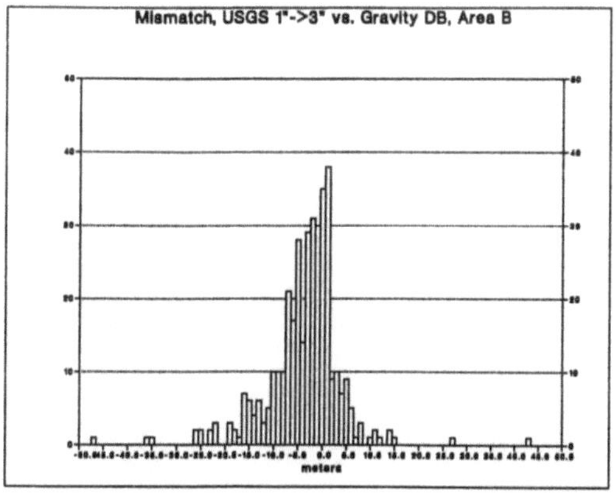

Fig. 3: Mismatch between USGS 30 meter DEM and point heights in the NGS Gravity database, Area B.

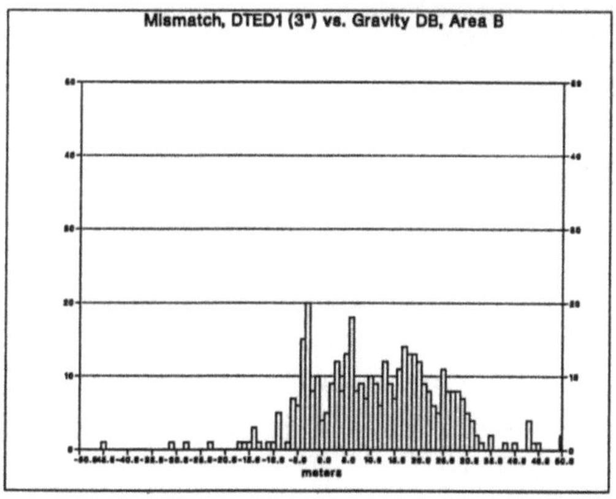

Fig. 4: Mismatch between NIMA Level 1 DTED and point heights in the NGS Gravity database, Area B.

Although these figures only represent two areas, they already show a clear indication that the USGS DEM data are superior to the NIMA DTED Level 1. Note the tighter histograms of USGS (Figures 1 and 3), compared to the more spread out ones of NIMA (Figures 2 and 4). This indicates a smaller standard deviation of the differences between USGS DEM heights and the point heights, and thus better agreement in general. It is also indicative of the lack of true 3 arcsecond data contained in the NIMA DTED Level 1 data. Additionally, note in Figure 4 the large (20 meter) shift of the histogram mean from zero. This is evidence of a 20 meter bias which exists in the NIMA DTED data of this region. Discussions with NIMA personnel indicate that the source of this bias is currently unknown. In general, therefore, it is seen that the USGS data seem to have a better agreement with existing point heights than DTED Level 1.

2.2 Terrain corrections from DEMs and their impact on the geoid

As a second test, terrain corrections (TCs) were computed from both DEMs, and compared to existing TCs based on the existing 30 arcsecond DEM, as well as compared to one another. The impacts of the various TCs on the geoid were also computed and compared. Table 2 shows the differences between various TC computations, and Table 3 shows the geoid impact of the differences seen in Table 2. In Table 3, the "Critical Distance" is the distance (in km) away from the maximum geoid impact, at which the geoid impact still exceeds 1 cm.

Table 2: Differences between various TC computations, Area A.

Category	USGS minus 30"	USGS minus NIMA
Average	1.4 mGals	0.7 mGals
Maximum	20 mGals	18mGals

Table 3: Geoid impact of differences between various TC computations, Area A.

Category	USGS minus 30"	USGS minus NIMA
Maximum	9.4 cm	5.8 cm
Critical distance	400 km	200 km

Certain conclusions can be drawn from these tables, and are further validated if one makes a detailed study of color plots (not shown in this paper) of the various data that are the foundation of these tables. First, it is clear that the larger average TCs of USGS relative to both the existing 30 arcsecond TCs as well as the NIMA Level 1 DTED TCs indicates that more "signal" or "roughness" exists in the USGS DEM. This is further evidence that the NIMA data are not completely portraying 3 arcsecond information, even though they are distributed with that horizontal resolution.

Secondly, these results were over a 1 x 1 degree area, but note that geoid impacts of switching from 30 arcsecond to 3 arcsecond data can yield geoid changes up to 400 km away from the center of Area A (which is outside of Area A). Thus there is the potential for geoid changes (at the 1+ cm level) to come not only from the data inside a 1x1 degree area itself, but also from the 8 surrounding 1 x 1 degree areas, causing constructive interference at the few decimeters level.

Having conducted these comparisons, it was clear that the USGS DEM would provide the best elevation data in the USA. Since these data are on a UTM grid, were on outdated datums (NAD27 and NGVD29) and contain some blunders, it was necessary to clean and re-grid the data, and transform onto modern datums (NAD 83 and NAVD 88) before being used on a large scale at NGS. Additionally, the incomplete (85% total coverage) nature of the available data meant that some DTED Level 1 data would also need to be used. During the Spring and Summer of 1999, NGS cleaned, combined, and regridded the USGS and NIMA data (giving priority to USGS data) into a new 1 arcsecond DEM covering the Northwest quarter of the United States, called NGSDEM99 (Smith and Roman, 1999). There are plans to expand this effort in 2000 and create a 1 arcsecond DEM for the entire United States.

3 New computational methods

Many classical approximations in the computation of terrain-induced gravity signals are being re-evaluated at NGS for their accuracy. A few are mentioned in this paper.

3.1 Spherical versus rectangular DEM prisms

When computing the gravitational attraction induced by the terrain, a planar approximation is often used. This approximation ignores the slight, but calculable, effects of the curvature of the Earth in two ways. The first, and more obvious, is the far-field spherical effect, which is the effect that the terrain dips below the horizon in a spherical co-ordinate frame, but not in planar. The second, more subtle effect, is that the shape of prisms is different in spherical versus planar co-ordinates, and this slight shape change has a noticeable impact nearest the station of interest. This is referred to as the near-field spherical effect and is illustrated in Figure 5.

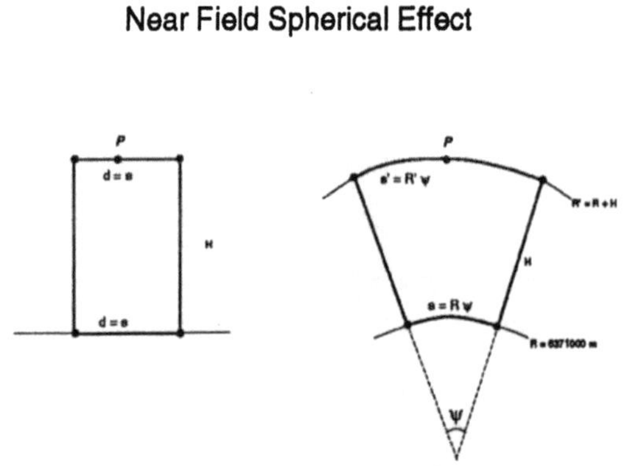

Fig. 5: Geometric illustration of the differences in near-field prisms between planar and spherical coordinates

From the simple geometry of the near-field prisms, an empirical difference was computed between the gravitational attraction of a rectangular prism and a spherical prism. This difference can be translated as one part of the error in computing Helmert anomalies when using the planar approximation. The empirical error seems to be almost entirely based on the height, H, of the prism and not on it's horizontal extent, ψ. The empirical function is shown in Figure 6:

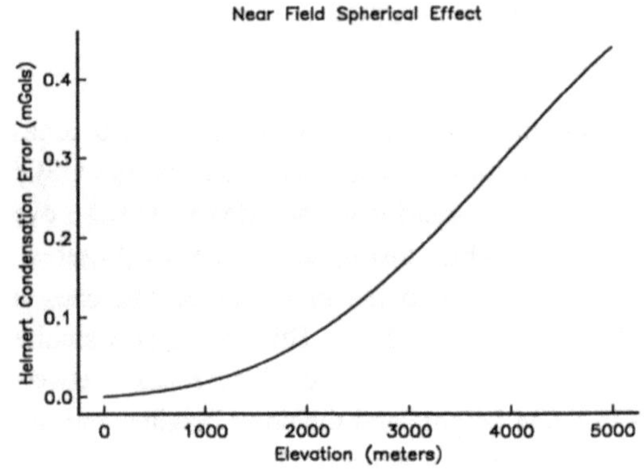

Fig. 6: Empirical error in Helmert anomaly computation due to the prism-shape error in planar coordinates

Although the Helmert anomaly error is relatively small, the effect is systematic and should be applied to every gridded DEM location where a TC is currently computed using planar co-ordinates. If this is done, this effect has a noticeable effect on the geoid, achieving over 7 cm in the Rocky Mountains of the United States. A map of this geoid impact is shown in Figure 7:

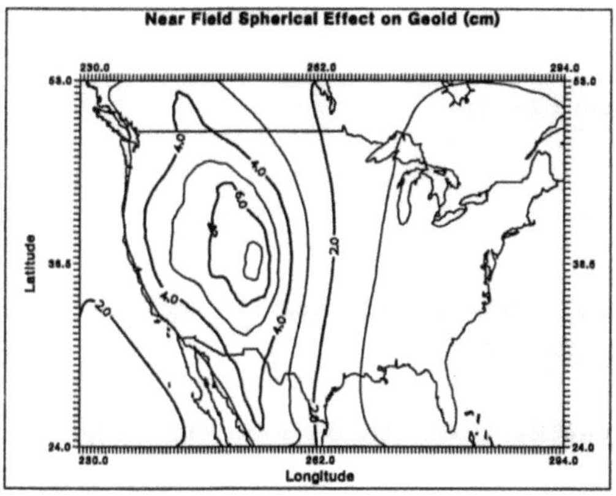

Fig. 7: Geoid impact (cm) of empirical errors in Helmert anomaly computation due to the prism-shape error in planar coordinates

3.2 Rigorous space domain computations

Because of the regularly gridded nature of most DEMs, computations done with them (such as gravitational attraction), are naturally suited to spectral methods such as FFTs (Schwarz, Sideris and Forsberg, 1990). However, transformation of data into the spectral domain can cause inaccuracies in computations, and therefore it is necessary to have highly accurate space-domain computations as a "golden rule" against which to check all future spectral computations. With this in mind, NGS performed an accuracy analysis, computing the gravitational signal induced by the removal of local terrain, and the restoration of that terrain as a 'condensed layer' (i.e. "Helmertization" of the terrain). In that study (Smith, Robertson and Milbert, 1999), it was found that high resolution DEM data was not necessary for large areas. In fact, a cut-off radius of 0.2 degrees was seen as acceptable in all forms of terrain for an accuracy of 0.01 mGals. Between 0.2 degrees and 4.0 degrees, high resolution DEMs may effectively be replaced with coarser DEMs (such as 30 arcseconds). Outside of 4 degrees, even coarser DEMs may be used (5 arcminutes) to compute the remaining global signal induced by topography.

4 Conclusions

The use of digital elevation models in the computation of geoid models is a long-established necessity. However, recently available DEMs, and the latest computer resources have made it possible, and worthwhile, to take steps beyond coarse DEMs with classical computational approximations. Two high-resolution DEMs, from USGS and NIMA were tested against one another, and the more accurate USGS data was used as the foundation for a new 1 arcsecond DEM (NGSDEM99) for the Northwest United States (Smith and Roman, 1999). This new DEM will be used in the next high-resolution geoid model for the United States, GEOID99 (scheduled for release September 1999; http://www.ngs.noaa.gov/GEOID/GEOID99/geoid99.html)

To supplement the new, accurate DEM data that is available, NGS has also been studying better ways of computing the terrain-induced gravity signal. Errors in old DEMs were only one reason why previous models were not of the best accuracy. The other reason has been the use of planar co-ordinates, and lack of rigorous checks on spectral computational methods. Future geoid models will make use of the latest techniques (Smith, Robertson and Milbert, 1999) to remove centimeters and even decimeters of lost accuracy in the geoid.

References

Schwarz, K.P., M.G. Sideris and R. Forsberg (1990). The use of FFT techniques in physical geodesy. *Geophysical Journal International*, Vol 100, pp. 485-514

Smith, D.A., D.G. Milbert (1999). The GEOID96 high resolution geoid height model for the United States, *Journal of Geodesy*, Vol 73, No. 5, pp 219-236.

Smith, D.A., D.S. Robertson, D.G. Milbert (1999). The gravitational attraction of local crustal masses in spherial coordinates, submitted to *Journal of Geodesy*.

Smith, D.A. and D.R. Roman (1999). A new high resolution DEM for the Northwest United States, submitted to *Surveying and Land Information Systems*.

Geoid modelling in coastal regions using airborne and satellite data: Case study in the Azores

M. J. Fernandes, L. Bastos
Faculdade de Ciências, Universidade do Porto
Observatório Astronómico, Monte da Virgem, 4430-146 Vila Nova de Gaia, Portugal

R. Forsberg, A. Olesen,
KMS, National Survey and Cadastre, Rentemestervej 8, DK-2400 Copenhagen NV, Denmark

F. Leite
Departamento de Matemática Aplicada F. C. T. Universidade de Coimbra
Apartado 3008, 3000 Coimbra, Portugal

Abstract. This study is concerned with the determination of a high resolution regional geoid in the area surrounding the Azores Islands, integrating multi-sensor airborne and marine observations with satellite derived and land gravity. It aims to prove the usefulness of airborne data in complementing existing marine and satellite derived gravity in coastal and insular regions, bridging the gap between land areas and the open ocean.

In the scope of the EU project AGMASCO (Airborne Geoid MApping System for Coastal Oceanography) several airborne gravimetric and altimetric campaigns have been performed. The new airborne and marine data used in this study have been collected in one of those campaigns, which took place in the Azores region, in October 1997. Profile and cross-over adjustments of the airborne to shipborne gravity suggest an accuracy of the airborne data of 4 to 5 mgal over wavelengths longer than 12 km.

Two geoid models have been computed using different data sets. The first was derived from all surface gravity (marine and land) merged with altimeter derived gravity anomalies. The second solution was computed from all previous data and airborne measured gravity. These solutions were called SURFACE and COMBINED respectively. Comparison of both solutions with the AZOMSS99 regional mean sea surface and with a set of GPS-levelling derived geoid undulations over the islands, shows that they agree very well. The difference between SURFACE and COMBINED solutions shows an average of 3 cm and a standard deviation of 4 cm. These results prove the ability of airborne data in complementing shipborne and satellite derived gravity for precise regional geoid determination.

Keywords. Geoid determination, airborne gravity.

1 Introduction

The Azores archipelago is composed by nine volcanic islands, located in the middle of the North Atlantic. This region possesses very rough topography and bathymetry, with important geological features related to seismic and volcanic activity, corresponding to high gradients in the gravity field. The study area focused in this study is the North Atlantic region surrounding the Azores islands ($36° \leq \varphi \leq 41°$, $-32° \leq \lambda \leq -24°$).

The accuracy of regional geoid solutions derived from surface gravity is strongly dependent on data distribution and accuracy. The marine gravity data available over the Azores region have an irregular distribution and variable accuracy. (Figure 1).

Regional geoids computed from surface data alone, will possess a inhomogeneous accuracy, dependent on data distribution. In the large wholes corresponding to areas without any surface gravity measurements, the computed solutions will fit the reference geopotential model used (e.g. EGM96, which has gross errors in the Azores region). The inclusion of satellite derived gravity anomalies has been shown to improve regional geoid solutions significantly, Fernandes et al (1999). However satellite altimetry lacks of sufficient accuracy close to land.

International Association of Geodesy Symposia, Vol. 121
Schwarz (ed.), Geodesy Beyond 2000 – The Challenges of the First Decade

Airborne gravimetry is a very attractive method for acquiring such data in coastal and insular regions, bridging the gap between land and open ocean areas where satellite derived gravity is more reliable.

Within the scope of the EU project AGMASCO, an observation campaign took place in the Azores in 1997. This included an airborne gravimetric and altimetric survey along several tracks in the NW-SE direction and over some satellite tracks (ERS2 and TOPEX/Poseidon) and a ship survey along some of the airborne NW-SE lines and two satellite tracks.

The purpose of this study is to prove the usefulness of airborne data in regional geoid determination and to compute a precise regional geoid in the Azores using all data available in the region.

2 AGMASCO - The Azores campaign

The Azores AGMASCO campaign took place in October 1997. The airborne survey was performed with a CASA Aviocar C212 of the Portuguese Airforce. The navigation system installed aboard the aircraft consisted in three GPS and one INS system. GPS reference stations were located in Flores, Faial, Terceira and S. Miguel islands.

Airborne measurements included gravimetry from a LaCoste & Romberg S-56 marine gravimeter and altimetry from an Optech 501 SX laser altimeter and from a radar altimeter prototype. The airborne profiles are shown in Figure 1. They were selected to cover the main features of interest in the area, forming a pattern of lines along the NW-SE direction, separated by about 20 km. Some profiles were chosen to coincide with some satellite tracks (ERS-2 and TOPEX/Poseidon).

Ground truth gravity data were collected on sea, along some of the airborne NW-SE lines and two satellite tracks (ERS2 and TOPEX/Poseidon), by the oceanographic Vessel R/V Håkon Mosby , belonging to the University of Bergen, Fernandes et. al (1998). Gravity measurements were also made on some of the islands.

3 Processing of airborne gravity

The quantity of interest for geoid computation is the free air anomaly Δg, which can be measured at aircraft height h, by a combination of gravity and GPS measurements, Olesen et al. (1997), Bastos et al. (1999):

$$\Delta g = g - \frac{\partial^2 h}{\partial t^2} + C_{eot} + C_{tilt} - \gamma_0 + \frac{\partial \gamma}{\partial h}(h - N) \qquad (1)$$

g - measured gravity
h -ellipsoidal height measured by GPS
C_{eot} - Eotvos Correction, Harlan (1968)
C_{tilt} - platform off-level corrections
γ_0 - normal gravity at the ellipsoid
N - geoid height from EGM96
$\frac{\partial \gamma}{\partial h}$ - gradient of normal gravity with height

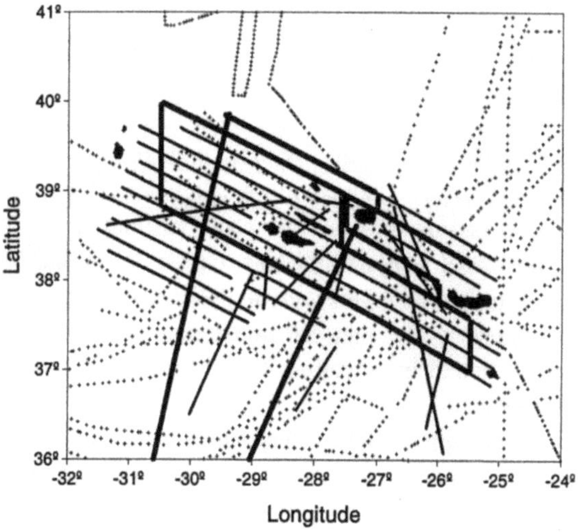

Fig. 1: Location of AGMASCO profiles: airborne (thin lines) and marine (thick lines). Small dots represent location of old data.

Upward continuation using a constant gradient of 0.3086 mgal/m has been used, which can be justified by the low flight elevation (nominally 1200 ft).

Data filtering plays an important role in airborne gravity processing to account for the difference in filtering inherent from the data and to remove the high frequency noise masking the gravity anomaly signal. The gravimeter data acquisition system uses a 1 sec. boxcar filter on internal 200 Hz data, whereas the inherent filtering of the accelerations derived from the GPS measurements depends on the GPS processing software, and the algorithm applied for the differentiation. This difference in filtering has little impact on the linear terms in the processing algorithm, because of the heavy final filtering. However the non linear terms, mainly represented by the tilt correction, are sensitive to the initial filtering.

Final filtering has been performed using a 3rd order Butterworth filter with a half power "cut-off" frequency ω_c (the frequency to which the filter attenuation factor is 0.5) corresponding to a period of 200 seconds, or a resolution (half wavelength) of 6 km. The frequency response function of the filter is:

$$F(\omega) = \frac{1}{1+\left(\frac{\omega}{\omega_c}\right)^6} \qquad (2)$$

where ω_c is the "cut-off" frequency.

The final track data were subjected to bias-only cross-over adjustment. It was found that the geometry of the tracks was not favourable to a free cross-over adjustment. A free adjustment of all airborne tracks conducted to a small rms after fit at the cross-over points, but each individual track could reveal a significant bias relative to marine or satellite derived gravity anomalies. Due to the poor geometry of the data set an alternative procedure was adopted by adjusting the airborne data to the marine data.

Data adjustment was performed in two steps:

First, the 10 tracks for which there were airborne and marine gravity measurements (coincident tracks) were adjusted separately. For each of these tracks, the airborne data was adjusted to the marine data, by removing the average difference between the two data types. The mean and rms of the differences between the marine and airborne data along these tracks was reduced from 7.3 mgal and 9.7 mgal to 0.0 mgal and 5.1 mgal respectively.

In the second step, the remaining airborne lines (non-coincident with marine tracks) were bias-only cross-over adjusted to the marine profiles, keeping the marine tracks fixed. This adjustment reduced the rms and maximum absolute value of the cross-over differences from 9.7 mgal and 28.2 mgal to 4.6 mgal and 10.7 mgal respectively.

The relatively high values of these residuals, comparatively to results obtained in other AGMASCO campaigns are believed to be due to the open ocean and weather conditions found in this campaign and the very irregular gravity field of the area. In addition, the large distances flown during this campaign (several hundreds of kilometres from the reference stations), conducted to larger errors in the GPS solutions, thus also affecting the gravity anomaly recovery.

Figure 2 shows the airborne gravity anomalies for the whole surveyed region. Some data had to be edited at the end of profiles due to filter edge effects and levelling of the gravimeter platform.

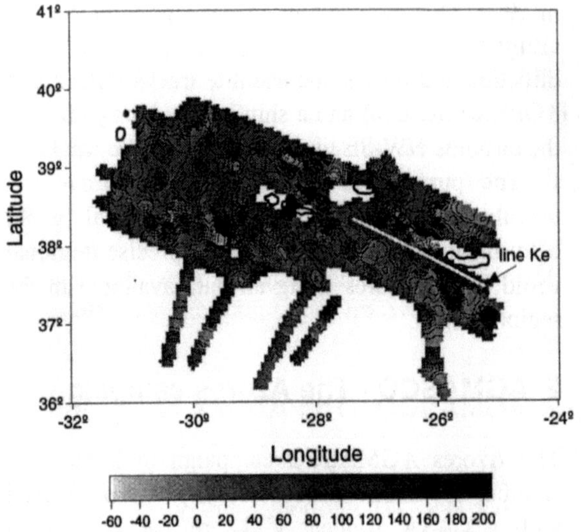

Fig. 2: Airborne free air anomalies. Contour interval is 10 mgal.

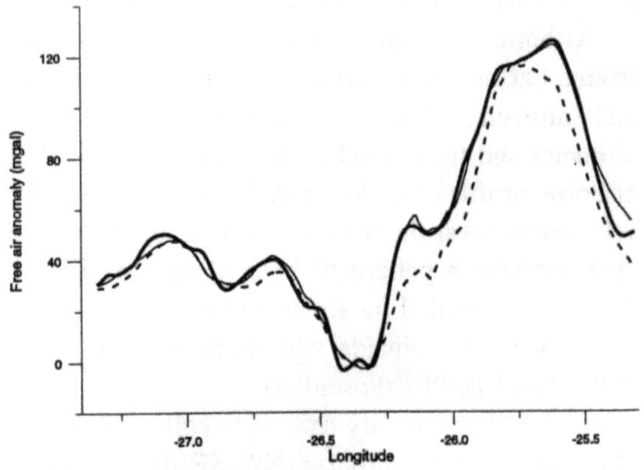

Fig. 3: Comparison of airborne (thick line), marine (thin line) and satellite (dashed line) along eastern part of line K (Line Ke). Location of profile is shown in Fig. 2.

In Figure 2 the full range of gravity anomalies, as measured by airborne gravimetry, can be depicted. These anomalies range from –62 mgal in deep sea depths to 213 mgal in some of the islands. Figure 3 shows a comparison of airborne, marine and satellite derived free air anomalies along an airborne/marine coincident profile after profile adjustment as described above. This is clear indication of the ability of airborne methods to measure the gravity anomaly in stable flight conditions.

4 Geoid computation

The spatial distribution of all gravity measurements available over the Azores region is shown in Figure 1. Marine gravity data provided by the international data banks are sparse, with large areas with no data at all. The accuracy of this old data set is difficult to access and is variable but usually not better than a few mgals. The 1997 AGMASCO marine data has an accuracy of about 1 mgal. 187 land points were also available from various campaigns that took place in the scope of Portuguese projects.

Due to the large holes present in the distribution of gravity data, in particular in the region north of the archipelago, gravity anomalies derived from satellite altimetry have been used to fill these large gaps. For this purpose, a set of gravity anomalies has been derived from a regional mean sea surface (MSS) model AZOMSS99, Fernandes (1999).

The gravity anomalies have been derived from the MSS heights using the remove restore method. The residual gravity anomalies Δg_{res} have been computed from residual heights Δh_{res} by inverse Stokes FFT techniques, Andersen and Knudsen (1998), Fernandes et al (1999). To reduce the effects of noise on the MSS heights a Wiener filter was applied with a "cut-off" wavelength of 25 km.

From the altimeter derived residual gravity anomalies a subset was selected covering the area where no surface or airborne gravity were available. This area was defined by creating an envelope around each surface/airborne measurement point with north-south and east-west radius of 0.15° and 0.18° respectively. Only altimeter derived observations outside this envelope were selected. The altimeter data set was called Δg_{altim}.

Since our aim is to study the influence of airborne data in geoid modelling, two data sets have been generated for geoid computation:

- a *SURFACE* data set, composed by all surface gravity observations (marine and land) and satellite derived anomalies
- a *COMBINED* data set – the *SURFACE* data set plus the airborne measurements.

For use with FFT methods a grid was built for each data set covering the study area, with a spacing of 0.05°x0.06°. The method used for merging and gridding the data are explained in Fernandes et al. (1999).

Using the two data sets described above two geoid solutions have been computed: a *SURFACE* and a *COMBINED* solution.

All geoid computations have been performed with the GRAVSOFT software package, Tscherning et al. (1992). The method used in the geoid computations was the remove-restore method. This splits the geoid undulation (N) into three components.

$$N = N_{EGM96} + N_{RTM} + N_{res} \tag{3}$$

where N_{EGM96} represents the contribution from EGM96 geopotential model, N_{RTM} is the residual terrain effect contribution and N_{res} the residual anomaly field after removal of the two previous contributions.

In a similar fashion the observed gravity anomalies are split into three components:

$$\Delta g_{obs} = \Delta g_{EGM96} + \Delta g_{RTM} + \Delta g_{res} \tag{4}$$

The computation of the residual contribution N_{res} has been performed by Fast Fourier Transform (FFT) methods. The geopotential model used in the computations is EGM96. The computations of the residual terrain model (RTM) contributions were made by prism integration, using a local digital elevation model derived from a local bathymetric (Lourenço et al. (1999)) and land altimetry.

The statistics of the gravity reduction for the combined solution are presented in Table 1 for reference. The first three raws refer to the combined data set (marine, land, airborne and satellite derived) before final gridding. Rows four to six show the individual contributions of each data type. Finally the last row shows refers to the combined data set after gridding. Note that these data sets are not directly comparable since they cover different spatial regions.

Using the methodology explained above, two geoid solutions were computed using each of the *SURFACE* and *COMBINED* data sets.

5 Results and discussion

The computed geoids were compared with eachother, with the AZOMSS99 mean sea surface and with a set of geoid undulations derived from GPS and levelling at 236 points on five islands. The results are shown in Tables 2 to 6. Comparisons relative to EGM96 are also presented for reference.

Table 1: Statistics of gravity reduction, *COMBINED* solution (in mgal)

	Average (m)	σ (m)	Min (m)	Max (m)
Δg_{obs} (all data)	31.57	24.29	-66.53	263.60
$\Delta g_{obs} - \Delta g_{EGM96}$	-0.49	17.77	-102.28	211.26
Δg_{res}	-0.36	11.56	-76.53	68.85
Δg_{res} (sea + land)	0.80	13.12	-61.02	68.85
Δg_{res} (airborne)	-1.90	14.16	-76.52	49.63
Δg_{res} (altimetry)	0.04	9.84	-74.00	51.10
Δg_{res} (all data gridded)	0.06	10.65	-73.47	51.07

Results show that all regional solutions represent a significant improvement over the global model EGM96.

The standard deviation of the differences between computed local geoids and the mean sea surface is about 9 cm in both solutions. The statistics of the differences between computed geoids and the GPS-levelling derived geoid undulations, shows that both solutions are similar over the islands. This result is not surprising since both models share the same data set over land. The standard deviation of these differences is less than 1 decimetre in three of the islands. It is however above the decimetre level in two of the islands, Terceira and S. Miguel. These results show that the gravity data available over the islands is not sufficient, in particular over S. Miguel. In this island the pattern of the residuals of both solutions relative to GPS/levelling derived geoid heights reveal that the error increases towards the edges of the island. This is due to the fact that this island has very high gravity anomalies and since the gravity coverage is sparse, the geoid near the coast is strongly influenced by the satellite derived gravity anomalies of the points around the island, which are much smaller.

The differences between the *SURFACE* and *COMBINED* solutions show an average of 3 cm and a standard deviation of 4 cm. The maximum absolute difference is 25 cm, on some points with large differences between the airborne and satellite derived gravity anomalies. Since there are no marine data in these locations and the method used to derived the satellite gravity anomalies underestimates, in general, the amplitudes of the shorter wavelengths, it is believed that these differences are more likely to be due to errors in the satellite derived anomalies than in the airborne data.

In these computations only FFT methods have been used. These have the advantage over least squares collocation of being very fast and efficient. The only drawback is that the FFT requires data on a grid, which has to be performed with care to avoid data distortions. The merge of the surface gravity and altimeter derived anomalies should be done using an appropriate method to reduce data discontinuities. This can be accomplished by application of low pass filter like the Wiener filter (Fernandes et. al. 1999).

Table 2: Differences between *SURFACE* and *COMBINED* solutions (in metres)

Average (m)	σ (m)	Min (m)	Max (m)
0.030	0.045	-0.289	0.230

Table 3: Differences between local geoid solutions and EGM96 (in metres)

	Average (m)	σ (m)	Min (m)	Max (m)
SURFACE	0.152	0.246	-0.857	1.965
COMBINED	0.122	0.245	-0.855	1.939

Table 4: Differences between model geoid heights and AZOMSS99 (in metres)

	Média (m)	σ (m)	Min (m)	Max (m)
EGM96	0.261	0.222	-0.792	2.044
SUPFACE	0.109	0.092	-0.345	0.554
COMBINED	0.138	0.087	-0.239	0.474

Table 5: Differences between geoid solution *COMBINED* and GPS-levelling derived geoid undulations (in metres)

	points	average (m)	σ (m)
S. Miguel	60	0.086	0.274
St. Maria	24	-0.051	0.055
Terceira	76	-0.298	0.136
S. Jorge	52	-0.380	0.052
Graciosa	26	-0.324	0.064

Table 6: Differences between geoid solution *SURFACE* and GPS-levelling derived geoid undulations (in metres)

	points	média (m)	σ (m)
S. Miguel	60	0.156	0.285
St. Maria	24	-0.075	0.053
Terceira	76	-0.375	0.135
S. Jorge	52	-0.499	0.052
Graciosa	26	-0.414	0.062

Results show that *SURFACE* and *COMBINED* solutions are similar. The geoid surface obtained with the *COMBINED* solution is shown in Figure 4.

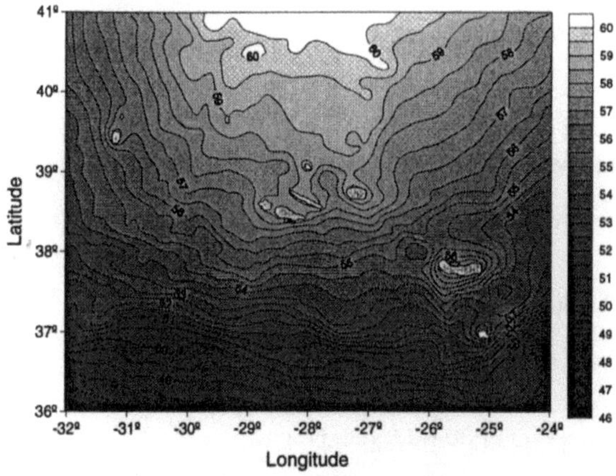

Fig. 4: Regional geoid, *COMBINED* solution

In conclusion, airborne data, subject to an appropriate processing, can be used in regional geoid computation in the same fashion as traditional marine gravity. In this processing data filtering and track bias removal are fundamental steps which determine the final accuracy of the data. Profile and cross-over adjustments of the airborne data to shipborne gravity suggest an accuracy of 4 to 5 mgal over wavelengths longer than 12 km. This result is not as good as obtained in the Skagerrak region, which was about 2 mgal. This is believed to be due to the larger variability and amplitude of the signals measured in the Azores compared to the Skagerrak sea. The experience acquired in the AGMASCO project also shows that the geometric configuration of the surveyed profiles is very important and it should be planned in such a way that a large number of cross-over points are obtained, not close to the track edges, where airborne data is less accurate.

In regions where surface gravity are sparse with irregular distribution and limited accuracy, as it happens in the Azores region, altimeter derived gravity anomalies are a powerful tool in the computation of regional geoids over ocean areas. However, due to the large altimeter footprint and to land contamination on some of the geophysical corrections, satellite altimetry cannot be used near land. It is in coastal and insular regions such as the Azores that airborne gravimetry can play an important role. The versatility and cost of airborne campaigns relative to shipborne missions makes airborne gravimetry a very attractive technique for collecting data for regional gravity field modelling.

Acknowledgements. This work has been funded by the European Commission, under the contract number MAS3-CT95-0014 and by JNICT (Junta Nacional de Investigação Científica e Tecnológica) under project PBICT/MAR/2261/95. This support is gratefully acknowledged.

References

Andersen, O., and P. Knudsen (1998). Global marine gravity field from the ERS-1 and Geosat geodetic mission altimetry, *J. Geophys. Res.* , Vol 103, No C4, pp. 8129-8137

Bastos, L, S. Cunha, R. Forsberg, A. Olesen, A. Gidskehaug, L. Timmen, U.Meyer (1999). The role of airborne gravimetry in gravity field modelling: experiences from the AGMASCO project - Presented at EGS-99, the Hague, Holland, April 19-23, 1999, Accepted for publication in *Physics and Chemistry of the Earth*.

Fernandes, M. J. (1999). Superfície média oceânica de alta resolução na zona Açores-Gibraltar usando altimetria dos satélites ERS, *Cartografia e Cadastro*, n 10, pp 3-12

Fernandes, M. J., A. Gidskehaug, D. Solheim, M. Mork, P. Jaccard, and J. Catalão (1998). Gravimetric and Hydrographic Campaign in the Azores Region, *1ª Assemb. Luso-Espanhola de Geodesia e Geofísica*, Almeria, Spain, February 9-13, 1998, to be published in the proceedings

Fernandes, M. J. L. Bastos, and J. Catalão (1999). The role of multi-mission ERS altimetry in the determination of the marine geoid in the Azores, Presented at INSMAP98, Melbourne, USA, Nov. 30 - Dec 4., 1998, submitted to *Marine Geodesy*.

Harlan, R. B. (1968). Eotvos corrections for airborne gravimetry, *J. of Geophys. Res.*, vol 73, pp. 4675-4679

Lourenço, N, J. M. Miranda, J. F. Luis, A. Ribeiro, L.A. Mendes Victor, Madeira, J & H. D. Needham (1999). Morpho-tectonic analysis of the Azores Volcanic Plateau from a new bathymetric compilation of the area, *Mar. Geophys. Res.*, in press.

Olesen, A. V., R. Forsberg, A. Gidskehaug (1997). Airborne gravimetry using the LaCoste & Romberg gravimeter – an error analysis In: M. E. Cannon and G. Lachapelle (eds.), *Proc. Int. Symp. on Kin. Systems in Geodesy Geomatics and Navigation*, Banff, Canada, June 3-6 1997, Publ. Univ. of Calgary, pp.613-618

Tsherning, CC., R. Forsberg, P. Knudsen (1992). Description of the GRAVSOFT package for geoid determination, *Proc. 1st Coninental. Workshop.. on Geoid in Europe*, Prague, pp. 327-334

Airborne gravity field surveying for oceanography, geology and geodesy - the experiences from AGMASCO

Ludger Timmen
GeoForschungsZentrum Potsdam, Telegrafenberg, D-14473 Potsdam, Germany
email: LudgerTimmen-Vermessung@t-online.de

Luisa Bastos
Univ. of Porto, Obs. Astronomico, Monte da Virgem, P-4430 V. N. Gaia, Portugal
email: lcbastos@oa.fc.up.pt

Rene Forsberg
National Survey and Cadastre, Rentemestervej 8, DK-2400 Copenhagen NV, Denmark
email: rf@kms.min.dk

Arne Gidskehaug
University of Bergen, Inst. of Solid Earth Physics, Allegaten 41, N-5007 Bergen, Norway
email: arneg@ifjf.uib.no

Uwe Meyer
Alfred Wegener Institute for Polar and Marine Research, Columbusstr., D-27568 Bremerhaven, Germany
email: umeyer@awi-bremerhaven.de

Abstract. Within the European AGMASCO project (Airborne Geoid Mapping System for Coastal Oceanography), cooperations with user groups from Oceanography, Geology and Geodesy have been arisen. The airplane based gravimetry/altimetry system provides a rapid surveying procedure which is specially employed in not easily available marine and land areas (e.g. coastal/shelf areas, polar regions). In Geodesy, it serves for the improvement and refinement of already existing geoids and of future satellite-only gravity field models. The availability of precise regional geoids allow the monitoring of the dynamic sea surface topography by satellite altimetry. For applications in geology and geotechnics, gravimetry is a pre-surveying method for exploration of energy resources.

The AGMASCO products such as gravity anomalies, sea surface heights and regional geoids (Skagerrak/1996, Fram Strait/1997, Azores/1997) will be discussed in this paper. The accuracy corresponds to the state-of-the-art of airborne gravimetry/altimetry.

Keywords. Airborne gravimetry/altimetry, coastal oceanography, EU project AGMASCO.

1 Introduction

The AGMASCO project ("Development of an Airborne Geoid Mapping System for Coastal Oceanography"), has been initiated by a consortium of 5 groups from 4 European nations under MAST-III contract MAS3-CT95-0014 of the Commission of the European Union. The project started on the 1st of January 1996 with a total run time of 38 month. The system has been designed as a supplement to the existing spaceborne and marine techniques.

To study the dynamic sea surface topography (SST), large progress has been achieved with the performance of satellite missions like ERS-1 and ERS-2. ENVISAT will give an additional push for oceanography. The monitoring of ocean currents from space by satellite altimetry requires a reliable separation of the SST from the geoid (≡ theoretical (equipotential) ocean surface). Therefore, the acquisition of gravity data is an important tasks, to derive a reliable "synthetic" geoid on regional scale in areas of interest. There are deficiencies especially in coastal areas (islands, shoals, local tides etc.), and in the polar regions (covered by sea ice, and only partly covered by satellite observations). In many cases the available data sets from space techniques or from older marine surveys do not satisfy the oceanographic goals for a "10-cm" geoid. The data density, their distribution and accuracy is not sufficient for an accurate geoid computation.

International Association of Geodesy Symposia, Vol. 121
Schwarz (ed.), Geodesy Beyond 2000 – The Challenges of the First Decade

The development of the AGMASCO airborne gravity system allows to overcome these problems for the most part. Because it is designed for economic rapid surveys over sea and over land, the system supports large and small scale tasks for oceanography as well as for geology and geophysics (e.g. resource exploration in off-shore areas) especially in coast-near ocean regions. By combining aerogravimetry with an airborne altimetry system in one aeroplane, the potential for simultaneous measurement of sea topography and geoid is available. With this system, flight pattern and flight parameters of the aircraft can be easily selected and changed, and the measurements can be performed right into the coastline. Because of the homogeneity of these data, they are well suited for assimilation with satellite data. A more detailed mapping of the geoid (with relatively large variations close to the coastline) and sea surface topography is possible. For monitoring time dependent variations of SST in special investigation areas, a merge of most accurate geoid information (±5 cm) with high resolution altimetry data from the future satellite mission ENVISAT (±1 . . 3 cm) appear to be very promising. The geoid can be assumed to be constant in time scales of some years (within the limits of the measurement uncertainties). Therefore a unique aerogravimetry campaign can serve to obtain full gain from a long-time satellite mission with a repeat cycle of a couple of weeks (ENVISAT: 35 days repeat).

GPS technology is one of the main components of the AGMASCO gravity/sea surface height determination system, see also Fig. 1. Gravity information is obtained from differencing accelerations

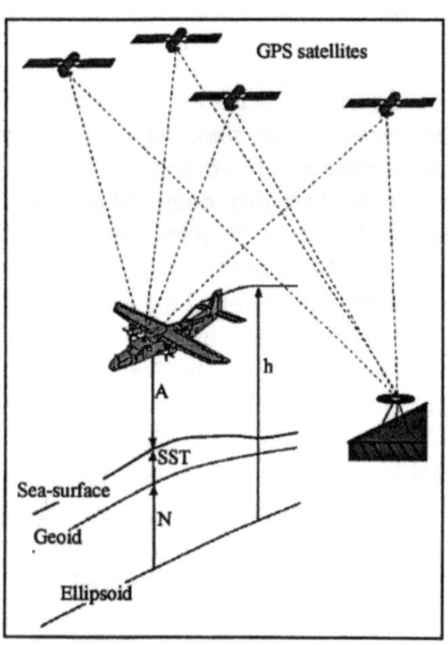

Fig. 1: The AGMASCO airborne geoid and sea surface mapping system for coastal oceanography

observed by a gravity meter (gravity plus aircraft accelerations) and derived from GPS observations (aircraft accelerations only). Simultaneously, the ocean surface is observed by altimeter measurements. By combining geoid, altimetry and GPS the sea surface topography is given through the equation:

$$SST = h - N - A$$

(where h: ellipsoidal height of aircraft, N: geoid height, A: aircraft height above sea surface).

The recent accuracy increase in airborne gravimetry is achieved due to large improvements in kinematic GPS data processing (cf Harrison et al. 1995, Klingelé et al. 1997, Olesen et al. 1997, Wei and Schwarz 1995, Xu et al. 1997). The progress becomes obvious in the gravity accuracy as well as in the spatial resolution of the gravity field which is nearly comparable to those obtained in marine gravity surveys. In addition, GPS combined with altimetry and attitude information allows the determination of the instantaneous sea surface height relative to a reference ellipsoid, e.g. WGS84.

The system involves the integration of available hardware components as well as of hard- and software which have been developed within the project. The employed gravity sensors are LaCoste and Romberg (LCR) model "S" air/sea gravity meters from the University of Bergen and from the Alfred Wegener Institute for Polar and Marine Research (AWI, Bremerhaven). Both instruments are specially upgraded by ZLS Corp. (Austin, Texas) for airborne application. To assess the performance of the system, new ground truth data (marine gravity and hydrographic data: ADCP (Acoustic Doppler Current Profiler), CTD (Conductivity, Temperature, Depth)) have been collected by R/V Håkon Mosby. The airborne and shipborne measurements along satellite tracks from ERS-2 and Topex/Poseidon allowed a comparison of the different techniques.

From AGMASCO two different airplanes (with experienced crews and other technical support) are available now: a Dornier 228 from AWI, Bremerhaven, and a CASA Aviocar C212 from the Portuguese air force.

2 Aero-gravimetry field campaigns and results

Two demonstration projects (Skagerrak/1996, Fram Strait/1997) and a full application project (Azores/1997) have been performed. The measurement and data processing methodology from the Skagerrak campaign has been applied to the Fram Strait and Azores survey. The aircraft flight velocities varied between 200 and 250 km/h (depends on wind) with altitudes of about 400 m and 250 m (Azores). The use of auto-pilot ensured an approximately constant engine speed and a horizontal aircraft orientation.

2.1 Skagerrak Airborne Gravity Campaign

Under the coordination of the National Survey and Cadastre (KMS) of Denmark, the Skagerrak airborne survey was done with a Dornier 228 aircraft ("Polar-4") belonging to the Alfred Wegener Institute. The airborne survey was carried out in the days September 13-21, 1996, followed by a 1-week ship cruise by the R/V Håkon Mosby (University of Bergen) collecting ground truth gravity and oceanographic data. The primary purpose of the measurements was to provide marine gravimetry for evaluating the airborne gravity measurements, but in addition oceanographic measurements were made (ADCP and some CTD casts), allowing independent determination of the sea surface topography.

Fig. 2 shows the location of the survey. The ship survey took place immediately following the airborne campaign, after moving the gravimeter from the airplane to the ship. On the "M" transect between Hanstholm, Denmark, and Kristianssand, Norway, a special effort was made to perform simultaneous airborne measurements and marine CTD/ADCP measurements, to allow a direct comparison of sea surface height measurements.

The altimetry allows an independent estimate of the vertical aircraft accelerations over sea areas, as well as allows a direct measurement of the height of the sea surface. The altimetry data was calibrated by runway overflights, with a detailed map of runway heights provided by local kinematic GPS surveys by car. To allow vertical pointing of the laser, and the correct lever arm transfer of GPS co-ordinates to the gravimeter sensor reference point, attitudes were determined by a Honeywell Lasernav INS. For most tracks the comparison of filtered GPS and altimeter accelerations agreed at an accuracy level of 1 mGal (10^{-5} m/s^2) or better, even for preliminary GPS solutions based on one reference station only.

We have attempted to determine the dynamic sea surface topography (SST) of Skagerrak during the experiment period by several methods: by GPS/laser airborne altimetry and TOPEX satellite altimetry (subtracting the "best" geoid model), by hydrodynamical models, interpolation of sea-level data from tide gauges, and - for the "M" track - by CTD/ADCP models. The hope was to get the latter hydrographic SST estimates to serve as "ground truth" for evaluating the SST estimation by airborne and satellite methods, and thus also get an improved error estimate of the marine geoid. A fully forced, baroclinic 3-D hydro-dynamical model was run for the month of September, 1996, by the Norwegian Meteorological Institute (NMI). The model run was a NMI variant of the Princeton Ocean Model (POM), termed ECOM3D. The model was forced by tides, wind, and atmospheric pressure, the freshwater run-off from rivers, and the flux of brackish waters in the Danish straits. The model domain covers the North Sea and the North Atlantic margins on a 20 km grid, and the region around Skagerrak on a 4 km grid. The results were given as hourly values of sea surface topography and average currents. The SST values obtained refer to an average ocean, and should thus only be used in a relative sense.

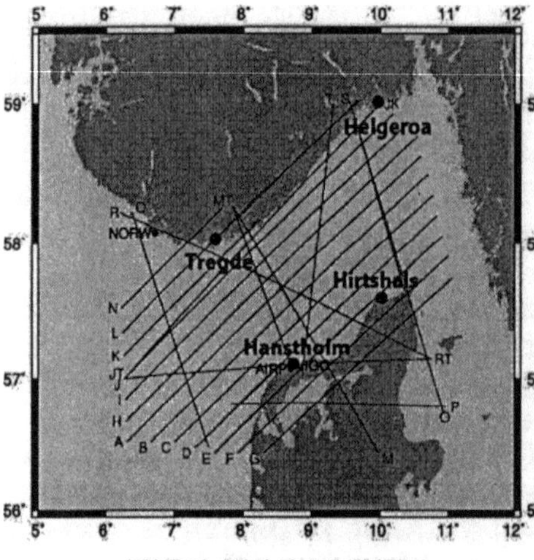

Fig. 2: Airplane flight pattern of the AGMASCO Skagerrak gravity survey

A very direct, albeit inaccurate, method to estimate the instantaneous sea-level is to interpolate actual sea surface heights from tide gauges. In the present case the four tide gauges shown in Fig. 2 (Hanstholm, Hirstshals, Tregde and Helgeroa/Nevlungshavn) have been used.

A cross-over analysis of all the processed gravity data have been performed, using an improved processing scheme compared to the results presented in Olesen et al. (1997). The cross-over errors for 54 cross-over points, before and after applying a bias correction to each track showed that the r.m.s. accuracy of the airborne data is better than 2 mGal. An independent test of data accuracy involves comparison to upward continued ground true data (R/V Håkon Mosby survey along 6 main aircraft tracks).

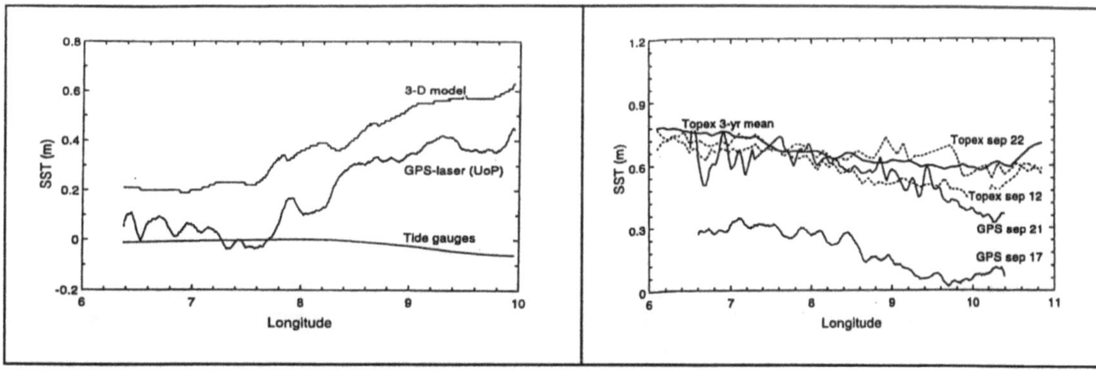

Fig. 3: (left) Example of GPS-laser SST measurements and the model results along profile J.
(right) Example of measured airborne SST values minus interpolated tide gauge data, profile A.

Considering the noise in the ship data and the upward continuation process, the comparison results support an r.m.s. airborne accuracy estimate around 2 mGal.

The Skagerrak geoid computation has been done using same methods as the current Nordic standard geoid NKG-96. The NKG-96 geoid solution has provided geoid fits better than 10 cm for GPS/levelling lines across Scandinavia, and is thus of very high quality. Adding the new AGMASCO data, the NKG geoid computations were repeated, using identical data grid spacings etc. The new geoid model shows nearly identical results to the NKG96 geoid, except for two areas with geoid discrepancies up to 15 cm. This is not surprising, given the fact that Skagerrak is covered with quite dense, albeit older, marine gravity data, which has entered the NKG96 solution. One can therefore say that the geoid computed from airborne and marine gravity data agrees very well, justifying the use of airborne data in other regions with insufficient marine gravity data coverage.

2.2 The Fram Strait survey

NORDGRAV'97 joins two surveys in one: an airborne geophysical survey over the Arctic Basin adjacent to the northern coast of Greenland including the Fram Strait, and an airborne approach to oceanography. Of basic interest was the first complete gravity mapping of the Morris Jesup Rise in an airborne reconnaissance type survey. The Fram Strait is a challenging area to test the AGMASCO system for it is a most important part of the major sub-polar current system of the northern hemisphere.

The NORDGRAV'97 project was initiated by the Alfred Wegener Institute for Polar and Marine Research. Major assistance during the project

Figure 3 (left) shows an example of the airborne SST measurement along the J-profile (subtracting measured SSH and geoid), compared to the results of the NMI 3-D hydrodynamic model, and the tide gauge interpolation. In this case the tide gauge interpolation and hydrodynamic model disagree strongly. A reasonably good fit is obtained between the airborne and the 3-D model SST. On most other tracks, however, no such agreement was found, and in those cases the airborne measurements tended to agree more with the interpolated tide gauge data than the hydrodynamic model. This was also the case for the TOPEX data. Figure 3 (right) shows the results of applying the tide gauge interpolated SST values to both TOPEX data and to two over-flights of profile "A". It is seen that the GPS results differ by about 0.5 m in bias, illustrating the problems of the GPS solutions (TOPEX SST data expected to be biased by about 70 cm due to the use of a different reference ellipsoid, the agreement with one of the GPS solutions must therefore be coincidental).

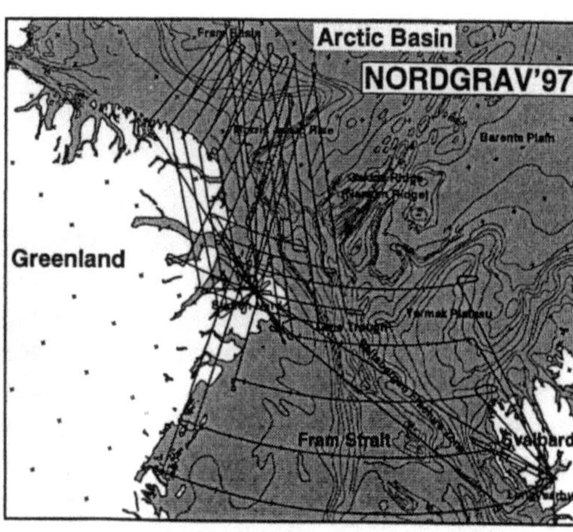

Fig. 4: Flight profiles covering the Fram Strait and the adjacent Arctic Basin

preparation and the field work was given by the National Survey and Cadastre of Denmark, the Norwegian Mapping Authority in Hønefoss and the University of Bergen, Institute of Solid Earth Physics (Norway). The AGMASCO part of the NORDGRAV mission covers the Fram Strait on six profiles following the latitudes 77°N up to 82°N in increments of one degree. Free air anomaly, bathymetry and derived Bouguer anomaly data were used as the basic data sets for modelling. For all acquired models the simple Bouguer anomaly was calculated by replacing the water column with a sediment cover of 2.4 g/cm^3 density to model the depth of the crust - mantle boundary, cf. Fig. 6.

2.3 The Azores Airborne Gravity Campaign

The Azores area was chosen because of the great interest in the application of a new technology to collect data in an area of extreme relevance for different earth sciences. There are very important geological features related to seismic and volcanic activity. This North Atlantic region is also of great oceanographic interest. It contains the main flow of the Azores current (between 34° - 28° W, 32° - 46° N). The AGMASCO system is especially suited for application in regions where the problem of geoid modelling and SST determination is more critical by the use of satellite altimetry methods.

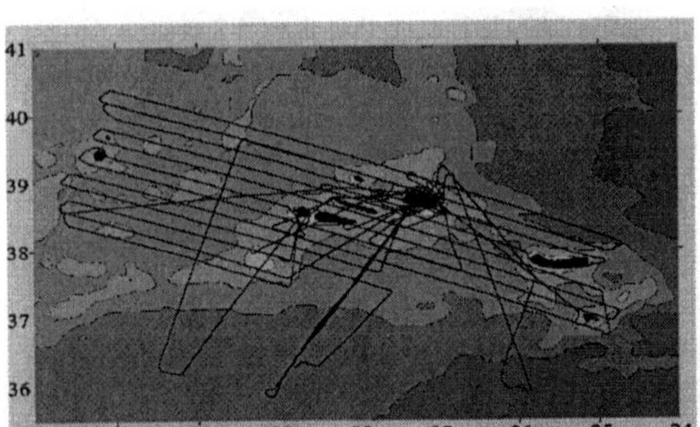

Fig. 5: AGMASCO airplane profiles, Azores'97

The Azores airborne survey was done with a CASA Aviocar C212, from the Geophysical Squadron of the Portuguese Airforce. For the sea measurements the oceanographic vessel of the University of Bergen, the R/V Håkon Mosby, made a dedicated cruise in the Azores region. The campaign was carried out during October 1997. The total flight time was 70 hours and the ship cruise lasted for two weeks. The land measurements included the establishment of GPS and gravity ties to known benchmarks on the islands. Reference GPS stations

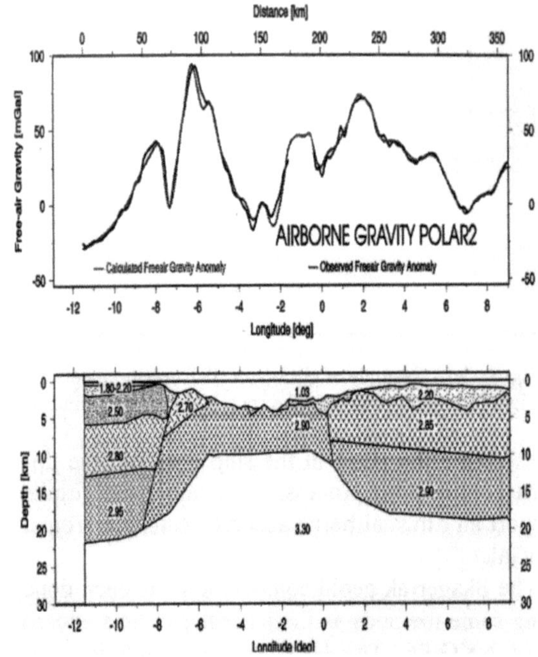

Fig. 6: Calculated and observed free air anomalies: 81°N (Greenland shelf – Lena Trough – Yermak Plateau)

were located at Flores, Faial, Terceira and S. Miguel islands.

The hardware installation made under the supervision of the UoP group involved the following devices: a gravimeter system based on a LaCoste & Romberg gravimeter (S-56) belonging to the Alfred Wegener Institute, Bremerhaven. Three GPS receivers were installed in the aircraft and a Litton 200 IMU from the Astronomical Observatory was used to provide attitude information.

The three different antennas onboard the aircraft were processed against one or more reference stations, yielding several solutions for each flight. Comparison of vertical accelerations derived from laser altimeter and from GPS height has been the main criterion for selecting between the different solutions. The disagreement between laser and GPS accelerations varied from 1.1 to 2.3 mGal after filtering. In the Skagerrak campaign these figures were 0.4 to 1.2 mGal. The less good agreement in the Azores campaign is believed to be due to the open ocean with its long waves, which will generate a signal in the altimetry data, even after filtering.

The gravity data available in the Azores region contain four main sets of data: marine gravity from BGI, DMA and NGDC; marine gravity data acquired during the AGMASCO campaign; airborne gravity acquired during the AGMASCO campaign; land gravity over the islands, partly acquired during the AGMASCO campaign and partly from previous projects.

The accuracy of these data sets is variable. The AGMASCO marine data after cross-over adjustment of the tracks, subject to bias only, has an r.m.s. and maximum cross-over difference of 0.8 mGal and 1.2 mGal respectively. The airborne data after a bias-only cross-over adjustment had an r.m.s. and maximum error of 3.5 mGal and 7.0 mGal. A joint adjustment of the new air and marine gravimetry, keeping the marine data fixed, has an r.m.s. and maximum cross-over difference of 4.1 mGal and 9.3 mGal respectively.

The computed geoid has been compared with a regional mean sea surface, AZOMSS98 (Fernandes, pers. comm.), after removal of the EGM96 SST, and with a set of geoid undulations derived from GPS and levelling at eight points on the islands. Using the computed geoid and the available MSS derived from satellite altimetry, the dynamic sea surface topography has been computed. The SST has been smoothed by keeping only wavelengths above 100 km approximately. The result is presented in Fig. 7.

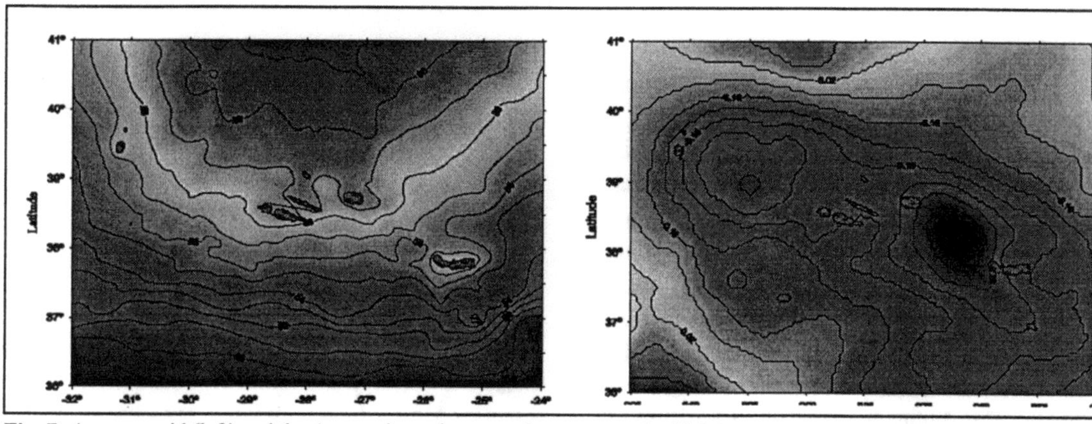

Fig. 7: Azores geoid (left) and the Azores dynamic sea surface topography (right)

3 Conclusions

The AGMASCO airborne survey experiments were successful in its primary goal: the determination of an accurate geoid model, based on airborne gravity. The airborne gravity results indicated errors at the 2 mGal (Skagerrak) and 3.5 mGal (Azores, Fram Strait) level r.m.s., at a resolution around 6 to 7 km. This means that airborne gravity has reached an accuracy level comparable not far from marine gravimetry. The geoid models computed from the airborne gravity match surface data solutions well, and estimated relative geoid errors for the Skagerrak region is at the 5 cm level. This means that the geoid model could be useful for detailed studies of dynamic sea surface topography. Given the economy of airborne gravimetry over ship-based gravimetry, it is therefore clear that the use of airborne gravimetry is an attractive alternative to map geoids in coastal regions.

Acknowledgements. This project is funded by the European Commission under the contract no. MAS3-CT95-0014. This support is gratefully acknowledged.

References

Harrison, J. C., MacQueen J. D., Rauhut, A. C. and Cruz, J. Y., 1995: The LCT Airborne Gravity System. Proc. IAG Symp. Airborne Gravity Field Determination. Special Report No. 60010, Dept. of Geomatics Engineering, Univ. of Calgary, 163-168.

Klingelé, E., Cocard, E., M. and Kahle, H.-G., 1997: Kinematic GPS as a source for airborne gravity reduction in the airborne gravity survey of Switzerland. J. Geophys. Res., 102, 7705-7715, 1997.

Olesen, A. V., Forsberg, R, and Gidskehaug, A., 1997: Airborne gravimetry using the LaCoste & Romberg gravimeter - an error analysis. In: M.E. Cannon and G. Lachapelle (Convenors): Proc. Int. Symp. Kin. Syst. in Geodesy, Geomatics and Navigation, Banff, Canada, 613-618.

Wei, M., Schwarz, K.P., 1995: Analysis of GPS-derived acceleration from airborne tests. Proc. IAG Symp. Airborne Gravity Field Determination. Special Report No. 60010, Dept. of Geomatics Engineering, Univ. of Calgary, 175-188.

Xu, G., Bastos, L. and Timmen, L., 1997: GPS kinematic positioning in AGMASCO campaigns - Strategic goals and numerical results. Proc. ION GPS-97, 1173-1183, Kansas City, USA.

A comparison of stable platform and strapdown airborne gravity

C.L. Glennie, K.P. Schwarz, A.M. Bruton
Department of Geomatics Engineering,
The University of Calgary, 2500 University Drive N.W., Calgary, Alberta, Canada, T2N 1N4

R. Forsberg, A.V. Olesen, K. Keller
KMS, National Survey and Cadastre, Rentemestervej 8, DK-2400 Copenhagen NV, Denmark

Abstract. To date, operational airborne gravity results have been obtained using either a damped two-axes stable platform gravimeter system such as the LaCoste and Romberg (LCR) S-model marine gravimeter or a strapdown inertial navigation system (SINS), both with comparable accuracies. In June of 1998 three flight tests were undertaken which tested a LCR gravimeter and the Honeywell Laseref III (LRF III) strapdown INS gravity system side-by-side in the same airplane. To our knowledge this was the first time such a comparison flight was undertaken. The flights occurred in Disko Bay, off the west coast of Greenland. Several of the flight lines were partly flown along existing shipborne gravity profiles to allow for an independent comparison of the results. The flight height was 300 m and the average flying speed was 70 m/s.

The results of the flight tests show that the gravity estimates from the two systems agree at the 2-3 mGal level, after the removal of a linear bias. This is shown in Figure 1. This small discrepancy is near the combined noise levels of the two systems. Also evident in Figure 1 is the fact that the estimates provided by both systems agree very well with the shipborne data that was available directly below the flight line. Tables 1 and 2 summarize the statistics for flight lines where both systems operated free of the power supply and hardware problems that affected both systems on some flight lines.

It appears that a combination of both systems would provide an airborne gravity survey system that would combine the excellent bias stability of the LCR gravimeter with the higher dynamic range and increased spatial resolution of the strapdown INS.

Keywords. Airborne gravimetry, gravimeter, strapdown inertial navigation system (SINS).

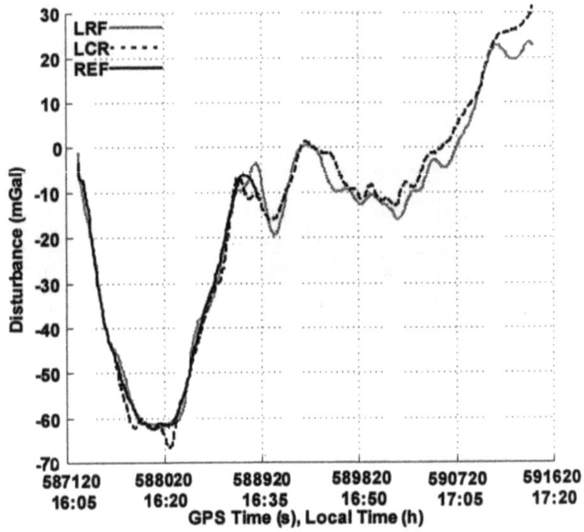

Fig. 1: The estimates from the LCR and LRF III SINS along a flight line of the survey. The filtering period is T_c = 200 seconds, corresponding to a half wavelength of 7 km.

Table 1: Comparison of gravity estimates from the LCR and LRF III SINS, in mGal (T_c = 200 sec)

Flight line	RMS	σ
A	2.4	1.7
B	3.0	2.1
C	1.4	1.1
G1	4.0	2.9

Table 2: Comparison of LCR and LRF III SINS with shipborne data, in mGal (T_c = 200 sec)

Flight Line	LCR	LRF III
A	1.7	2.0
G1	2.3	3.8

Currently under review for publication in the Journal of Geodesy.

International Association of Geodesy Symposia, Vol. 121
Schwarz (ed.), Geodesy Beyond 2000 – The Challenges of the First Decade

The NRL airborne geophysics program

J.M. Brozena, V.A. Childers
Naval Research Laboratory,
Code 7421, 4555 Overlook Ave. SW, Washington, DC 20375-5350

Abstract. The Naval Research Laboratory (NRL) has been conducting large area, reconnaissance aerogeophysical surveys in the Arctic since 1991. In projects funded by the National Imagery and Mapping Agency, the Office of Naval Research, and the National Science Foundation, NRL has surveyed all of Greenland and nearly two-thirds of the Arctic Ocean basin during the course of eight field seasons. Simultaneously, NRL has developed system and methodology improvements for airborne gravimetry. These improvements have addressed all aspects of the survey from instrumentation to aircraft operation to data processing. In this paper we address both the Arctic survey and the system improvement aspects of our aerogeophysical program.

Keywords. Airborne gravimetry, airborne geophysics, potential fields of the Arctic Ocean.

1 Introduction

Unlike all other gravity measurement methods, airborne gravimetry has the potential to recover an accurate gravity field at any place on Earth. With the development of precise Global Positioning System (GPS) aircraft positioning, airborne gravimetry can measure the field over any continent, coastal area, or ocean within range of aircraft operations. In the Arctic, airborne surveying has proven an effective method for collecting gravity and other geophysical data. During the course of eight field seasons, NRL has surveyed all of Greenland and nearly two-thirds of the Arctic Ocean. In addition, a host of smaller scale, higher resolution airborne geophysical surveys have been conducted in the Arctic using smaller twin engine aircraft by Danish, Norwegian, German, and Canadian researchers. (Forsberg et al., (1999); Timmen et al., (1998); Forsyth et al., (1994))

Simultaneously, NRL has made a large-scale effort at system and methodology improvements in gravity measurement. These improvements have addressed all aspects of the survey from instrumentation to aircraft operation to data processing. In this paper we address both the Arctic survey and the system improvement aspects of our aerogeophysical program.

2 The NRL aerogeophysical program

The objective of the aerogeophysical program is to measure gravity, magnetics, and sea-surface topography from aircraft for studies of geophysics and geodesy. In general, NRL has specialized in the broad area, medium resolution survey over remote regions where little or no data exist. The range of the NRL survey capability is maximized by the use of the U.S. Navy P-3 Orion aircraft. Designed for submarine surveillance, the P-3 can cover up to 6,000 km in a single flight at low altitude.

The P-3 is a large 4-engine, turboprop aircraft with a fuselage large enough to accomodate many instruments. Two LaCoste and Romberg S-model marine gravimeters, modified for the airborne environment, are located at the center of the aircraft's rotation. Scalar magnetics measurement is made by an EG&G proton precession magnetometer located in the nonmagnetic tail boom. Radar altimetry is provided by a 10 GHz radar pulsed at 10 KHz, then averaged to 10 Hz. Aircraft attitude information is provided by a Litton 72 inertial navigation system. Real-time navigation information is obtained from the inertial system updated with C/A code GPS positions. An autopilot holds the aircraft altitude fixed to a given barometric pressure. Post-mission, three dimensional aircraft positioning is calculated from GPS carrier-phase data collected simultaneously by

International Association of Geodesy Symposia, Vol. 121
Schwarz (ed.), Geodesy Beyond 2000 – The Challenges of the First Decade

Ashtech Z-12 dual-frequency receivers in the aircraft and at a base station using the XOMNI software (Ball et al., (1995)).

2.1 The NRL arctic program

The goal of the NRL Arctic program is to provide continuous measurement of gravity and magnetics across the Arctic Ocean to help understand the structure and tectonic evolution of the region. The surveys also are designed to overlap and densify historical aeromagnetic measurements in the region so that these historical data can be adjusted. In projects funded by the National Imagery and Mapping Agency, the Office of Naval Research, and the National Science Foundation, more than 200,000 line km of measurements have been made over the Arctic Ocean.

Surveying was conducted out of Thule, Greenland (1992), Prudhoe Bay, AK (1994), Barrow, AK (1995, 1996, and 1997), and most recently out of Svalbard (1998 and 1999) (Figure 1). The 1992 through 1997 surveys were flown over the Amerasia Basin, and the 1998 and 1999 surveys were flown in the Eurasia Basin. The 1994, 1995, 1998, and 1999 seasons cover portions of the polar gap in the ERS 1 and 2 marine gravity field calculated from satellite altimetry. The principal data lines are spaced every 18 km (10 nautical mi) with cross tracks spaced every 75-130 km (40-70 nautical mi). For two surveys, 1996 and 1997, the data line spacing was reduced to 10-14 km (6-8 nautical mi) to improve the resolution of the recovered gravity field over high gradient areas of the southern Canada Basin and the Chukchi Plateau. Nominal altitude is 600 m and speed is 250 kts (~450 km/hr).

Large area surveys trade off some resolution and accuracy in exchange for the broad coverage. Such surveys require high aircraft speeds and wide line spacing. Resolution of the gravity anomalies along track is a function of the lowpass filtering employed and the speed of the aircraft. The half-power points of the lowpass filters used vary from 23.5 to 30 km full wavelength, yielding an along-track resolution of approximately 12 - 15 km half-wavelength. Resolution in the cross track direction is defined by the data line spacing, or about 18 km half-

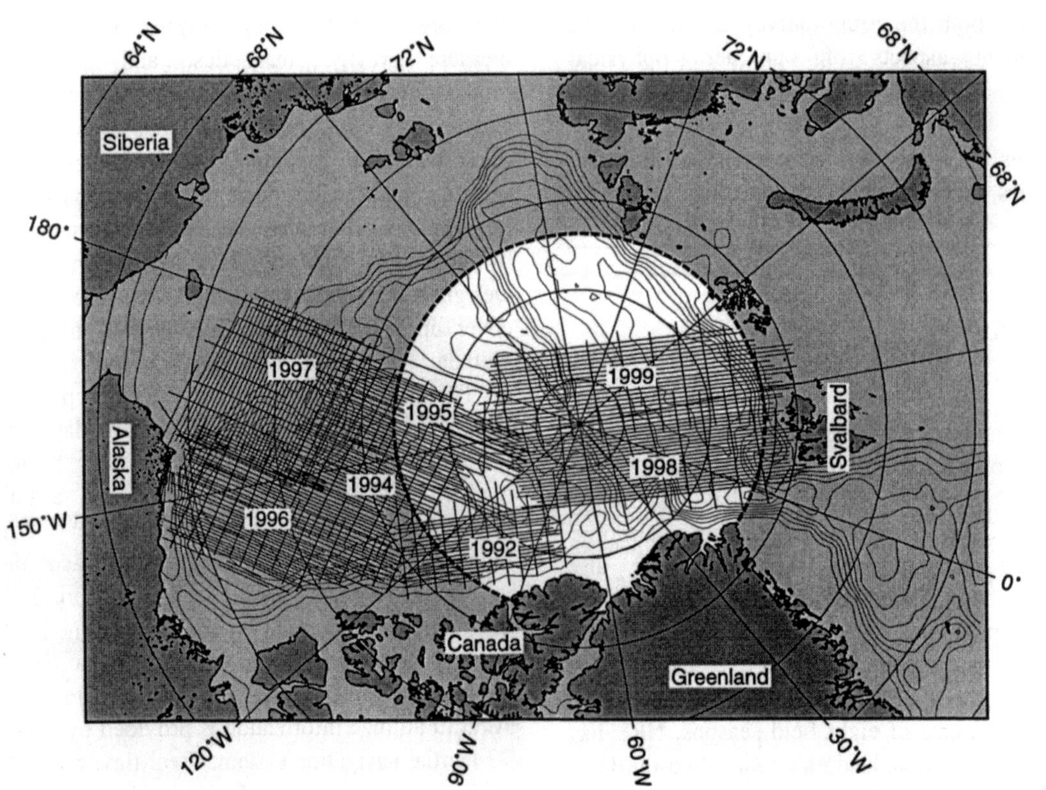

Fig. 1: Tracklines for the seven NRL Arctic field surveys are shown, identified by year. The gray shaded region shows the coverage of the ERS 1/2 satellite marine gravity field with the heavy dashed line showing the northernmost extent of its coverage.

wavelength. Accuracy is also decreased with increased survey speeds. Some errors scale with velocity, so increased speed requires more filtering which results in more spatial averaging. The accuracy of the gravity measurements for each survey, as determined by the size of the rms crossover error, is in the 2-3 mGal range. Our 1997 Chukchi Borderlands survey had the best internal comparison statistics, with intersection errors of 2.0 mGal rms.

3 Aerogravimetry improvements

3.1 How airborne gravity is measured

Isolation of the target gravity signal from the effects of aircraft motion is the primary challenge in airborne gravimetry. A gravimeter is an accelerometer that is kept in alignment with the local vertical by a two-axis gyro stabilized platform. To measure gravity aboard an aircraft, acceleration measured along the local vertical by the gravimeters must be differenced with the vertical acceleration of the aircraft. The Eötvös correction corrects for the effects of measuring from a moving platform over a rotating earth (Harlan, (1968)). The resultant accelerations are then differenced with the normal gravity field and corrected with the free-air correction, to yield the free-air anomaly at aircraft altitude (Telford et al., (1990)). The free-air anomaly is further corrected for offleveling errors that result when horizontal accelerations from aircraft trajectory drive the gyro-stabilized gravimeter platform offlevel (Peters and Brozena, (1995)). Final noise reduction in the free-air anomaly is achieved through a cosine taper lowpass filter applied in the frequency domain. This filter is tailored to best remove noise while optimizing the signal based upon aircraft speed and alititude and noise characteristics of the survey (Childers et al., (1999)). After all free-air anomaly profiles from a survey are calculated, a least-squares network adjustment is applied to the survey to minimize misties at intersections (Peters and Brozena, (1988)).

3.2 Noise sources in the data

Remaining noise in the final free-air anomaly can be attributed to two sources: error inherent in the technique and measurement error. The inherent error is signal lost to upward continuation effects and to lowpass filtering, both of which result in the preferential attenuation of the shorter wavelength anomalies. Downward continuation can restore some of the signal lost to measurement at aircraft altitude, but can be unstable and substantially amplify noise. Substantial lowpass filtering is required to attenuate the high frequency noise in the measurement resulting from aircraft motion. Unfortunately, some of the observable signal at the aircraft altitude is lost to these heavy filters. In fact, the filter plays the limiting role in determining the resolution of the airborne measurement along track.

Measurement noise can be introduced in the aircraft acceleration determination and by the gravimeter. GPS positions provide the primary source of aircraft vertical accelerations. Accelerations from these positions can be noisy for a variety of reasons. Positioning error scales with base line distance, although these errors tend to add linear trends to the altitude that are lost in the differencing process. Ionospheric disturbances at high latitudes can create significant noise, even in the dual frequency solution designed to minimize these effects. Changes in the visible GPS constellation tend to cause "steps" in the altitude solution that translate into a dipole-shaped acceleration error of a few mGal.

Alternatively, accelerations can be derived from radar altimetry measurements. Short wavelength errors are common in the radar altitudes as a result of sea ice or surface waves, although these generally are removed by the lowpass filtering. As a source of acceleration, there is an inherent error introduced by the undulation of the ocean surface with the geoid. The aircraft accelerations must be referenced to the ellipsoid, and geoidal undulations introduce spurious accelerations.

Gravimeter error occurs primarily as a result of the behavior of the gyro-stabilized platform and the condition of the sensor. Healthy gyros and accelerometers and their proper alignment are critical to optimal platform performance. Even a meter in peak condition will introduce errors in response to high aircraft dynamics. Long period horizontal accelerations drive the platform offlevel, causing some of the horizontal accelerations to map into the vertical. The offlevel correction repairs some but not all of these errors.

Noise also can be introduced in the data processing methodology. The lowpass filter used in the processing plays an important role in the resultant free-air anomaly. In addition to attenuating signal with the noise, the filter can also distort the shape of the anomalies. In fact, some distortion is inevitable, given the amount of noise attenuation required, so that one must choose the type of distortion that is acceptable for a given application (Childers, et al., (1999)).

3.3 Methodology improvements

Our efforts in methodology improvements have been facilitated by two opportunities for comparison with high-quality sea surface gravity measurements. In the eastern Pacific off the coast of San Diego, we overflew several high-quality marine gravity profiles. Also, our 1996 survey in the southern Canada Basin overlaps with high-quality ice surface gravity measurements made by the Geological Survey of Canada (GSC). These two opportunities for "ground truth" comparison have allowed us to test various methodology improvements.

To improve our aircraft vertical acceleration determination, we have a two-pronged approach. First, we check for "steps" introduced into the GPS solution as a result of changes in the GPS satellite constellation. An automated check for steps is followed with a manual check by comparing the GPS vertical acceleration with the beam velocity from the gravimeter. Once a step in the altitude is found, the remainder of the time series is shifted to eliminate the step. This procedure removes problems in the derived accelerations, but leaves a time series unsuited for calculating the free-air correction.

A second approach is also useful when surveying over oceanic areas including sea-ice covered regions such as the Arctic. We first process all the tracks using GPS accelerations and then calculate a gravimetric geoid for the region. Then, vertical accelerations derived from the radar altimeter are summed with the accelerations generated from the geoid sampled along the profile to create a nearly independent determination of aircraft acceleration. This acceleration yields a less noisy free-air anomaly, although some peaks may be underestimated if the geoid is not sufficiently sampled by the survey (Figure 2). Errors sometimes can be reduced further by averaging the free-air anomalies created by these two methods. This approach is particularly helpful in high-latitude surveys where long base lines and ionospheric interference can make the GPS accelerations too noisy.

To reduce gravimeter errors, we check our gyro alignment before surveying. We also try to maintain a sufficient supply of spare parts so that

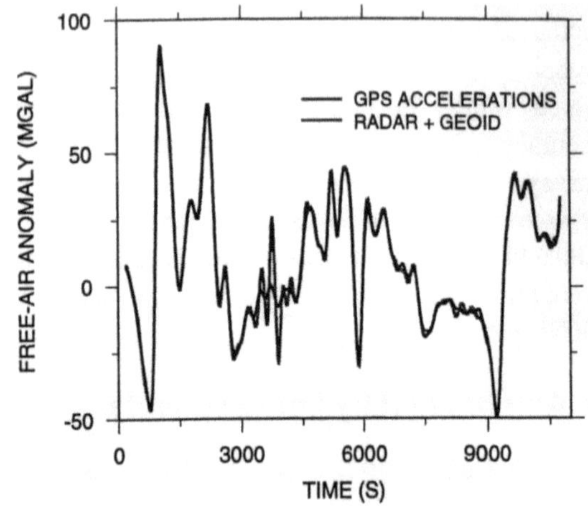

Fig. 2: Free-air anomaly for a profile from the 1999 survey calculated with both GPS accelerations and the "radar plus geoid" accelerations. Note that the radar plus geoid method loses some amplitude on the peaks but has less noise.

repairs can be made quickly in the field at the first sign of a problem.

Another improvement in the gravimeter measurement has been made in our offlevel correction. The formula for the offlevel correction is (Peters and Brozena, (1995)):

$$c = (a_x^2 + a_l^2 + a_e^2 - a_n^2)/2g_m \quad (1)$$

where a_x and a_l are the cross and long accelerometer output from the meter, a_e and a_n are the east and north components of aircraft acceleration from GPS, and g_m is the measured gravity. Previously, we had filtered these accelerations with the lowpass filter used for the final processing prior to calculating the offlevel correction. We have since found a significant improvement from matching the response of the filter to that of the stable platform (Figure 3). We find that some filtering is required, because measurement noise in the accelerometers or in the GPS accelerations is rectified when squared by the correction. The filter we use begins its frequency rolloff at 0.05 Hz.

In the realm of processing, we first attempt to match the lowpass filter to the survey. Our goal in filter design is to maximize the bandwidth of the gravity signal while keeping noise at acceptable levels. The amount of filtering required depends upon the signal-to-noise characteristics of the particular survey, as determined by the power in the gravity field and the measurement noise levels. If the gravity field has substantial power at short wavelengths, we can increase the cut-off frequency

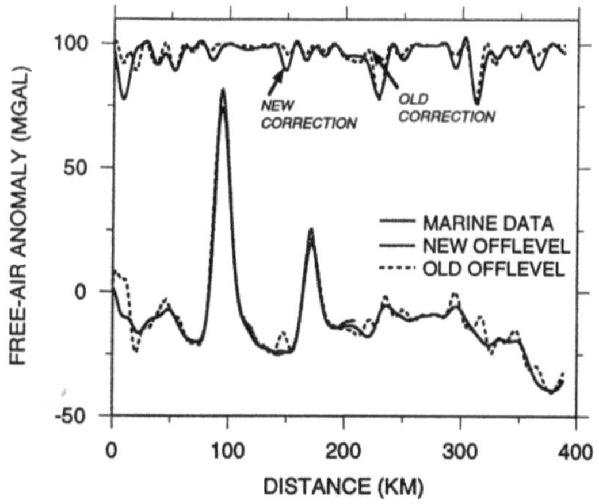

Fig. 3: Free-air anomaly for East Pacific profile calculated with improved offlevel correction. Offlevel corrections are displayed at top shifted by 100 mGal for visibility. The gray line is the "ground truth" free-air anomaly. The solid line shows the offlevel correction when accelerations are filtered to match the platform response, dashed line shows offlevel correction with accelerations "overfiltered".

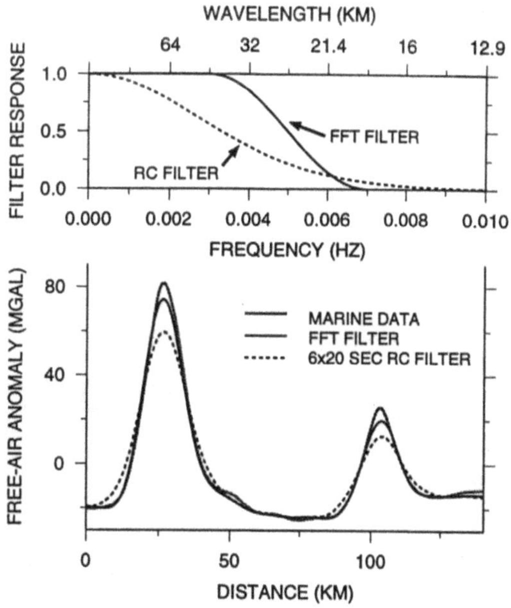

Fig. 4: Airborne data along an East Pacific profile compared to "ground truth" marine data. The FFT filter designed to pass more signal unattenuated in the lowest frequencies recovers more anomaly amplitude over the seamounts.

of the filter and pass more signal. Figure 4 shows the signal gained by switching from the 6x20 sec RC filter to one designed in the frequency domain to maximize our signal.

Another improvement in processing methodology is achieved by averaging the output of the two gravimeters. If both meters are performing well, we tend to gain some noise cancellation through averaging as the meters respond slightly differently to the aircraft motion. This generally is only a second-order improvement, but worth the effort if available.

Noise as a result of inherent error is expressed as a loss of amplitude of shorter wavelength anomalies while measurement noise tends to be expressed as oscillatory errors of a few mGal superimposed upon the gravity signal. Our various methods of reducing measurement noise have, in general, allowed us to lessen our lowpass filtering to allow more bandwidth in the resulting free-air anomaly.

There is a noticable correlation between changes in heading and/or along-track velocity and errors in the free-air anomaly. As always in airborne gravimetry, we find that the best strategy is to fly smoothly at constant velocity, and minimize course changes.

4 Conclusions

The NRL airborne geophysics program has simultaneously conducted a large area aerogeophysical survey in the Arctic and developed system and methodology improvements for airborne gravimetry. As of August 1999, NRL has now measured gravity and magnetics over a region in the Arctic Ocean of more than 3 million km^2.

Substantial improvements have been realized in the NRL aerogravity operation. The new techniques, especially the filter design and offlevel correction improvements, have resulted in a 40% improvement in the rms correspondence of the airborne data with the East Pacific marine data in our comparison. Noise reduction in vertical accelerations, gravimeter issues, and processing techniques have allowed us to lessen our lowpass filtering to allow more bandwidth in the resulting free-air anomaly for a given level of error allowed in the final map.

References

Forsberg, R., A. Olesen, and K. Keller, (1999). Airborne Gravity Survey of the North Greenland Shelf 1998. *Kort & Matrikelstyrelsen Technical Report no. 10.*

Timmen, L., G. Boedecker, and U. Meyer, (1998). Flugzeuggestutzte Vermessung des Erdschwerefeldses Zeitschrift fuer Vermessungswesen. *Jahrgang*, 11, pp. 378-384.

Forsyth, D.A., M. Argyle, A. Okulitch, and H.P. Trettin, (1994). New seismic, magnetic and gravity constraints on the crustal structure of the Lincoln Sea continent-ocean transition. *Canadian Journal of Earth Science*, 31, pp. 905-918.

Ball, D.G., J.L. Jarvis, J.M. Brozena, and M.F. Peters, (1995). XOMNI User and Technical Documentation. *Naval Research Laboratory Technical Report no. NRL/MR/7420--95-7774.*

Harlan, R.B., (1968). Eötvös corrections for airborne gravimetry. *Journal of Geophysical Research*, 73, pp. 4675-4679.

Telford, W.M., L.P. Geldart, and R.E. Sheriff, (1990). *Applied geophysics*. New York, NY, Cambridge Univ. Press, 770 p.

Peters, M.F., and J.M. Brozena, (1995). Methods to improve existing shipboard gravimeters for airborne gravimetry, *International Symposium on Kinematic Systems in Geodesy, Geomatics, and Navigation*: Boulder, CO, IUGG, pp. 39-46.

Childers, V.A., R.E. Bell, and J.M. Brozena, (1999). Airborne gravimetry: an investigation of filtering. *Geophysics*, 64, pp. 61-69.

Peters, M.F., and J.M. Brozena, (1988). Constraint criteria for adjustment of potential field surveys. *Geophysics*, 53, pp. 1601-1604.

On the modeling of long wavelength systematic errors in surface gravimetric data

Nikolaos K. Pavlis
Raytheon ITSS Corporation
7701 Greenbelt Road, Greenbelt, MD 20770, USA
email: npavlis@geodesy2.gsfc.nasa.gov

Abstract. A satellite-only model and a set of 1°x1° area-mean terrestrial gravity anomalies were used to estimate simultaneously geopotential coefficients to degree and order 70 and a set of spherical harmonic coefficients representing regional systematic errors in the gravity data. Several test solutions were developed whereby the weighting of the gravity data and the maximum degree and order of the systematic bias coefficient set were varied. The results were evaluated using both internal consistency statistics (e.g., *a posteriori* signal and error statistics, calibration factors) and comparisons with independent data (orbit fits, comparisons with GPS/leveling-derived geoid undulations). The global RMS error correlation between geopotential and bias coefficients was 6.7% for an unconstrained bias expansion to degree 20, and 7.8% for an expansion to degree 25, where an *a priori* constraint on the bias parameters was also employed. The bias expansion to degree 25 resulted in a slight degradation of the results from comparisons with GPS/leveling-derived geoid undulations or height anomalies. Optimization of this technique requires additional tests, especially with regard to terrestrial data weighting issues. Future satellite missions (e.g., CHAMP, GRACE and GOCE) will permit bias recovery at finer resolution, thereby improving the overall effectiveness of this technique.

Keywords. Terrestrial gravity anomaly data, systematic errors, satellite-only geopotential models, combination geopotential models.

1 Introduction

Terrestrial gravity data provide a valuable source of short wavelength gravitational information within global geopotential solutions. Over land areas these are the only data presently available to support short wavelength geopotential modeling. Over ocean areas, accurate marine gravimetry can aid the separation between geoid undulation and Dynamic Ocean Topography (DOT) over medium and short wavelengths, given accurate sea surface height data from satellite altimetry. The introduction of surface gravity data into combined global geopotential solutions is a challenging task, complicated primarily by long wavelength inconsistencies between the gravitational signal sensed by tracking data and that implied by surface gravimetry. These long wavelength inconsistencies are attributed primarily to regional biases in the surface gravity data. Orbit fit tests that are sensitive to the long wavelength ($n \leq 36$) accuracy of a geopotential model demonstrate clearly the superiority of satellite-derived over surface gravity-derived models within that degree range. The specific origin of these systematic errors is not clearly understood, although several possible causes have been identified (Heck, 1990). Mainville and Rapp (1985) examined these long wavelength inconsistencies, and concluded that over several regions these were associated with the presence of "geophysically predicted" anomalies in the terrestrial data that were available at that time. Pavlis (1998a) revisited this problem, using recent terrestrial and satellite information, and showed that despite significant advances in the coverage and accuracy of the terrestrial gravity data, some of these regional biases persist. Pavlis (ibid.) found that the currently available marine gravity data contribute ~50% of the total geoid undulation difference observed between satellite-only and surface gravity-only models, up to degree 20. In addition, the observed biases are significantly larger than those expected from the postulated models of Laskowski (1983) that approximate vertical datum inconsistency effects.

These regional biases in surface gravimetry imply the presence of correlated errors in these data. However, formation of a full error covariance matrix to accompany these data is problematic primarily because the error covariance properties of the data are not (well) known, but also because the size of such a matrix presents a computational

International Association of Geodesy Symposia, Vol. 121
Schwarz (ed.), Geodesy Beyond 2000 – The Challenges of the First Decade

challenge. To compensate for these errors and preserve the long wavelength integrity of combined solutions, it has been common practice to increase the error variances of the gravity data, while retaining a diagonal error covariance matrix. In addition to down weighting, some combination solutions exclude terrestrial gravity data from contributing to the definition of the very long wavelength ($n \leq 5$) portion of the model. These practices have their limitations as it is discussed by Pavlis in (Lemoine et al., 1998a, p. 7-24). Pavlis (1998b) proposed an alternative approach for the treatment of these effects, whereby one attempts to separate the long wavelength systematic errors (in the form of a low degree surface spherical harmonic expansion), from the valid gravitational signal present in the terrestrial gravity data. This can be accomplished in combination solutions where satellite tracking information and surface gravity data are adjusted simultaneously. The preliminary analysis reported by Pavlis (1998b) is pursued further here.

2 Methodology

In the conventional approach (see e.g., Pavlis, 1988), observation equations for the terrestrial gravity anomalies are written as:

$$v = \overline{\Delta g}_{signal} - \overline{\Delta g}_{obs.} \ , \quad (1)$$

where v is the residual. In the current approach observation equations take the form:

$$v' = \overline{\Delta g}_{signal} + \overline{\Delta g}_{bias} - \overline{\Delta g}_{obs.} \ , \quad (2)$$

where (see Pavlis, 1988 for notation definitions):

$$\overline{\Delta g}_{signal} = \frac{1}{\Delta\sigma}\frac{GM}{r^2}\sum_{n=2}^{N}(n-1)\left(\frac{a}{r}\right)^n \sum_{m=-n}^{n}\overline{C}_{nm}\overline{IY}_{nm} \quad (3)$$

$$\overline{\Delta g}_{bias} = \frac{1}{\Delta\sigma}\frac{GM}{r^2}\sum_{k=0}^{K} f_k \sum_{l=-k}^{k}\overline{B}_{kl}\overline{IY}_{kl} \quad (4)$$

$$f_k = \begin{cases} 1 & if\, k \leq 1 \\ k-1 & if\, k > 1 \end{cases} . \quad (5)$$

Notice that:

$$v \Leftrightarrow v' - \overline{\Delta g}_{bias} \ . \quad (6)$$

Obviously from terrestrial gravity anomaly data alone one cannot estimate simultaneously both $\overline{C}_{nm}$ and $\overline{B}_{kl}$ coefficients, unless $K \leq 1$. However, normal equations formed from the terrestrial data for the $\overline{C}_{nm}$ and $\overline{B}_{kl}$ parameters can be combined with corresponding normal equations obtained from satellite tracking data. The latter contribute information pertaining only to the $\overline{C}_{nm}$ terms, and thus permit the separation of the potential coefficient parameters ($\overline{C}_{nm}$) from the spherical harmonic coefficients representing the regional biases in the gravity anomaly data ($\overline{B}_{kl}$). The principle of this technique is quite analogous to the simultaneous estimation of potential coefficients and spherical harmonic coefficients of the DOT within combination solutions incorporating range data from satellite altimetry.

In the context of the present study the discrimination between the bias term ($\overline{\Delta g}_{bias}$) and the residual ($v'$) is only a matter of spatial frequency content, i.e., bias is that part of the residual anomaly field that exhibits spatial coherence. This implies that within a combination solution (with satellite tracking information), if the bias term can be separated entirely from the gravity anomaly signal term, one should arrive at the same estimate of $\overline{\Delta g}_{signal}$ (and of $\overline{C}_{nm}$) using either the form (1) or the form (2) of the gravity anomaly observation equations. In this (ideal) case the correspondence (6) becomes an equality (notice however that the error estimates of $\overline{C}_{nm}$ will depend on the form of observation equations used). Separation of signal from bias terms depends on the maximum degree of the bias expansion (K), the error properties of the satellite-only model, the magnitude and frequency content of the bias term itself, and the weight assigned to $\overline{\Delta g}_{obs.}$ relative to the satellite tracking information within the combination solution adjustment. Increasing the weight of $\overline{\Delta g}_{obs.}$ decreases the formal *a posteriori* errors of signal and bias anomaly terms but increases their error correlation. Infinite weight of $\overline{\Delta g}_{obs.}$ implies the total absence of satellite tracking information, and renders the problem singular (for $K > 1$). In contrast, decreasing the weight of $\overline{\Delta g}_{obs.}$ implies a "looser" constraint on the sum of signal and bias terms, and may result in positive error correlation for the estimated values of these terms (ideally the correlation should always be negative).

Table 1: Test solution statistics

Name	Kmax	Average Calibr. Factor	RMS Δg Signal (mGal)	RMS Δg Bias (mGal)	RMS Residual (mGal)
s004	0	1.013	18.74	0.03	5.09
s006	20	1.015	18.75	2.28	4.70
s012	25	1.055	19.02	2.15	4.61
s017 (‡)	25	1.003	18.75	1.94	4.62

(‡) A priori constraint $RMS(k) = 5.35 \times 10^{-7} / k^{1.63}$ used for the bias parameters.

The representation chosen here for the regional anomaly biases is obviously not the only possible choice. Discrete values (e.g., over 5° cells) or low order polynomials representing these biases over various regions could also have been used. However, the chosen representation has the considerable advantage that the normal equations for the parameter vector consisting of $\overline{C}_{nm}$ and $\overline{B}_{kl}$ can be formed with great efficiency, compared to those corresponding to the alternative representations. The disadvantage of the low degree spherical harmonic expansion used here, is that it may be "smearing" an existing regional bias to neighboring areas, that in reality should not be affected. In this regard, future geopotential mapping missions (especially GRACE and GOCE) will benefit the effectiveness of this technique, because they are expected to allow for much higher degree expansions for the biases, than those described here.

3 Data used and numerical tests

The satellite-only model used here is designated PGS7609 (Lemoine et al., 1998b). It represents an incremental improvement over EGM96S, mainly through the addition of TDRSS tracking data. The 1°x1° surface gravity data used are those used in EGM96 (Lemoine et al., 1998a, section 3.5). Using these data in combination, several test solutions were performed (notice that satellite altimeter data are not included in any of these test models). Four of them are discussed next:

- s004 is a solution where surface gravimetry was treated exactly as in the development of EGM96, and thus constitutes a "benchmark" for the evaluation of the bias recovery method.
- s006 uses the same weights as s004, but estimates also bias coefficients to degree 20.
- s012 is as s006 but biases are estimated to degree 25, and **land** data weights are increased by a factor of approximately 4.
- s017 uses the same weights as s004, estimates biases to degree 25, but uses also an *a priori* constraint on the bias parameters.

Table 1 provides some statistics for these solutions, which are mainly internal consistency indicators. Figure 1 shows the spectra of several coefficient sets. Comparisons with independent data that can gauge the performance of a test model included orbit fit tests (Table 2), and comparisons with GPS/leveling derived geoid undulations or height anomalies (Table 3). While all four solutions perform similarly, the bias recovery tends to improve slightly (compared to s004) the orbit fits for high-altitude spacecraft, and degrade the fits for low altitude ones. The GPS/leveling comparisons degraded slightly when the bias terms were extended beyond degree 20. These tests indicate that careful choice of data weights, K , and the use of *a priori* constraint is required to optimize the performance of this technique. The gravity anomaly biases (to degree 25) estimated in solution s017 are shown in Figure 2. Figure 3 shows the geoid

Table 2: Orbit fit results - RMS of fit in (cm) for range (r) and (mm/s) for range-rate (r-r) data

Model	PGS7609	s004	s006	s012	s017
LAGEOS(‡)	2.98	3.15	3.14	3.16	3.13
LAGEOS 2	3.96	3.91	3.90	3.91	3.91
Starlette	9.77	9.93	9.92	10.01	9.90
Ajisai	4.98	5.11	5.08	5.12	5.09
Stella	9.99	8.78	8.90	8.93	8.81
GFZ-1	166.60	126.51	121.76	152.08	123.77
GFZ-1 (‡)	160.61	95.80	96.48	106.14	96.18
TRMM (r)	239.56	218.85	223.77	233.32	223.20
TRMM (r-r)	4.61	4.42	4.56	4.79	4.55

Table 3: GPS/leveling minus model geoid undulation statistics. All models complete to Nmax=360 (EGM96 used for n>70). Total sites used are 9307. Units are cm.

	s004		s006		s012		s017		EGM96	
Area	Mean	S. Dev.	Mean	S. Dev.	Mean	S. Dev.	Mean	S. Dev.	Mean	S. Dev.
Australia	1.9	46.3	3.7	46.4	0.1	48.2	3.1	47.6	-1.2	46.4
Baltic Region	-57.7	31.5	-57.9	32.6	-61.5	33.9	-58.4	32.0	-57.1	26.0
Canada	-102.8	40.1	-102.7	40.4	-102.0	41.2	-103.1	39.8	-102.9	37.3
Czech Republic	-60.2	22.6	-57.3	22.7	-63.5	24.8	-58.0	22.8	-65.0	21.9
N-S Euro. Trav.	12.5	31.4	13.4	31.1	13.7	31.5	12.5	30.8	8.4	31.9
France	-111.9	37.9	-109.4	36.5	-111.5	41.6	-110.3	37.3	-116.9	35.4
Germany	-52.2	12.6	-49.6	12.7	-56.6	13.5	-51.9	12.7	-60.6	13.5
Hungary	-94.3	13.6	-101.5	14.9	-113.6	17.6	-102.8	15.3	-73.6	13.7
USA	-93.8	40.8	-93.6	41.0	-94.6	43.1	-94.2	42.6	-97.5	39.0
Mean		39.1		39.2		41.4		40.2		37.2

undulation effects that are implied by the biases estimated in the solution s017. These effects have a global RMS value of 3.1 m, and would be present (approximately) in a solution based **only** on surface gravity data. The geoid undulation difference s017 minus s004, up to degree 25 is shown in Figure 4. This difference has a global RMS value of 6.3 cm. Figure 4 has a very distinct signature over the tropics; a similar signature was present in the geoid difference between JGM3 and JGM2. Figure 4 suggests that systematic biases may have been (at least partially) responsible for geoid error present in JGM2 over the tropics, which was reduced in JGM3, through down weighting of surface gravity anomaly data.

4 Summary and conclusions

The investigation reported here indicates that modeling of surface gravity biases (as a spherical harmonic expansion), within solutions that combine satellite tracking and surface gravity data, is possible and the separation of signal and bias terms can be accomplished with presently available data to (approximately) degree 25. Additional tests and comparisons are needed however in order to optimize the performance of this technique. Such tests need to consider carefully issues of data weighting. The impact of satellite altimeter data (in the form of ranges), in the solution needs also to be investigated.

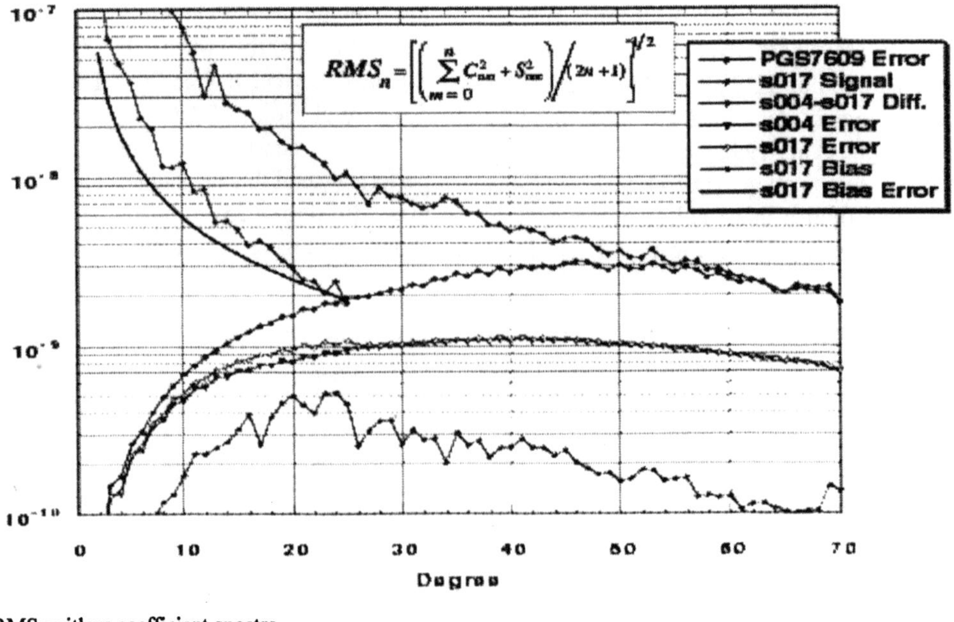

Fig. 1: RMS unitless coefficient spectra

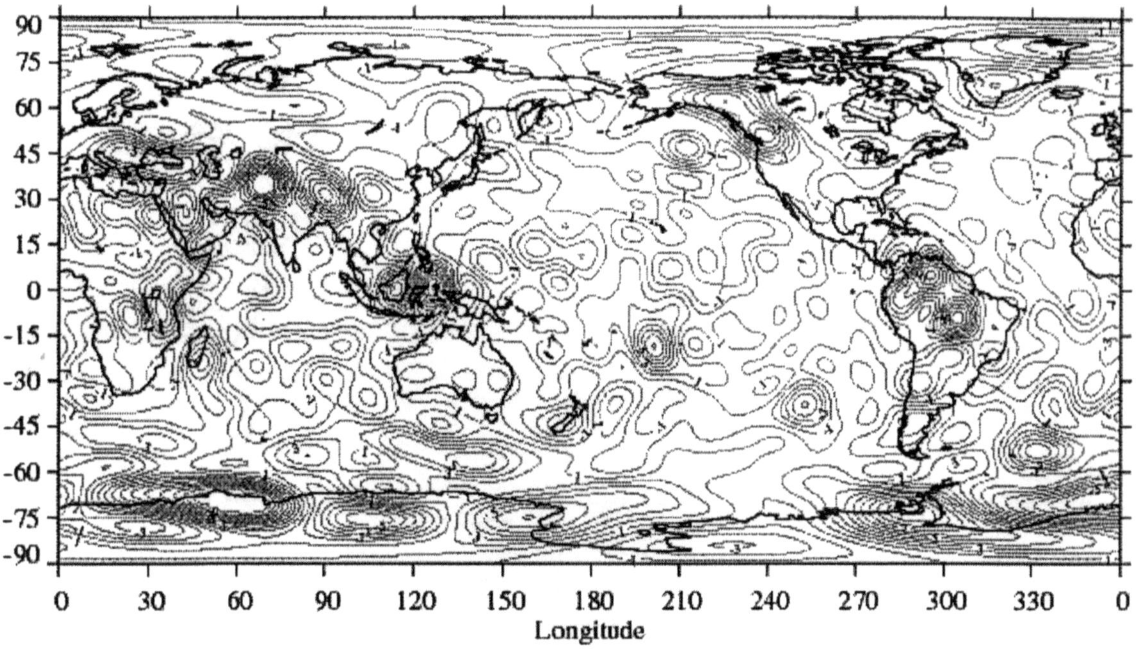

Fig. 2: Systematic gravity anomaly biases to degree 25 from solution s017. Contour interval is 1 mGal.

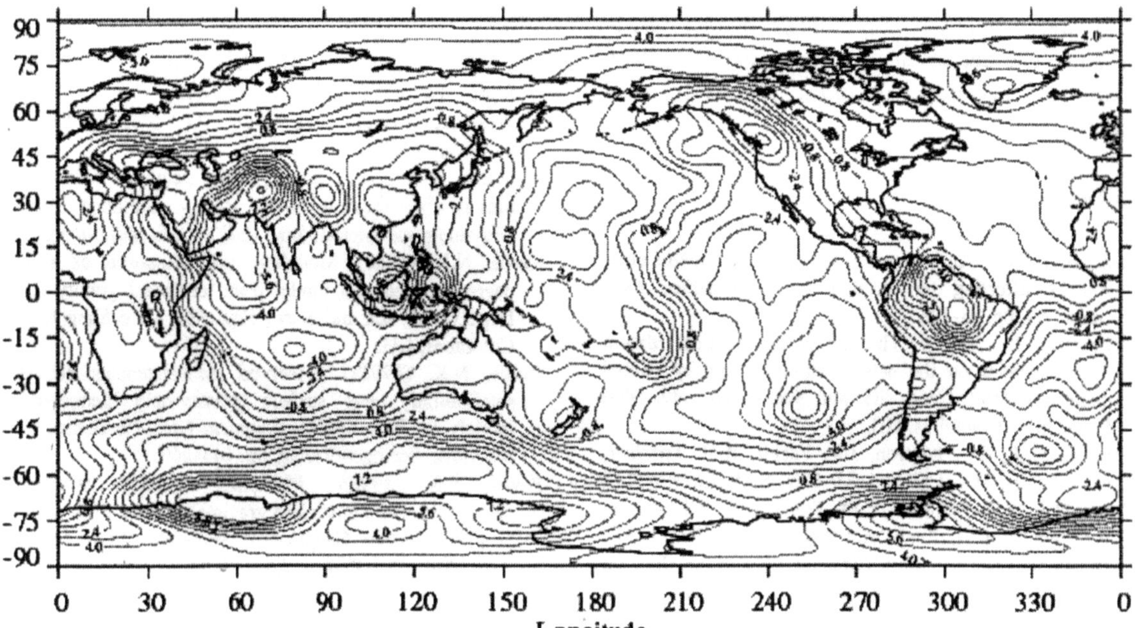

Fig. 3: Geoid undulation effects (to degree 25) implied by anomaly biases estimated in solution s017. Contour interval is 0.8 m.

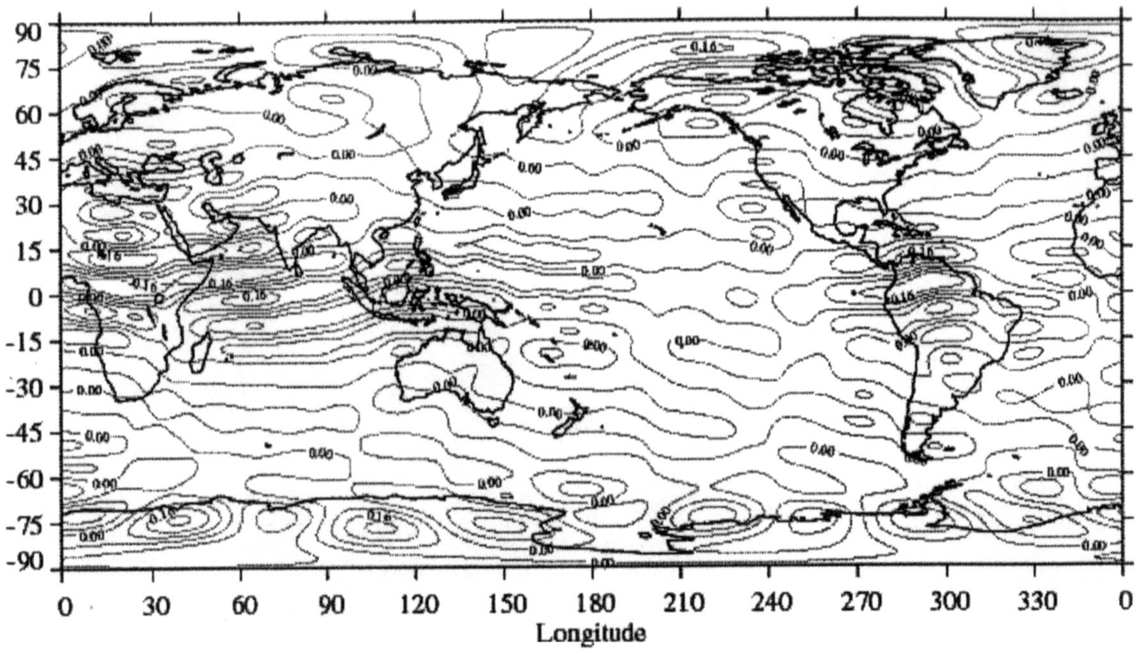

Fig. 4 Geoid undulation difference s017-s004 to degree 25. Contour interval is 0.08 m.

Acknowledgments. This research was supported in part by NIMA through contract NMA202-98-C-1021, and in part by NASA through contract NAS5-32352. The author thanks Chris Cox (Raytheon ITSS) who performed the orbit fit tests reported here.

References

Heck, B. (1990). An evaluation of some systematic error sources affecting terrestrial gravity anomalies. *Bull. Geod.*, 64, pp. 88-108.

Laskowski, P. (1983). The effect of vertical datum inconsistencies on the determination of gravity related quantities. Dept. Geod. Sci. and Surv., *Rep. 349*, Ohio State Univ., Columbus, OH.

Lemoine, F.G., S.C. Kenyon, J.K. Factor, R.G. Trimmer, N.K. Pavlis, D.S. Chinn, C.M. Cox, S.M. Klosko, S.B. Luthcke, M.H. Torrence, Y.M. Wang, R.G. Williamson, E.C. Pavlis, R.H. Rapp, and T.R. Olson (1998a). The Development of the Joint NASA GSFC and the National Imagery and Mapping Agency (NIMA) Geopotential Model EGM96. *NASA/TP-1998-206861*. Goddard Space Flight Center, Greenbelt, MD.

Lemoine, F.G., C.M. Cox, D.S. Chinn, M.H. Torrence, N.K. Pavlis, Y.M. Wang, R.G. Williamson, and E.C. Pavlis (1998b). Gravitational Models Including TDRSS Data from EP/EUVE, RXTE, CGRO, and TRMM. Presentation at the 1998 Spring AGU Meeting, Boston, MA, May 26-29.

Mainville, A. and R.H. Rapp (1985). Detection of regional bias in 1°x1° mean terrestrial gravity anomalies. Bureau Gravimetrique International, *Bull. d' Information*, No. 57.

Pavlis, N.K. (1988). Modeling and estimation of a low degree geopotential model from terrestrial gravity data. Dept. Geod. Sci. and Surv., *Rep. 386*, Ohio State Univ., Columbus, OH.

Pavlis, N.K. (1998a). Observed inconsistencies between satellite-only and surface gravity-only geopotential models. In: *Proc. Of IAG Symposia Vol. 119: Geodesy on the Move*, Springer-Verlag, pp. 144-149.

Pavlis, N.K. (1998b). Modeling of long wavelength systematic errors in surface gravimetric data. Presentation at the European Geophysical Society's XXIII General Assembly, Nice, France, April 20-24.

Investigation of different methods for the combination of gravity and GPS/levelling data

H. Denker, W. Torge, G. Wenzel
Institut für Erdmessung (IfE), Universität Hannover, Schneiderberg 50, D-30167 Hannover, Germany

J. Ihde, U. Schirmer
Bundesamt für Kartographie und Geodäsie, Außenstelle Leipzig, Karl-Rothe-Straße 10-14, D-04105 Leipzig, Germany

Abstract. Two different methods for the combined computation of the quasigeoid are compared in a test area in Germany. Both methods are based on the remove-restore technique and use the global geopotential model EGM96, point gravity data with a spacing of a few km, a digital terrain model and GPS/levelling control points (with a spacing of about 25 km).

In method I the global model is combined first with the gravity and terrain data using the least squares spectral combination technique with integral formulas. The resulting height anomalies are given in a 1.0′x 1.5′grid. Then a smooth corrector surface is developed from the GPS/levelling data by least squares collocation, using a signal and a trend component.

The second method (II) is based on a common adjustment of the EGM96 reduced gravity and height anomaly observations using point masses and appropriate weight relations. The point masses are arranged at a depth of 10 km, 30 km and 200 km, and in hilly areas also at a depth of 5 km.

Both techniques are compared from the methodological and numerical point of view. The results are discussed and show an agreement at the cm level.

Keywords. Geoid, levelling, gravity, GPS.

1 Introduction

The Bundesamt für Kartographie und Geodäsie (BKG) and the State Survey Offices have organised the observation of precise GPS/levelling control points in Germany, following a BKG proposal from 1992. The average spacing of the control points is about 25 km. The ellipsoidal GPS heights are referring to the ETRS89 reference system, while the levelled heights are given as normal heights in the DHHN92 system. The main objective of these GPS/levelling control points is to serve for the computation of a new quasigeoid model for Germany, in connection with high resolution gravity and terrain data. The BKG and the Institut für Erdmessung (IfE) agreed to cooperate on the computation of this new combined quasigeoid, which shall become a standard for the transformation of heights between the ETRS89 and the DHHN92 height system. The present paper describes some first results based on two different combination procedures for a test area in East Germany, as at present all required data sets are only available for this subarea (with a size of about 100,000 km^2).

2 Data description

For the quasigeoid determination 4 groups of data are available in the test area (50°N - 55°N, 9°E - 16°E):

- height anomalies from GPS and levelling (ζ_{GPS}),
- terrestrial gravity anomalies (Δg),
- digital terrain models (*DTM*),
- geopotential models (*GPM*).

Height anomalies from GPS and levelling are available for a total of 196 points (see Fig. 1). The GPS observations were always made in two sessions of 24 hours. The average point-to-point distance is about 25 km. The computation of the GPS heights, h_{ETRS}, is based on the European Terrestrial Reference System 1989 (ETRS89). Normal heights, H^N_{DHHN}, referring to the DHHN92 height system, were determined for all GPS stations by precise levelling. Considering the accuracy of the levelling heights and of the ellipsoidal heights from GPS, the accuracy of the GPS/levelling quasigeoid heights,

International Association of Geodesy Symposia, Vol. 121
Schwarz (ed.), Geodesy Beyond 2000 – The Challenges of the First Decade

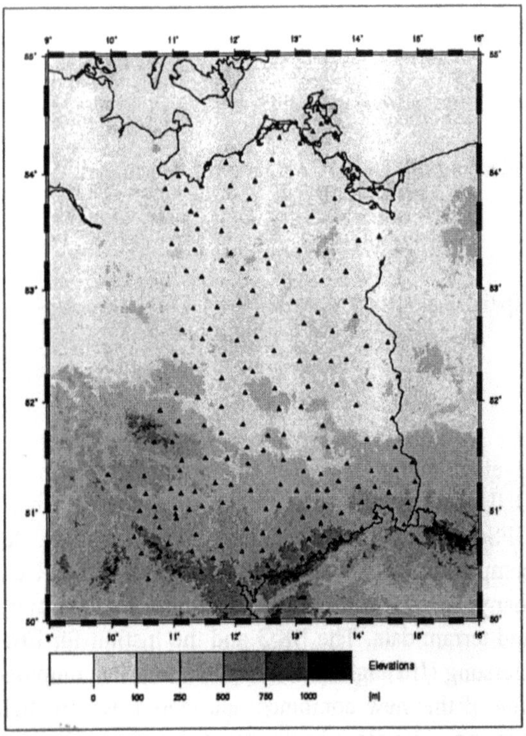

Fig. 1: Locations of GPS/levelling stations and digital terrain model

$\zeta_{GPS} = h_{ETRS} - H^{N}_{DHHN}$, is estimated as ± 0.015 m.

Furthermore, for the test area there are more than 70,000 point gravity values available. Outside of the test area there are additional point and mean gravity anomalies available from the IfE and BKG data base. Fig. 2 displays the locations of the gravity data from the IfE data base.

Finally, a high resolution digital terrain model with an original block size of about 30 m as well as the geopotential model EGM96 (Lemoine et al. 1996) were utilized.

3 Combination method I

In 1997, the high resolution European gravimetric (quasi)geoid model EGG97 was computed at the Institut für Erdmessung (IfE), University of Hannover, Germany, operating as the computing centre of the International Association of Geodesy (IAG) Subcommission for the Geoid in Europe (Denker and Torge 1997). The EGG97 model was computed in a 1.0′ x 1.5′ grid and combines about 2.7 million terrestrial gravity data, 700 million terrain data and the spherical harmonic model EGM96 from NASA/NIMA (Lemoine et al. 1996).

The mathematical modelling is based on the spectral combination technique in connection with a remove-restore procedure. Formal error estimates of the resulting height anomalies were derived on the basis of corresponding degree variances. Based on a ± 1 mgal correlated noise for the gravity data, the standard deviations of the height anomalies are ± 0.064 m, while the standard deviations for height anomaly differences are ± 0.039 m over 100 km and ± 0.076 m over 1000 km distance, respectively. The analysis also shows that the major error contribution is coming from the spectral band below degree l=360, suggesting that the EGG97 error is predominantly long-wavelength (Denker 1998). This finding is also confirmed by intercomparisons with GPS/levelling data, showing long to medium wavelength discrepancies (see *Denker* 1998, *Milbert* 1995).

This circumstance opens the possibility to develop an empirical corrector surface which relates the given gravimetric quasigeoid model to the reference system of GPS and levelling heights (*Milbert* 1995, *Denker* 1998). It must be understood that such a corrector surface will incorporate systematic errors from ellipsoidal, levelling, and geoidal sources. Modelling of the corrector surface begins by forming residuals in the sense of

$$\zeta_{GPS} - \zeta_{EGG97} = \left(h_{GPS} - H^{N}\right) - \zeta_{EGG97} = l = t + s + n, \quad (1)$$

where ζ_{GPS} is the GPS/levelling quasigeoid undulation, computed as the difference of the ellipsoidal height from GPS, h_{GPS}, and the normal height from levelling, H^{N}, ζ_{EGG97} is the quasigeoid undulation from the gravimetric model EGG97, and l are the raw residuals, which are considered as a trend (t), signal (s), and noise (n) component in a least-squares collocation model.

The trend component (t) is modelled by a 3-parameter datum shift in the form

$$t = \cos\varphi\cos\lambda\Delta X + \cos\varphi\sin\lambda\Delta Y + \sin\varphi\Delta Z, \quad (2)$$

with the ellipsoidal latitude and longitude φ and λ, and the datum shift constants $\Delta X, \Delta Y, \Delta Z$. Instead of $\Delta X, \Delta Y, \Delta Z$ one can also introduce changes in the ellipsoidal coordinates of an initial point, which can be interpreted as a height bias and tilts in NS and WE direction. For the present test area the magnitude of the tilt is 0.18 ppm at an azimuth of 175°.

After computing the trend parameters, an empirical covariance function of the de-trended residuals (observations), l-t, was computed and modelled by a simple mathematical function (see Fig. 3).

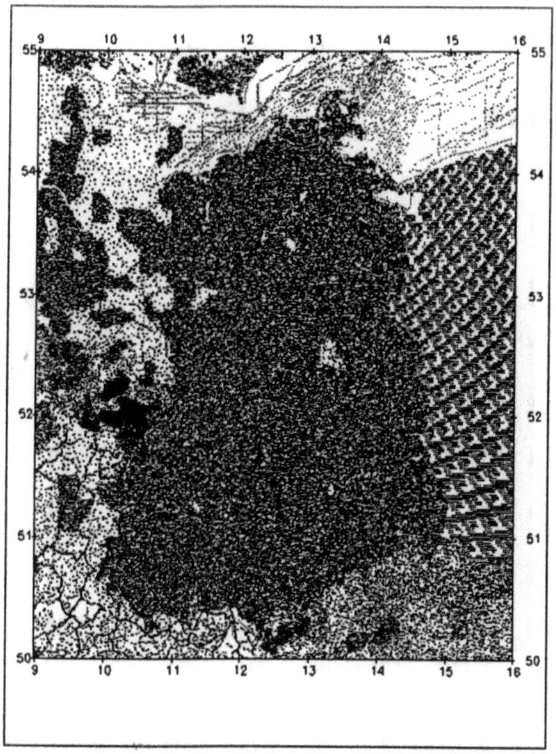

Fig. 2: Locations of gravity stations

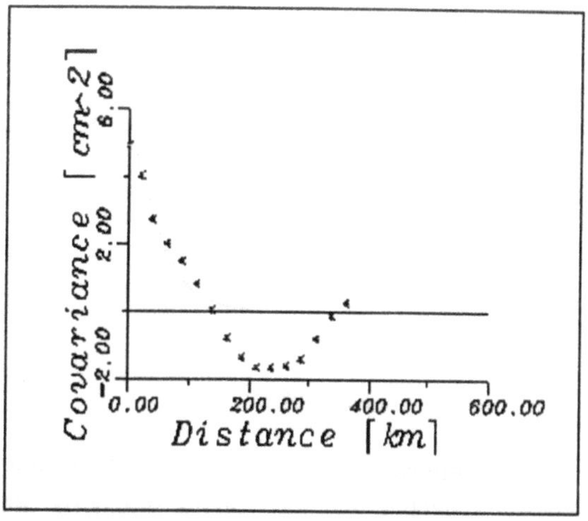

Fig. 3: Empirical covariance function (signal)

Table 1: Statistics of the individual model components for method I (computed in 196 GPS/Levelling points)

Parameter	Mean	Std. Dev.	Min.	Max.
$\zeta_{GPS} - \zeta_{EGG97}$	0.023	0.033	-0.044	+0.088
$\zeta_{GPS} - (\zeta_{EGG97} + t)$	0.000	0.022	-0.046	+0.098
$\zeta_{GPS} - (\zeta_{EGG97} + t + s)$	0.000	0.006	-0.022	+0.027
Trend t	0.023	0.024	-0.027	+0.060
Signal s	0.000	0.020	-0.033	+0.071
Corr. Surf. (t+s)	0.023	0.031	-0.033	+0.082

We used a second order Markov covariance model in the form

$$Cov(s) = C_0 (1 + s / \alpha) \exp(-s / \alpha), \qquad (3)$$

where s is the distance, C_0 is the signal variance, and α is a parameter that describes the characteristic length of the covariance function. After fixing the signal and error covariance models (± 0.022 m signal standard deviation, 50 km signal correlation length, ± 0.015 m uncorrelated noise), the signal component can be computed in an arbitrary station P by the formula

$$\hat{s} = \boldsymbol{c}_P^T (\boldsymbol{C} + \boldsymbol{D})^{-1} (\boldsymbol{l} - \boldsymbol{t}). \qquad (4)$$

In Eq. (4) $\hat{s}$ is the predicted signal in station P, $\boldsymbol{C}$ is a matrix containing the signal covariances between the observations, $\boldsymbol{D}$ is the noise covariance matrix, and the vector $\boldsymbol{c}_P$ contains the signal covariances between the predicted signal and the observations. Finally, the predicted signal and the trend component are added to the original gravimetric quasigeoid (EGG97), yielding the corrected (improved) geoidal surface (denoted as EGG97C) in the form

$$\zeta_{EGG97}^{corr} = \zeta_{EGG97} + t + \hat{s}. \qquad (5)$$

The corrector surface, i.e. t+s, is shown in Fig. 4. The above described technique to combine a gravimetric geoid/quasigeoid with GPS/levelling can be regarded as a stepwise solution, similar to stepwise collocation, where in the first step the gravity and terrain data are combined with the global model, while in the second step the GPS/levelling data are added on the basis of empirical covariance modelling.

The statistics of all relevant model components were computed in the 196 GPS/levelling stations and are presented in Table 1. The raw residuals $\zeta_{GPS} - \zeta_{EGG97}$ according to Eq. (1) show a mean value of 0.023 m and a standard deviation of ± 0.033 m. The detrended residuals $(\zeta_{GPS} - \zeta_{EGG97} - t)$ show a standard deviation of ± 0.022 m with maximum values of 0.098 m. The largest values are located in the north around the island of Rügen. The predicted signal has a standard deviation of ± 0.020 m with maximum values up to 0.071 m. The largest values are again found in the north. Further interpretation of the signal component is difficult, because it contains effects from GPS, levelling and the gravimetric quasigeoid. The residuals about the predictions, i.e. $\zeta_{GPS} - \zeta_{EGG97} - t - s$, are shown in Fig. 5. The Rms value was found to be ± 0.006 m, being significantly smaller than the assigned data noise. The remaining maximum discrepancies are only ± 0.027 m. This documents the efficiency of the procedure.

4 Combination method II

In this combination method, the GPS/levelling quasigeoid heights and the gravity anomalies are introduced as observations in an adjustment of point masses. The adjustment procedure is also based on a remove-restore technique, where the observations are reduced for the long-wavelength effect of the geopotential model EGM96 (ζ_{GPM}, Δg_{GPM}). Previous investigations showed that short-wavelength effects from a digital terrain model were not giving a significant improvement for the current setup of the point mass adjustment. Thus terrain effects were neglected in the point mass modelling to date.

The residual gravity anomalies $\Delta g'$ and height anomalies ζ', that are used in the point mass adjustment, are thus defined as :

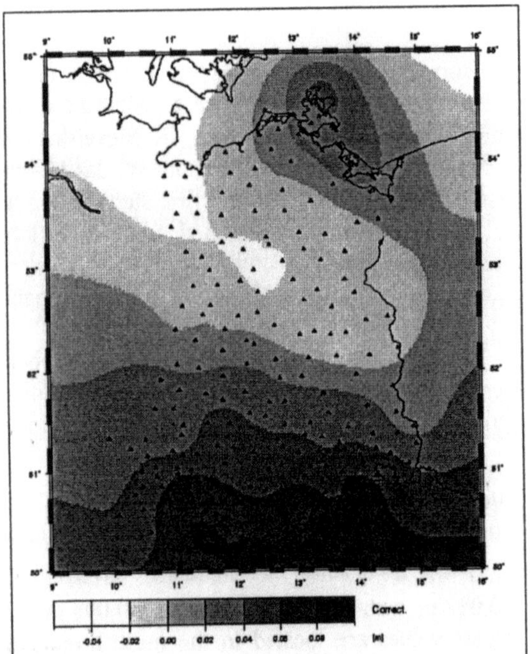

Fig. 4: EGG97 corrector surface (method I)

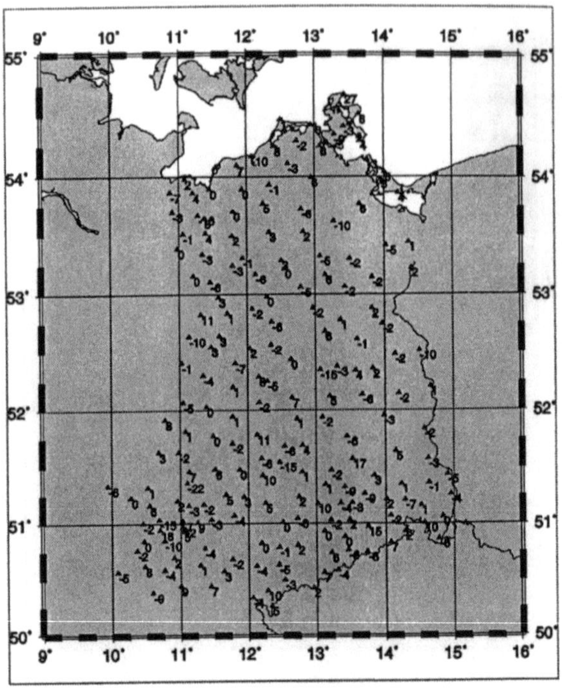

Fig. 5: Residuals for method I (in mm) in the 196 GPS/levelling stations

$$\Delta g' = \Delta g - \Delta g_{GPM}, \tag{6}$$

$$\zeta' = \zeta_{GPS} - \zeta_{GPM}. \tag{7}$$

The relations between the masses m and the residual height and gravity anomalies (observation equations) are as follows:

$$\Delta g' = G \sum \frac{m(H - H')}{d^3}, \tag{8}$$

$$\zeta' = \frac{G}{\gamma_0} \sum \frac{m}{d}. \tag{9}$$

In these equations G is the Newton's gravitational constant and d is the distance.

The accuracy of both observation types is considered in a weight matrix. As a priori accuracies, ± 0.015 m are introduced for the height anomalies and ± 1 mgal for the gravity anomalies. Here it should be noted that in the adjustment the gravity anomalies are introduced as mean values with a block size of 2 km and 5 km. The total number of mean values is 25700.

Investigations showed that a hierarchical arrangement of the point masses at different depths leads to optimal results (ratio of observations, unknowns and accuracy). The point masses, arranged at a depth of 10 km and with a distance of

0.1°x 0.15° shall approximate mainly the short-wavelength parts of the quasigeoid. The point masses at a depth of 30 km with a distance of 0.2°x 0.3° cover basically those frequencies which are determined by the height anomalies from GPS and levelling. 8 point masses at a depth of 200 km shall compensate the long wavelength and slope influences. In hilly areas additional point masses with a distance of 0.05° x 0.075° are arranged at a depth of 5 km (Fig. 6).

The quasigeoid model from this computation is denoted as BKG98. The Rms residual of the GPS/levelling derived height anomalies is ± 0.011 m. For the northern flat country area the Rms value is ± 0.010 m, while the value for the southern hilly area is ± 0.012 m, respectively. The individual residuals are also shown in Fig. 7. The Rms residual of the gravity anomaly is ± 2.2 mgal.

5 Comparison of methods and discussion

The combined quasigeoid solutions from method I (EGG97C) and method II (BKG98) were intercompared in the area covered by the GPS/levelling stations. The differences between EGG97C and BKG98 are displayed in Fig. 8. The Rms difference is ± 0.010 m. The maximum differences are located close to the boundary of the comparison area (see Fig. 8), the main reason being that the BKG98 solution does not include detailed gravity data

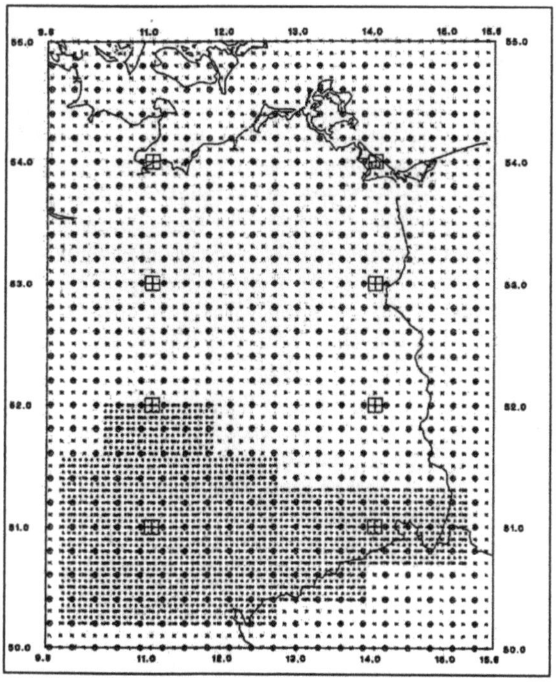

Fig. 6: Locations of point masses (method II)

outside the area covered by the GPS/levelling stations. Therefore, the comparison was repeated with a 10 km border area being excluded. For this case the Rms difference is ± 0.009 m with maximum values of 0.046 m. In the inner area the maximum differences of about 0.045 m are found around 52°N and 12°W. Especially in this area, the differences show structures that are correlated with the location of the point masses. It should also be noted that in this region the dense point mass grid (5 km) ends (see Fig. 6). In the future this phenomena needs to be studied in more detail. Furthermore, the differences were also analysed along profiles and spectra were computed. The spectra show peaks at half wavelengths of about 10 km and 20 km, which corresponds to the grid spacing used in the point mass modelling.

In another analysis the residuals from both methods (see Fig. 5 and 7) were studied. The correlation between the residuals from both methods is about 65 %. This shows that both methods have the same tendency in the GPS/levelling stations, and larger residuals may also indicate small height errors in these stations.

A strong test of the used combination procedures would be to inter-compare the two combined quasigeoid solutions with an independent GPS/levelling data set. However, as at present such a data set is not available, additional solutions were derived using only about one half of the GPS/levelling control points in the computation of the combined quasigeoid models, while the remaining stations were used for a comparison only. This was done for both combination procedures, and the corresponding solutions, based on only 94 GPS/levelling points, are denoted as EGG97C (B) and BKG98 (B), respectively.

The Rms residuals for the previously described solutions and the corresponding B solutions are given in Table 2. For the B solutions, the statistics are given for the 94 stations used in the development of the corresponding solutions as well as for the remaining 102 stations used only for the evaluation. The table shows a slight increase of the residuals in the independent comparison stations, but the results are still very satisfactory.

Table 2: Statistics of the residuals in GPS/levelling points

Solution		Mean	Std. Dev.	Min.	Max
BKG98	(196)	0.000	0.011	-0.031	+0.031
EGG97C	(196)	0.000	0.006	-0.022	+0.027
BKG98 (B)	(94)	0.001	0.013	-0.040	+0.039
	(102)	0.001	0.015	-0.033	+0.034
EGG97C (B)	(94)	0.000	0.005	-0.013	+0.025
	(102)	0.001	0.011	-0.033	+0.024

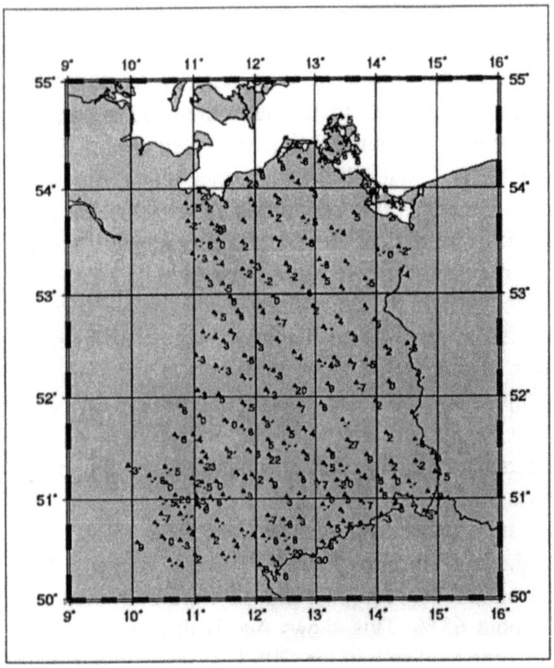

Fig. 7: Residuals for method II (in mm) in the 196 GPS/ levelling stations

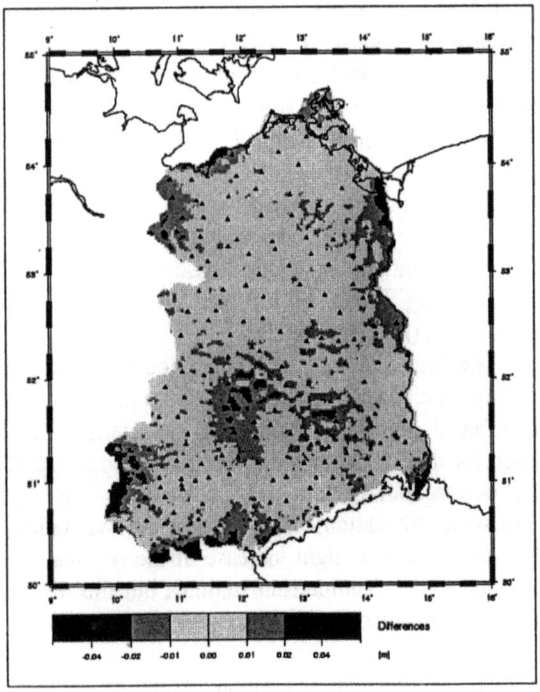

Fig. 8: Height anomaly differences EGG97C - BKG98

6 Conclusions

The two investigated methods to combine GPS/levelling data with gravimetric data are based on a totally different concept. Both methods show a satisfactory agreement of ± 0.01 m (Rms). The maximum differences of about 0.045 m are located in the inner area and show a correlation with the location of the point masses, indicating that the point mass modelling should be improved. The residuals in the GPS/levelling control points from both combination procedures show a high correlation (65 %), indicating that small height errors (resulting from inaccurate centering data, different observation epochs of the GPS and levelling data, etc.) might still exist. Before doing the final computations for the entire area of Germany, the existing problems have to be further studied and clarified.

References

Denker, H. (1998): Evaluation and improvement of the EGG97 quasigeoid model for Europe by GPS and levelling data. Second Cont. Workshop on the Geoid in Europe, Rep. Finn. Geod. Inst. 98:4, 53-61, Masala, 1998.

Denker, H., W. Torge (1997): The European gravimetric quasigeoid EGG97 - An IAG supported continental enterprise. In: R. Forsberg et al. (eds.): Geodesy on the Move, IAG Symp., Vol. 119, 245-254, Springer, Berlin, Heidelberg, New York, 1998.

Ihde, J. (1995): Geoid Determination by GPS and Levelling. In: *International Association of Geodesy. Symposia 113, Gravity and Geoid.* Springer-Verlag Berlin, Heidelberg

Ihde, J. , Schirmer, U., Stefani, F., Töppe, F. (1998): Geoid Modelling with Point Masses. In: *Proceedings of the Second Continental Workshop on the Geoid in Europe, Reports of the Finnish Geodetic Institute.* Budapest, Hungary, March 10-14, 1998

Lemoine, F. G. et al. (1996): The Development of the NASA GSFC and NIMA Joint Geopotential Model. In: *Proceedings of the International Symposium on Gravity, Geoid, and Marine Geodesy, (GRAGEOMAR 1996).* The University of Tokyo, Tokyo, Japan, September 30 - October 5, 1996

Milbert, D.G. (1995): Improvement of a high resolution geoid height model in the United States by GPS height on NAVD 88 benchmarks. Bull. d'Informations 77 and IGeS Bull. 4, Special Issue, New Geoids in the World, 13-16, Milan, Toulouse.

The regional geopotential model to degree and order 720 in China

Yang Lu, H.T. Hsu, F.Z. Jiang
Laboratory of Dynamical Geodesy,
Institute of Geodesy and Geophysical,
Chinese Academy of Sciences, 54 Xudong Road, 430077 Wuhan, Hubei, P.R.China

Abstract. The spherical harmonic expansion is a powerful method to describe local and global gravity field in the frequency domain. In principle, the resolution and precision of gravity field expressed by the spherical harmonic model are proportional to the degree and order of model expressed, for example, to degree 720 that has a resolution about 28 km. For this reason a higher degree geopotential model is of the advantage for us. It is possible to develop the regional model to degree and order 720 although that seems more difficult for global model in terms of the resolution and precision of gravity data over global area. On the basis of the tailored method, we have developed the regional higher resolution geopotential model IGG97LB to degree and order 720 suitable to the mainland of China. The gravity anomalies computed from the new model can achieve the mean square error of ±8.8 mGal. The geoid undulations from model have a mean square error ±0.67 m compared with height anomalies derived from GPS and levelling in the research area. The mean square errors of gravity anomalies and geoid undulations are the same as that computed from reference model EGM96 in global area without China. A clear linear relation between gravity anomalies and geoid undulations on short wavelength parts both computed from the new model has been found in the research area.

Keywords. Regional geopotential model.

---

1 Introduction

The geopotential model can be expressed by the spherical harmonics expansions, which represent the complex Earth's gravity field in the frequency domain by means of analysis method. Moreover, the resolution and precision of gravity field expressed by spherical harmonic model are proportional to degree and order of model expressed. From the geopotential model we can easily obtain any gravity field quantity in different resolutions as we want such as geoid undulations, gravity anomalies and vertical deflections and so on.

Only the long wavelength parts of the gravity field can be determined from satellite to satellite tracking. Combining the long wavelength parts with terrestrial gravity and altimeter data, which reveal the short wavelength character, produced a complete geopotential model. During the past ten years, a number of the high resolution and accurate global geopotential models have been developed to degree and order 360, for example OSU86 and OSU91A1F (Rapp and Cruz (1986), Rapp and Pavlis (1990), Rapp et al. (1991)). To the best of author's knowledge N.K.Pavlis had developed a higher resolution global geopotential model to degree and order 500. In particular, the currently most accurate high resolution global geopotential model EGM96 (Lemoine et al. (1996)) has been computed by a collaboration of NASA, US National Imagery and Mapping Agency (NIMA) and Ohio State University (OSU). An international group under the umbrella of IAG has validated EGM96 model.

In the near future, the research of geopotential model will focus on the global higher degree geopotential model such as to degree and order 720. Unfortunately, it seems more difficult to compute such a global high degree geopotential model. The main limitations are lack of homogeneous data in space, time and precision within global area. However, it is possible to develop a higher degree regional geopotential model in terms of tailored method (G.Wenzel, (1998)).

This paper intends to develop the higher degree geopotential model IGG97LB to degree and order 720 tailored to the gravity data in the mainland of China.

International Association of Geodesy Symposia, Vol. 121
Schwarz (ed.), Geodesy Beyond 2000 – The Challenges of the First Decade

2 Gravity data processing

The research area is on a 25°x55° region in the mainland of China. We chose EGM96 model as the reference model. The gravity anomalies computed from EGM96 model have been used outside the research area and some blocks lacking observed value. All used free air gravity anomalies have been reduced to IGSN71 system.

To compute a new model to degree and order 720, we need a dataset of gravity anomalies on a 15'x15' grid. Since the original dataset of gravity anomalies come from many ways, there may be some rough errors and discrepancies. Before using this dataset, a pre-treatment should be made.

There are several ways to find rough errors. We applied rejection technique of standard deviation in a statistical-probability sense. For this purpose, the research area is divided into many blocks (5'x5' each). We evaluated residual gravity anomalies between the observed and the reference gravity anomalies computed from model (i.e. EGM96), and evaluated standard deviations of residual anomalies in each block, gives a criterion to reject rough errors, in this case three times standard deviations.

If the differences between the observed and the mean value of residual anomalies in a block are greater than three times standard deviations, the observed will be rejected. Repeating the above procedure, will be generated the required data in each block. Implicitly in this process we will obtain the mean of residual anomalies on a grid (5'x5'). Combining the residual anomalies and reference anomalies, we have a set of 130479 mean 5'x5' free air gravity anomalies in the research area. Furthermore we have a set of 13018 mean 15'x15' gravity anomalies used to compute the new geopotential model IGG97LB.

3 Computation of geopotential model IGG97LB to degree 720

The geopotential model IGG97LB complete to degree and order 720 has been computed in October 1997, and has been used to descriptive earth crust interior tectonic and dynamic feature in northern China (Lu and Jiang (1997)).

The coefficients of IGG97LB up to degree 36 have been taken from geopotential model EGM96; the coefficients from degree 37 to degree 720 have been computed from the mean 15'x15' gravity anomalies by the tailored method (H.T.Hsu and Y.Lu, 1995). As initial approximate model, global geopotential model EGM96 was used to degree 360.

In the idea behind, the tailored method (Weber and Zomorrodian (1988), Basic (1989), Hsu and Lu (1995)) there are discrepancies, called regional residual gravity field, between the regional observed gravity field and the reference field that obtained from global potential model. The main reasons of these are as follows. First, the initial global model (for example OSU91A1F, EGM96 etc.) is a model in a global-average sense, and is not very accurate at the shorter wavelength. Second, the dataset used in the solution of model is incomplete and the omission errors of model can not be avoided. Therefore we can use higher resolution and accuracy local gravity data to improve a global model to higher degree and order. The model can fit to local gravity field well.

To some extent, so called tailored method is a 'remove-restore' method. First, removing reference gravity anomalies (i.e. EGM96) from observed gravity anomalies produces regional residual gravity anomalies, which are errors of the anomalies computed from EGM96 model shown in Fig.1. Second, the corrections for new coefficients are calculated from residual gravity anomalies. Finally, restoring complete model from the reference model and the corrections calculated for coefficients generates the new model.

In the tailored method (Hsu and Lu (1995)), for gravity anomaly, we developed following formulas to compute the corrections of new coefficients. These formulas based on the solutions by Rapp et al. (1991) in terms of the ellipsoidal coordinate.

$$\Delta\overline{C}_{nm} = \frac{1}{4a\pi\gamma}\sum_{t=1}^{T} r_t^{E} \sum_{k=0}^{s} \frac{L_{nmk}}{\overline{S}_{n-2k,|m|}(\frac{b}{E})} \bullet$$

$$\frac{I\overline{P}_{n-2k,|m|}}{(n-2k-1)q^{t}_{n-2k}}\,\delta\Delta\overline{g}_t \begin{cases} IC_t & if \quad m \geq 0 \\ IS_t & if \quad m < 0 \end{cases} \tag{1}$$

where

$$I\overline{P}_{n|m|} = \int_{\delta_i}^{\delta_{i+1}} \overline{P}_{n|m|}(\cos\delta)\sin\delta d\delta$$

$$IC_j = \int_{\lambda_j}^{\lambda_{j+1}} \cos\lambda d\lambda$$

$$IS_j = \int_{\lambda_j}^{\lambda_{j+1}} \sin\lambda\, d\lambda$$

where $\Delta\overline{C}_{nm}$ is the corrections of geopotential coefficients caused by regional residual field, $\delta\Delta g$ is residual gravity anomaly, L_{nmk} is defined by Gleason (1988), $\overline{S}_{nm}$ is related to the associated Legendre function of the second kind (Gleason, (1988)), q_n is smoothing factor, s is the greatest integer less than or equal to $\frac{1}{2}(n-|m|)$, and T is the total amount blocks in local area.

The new coefficients of potential model become

$$\overline{C}_{nm} = \overline{C}^s_{nm} + \Delta\overline{C}_{nm}W_n \qquad for \quad n,m = 2-360 \quad (2)$$

$$\overline{C}_{nm} = \overline{C}^s_{nm} + \Delta\overline{C}_{nm} \qquad for \quad n,m = 361-720 \quad (3)$$

where $\overline{C}^s_{nm}$ is geopotential coefficients of initial model, W_n is the weight functions.

The weight function W_n was selected as follows (Basic (1989)),

$$W_n = \begin{cases} 0 & n < N_{min} \\ (n-N_{min})/(N_{max}-N_{min}) & N_{min} \le n \le N_{max} \\ 1 & N_{max} < n \end{cases} \quad (4)$$

where N_{min} is the lowest degree that should be kept in initial model and N_{max} is the starting degree that should be fully corrected.

From above, we may obtain the regional geopotential model IGG97LB to degree 720. In IGG97LB model, the reference ellipsoid is the same as in EGM96. We have

$$GM = 3986004.415E + 8\, m^3/s^2$$
$$a = 6378136.3 \qquad m.$$

4 Comparisons of results from models

4.1 Comparisons with observed gravity anomalies

The observed gravity anomalies have been compared with that computed from different geopotential models in the research area. The results compared are summarised in Table 1. We can see an obvious improvement by a factor of 2.8 for the new model IGG97LB over model EGM96 on a 15'x15' grid. We can also see the accuracy by IGG97LB model to degree 720 is better than that to degree 360.

Table 1: Comparison of the observed gravity anomalies with computed that from models in the mainland of China, in mGal.

Model	OSU91A	EGM96	IGG97LB	
Max. Degree	360	360	360	720
Resolution	15	15'	15'	15'
Number	13018	13018	13018	13018
Mean diff.	4.2	4.0	4.1	4.1
Rms diff.	±25.6	±24.3	±13.8	±8.8
Min. Diff.	-114	-263	-115	-64
Max. Diff.	170	213	97	67

The statistical results reported that the absolute magnitude of errors, which less than 10.0 mGal, account for 78.2 percent of the total for the IGG97LB and 45.7 percent for the EGM96.

Table 2 shows the comparisons of the gravity anomalies and geoid undulations both of these computed from IGG97LB and EGM96 model in global area without China on a 60'x60' grid. The both of RMS difference of gravity anomalies and geoid undulations are very small. That means the differences of the model IGG97LB and EGM96 are so small as to be negligible in global area without China. The new model to degree 720 is the same as that to degree 360.

Table 2: Comparisons of the gravity anomalies and geoid undulations computed from IGG97LB and EGM96 in global area without China area, on a 60'×60' grid (number: 63840)

Model	EGM96-IGG97LB			
	Anomaly (mGal)		undulation (m)	
Max. Degree	360	720	360	720
Mean diff.	-0.0	-0.0	-0.00	-0.00
Rms diff.	±0.6	±0.6	±0.06	±0.06
Min. Diff.	-27.2	-23.8	-2.75	-2.35
Max. Diff.	52.7	56.5	2.43	2.81

We have noted there are some big differences between the IGG97LB and the EGM96, for example the differences of geoid undulation achieve

2.8m. We found these differences are located nearby the research area, and quickly reduced from this area outward. The absolute magnitude of difference, which greater than 1.0mGal, account for 1.7 percent of total for gravity anomaly and that, which greater than 0.1m, account for 3.3 percent for undulation.

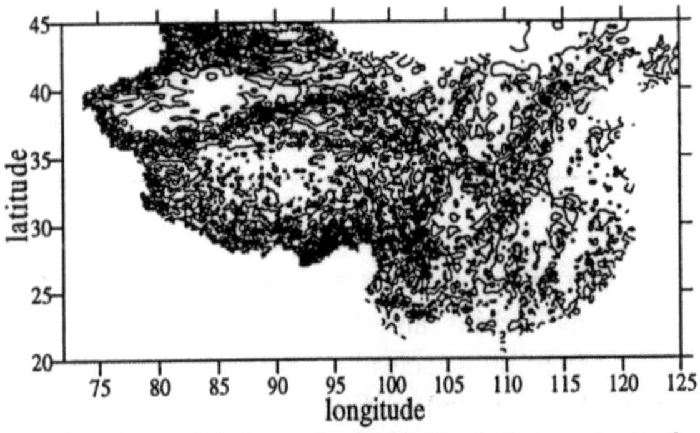

Fig. 1: The residual gravity anomalies between observed and EGM96 model in the research area of China (on a 15'×15' grid, c.i. =20 mGal)

What we can learn from above are that the new model based on the tailored method fit only to the local research area, and it would be a negative effect to nearby the local area. The reasons of this are the new model IGG97LB referenced to the EGM96 and the gravity anomalies outside China area were not used for the new model.

Fig.1 and Fig.2 illustrate the difference of gravity anomalies between the observed anomalies and that computed from EGM96 model to degree 360 and from IGG97LB model to degree 720 on a 15'x15' grid. In this case, the differences are the errors of EGM96 and IGG97LB, respectively.
Comparing Fig.1 and Fig.2, we can see that the errors in IGG97LB (i.e. Fig.2) much smaller and more smooth than that in EGM96 (i.e. Fig.1).

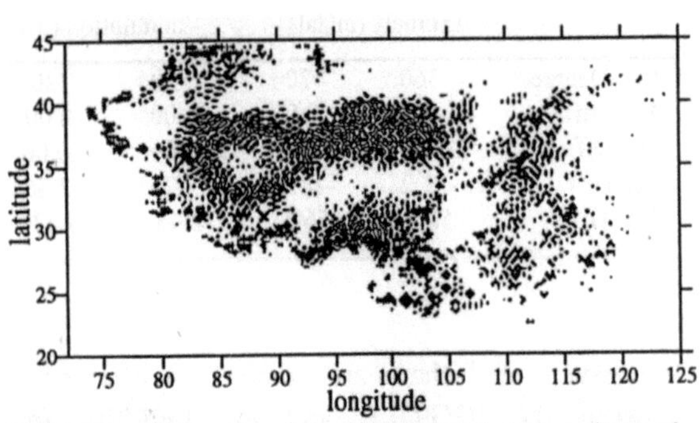

Fig. 2: The differences of gravity anomaly between observed and IGG97LB model (to degree 720) in the research area of China (on a 15'×15' grid, c.i.=20 mGal)

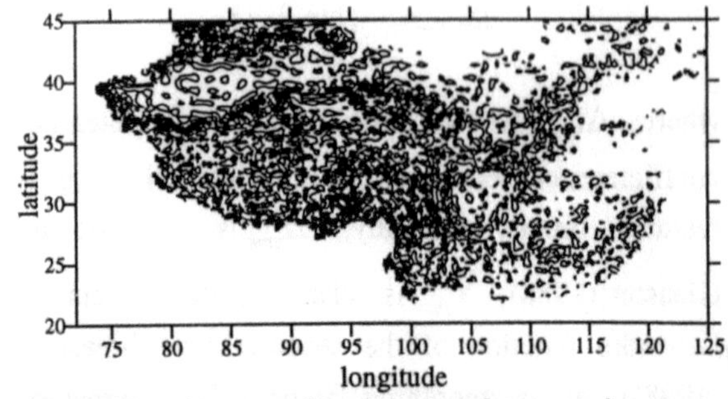

Fig. 3: The short wavelength parts of topography on a resolution from 30' to 15' in the research area of China (c.i.=300 m)

It indicates that the accuracy of gravity anomalies recovered by IGG97LB model is distinctly improved in the mainland of China. Fig.3 shows the short wavelength parts of the topography on a resolution about 28~55 km (i.e. 15'~30'). It is taken by subtracting the model heights, which computed from a regional model of the Earth's topography to degree 360 on a 15'x15' grid, from the observed heights. These parts represent the local variation of topography. In addition, we illustrate the short wavelength parts of gravity anomaly and geoid undulation in Fig.4 and Fig.5, respectively. Both of these are computed from IGG97LB model from degree 361 to 720 and they represent the local variations of gravity field. We can see that Fig.4 and Fig.5 are similar to Fig.3 in feature. To some extent, the variations of gravity anomalies and geoid undulations are corresponding to the variations of topography.

The parts showed in Fig.4 and Fig.5 both are the omission errors of the model to degree 360 on a resolution about 28~55 km. The standard deviations are ±10.0mGal for former and ±0.13 m for latter. The maximum errors are 77.0 mGal and 0.9 m, respectively.

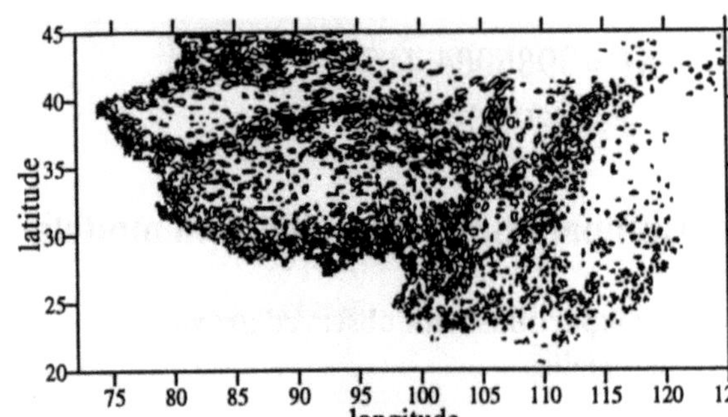

Fig. 4: The short wavelength parts of gravity anomalies computed from IGG97LB model from degree 361 to 720 in the research area of China (c.i.=15 mGal)

It could be safely said the short wavelength parts of gravity anomalies and geoid undulations, both computed from IGG97LB model from degree 361 to 720, reveal the information of gravity field on the resolution about 28~55 km. In other words, the IGG97LB model to degree 720 can be used to complement the parts of higher frequency of gravity field in contrast the model to degree 360.

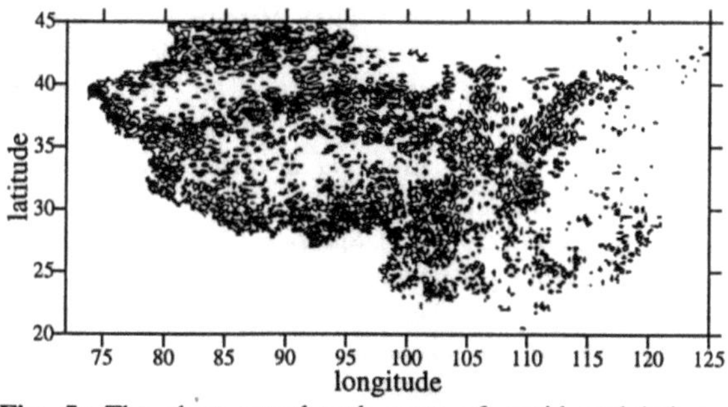

Fig. 5: The short wavelength parts of geoid undulation computed from IGG97LB model from degree 361 to 720 in the research area of China (c.i.=0.2m)

It is well known that there is a linear relation between gravity anomalies and geoid undulations. From Fig.4 and Fig.5, we have found a very clear linear relation on the short wavelength parts plotted in Fig.6. We have the linear relation formula as follows,

$$\Delta N = 1.28387 \times \delta\Delta g - 0.004 \quad cm \qquad (5)$$

where ΔN is residual geoid undulation and $\delta\Delta g$ is residual gravity anomaly in mGal.

The mean value of linear correlation coefficient is 0.98. That means we could easily simulate the ultra-short wavelength parts of geoid undulation from residual gravity anomalies with this linear relation. The mean fitting error is ±2.6 cm. The result simulated was given in the last column of Table 3.

4.2 Comparisons with height anomalies from GPS/Levelling

As well known, the height anomaly derived from GPS/Levelling is a geometric quantity. It is independent of the gravity anomalies.

We have 78 height anomalies from GPS network in the mainland of China. The models have been validated using GPS and levelling derived height anomalies. The comparisons described have clearly demonstrated the improvement of the new model over the global models.

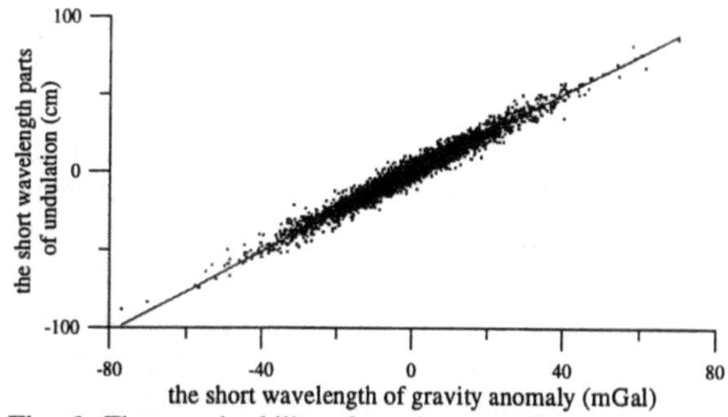

Fig. 6: The correlatability of gravity anomalies and geoid undulations on short wavelength in the research area of China

Table 3: Results from the comparison of models and height anomalies, in metres

Model	OSU91	EGM96	IGG97LB		simulate
Degree	360	360	360	720	
Sta.Dev.	1.50	1.01	0.71	0.67	0.61
Min.diff	-5.63	-2.26	-1.80	-1.66	-1.66
Max.diff	2.65	1.42	1.01	1.16	0.93

The results compared are given in Table 3. The height anomalies computed from new model IGG97LB can achieve the mean errors of ±0.67 m. The maximum error is 1.66 m. The IGG97LB to degree 720 solution is better than that to degree 360, but it is only few centimetres. It is in keeping with the spectral characteristic of geoid undulation.

The OSU91A1F solution is above two times poorer than the new model. The EGM96 solution is better than the OSU91 for the reason that the gravity data on the grid from China haven been used to compute the EGM96.

The last column in the Table 3 given the result compared of the height anomaly from GPS/Levelling with that from simulation, which calculating from the above linear formula and adding to the IGG97LB model to degree 720. We have an improvement of some centimetres.

5 Some Remarks

Using the tailored method, we have computed the higher-resolution regional geopotential model IGG97LB to degree and order 720. The new model can fit to the gravity field in the mainland of China well. Its resolution of the gravity field represented gives about 28 km. The new potential model to degree 720 can effectively complement high frequency information of regional gravity field as compared with model to degree 360, and the new model could simulate the ultra-short wavelength parts of geoid undulation from a linear relation.

The results from this work have shown that the higher-resolution geopotential model (for example to degree 720) can be developed in local area using regional high resolution and accuracy gravity data based on the 'tailored' method. At present, to attain this goal is more safely than to develop a global higher degree model for the resolution and accuracy of gravity field data.

Finally, one must note that this new model is based on the combination of a global model and local gravity data. It is improved only within the local area.

Acknowledgements. This work was supported in part by the National Science Foundation of China (Project No.49674209 and No.49634140). Many thanks to Prof. R.H.Rapp for the model OSU91A. Many thanks to Prof. F.G.Lemoine et al. For the model EGM96.

References

Basic, T., (1989). Untersuchungen zur regionalen Geoidbestimmung mit 'dm' Genauigkeit, Dr.-Ing. Dissertation, *Wissenschaftliche Arbeiten der Vermessungswesen der Universitaete Hannover*, Nr.157, Hannover.

Gleason, D.M., (1988). Comparing ellipsoidal corrections to the transformation between the geopotential's spherical and ellipsoidal spectrums. *Manusc.Geod*, 3, pp.14-129.

Hsu, H.T. and Y.Lu, (1995). The regional geopotential model in China, *Bolletino di Geodesia e scienze Affini*, N.2, pp.61-175.

Lemoine, F.G., D.E.Smith, L.Kunz, R.Smith, E.C.Pavlis, N.K.Pavlis, S.M.Klosko, D.S.Chinn, M.H.Torrence, R.G.Williamson, C.M.Cox, K.E.Rachlin, Y.M.Wang, S.C.Kenyon, R.Salman, R.Trimmer, R.H.Rpp and R.S.Neren, (1996). The development of the NASA GSFC and NIMA joint geopotential model, *International Symposium Gravity, Geoid and Marine Geodesy, International Association of Geodesy Symposia*, Tokyo, Japan, September 30-October 5, vol.117, pp.461-469.

Lu, Yang and F.Z.Jiang, (1997). High-resolution gravity model and interior tectonic and dynamic feature in northern China, Proceedings, *IUGG IAG, International Symposium on Current Crustal Movement and Hazard Reduction in east Asia and Southeast Asia*, Wuhan, November 18-25, pp.396-404.

Rapp, R. and J.Cruz, (1986). Spherical harmonic expansions of the Earth's gravitational potential to degree 360 using 30' mean anomalies. *The Ohio State University, Department of Geodetic Science and Surveying*, Columbus/Ohio, Report no.376.

Rapp, R. and N.K.Pavlis, (1990). The development and analysis of geopotential coefficient models to spherical harmonic degree 360. *Journal of Geophysical Research*, 95(b13), pp.21885-21911.

Rapp, R. Y.Wang and N.K.Pavlis, (1991). The Ohio State 1991 geopotential and sea surface topography harmonic coefficient models. *The Ohio State University, Department of Geodetic Science and Surveying*, Columbus/Ohio, Report no.410.

Weber, G. And H.Zomorrodian, (1988). Regional geopotential model improvement for the Iranian geoid determination, *Bulletin Geodesique*, Vol.62, pp.125-141.

Wenzel, G., (1998). Ultra high degree geopotential model GPM3E97A to degree and order 1800 tailored to Europe, *Proceedings of the 2nd Continental Workshop on the Geoid in Europe*, Budapest/Hungary, March 10-14, pp.77-80.

Gravity field and geoid for Japan

Y. Kuroishi
Space Geodesy Laboratory,
Geographical Survey Institute, 1 Kitasato, Tsukuba, Ibaraki, 305-0811, Japan

Abstract. The latest gravimetric geoid of Japan, JGEOID98 was evaluated, focussed on the long-wavelength distortion. The network adjustment applied to the ship gravity data was considered to yield the distortion in the gravity field.

A recent global marine gravity model from altimetry data, KMS98 was assessed over Japan, demonstrating significant improvement in long wavelength. A preliminary test to use KMS98 for correcting JGEOID98 was made with a simple filtering approach and the results show substantial improvement in fixing the long-wavelength errors. The distortion in JGEOID98 was estimated to be on the order of ! 10 mgal.

KMS98 worsened the geoid in short wavelength, having some problematic areas near the coasts. We need to develop an optimal way to incorporate altimeter data in the geoid determination over Japan.

Keywords. Gravimetric geoid, gravity field.

1 Introduction

The Japanese Islands are located in a trench and island-arc region and the topography and accordingly the gravity field shows complicated undulations. To determine the geoid in this area accurately, we must recover the gravity field precisely even in the surrounding seas.

The Kuroshio Current, one of the strongest oceanic currents in the world flows along the south coasts of Japan. We should be careful when handling altimeter data for geoid determination.

Kuroishi (1995) determined a gravimetric geoid of Japan, JGEOID93 from land and ship gravity data by the 2D-FFT method for Stokes integration (Strang van Hees, 1990). He applied a simultaneous net adjustment of crossover errors of the ship data and obtained the model at a precision better than 10 cm in short wavelength. Trend errors of 2-4 ppm in the east-west direction were found in comparison with GPS/leveling geoid undulations, suggesting long-wavelength errors.

The author has been working on the improvement over JGEOID93 (e.g. Kuroishi, 1999a, 1999b). Newly obtained ship gravity data of wider coverage were used with the net adjustment and a bias fitting to a global reference geopotential model, EGM96 (Lemoine et al., 1997). In order to perform the strict computation of the generalized Stokes equation, the 1D-FFT method (Haagmans et al., 1993) was introduced and the latest model, JGEOID98 was obtained (Kuroishi, 1999b).

Comparing with the nation-wide GSP/leveling network, Kuroishi (1999b) showed the improvement by JGEOID98 over JGEOID93 by 28 % in short wavelength and reduction of tilt from 2.2 to 1.4 ppm in long wavelength. It was pointed out that the net adjustment of ship data yielded the distortion of the gravity field in long wavelength.

Kuroishi (1998) compared the JGEOID98 gravity field with the altimeter-derived global marine gravity model by Sandwell and Smith (1997). The altimeter model was found to contain several tens of mgal errors along the Japanese coasts and could not be easily used for correcting the JGEOID98 gravity field.

In this paper, we review the accuracy of JGEOID98, focussing on the distortion in long wavelength. And the evaluation of the performance of a recent global marine gravity grid, KMS98 (Andersen and Knudsen, 1998b) is made in the geoid determination over Japan. Then a preliminary test is carried out for correcting the long-wavelength errors in JGEOID98 with KMS98.

2 Latest gravimetric geoid of Japan, JGEOID98, and its gravity field model

2.1 Ship data adjustment

The latest gravimetric geoid model of Japan, JGEOID98 was determined by the 1D-FFT method

International Association of Geodesy Symposia, Vol. 121
Schwarz (ed.), Geodesy Beyond 2000 – The Challenges of the First Decade

in remove and restore manner from land and ship gravity data collected by the Bureau Gravimetrique International. Detailed description on the preprocessing of the ship data is given in Kuroishi (1999b). In this section the main points regarding the fixing of the gravity field in long wavelength are summarized.

After the gravity system was unified and blunder errors were removed, a net adjustment of crossover errors (COEs) was applied to all the data cruises with a linear drift model in time. Then all the data were fitted to the EGM96 gravity field to take the bias out. Fig. 1 shows the distribution of all the gravity data.

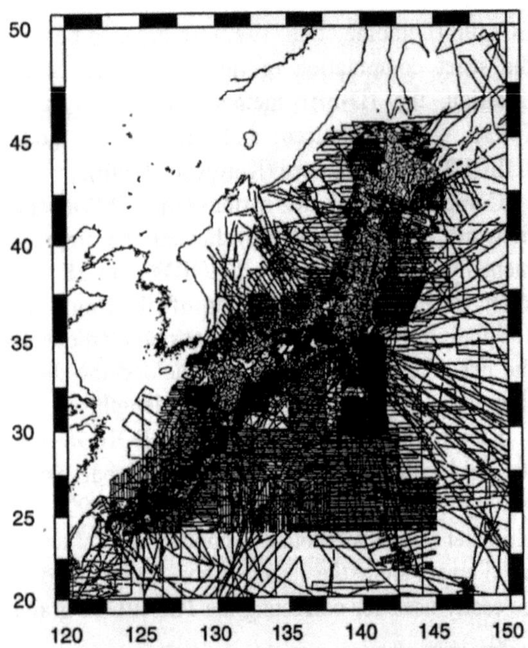

Fig. 1: Surface gravity data coverage for JGEOID98 (Kuroishi, 1999b)

No reference cruises were explicitly selected in the adjustment, but eventually three cruises were playing an anchoring role because of the smallness of internal and external COE point numbers. The locations of the data in the three cruises are presented in Fig. 2. Each cruise contains only two external COE points. The mean of the external COEs changed by + 5.1, + 7.1 and + 2.8 mgal for the cruises of USSR, Japan Hydrographic Department and Hawaii Institute of Geophysics, respectively. The gravity field model of JGEOID98 is likely to be distorted in long wavelength by such amount.

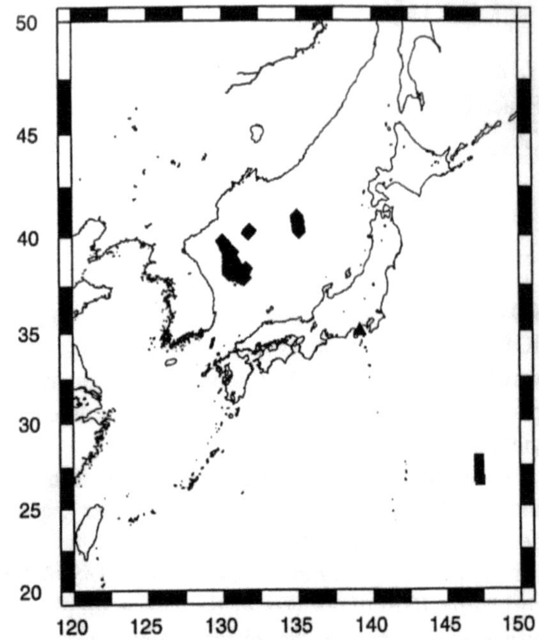

Fig. 2: Locations of three eventual reference cruises. Diamonds, triangles and squares correspond to cruises conducted by USSR in 1969, Japan Hydrographic Department in 1989 and Hawaii Institute of Geophysics in 1970, respectively.

2.2 Geoid determination method

The geoid was determined from refined Faye anomalies on a 3′% 3′ grid by the 1D-FFT method for the generalized Stokes integral with 100 % zero-padding in the longitudinal direction. EGM96 was employed as the global reference model and the scaled geopotential coefficients (Smith and Milbert, 1997) were used in the computation. The normal gravity and the atmospheric correction were evaluated up to the second order of heights and the indirect effect was evaluated by Grushinsky's formula (Wichiencharoen, 1982) with the normal density of 2.67 %10^3kg/m^3.

JGEOID98 is given in the non-tidal system. The geocentric gravity-mass constant of EGM96 and the geopotential on the geoid of GRS80 were used to scale the absolute values. For detailed description, see Kuroishi (1999b).

2.3 Evaluation of geoid by comparing with nation-wide GPS/leveling network

Kuroishi (1999b) assessed the accuracy of the geoid models, JGEOID93 and JGEOID98 externally with five local and one nation-wide GPS/leveling networks. JGEOID98 shows significant improvement over JGEOID93 in both long and short wavelength.

Relative accuracy plots for the two models before and after trend fit are given in Fig. 3. It is clearly demonstrated that the post-fit accuracy of JGEOID98 is better than that of JGEOID93 over almost all ranges of baseline distances. This suggests that the improved information on the gravity field at sea enhances the geoid precision in short wavelength on land.

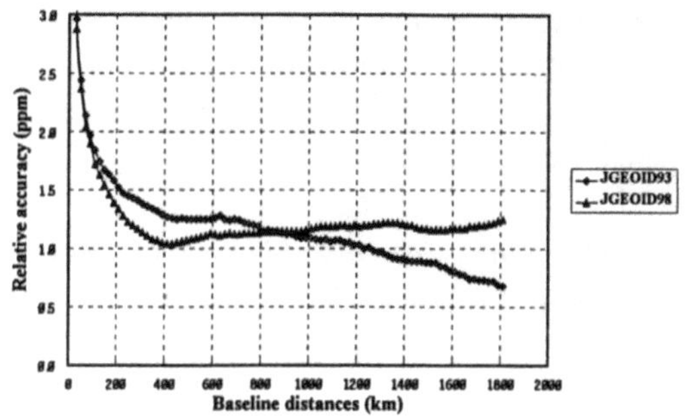

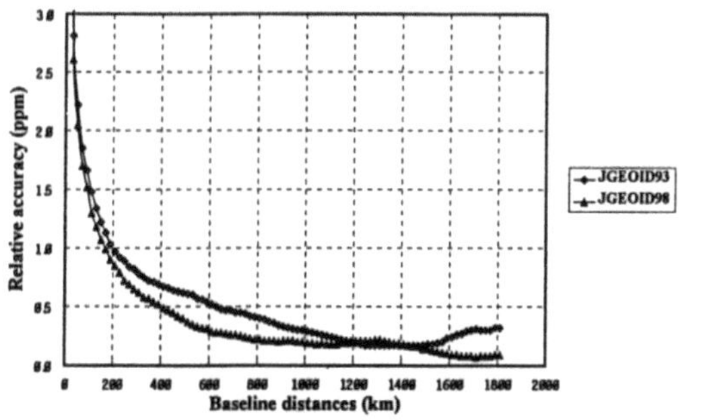

Fig. 3: Relative accuracy of JGEOID93 and JGEOID98 w.r.t. nation-wide GPS/leveling network: (upper) raw and (bottom) de-trended results.

Although the tilt of JGEOID98 with respect to the nationwide network is substantially smaller than that of JGEOID93, the pre-fit accuracy of JGEOID98 increases as baseline distances become larger and exceeds that of JGEOID93 over distances of 900 km or longer. This is because the tilt of JGEOID98 is in the north-south direction, which is in accordance with the extension of the Japanese islands. The direction does not agree with that of EGM96. The tilt should, therefore, be attributed to the long-wavelength distortion of the gravity field model for JGEOID98 due to the net adjustment processing.

3 Comparison with altimeter-derived global marine gravity grid, KMS98

3.1 Performance of KMS98 in the geoid determination over Japan

Andersen et al. (1998) released the latest global marine gravity grid, KMS98 from ERS-1 and Geosat altimetry. Their method to recover the gravity field yielded significant improvement over the gravity grid by Sandwell and Smith (1997) in terms of bias and deviations (Andersen and Knudsen, 1998).

KMS98 was combined with the same land gravity data as used in JGEOID98. The geoid was determined from them by the same method as applied to JGEOID98 and is called here as JKMS98. The similar process was also made for the Sandwell and Smith (1997) grid and the corresponding geoid is designated here as S&S.

Comparisons with the GPS/leveling networks were made for evaluating the performance of the geoid models. Table 1 lists the statistics of the differences between the gravimetric geoids and GPS/leveling geoid undulations.

Table 1: Statistics of differences between gravimetric geoids and GPS/leveling geoid undulations in cm. Top Shows means and bottom the standard deviations about the mean.

Network	EGM96	JGEOID98	S&S	JKMS98
National	8.89	84.84	149.06	52.12
SW Japan	- 132.29	- 89.13	37.16	- 66.79
Kinki	- 138.97	- 61.35	48.95	- 42.80
Shizuoka	- 97.88	- 31.60	53.80	5.20

Network	EGM96	JGEOID98	S&S	JKMS98
National	38.48	66.65	79.98	20.09
SW Japan	26.85	35.12	41.15	23.82
Kinki	24.83	25.85	45.57	9.79
Shizuoka	42.02	8.18	18.62	20.08

JKMS98 shows the best performance over the other models except Shizuoka network, which includes the highest peak of Japan, Mt. Fuji and is located north to Izu Peninsula bounded by Suruga and Sagami Troughs. De-trending of the differences reveals that JKMS98 has the smallest tilt of 0.5 ppm with a standard deviation of 19 cm for the nation-wide net. It is, therefore, expected to reduce the long-wavelength distortion in JGEOID98 effectively by using KMS98.

3.2 Comparison of gravity field

The JGEOID98 gravity field is compared with KMS98 (Fig. 4). Near the coasts and at shallow waters, such as the Inland Sea (Seto-naikai), Toyama (middle of the coasts facing the Sea of Japan), Suruga and Sagami Bays, are found wide zones of big differences. Most of those, especially around Izu Peninsula, should be attributed to the errors in KMS98.

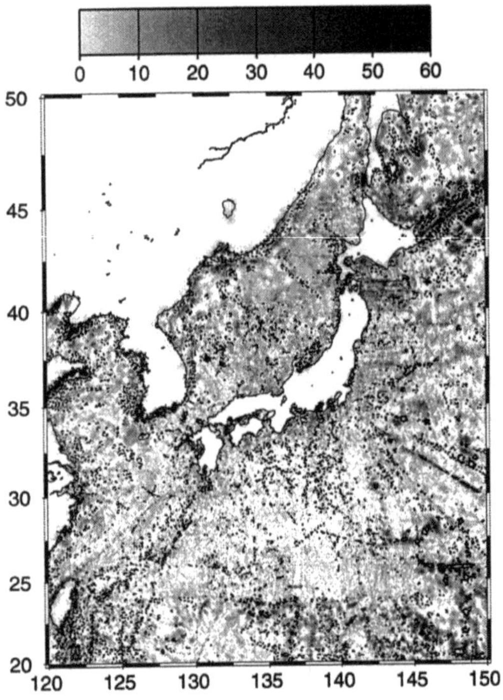

Fig. 4: Gravity differences between JGEOID98 and KMS98. Positive values are shown by gray shading and negative with Contours in 10 mgal Interval. Numerals in mgal.

There exist some areas of a few tens of mgal differences in very short wavelength. Where the ship data coverage is sparse, KMS98 recovers fine undulations of the gravity field associated with the geological structures such as sea mounts. For other areas in good ship data coverage, JGEOID98 resolves the finer undulations than KMS98.

In long wavelength, a few tens of mgal differences exist, for example, south off Japan. These differences would cause the errors of JGEOID98 in long wavelength. We will consider using the differences for correcting the possible distortion in JGEOID98 in the next session.

4 Preliminary correction with KMS98 to JGEOID98

4.1 Filtering of gravity discrepancies

In this chapter, we will make a preliminary test to reduce the errors of JGEOID98 in long wavelength with KMS98. Under the consideration that the adjusted ship gravity data resolved short-wavelength undulations of the gravity field better than the altimetry-derived model, high-cut filtering of the gravity differences is considered for that purpose.

As one of the simplest methods, we use a cosine-tapered weighting function in frequency domain. Fig. 5 shows the basic shape of the filter function. We must define two filter parameters, *f*stop and *f*start, whose reciprocals are called here the terminating and the full-weighting wavelengths, respectively.

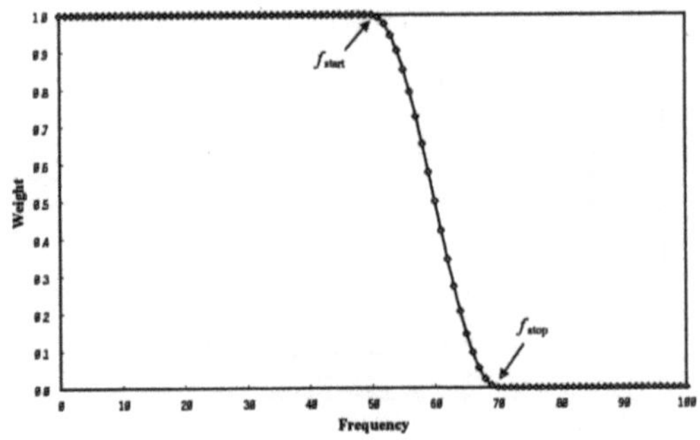

Fig. 5: High-cut filter in frequency domain

Two functions are considered as test cases. In one filter *A*, 1 and 2 arc-degrees are assigned to the terminating and full-weighting wavelengths: in the other one *B*, those are 2 and 5 arc-degrees.

The resulting gravity difference models are shown in Fig. 6. The differences range from – 38 to + 30 mgal for model *A* and from – 13 to + 10 mgal for model *B*. These models with the opposite sign are used as correction surface to the JGEOID98 gravity field.

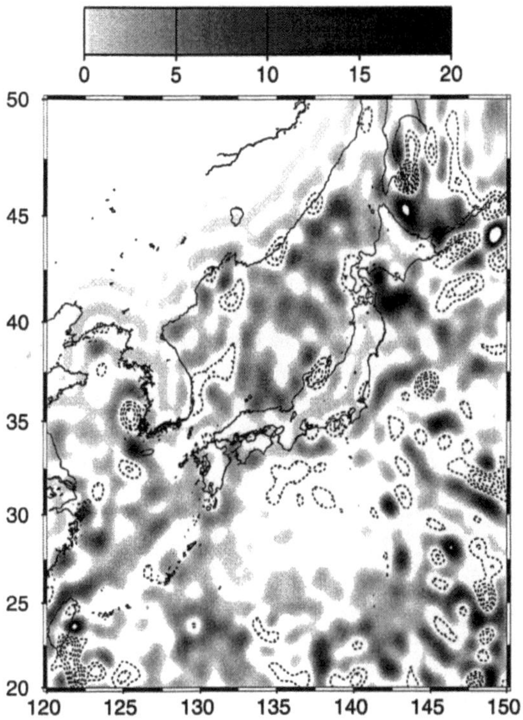

Fig. 6: Filtered gravity differences (upper) for model *A* and (bottom) for *B*. Positive values are shown by gray shading and negative with contours in 10 mgal interval. Numerals in mgal.

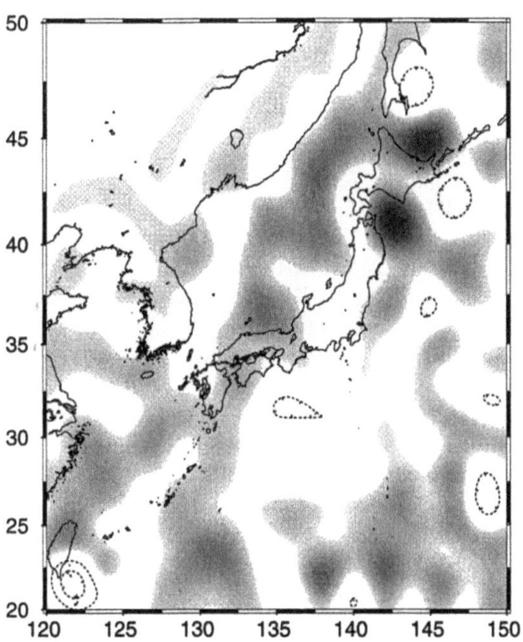

Fig. 7 Tilts of gravimetric geoids w.r.t. GPS/leveling geoid undulations.

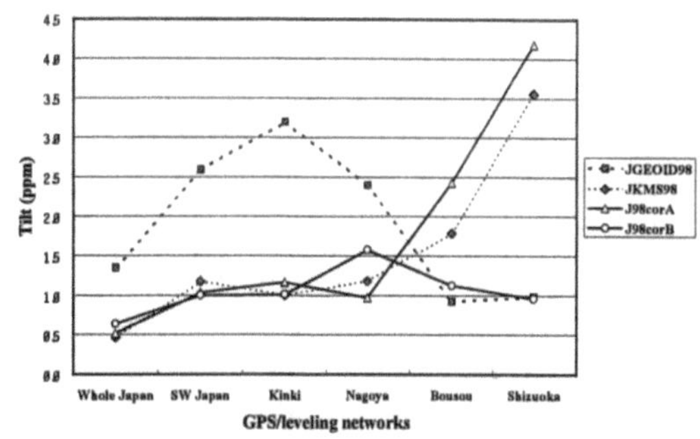

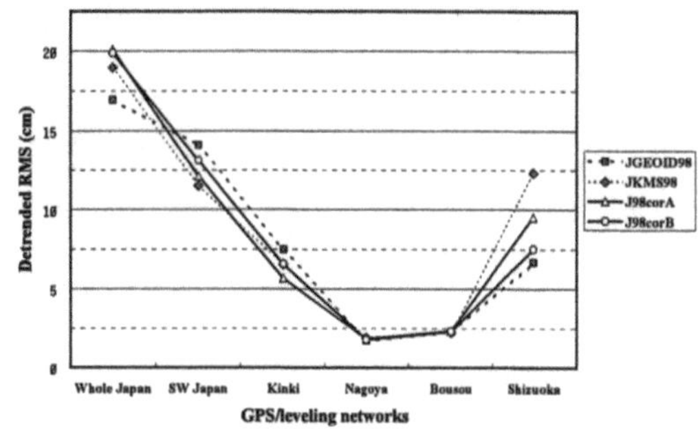

Fig. 8: De-trended RMS differences between gravimetric geoids and GPS/leveling geoid undulations

4.2 Evaluation with the nation-wide GPS/leveling network

The effectiveness of the two models *A* and *B* as correctors is evaluated by comparing the resulting geoid models (named as J98corA and J98corB, respectively) with GPS/leveling geoid undulations. Tilts and de-trended RMS differences are computed for each GPS/leveling network and shown in Fig. 7 and Fig. 8, respectively.

Noticeable differences in the tilts are found in Boso and Shizuoka Networks, where J98corA gives larger values than J98corB. In the two networks, JKMS98 results in much bigger tilts than JGEOID98. As we discussed in the chapter 3, KMS98 should have some problematic areas near Izu Peninsula and worsen the geoid determination around there.

The problematic zones are relatively narrow as seen in Fig. 6. The model B, therefore, does not have substantial effects on the geoid determination, in contrast to the model A.

For the remaining networks, there is no major difference between J98corA and J98corB, and either model shows major improvement over JGEOID98. We, therefore, conclude that the model

B is superior to the model A as the corrector and the magnitude of the distortion would be on the order of ! 10 mgal. This is compatible with the changes in the mean of the external COEs of the eventual reference cruises.

Even with this simple approach, KMS98 successfully constrained the gravity field over Japan. The effectiveness of KMS98 is confirmed in the correction of JGEOID98.

For the nation-wide network, the corrections worsen the short-wavelength accuracy not only over JGEOID98 but also JKMS98. In addition to the results for Boso and Shizuoka Networks, this suggests that the accuracy of the adjusted ship gravity data is significantly higher than that of the altimeter-derived gravity near the coasts, especially at highly undulated regions. We need to develop an optimal way to incorporate altimeter data in the geoid determination over Japan.

5 Conclusions

The latest gravimetric geoid of Japan, JGEOID98 and its gravity field were evaluated, focussed on the long wavelength distortion. Comparisons with JGEOID93 over the GPS/leveling networks indicate that JGEOID98 is likely to be distorted in the network adjustment of COEs. Looked into the changes of the means of the external COEs of the eventual reference cruises, the distortion in the gravity field would be on an order of several mgal or bigger.

The performance of a recent global marine gravity model from altimetry data, KMS98 was assessed over the Japanese Islands. The model shows significant improvement over the existing ones especially in long wavelength. The model is expected to play a major role in constraining the geoid determination in long wavelength.

A preliminary test to use KMS98 for correcting JGEOID98 was made with a simple filtering approach. A high-cut filter of the cosine-tapered shape at a wavelength of several arc-degrees demonstrates substantial improvement in fixing the long wavelength errors. The distortion of the gravity field was estimated to be on the order of ! 10 mgal. The effectiveness of KMS98 is confirmed in the improvement of geoid determination over Japan.

KMS98 worsened the geoid in short wavelength and is considered to have some problematic areas near the coasts, especially at highly undulated regions. Ship gravimetry should be still superior in short-wavelength resolution to altimetry. We need to develop an optimal way to incorporate altimeter data in the geoid determination over Japan.

References

Andersen, OB, P Knudsen (1998) Global Marine Gravity Field from the ERS-1 and Geosat Geodetic Mission Altimetry. J Geophys Res 103(C4): 8129-8137

Andersen, OB, P Knudsen, S Kenyon, R Trimmer (1998) The KMS98 Global Marine Gravity Field. Abstracts 2nd Joint Meeting of the International Gravity Commission and the International Geoid Commission, Trieste, Italy, September 7-12, 1998: 46

Haagmans, R, E de Min, M van Gelderen (1993) Fast Evaluation of Convolution Integrals on the Sphere Using 1D-FFT and a Comparison with Existing Methods for Stokes' Integral. Maniscr Geod, 18(5): 227-241

Kuroishi, Y (1995) Precise Gravimetric Determination of Geoid in the Vicinity of Japan. Bull Geographical Survey Inst 41:1-93

Kuroishi, Y (1998) Determination of Gravimetric Geoid of Japan and Its Improvement. Proc Symp Ocean-Earth Dynamics and Satellite Altimetry, Ocean Res Inst, Univ Tokyo, November 11-12, 1997: 119-128

Kuroishi, Y (1999a) Improvement of Japanese Geoid with 1D-FFT Method and Its Comparison with Altimetry-derived Geoid. In: Proc. of 2nd Joint Meeting of the International Gravity Commission and the International Geoid Commission, Trieste, Italy, September 7-12, 1998 in press.

Kuroishi, Y (1999b) Improved Gravimetric Geoid for Japan, JGEOID98. Paper submitted to J Geodesy.

Lemoine, FG, DE Smith, L Kunz, R Smith, EC Pavlis, NK Pavlis, SM Klosko, DS Chinn, MH Torrence, RG Williamson, CM Cox, KE Rachlin, YM Wang, SC Kenyon, R Salman, R Trimmer, RH Rapp, RS Nerem (1997) In Segawa J et al.(eds.): Proc. of the IAG Symp. No 117 Geoid and Marine Geodesy. Tokyo, Japan, September 30- Oct. 5, 1996, Springer Verlag, 461-469

Sandwell, DT, WHF Smith (1997) Marine Gravity Anomaly from Geosat and ERS 1 Satellite Altimetry. J Geophys Res 102 (B5): 10039-10054

Smith, DA, DG Milbert (1997) Evaluation of Preliminary Models of the Geopotential in the United States. International Geoid Service Bull 6, DIIAR- Politecnico di Milano, Milano, Italy: 7-32

Strang van Hees, G (1990) Stokes Formula Using Fast Fourier Techniques. Manuscr Geod 15: 235-239

Wichiencharoen C (1982) The Indirect Effects on the Computation of Geoid Undulations. Rep 336, Dep Geod Sci Surv, The Ohio State University, Columbus

The dual sphere superconducting gravimeter GWR CD029 at Frankfurt a.M. and Wettzell – first results and calibration

M. Harnisch, G. Harnisch, I. Nowak, B. Richter, P. Wolf
Bundesamt für Kartographie und Geodäsie, Richard-Strauss-Allee 11, D-60598 Frankfurt am Main, Germany

Abstract. The first dual sphere superconducting gravimeter CD029 is operating since November 1998, at first at the test site Frankfurt a.M., further permanently at Wettzell (Bavarian Forest, Germany). The GWR dual sphere superconducting gravimeter, developed by GWR Instruments, Inc., San Diego, California, USA, enables enhanced consistencies in data and drift monitoring to achieve a high resolution and reliability for studying temporal gravity variations. The dual sphere system is equivalent to two single instruments, but the technical expenditure and the costs are considerably reduced. By differencing the signals of both systems even micro steps can be detected. The first experiences with the CD029 show, that it meets the expectations. Under regular conditions noise and drift rate are very small. No steps were observed at Wettzell down to a level of some nm s^{-2}, besides those occurred during maintenance activities. The scale factors of both systems of the C029 have been estimated by the acceleration method and by comparisons with absolute gravity measurements.

Keywords. Superconducting gravimeter, calibration, Earth tides

1 Introduction

The high quality of gravity registrations with superconducting Gravimeters (SG) are demonstrated by time series derived by various authors the last 15 years. However, even with the high quality of the time series there are still gaps, offsets and noise in the registrations due to strong environmental effects like earthquakes, man-made disturbances from instrument maintenance and liquid helium transfer, and limitations in the long term stability of the electronic components.

During the 13th International Symposium on Earth Tides, Brussels 1997, B. RICHTER and R. WARBURTON announced the dual sphere superconducting gravimeter (DSG) as a new type of superconducting gravimeters [2]. Between July 28 and October 19, 1998 the CD029 was installed for a test run at Frankfurt a.M. Since November 4, 1998 it replaces the older SG103 at Wettzell (Bavarian Forest, Germany). Based on data from the long-term measurements at Wettzell and test measurements at Frankfurt a.M. some first results and experiences with this new type of gravimeters are presented.

2 Some technical details of the CD029

The DSG was developed with the aim of enhanced reliability of the recorded data and especially of a better drift monitoring.

The DSG is equivalent to two single systems combined in one instrument where a second sensor is placed approximately 20 cm above the first. Each gravity sensor contains a hollow niobium sphere of 1 inch in diameter whose vertical displacement is measured by a capacitance bridge.

The magnetic coils, feedback coils, and thermal switches are wound and mounted on a single cylinder of copper. The Cu block is temperature regulated using a Ge thermometer thermally anchored at the midpoint of the block and a resistive heater placed nearby. In this configuration, the conductive heat flow between the sensor and the helium gas goes through the temperature-controlled midpoint of the sensor. This minimises thermal gradients in the sensor.

The masses of the test spheres, magnetic shielding, coil winding and machining differences all produce magnetic asymmetries that differ between the two sensors in the DSG. These affect both the coefficient and null point of the individual sensors, so that the individual sensors have different tilt minima. To solve the alignment and mode problems, two sets of concentric side coils were added to the instrument.

The DSG operates with two sets of gravity electronics, the one dedicated to Unit 1 (the lower system) and the other to Unit 2 (the upper system). So each of the two sensors can be operated as a single sphere SG. The outputs of these electronics are filtered and then sampled by a data acquisition

International Association of Geodesy Symposia, Vol. 121
Schwarz (ed.), Geodesy Beyond 2000 – The Challenges of the First Decade

system in the standard manner used with the single sphere SG.

Another essential feature of the DSG is the ultra long time Dewar refrigeration system. The neck is modified to interface with a two stage 4 °K cryocooler capable of obtaining temperatures below the vaporisation point of liquid helium. Normally the lower stage is operated below the vaporisation temperature of helium. so that the boiled off helium gas is re-condensed and drips back into the storage belly. Therefore, during normal operation the system consumes no liquid helium and will operate indefinitely.

Some problems arise from the accidental formation of ice in the neck of the Dewar. Due to the mechanical contact between cold head and Dewar disturbances in the gravity residuals become evident. The ice must be blown out by warm helium gas. During this rigorous procedure steps may occur.

Some further technical details of the DSG are given in [2].

3 The difference signal

A characteristic feature of the dual sphere gravimeter is the direct comparison of the gravity variations measured by both systems. Precise identification of offsets (both time location and magnitude) is made possible by the fact that most geophysical signals have common mode and are removed in the difference signal. But the difference signal is sensitive to all influences acting on one of the sensors mostly introduced by instrumental anomalies.

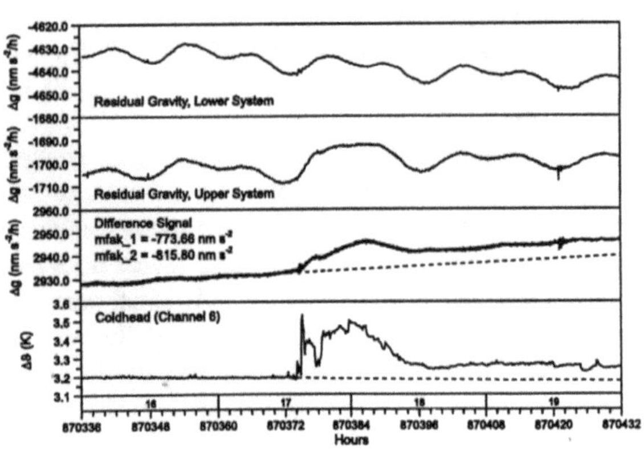

Fig. 1: Malfunction of the cold head 16.-19.4.1999. The disturbances of the residual gravity are masked by residual tidal influences due to an unsuited (older) set of tidal parameters. They are clearly detected by the difference signal. Also a systematic shift of the difference signal is to be seen.

As an example Figure 1 shows the influence of a malfunction of the cold head on the recorded gravity variations. The effect shown in the gravity residuals for the single systems is nearly totally masked by insufficient eliminated tides. However, the disturbance becomes evident in the difference signal and it is easy to find out a correlation to the cold head temperature (cold head malfunction). Moreover the difference signal after the disturbance has a systematic offset of about 6 nm s^{-2}. That means, that at least one of the systems is disturbed or both are influenced by offsets with a difference of this amount.

The most cited application of the difference signal is the detection of steps. This aspect is discussed in more detail in paragraph 5.

4 Noise and drift rate

The noise of the CD029 is very low. This may be seen already from comparisons of the residual gravity curves of different gravimeters (see e.g. Figure 4). A more objective way is the estimation of the standard deviation m_0 of the residual gravity over short periods. Even during the test measurements at Frankfurt a.M., which were carried out under very unfavourable environmental conditions, m_0-values of 0.44 (lower system) and 0.59 nm s^{-2} (upper system) were found, later at Wettzell they diminished to 0.21 (lower system) and 0.19 nm s^{-2} (upper system). The corresponding values of the SG103 at Bad Homburg (25.3.99) and of the C023 at Medicina (17.-18.1.99) were 0.97 and 0.38 nm s^{-2}.

The drift rates during the different operation periods of the CD029 are summarised in Figure 2.

At Frankfurt a.M. the drift rates of both systems were very small. Only weak exponential constituents occurred. Decay times of 41.7 and 21.1 days for the lower and the upper system respectively were found. After subtracting these exponential constituents linear drift rates of (349.7 ± 1.7) nm s^{-2}/year and (413.9 ± 1.5) nm s^{-2}/year remained.

A quite different drift behaviour came out after the CD029 was transported to Wettzell with the sensors kept at liquid helium temperature. Nevertheless both systems started with very large drift rates of about 64.6 nm s^{-2}/day (lower system) and 23.8 nm s^{-2}/day (upper system). After about 2 weeks the upper system reached steady values close to a zero drift while the very strong drift of the lower system continued. As in other cases the drift is composed of a linear and an exponential component. The decay constant of 25.2 days corresponds to the decay time during the test measurements. However, due to a very large initial disturbance of about 1000 nm s^{-2} also after more than 3 months a significant exponential constituent could be observed. After correcting the exponential constituent of the lower system linear drift rates of (-1739.2 ± 2.6) nm s^{-2}/year and (+93.8 ± 4.4) nm s^{-2}/year were found.

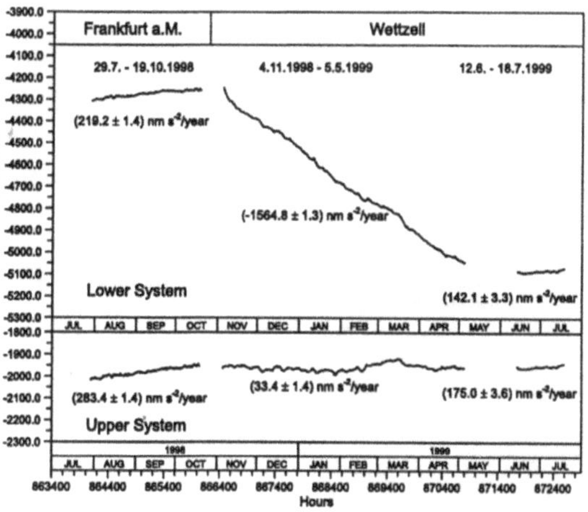

Fig. 2: Drift rates of both systems of the CD029. First section: test measurements at Frankfurt a.M., in both systems exponential constituent eliminated. Second section: Wettzell, exponential constituent eliminated (only in the lower system). Third section: after re-initialisation, no exponential constituent. – In all graphs the influence of polar motions is subtracted.

Due to an unexpected total loss of helium the registrations at Wettzell ended abruptly in the beginning of May 1999. After a break of about six weeks the CD029 was set in operation again. After cooling down and reinitialization it started with encouraging low drift rates of -308.8 nm s^{-2}/year of the lower and -497.5 nm s^{-2}/year of the upper system. No exponential constituents appeared. This drift behaviour corresponds to that during the first testing period at Frankfurt a.M. Presently there is no explanation for the abnormal drift behaviour during the first set up period at Wettzell.

5 Steps

During the test measurements with the CD029 at Frankfurt a.M. altogether five steps could be detected in the difference signal. An example is given in Figure 3. Clearly it can be seen, that in this way also under unfavourable environmental conditions steps smaller than a few nm s^{-2} may be detected with high significance. During the measurements at Wettzell no steps were observed exceeding this very low threshold. In conclusion it seems that the gravity signal of modern SG is not seriously affected by steps (apart from offsets due to failures of the diverse peripheral instruments or extensive maintenance activities). Especially the danger is minimised, that step accumulations affect the long-term behaviour of the gravimeter [4].

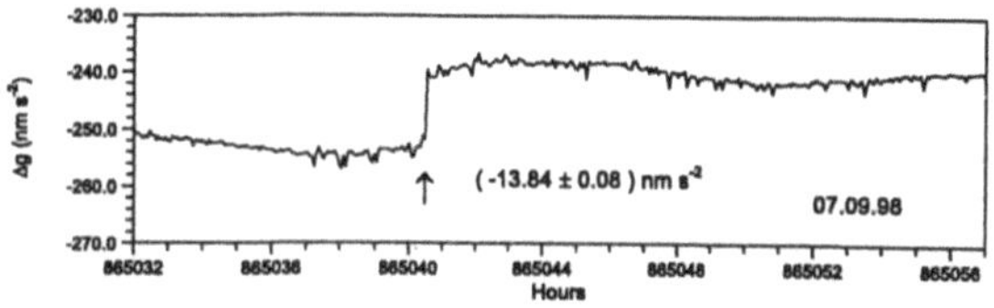

Fig. 3: Example of a step in the difference signal of the CD029 during the test measurements at Frankfurt a.M. Definitively only this step (the largest one) had to be corrected.

6 Tidal analysis (ETERNA)

The data recorded at Wettzell from the beginning of the series in November 1998 till May 1999 were analysed both with and without filtering. The short data series of 161 days allows only subdivisions into 13 and 16 wave groups. Roughly speaking the results are equivalent to those of the SG103 at the same station, based on a series of more than two years [6].

Table 1: Tidal parameters of the greatest tides from the gravimeters CD029 (161 days) and SG103 (864 days). Analyses without filtering.

Instr.	O1	K1	M2	S2	m0
δ-factors:					
CD029	1.14883	1.13563	1.18477	1.18345	13.852
Lower S.	±.00058	±.00035	±.00018	±.00039	
CD029	1.14897	1.13587	1.18498	1.18394	16.295
Upper S.	±.00054	±.00033	±.00019	±.00039	
SG103	1.14388	1.13077	1.18050	1.17913	25.950
	±.00060	±.00036	±.00028	±.00061	
phases:					
CD029	0.0905	0.2093	1.3701	0.2751	
Lower S.	±.0288	±.0178	±.0089	±.0190	
CD029	0.1008	0.2313	1.4394	0.3332	
Upper S.	±.0271	±.0167	±.0090	±.0191	
SG103	0.1317	0.2237	1.4850	0.4332	
	±.0299	±.0183	±.0134	±.0297	

All error bars of the δ-factors are in the same order, though the standard deviation m_0 for the SG103 is more then two times larger than those for both systems of the CD029. The results of both systems of the CD029 agree very well due to the consistent scale factors derived from the internal comparison by means of the difference signal. But if the δ-factors of the CD029 and the SG103 are compared, significant systematic differences become evident, characterised by a constant ratio. These differences are due to uncertainties of the scale factors used for the analyses, especially in the case of the SG103.

Contrary to the δ-factors of the CD029 the phases κ differ significantly. The differences are in the order of 5 s. This result is confirmed by the investigations of the difference signal. Meanwhile it becomes evident, that this difference is caused by an asynchronous readout of the digital voltmeters. Such an error may only be found by comparison of the results of two instruments.

7 Environmental influences

The higher accuracy and reliability of modern SGs demand a more careful consideration of environmental influences.

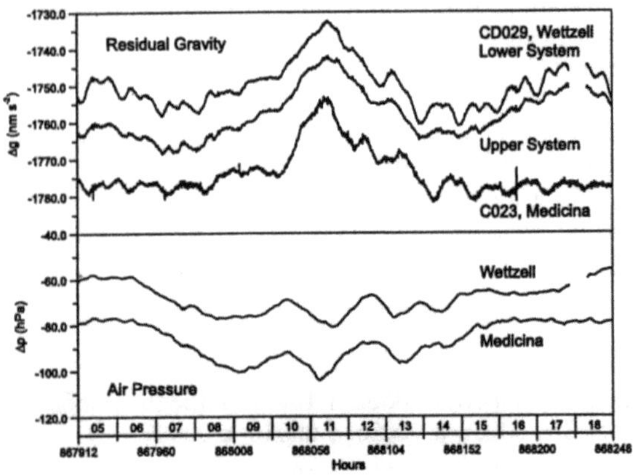

Fig. 4: Residual air pressure influences at Wettzell and Medicina (5. - 18.1.1999)

It is common use to correct the influence of air pressure variations by means of a linear regression model. The local air pressure is used and the regression coefficient is estimated together with the tidal parameters during the tidal analysis. Though all gravimeter data are corrected in this way, residual disturbances may remain. As an example Figure 4 shows nearly isochronous anomalies of the gravity residuals at Wettzell/Germany and Medicina/Italy (400 km apart), which obviously are produced by insufficient air pressure corrections.

To an increasing extend also hydrological influences (precipitation, soil moisture, ground water) are taken into consideration. The seasonal variations of these influences may disturb all investigations of long-term effects [4,5]. – Figure 5 shows precipitation and measured variations of the groundwater level along with the modelled gravity effect of the precipitation and the measured gravity variations. It should be noticed, that also on top of mountains as it is the case at Wettzell considerable variations of the hydrological regime may occur.

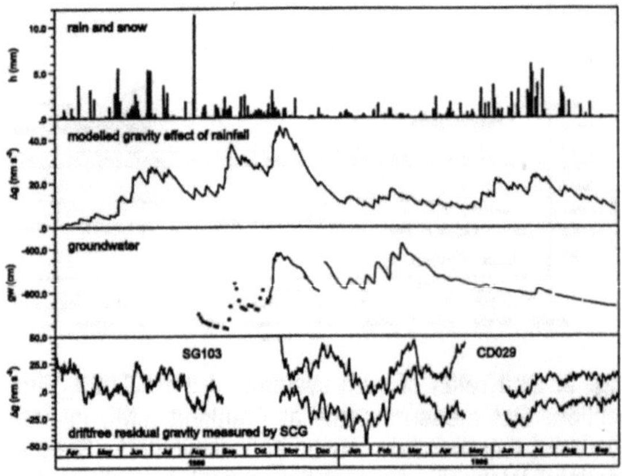

Fig. 5: Hydrological Influences at Wettzell, April 1998 – Sept. 1999. From top to bottom: Precipitation (summed up over 15 minutes). Modelled gravity effect of precipitation. Measured groundwater variations (small circles denote single measurements "by hand"). Measured gravity variations (residual gravity of gravimeters SG103 and both systems of CD029, exponential constituents, linear drift and gravity effect of polar motions eliminated).

For the estimation of the gravity effect of rainfall the simple exponential model

$$\Delta g_{i,rain} = 2 \pi G \rho r_j (1-\exp(-(i-j)/\tau_1)) \exp(-(i-j)/\tau_2)$$

was used with G = gravitational constant, ρ = density contrast (1 g cm^{-3}), r = precipitation and the empirical time constants τ_1 = 4 hours and τ_2 = 91 days [7,8]. Clearly a correspondence of the modelled gravity effect of rainfall and of the measured variations of the groundwater level is evident, especially between April and December 1998. Also a correspondence between groundwater variations and measured gravity variations is indicated (most clearly between November 1998 and April 1999). However, the relation is not stable over the whole observation period. But this is not to be expected because of the very complicated geological and hydrological situation in the environment of the

Wettzell station. Therefore more data must be gathered and complex models have to be developed before reliable corrections can be derived.

8 Scale factors of the CD029

At the beginning only scale factors could be used, which were derived from direct comparisons of the measured signal with the theoretical tidal model. Later, when time series became available, long enough for reliable tidal analysis, also single spectral components of the tidal signal could be compared (e.g. O1). It should be noted, that all these estimations are based on tidal registrations with the LCR-gravimeter ET-15, carried out at Bad Homburg between December 1983 and July 1984 in parallel with the TT40 [1].

Meanwhile the CD029 could be calibrated also directly by comparisons with absolute gravity measurements as well as by artificial accelerations, derived from small sinusoidal vertical displacements of the gravimeter ("platform calibration", "Frankfurt Calibration System").

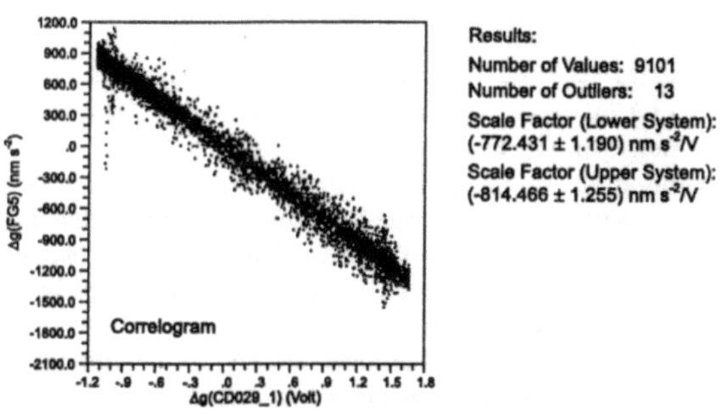

Fig. 6: Calibration of the CD029 by comparison with the absolute gravimeter FG5-101, Wettzell, 9 – 11.8. 1999.

A great problem of the calibration by comparison with absolute measurements are the very different accuracies of both systems to be compared (Figure 6). However, if at first all extreme outliers are eliminated, accuracies in the order of 1 nm s^{-2} may be reached. This level of accuracies seems to be the limit of this calibration method.

In Table 2 the results of 2 calibration campaigns are summarised. The mean values of both comparisons agree very well with the result of the platform calibration.

The acceleration experiment (Table 3) was carried out on October 22 and 23, 1998 at Frankfurt a.M. In addition to the scale factors of both systems also the phase shift of the tide filter could be estimated. It should be noticed that at Frankfurt a.M. the environmental conditions were not the best ones to carry out high precision measurements. Under more favourable conditions a higher accuracy can be expected as it is known from earlier acceleration experiments [3].

Table 2: Calibration of the CD029 by comparison with absolute gravity measurements

Date	Scale Factor nm s^{-2}/V	Δgmax nm s^{-2}	N	Absolute Gravimeter
15.-16.12.98	-771.495 ± 3.293	1663.5	3550	FG5-101 BKG
	-813.477 ± 3.472			
9.-11.8.99	-772.431 ±.1.190	2163.79	9101	FG5-101 BKG
	-814.466 ± 1.255	2163.82		

The relationship of the scale factors of both gravimeter systems to each other may be checked by means of the difference signal. If they are not consistent, the expected flat shape of the difference signal is disturbed by residuals of the tidal gravity variations as may be seen from the uppermost graph of Figure 7. If at first the scale factor of the upper system is varied and in a second step also a small phase difference between both systems is taken into account, the residual gravity curve becomes more and more flat as may be seen from the lower graphs of Figure 7. From this example it may be concluded, that from the viewpoint of the difference signal the calibration accuracy must be at least ±0.1 nm s^{-2}/V and the phase difference ±0.1°.

Table 3: Calibration of the CD029 by artificial accelerations

Date	Lower System nm s^{-2} /V	Upper System nm s^{-2} /V	Remarks
22.10.1998	-773.41 ± 0.36	-816.79 ± 0.33	varying room temperature
23.10.1998	-773.91 ± 0.16	-816.65 ± 0.03	stable environmental conditions
Mean	-773.66 ± 0.2	-816.72 ± 0.2	
Phase shift (s)	44.1 ± 0.1	43.7 ± 0.1	up to now 38.0 s were used

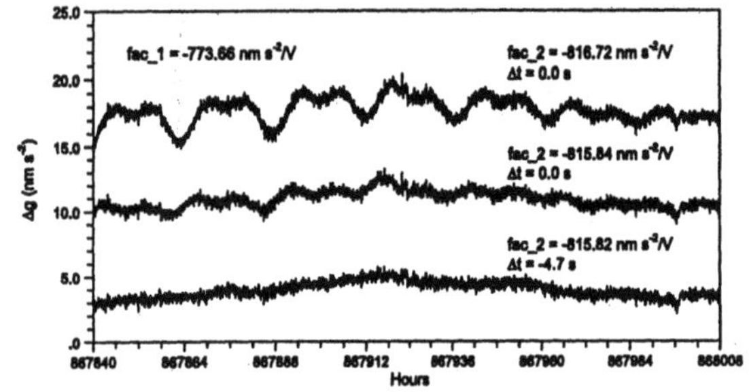

Fig. 7: Influence of the scale factors on the difference signal (2.-8.1.1999). Top graph: All calibration parameters taken from the platform calibration (October 1998). Middle graph: Calibration factor of the upper system adjusted. Bottom graph: Calibration factor of the upper system and phase difference adjusted.

The sensitivity of the difference signal to errors of the calibration factors may be used to derive a consistent set of calibration parameters. Therefore the scale factor of the lower system resulting from the platform calibration was used as reference value and a trial and error procedure was started assuming a zero phase difference between both systems. Then the scale factor of the upper system and the phase difference were varied until the best fit was reached (described by the mean square error m_0), i.e. until the undulations of the difference signal were minimised. The result is given in Table 4. These values are consistent between each other but not necessarily without any errors. The problem of a precise absolute calibration may not be resolved yet because from the point of view of the difference signal also the method of the platform calibration is not accurate enough for this decision.

Table 4: Consistent set of calibration parameters of the CD029

	Lower System	Upper System	Unit
Scale factor	-773.66 ± 0.2	-815.82	nm s^{-2}/V
Phase	43.9 ± 0.1	39.2	s

9 Final remarks

First experiences with the new dual sphere gravimeter CD029 confirm the high quality and the high accuracy of this new type of superconducting gravimeters. In general the noise of the tidal signal and the drift rate are very low. The very strong drift behaviour during the first operation period at Wettzell has to be considered as an exception. During the same time no steps occurred. The difference signal which compares the gravity values of both systems is a very helpful tool for the monitoring of the instrumental drift and the detection of disturbances of different kind. It also may be used to derive a consistent set of calibration factors and the phase lag between the two systems.

References

Richter, B. (1987) [1]. Das supraleitende Gravimeter. *Dt. Geodät. Kommiss.*, R. C, H. 329, Inst. Angew. Geodäsie, Frankfurt a.M., 126 p.

Richter, B. and R.J. Warburton (1998) [2]. A New Generation of Superconducting Gravimeters. *Proc. 13th Int. Symp. Earth Tides, Brussels 1997.* Obs. Royal Belgique, Sér. Géophys., Brussels, pp.545 - 555.

Richter, B., H. Wilmes, I. Nowak and P. Wolf (1995) [3]. Calibration of a Cryogenic Gravimeter (SCG TT60) by Artificial Accelerations and Comparisons with Absolute Measurements. Poster Presentation at the General Assembly IUGG Boulder, unpublished.

Harnisch, M. and G. Harnisch (1995) [4]. Processing of the data from two superconducting gravimeters, recorded in 1990 - 1991 at Richmond (Miami, Florida). Some problems and results. Working Group Bonn 1994. *Marées Terrestres Bull. Inf.*, Bruxelles 122, pp.9141 - 9147.

Harnisch, M., G. Harnisch, B. Richter and W. Schwahn (1998). [5] Estimation of polar motion effects from time series recorded by superconducting gravimeters. *Proc. 13th Int. Symp. Earth Tides, Brussels 1997.* Obs. Royal Belgique, Sér. Géophys, Brussels, pp.511 - 518.

Harnisch, M., G. Harnisch, H. Jurczyk and H. Wilmes (1999) [6]. 889 Days of Registrations with the Superconducting Gravimeter SG103 at Wettzell (Germany). *Proc. Second GGP Workshop Munsbach Castle (Luxembourg) 24-26 March,* in preparation.

Harnisch, M. and G. Harnisch (1999) [7]. Hydrological Influences in the Registrations of Superconducting Gravimeters and Ways to their Elimination. Workshop Jena 1998. *Marées Terrestres Bull. Inf.,* Bruxelles 131, pp.10161 - 10170.

Crossley, D.J., Su Xu and T. van Dam (1998) [8]. Comprehensive Analysis of 2 years of SG Data from Table Mountain, Colorado. . *Proc. 13th Int. Symp. Earth Tides, Brussels 1997.* Obs. Royal Belgique, Sér. Géophys, Brussels, pp.659 - 668.

PART 3

Advances in Theory and Numerical Techniques

Bernhard Heck
Petr Holota

Direct methods in physical geodesy

P. Holota
Research Institute of Geodesy, Topography and Cartography, 250 66 Zdiby 98, Praha-východ, Czech Republic

Abstract. General principles of direct and variational methods are discussed first, together with basic functional-analytic tools, especially Sobolev's weight space. An interpretation of Neumann's problem as a minimization problem for a quadratic functional is approached as an example. The focus, however, is on the linear gravimetric boundary value problem and a successive rectification of an oblique derivative in the respective boundary condition for the disturbing potential. The convergence and a tie of this concept to minimization principles is discussed. Finally, an interpretation in terms of function bases is shown.

Keywords. Boundary value problems, numerical methods, Sobolev's space, Dirichlet's principle, minimum of a quadratic functional.

---

1 Introduction

The principle objective of this paper is to discuss an approach to the solution of boundary value problems in the determination of the external gravity field of the Earth which is not standard in geodetic literature. We, therefore, start with somewhat more general aspects.

Problems of mathematical physics usually result in partial differential equations which have to be integrated under given initial or boundary conditions. For applications it is important to express numerical values of the respective solutions. Probably, methods best adapted to this goal are the so-called direct methods. According to Sobolev (1954) under these methods one can understand such approximate methods for the solution of problems in the theory of differential and integral equations that transform these problems in systems of algebraic equations, cf. also Michlin (1970) or Rektorys (1974).

We often meet a fact that is of principal importance. In many cases we can replace the integration of a differential equation by an equivalent problem of getting a function that minimizes some integral. Thus the integration of the respective differential equation can be replaced by an equivalent variational problem. Methods that enable to interpret the integration of a differential equation as an equivalent variational problem are usually called variational methods.

In the history the first use of the variational methods was in the form of *Dirichlet's principle.* According to this principle among functions which attain given values on the boundary $\partial\Omega$ of a domain Ω that and only that function is harmonic in Ω that minimizes the so-called *Dirichlet's integral*, see Michlin (1970) or Rektorys (1974). Dirichlet's principle was extensively used by Riemann, but critically commented by Weierstrass and later also by Hadamard. Only in the beginning of the 20th century in connection with Hilbert's works the principle got a new interest. Hilbert showed that the justification of Dirichlet's principle is much deeper and essentially associated with the notion of the completeness of the metric space.

Return now to our geodetic applications. First we give some auxiliary definitions.

2 Basic notations and tools

In literature the direct methods are usually considered for bounded solution domains. Our aim, however, is to discuss problems associated with the determination of the external gravity field of the Earth. Therefore, we have to consider boundary value problems for an unbounded solution domain Ω. First we will construct the basic apparatus as an analogue to that discussed (for bounded domains) e.g. in Nečas (1967) and Rek-

International Association of Geodesy Symposia, Vol. 121
Schwarz (ed.), Geodesy Beyond 2000 – The Challenges of the First Decade

torys (1974).

Let $\varepsilon(\bar{\Omega})$ be a space of functions having derivatives of all orders continuous in Ω and continuously extendable to the closure $\bar{\Omega}$ of Ω. The functions from $\varepsilon(\bar{\Omega})$ are supposed to equal zero in a neighbourhood of infinity.

In $\varepsilon(\bar{\Omega})$, supposing that $|x| > 0$ for all $x \in \Omega$, we define an inner product by

$$(u,v)_1 \equiv \int_\Omega \frac{uv}{|x|^2}\,dx + \sum_{i=1}^{3} \int_\Omega \frac{\partial u}{\partial x_i}\frac{\partial v}{\partial x_i}\,dx \quad (1)$$

where x_i, $i = 1,2,3$, are rectangular Cartesian coordinates in Euclidean three-dimensional space $\boldsymbol{R}^3$. This product induces the norm

$$(u,u)_1^{1/2} \equiv \|u\|_1 \quad (2)$$

The completion of $\varepsilon(\bar{\Omega})$ in this norm represents a *Sobolev weight space*. We will denoted it by $W_2^{(1)}$. Roughly speaking, $W_2^{(1)}$ is a space of functions that are square integrable on Ω under the weight $|x|^{-2}$ and have derivatives of the first order in a certain generalized sense. We could also show that $W_2^{(1)}$ cannot contain functions with discontinuities as e.g. a jump. On the other hand harmonic functions with their characteristic regularity at infinity belong to $W_2^{(1)}$.

We suppose that also the boundary $\partial\Omega$ of Ω has a clearly defined regularity. Putting $\Omega' = \boldsymbol{R}^3 - \bar{\Omega}$, we suppose that Ω' is a domain with *Lipschitz' boundary*, see Nečas (1967), Rektorys (1974) or Kufner et al. (1977). Lipschitz' boundary is already general enough to represent (under a mild idealization) the topography of the Earth. Domains with Lipschitz' boundary are e.g. the sphere, ellipsoid, cube, polyhedron as well as more general domains with smooth or piecewise smooth boundaries. Note, however, that among domains with Lipschitz' boundary one cannot range those having singularities like highly sharp edges, vertices or turning points analogous to the two dimensional case.

Lipschitz' boundary has an outer (inner) normal almost everywhere. The proof can be found in Nečas (1967) or Kufner et al. (1977). This important property will enable us to use the famous Green's theorem. In the sequel we will denote the unit outer normal of $\partial\Omega \equiv \partial\Omega'$ by $\boldsymbol{n}$.

It is obvious that every function from $\varepsilon(\bar{\Omega})$ has uniquely defined values on $\partial\Omega$. Let u be a function from $\varepsilon(\bar{\Omega})$. A restriction of $u(x)$ for $x \in \partial\Omega$ is usually referred to as the trace of u on $\partial\Omega$. Evidently, the trace of u is continuous and square integrable on $\partial\Omega$. In case of bounded domains the possibility to extend the notion of the trace to all functions from a Sobolev space is demonstrated in Nečas (1967), Kufner et al. (1977) or Nečas and Hlaváček (1981). For our unbounded domain Ω the extension of the notion of the trace to all functions from $W_2^{(1)}$ can be constructed in a similar way. However, here and further on we will confine ourselves to domains Ω such that Ω' is also *starshaped at the origin.*

Denoting by $L_2(\partial\Omega)$ the space of square integrable functions on $\partial\Omega$ and by $\langle\,,\rangle$ the scalar product of two vectors in $\boldsymbol{R}^3$, we can formulate our assertion in the form of the following

Theorem 1 (*Trace theorem*). Let Ω be an unbounded domain and $\Omega' = \boldsymbol{R}^3 - \bar{\Omega}$ be a starshaped domain at the origin with the Lipschitz boundary such that $\langle x, \boldsymbol{n}\rangle > 0$ for almost all $x \in \partial\Omega$. Then there exists a continuous linear mapping Z from $W_2^{(1)}$ into $L_2(\partial\Omega)$ such that $Zu = u$ for $u \in \varepsilon(\bar{\Omega})$ and that for all $u \in W_2^{(1)}$

$$\|Zu\|_{L_2(\partial\Omega)} = \|u\|_{L_2(\partial\Omega)} = \left(\int_{\partial\Omega} u^2\,dS\right)^{1/2}$$

$$\leq \beta\left(\int_\Omega |\boldsymbol{grad}\,u|^2 dx\right)^{1/2} \quad (3)$$

where β is a positive constant: $\beta = 1/\sqrt{b}$ and

$$b = \inf_{\partial\Omega}\left\langle \frac{x}{|x|^2}, \boldsymbol{n}\right\rangle = \inf_{\partial\Omega}\left[\frac{\cos(x,\boldsymbol{n})}{|x|}\right] \quad (4)$$

Zu is called the trace of the function u on $\partial\Omega$. (Usually it is also denoted by u.) Moreover, from eq. (3) it is immediately obvious that

$$\|u\|_{L_2(\partial\Omega)} \leq \beta\,\|u\|_1 \quad (5)$$

For *Proof* see Holota (1997).

Finally, we can also show that for $u \in W_2^{(1)}$

$$\|u\| = \left(\int_\Omega |\boldsymbol{grad}\,u|^2 dx\right)^{1/2} \quad (6)$$

has all the characteristic properties of a functional norm and that even

$$\int_\Omega \frac{u^2}{|x|^2}\,dx \leq 4\int_\Omega |\boldsymbol{grad}\,u|^2 dx \quad (7)$$

holds for all $u \in W_2^{(1)}$. The respective proof can be found in (Holota, 1997). In consequence one can easily deduce from eq. (7) that for $u \in W_2^{(1)}$

$$\|u\| \leq \|u\|_1 \leq \sqrt{5}\,\|u\| \quad (8)$$

Thus $||u||_1$ and $||u||$ are equivalent norms.

3 Neumann's problem and a quadratic functional

Put first

$$A(u,v) = \sum_{i=1}^{3} \int_{\Omega} \frac{\partial u}{\partial x_i} \frac{\partial v}{\partial x_i} \, dx \tag{9}$$

which is a bilinear form defined on the Cartesian product $W_2^{(1)} \times W_2^{(1)}$ and consider the following quadratic functional

$$\Phi(u) = A(u,u) - 2 \int_{\partial\Omega} uf \, dS \tag{10}$$

defined on $W_2^{(1)}$, where $f \in L_2(\partial\Omega)$.

From inequality (8) we immediately see that

$$A(u,u) = ||u||^2 \geq \frac{1}{5} ||u||_1^2 \tag{11}$$

Similarly, using Schwarz's inequality we obtain

$$\int_{\partial\Omega} uf \, dS \leq ||u||_{L_2(\partial\Omega)} ||f||_{L_2(\partial\Omega)} \tag{12}$$

so that from the trace theorem above and especially from inequality (5) we have

$$\int_{\partial\Omega} uf \, dS \leq \beta ||f||_{L_2(\partial\Omega)} ||u||_1 \tag{13}$$

Hence it is clear that for $v \in W_2^{(1)}$

$$\lim \Phi(u) \to +\infty \quad \text{as} \quad ||u||_1 \to \infty \tag{14}$$

Recall that in a usual sense this property means that Φ is a coercive functional on $W_2^{(1)}$.

Take now an arbitrary $v \in W_2^{(1)}$ and form the function $\Phi(u+tv)$ of $t \in (-\infty, \infty)$. We can easily compute that

$$\frac{d}{dt} \Phi(u+tv)\bigg|_{t=0} = 2 \left[A(u,v) - \int_{\partial\Omega} vf \, dS \right] \tag{15}$$

Moreover, writing v instead of u in inequality (13) and taking into consideration that

$$A(u,v) \leq ||u||_1 ||v||_1 \tag{16}$$

which in combination with inequality (8) is a consequence of Lagrange's identity, we see that

$$D\Phi(u,v) \equiv \frac{d}{dt} \Phi(u+tv)\bigg|_{t=0} \tag{17}$$

is a bounded linear functional of the variable v (usually called Gâteaux' differential of Φ at the point u.) Inspecting now eq. (15), we can easily deduce that

$$D\Phi(u+v,v) - D\Phi(u,v) = 2A(v,v) \geq \frac{2}{5} ||v||_1^2$$

in view of inequality (11). Hence

$$D\Phi(u+v,v) - D\Phi(u,v) \geq 0 \tag{18}$$

for all $u, v \in W_2^{(1)}$.

The coercivness of the functional Φ and inequality (18) now enable us to use general results in the theory of non-linear functionals and to conclude that Φ attains its minimum in $W_2^{(1)}$. The proof results from the solution of an abstract variational problem as e.g. in Nečas and Hlavaček (1981).

On the contrary suppose that at a point $u \in W_2^{(1)}$ the functional Φ has its local minimum. In consequence $\Phi(u+tv)$ as a function of t attains its minimum for $t = 0$. Thus necessarily

$$\frac{d}{dt} \Phi(u+tv)\bigg|_{t=0} = 0 \tag{19}$$

for all $v \in W_2^{(1)}$ and it follows from eq. (15) that

$$A(u,v) = \int_{\partial\Omega} vf \, dS \tag{20}$$

holds for all $v \in W_2^{(1)}$. In the terminology of calculus of variations eq. (20) represents Euler's necessary condition for our functional Φ to have a minimum at the point u.

The last result can be given a classical interpretation, provided that u is sufficiently regular. In this case, using Green's theorem, we have

$$A(u,v) = -\int_{\Omega} v\Delta u \, dx - \int_{\partial\Omega} v \frac{\partial u}{\partial n} \, dS \tag{21}$$

where Δ means Laplace's operator and $\partial/\partial n$ denotes the derivative in the direction of the unit normal $\boldsymbol{n}$. Thus it follows from eq. (20) that

$$\int_{\Omega} v\Delta u \, dx + \int_{\partial\Omega} v \left(\frac{\partial u}{\partial n} + f \right) dS = 0 \tag{22}$$

is valid for all $v \in W_2^{(1)}$. Supposing in addition that also f and $\partial\Omega$ are sufficiently smooth and applying the usual reasoning to integrals from continuous functions, we obtain from eq. (22) that u has to satisfy the following equations

$$\Delta u = 0 \quad \text{in} \quad \Omega \tag{23}$$

and

$$\frac{\partial u}{\partial n} = -f \quad \text{on} \quad \partial\Omega \tag{24}$$

In other words u has to be a solution of Neumann's boundary value problem.

4 Linear gravimetric boundary value problem

We will consider our rectangular Cartesian coordinates x_i, $i = 1, 2, 3$, again and place their origin into the center of gravity of the Earth. In addition we will identify with W and U the gravity and a standard (normal) potential of the Earth. Under this notation $\boldsymbol{grad}\, W$ is the gravity vector and its length $g = |\boldsymbol{grad}\, W|$ is the measured gravity. By analogy we put $\gamma = |\boldsymbol{grad}\, U|$ for the normal gravity. Finally, $T(\boldsymbol{x}) = W(\boldsymbol{x}) - U(\boldsymbol{x})$ is the disturbing potential and $\delta g(\boldsymbol{x}) = g(\boldsymbol{x}) - \gamma(\boldsymbol{x})$ is known as the gravity disturbance. We assume that g is corrected for gravitational interaction with the Moon, the Sun and the planets, for the precession and nutation of the Earth and so on.

The linear gravimetric boundary value problem is an oblique derivative problem, cf. Koch and Pope (1972), Bjerhammar and Svensson (1983) or Grafarend (1989). Its solution domain is the exterior of the Earth. We will denote it by Ω. The problem may be formulated as follows:

$$\Delta T = 0 \quad \text{in} \quad \Omega \tag{25}$$

$$\langle \boldsymbol{s}, \boldsymbol{grad}\, T\rangle = -\delta g \quad \text{on} \quad \partial\Omega \tag{26}$$

where $\boldsymbol{s} = -(1/\gamma)\,\boldsymbol{grad}\, U$. Moreover, T is assumed regular at infinity, i.e.,

$$T = \mathcal{O}(|\boldsymbol{x}|^{-1}) \quad \text{as} \quad \boldsymbol{x} \to \infty \tag{27}$$

The concept of this paper and the idea of the last section lead us to a question whether it is possible to formulate a minimization problem that would lead (possibly in combination with additional steps like iterations) to the solution of an oblique derivative boundary value problem as that given by eqs. (25) and (26).

First, however, we try to formulate an oblique derivative boundary value problem in terms of an integral identity formally of a similar structure as Euler's condition mentioned in the last section, cf. eq. (20). The crucial problem is how to define the bilinear form $A(u, v)$ in this case. Following Holota (1997), we can verify that

$$A(u,v) = \int_\Omega \langle \boldsymbol{grad}\, u, \boldsymbol{grad}\, v\rangle\, d\boldsymbol{x}$$

$$- \int_\Omega \langle \boldsymbol{grad}\, v, \boldsymbol{a} \times \boldsymbol{grad}\, u\rangle\, d\boldsymbol{x}$$

$$- \int_\Omega v\, \langle \boldsymbol{curl}\, \boldsymbol{a}, \boldsymbol{grad}\, u\rangle\, d\boldsymbol{x} \tag{28}$$

serves our purpose, provided that $\boldsymbol{a} = (a_1, a_2, a_3)$ is a vector field such that the components a_i and also $|\boldsymbol{x}|(\boldsymbol{curl}\, \boldsymbol{a})_i$, $i = 1, 2, 3$, are Lebesgue-measurable functions defined and bounded almost everywhere on Ω. [This typically means that they belong to the space $L_\infty(\Omega)$.]

Indeed, supposing that a_i and u are sufficiently smooth and using Green's theorem, we arrive at

$$A(v,u) = -\int_\Omega v\, \Delta u\, d\boldsymbol{x}$$

$$- \int_{\partial\Omega} v\, \langle \boldsymbol{n} + \boldsymbol{a} \times \boldsymbol{n}, \boldsymbol{grad}\, u\rangle\, dS \tag{29}$$

Thus it follows from eq. (20) that

$$\int_\Omega v\, \Delta u\, d\boldsymbol{x}$$

$$+ \int_{\partial\Omega} v\, [\langle \boldsymbol{n} + \boldsymbol{a} \times \boldsymbol{n}, \boldsymbol{grad}\, u\rangle + f]\, dS = 0 \tag{30}$$

holds for all $v \in W_2^{(1)}$. Finally, supposing that also f and $\partial\Omega$ are smooth enough and applying the usual reasoning to integrals from continuous functions again, we easily deduce that

$$\Delta u = 0 \quad \text{in} \quad \Omega \tag{31}$$

$$\langle \boldsymbol{\sigma}, \boldsymbol{grad}\, u\rangle = -f \quad \text{on} \quad \partial\Omega \tag{32}$$

where the vector

$$\boldsymbol{\sigma} = \boldsymbol{n} + \boldsymbol{a} \times \boldsymbol{n} \tag{33}$$

is oriented towards the exterior of the domain Ω' and never tangential to its boundary since

$$\langle \boldsymbol{\sigma}, \boldsymbol{n}\rangle = 1 \tag{34}$$

Eqs. (31) and (32) represent a classical oblique derivative boundary value problem. Note also that u satisfies an asymptotic condition at infinity which expresses the same degree of regularity as eq. (27). This follows from the fact that $u \in W_2^{(1)}$. The tie of the problem to that given by eqs. (25) and (26) is clear. It is enough to put

$$\boldsymbol{\sigma} = \frac{\boldsymbol{s}}{\langle \boldsymbol{s}, \boldsymbol{n}\rangle} = \frac{\boldsymbol{grad}\, U}{\langle \boldsymbol{n}, \boldsymbol{grad}\, U\rangle} \quad \text{and} \tag{35}$$

$$f = \frac{\gamma}{\langle \boldsymbol{n}, \boldsymbol{grad}\, U \rangle} \delta g = \gamma \left(\frac{\partial U}{\partial n} \right)^{-1} \delta g \qquad (36)$$

5 Successive approximations

Inspecting $A(u,v)$, we immediately see that for an oblique derivative boundary value problem $A(u,v)$ is a non-symmetric bilinear form. We could verify easily that for this fact it is not possible to construct a quadratic functional such that its minimization would lead to the solution of the boundary value problem under consideration.

Here we try to formulate a sequence of problems instead that will allow an application of minimization principles. For this purpose put

$$A_1(u,v) = \int_\Omega \langle \boldsymbol{grad}\, u, \boldsymbol{grad}\, v \rangle \, dx \qquad (37)$$

and

$$A_2(u,v) = \int_\Omega \langle \boldsymbol{grad}\, v, \boldsymbol{a} \times \boldsymbol{grad}\, u \rangle \, dx$$

$$+ \int_\Omega v \, \langle \boldsymbol{curl}\, \boldsymbol{a}, \boldsymbol{grad}\, u \rangle \, dx \qquad (38)$$

so that $A(u,v) = A_1(u,v) - A_2(u,v)$.

Our aim is now to examine the sequence of functions $\{u_m\}_{m=0}^{\infty}$ defined by the following equations

$$A_1(u_{m+1}, v) = \int_{\partial\Omega} v f \, dS + A_2(u_m, v) \qquad (39)$$

that are assumed to hold for all $v \in W_2^{(1)}$ and $m = 0, 1, \dots, \infty$. Putting in addition

$$F_m(v) = \int_{\partial\Omega} v f \, dS + A_2(u_m, v) \qquad (40)$$

we see that for u_m fixed $F_m(v)$ is a functional defined on $W_2^{(1)}$. Moreover, $F_m(v)$ is a bounded functional for $m = 0, 1, \dots, \infty$. Indeed, using Schwarz's inequality again, we can compute that

$$A_2(u_m, v) \le \left(\int_\Omega |\boldsymbol{a}|^2 |\boldsymbol{grad}\, u_m|^2 dx \right)^{1/2} \|v\|_1$$

$$+ \left(\int_\Omega \langle |\boldsymbol{x}| \boldsymbol{curl}\, \boldsymbol{a}, \boldsymbol{grad}\, u_m \rangle^2 dx \right)^{1/2} \|v\|_1 \qquad (41)$$

This in combination with inequality (13) yields

$$|F_m(v)| \le C_m \|v\|_1 \qquad (42)$$

where

$$C_m = \beta \|f\|_{L_2(\partial\Omega)} + \left(\int_\Omega |\boldsymbol{a}|^2 |\boldsymbol{grad}\, u_m|^2 dx \right)^{1/2}$$

$$+ \left(\int_\Omega \langle |\boldsymbol{x}| \boldsymbol{curl}\, \boldsymbol{a}, \boldsymbol{grad}\, u_m \rangle^2 dx \right)^{1/2} \qquad (43)$$

so that obviously $\|F_m\| \le C_m$, where $\|F_m\|$ denotes the norm of the functional F_m and we see that F_m is a bounded functional.

Recalling in addition inequalities (11) and (16) valid for the bilinear form given by eq. (9) we immediately see that for all $u, v \in W_2^{(1)}$

$$A_1(u,v) \le \|u\|_1 \|v\|_1 \quad \text{and} \qquad (44)$$

$$A_1(v,v) \ge \alpha \|v\|_1^2 , \quad \alpha = \frac{1}{5} \qquad (45)$$

i.e., $A_1(u,v)$ is a bounded and $W_2^{(1)}$-elliptic bilinear form.

These properties of $A_1(u,v)$ make it now possible to apply Lax-Milgram's theorem which has a key position in the theory of the weak solution, see Nečas (1997), Rektorys (1994) or Bers et al. (1964). Together with the boundedness of the functional F_m it allows us to write

$$A_1(u^*, v) = F_m(v) , \quad v \in W_2^{(1)} \qquad (46)$$

where the element $u^* \in W_2^{(1)}$ is uniquely given by the functional F_m. Writing u_{m+1} instead of u^*, we immediately see that eq. (39) has a unique solution for all $m = 0, 1, \dots, \infty$. Moreover, it also follows from Lax-Milgram's theorem that

$$\|u_{m+1}\|_1 \le \frac{1}{\alpha} \|F_m\| \qquad (47)$$

The open question still is whether the sequence of the solutions $\{u_m\}_{m=0}^{\infty}$ has a limit. For this purpose we first deduce from eqs. (39) that

$$A_1(u_{m+2} - u_{m+1}, v) = A_2(u_{m+1} - u_m, v) \quad (48)$$

holds for all $v \in W_2^{(1)}$. For $u_{m+1} - u_m$ fixed $A_2(u_{m+1} - u_m, v)$ is a bounded linear functional of the variable v. Let $\|A_2\|$ denotes its norm. From the results above we then easily obtain that

$$\|A_2\| \le C_a \qquad (49)$$

where

$$C_a = \left[\int_\Omega |\boldsymbol{a}|^2 |\boldsymbol{grad}\, (u_{m+1} - u_m)|^2 dx \right]^{1/2}$$

$$+ \left[\int_\Omega \langle |\boldsymbol{x}| \boldsymbol{curl}\, \boldsymbol{a}, \boldsymbol{grad}\, (u_{m+1} - u_m) \rangle^2 dx \right]^{1/2}$$

Thus by analogy to inequality (47) we have

$$\|u_{m+2} - u_{m+1}\|_1 \leq \frac{1}{\alpha} C_a = 5C_a \qquad (50)$$

and it remains to derive an estimate of C_a in terms of $\|u_{m+1} - u_m\|_1$. Put, therefore,

$$K = \operatorname*{supess}_{\Omega} \left[|\boldsymbol{a}| \right] + \operatorname*{supess}_{\Omega} \left[|\boldsymbol{x}| |\boldsymbol{curl}\, \boldsymbol{a}| \right] \qquad (51)$$

where "suppess" means an essentially supreme value, so that in view of the mean value theorem

$$C_a = \widetilde{K} \, \|u_{m+1} - u_m\|_1 \qquad (52)$$

where $\widetilde{K}$ is a number such that $0 \leq \widetilde{K} \leq K$. Hence it is obvious that

$$\|u_{m+2} - u_{m+1}\|_1 \leq 5\widetilde{K} \, \|u_{m+1} - u_m\|_1 \qquad (53)$$

and in consequence we have

$$\|u_{m+2} - u_{m+1}\|_1 \leq (5\widetilde{K})^{(m+1)} \, \|u_1 - u_0\|_1 \qquad (54)$$

Clearly, in case that

$$5\widetilde{K} < 1 \qquad (55)$$

this inequality shows that $\{u_m\}_{m=0}^{\infty}$ is a Cauchy sequence. Because $W_2^{(1)}$ is a complete space, we can then conclude that the sequence converges to a function $u \in W_2^{(1)}$ in the norm $\| \, . \, \|_1$.

6 Convergence condition

Inequality (55) represents a sufficient condition for the iteration procedure above to converge. It is a certain bound for $|\boldsymbol{a}|$ and $|\boldsymbol{x}||\boldsymbol{curl}\,\boldsymbol{a}|$. Thus the question is how the vector $\boldsymbol{a}$ is actually defined in Ω. We know that its trace on $\partial\Omega$ has to satisfy eq. (33) with $\boldsymbol{\sigma}$ given, i.e.,

$$\boldsymbol{a} \times \boldsymbol{n} = \boldsymbol{\sigma} - \boldsymbol{n} \quad \text{on} \quad \partial\Omega \qquad (56)$$

It is a linear system for the traces of the components a_i, $i = 1, 2, 3$, of the vector $\boldsymbol{a}$ on $\partial\Omega$. We can easily verify that the equation is met by

$$\boldsymbol{a} = \boldsymbol{n} \times \boldsymbol{\sigma} \qquad (57)$$

However, the system is singular and together with $\boldsymbol{a}$ also $\boldsymbol{a} + \omega\boldsymbol{n}$ represents its solution for any factor ω. Using this liberty, we take $\boldsymbol{a}$ given by eq. (57) for simplicity reason. This enables us to define $\boldsymbol{a}$ inside Ω. Indeed, in case of our linear gravimetric boundary value problem eq. (35) defines $\boldsymbol{\sigma}$ not only on $\partial\Omega$, but also in Ω. Similarly, we can think that Q is another scalar function of position in $\bar{\Omega}$ that has the necessary degree of regularity and we will suppose that one of it level surfaces, $Q(\boldsymbol{x}) = const.$, coincides with $\partial\Omega$. A function like this exists. It can be obtained, e.g., as a solution of Dirichlet's problem for Laplaces's equation. Obviously,

$$\boldsymbol{n} = \frac{\boldsymbol{grad}\, Q}{|\boldsymbol{grad}\, Q|} \qquad (58)$$

then defines $\boldsymbol{n}$ in $\bar{\Omega}$.

Because U and Q are harmonic functions in an unbounded domain that are also regular at infinity their derivatives have for $\boldsymbol{x} \to \infty$ the following asymptotic behaviour:

$$\frac{\partial U}{\partial x_i} = \mathcal{O}(|\boldsymbol{x}|^{-2}) \quad \text{and} \quad \frac{\partial Q}{\partial x_i} = \mathcal{O}(|\boldsymbol{x}|^{-2}) \qquad (59)$$

$i = 1, 2, 3$. In consequence we can derive that

$$|\boldsymbol{a}| = \mathcal{O}(|\boldsymbol{x}|^{-2}) \quad \text{as} \quad \boldsymbol{x} \to \infty \qquad (60)$$

For $|\boldsymbol{x}||\boldsymbol{curl}\,\boldsymbol{a}|$ the similar conclusion need somewhat more computation. First we obtain that

$$(\boldsymbol{curl}\,\boldsymbol{a})_i = n_i \operatorname{div} \boldsymbol{\sigma} - \sigma_i \operatorname{div} \boldsymbol{n}$$

$$+ \langle \boldsymbol{\sigma}, \boldsymbol{grad}\, n_i \rangle - \langle \boldsymbol{n}, \boldsymbol{grad}\, \sigma_i \rangle \qquad (61)$$

$i = 1, 2, 3$. This equation can be interpreted geometrically, see Holota (1997). Using this interpretation we can then conclude that

$$|\boldsymbol{x}||\boldsymbol{curl}\,\boldsymbol{a}| = \mathcal{O}(|\boldsymbol{x}|^{-2}) \quad \text{as} \quad \boldsymbol{x} \to \infty \qquad (62)$$

Considering finally both the eqs. (60) and (62), we can expect that $\widetilde{K}$ will be much smaller than K which gives us also more chance to expect that inequality (55) holds for cases that are not too far from reality. A final answer of quantitative nature needs, however, still some investigation.

7 Interpretation of an asymmetric part

We start with eq. (38). Assuming that the components a_i of the vector $\boldsymbol{a}$ as well as the solution u are sufficiently smooth and using Green's theorem, we obtain

$$A_2(u_m, v) = -\int_{\partial\Omega} v \, \langle \boldsymbol{a} \times \boldsymbol{n}, \boldsymbol{grad}\, u_m \rangle \, dS \qquad (63)$$

cf. eq. (30). Thus, recalling eq. (56), we have

$$A_2(u_m, v) = -\int_{\partial\Omega} v \, \langle \boldsymbol{\sigma} - \boldsymbol{n}, \boldsymbol{grad}\, u_m \rangle \, dS$$

$$= -\int_{\partial\Omega} v|\boldsymbol{\sigma} - \boldsymbol{n}| \frac{\partial u_m}{\partial t} dS \tag{64}$$

where $|\boldsymbol{\sigma}-\boldsymbol{n}| = \sqrt{|\boldsymbol{\sigma}|^2 - 1} = \tan(\boldsymbol{\sigma}, \boldsymbol{n})$ as we can verify by direct computation and $\partial/\partial t$ denotes the derivative in the direction of the vector

$$\boldsymbol{t} = \frac{\boldsymbol{\sigma} - \boldsymbol{n}}{|\boldsymbol{\sigma} - \boldsymbol{n}|} \tag{65}$$

which obviously is tangential to $\partial\Omega$.

In our iteration procedure given by eq. (39) it means that we can modify it as follows:

$$A_1(u_{m+1}, v) = \int_{\partial\Omega} v f_m \, dS \tag{66}$$

for all $v \in W_2^{(1)}$ while

$$f_m = f - \frac{\partial u_m}{\partial t} \tan(\boldsymbol{\sigma}, \boldsymbol{n}) \tag{67}$$

For our geodetic case the correction term that appears in eq. (67) can obviously be expressed as

$$\frac{\partial u_m}{\partial t} = \frac{\partial u_m}{\partial \sigma} \cos(\sigma, t) + \frac{\partial u_m}{\partial \tau} \cos(\tau, t) \tag{68}$$

where in the particular boundary point of consideration $\boldsymbol{x} \in \partial\Omega$ the derivative $\partial u_m/\partial\tau$ is a component of $\boldsymbol{grad}\, u_m$ in the tangential plane of the level surface $U = const.$ of the model gravity potential passing through the point $\boldsymbol{x}$. Because $\langle \boldsymbol{\sigma}, \boldsymbol{n} \rangle = 1$ we also have

$$\frac{\partial u_m}{\partial t} = \frac{1}{|\sigma|} \left[\langle \boldsymbol{\sigma}, \boldsymbol{grad}\, u_m \rangle \sin(\sigma, n) + \frac{\partial u_m}{\partial \tau} \right] \tag{69}$$

Remark. In practice under our smoothness assumptions we can compute this expression approximately by means of f and the components ξ and η of the deflection of the vertical. For $\langle \boldsymbol{\sigma}, \boldsymbol{grad}\, u_m \rangle$ we can approximately take $\langle \boldsymbol{\sigma}, \boldsymbol{grad}\, u \rangle = f$ and similarly for $\partial u_m/\partial\tau$ we can take $\partial u/\partial\tau$ which according to Heiskanen and Moritz (1967) can be expressed as

$$\frac{\partial u}{\partial \tau} = \gamma(\xi \tan\beta_1 + \eta \tan\beta_2) \tan^{-1}(\boldsymbol{\sigma}, \boldsymbol{n}) \tag{70}$$

where β_1, β_2 are the angles of inclination of the north-south and east-west profiles of $\partial\Omega$.

8 A final remark

Formally, eq. (66) has the same structure as eq. (20). This, of course, means that every iteration step may be interpreted as a minimization problem. In addition we also see that in our iteration procedure we rectify the oblique derivative and that within individual iteration steps we treat the problem as a Neumann's problem (in a generalized setting).

However, it is worth mentioning that in our generalized formulation u_{m+1} is defined as an element among an excessively great multitude of functions. This is, but an artificial problem only, cf. Holota (1998a,b). It is enough to consider a space $H_2^{(1)}(\Omega)$ of those functions of $W_2^{(1)}(\Omega)$ which are harmonic in Ω and reformulate our definition, i.e. to look for $u_{m+1} \in H_2^{(1)}(\Omega)$ such that eq. (66) holds for all $v \in H_2^{(1)}(\Omega)$.

The simple structure of the bilinear form $A_1(u, v)$ makes it also possible to interpret the problem numerically. Indeed, taking into consideration the practice applied in geodesy, we agree that it is convenient to use Runge's property of Laplace's equation and to work with a space of functions which are harmonic outside a domain B completely embedded in the body of the Earth. Putting $B' = \boldsymbol{R}^3 - \bar{B}$, we will denote this space by $\mathcal{H}(B')$.

Following Neyman (1979) and an analogue to his reasoning related to Bjerhammar's sphere, one can show that in terms of the norm $\|\,.\,\|_1$ the space $\mathcal{H}(B')$ is dense in $H_2^{(1)}(\Omega)$. This enables us to approximate u_{m+1} by means of

$$u^{(n)} = \sum_{i=0}^{n} c_i^{(n)} v_i \tag{71}$$

where $c_i^{(n)}$ are numerical coefficients and v_i are members of a function base of $\mathcal{H}(B')$. Moreover, eq. (66) offers a natural starting point for a numerical interpretation of the problem. Indeed, for the coefficients $c_i^{(n)}$ we can immediately write Galerkin's system

$$\sum_{i=0}^{n} c_i^{(n)} A_1(v_i, v_j) = \int_{\partial\Omega} v_j f_m \, dS \tag{72}$$

where $j = 0, \ldots, n$. For v_i, $i = 0, \ldots, n$ let us take here e.g. a set of elementary potentials

$$v_i(\boldsymbol{x}) = \frac{1}{|\boldsymbol{x} - \boldsymbol{y}_i|}, \quad \boldsymbol{y}_i \in B \tag{73}$$

and add some comments related to the computation of the elements $A_1(v_i, v_j)$. Recall that the domain Ω' is assumed star-shaped at the origin and define $\boldsymbol{y}_i'$ as a point of intersection of $\partial\Omega$ with a radial ray passing through the point $\boldsymbol{y}_i$. For $i, j = 0, \ldots, n$ put $R_{ij} = (|\boldsymbol{y}_i'| + |\boldsymbol{y}_j'|)/2$ or, alternatively, $R_{ij} = \max_{i,j}[\,|\boldsymbol{y}_i'|, |\boldsymbol{y}_j'|\,]$ in case that

$(|y_i'|+|y_j'|)/2 \leq \max_{i,j}[\,|y_i|,|y_j|\,]$. Now, writing R instead of R_{ij} for simplicity reasons, we can try to approximate $A_1(v_i,v_j)$ by means of

$$A_R(v_i,v_j)=\sum_{k=1}^{3}\int_{S_R}\frac{\partial v_i}{\partial x_k}\frac{\partial v_j}{\partial x_k}\,dx \qquad (74)$$

where $S_R \equiv \{x \in \boldsymbol{R}^3; |x| > R\}$. ($R$ can be also defined in an average sense so that in a surrounding of y_i' and y_j' the sphere ∂S_R approximates $\partial\Omega$.) The diagonal terms can be computed by means of a standard integration. We obtain

$$A_R(v_i,v_i)=\frac{2\pi R}{R^2-|y_i|^2}+\frac{\pi}{|y_i|}\ln\frac{R+|y_i|}{R-|y_i|} \qquad (75)$$

For the off-diagonal terms the computation is a bit more complicated. The result is as follows:

$$A_R(v_i,v_j)=\frac{2\pi}{RL}$$

$$+\frac{5\pi}{Rz}\ln\frac{L+z-\cos\psi_{ij}}{1-\cos\psi_{ij}}+\frac{2\pi}{R}\delta S \qquad (76)$$

where

$$L=\sqrt{1-2z\cos\psi_{ij}+z^2}, \quad z=\frac{|y_i||y_j|}{R^2} \qquad (77)$$

$$\delta S=\frac{1}{2}\sum_{n=0}^{\infty}\frac{1}{2n^2+3n+1}z^nP_n(\cos\psi_{ij}) \qquad (78)$$

and ψ_{ij} is the angle between the vectors y_i and y_j, cf. Holota (1998b). Note that in eq. (78) the magnitude of individual terms decreases very quickly. Therefore, it is enough to sum up its first terms only to guarantee the sufficient accuracy.

Acknowledgements. The work on this paper was supported by the Grant Agency of the Czech Republic through Grant No. 205/99/0833. The presentation of the paper at the 22nd General Assembly of the IUGG (Birmingham, 1999) was sponsored also by the Ministry of Education, Youth and Sports of the Czech Republic through Grant No. LA 015(1999). Both these supports are gratefully acknowledged.

References

Bers L., John F., and Schechter M. (1964). *Partial differential equations*, John Wiley and Sons, Inc., New York – London – Sydney.

Bjerhammar A., and Svensson L. (1983). On the geodetic boundary-value problem for a fixed boundary surface – A satellite approach. *Bull. Géod. 57*, pp. 382-393.

Grafarend E.W. (1989). The geoid and the gravimetric boundary-value problem. *Report No. 18 from the Dept. of Geod.*, The Royal Inst. of Technology, Stockholm.

Heiskanen, W.A., and H. Moritz. (1967). *Physical geodesy*, W.H. Freeman and Company, San Francisco and London.

Holota, P. (1997). Coerciveness of the linear gravimetric boundary-value problem and a geometrical interpretation, *Journal of Geodesy, 71*, pp. 640-651.

Holota P. (1998a). Variational methods and subsidiary conditions for geoid determination. In: Vermeer, M. and Ádám, J. (eds.), Second Continental Workshop on the Geoid in Europe, Budapest, Hungary, March 10-14, 1998, Proceedings, *Reports of the Finnish Geodetic Inst. No. 98:4*, Masala, 1998, pp. 99-105.

Holota P. (1998b). Variational methods in geoid determination and function bases. Presented at the 23rd General Assembly of the European Geophysical Society (Symposium G11), Nice, France 20-24 April 1998. *Physics and Chemistry of the Earth*, Part A: Solid Earth and Geodesy - Recent Advances in Precise Geoid Determination Methodology, Vol. 24, No. 1, pp. 3-14, 1999.

Koch K.R., and Pope A.J. (1972). Uniqueness and existence for the geodetic boundary-value problem using the known surface of the Earth. *Bull. Geod. 106*, pp. 467-476.

Kufner A., John O., and Fučík S. (1977). *Function spaces.* Academia, Prague.

Michlin, A.G. (1970). *Variational methods in mathematical physics*, Nauka Publisher, Moscow (in Russian); also in Slovakian: Alpha, Bratislava 1974.

Neyman Yu.M. (1979). *A variational method in physical geodesy*, Nedra Publishers, Moscow (in Russian).

Nečas, J. (1967). *Les méthodes directes en théorie des équations elliptiques*, Academia, Prague.

Nečas, J., and I. Hlaváček (1981). *Mathematical theory of elastic and elasto-plastic bodies: An introduction*, Elsevier Sci. Publ. Company, Amterdam-Oxford-New York.

Rektorys, K. (1974). *Variační metody v inženýrských problémech a v problémech matematické fyziky*, SNTL Publishers of Technical Literature, Prague 1974; also in English: *Variational methods*, Reidel Co., Dordrecht-Boston, 1977.

Sobolev, S.L. (1954). *Equations of mathematical physics*, Moscow (in Russian).

A general least-squares solution of the geodetic boundary value problem

Martin van Gelderen
Delft Institute for Earth-Oriented Space Research,
Delft University of Technology, Thijsseweg 11, NL-2629 JA Delft

Reiner Rummel
Institut für Astronomische und Physikalische Geodäsie,
Munich University of Technology, Arcisstrasse 21, D-80290 Munich

Abstract. A general least squares solution is derived for a whole class of GBVPs in spherical, constant radius approximation by means of a transformation to the spectrum. For a number of GBVPs the kernels providing the solution can be exactly computed and are shown here as examples.

Keywords. Geodetic boundary value problems.

1 Introduction

In (Rummel and Teunissen, 1986) and in (Rummel et al., 1989) a solution of the GBVP in spherical, constant radius approximation was presented for various types of observations; uniquely and over-determined. In a later publication (Rummel and van Gelderen, 1992) the relation between the gradiometer observations, the second derivatives of the gravitational potential, and the disturbing potential was demonstrated with the use of vector and tensor spherical harmonics. Here we will show how a GBVP in the same degree of approximation can be solved with more general types of observations. The basic condition for this to be applicable is that the spectral relation between the disturbing potential and the observable only depends on the degree of their spherical harmonic expansions.

The solutions by e.g. (Sacerdote and Sansó, 1985) and (Thong and Grafarend, 1985) for some GBVPs are obtained by a completely different approach but are identical to ours for the GBVP in spherical, constant radius approximation.

2 Observation equation

All observations of the linearised GBVP with unknown boundary surface can be written as a linear combination of a position dependency and a differential operator applied to the disturbing potential:

$$\mathbf{f}(\mathbf{P}) = a\mathbf{x}(\mathbf{P}) + \mathbf{D}T \tag{1}$$

with

$\mathbf{f}(\mathbf{P})$	observable boundary function (scalar, vector or tensor) minus its approximate value, e.g. from the ellipsoid
a	constant scalar ($a \in \mathcal{R}$)
$\mathbf{x}(\mathbf{P})$	position displacement function (scalar, vector or tensor) between actual and given approximate boundary surface, e.g. telluroid
$\mathbf{D}$	linear differential operator
T	disturbing potential.

When necessary, we write the vector functions in components with respect to a local Cartesian frame: x North, y East and z up. All functions that can be of vector or tensor type are printed bold.

All quantities will be normalised to the unit sphere without a dimension in order to get compact expressions. The inner product on the unit sphere Σ will be denoted by

$$<\mathbf{f}, \mathbf{g}> = \int_{\Sigma} \mathbf{f} \cdot \mathbf{g} \, d\sigma \tag{2}$$

where $\mathbf{f}$ and $\mathbf{g}$ are two scalar, vector or tensor functions on the unit sphere and $\mathbf{f} \cdot \mathbf{g}$ the standard Cartesian inner product between them.

International Association of Geodesy Symposia, Vol. 121
Schwarz (ed.), Geodesy Beyond 2000 – The Challenges of the First Decade

Table 1: The operators most relevant for GBVPs. The derivatives are written w.r.t. local Cartesian coordinates

OPERATOR	EIGEN/SINGULAR VALUE
∂_z	$(\ell+1)$
$\{\partial_x,\partial_y\}$	$\sqrt{\ell(\ell+1)}$
∂_{zz}	$(\ell+1)(\ell+2)$
$\{\partial_{xz},\partial_{yz}\}$	$(\ell+1)\sqrt{\ell(\ell+1)}$
$\{\partial_{xx}-\partial_{yy},2\partial_{xy}\}$	$\sqrt{(\ell-1)\ell(\ell+1)(\ell+2)}$
Poisson	$\left(\frac{R}{r}\right)^{\ell+1}$

Only operators $\mathbf{D}$ will be allowed such that $\mathbf{D}^{+}\mathbf{D}$ diagonalizes on the basis of spherical harmonic functions $\frac{1}{r^{\ell+1}}Y_{\ell m}(P)\equiv\mathcal{Y}_{\ell m}(P)$ with weights only dependent on the degree:

$$<\mathbf{D}\mathcal{Y}_{\ell m},\mathbf{D}\mathcal{Y}_{\ell' m'}>=\lambda_\ell^2\delta_{\ell,\ell'}\delta_{m,m'}. \quad (3)$$

If T and $\mathbf{f}$ are square integrable on the sphere we can write

$$T(P)=\sum_{\ell m}c_{\ell m}\mathcal{Y}_{\ell m}(P) \quad (4)$$

$$\mathbf{f}(\mathbf{P}) \;=\sum_{\ell m}f_{\ell m}\frac{1}{\lambda_\ell}\mathbf{D}\mathcal{Y}_{\ell m}(\mathbf{P}) \quad (5)$$

with the potential coefficients $c_{\ell m}$ and the scalar coefficients of $\mathbf{f}$:

$$c_{\ell m}=<T,\mathcal{Y}_{\ell m}>\;,\; f_{\ell m}=\frac{1}{\lambda_\ell}<f,\mathbf{D}\mathcal{Y}_{\ell m}>. \quad (6)$$

Observation equation (1) can now be transformed to the spectral domain on the sphere:

$$f_{\ell m}=ax_{\ell m}+\lambda_\ell c_{\ell m} \qquad \forall(\ell,m). \quad (7)$$

The advantage of a spectral formulation is that the observation equation is now in pure diagonal form which simplifies the solution of the GBVP. No limits were given for the summation of the spectral components. Generally the lower limit for the degree will be zero, one or two, depending on the type of operator $\mathbf{D}$. The upper limit is infinity for a general L^2 function but will be finite for practical applications.

The operators $\mathbf{D}$ which are the most relevant for GBVPs are listed in Table (1).

2.1 General solution of the overdetermined GBVP

Since the observation equations are completely decoupled for each degree-order combination, the least squares solution of a combination of observations can be computed coefficient by coefficient. If we have the observation types $\mathbf{f}^1,\mathbf{f}^2,\ldots$ the observation equations are

$$\begin{pmatrix} f_{\ell m}^{i=1} \\ f_{\ell m}^{i=2} \\ \vdots \end{pmatrix}=\begin{pmatrix} a^{i=1} & \lambda_\ell^{i=1} \\ a^{i=2} & \lambda_\ell^{i=2} \\ \vdots & \vdots \end{pmatrix}\begin{pmatrix} x_{\ell m} \\ c_{\ell m}\end{pmatrix}. \quad (8)$$

To each of the observation types can be assigned a weight which may depend on the degree. Then the weight matrix becomes

$$P=\begin{pmatrix} p_\ell^{i=1} & 0 & \cdots \\ 0 & p_\ell^{i=2} & \cdots \\ \vdots & \vdots & \cdots \end{pmatrix}. \quad (9)$$

The implication of this choice of the weight matrix for the possible stochastic models we can use for the observation data will be presented in a future publication. The normal matrix then becomes

$$N=\begin{pmatrix} N_{11} & N_{12} \\ \cdot & N_{22}\end{pmatrix}=$$

$$\begin{pmatrix} \sum_i(a^i)^2p_\ell^i & \sum_i a^i\lambda_\ell^i p_\ell^i \\ \cdot & \sum_i(\lambda_\ell^i)^2p_\ell^i \end{pmatrix}. \quad (10)$$

The summation $\sum_i$ should be read as the sum over all observation types. The inverse normal matrix is

$$Q\equiv N^{-1}=$$

$$\frac{1}{E_\ell}\begin{pmatrix} \sum_i(\lambda_\ell^i)^2p_\ell^i & -\sum_i a^i\lambda_\ell^i p_\ell^i \\ \cdot & \sum_i(a^i)^2p_\ell^i \end{pmatrix}. \quad (11)$$

where, after some re-arranging,

$$E_\ell=\sum_{i,j>i}p_\ell^ip_\ell^j(\lambda_\ell^ia^j-\lambda_\ell^ja^i)^2. \quad (12)$$

The least squares solution is therefore

$$\hat{x}_{\ell m}=Q_{11}\sum_i a^ip_\ell^if_{\ell m}^i+Q_{12}\sum_i\lambda_\ell^ip_\ell^if_{\ell m}^i=$$

$$\frac{1}{E_\ell}\sum_i\left(\sum_j(a^i\lambda_\ell^j-a^j\lambda_\ell^i)p_\ell^j\lambda_\ell^j\right)p_\ell^if_{\ell m}^i=$$

$$\frac{1}{E_\ell}\sum_{i,j>i}(a^i\lambda_\ell^j-a^j\lambda_\ell^i)p_\ell^ip_\ell^j(\lambda_\ell^jf_{\ell m}^i-\lambda_\ell^if_{\ell m}^j) \quad (13)$$

and

$$\hat{c}_{\ell m} = Q_{12} \sum_i a^i p_\ell^i f_{\ell m}^i + Q_{22} \sum_i \lambda_\ell^i p_\ell^i f_{\ell m}^i =$$

$$\frac{1}{E_\ell} \sum_i \left(\sum_j (a^j \lambda_\ell^i - a^i \lambda_\ell^j) p_\ell^j a^j \right) p_\ell^i f_{\ell m}^i =$$

$$\frac{1}{E_\ell} \sum_{i,j>i} (a^j \lambda_\ell^i - a^i \lambda_\ell^j) p_\ell^i p_\ell^j (a^j f_{\ell m}^i - a^i f_{\ell m}^j) \,. \quad (14)$$

The solution for T in the space domain is, with (6 and 14),

$$\hat{T}(P) = \sum_{\ell m} \hat{c}_{\ell m} \mathcal{Y}_{\ell m}(P)$$

$$= \sum_{\ell m} \frac{1}{E_\ell} \sum_{i,j} a^j p_\ell^i p_\ell^j (a^j \lambda_\ell^i - a^i \lambda_\ell^j) \,\cdot$$

$$<\mathbf{f}^i(\mathbf{Q}), \mathbf{D}_\mathbf{Q}^i \frac{1}{\lambda_\ell^i} \mathcal{Y}_{\ell m}(\mathbf{Q})> \; \mathcal{Y}_{\ell m}(\mathbf{P})$$

$$= \sum_i <\mathbf{f}^i(\mathbf{Q}), \sum_j \mathbf{a}^j \mathbf{K}^{i,j}(\psi_{\mathbf{PQ}})>_\mathbf{Q} \,. \quad (15)$$

The kernel function is

$$\mathbf{K}^{i,j}(\psi) = \sum_{\ell \mathbf{m}} \kappa_\ell^{i,j} \mathbf{D}_\mathbf{Q}^i \mathcal{Y}_{\ell \mathbf{m}}(\mathbf{Q}) \mathcal{Y}_{\ell \mathbf{m}}(\mathbf{P}) \quad (16)$$

with

$$\kappa_\ell^{i,j} = \frac{a^j \lambda_\ell^i - a^i \lambda_\ell^j}{\lambda_\ell^i E_\ell} p_\ell^i p_\ell^j. \quad (17)$$

Some observations, such as disturbances, do not depend on the position. For them a slightly different solution is obtained. The observation equation reduces to

$$f_{\ell m} = \lambda_\ell c_{\ell m}. \quad (18)$$

Leaving out the intermediate steps, we obtain

$$Q = \frac{1}{\sum_i p_\ell^i (\lambda_\ell^i)^2} \,, \; \hat{c}_{\ell m} = \frac{\sum_i p_\ell^i \lambda_\ell^i f_{\ell m}^i}{\sum_i p_\ell^i (\lambda_\ell^i)^2} \quad (19)$$

and

$$\hat{T}(P) = \sum_i <\mathbf{K}^i(\psi_{\mathbf{PQ}}), \mathbf{f}(\mathbf{Q})>_\mathbf{Q} \quad (20)$$

with

$$\mathbf{K}^i(\psi) = \sum_{\ell \mathbf{m}} \kappa_\ell^{i,j} \mathbf{D}_\mathbf{Q}^i \mathcal{Y}_{\ell \mathbf{m}}(\mathbf{Q}) \mathcal{Y}_{\ell \mathbf{m}}(\mathbf{P}) \quad (21)$$

with

$$\kappa_\ell^{i,j} = \frac{p_\ell^i}{\sum_i p_\ell^i (\lambda_\ell^i)^2}. \quad (22)$$

The inner products in (15) and (20) can be rewritten to simple surface integrals if we look at specific operators. For operators $\mathbf{D}$ of vertical type (∂_z or ∂_{zz}) we have

$$\mathbf{D}_\mathbf{Q}^i \mathcal{Y}_{\ell \mathbf{m}}(\mathbf{Q}) = \lambda_\ell^i \mathcal{Y}_{\ell \mathbf{m}}(\mathbf{Q}). \quad (23)$$

Equations (16) and (21) can then be rewritten as

$$K^{i,j}(\psi) = \sum_\ell (2\ell + 1) \left[\frac{a^j \lambda_\ell^i - a^i \lambda_\ell^j}{E_\ell} p_\ell^i p_\ell^j \right] \cdot$$

$$P_\ell(\cos\psi), \quad (24)$$

$$K^i(\psi) = \sum_{\ell m} (2\ell + 1) \left[\frac{p_\ell^i \lambda_\ell^i}{\sum_i p_\ell^i (\lambda_\ell^i)^2} \right] \cdot$$

$$P_\ell(\cos\psi) \,. \quad (25)$$

and for (15) and (20) we can use

$$<\mathbf{K}(\psi_{\mathbf{PQ}}), \mathbf{f}(\mathbf{Q})> = \int_\Sigma \mathbf{K}(\psi_{\mathbf{PQ}}) \mathbf{f}(\mathbf{Q}) d\sigma_\mathbf{Q}. \quad (26)$$

If so desired, (15) can be written as

$$\hat{T}(P) = \sum_{i,j>i} <K^{i,j}(\psi_{PQ}), a^j \mathbf{f}^i(\mathbf{Q}) -$$

$$a^i \mathbf{f}^j(\mathbf{Q})>_\mathbf{Q}. \quad (27)$$

For operators $\mathbf{D}$ of horizontal type the inner product is not only an integration over Σ but also a vectorial inner product. For $\mathbf{D} = \{\partial_\mathbf{x}, \partial_\mathbf{y}\}$ we have

$$<\mathbf{K}(\psi_{\mathbf{PQ}}), \mathbf{f}(\mathbf{Q})>_\mathbf{Q} =$$

$$\int_\Sigma \left[K_x(\psi_{PQ}) f_x(Q) + K_y(\psi_{PQ}) f_y(Q) \right] d\sigma_Q. \quad (28)$$

From spherical trigonometry (α_Q: the azimuth at Q towards P, ψ: the spherical distance between P and Q) we have

$$\begin{pmatrix} \partial_x \\ \partial_y \end{pmatrix}_Q =$$

$$\begin{pmatrix} -\cos\alpha_Q & \sin\alpha_Q \\ -\sin\alpha_Q & -\cos\alpha_Q \end{pmatrix} \begin{pmatrix} \partial_\psi \\ \sin^{-1}\psi\partial_{\alpha_P} \end{pmatrix} \quad (29)$$

which yields for the integral solution

$$\int_\Sigma K_\psi(\psi_{PQ})\,[\cos\alpha_Q f_x(Q)+$$

$$\sin\alpha_Q f_y(Q)]\; d\sigma_Q; \quad (30)$$

with (cf. 16 and 21)

$$K_\psi(\psi) = -\partial_\psi \sum_\ell (2\ell+1)\kappa_\ell P_\ell(\cos\psi). \quad (31)$$

Likewise we have for the surface operator $\{\partial_{xx} - \partial_{yy}, 2\partial_{xy}\}$

$$<\mathbf{K}(\psi_{PQ}), \mathbf{f}(Q)>_Q =$$

$$\int_\Sigma [(K_{xx} - K_{yy})(\psi_{PQ})(f_{xx} - f_{yy})(Q) +$$

$$4K_{xy}(\psi_{PQ})f_{xy}(Q)]\; d\sigma_Q \quad (32)$$

which can be rewritten as

$$\int_\Sigma K_{\psi\psi}(\psi_{PQ})\,[\cos 2\alpha_Q(f_{xx} - f_{yy})(Q)-$$

$$\sin 2\alpha_Q(2f_{xy})(Q)]\; d\sigma_Q. \quad (33)$$

with

$$K_{\psi\psi}(\psi) = (\partial_{\psi\psi} - \cot\psi\partial_\psi)$$

$$\sum_\ell (2\ell+1)\kappa_\ell P_\ell(\cos\psi). \quad (34)$$

2.2 Closed expressions for the kernels

All the integral kernels $K(\psi)$ in the solution of the GBVP were written as series expansions in Legendre polynomials (16 and 21). By the splitting of the fraction in the series most of them can be written as a linear combination of the following functions (this works at least for the kernels for the unique GBVPs, for GBVP of overdeterminded type the series can be more complicated and a closed expression difficult to find)

$$F_A(\psi) = \sum_{n=2}^{\infty} \frac{1}{n+A} P_n(\cos\psi) \quad (35)$$

$$G_0(\psi) = \sum_{n=2}^{\infty} P_n(\psi)\,. \quad (36)$$

With (Moritz, 1980, Chap. 22)

$$F_{-1}(\psi) = 1 - 2\sin\frac{\psi}{2} - \cos\psi -$$

$$\cos\psi \ln(\sin\frac{\psi}{2} + \sin^2\frac{\psi}{2}) \quad (37)$$

$$F_0(\psi) = -\cos\psi - \ln(\sin\frac{\psi}{2} + \sin^2\frac{\psi}{2}) \quad (38)$$

$$F_1(\psi) = -1 - \tfrac{1}{2}\cos\psi + \ln(1 + \frac{1}{\sin\frac{\psi}{2}}) \quad (39)$$

$$F_2(\psi) = -\tfrac{3}{2} - \tfrac{1}{3}\cos\psi + 2\sin\frac{\psi}{2}$$

$$+\cos\psi \ln(1 + \frac{1}{\sin\frac{\psi}{2}}) \quad (40)$$

$$F_4(\psi) = \tfrac{5}{12} - \tfrac{1}{5}\cos\psi - \tfrac{5}{2}\cos^2\psi$$

$$+(5\cos^2\psi + \tfrac{5}{3}\cos\psi - \tfrac{2}{3})\sin\frac{\psi}{2}$$

$$+\tfrac{1}{2}\cos\psi(5\cos^2\psi - 3)\ln(1 + \frac{1}{\sin^2\frac{\psi}{2}}) \quad (41)$$

$$G_0(\psi) = -1 - \cos\psi - \frac{1}{2\sin\frac{\psi}{2}}\,. \quad (42)$$

3 Examples of GBVPs

With the general solution presented in the last section, some (free) GBVPs will be worked out here. Other GBVPs can be easily derived from either the general expression (13 - 15) or by specializing the solutions given below; i.e. by setting the weights of absent observables to zero.

3.1 Vertical GBVP

We have four observables which only depend on the vertical position displacement: the potential differences dW, the gravity anomaly dg, the vertical gradient of gravity $d\Gamma_{zz}$ and the vertical position itself (observed by e.g. GPS). When all observation equations are normalized, see (Rummel et al., 1989) the system becomes

$$\begin{pmatrix} dz_{\ell m} \\ dW_{\ell m} \\ dg_{\ell m} \\ d\Gamma_{\ell m} \end{pmatrix}$$

$$\begin{pmatrix} 1 & 0 \\ -1 & 1 \\ -2 & (\ell+1) \\ -6 & (\ell+1)(\ell+2) \end{pmatrix} \begin{pmatrix} dz_{\ell m} \\ c_{\ell m} \end{pmatrix} . \qquad (43)$$

The auxiliary function E_ℓ is (12):

$$E_\ell = p_\ell^W p_\ell^g (\ell-1)^2 + p_\ell^W p_\ell^\Gamma (\ell-1)^2(\ell+4)^2 +$$

$$4 p_\ell^g p_\ell^\Gamma (\ell-1)^2(\ell+1)^2 + p_\ell^g p_\ell^z (\ell+1)^2 +$$

$$p_\ell^\Gamma p_\ell^z (\ell+1)^2(\ell+2)^2 \qquad (44)$$

and the $c_{\ell m}$-part of the inverse normal matrix (11)

$$Q_{22} = \frac{p_\ell^z + p_\ell^W + 4p_\ell^g + 36 p_\ell^\Gamma}{E_\ell} . \qquad (45)$$

The solution for the potential coefficients is (14):

$$\hat{c}_{\ell m} = \frac{1}{E_\ell} \left[(\ell-1) p_\ell^W p_\ell^g (dg - 2dW)_{\ell m} + \right.$$

$$(\ell-1)(\ell+4) p_\ell^W p_\ell^\Gamma (d\Gamma - 6dW)_{\ell m} +$$

$$p_\ell^z p_\ell^W (dz + dW)_{\ell m} +$$

$$4(\ell-1)(\ell+1) p_\ell^g p_\ell^\Gamma (d\Gamma - 3dg)_{\ell m} +$$

$$(\ell+1) p_\ell^z p_\ell^g (2dz + dg)_{\ell m} +$$

$$\left. (\ell+1)(\ell+2) p_\ell^z p_\ell^\Gamma (6dz + d\Gamma)_{\ell m} \right] \qquad (46)$$

and for the vertical position (13):

$$\hat{z}_{\ell m} = \frac{1}{E_\ell} \left[(\ell-1) p_\ell^W p_\ell^g ((\ell+1) dW_{\ell m} - dg_{\ell m}) + \right.$$

$$(\ell-1)(\ell+4) p_\ell^W p_\ell^\Gamma ((\ell+1)(\ell+2) dW_{\ell m} - d\Gamma_{\ell m}) +$$

$$2(\ell-1)(\ell+1)^2 p_\ell^g p_\ell^\Gamma ((\ell+2) dg_{\ell m} - d\Gamma_{\ell m}) +$$

$$\left. (p_\ell^W + (\ell+1)^2 p_\ell^g + (\ell+1)^2(\ell+2)^2 p_\ell^\Gamma) p_\ell^z dz_{\ell m} \right] . \qquad (47)$$

The space domain solution is (26)

$$\hat{T}(P) = \int_\Sigma K^{W,z}(\psi_{PQ}) \left[dW + dz\right]_Q d\sigma$$

$$+ \int_\Sigma K^{g,z}(\psi_{PQ}) \left[dg + 2dz\right]_Q d\sigma$$

$$+ \int_\Sigma K^{\Gamma,z}(\psi_{PQ}) \left[d\Gamma + 6dz\right]_Q d\sigma$$

$$+ \int_\Sigma K^{g,W}(\psi_{PQ}) \left[dg - 2dW\right]_Q d\sigma$$

$$+ \int_\Sigma K^{\Gamma,W}(\psi_{PQ}) \left[d\Gamma - 6dW\right]_Q d\sigma$$

$$+ \int_\Sigma K^{\Gamma,g}(\psi_{PQ}) \left[d\Gamma - 3dg\right]_Q d\sigma \qquad (48)$$

with the kernels (16)

$$K^{W,z}(\psi) = \sum_\ell (2\ell+1) \frac{1}{E_\ell} p_\ell^W p_\ell^z P_\ell(\cos\psi)$$

$$K^{g,z}(\psi) = \sum_\ell (2\ell+1) \frac{(\ell+1)}{E_\ell} p_\ell^g p_\ell^z P_\ell(\cos\psi)$$

$$K^{\Gamma,z}(\psi) = \sum_\ell (2\ell+1) \frac{(\ell+1)(\ell+2)}{E_\ell} p_\ell^\Gamma p_\ell^z$$

$$P_\ell(\cos\psi)$$

$$K^{g,W}(\psi) = \sum_\ell (2\ell+1) \frac{(\ell-1)}{E_\ell} p_\ell^g p_\ell^W P_\ell(\cos\psi)$$

$$K^{\Gamma,W}(\psi) = \sum_\ell (2\ell+1) \frac{(\ell-1)(\ell+4)}{E_\ell} p_\ell^\Gamma p_\ell^W$$

$$P_\ell(\cos\psi)$$

$$K^{\Gamma,g}(\psi)=\sum_\ell(2\ell+1)\frac{(\ell-1)(\ell+1)}{E_\ell}p_\ell^\Gamma p_\ell^g$$

$$P_\ell(\cos\psi)\,. \tag{49}$$

3.2 Horizontal GBVP

The horizontal components of the gravity vector, the anomalies $d\Phi$ and $\cos\phi d\Lambda$, the gradients $d\Gamma_{xz}$ and $d\Gamma_{yz}$ and the horizontal coordinates themselves only depend on the horizontal position displacement:

$$\begin{pmatrix} \mathbf{dx}_{\ell m}\\ \mathbf{d\Phi}_{\ell m}\\ \mathbf{d\Gamma}_{\ell m}\end{pmatrix}=$$

$$\begin{pmatrix} 1 & 0\\ 1 & \sqrt{\ell(\ell+1)}\\ 3 & (\ell+2)\sqrt{\ell(\ell+1)}\end{pmatrix}\begin{pmatrix} \mathbf{dx}_{\ell m}\\ c_{\ell m}\end{pmatrix} \tag{50}$$

with

$$\begin{pmatrix} \mathbf{dx}_{\ell m}\\ \mathbf{d\Phi}_{\ell m}\\ \mathbf{d\Gamma}_{\ell m}\end{pmatrix}=\begin{pmatrix} \{dx,dy\}_{\ell m}\\ \{d\Phi,\cos\phi d\Lambda\}_{\ell m}\\ \{d\Gamma_{xz},d\Gamma_{yz}\}_{\ell m}\end{pmatrix}. \tag{51}$$

Then (we write F_ℓ instead of E_ℓ)

$$F_\ell=p_\ell^x p_\ell^\phi \ell(\ell+1)+p_\ell^x p_\ell^\Gamma(\ell+2)^2\ell(\ell+1)+$$

$$p_\ell^\phi p_\ell^\Gamma(\ell-1)^2\ell(\ell+1) \tag{52}$$

and the $c_{\ell m}$-part of the inverse normal matrix is

$$Q_{22}=\frac{p_\ell^x+p_\ell^\phi+9p_\ell^\Gamma}{F_\ell}. \tag{53}$$

The least squares estimate of the unknowns is

$$\hat{c}_{\ell m}=\frac{1}{F_\ell}\Big[\sqrt{\ell(\ell+1)}p_\ell^x p_\ell^\phi(\mathbf{d\Phi}-\mathbf{dx})_{\ell m}+$$

$$(\ell+2)\sqrt{\ell(\ell+1)}p_\ell^x p_\ell^\Gamma(\mathbf{d\Gamma}-\mathbf{3dx})_{\ell m}+$$

$$(\ell-1)\sqrt{\ell(\ell+1)}(\mathbf{d\Gamma}-\mathbf{3d\Phi})_{\ell m}\Big] \tag{54}$$

and

$$\hat{\mathbf{x}}_{\ell m}=\frac{1}{F_\ell}\Big[\ell(\ell+1)(p_\ell^\phi+(\ell+2)^2p_\ell^\Gamma)p_\ell^x\mathbf{dx}_{\ell m}+$$

$$(\ell-1)\ell(\ell+1)((\ell+2)\mathbf{d\Phi}-\mathbf{d\Gamma})_{\ell m})]\,. \tag{55}$$

For the space domain solution we write with (30):

$$\hat{T}(P)=\int_\Sigma \partial_\psi K^{\phi,x}(\psi_{PQ})\Big(\cos\alpha_Q\,[dx-d\Phi]_Q+$$

$$\sin\alpha_Q\,[dy-\cos\phi d\Lambda]_Q\Big)\,d\sigma_Q+$$

$$\int_\Sigma \partial_\psi K^{\Gamma,x}(\psi_{PQ})\Big(\cos\alpha_Q\,[3dx-d\Gamma_{xz}]_Q+$$

$$\sin\alpha_Q\,[3dy-d\Gamma_{xy}]_Q\Big)\,d\sigma_Q+$$

$$\int_\Sigma \partial_\psi K^{\Gamma,\phi}(\psi_{PQ})\Big(\cos\alpha_Q\,[3d\Phi-d\Gamma_{xz}]_Q+$$

$$\sin\alpha_Q\,[3\cos\phi d\Lambda-d\Gamma_{yz}]_Q\Big)\,d\sigma_Q \tag{56}$$

with the kernels

$$K^{\phi,x}(\psi)\;=\sum_\ell(2\ell+1)\tfrac{1}{F_\ell}p_\ell^\phi p_\ell^x P_\ell(\cos\psi)$$

$$K^{\Gamma,x}(\psi)\;=\sum_\ell(2\ell+1)\frac{(\ell+1)}{F_\ell}p_\ell^\Gamma p_\ell^x P_\ell(\cos\psi)$$

$$K^{\Gamma,\phi}(\psi)\;=\sum_\ell(2\ell+1)\frac{(\ell+1)(\ell+2)}{F_\ell}p_\ell^\Gamma p_\ell^\phi$$

$$P_\ell(\cos\psi)\,. \tag{57}$$

3.3 GBVP of torsion balance type

Here we have only one observable of the free GBVP. It is independent of the position, at least in the approximation we are working in, the pure horizontal part of the tensor of second derivatives:

$$\mathbf{d\Gamma}_{\ell m}=\{\mathbf{\Gamma_{xx}}-\mathbf{\Gamma_{yy}},\mathbf{2\Gamma_{xy}}\}_{\ell m}=$$

$$\sqrt{(\ell-1)\ell(\ell+1)(\ell+2)}\,c_{\ell m}. \tag{58}$$

We write

$$G_\ell=(p_\ell^\Gamma)^2(\ell-1)\ell(\ell+1)(\ell+2) \tag{59}$$

and the Q-matrix becomes

$$Q_{22} = \frac{p_\ell^\Gamma}{G_\ell}. \qquad (60)$$

The solution for the potential coefficients is

$$\hat{c}_{\ell m} = \frac{p_\ell^\Gamma}{G_\ell}\sqrt{(\ell-1)\ell(\ell+1)(\ell+2)}\,d\Gamma_{\ell m} =$$

$$\frac{1}{\sqrt{(\ell-1)\ell(\ell+1)(\ell+2)}}d\Gamma_{\ell m} \qquad (61)$$

with the space domain solution (33)

$$\int_\Sigma (\partial_{\psi\psi} - \cot\psi\partial_\psi)K(\psi_{PQ})\cdot$$

$$(\cos 2\alpha_Q\,[\Gamma_{xx} - \Gamma_{yy}]\,(Q) -$$

$$\sin 2\alpha_Q\,[2\Gamma_{xy}]\,(Q))\,d\sigma_Q \qquad (62)$$

with

$$K(\psi) = \sum_\ell \frac{2\ell+1}{\sqrt{(\ell-1)\ell(\ell+1)(\ell+2)}}$$

$$P_\ell(\cos\psi)\ . \qquad (63)$$

3.4 Kernels of some determined GBVPs

Specializing the solutions above we obtain the solution of various GBVPs. The kernels for the uniquely determined GBVPs are listed in Table (2). For the kernels for the horizontal GBVPs the appropriate derivative has been applied ($-\partial_\psi$ or $\partial_{\psi\psi} - \cot\psi\partial_\psi$). The summations over the degree were taken from degree two up to infinity.

3.5 Gradiometric GBVP

Five independent components build the tensor of second derivatives of the potential. They can be grouped into three combinations (Rummel and van Gelderen, 1992):

$$\begin{pmatrix} \Gamma^0 \\ \Gamma^1 \\ \Gamma^2 \end{pmatrix} = \begin{pmatrix} \Gamma_{zz} \\ \{\Gamma_{xz}, \Gamma_{yz}\} \\ \{\Gamma_{xx} - \Gamma_{yy}, 2\Gamma_{xy}\} \end{pmatrix}. \qquad (64)$$

Omitting the condition equations for Γ^1 and Γ^2 the model for the fixed gradiometric GBVP for observations at altitude h are

$$\begin{pmatrix} \Gamma^0_{\ell m} \\ \Gamma^1_{\ell m} \\ \Gamma^2_{\ell m} \end{pmatrix} =$$

$$\begin{pmatrix} \sigma^{\ell+3}(\ell+1)(\ell+2) \\ \sigma^{\ell+3}(\ell+2)\sqrt{\ell(\ell+1)} \\ \sigma^{\ell+3}\sqrt{(\ell-1)\ell(\ell+1)(\ell+2)} \end{pmatrix} c_{\ell m} \qquad (65)$$

where $\sigma = R/(R+h)$; R the mean earth radius. Then

$$Q = \frac{1}{H_\ell} \qquad (66)$$

with

$$H_\ell = \sigma^{2\ell+6}(\ell+1)(\ell+2)(p_\ell^0(\ell+1)(\ell+2)+$$

$$p_\ell^1\ell(\ell+2) + p_\ell^2(\ell-1)\ell), \qquad (67)$$

$$\hat{c}_{\ell m} = \frac{\sigma^{\ell+3}}{H_\ell}\Big((\ell+1)(\ell+2)p_\ell^0 d\Gamma^0_{\ell m}+$$

$$(\ell+2)\sqrt{\ell(\ell+1)}p_\ell^1 d\Gamma^1_{\ell m}+$$

$$\sqrt{(\ell-1)\ell(\ell+1)(\ell+2)}p_\ell^2 d\Gamma^2_{\ell m}\Big), \qquad (68)$$

and finally

$$\hat{T}(P) = \int_\Sigma (K^0(\psi_{PQ})d\Gamma_{zz}(Q) -$$

$$\partial_\psi K^1(\psi_{PQ})\,[\cos\alpha_Q d\Gamma_{xz}(Q) + \sin\alpha_Q d\Gamma_{yz}(Q)]$$

$$+(\partial_{\psi\psi} - \cot\psi\partial_\psi)K^2(\psi_{PQ})\cdot$$

$$[\cos 2\alpha_Q(d\Gamma_{xx} - d\Gamma_{yy})(Q) -$$

$$\sin 2\alpha_Q(2d\Gamma_{xy})(Q)])\cdot d\sigma_Q \qquad (69)$$

with

$$K^0(\psi) = \sum_\ell (2\ell+1)\frac{p_\ell^0\sigma^{\ell+3}(\ell+1)(\ell+2)}{H_\ell}P_\ell(\cos\psi)$$

$$K^1(\psi) = \sum_\ell (2\ell+1)\frac{p_\ell^1\sigma^{\ell+3}(\ell+2)\sqrt{\ell(\ell+1)}}{H_\ell}P_\ell(\cos\psi)$$

$$K^2(\psi) = \sum_\ell (2\ell+1)\frac{p_\ell^2\sigma^{\ell+3}\sqrt{(\ell-1)\ell(\ell+1)(\ell+2)}}{H_\ell}.$$

$$P_\ell(\cos\psi). \qquad (70)$$

Table 2: The singular values and the corresponding integral kernels for the uniquely determined GBVPs. For the kernels of the horizontal GBVPs the appropriate derivative of the kernel is given. The summation of the series is taken from degree two up to infinity.

Observations	Singular value	Kernel
$\{W, g\}$	$\frac{1}{n-1}$	$1 - 5\cos\psi + \frac{1}{\sin\frac{\psi}{2}} - 6\sin\frac{\psi}{2} - 3\cos\psi\ln(\sin\frac{\psi}{2} + \sin^2\frac{\psi}{2})$
$\{W, \Gamma\}$	$\frac{1}{(n-1)(n+4)}$	$\frac{71}{60} - \frac{22}{25}\cos\psi - \frac{7}{2}\cos^2\psi + (7\cos^2\psi + \frac{7}{3}\cos\psi - \frac{32}{15})\sin\frac{\psi}{2}$ $-\frac{3}{5}\cos\psi\ln(\sin\frac{\psi}{2} + \sin^2\frac{\psi}{2}) + \frac{7}{10}\cos\psi(5\cos^2\psi - 3)\ln(1 + \frac{1}{\sin\frac{\psi}{2}})$
$\{W, z\}$	1	$\delta(\psi)$
$\{g, \Gamma\}$	$\frac{1}{(n-1)(n+1)}$	$1 - 3\sin\frac{\psi}{2} - \frac{7}{4}\cos\psi - (\frac{3}{2}\cos\psi + \frac{1}{2})\ln\sin\frac{\psi}{2}$ $-\frac{1}{2}(3\cos\psi - 1)\ln(1 + \sin\frac{\psi}{2})$
$\{g, z\}$	$\frac{1}{n+1}$	$-1 - \frac{3}{2}\cos\psi + \frac{1}{\sin\frac{\psi}{2}} + \ln\frac{\sin\frac{\psi}{2}}{1+\sin\frac{\psi}{2}}$
$\{\Gamma, z\}$	$\frac{1}{(n+1)(n+2)}$	$-\frac{7}{2} - \frac{1}{2}\cos\psi + 6\sin\frac{\psi}{2} + (3\cos\psi - 1)\ln(1 + \frac{1}{\sin\frac{\psi}{2}})$
$\{x, \phi\}$	$\frac{1}{n(n+1)}$	$\frac{3}{2}\sin\psi - \cot\frac{\psi}{2}$
$\{x, \Gamma\}$	$\frac{1}{n(n+1)(n+2)}$	$-\frac{1}{4}\sin\psi - \frac{3}{2}\cos\frac{\psi}{2} + \frac{3}{2}\sin\psi\ln(1 + \frac{1}{\sin\frac{\psi}{2}}) + \frac{1}{2}\cos\frac{\psi}{2}\frac{3\sin^2\frac{\psi}{2} - 1}{1+\sin\frac{\psi}{2}}$
$\{\Gamma, \phi\}$	$\frac{1}{(n-1)n(n+1)}$	$\frac{7}{4}\sin\psi - 3\cos\frac{\psi}{2} + \frac{3}{2}\sin\psi\ln(\sin\frac{\psi}{2} + \sin^2\frac{\psi}{2}) + \frac{1-\sin\frac{\psi}{2}}{\cos\frac{\psi}{2}}$
$\{\Gamma_{xx} - \Gamma_{yy}, 2\Gamma_{xy}\}$	$\frac{1}{(n-1)n(n+1)(n+2)}$	$\frac{1}{2} + \frac{1}{2}\cos\psi$

4 Conclusions and outlook

A general solution for the GBVP in spherical, constant radius approximation was derived by using a least squares procedure in the spectral domain. As long as the relation between observable and disturbing potential diagonolizes in the spectral domain, all types of observations can be incorporated into this framework.

In future work more attention will be given to the stochastic properties of the boundary data and to the nature of the operators **D**.

References

Moritz, H. (1980). *Advanced physical geodesy*. Wichmann, Karlsruhe.

Rummel, R. and Teunissen, P.J.G. (1986). Geodetic boundary value problem and linear inference. In: *International Symposium Figure and Dynamics of the Earth, Moon, and Planetes*: 227-264, Prague.

Rummel, R., and van Gelderen, M. (1992). Spectral analysis of the full gravity tensor. *Geophys. J. Int.*, 111(1): 159-169.

Rummel, R., Teunissen, P.J.G, and van Gelderen, M. (1989). Unique and overdetermined geodetic boundary value problems by least squares. *Bull. Géod.*, 63: 1-33.

Sacerdote, F. and Sansò, F. (1985). Overdetermined boundary value problems in physical geodesy. *Manuscr. Geod.*, 10: 195-207.

Thong, N. and Grafarend, E.W. (1985). A spheroidal harmonic model for the terrestrial gravitational field. *Manuscr. Geod.*, 14: 285-304.

On an O(N) algorithm for the solution of geodetic boundary value problems

R. Klees, M. van Gelderen
Delft Institute for Earth-Oriented Space Research (DEOS),
Faculty of Civil Engineering and Geosciences,
Delft University of Technology, Thijsseweg 11, NL-2629 JA Delft, The Netherlands

Abstract. We study a fast algorithm for the solution of geodetic boundary value problems. The algorithm uses basis functions that ideally localize in space. It can handle any smooth enough boundary surface and does not require spherical and constant radius approximation. It solves a problem with N unknowns in $O(N)$ operations up to some logarithmic terms. A priori given satellite models can easily be taken into account without degrading the performance. Some numerical experiments based on a synthetic earth model show that the algorithm is suited for ultra-high resolution global gravity field recovery from terrestrial data on any hardware platform including PC's. For $N = 65538$ unknowns the matrix assembly takes less than 1 hour, and the solution of the linear system of equations using GMRES without any preconditioning takes little more than 1 hour. The accuracy obtained so far is not satisfactory yet and needs further investigation.

Keywords. Geodetic boundary value problems, fast numerical algorithm, panel clustering, multipole methods.

1 Introduction

(Klees et al., 1998a) proposed a fast algorithm, which theoretically allows to solve integral equations of the second kind defined on sufficiently smooth boundary surfaces in about $O(N)$ operations, where N denotes the number of unknowns. Our objective is to investigate the performance of the algorithm for the solution of geodetic boundary value problems in terms of computer time, storage requirements, and accuracy. The algorithm requires that the geodetic boundary value problem is first transformed into an integral equation of the second kind, which is always possible, cf. (Klees, 1997; Klees et al., 1998b). A numerical experiment is defined for a synthetic earth model, cf. (Haagmans, 1999), which has spectral characteristics similar to the Earth in terms of gravity field and topography. The determination of the disturbing potential outside the topography is formulated as a Robin problem for the Laplace operator.

The outline of the paper is the following: We start with the formulation of the mathematical model, which forms the basis of the fast algorithm. It is in fact the weak form of an integral equation over the boundary of the domain of harmonicity, which may be for instance the earth's surface or the telluroid, depending on the type of boundary value problem we consider. Then, we briefly discuss the main features of the fast algorithm. Finally, we describe the numerical experiment and discuss the results in terms of efficiency and accuracy.

2 The mathematical model

Our starting point is the linearized geodetic boundary value problem, which aims at the determination of the disturbing potential from given data on the boundary surface:

$$\begin{aligned} \Delta T(x) &= 0, \quad x \in \text{ext}\ \Gamma, \\ BT(x) &= f(x), x \in \Gamma, \\ T(x) &\to 0, \quad |x| \to \infty. \end{aligned} \tag{1}$$

T denotes the disturbing potential, B a differential operator of order smaller than two, and f the given boundary data. We assume the usual radiation condition. More general conditions are discussed for instance in (Lehmann and Kless, 1999). Generalization of the algorithm, which can handle these conditions are presented in (Klees et al., 1998a). The type of boundary condition in (1) contains as special cases all rel-

International Association of Geodesy Symposia, Vol. 121
Schwarz (ed.), Geodesy Beyond 2000 – The Challenges of the First Decade

evant boundary conditions to linearized geodetic boundary value problems, e.g. the Stokes and Molodensky boundary conditions with or without spherical and constant radius approximation, or the fixed gravimetric boundary value problem. Moreover, the extension to mixed boundary value problems is straightforward.

Next, we represent the disturbing potential in terms of a single layer potential with density defined on the boundary surface Γ:

$$T(x) = \int_{y \in \Gamma} k(x-y)\, u(y)\, d\Gamma(y), \; x \in \text{ext}\, \Gamma. \quad (2)$$

Observing the boundary condition and taking the limit to the boundary surface yields a linear operator equation relating the unknown single layer density u to the observations f:

$$(Au)(x) = f(x), \qquad x \in \Gamma. \quad (3)$$

The linear operator A is defined by

$$(Au)(x) := \lambda(x)u(x) - \int_{y \in \Gamma} k(x,y)\, u(y)\, d\Gamma(y), \qquad x \in \Gamma. \quad (4)$$

λ is a given function on Γ depending on the jump relation associated with the differential operator B and the single layer operator, and k is the kernel. Obviously, A is a linear integral operator of the second kind.

The algorithm is based on the weak form of the operator equation (3):

$$u \in L^2(\Gamma): \; \langle Au, v\rangle = \langle f, v\rangle, \quad \forall v \in L^2(\Gamma), \; (5)$$

where $\langle \cdot, \cdot \rangle$ denotes the $L^2(\Gamma)$-inner product. The approximate solution u_N of the weak form is sought in a subspace $V_N \in L^2(\Gamma)$ of finite dimension N. It is uniquely determined by demanding that the residuum $Au_N - f$ is orthogonal to the solution space, which is equivalent to

$$u_N \in V_N: \; \langle Au_N, v\rangle = \langle f, v\rangle, \quad \forall v \in V_N. \; (6)$$

Equation (6) defines a linear system of equations for the coefficients of the single layer density w.r.t. the basis $\{b_i : i = 1, \ldots, N\}$ of the solution space V_N:

$$\mathbf{A}\mathbf{u} = \mathbf{f}, \quad (7)$$

with

$$\mathbf{A} = (a_{ij})_{i,j=1,\ldots N}, \quad a_{ij} = \langle b_i, Ab_j\rangle \quad (8)$$
$$\mathbf{f} = (f_i)_{i=1,\ldots N}, \quad f_i = \langle b_i, f\rangle. \quad (9)$$

The vector $\mathbf{u}$ contains the coefficients of u_N w.r.t. the basis $\{b_i : i = 1, \ldots, N\}$. The stiffness matrix $\mathbf{A}$ has dimension $N \times N$ and is full. The reason lies in the fact that the kernel function $k(x, y)$ links every point $x \in \Gamma$ to every point $y \in \Gamma$. Hence, any numerical scheme is of the order $O(N^2)$. However, the application of this technique to geodetic boundary value problems requires the reduction of the complexity of $\mathbf{u} \to \mathbf{Au}$ from $O(N^2)$ to essentially $O(N)$ since N is large. This reduction should be possible for general boundary surfaces Γ and for any kernel k related to geodetic problems.

3 The fast algorithm

The fast algorithm is discussed in details in (Klees et al., 1999a), see also (Lage, 1999). We only want to summarize the main lines. The basic idea of the fast algorithm is in fact the approximation of the kernel k by a degenerate kernel k_m, that is, through the sum of a finite number of products of functions of only x and functions of only y:

$$k(x,y) \approx k_m(x, y; c_\sigma, c_\tau) = \sum_{(\mu,\nu)\in I_m} \kappa_{\mu\nu}(c_\sigma, c_\tau) X_\mu(x; c_\sigma) Y_\nu(y; c_\tau). \quad (10)$$

I_m denotes an index set and $m \in \mathcal{N}_0$; c_σ and c_τ are points in $\mathcal{R}^3$. The degenerate kernel approximation allows to separate the x-integration (due to the $L^2(\Gamma)$ scalar product of the weak formulation) from the y-integration (due to the integral operator), which reduces the amount of work substantially. Since the kernel k satisfies for some $s \geq 0$ and for all $\alpha, \beta \in \mathcal{N}_0^3$

$$|D_x^\alpha D_y^\beta k(x,y)| \leq C(\alpha,\beta) \cdot |x-y|^{-(s+|\alpha|+|\beta|)}, \quad (11)$$

that is, it decays very fast as function of the distance between x and y, we can control the approximation error $|k - k_m|$ as function of the distance between any pair of points $x, y \in \mathcal{R}^3$. That is, for given $c_\sigma, c_\tau \in \mathcal{R}^3$ and given $m \in \mathcal{N}$ and $0 < \eta < 1$ it holds for all $x, y \in \mathcal{R}^3$ and $|y - c_\tau| + |x - c_\sigma| \leq \eta |c_\sigma - c_\tau|$:

$$|k(x,y) - k_m(x,y)| \leq C_\eta \eta^m \, |k(x,y)| \quad (12)$$

There are several possibilities how to choose k_m, e.g. Taylor expansion, multipole expansion or Chebyshev expansion. In our experiments we use a truncated multipole expansion. Then, the functions X_μ and Y_ν are homogeneous harmonic polynomials of positive degree and the expansion coefficients $\kappa_{\mu\nu}$ are homogeneous harmonic polynomials of negative degree.

The space V_N is commonly a finite element space of piecewise polynomials on a triangulation of Γ. The realization of $\mathbf{u} \to \mathbf{Au}$ involves the computation of integrals over every pair of triangles (panels). If the panels are well separated we can form union of panels (clusters) and use the degenerate kernel approximation to approximate their contribution to the realization of $\mathbf{u} \to \mathbf{Au}$. This is equivalent to the splitting of the stiffness matrix $\mathbf{A}$ into the *near-field part* $\mathbf{N}$ and the *far-field part* $\mathbf{F}$, and to approximate the far-field part by replacing k by k_m:

$$\mathbf{A} = \mathbf{N} + \mathbf{F} \approx \mathbf{N} + \tilde{\mathbf{F}}, \tag{13}$$

$$\tilde{\mathbf{F}} = \sum_{(\sigma,\tau)} \mathbf{X}_\sigma^\mathbf{T} \mathbf{F}_{\sigma\tau} \mathbf{Y}_\tau. \tag{14}$$

The $N \times N$ near-field matrix $\mathbf{N}$ should only contain $O(N)$ non-zero entries. The degenerate kernel approximation allows to represent the approximate far-field matrix $\tilde{\mathbf{F}}$ by a finite sum of the products of three types of matrices $\mathbf{X}_\sigma, \mathbf{F}_{\sigma\tau}$, and $\mathbf{Y}_\tau$. The entries of the matrices $\mathbf{X}_\sigma$ involve only integrations over x, those of $\mathbf{Y}_\tau$ only over y, and the matrices $\mathbf{F}_{\sigma\tau}$ contain the coefficients $\kappa_{\mu,\nu}$. The realization $\mathbf{u} \to \mathbf{Nu}$ involves integrals over pairs of panels and is computed as usual. The realization $\mathbf{u} \to \tilde{\mathbf{F}}\mathbf{u}$, however, is done by

$$\tilde{\mathbf{F}}\mathbf{u} = \sum_{\sigma,\tau} \mathbf{X}_\sigma(\mathbf{F}_{\sigma\tau}(\mathbf{Y}_\tau \mathbf{u})), \tag{15}$$

i.e., without direct access to the entries of $\tilde{\mathbf{F}}$. In particular it does not require to form the matrices $\mathbf{X}_\sigma, \mathbf{F}_{\sigma\tau}, \mathbf{Y}_\tau$ explicitly. Due to the existence of shift operators for multipole expansions, cf. (Greengard and Rokhlin, 1997) we only have to assemble the matrices $\mathbf{X}_\sigma$ and $\mathbf{Y}_\tau$ for σ and τ, respectively, being panels. These matrices, however, are sparse with $O(m^2)$ entries (Greengard and Rokhlin, 1997).

The sparseness of the near-field matrix $\mathbf{N}$ and the fast evaluation of $\tilde{\mathbf{F}}\mathbf{u}$ depend on how the clusters are formed. The clusters we use are organized in a tree. The panels are the leaves of the tree, and the whole boundary surface Γ is the root of the tree. First, we subdivide the root into four about equally large sets. Then, the subdivision is recursively continued until the subsets contain $O(1)$ panels. The set of all clusters obtained in that way forms a hierarchical subdivision of the boundary surface. Next, we can form pairs of clusters (σ, τ), the so-called partition of $\Gamma \times \Gamma$. If

$$r_\sigma + r_\tau \leq \eta |c_\sigma - c_\tau|, \qquad 0 < \eta < 1, \tag{16}$$

where r_σ, r_τ denote the radii of the clusters σ, τ and c_σ, c_τ the cluster centers, the clusterpair σ, τ belongs to the far-field of the partition. Otherwise, the clusters are subdivided into smaller clusters by running one level deeper in the tree. If the finest level is reached, the level of the panels, the remaining pairs of panels define the near-field of the partition. The clusterpairs in the far-field define the realization $\mathbf{u} \to \tilde{\mathbf{F}}\mathbf{u}$, the pairs of panels in the near-field define the realization $\mathbf{u} \to \mathbf{Nu}$. The approximate linear system of equations $(\mathbf{N} + \tilde{\mathbf{F}})\mathbf{u} = \mathbf{f}$ is then solved using e.g. the Generalized Minimum Residuum (GMRES) method (Saad and Schultz, 1986).

The complexity of the fast algorithm can be expressed in terms of the dimension N of the solution space V_N and the order m of the multipole expansion, cf. (Hackbusch and Novak, 1989). The organization of the clusters in a tree guarantees that the $N \times N$ near-field matrix $\mathbf{N}$ is sparse and contains only $O(N)$ entries. Moreover, the approximate far-field computation requires $O(Nm^4)$ operations and the storage requirements are of order $O(Nm^2)$. In order to ensure that the error of the far-field approximation is asymptotically equal to the order of the discretization error, we have to choose $m = O(\log N)$.

Table 1: Number of elements and minimum resolution for each level

LEVEL	TRIANGLES	NODES	RES. (deg.)
4	2048	1026	8.7
5	8192	4098	4.4
6	32768	16386	2.2
7	131072	65538	1.1

4 The numerical experiment

In order to study the performance of the algorithm some numerical test computations were carried out. The boundary value problem we used is an artificial Robin problem which does not relate to a physical problem. Moreover, for simplicity dimensionless quantities are used. The

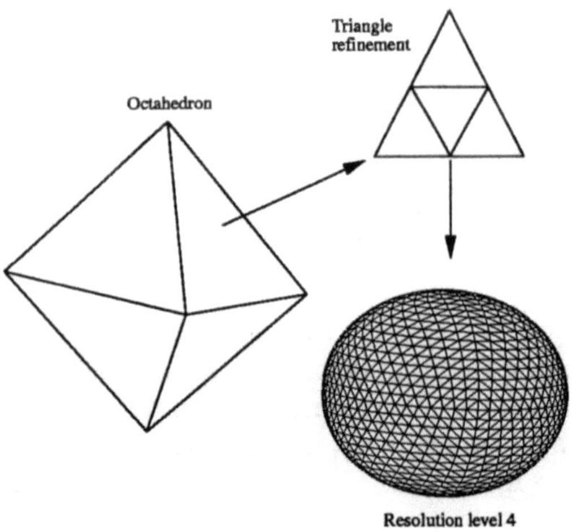

Fig. 1: The panel subdivision

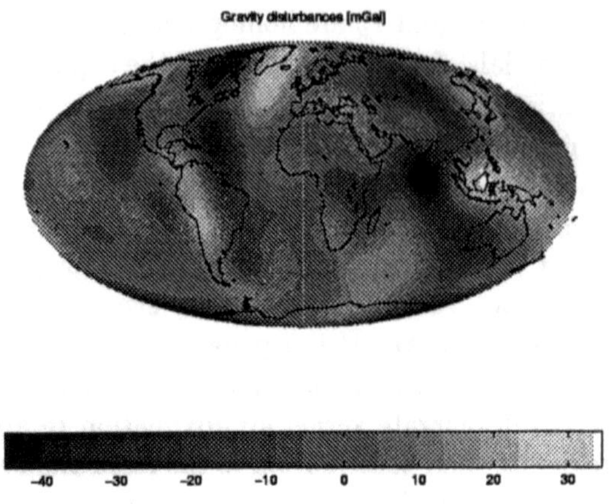

Fig. 2: The boundary data (without zero-degree term)

boundary value problem is:

$$\begin{aligned} \Delta T(x) &= 0, \quad x \in \text{ext } \Gamma, \\ -T(x) + \frac{\partial T}{\partial n} &= f(x), x \in \Gamma, \\ T(x) &\to 0, \quad |x| \to \infty. \end{aligned} \tag{17}$$

For the boundary Γ a sphere was taken. Its triangulation was generated by using an octahedron for approximation level 0 and projecting the corners onto the sphere. For the higher levels each triangle was subsequently subdivided into 4 congruent subtriangles by halving the sides; see figure 1. For an overview of the number of panels etc. see table 1.

A synthetic gravity field was generated with a maximum resolution of approximately 750 km (Haagmans, 1999). See figure 2 for a picture of the boundary data for the BVP computed from this field. This field resembles as closely as possible the true gravity field of the earth but filtered such that its content can be represented by the sampling points of the level 4 triangulation, which have a maximal distance of 8.7 degrees.

5 Results and discussion

We carried out several test runs in order to study the performance of the fast algorithm. For comparison some runs were made with the standard BEM algorithm; i.e., the stiffness matrix in equation (7) was computed directly without kernel approximation. The linear system of equations were solved using GMRES without preconditioning or restart. The order m of the cluster expansion (10) was fixed to $m = 3$ or $m = 6$.

The experiments were done on a HP-C180 workstation with 512Mb internal memory. Due to memory constraints the simulations were only carried out up to level 5 for the standard BEM and up to level 6 for the fast algorithm. The results for higher levels were obtained by extrapolation. They are indicated in the figures by dashed lines. The maximum distance between the data points at level six is about 2.2 degree. Since piecewise linear polynomials have been used as trial functions, the maximum resolution the trial space provides is about one level higher. Therefore, a level-7 solution represents similar details as a spherical harmonic expansions up to degree and order 360.

The performance of the fast algorithm shows up in the CPU time for the matrix assembly and the solution of the linear system of equations when compared with the CPU time for the standard BEM algorithm. In figure 3 the CPU time for the matrix assembly is shown. Note the logarithmic scale of the vertical axis. For level 7 we estimate a speed up of about 3 orders of magnitude! The computations were done with $m = 3$. Higher values of m will slightly increase the CPU time (Klees et al., 1999a; Lage, 1999).

Figure 4 shows the CPU time for the solution of the linear system of equations. There is no difference between the standard BEM and the fast algorithm. About 30 iterations were necessary in all tests, independent of the number of unknowns. This indicates the good conditioning of the stiffness matrix for the standard BEM and the fast algorithm.

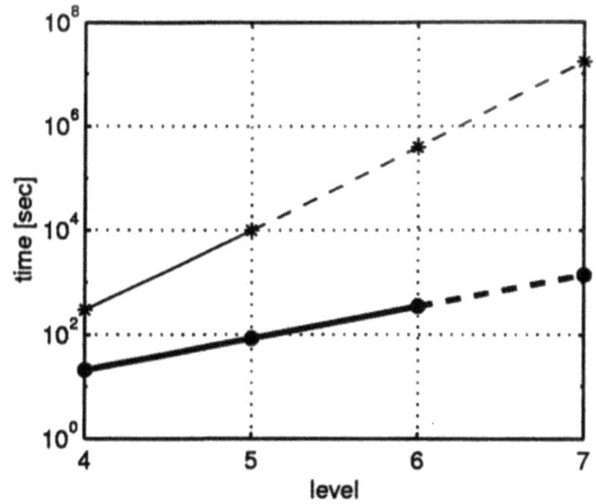

Fig. 3: The CPU time for stiffness matrix assembly (bold: fast algorithm, m=3; normal: standard BEM algorithm)

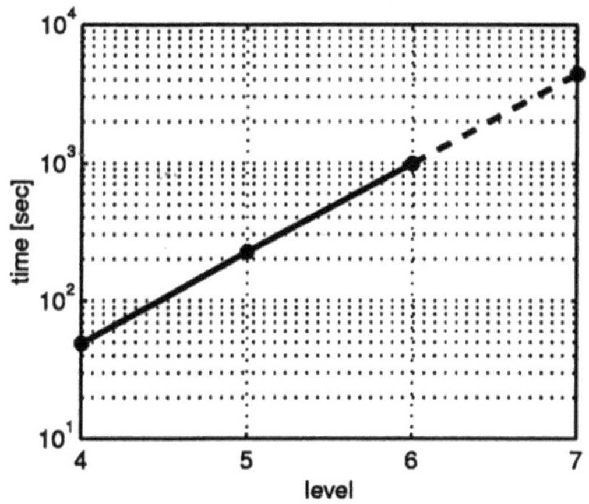

Fig. 4: The CPU time for the solution of the linear system of equations using GMRES

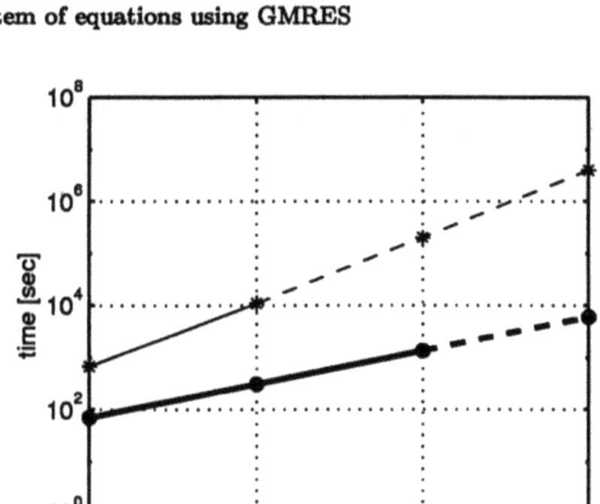

Fig. 5: The total CPU time for the solution of the BVP (bold: fast algorithm; normal: standard BEM algorithm)

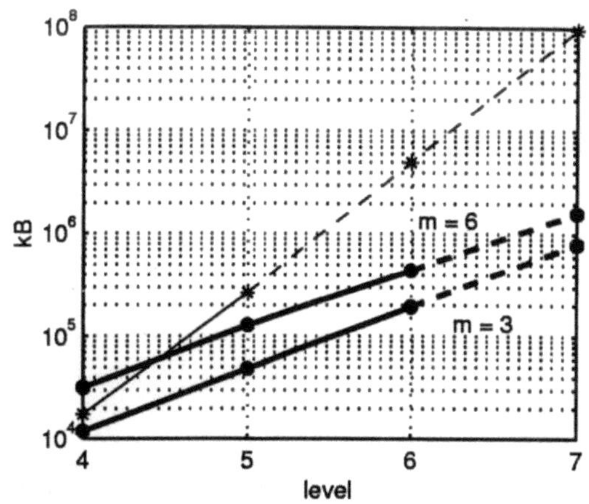

Fig. 6: Storage requirements (bold: fast algorithm; normal: standard BEM algorithm)

In figure 5 the total CPU time, matrix assembly plus solution of the linear system, is shown. It can be clearly seen that the fast algorithm is over two orders of magnitude faster than the standard BEM algorithm.

The storage requirements for the standard BEM algorithm increase very fast with higher levels. Due to the efficient approximation of the matrix-vector product **Fu**, which is the basic operation in the GMRES algorithm, much memory can be saved with the fast algorithm: up to two orders of magnitude for level 7, depending on the order m of the cluster expansion; see figure 6. The results shown for $m = 3$ and $m = 6$ indicate that the storage requirements increase with increasing degree of the multipole expansion. Theoretically, the increase is proportional to m^2, which may cause a serious problem for large m. The choice of m depends on the dimension N of the solution space, i.e. we have to choose $m = O(\log N)$ in order to keep the error due to the degenerate kernel approximation below the error of the Galerkin discretization. The optimal choice of m for given N needs further investigations.

For practical applications the accuracy of the approach is an important criterium. With accuracy we mean the magnitude of the errors of functionals of the geopotential field in points outside or on the boundary surface. The accuracy depends, however, on many parameters, the error due to the kernel approximation in the far-field is just one example, and by far not the most important one. Other error sources such as discretization error, numerical integration error, and boundary surface approximation error

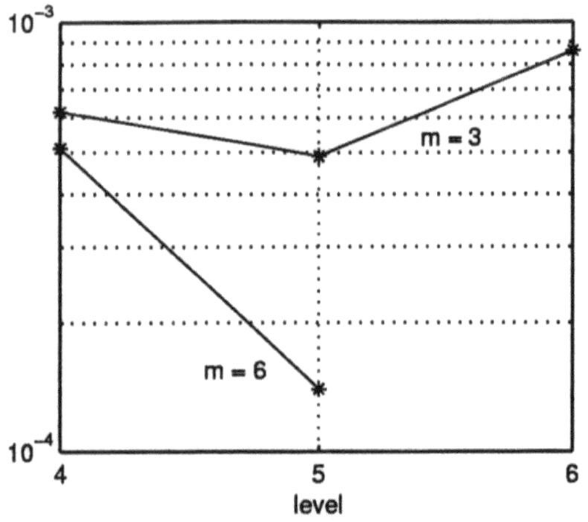

Fig. 7: Relative rms potential errors in a number of control points at 15 km height above the boundary surface

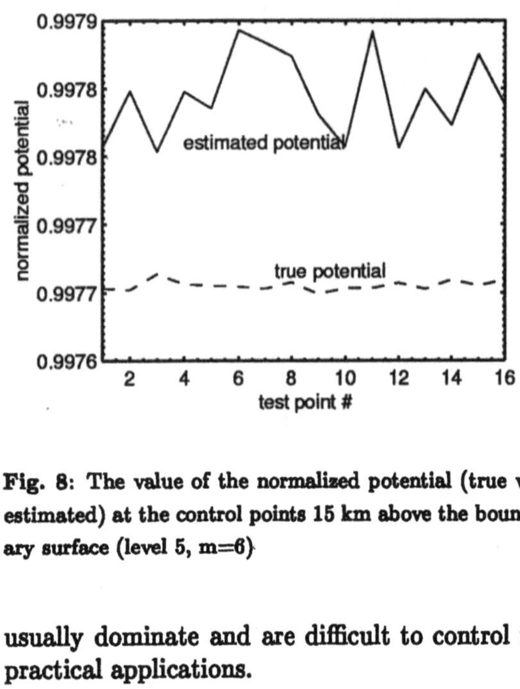

Fig. 8: The value of the normalized potential (true vs. estimated) at the control points 15 km above the boundary surface (level 5, m=6)

usually dominate and are difficult to control in practical applications.

In figure 7 the relative accuracy of the solution, i.e. with respect to the true potential, evaluated at a height of 15 km above the boundary surface, is shown. We observe that for $m = 3$ we do not get a steady improvement by augmenting the resolution. Similar has also been reported by (Kleess et al., 1998a; Lage, 1999). It is likely that for low m the error of the kernel approximation dominates the error budget. We expect that for higher m the error function behaves monotone. Figure 7 also indicates that a higher order $m = 6$ of the cluster expansion clearly improves the results compared to $m = 3$; unfortunately we were not able to compute a solution at level 6 with $m = 6$ or to increase m significantly due to memory constraints. The accuracy of our solution is reasonable but still not very satisfactory. A $10^{-3} - 10^{-4}$ relative error means that only the zero-degree term can be estimated, the much smaller high resolution components of the geopotential field are hardly resolved at all. See figure 8 for a picture of the error in a number of test points. Clearly the solution obtained is biased. So far, no satisfactory explanation was found for this behaviour.

6 Conclusions

The numerical experiments confirm that the algorithm is in fact essentially of order $O(N)$. The speed up to compared to the standard BEM algorithm growths exponentially with the number of unknowns. For level 7, which corresponds to a degree and order 360 spherical harmonic expansions, the gain is about three orders of magnitude. Therefore, high-resolution global and regional gravity field recovery is feasible with the fast algorithm on workstations and PC's. The memory requirements are stronger. Current applications require about $1 - 2$ GB RAM. In geodetic applications the order m of the cluster expansion is a major concern. Low values may result in approximation errors that dominate other error sources. More studies are needed in order to figure out the optimal choice of m for a given resolution. The accuracy obtained so far is not satisfactory yet. The bias observed in the solution requires further investigations.

Acknowledgements. Roger Haagmans developed and implemented the synthetic earth model we used in our test computations. His support is gratefully acknowledged.

References

Greengard, L. and Rokhlin, V. (1997). A new version of the fast multipole method for the Laplace equation in three dimensions. *Acta Numerica*, 6: 229-269.

Haagmans, R. (1999). A synthetic earth model for use in geodesy. Submitted to Journal of Geodesy.

Hackbusch, W. and Novak, Z. (1989). On the fast matrix multiplication in the BEM by panel clustering. *Numer. Math*, 54: 463-491.

Klees, R. (1997). Topics on the boundary value problems. In: F. Sansò and R. Rummel (eds.) *Geo-*

detic boundary value problems in view of the one centimeter geoid, Lecture Notes in Earth Sciences 65: 482-531, Springer, Berlin.

Klees, R., Lage, C., and Schwab, C. (1998a). Fast numerical solution of the vector Molodenski problem. *DEOS Progress Letter*, 98.1: 31-42.

Klees, R., Ritter, S., and Lehmann, R. (1998b). Integral equation formulations for geodetic mixed boundary value problems. *DEOS Progress Letter*, 98.2: 1-14.

Klees, R., Lage, C., and Schwab, C. (1999a). Research Report, Seminar für Angewandte Mathematik, ETH Zürich.

Klees, R., Lage, C., and Schwab (1999b). Fast numerical solution of the vector Molodenski problem. In: *Proc. IV Hotine-Marussi Symposium on Mathematical Geodesy*, Trento, Italy.

Lage, C. (1999). *Advanced boundary element algorithms*. Research Report 99-11, Seminar für Angewandte Mathematik, ETH Zürich.

Lehmann, R. and Klees, R. (1999). Numerical solution of geodetic boundary value problems using a global reference field. Accepted for publication in Journal of Geodesy.

Saad, Y. and Schultz, M. (1986). GMRES: a generalized minimal residual algorithm for solving nonsymmetric linear systems. *SIAM J. Scientific and Statistical Computing* 7: 856-869.

The multigrid method for satellite gravity field recovery

J. Kusche, S. Rudolph
Institute of Theoretical Geodesy,
University of Bonn, Nußallee 17, 53115 Bonn, Germany

Abstract. Dedicated SST- or gradiometry missions like CHAMP, GRACE and GOCE will provide gravity field information of unprecedented resolution and precision. It has been recognized that better gravity field models and estimates of the geoid are useful for a wide range of research and application, including ocean circulation and climate change studies, physics of the earth's interior and height datum connection and unification (ESA 1996, NRC 1997). The computation of these models will require the solution of large normal equation systems, especially if "brute force" approaches are applied. Evidently there is a need for fast solvers. The multigrid method (MGM) is not only an extremely fast iterative solution technique, it yields a well-defined sequence of coarser approximations as a by-product to the final gravity field solution. We investigate the application of MGM to satellite gravity field recovery using space-localizing kernel functions, for theoretical as well as numerical aspects.

Keywords. Satellite gravity recovery, multigrid method, fast solvers, regularization.

1 Mathematical setting

The mathematical setting for gravity field recovery, as far as considered in this paper, is as follows:

We assume the geopotential $T \in H$, where H is a Hilbert space equipped with a reproducing kernel. The reproducing kernel may be identified with a covariance function of T describing the state of knowledge of the geopotential power density. A linear operator $\mathcal{A}_{(n)} : H \to \mathbb{R}^n$ is supposed to map T onto n satellite observations l_i of SST or gradiometry type,

$$l_i + \epsilon_i = \mathcal{A}_i T = (A_i, T)_H , \qquad i = 1..n . \quad (1)$$

Here, $A_i \in H$ are Riesz's representers of observation functionals. The normal equations read

$$\mathcal{A}^*_{(n)} \mathcal{A}_{(n)} T = \mathcal{A}^*_{(n)} l_{(n)} , \quad (2)$$

where $\mathcal{A}^*_{(n)}$ is the adjoint operator. Due to the ill-posed (ill-conditioned) nature of the downward continuation process, one usually regularizes the problem. Applying the Tykhonov regularization with parameter α, the normals take the form

$$\left(\mathcal{A}^*_{(n)} \mathcal{A}_{(n)} + \alpha \mathcal{I}\right) T = \mathcal{A}^*_{(n)} l_{(n)} , \quad (3)$$

where $\mathcal{I}$ is the identity operator. For numerical purposes, these equations have to be discretized. The most popular technique is Galerkin least-squares projection. If we look for a solution in a finite-dimensional subspace $H_j = \text{span}\{\Phi_i, i = 1..u_j\}$, the normal equations for the approximation coefficients finally yield

$$(A^T A + \alpha M)\, \chi = A^T l \quad (4)$$

with Gramian matrix $M_{ij} = (\Phi_i, \Phi_j)_H$ and design matrix $A_{ij} = \mathcal{A}_i \Phi_j = (A_i, \Phi_j)_H$. Common choices for the basis of the approximation subspaces are the spherical harmonics $\Phi_j = Y_{jk}, k = -j \ldots j$, the $u_j = n$ representers of the observation functionals $\Phi_j = A_j$ (i.e. in least-squares collocation), or a predefined system of radial base functions (i.e. harmonic localizing kernels) $\Phi_j = \sum_0^\infty \frac{2n+1}{4\pi}\, \varphi_n\, P_n(\cdot, q_j)$.

The equation system (4), at least in global applications, contains several thousands of unknowns. Besides, the optimal value of α is not known a-priori and has to be determined from parameter choice rules like generalized cross-validation or the L-curve method. Then (4) has to be solved several times and for condition numbers varying over several orders of magnitude. Fast iterative solvers are capable to handle the problem within acceptable computation time. A linear iteration of (3) reads

$$T^{k+1} = T^k + \mathcal{R}^\alpha \Big(\mathcal{A}^*_{(n)} l_{(n)} - (\mathcal{A}^*_{(n)} \mathcal{A}_{(n)} + \alpha \mathcal{I})\, T^k\Big) = (\mathcal{I} - \alpha \mathcal{R}^\alpha)\, T^k + \mathcal{R}^\alpha \mathcal{A}^*_{(n)} (l_{(n)} - \mathcal{A}_{(n)} T^k) . \quad (5)$$

International Association of Geodesy Symposia, Vol. 121
Schwarz (ed.), Geodesy Beyond 2000 – The Challenges of the First Decade

where $\mathcal{R}^\alpha$ is an easy computable approximation of the operator inverse to (3) and

$$||\mathcal{I} - \mathcal{R}^\alpha(\mathcal{A}^*_{(n)}\mathcal{A}_{(n)} + \alpha\mathcal{I})|| = \kappa < 1 \quad (6)$$

must be guaranteed to ensure the convergence of (5). But the actual convergence rate of iterative techniques strongly depends on the individual structure of the equation system under consideration, which, on the other hand, depends on the chosen base functions as well as the satellite data distribution and the numbering scheme of the unknowns. Considering the case of space-localizing functions, unfortunately the normal system (4) lacks a special block structure (suitable for block Jacobi techniques) as well as sparsity.

Fast solvers are preconditioned conjugate gradient algorithms (PCCGA) and multigrid methods (MGM). The application of PCCGA and block-Jacobi techniques to gravity recovery from SST and SGG is discussed by Schuh et al. (1996), where spherical harmonics are used as basis. MGM algorithms cannot be applied as black-box algorithm, but when adopted to the problem they may run faster than PCCGA. Moreover, as a by-product they yield a well-defined sequence of coarser approximations to the final solutions in subspaces $H_1 \subset \cdots \subset H_{j-1} \subset H_j$.

2 The multigrid method (MGM)

The multigrid approach has been developed originally during the sixties for the iterative solution of discrete elliptic boundary value problems. MGM iterations belong to the class of fastest iterations because their convergence rate is independent of the discretization width. Introductory texts are Bramble (1993), Hackbusch (1995) or Braess (1996). In Xu (1992) the multigrid method is presented within a large group of iterative techniques from a unified point of view.

The principle of MGM is simple: Approximate solutions with *smooth errors* are obtained very efficiently by applying standard *relaxation methods* like Richardson iteration, Jacobi overrelaxation (JOR), successive overrelaxation (SOR), symmetric SOR (SSOR), or block versions of these methods. Here smoothness means that the short wavelenghts of the errors are reasonable damped, where the error is defined with respect to the discrete solution of the problem. Because of the error smoothness, corrections of these approximations can be calculated cheaply on *coarser grids*. This basic idea can be used recursively employing coarser and coarser grids. Only on the coarsest grid a direct solution must be computed. The *V-cycle-algorithm* for the solution of the discretized equation $\mathcal{L}_j : H_j \to H_j$

$$\mathcal{L}_j v_j = f_j \quad (7)$$

is as follows:

V-cycle algorithm

Step 1. Pre-smoothing

$$v_j^{k+1/3} = v_j^k + \mathcal{S}_j(f_j - \mathcal{L}_j\, v_j^k) \quad (8)$$

Step 2. Coarse-Grid Correction

$$v_j^{k+2/3} = v_j^{k+1/3} + \mathcal{R}_{j-1}\mathcal{Q}_{j-1}(f_j - \mathcal{L}_j\, v_j^{k+1/3})$$

Step 3. Post-smoothing

$$\begin{aligned} v_j^{k+1} &= v_j^{k+2/3} + \mathcal{S}_j(f_j - \mathcal{L}_j\, v_j^{k+2/3}) \\ &= v_j^k + \mathcal{R}_j(f_j - \mathcal{L}_j\, v_j^k) \end{aligned}$$

Exact Solution in H_1

$$\mathcal{R}_1 = (\mathcal{L}_1)^{-1}$$

Here k is the iteration index and j denotes the grid index or discretization level. It should be noted that the iteration operator $\mathcal{R}_j$ is defined recursively by $\mathcal{R}_{j-1}$. $\mathcal{S}_j$ is a relaxation or smoothing operator and $\mathcal{Q}_{j-1}$ denotes projection of the defect or residual equation onto coarser space.

In matrix-vector formulation (7) reads $\boldsymbol{L}\boldsymbol{v} = \boldsymbol{f}$. Within the algorithm (8), a sequence of auxiliary problems $\boldsymbol{L}_{i-1}\delta\boldsymbol{v}_{i-1} = \boldsymbol{d}_{i-1}$ is solved instead (the iteration index is omitted here). One has to implement intergrid transfers

$$\begin{aligned} \boldsymbol{L}_{i-1} &= \boldsymbol{R}\boldsymbol{L}_i\boldsymbol{P} \\ \boldsymbol{d}_{i-1} &= \boldsymbol{R}\boldsymbol{d}_i = \boldsymbol{R}(\boldsymbol{f}_i - \boldsymbol{L}_i\boldsymbol{v}_i) \\ \delta\boldsymbol{v}_i &= \boldsymbol{P}\,\delta\boldsymbol{v}_{i-1}\,. \end{aligned} \quad (9)$$

The *restriction matrix* $\boldsymbol{R}$ and the *prolongation matrix* $\boldsymbol{P}$ are crucial for the efficiency of the algorithm. Usually they are sparse and need not be stored as full matrices. The multigrid iteration process can be further accelerated by means of the following concepts:

Nested iteration. Good initial approximations v_i^0 on *each grid* $H_i, i = 1..j$ can be found by performing a few multigrid steps, beginning with v_1^0.

Stopping rule. The MGM iteration may be safely stopped when the iteration error $||v_j^k - v_j||$

is smaller than the discretization error $||v_j - v||$. An estimate for the iteration error is

$$||v_j^k - v_j|| \leq \frac{\kappa}{1-\kappa} ||v_j^k - v_j^{k-1}|| \tag{10}$$

where κ is the contraction number of the iteration process.

3 Iterative solution of regularized satellite normal equations by MGM

For normal equations resulting from Tykhonov regularization, a simple smoother of Richardson type can be given by

$$\mathcal{S}_j = \alpha^{-1}\mathcal{I}_j \,, \tag{11}$$

see Maaß and Riederer (1996). In matrix formulation this requires the computation of $\boldsymbol{M}^{-1}$, which may be well approximated by $\boldsymbol{I}$ provided the base functions are normalized with respect to H. However, convergence of (11) cannot be guaranteed for arbitrary small α.

Since we would like to use a *nested sequence* $H_{j-1} \subset H_j$ of approximation subspaces, there must exist coefficients c_{ik} such that

$$\Phi_{i,j-1} = \sum_k c_{ik}\, \Phi_{k,j} \tag{12}$$

holds for the base functions $\Phi_{i,j-1}, \Phi_{i,j}$. Using the *canonical* prolongation and restriction

$$\boldsymbol{R}_{ik} = c_{ik} \qquad \boldsymbol{P} = \boldsymbol{R}^T \tag{13}$$

it is clear that

$$\chi_{j-1}^{k+1} = \chi_{j-1}^k + \delta\chi_{j-1} \tag{14}$$

converges towards the solution of the original problem (3) when discretized in H_{j-1}. Here, $\delta\chi_{j-1}$ solves the defect equation in H_{j-1}

$$\boldsymbol{R}(\boldsymbol{A}^T\boldsymbol{A} + \alpha\boldsymbol{M})\boldsymbol{R}^T\delta\chi_{j-1} = \boldsymbol{R}\boldsymbol{d}_j \,. \tag{15}$$

In the final approximation space H_j

$$\chi_j^{k+1} = \chi_j^k + \boldsymbol{R}^T\delta\chi_{j-1} \tag{16}$$

reasonably approximates the solution of (4), provided that the space H_{j-1} as well as the c_{ik} are "well-chosen".

4 Spherical context and relations to other multilevel techniques

Now we turn to a *spherical setting*, assuming space-localizing kernel functions

$$\Phi_{k,j} = \Phi_j(\cdot, q_k) = \sum_{n=0}^{\infty} \frac{2n+1}{4\pi}\, \varphi_n^j\, P_n(\cdot, q_k) \tag{17}$$

as basis in the approximation space H_j. P_n are the Legendre polynomials and $q_k \in \Omega$, where Ω is a suitable chosen Bjerhammar sphere. It seems natural to assume that the restriction coefficients originate from a spherical radialsymmetric function

$$c_{ik} = c(q_i, q_k) = \sum_{n=0}^{\infty} \frac{2n+1}{4\pi}\, c_n\, P_n(q_i, q_k) \tag{18}$$

which should be of local support in order to minimize the computational burden. Nevertheless, the base functions $\Phi_{k,i}, i < j$ defined on the coarser spaces are in general not radialsymmetric.

Assuming large u_j, the restriction equation (12) may be regarded as an approximate convolution integral

$$\begin{aligned} \Phi_{i,j-1} &\approx \int_\Omega c(q_i, q')\, \Phi_{j-1}(\cdot, q')\, d\omega' \\ &= \sum_{n=0}^{\infty} \frac{2n+1}{4\pi}\, c_n\, \varphi_n^j\, P_n(\cdot, q_i)\,. \end{aligned} \tag{19}$$

For the coarsest grid, allowing different restriction functions on each intergrid transfer, we have

$$\begin{aligned} \Phi_{i,1} &\approx c^1 * \cdots * c^{j-1} * \Phi_{i,j} \\ &= \sum_{n=0}^{\infty} \frac{2n+1}{4\pi}\, c_n^1 \cdots c_n^{j-1}\, \varphi_n^j\, P_n(\cdot, q_i)\,. \end{aligned} \tag{20}$$

Clearly, the base functions on each grid tend towards a low-pass filtered version of the fine-grid kernel function. Vice versa, $\Phi_{i,j}$ can be written as a sum of low-pass and band-pass versions

$$\Phi_{i,j} \approx \sum_{l=1}^{j} d^l * \Phi_{i,j} \tag{21}$$

where

$$\begin{aligned} d^1 &= c^1 * \cdots * c^{j-1} \\ d^l &= (1 - c^{l-1}) * c^l \cdots * c^{j-1} \qquad 1 < l < j\,, \end{aligned} \tag{22}$$

thus resembling the constructions of certain spherical wavelets. This can be used to obtain desired effects:

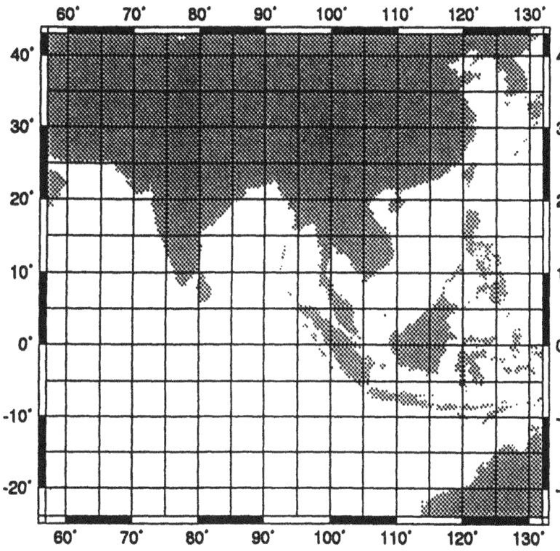

Fig. 1: Area under consideration

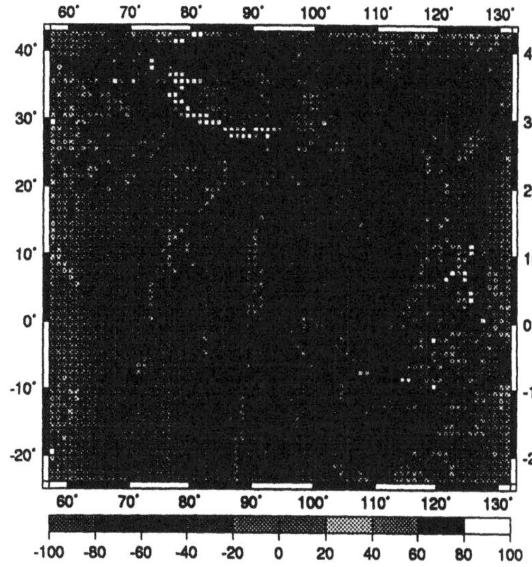

Fig. 2: (Pseudo–) true gravity anomalies

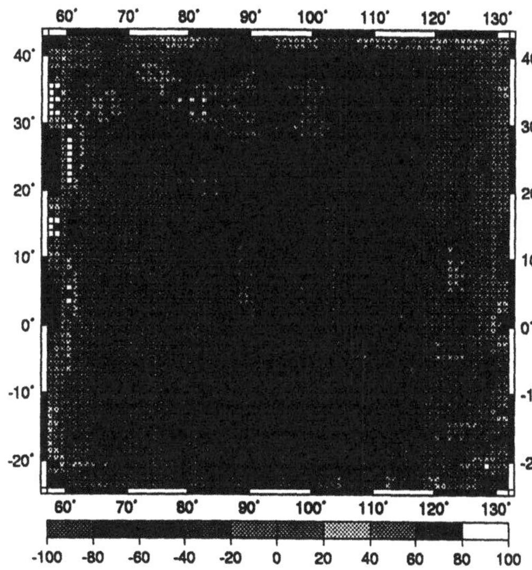

Fig. 3: Initial approximation

Example: Poisson-kernel restriction. Choose a sequence $0 < \sigma^1 < \cdots < \sigma^{j-1} < 1$. With

$$\begin{aligned} c_n^{j-1} &= (\sigma^{j-1})^n \\ c_n^{j-2} &= (\sigma^{j-2}/\sigma^{j-1})^n \\ &\vdots \\ c_n^1 &= (\sigma^1/\sigma^2)^n \end{aligned} \tag{23}$$

the $\Phi_{k,i}, k < j$ appear as damped versions of $\Phi_{k,j}$, related by a specific depth ratio σ^{j-l}. The restriction functions (18) are Poisson kernels with the choice (23).

5 Numerical example

A numerical example should demonstrate the application of the MGM technique to satellite gravity recovery. Satellite data and normal matrix have been created within a simulated experiment of GRACE type, i.e. a low–low high–precision SST mission, based on the EGM96 spherical harmonic geopotential model. An analysis period of 31 days was chosen. Intersatellite range–rate measurements have been generated at a sampling rate of 0.2Hz and corrupted with gaussian white noise of $\sigma = 1\mu m/s$. This means that a total number of 535,000 simulated observations has been processed. The methology is presented in Kusche et al (1998).

The area under consideration was chosen to be $[57°..132°] \times [-24°..43°]$, see Figure 1. A set of 5,025 $1° \times 1°$ mean anomaly blocks, computed from the EGM96 model complete up to degree and order 360 and reduced by a low–degree reference model, is shown in Figure 2.

In this example one of the simplest MGM algorithms was implemented: the V–cycle algorithm (8) using only two grids. The coarse grid was obtained by standard coarsening, i.e. as a $2° \times 2°$ grid. A restriction function was designed in a way that (12) acts as a 9–point stencil. This means $c(q_i, q_k)$ vanishes if the distance between q_i and q_k exceeds $\sim 150km$. Figure 3 illustrates the initial approximation obtained from the coarse–grid solution of (4).

Convergence rates, obtained with two different relaxation methods, are given in Figure 4. Here JOR denotes the Jacobi overrelaxation smoother whereas ALPHA means the Richardson smoother (11). Estimated as well as true rates fit very good together.

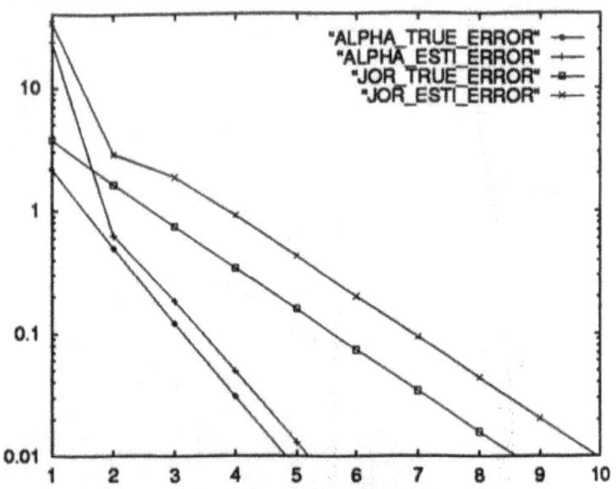

Fig. 4: Convergence rates (rms of estimated (10) as well as true iteration error [mGal] vs. cycle)

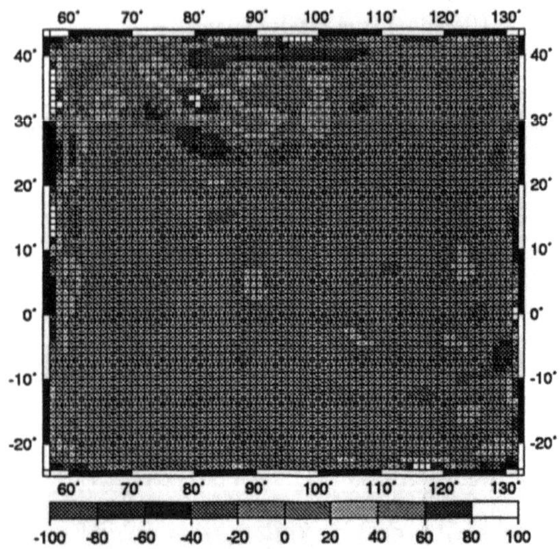

Fig. 5: Final approximation

It should be noted that the actual convergence rate strongly depends on the regularization parameter. It is obvious that the simple Richardson smoother (11) works fine within the two–grid algorithm and its convergence rate is superior to the more elaborated JOR smoother. This, however, is valid only if the regularization parameter is "well–chosen", for example already near the corner of the L–curve. With JOR the two–grid algorithm also converges if α is a coarse guess.

The iterated approximation after 7 cycles is given in Figure 5, where the estimated iteration error is far below $0.1mGal$. When compared to the pseudo–true field (Figure 2) one observes that the main features are well–detected. Clearly one cannot hope to recover the gravity field at $1° \times 1°$ resolution from a GRACE–type mission with satisfying S/N ratio. Especially ocean ridge features of alternating sign remain invisible within this experiment. But the important result is that a suitable approximation is available after very few iteration cycles, this means by far less than 1% computation time when compared to a direct solver.

6 Outlook

We found that the MGM offers a flexible tool to produce fast solutions for regularized satellite normal equations, which will arise from the dedicated gravity missions of the next decade. We expect that global high–resolution solutions, based on space–localizing base functions, will be possible within acceptable computation time. Further investigations will concern the convergence behaviour of the algorithm, the self–regularizing properties of the iterative schemes as well as a refinement of the smoothing operator with regard to the individual regularization parameter.

Acknowledgement. This work was supported in part (S. R.) by the Deutsche Forschungsgemeinschaft under grant IL17/5–1.

References

Braess, D. (1996). *Finite Elemente.* Springer, Berlin

Bramble, J. H. (1993). *Multigrid Methods.* Pitman Research Notes in Mathematics Series, Harlow

ESA (1996). *Gravity Field and Steady–State Ocean Circulation Mission.* ESA Publications Division, Reports for Assessment, Noordwijk

Hackbusch, W. (1985). *Multi–Grid Methods and Applications.* Springer Series in Computational Mathematics, Berlin

Kusche, J., Ilk, K. H. and Rudolph, S. *Two–step data analysis for future satellite gravity field solutions: A simulation study.* Proceedings of the 2nd Joint Meeting of IGC/IGeC at Trieste, Italy, 7–12 September 1998, to appear in Bollettino Geofisica

Maaß, P. and A. Riederer (1996). *Wavelet-accelerated Tykhonov-Regularization with Applications*, preprint, University of Potsdam

NRC (1997). *Satellite Gravity and the Geosphere.* National Academy Press, Washigton D.C.

Schuh, W.–D., Sünkel, H., Hausleitner, W. and Höck, E. (1996) *Refinement of Iterative Procedures for the Reduction of Spaceborne Gravimetry Data.* Study of advanced reduction methods for spaceborn gravimetry data, and of data combination with geophysical parameters, CIGAR IV Final Rep., ESTEC/JP/95-4-137/MS/nr

Xu, J. (1992). *Iterative methods by Space decomposition and Subspace Correction.* SIAM Review, 34(4): 581-613

Numerical realization of a new iteration procedure for the recovery of potential coefficients

M.S. Petrovskaya, A.N. Vershkov
Main (Pulkovo) Astronomical Observatory of Russian Academy of Sciences
Pulkovskoe Shosse 65, St. Petersburg, 196140, Russia; e-mail: petrovsk@gao.spb.ru

N.K. Pavlis
Raytheon STX Corporation, 7701 Greenbelt Road Suite 400, Greenbelt MD 20770, USA
e-mail: npavlis@geodesy2.gsfc.nasa.gov

Abstract. The application of the standard iteration procedure, developed by Pellinen, Rapp and Cruz for recovering the spherical harmonic coefficients $\bar{C}_{n,m}$ of the earth's potential, reveals their «exotic» behavior, as a consequence of the unbounded increment of the ellipsoidal correction terms with increasing the degree n. The new iteration procedure, proposed by Petrovskaya (1999), is more appropriate for evaluating high degree potential coefficients. In the present paper the efficiency of this procedure is studied numerically for $n \leq 358$. As the input data, the same surface gravity anomaly is used as was applied for constructing EGM96 geopotential model. From the coefficients $\bar{C}_{n,m}$ derived by the standard and new iteration procedures the corresponding gravity anomaly degree variances are evaluated. The similar variances are also estimated for $\bar{C}_{n,m}$ obtained by applying Jekeli's ellipsoidal harmonic approach for deriving the spherical harmonic potential coefficients. It appears that the variances corresponding to the new iteration procedure are close to the ones derived by Jekeli's approach, as opposed to the standard iteration procedure. Several possible applications of the derived solution for $\bar{C}_{n,m}$ are discussed.

Keywords. Gravitational potential coefficients, gravimetric boundary value problem solution, iteration procedure.

1 Introduction

When constructing the geopotential models on the base of surface gravity data Δg two main analytical approaches are applied. In the standard approach, developed in particular by Pellinen (1982), Rapp and Cruz (1986), the surface gravimetric data are reduced to the external sphere Σ of the minimal radius and then a relation is derived between the potential coefficients $\bar{C}_{n,m}$ and the spectral characteristics of Δg. This relation is solved with respect to $\bar{C}_{n,m}$ by an iteration procedure. An alternative approach was proposed by Jekeli (1981, 1988). It is based on reducing the gravity data to an external ellipsoid surface, expanding the earth's potential on/outside it in the ellipsoidal harmonic series and then transferring by analytical relations from the ellipsoid harmonic coefficients to the spherical harmonic ones, $\bar{C}_{n,m}$. This solution is more strict as compared to the above iteration one due to taking into account higher degrees of e^2. At the same time, the ellipsoidal harmonic approach is much more complicated because the relations between the spherical and ellipsoidal harmonic coefficients represent very slowly convergent hypergeometric series. Besides, small corrections are left in the boundary value (BV) equation in form of the infinite series, depending on the unknown coefficients $\bar{C}_{n,m}$, while in the iteration procedure there are no such series and only some constants are slightly modified. Jekeli's approach was used for constructing the geopotential models OSU91 (Rapp and Pavlis, 1990) and EGM96 (Lemoine et al., 1998).

It is shown analytically by Petrovskaya (1999) that the standard iteration procedure is ineffective for evaluating the potential coefficients of high degree and order. With enlarging the degree n in the solution for $\bar{C}_{n,m}$ the ellipsoidal correction exceeds the spherical approximation and tends to infinity by the absolute value with $n \to \infty$. In (ibid.) another iteration solution was proposed for evaluating $\bar{C}_{n,m}$, derived from the same BV equation. The corresponding solution for $\bar{C}_{n,m}$ behaves quite «natural»: the principal term (spherical

International Association of Geodesy Symposia, Vol. 121
Schwarz (ed.), Geodesy Beyond 2000 – The Challenges of the First Decade

approximation) is always dominating as compared to the ellipsoidal correction.

In the present paper a detailed numerical analysis is carried out for estimating the efficiency of different iteration procedures. From the standard and new iteration solutions for the coefficients $\overline{C}_{n,m}$ the gravity anomaly degree variances are evaluated and compared with the ones derived from the same gravity data by Jekeli's approach.

2 Standard and new iteration procedures

In the present paper a more exact *BV* relation is used than was applied in (Rapp and Cruz, 1986) and (Petrovskaya, 1999). It was derived by Jekeli (1981) and Pavlis (1988) and utilized for constructing the geopotential models OSU91 (Rapp and Cruz, 1990) and EGM96 (Lemoine *et al.*, 1998). This equation includes, besides the ellipsoidal corrections ε_h and ε_γ, an additional correction ε_p, representing the effect of the difference between the gravity anomaly and the isozenital projection of the gravity anomaly vector. (As was indicated by Pavlis (1988, pp. 44-45), the corrections ε_h and ε_p are of the same magnitude). From this equation the following basic relation is derived between the potential coefficients and the ones for the gravity anomaly:

$$\overline{C}_{n,m} = \overline{C}^{(0)}_{n,m} - e^2(n-1)q_{nm}\overline{C}_{n,m} - e^2(n-1)\beta_{nm}\overline{C}_{n-2,m} - \\ - e^2(n-1)\gamma_{nm}\overline{C}_{n+2,m}, \; n = 2,3,\ldots, |m| \le n \quad (1)$$

where

$$\overline{C}^{(0)}_{n,m} = \frac{a^2}{4\pi GM(n-1)}\int_\sigma \Delta g_E \overline{Y}_{n,m}(\theta,\lambda)d\sigma, \\ \Delta g_E = \Delta g - H\frac{\partial \Delta g}{\partial H}. \quad (2)$$

Here GM is the gravitational constant multiplied by the earth's mass; a, e are the semi-major axis and the first eccentricity of the normal ellipsoid; θ, λ are the polar angle and longitude of a point under consideration; $\overline{Y}_{n,m}(\theta,\lambda)$ are the fully normalized spherical functions; H is the normal (or orthometric) height; σ is the unit sphere. By $\overline{C}_{n,m}$ the fully normalized harmonic coefficients of the earth's disturbing potential are designated, scaled with respect to a. By q_{nm}, β_{nm} and γ_{nm} the bounded numerical constants are designated.

The quantity $\overline{C}^{(0)}_{n,m}$ in (1) – (2) is induced by the spherical harmonic coefficient for Δg_E, representing the surface gravity anomaly Δg reduced to the ellipsoid surface. It can be also treated as the spherical approximation for the «disturbed» coefficient $\overline{C}_{n,m}$.

The equation (1) can be solved with respect to $\overline{C}_{n,m}$ in different ways. For instance, the following iteration procedure can be applied:

$$\overline{C}^{(k)}_{n,m} = \overline{C}^{(0)}_{n,m} - e^2(n-1)q_{nm}\overline{C}^{(k-1)}_{n,m} - e^2(n-1)\times \\ \times\beta_{nm}\overline{C}^{(k-1)}_{n-2,m} - e^2(n-1)\gamma_{nm}\overline{C}^{(k-1)}_{n+2,m}, \; k = 1,2,3,\ldots \quad (3)$$

where k is the number of iteration.

At $k = 1$ it gives (after omitting the upper index in $\overline{C}^{(1)}_{n,m}$):

$$\overline{C}_{n,m} = \alpha_{nm}\overline{C}^{(0)}_{n,m} - e^2(n-1)\beta_{nm}\overline{C}^{(0)}_{n-2,m} - \\ - e^2(n-1)\gamma_{nm}\overline{C}^{(0)}_{n+2,m} \quad (4)$$

where

$$\alpha_{nm} = 1 - e^2(n-1)q_{nm}.$$

Formula (3) defines the standard iteration procedure and (4) gives the solution in the first approximation. Since $q_{nm} \approx 1/4$ then the principle ellipsoidal correction terms in (3) and (4) increase infinitely by the absolute value with $n \to \infty$.

From the same basic relation (1) – (2), by means of its elementary transformation, another iteration formula was derived in (ibid.) in the following way. In (1) the terms depending on $\overline{C}_{n,m}$ from the left and right hand sides were combined together. After dividing the resulting equation by the obtained coefficient of $\overline{C}_{n,m}$ the new iteration procedure was developed:

$$\overline{C}^{(k)}_{n,m} = a_{nm}\overline{C}^{(0)}_{n,m} - e^2(n-1)b_{nm}\overline{C}^{(k-1)}_{n-2,m} - \\ - e^2(n-1)c_{nm}\overline{C}^{(k-1)}_{n+2,m}, \; k = 1,2,\ldots$$

The following initial approximation is taken, instead of (2) for the iteration procedure (3):

$$\overline{C}^{(1)}_{n,m} = a_{nm}\overline{C}^{(0)}_{n,m}. \quad (5)$$

At k=2 one has

$$\overline{C}_{n,m}=\overline{C}^{(1)}_{n,m}-e^2(n-1)\,b_{nm}\overline{C}^{(1)}_{n-2,m}-$$
$$-e^2(n-1)\,c_{nm}\overline{C}^{(1)}_{n+2,m} \qquad (6)$$

where the upper index of $\overline{C}^{(2)}_{n,m}$ is omitted.

In (5) and (6) the constants a_{nm}, b_{nm} and c_{nm} are as follows:

$$a_{nm}=\left[1+e^2(n-1)\,q_{nm}\right]^{-1},$$
$$b_{nm}=a_{nm}\,p_{nm}u_{nm},$$
$$c_{nm}=a_{nm}\,r_{nm}v_{nm}.$$

Here

$$p_{nm}=\frac{(n^2+n-2)(n-\overline{m})(n-\overline{m}-1)}{2(n-1)^2(2n-3)(2n-1)};\ \overline{m}=|m|,\ n=4,5,...$$

$$p_{nm}=0,\ n=2,3,$$

$$q_{nm}=\frac{4n(n+1)-4m^2-23}{16(n-1)^2}+\frac{5(4m^2-1)}{16(n-1)^2(2n+3)(2n-1)},$$

$$r_{nm}=\frac{(n^2+n-2)(n+\overline{m}+2)(n+\overline{m}+1)}{2(n-1)^2(2n+3)(2n+5)},$$

$$u^2_{nm}=\frac{(2n-3)(n+\overline{m}-1)(n+\overline{m})}{(2n+1)(n-\overline{m})(n-\overline{m}-1)};\ u_{nm}=0,\ \overline{m}=n-1,n,$$

$$v^2_{nm}=\frac{(2n+5)(n-\overline{m}+1)(n-\overline{m}+2)}{(2n+1)(n+\overline{m}+1)(n+\overline{m}+2)}.$$

All the coefficients of $\overline{C}^{(1)}_{p,m}$ ($p=n$, n-2, n+2) on the right hand side of (6) are bounded by the absolute value for any n and m, as opposed to those of $\overline{C}^{(0)}_{p,m}$ in (4).

Formulas (5) and (6) provide the accepted approximation for the new iteration solution.

3 Numerical tests for estimating the efficiencies of the iteration procedures

For revealing the efficiency of the new iteration procedure as compared with the standard one we compare both sets of the corresponding spherical harmonic coefficients with $\overline{C}_{n,m}$ derived by the ellipsoidal harmonic approach (Jekeli, 1981, 1988). The calculations are performed proceeding from a more simple *BV* relation, than was used for deriving the basic formula (1). In it the ellipsoidal corrections are omitted which are much smaller than the principal ellipsoidal correction term. The reduced equation differs from (1) only by slightly different values of the numerical coefficients. Such approximation is sufficient for discovering the main features of the discussed sets of the potential coefficients.

To evaluate the spherical approximation coefficients defined in (2) (that is Δg_E coefficients) the same global gravity data are used as was applied for constructing the geopotential model EGM96 (Lemoine et al, 1998). The most important of them are the gravity measurements on the earth's surface, the satellite altimeter and satellite tracking data.

Three sets of the potential coefficients, $\overline{C}^{\,k}_{n,m}$ ($k=1,2,3$), are calculated, corresponding to the standard iteration procedure ($k=1$), the new iteration procedure ($k=2$) and the ellipsoidal harmonic approach ($k=3$). Then three sets the corresponding gravity anomaly degree variances $c^{(k)}_n$ ($k=1,2,3$) are evaluated by the formula

$$c^{(k)}_n=\left(\frac{GM}{a^2}\right)^2(n-1)^2\sum_{m=-n}^{n}\left[\overline{C}^{(k)}_{n,m}\right]^2,\ k=1,2,3. \qquad (7)$$

Two sets of the degree variances for differences are calculated:

$$\Delta^{(k)}_n=\left(\frac{GM}{a^2}\right)^2(n-1)^2\sum_{m=-n}^{n}\left[\overline{C}^{(3)}_{n,m}-\overline{C}^{(k)}_{n,m}\right]^2,\ k=1,2. \qquad (8)$$

The corresponding differences in percentages are also estimated:

$$\delta^{(k)}_n=\sqrt{\frac{\Delta^{(k)}_n}{c^{(3)}_n}}\times 100,\ k=1,2. \qquad (9)$$

Table 1: Degree variances for differences (δ_n %)

N	Standard iteration max (2 ≤ n ≤ N)	New iteration max (2 ≤ n ≤ N)
70	0.87 %	0.65 %
100	1.81	1.09
150	4.28	2.35
200	8.78	4.05
250	13.52	5.53
300	22.76	7.71
358	33.14	9.91

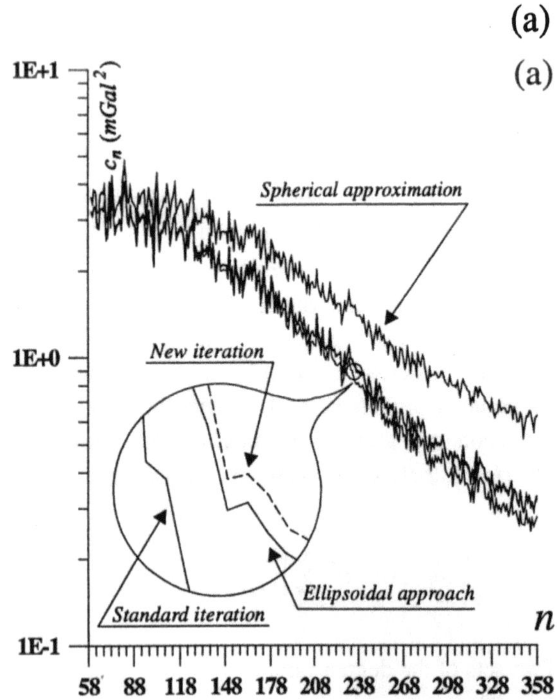

Fig. 1a: Gravity anomaly degree variances over the whole interval

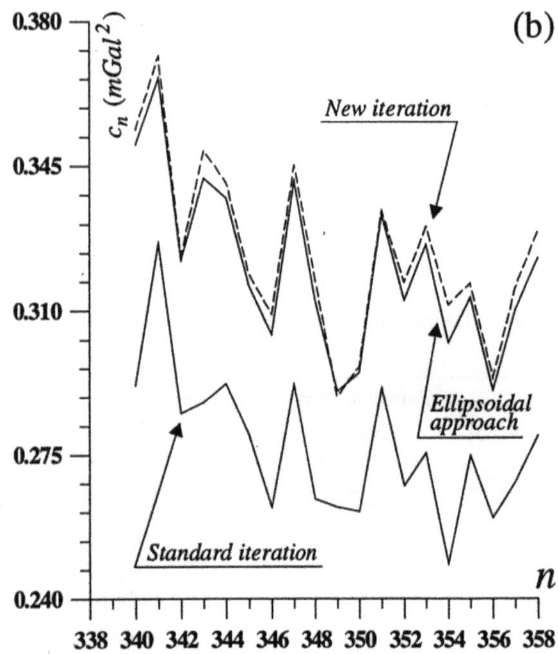

Fig. 1b: Gravity anomaly degree variances in the end of interval

The quantities (7) – (9) are presented in Fig. 1–2 and Table 1. It can be seen that the application of the new iteration procedure shows rather good results. In particular, for $n \leq 100$ this simple procedure gives practically the same result as is derived by the complicated ellipsoidal harmonic approach.

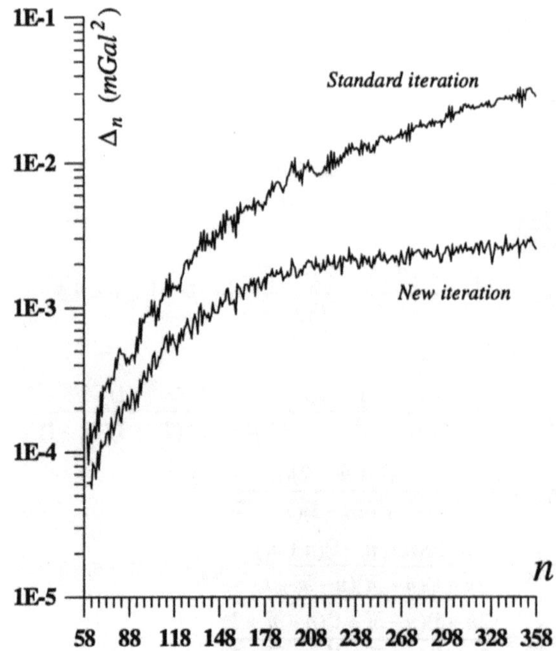

Fig. 2: Degree variances for the differences

There can be two main reasons for appearing the above differences between the potential coefficients evaluated by the new iteration procedure and by the ellipsoidal harmonic approach.

One of them may be caused by taking into account in the former the ellipsoidal corrections only of the order e^2, as opposed to the latter. The second reason is not so evident and its effect can not be easily estimated. This effect may be caused by applying different latitudes in the integral formulas in the iteration and Jekeli's procedures. In the iteration procedure the integration is performed over the sphere, with respect to the geocentric co-latitude θ and the longitude λ. Therefore the corresponding optimal equi-angular gridding should be carried out with respect to the angular variables θ and λ. In difference to it, in the ellipsoidal harmonic approach the integration is performed with respect to another co-latitude, the reduced one (δ), and the corresponding optimal gridding would be with respect to δ and λ. In practice, instead of gridding by θ or δ over the spheres, the discretization in the integral formulas is performed over the ellipsoid surface with respect to the geodetic latitude β. It may result in appearing different sampling errors of discretization in the iteration and ellipsoidal harmonic solutions, especially for large values of n.

4 Conclusions

The very simple analytical expression for the coefficients $\bar{C}_{n,m}$, derived by the new iteration procedure, and the closeness of this set of coefficients to the one for the ellipsoidal harmonic approach enable the possibility of different applications of the derived formulas when constructing geopotential models or evaluating various transformants of the gravity field. Some of them are mentioned below.

From relation (1) – (2), connecting $\bar{C}_{n,m}$ and the spherical harmonic coefficients for the gravity anomaly Δg_E on the ellipsoid surface, a BV relation can be derived, connecting the potential coefficients and the full quantity Δg_E. It can be solved by the LS procedure similarly to the usual BV relation (Lemoine et al., 1998) in which the truncated solid harmonic series is used. The new BV relation will be more simple because in it the coefficients of the spherical functions $\bar{Y}_{n,m}(\theta,\lambda)$ will be constants, instead of depending on the factors r_e^{-n-3}. Such relation can be processed by LS adjustment approach for $n \le 70$ on the base of Δg_E data over land areas, simultaneously with the observation equations based on satellite data.

It will be of interest to compare for $n \le 70$ the LS iteration solution for $\bar{C}_{n,m}$ with the above integral iteration solution, both based on the same model of the BV equation. Then these two solutions can be compared with the one derived by LS approach from the standard BV equation (ibid.). Finally, these three solutions can be compared with the most strict and exact (theoretically) Jekeli's solution for $\bar{C}_{n,m}$.

The coefficients $\bar{C}_{n,m}$ evaluated by the new iteration procedure can be used as a priori information when solving the observation equations by LS procedure on the base of the gravity data for large values of n and m. The same potential coefficients can be applied to filter out the higher degree harmonics in the observation equations, in order to reduce the aliasing effect in the solution. The new iteration solution can be also applied for computation of the series expressions for the non-harmonic correction terms when evaluating the potential coefficients by the ellipsoidal harmonic approach.

The present solution takes into account the ellipsoidal corrections only of the order e^2 in the BV relation. For constructing super-high-degree expansions of the geopotential further efforts are needed for deriving an analytical expression for $\bar{C}_{n,m}$ depending on high order degrees of e.

Acknowledgment. This research was supported by the Russian Ministry of Science (Project No. 1.7.3.1).

References

Jekeli, C. (1981). The downward continuation to the earth's surface of truncated spherical and ellipsoidal harmonic series of the gravity and height anomaly. Dept. Geod. Sci., Ohio State Univ., *Rep. 323*.

Jekeli, C. (1988). The exact transformation between ellipsoidal and spherical harmonic expansions. *Manuscripta Geodaetica*, Vol. 13, No 2, pp. 106 – 113.

Lemoine, F.G., S.C. Kenyon, J.K. Factor, R.G. Trimmer, N.K. Pavlis, D.S. Chinn, C.M. Cox, S.M. Klosko, S.B. Luthcke, M.H. Torrence, Y.M. Wang, R.G. Williamson, E.C. Pavlis, R.H. Rapp, T.R. Olson (1998). Geopotential Model EGM96. *NASA/TP-1998-206861*. Goddard Space Flight Center, Greenbelt.

Pellinen, L.P. (1982). Effects of the earth's ellipticity on solving geodetic boundary value problem. *Bollettino di Geodesia Scienze Affini*, Vol. 41, No 1, pp. 89 – 103.

Petrovskaya, M.S. (1999). Iteration procedure for evaluating high degree potential coefficients from gravity data. In: *Proc. of International Association of Geodesy Symposia (IV Hotine-Marussi Symposium on Mathematical Geodesy, Trento, Italy, 1988)*, in print.

Rapp, R.H. and J.Y. Cruz (1986). Spherical harmonic expansions of the earth's gravitational potential to degree 360 using 30' mean anomalies. Dept. Geod Sci., Ohio State Univ., *Rep. 376*.

Rapp, R.H. and N.K. Pavlis (1990). The development and analysis of geopotential coefficient models to spherical harmonic degree 360. *Journal of Geophysical Research*, Vol. 95, B13, pp. 21885 – 21911.

Improved analytical approximations of the Earth's gravitational field

V.N. Strakhov
United Institute of Physics of the Earth, Russ. Acad. of Sciences, B.Gruzinskaya 10, 123810 Moscow, Russia

U. Schäfer
BKG, Federal Agency of Cartography and Geodesy, Potsdam branch, PF 606008, 14473 Potsdam, Germany

A.V. Strakhov
Computing Center of the Russ. Acad. of Sciences, Vavilova street 40, 117967 Moscow, Russia

Abstract. We present a new approach suited to solve efficiently large systems of linear algebraic equations (SLEs) having full matrices with up to hundreds of thousands of unknowns. This approach is perfectly new in numerical algebra: it reduces an SLE to one equation with one unknown. We consider large SLEs that arise in establishing linear models of the Earth's gravitational field by using directly the field data, given at the observation points on the physical surface of the Earth and/or in outer space. Numerical tests with synthetic and real gravity data demonstrate the power of this approach w.r.t. accuracy, stability and efficiency of the solution, making it a promising tool for obtaining high-grade analytical approximations of the Earth's gravitational field on a regional as well as on a global scale. A very first practical application of the new approach yields low-degree regional gravity field models for an East European region, that covers more than 4 mill.km^2, with an outstanding prediction power at independent gravity control points.

Keywords. System of linear equations, gravitational field, analytical approximation, regional model

1 Introduction

One of the burning problems in geophysics and geodesy is to derive linear analytical models of the Earth's gravitational field with a high spatial resolution that are consistent with the observations of various field elements at the Earth's surface and in outer space. From a geodetic point of view gravity field modeling requires to keep pace and to be consistent with the precision of 10^{-9} already obtained in space positioning and for Earth rotation parameters.

For modeling the anomalous potential it is meaningful to apply linear approximations since they allow to find other field elements (e.g. higher derivatives) by means of elementary operations.

From a methodological point of view it is essential to make direct use of the observation values at those points in space where they have been measured, without any reductions to certain reference surfaces. This is also important from a pure practical point of view since the data of impending satellite / airborne gravimetric missions will be scattered over a wide range of altitudes / heights. Prior procedures commonly used to reduce the observed data to a certain reference surface may strongly predefine the solution, i.e. the model one is going to determine.

Hence, creating high resolution global or regional gravitational field models and avoiding predefining hypotheses a challenge of the next decade is to make direct use of the 3D spatiality of B radially scattered B observation points. This leads to the problem of solving huge SLEs, where the respective matrix is full and the number of unknowns may reach the order of 10^6 for models with L>360 in terms of spherical harmonics. This problem is difficult to handle even in the ages of modern supercomputers. In consequence, nowadays the SLEs that are used for the derivation of high resolution models are mostly simplified; the respective matrices are not full, cf. e.g. the EGM96 (Lemoine et al. 1998) or the GFZ97 (Gruber et al. 1997). In both cases is L=360 and the solutions, i.e. the spherical harmonic coefficients, have been obtained by combining a solution of a complete low-degree SLE and of a block-diagonal solution for the higher degrees to be solved independently.

This paper raises the issue to improve analytical models of the Earth's gravitational field by solving directly, accurately and efficiently large SLEs that have full matrices. We propose to solve this challenging task by means of a new approach called STRED (Strakhov's reduction method).

International Association of Geodesy Symposia, Vol. 121
Schwarz (ed.), Geodesy Beyond 2000 – The Challenges of the First Decade

2 Set up of the SLE using SNAP

For approximating the gravitational potential we use a representation called SNAP (Strakhov's New Analytical APproximation), valid for functions that are harmonic outside a unit sphere. A detailed description is given in Strakhov et al. (1995, 1997, 1998). The SNAP representation is equivalent to an approximation by truncated series of spherical functions but has a number of numerical advantages. Both approximations may be derived from Whittaker's classical integral representation ($x = (x_1, x_2, x_3) = (r, \varphi, \theta)$):

$$V(x) = \frac{1}{2\pi r}\int_0^{2\pi} F(\psi, t(x,\psi))\, d\psi \qquad (1)$$

by specifying the integrand $F(\psi,t)$. In case of the SNAP representation we made the following choice:

$$V(x) \approx V_N(x) = \frac{1}{r(L+1)}\sum_{S=0}^{L} F_S(x) \qquad (2)$$

where

$$F_S(x) = \operatorname{Re}\left\{\sum_{\nu=0}^{m} a_\nu^S \left(t_S(x)\right)^\nu\right\}, \quad 0 < m \le 2L \qquad (3)$$

with the SNAP coefficients $a_\nu^{(S)} = p_\nu^{(S)} + i q_\nu^{(S)};\ q_0^{(S)} \equiv 0$ and

$$t_S(x) = \frac{\cos\theta + i\sin\theta\cos(\varphi - \psi_S)}{r},$$

$$\psi_S = \frac{2S+1}{2(L+1)}\pi \qquad (4)$$

Higher derivatives of the gravitational potential may be obtained by differentiating equation (2).

Equating the observed gravity field values with a SNAP representation of type (2)-(4) yields an SLE for the determination of the SNAP coefficients.

One has to determine 260 281 SNAP coefficients in case that in (2) the degree L is 360; for L=48 the number M of unknown coefficients is 4753.

It had been shown (Strakhov et al. 1997) that stable solutions of these large SLEs may be obtained using special iteration methods – so-called methods of *successive polynomial multiplication* (SPM). Using SPM methods we were able to derive regional gravity models for Central Europe which are complete up to L=1530. These models show a better approximation accuracy in this particular region than today's global spherical harmonic models (Strakhov et al. 1999). One drawback of these iteration methods is the enormous computing effort, requiring supercomputer facilities to solve the SLEs within a reasonable time span. In comparison to SPM methods the approach presented below allows to solve SLEs much faster using an up-to-date personal computer.

3 STRED – a new approach to solve SLEs

For generating the SLEs we simply equate the analytical representation with the observed data (chap. 2). Normally, one gets systems with rectangular matrices with either N>M – mainly in case of global models – or N<M (N – is the number of equations; M – is the number of unknowns).

That means, we consider the problem to obtain stable approximate solutions of the following SLE:

$$Ax = f_\delta = f + \delta f \qquad (5)$$

where A is a N×M-matrix with real coefficients a_{ij}, $1 \le i \le N$, $1 \le j \le M$; x is the M-vector that is to be determined; f_δ denotes the given field element, f is the N-vector describing the signal, δf characterizes the N-vector of the noise, resp. of the errors.

In case of N>M the rectangular SLE can be converted by means of First Gauss' transformation into a normal equation system with a quadratic symmetric matrix. Here we assume that sufficient main storage is available for loading the matrix in the computer.

For N<M one also gets an SLE with a quadratic symmetric matrix using 2nd Gauss' transformation. In both cases the matrices are positiv semi-definite.

The basic idea of the new approach – recently proposed by V.N.Strakhov – is to reduce the solution of SLE (5) to the solution of merely one regularized equation with one unknown, or more sophisticated, to reduce the determination of a sequence of approximate solutions of the SLE (5) to the solution of a sequence of regularized linear equations with one unknown using the so-called auto-regularization (Strakhov and Teterin, 1991). Further this approach is named *St*rakhov's *red*uction (STRED). Below it is outlined for the case N>M.

In frame of the STRED approach the following main constructive ideas are used.

The *first idea* consists in transforming the SLE (5) to a system, named *system in canonical form*:

$$Bz = \kappa_\delta \check{e} \qquad (6)$$

where κ_δ is a constant that depends on vector f_δ, once determined it will be fixated; $\check{e}$ is a M-vector that looks like

$$\breve{e} = \left\| \begin{matrix} 0 \\ \vdots \\ 0 \\ \cdots \\ 1 \end{matrix} \right\| \begin{matrix} \uparrow \\ M-1 \\ \downarrow \\ \cdots\cdots \\ 1 \end{matrix} \tag{7}$$

and **B** denotes a M×M-matrix of type:

$$B = \hat{A}^T \hat{A} \tag{8}$$

with

$$\hat{A} = A\,U\,, \qquad z = U^T x \tag{9}$$

where **U** is a matrix with orthogonality of columns. **U** may be represented as a product of elementary plane rotation matrices

$$U = \prod_{r=1}^{M-1} T_{r,M}(\varphi_r)\,, \tag{10}$$

whose parameters are selected by the condition:

$$\left(\hat{a}^{(i)}, f_\delta\right) = 0\,, \quad i = 1,2,\ldots,M-1, \tag{11}$$

i.e. there are M-1 conditions for selecting φ_r. The $\hat{a}^{(i)}$ are column vectors of the N×M-matrix **Â**.

In accordance with the definition of matrices **Â** and **B** one gets in (6)

$$\kappa_\delta = \left(\hat{a}^{(M)}, f_\delta\right) \neq 0 \tag{12}$$

That means, in the canonical form of system (6) the vector of the right-hand side possesses only one non-zero component – the last one.

The *Second idea* in frame of the STRED approach comprises two consecutive orthogonal transformations of the SLE (6) (cf. eq. (18) and (20) below). These orthogonal transformations do not change the vector of the right-hand side, but convert system (6) into an SLE with a lower tridiagonal matrix $\mathfrak{L}$, that is equivalent to (6):

$$\mathfrak{L}\,w = \kappa_\delta \breve{e} \tag{13}$$

For the elements l_{ij} of matrix $\mathfrak{L}$ we have

$$l_{ij} = 0\,, \qquad i-j<0\,,\ i-j \geq 3 \tag{14}$$

and therefore due to (7) the SLE (13) is equivalent to an equation with one unknown, because one can set

$$w = \left\| \begin{matrix} 0 \\ \vdots \\ 0 \\ \cdots \\ w_M \end{matrix} \right\| \begin{matrix} \uparrow \\ M-1 \\ \downarrow \\ \cdots\cdots \\ 1 \end{matrix}\,, \qquad l_{M,M}\, w_M = \kappa_\delta \tag{15}$$

In the presence of errors δf at the right-hand side of the original SLE (5) it is necessary to regularize system (15). This is due to the fact that κ_δ is a constant determined from this right-hand side of (5) and $l_{M,M}$ contains rounding errors of calculation. Hence, we get the following regularized system:

$$\left(l_{M,M} + \frac{\alpha}{l_{M,M}} \right) w_{M,\alpha} = \kappa_\delta\,, \tag{16}$$

with the parameter of regularization α, given by

$$\alpha > -l^2_{M,M}\,, \tag{17}$$

where α may be: $\alpha>0$ as well as $\alpha<0$. In order to define a set of approximate solutions of system (5) the selection principle for the values of α is evident.

It remains to describe the conversion of the system (6)-(12) into the SLE (13).

We have first an orthogonal transformation of matrix **B**:

$$T = V^T B V\,, \tag{18}$$

where **V** is a column-wise orthogonal M×M-matrix that may be represented as the product of plane elementary rotation matrices. These matrices may be chosen in such a way that **T** will be a symmetric, tridiagonal matrix (Voevodin and Kusnetsov, 1984). In doing so we get an orthogonal transformation of the independent variables

$$u = V^T z, \qquad z = V u \tag{19}$$

Then a second orthogonal transformation is applied

$$T\Phi = \mathfrak{L}\,, \quad w = \Phi^T z\,, \quad z = \Phi w\,, \tag{20}$$

where the column-wise orthogonal matrix **Φ** may be represented as the multiplication of matrices of elementary plane rotation

$$\Phi = \prod_{k=1}^{M-1} T_{k,k+1}(\varphi_k)\,, \tag{21}$$

The parameters of these rotation matrices may be defined by successive annihilation of matrix' **T** super-diagonal elements starting with the element at position (1,2).

These first two ideas, i.e. the initial reduction of the original SLE (5) to a system in canonical form (6) and afterwards the transition from (6) to the system (13)-(14) are the essentials of the STRED approach. However, the numerical results obtained so far indicate that for large values of M (conditionally $M>10^3$) the transition from system (6) first to system

$$T u = \kappa_\delta \check{e} , \qquad (22)$$

and then to the SLE (13)-(14) is connected with a considerable accumulation of rounding errors. These rounding errors often do not allow to get the approximate solutions x_α

$$x_\alpha = V\left\{\Phi\{w_\alpha\}\right\} \qquad (23)$$

with the necessary accuracy, where

$$w_\alpha = \left|\begin{array}{c} 0 \\ \vdots \\ 0 \\ \cdots\cdots \\ w_{M,\alpha} \end{array}\right| \begin{array}{c} \uparrow \\ M-1 \\ \downarrow \\ \cdots\cdots \\ 1 \end{array} , \qquad (24)$$

and $w_{M,\alpha}$ is defined from the solution of eq. (16).

Therefore, in case of practical applications where M is quite a large number, it seems appropriate to make use of auto-regularization algorithms (Strakhov and Teterin 1991). The usage of auto-regularization can be regarded as the *third constructive idea* in connection with the proposed STRED approach. Due to the lack of space it is not possible to outline here the idea of auto-regularization in more detail.

In the frame of the STRED approach there is a *fourth constructive idea* [1]:
By means of well-known methods (e.g. Lavrentiev's method of regularization; Cholesky's generalized decomposition) the matrix (6) may be regularized and represented as the product of a lower and a upper trigonal matrices, realizing a transition from (6) to the following SLE:

$$L^{(reg)} L^{(reg),T} z^{(reg)} = \kappa_\delta \check{e} . \qquad (25)$$

The solution of (25) may be achieved by successively solving the two systems of trigonal matrices:

$$\begin{aligned} L^{(reg)} v^{(reg)} &= \kappa_\delta \check{e} , \\ L^{(reg),T} z^{(reg)} &= v^{(reg)} . \end{aligned} \qquad (26)$$

Solving the first system leads to one equation with one unknown:

$$v^{(reg)} = \left|\begin{array}{c} 0 \\ \vdots \\ 0 \\ \cdots\cdots \\ v^{(reg)}_M \end{array}\right| \begin{array}{c} \uparrow \\ M-1 \\ \downarrow \\ \cdots\cdots \\ 1 \end{array} , \qquad (27)$$

[1] From the viewpoint of practical implementation this idea is currently the most developed one. The respective program code was written by A.V.Strakhov and a number of synthetic and practical numerical examples was calculated (Chap. 4).

where $v^{(reg)}_M$ has to be find from equation

$$l^{(reg)}_{M,M} \, v^{(reg)}_M = \kappa_\delta \qquad (28)$$

At the same time we get for the second system in (26)

$$L^{(reg),T} z^{(reg)} = \frac{\kappa_\delta}{l^{(reg)}_{M.M}} \check{e} . \qquad (29)$$

One can show, that system (29) may be transformed into a system in canonical form with an upper trigonal matrix. That means, that the vector at the right-hand side has purely one non-zero component – the first one. This transformation may be performed by subsequent application of the following two orthogonal transformations:

$$\Psi^T L^{(reg),T} P = H^{(reg)}, \quad H^{(reg)} W = R^{(reg)}, \qquad (30)$$

where matrices **P**, **Ψ**, **W** are column-wise orthogonal and representable in form of products of elementary matrices of plane rotation. **P** is a permutation matrix. The upper Hessenberg matrix $\mathbf{H}^{(reg)}$ and the upper tridiagonal matrix $\mathbf{R}^{(reg)}$ are both M×M-matrices. This transformation gives

$$R^{(reg)} u^{(reg)} = \kappa_\delta \hat{e} , \qquad (31)$$

with

$$\hat{e} = \left|\begin{array}{c} 1 \\ \cdots \\ 0 \\ \vdots \\ 0 \end{array}\right| \begin{array}{c} 1 \\ \cdots\cdots \\ \uparrow \\ M-1 \\ \downarrow \end{array} . \qquad (32)$$

The solution of the system (31)-(32) leads clearly to the problem to solve one linear equation with one unknown. The remaining steps (i.e. back transformation/substitution, evaluation of deviations, etc.) should be evident to the reader.

4 Numerical experiments

4.1 Synthetic models

To evaluate the capabilities of the STRED approach we carried out numerical tests with synthetic data. We assumed vectors x to be given and simulated the matrix **A** and the noise, resp. the errors δf in the right-hand side of (1). We replaced **A** by $\mathbf{A_Q}$:

$$\begin{aligned} A_Q \, x &= f_\delta = f + \delta f, \\ A_Q &= Q H + (1-Q) T, \quad 0 \le Q < 1 \end{aligned} \qquad (33)$$

with two matrices **H** and **T** of dimension N×M. Their matrix elements were chosen as follows:

$$h_{ij} = \frac{1}{i+j-1}, \qquad \begin{array}{l} 1 \le i \le 1000 = N \\ 1 \le j \le 250 = M \end{array}, \tag{34}$$

$$t_{ij} = \frac{n}{(i-j)^2 + n^2}, \qquad \begin{array}{l} 1 \le i \le 1000 = N \\ 1 \le j \le 250 = M \end{array}, \; n = \frac{1}{2}. \tag{35}$$

We made the numerical tests for different vectors x,

$$\begin{array}{ll} \text{vector 1:} & x_i \equiv 1, \\ \text{vector 2:} & x_i = \dfrac{i}{M}, \\ \text{vector 3:} & x_i = (-1)^i \dfrac{i}{M}. \end{array} \tag{36}$$

Vector f in (33) was computed by using one of the vectors x from (36) multiplied with $\mathbf{A}_Q$, whereas vector δf was created by a random number generator:

$$\delta f_i = \gamma_c \varepsilon_i, \tag{37}$$

where γ_c was determined from given values of c^2:

$$c^2 = \frac{\| \delta f \|_E^2}{\| f_\delta \|_E^2}. \tag{38}$$

We used the two values: $c^2 = 10^{-3}$ and $c^2 = 10^{-5}$.

The established SLEs were solved for various values of Q and c^2 by the STRED approach and by two other regularization methods: Lavrentiev's regularization method (LRM) and by a regularized version of Cholesky's decomposition method (CDM). The last two methods are well-known. Therefore, their description may be omitted.

Results of this comparison are given in Table 1. The parameter η describes the quality of the solution:

$$\eta = \frac{\| x^{true} - x^{calc} \|_E}{\| x^{true} \|_E} \tag{39}$$

Table 1: Quality of solutions of SLE (33) evaluated by parameter η for different methods

Q=0.1

Method	x-vector 1		x-vector 2		x-vector 3	
	$c^2=10^{-3}$	$c^2=10^{-5}$	$c^2=10^{-3}$	$c^2=10^{-5}$	$c^2=10^{-3}$	$c^2=10^{-5}$
LRM	0.126	0.015	0.125	0.015	0.181	0.019
CDM	0.143	0.016	0.137	0.015	0.153	0.017
STRED	0.028	0.003	0.028	0.003	0.040	0.004

Q=0.75

Method	x-vector 1		x-vector 2		x-vector 3	
	$c^2=10^{-3}$	$c^2=10^{-5}$	$c^2=10^{-3}$	$c^2=10^{-5}$	$c^2=10^{-3}$	$c^2=10^{-5}$
LRM	0.322	0.074	0.340	0.048	0.466	0.071
CDM	0.547	0.080	0.325	0.048	0.513	0,071
STRED	0.060	0.007	0.049	0.005	0.090	0.009

It is striking, that for all values of Q and c^2, and for all three synthetic vectors the STRED method gives the best approximation.

4. 2 Gravity models for Eastern Europe derived by means of the STRED approach

To demonstrate the power of the STRED approach in case of real data we derived analytical models of the Earth's gravity field using gridded gravity anomaly values from 15 360 compartments 10' by 15' in the area enclosed between 50° to 68° N and 20° to 60° E (see Fig. 1). This data was placed at our disposal by G.V.Demianov (Personal communication, 1998).

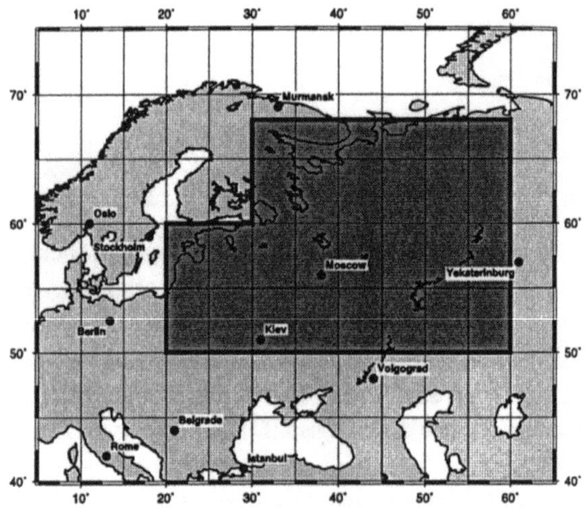

Fig. 1: The area of used gravity anomaly data

From the totality of 15 360 points we formed four subsets #1, #2, #3, and #4, each of them consists of 3840 anomaly data with a regular point spacing of 20' by 30'. These sets were selected by the following rule starting in the northmost row at the westernmost point of the area marked in Fig.1:

1212121···
3434343···
1212121···
3434343···
⋮

Then we grouped together three of the four data sets (11 520 values), leaving apart the remaining set for control purposes. This procedure yields by permutation four data sets #123, #234, #341, #412 and four control data sets #4, #1, #2, #3, respectively.

The four data sets – each of 11 520 points – were used to derive four different analytical SNAP models of the gravity anomaly field for the particular region indicated in Fig. 1 by means of the STRED approach. In case of L=48 the number M of unknown SNAP coefficients to be determined is (L+1)(2L+1)=4753, yielding a matrix **A** with N>M, i.e. 11 520 rows by 4753 columns.

The resulting approximation accuracy (observed minus modelled gravity values) are given in Table 2.

It is remarkable, that the fit of the mean is below the mikrogal level and that in case of all four SNAP models derived for this East European region the

RMS values differ within a few mikrogals only.

To get an independent check of the models' quality each solution, i.e. each of the four derived SNAP coefficient sets, was used to synthesize the anomaly values in the respective 3840 control points that have not been used in deriving the models. The prediction power of the individual SNAP models in the corresponding control data set (observed minus predicted) is documented in Table 3. Although the approximation/recovery of the gravity mean of all independent control points slightly rises up to a few mikrogals, there is no noticeable rise w.r.t. the RMS.

In Table 4 and 5 we present the accuracy obtained in the same control data sets with the global models EGM96 (Lemoine et al. 1998) and GFZ97 (Gruber et al. 1997), respectively. Both global models are truncated to degree and order L=48. This allows a raw comparison with our results which confirms the high quality of the regional SNAP models for this region.

Table 2: Fit accuracy of four SNAP gravity models (L=48) derived by means of the STRED approach using 11 520 gravity anomaly values

Parameter [mgal]	model #123	model #234	model #341	model #412
Mean	-0.000	-0.000	-0.000	-0.000
RMS	12.129	12.157	12.121	12.123
Minimum	-52.749	-53.104	-51.110	-54.193
Maximum	79.586	79.345	75.634	79.452

The computing time for each solution is about 16 hours ona modern PC Pentium II in single user mode.

Table 3: Approximation accuracy of SNAP gravity models at 3840 independent control points (L=48)

Parameter [mgal]	#123 in #4	#234 in #1	#341 in #2	#412 in #3
Mean	-0.035	0.064	-0.021	0.014
RMS	12.211	12.120	12.234	12.233
Minimum	-48.740	-48.334	-52.030	-51.069
Maximum	70.138	75.620	79.391	63.880

Table 4: Approximation accuracy of EGM96 gravity model (truncated at L=48) at sets of 3840 control points

Parameter [mgal]	EGM96 in #4	EGM96 in #1	EGM96 in #2	EGM96 in #3
Mean	1.325	1.269	1.422	1.170
RMS	17.819	17.602	17.847	17.666
Minimum	-56.127	-56.329	-58.933	-58.303
Maximum	94.994	105.45	106.82	94.986

Table 5. Approximation accuracy of GFZ97 gravity model (truncated at L=48) at sets of 3840 control points

Parameter [mgal]	GFZ97 in #4	GFZ97 in #1	GFZ97 in #2	GFZ97 in #3
Mean	1.457	1.394	1.547	1.303
RMS	17.871	17.660	17.901	17.722
Minimum	-57.618	-55.983	-57.838	-57.559
Maximum	96.618	106.57	106.28	96.583

5 Conclusions

The presented STRED approach – applied to the problem of gravity field modeling – allows to obtain high-grade analytical models of the Earth's gravity field as stable approximate solutions of large SLEs with several thousands of unknowns. On top of that this approach seems to be perfectly new in numerical algebra, allowing to reduce huge SLEs to one regularized equation with one unknown.

The numerical simulation tests with well-defined synthetic models show a significant better performance of the STRED approach in comparison to other existing methods for solving large SLE's.

SNAP models derived from gravity data of Eastern Europe are practically bias-free with an outstanding prediction power at independent control points.

Further studies are needed in order to implement the approach to underdetermined systems with $N<M$. More investigations are required also w.r.t. the suppression of rounding errors, especially when $L \gg 48$.

References

Gruber, Th., A. Bode, Ch. Reigber, P. Schwintzer (1997). D-PAF Global Earth Gravity Models Based on ERS. Proceedings of 3rd ERS Symposium: Space at the service of our environment, ESA SP-414 Vol.III, , 1661-1668.

Lemoine, F.G., S.C. Kenyon et al. (1998). TheDevelopment of the Joint NASA GSFC and the National Imagery and Mapping Agency (NIMA) Geopotential Model EGM96. Greenbelt, NASA/TP—1998–206861, 575 pages.

Strakhov, V.N., D.E. Teterin (1991). Linear Transformations of Gravitational and Magnetic Anomalies in case of Multi-element Surveys and Arbitrarily Observation Grids (in Russian). *Doklady AN SSSR*, vol. 318, No. 3, pp.572-576.

Strakhov, V.N., U. Schäfer, A.V. Strakhov, A.I. Luchitsky, D.E. Teterin (1995). A new Approach to Approximate the Earth's Gravity Field. In: Sünkel, H.,I. Marson (eds.) Proc. of IAG Symposium No. 113, 'Gravity and Geoid', Graz, Austria, Sep 11-17, 1994, Springer, Berlin Heidelberg New York, pp 225-237.

Strakhov, V.N., U. Schäfer, A.V. Strakhov, D.E. Teterin (1997). Ein neuer Ansatz zur Approximation des Gravitationsfeldes der Erde. Interim-Report, Projekt INTAS-93-1779, Institut für Angewandte Geodäsie, Potsdam, 382 S.

Strakhov, V.N., U. Schäfer, A.V. Strakhov, (1998). Neue lineare Approximationen linearer Elemente des Gravitationsfeldes der Erde. In: Freeden, W. (ed.) Progress in Geodetic Science. Shaker, Aachen, pp 315-322.

Strakhov, V.N., U. Schäfer, A.V. Strakhov, (1999). Regional Gravitational Models Derived by Means of the SNAP Method. IUGG99, XXII General Assembly, Symposium G3, Birmingham, UK, July 19-30, 1999, Poster G3/E/17, Abstracts A.433.

Voevodin, V.V., Yu.A. Kusnetsov (1984). Matrices and Calculations. Nauka, Moscow, 318 pp (in Russian).

Sparse preconditioners of Gram's matrices in the conjugate gradient method

G. Moreaux
University of Copenhagen, Department of Geophysics,
Juliane Maries Vej 30, DK-2100 Copenhagen O, Denmark (E-mail: gm@gfy.ku.dk)

Abstract. When harmonic spherical splines are used to interpolate and predict discretely given data we are confronted with the problem of solving symmetric positive-definite systems involving as many equations as the number of data. Due to harmonicity, these systems are full and thus iterative methods like the conjugate gradient method should be prefered to direct methods such as the Cholesky factorization for large data sets. In order to speed-up classic iterative solvers, we present a class of sparse symmetric positive definite preconditioners which hold for any minimization norm, i.e. for any harmonic reproducing kernel, and for any data distribution. Numerical results are shown and demonstrate the efficiency of our method.

Keywords. Reproducing kernels, positive definite functions, iterative methods.

1 Introduction

Spherical splines have been successfully used in geodesy to interpolate and to predict data discretely given (Freeden and Reuter (1983), Freeden et al. (1997), Moreaux et al. (1999a)). Such a technique has the main drawback that it requires the solution of full symmetric positive definite linear systems involving as many equations as the number of data, the left hand side matrices being Gram-matrices of reproducing kernels associated with the minimization norm. Thus, since modern satellites provide a huge amount of observational data, direct solvers are unpractical and then iterative methods should be used. However iterative methods like the conjugate gradient method require at least one matrix-vector product in each iteration step the use of sparse preconditioners should be investigated to speed up these methods.

The system matrices being Gram-matrices of harmonic kernels and as in most applications there exists an appropriate distance of separation beyond which the kernel values, i.e. the matrix entries, are negligibly small, a legitime thinking to obtain sparse preconditioners could be to first construct locally supported positive definite approximations of the kernels and use its Gram-matrices as preconditioners of classic iterative solvers.

Following the concept of "finite covariance functions" of Sansò and Schuh (1987), Moreaux et al. (1999b) developed three different methods to get locally supported positive definite approximations of any harmonic kernel and shown that in both terms of errors and CPU-time the one should be prefered to the others. Therefore this paper is focussed on the first method for which we begin by showing that closed expressions of the truncated Legendre coefficients are available for some minimization norms, and then we go further into its use as preconditioning of the conjugate gradient method.

This article is organized as follows: after some preliminary facts given at the beginning of the Section 2, we recall the theoretical aspects of the first method developed in Moreaux et al. (1999b). Then, in Section 3, we give closed expressions of Legendre truncation coefficients. In Section 4, we first show how these locally supported functions approximate a given harmonic reproducing kernel and then we also investigate the efficiency of using sparse Gram-matrices as preconditioners of the conjugate gradient method for the resolution of different linear systems involving the same kernel. After that, conclusions are drawn in the fifth section.

2 Reproducing kernels

For any point $P \in \mathcal{R}^3$ different from the origin

International Association of Geodesy Symposia, Vol. 121
Schwarz (ed.), Geodesy Beyond 2000 – The Challenges of the First Decade

we may write $P = r_P \xi_P$ where $\xi_P \in \Omega_1 = \{P \in \mathcal{R}^3 \mid r_P = 1\}$ (unit sphere). To be consistent we define $\Omega_R = \{P \in \mathcal{R}^3 \mid r_P = R\}$ to be the sphere of radius R centered at the origin, denote by $\Omega_R^e = \{P \in \mathcal{R}^3 \mid r_P > R\}$ to be the outer space of Ω_R and set $\overline{\Omega_R^e} = \Omega_R^e \cup \Omega_R$.

Given an integer N and a real sequence $\{\kappa_n\}$ satisfying

$$\sum_{n=N}^{+\infty} \frac{2n+1}{4\pi} |\kappa_n| < \infty , \tag{1}$$

let us consider the harmonic reproducing kernel K defined on $\overline{\Omega_R^e} \times \overline{\Omega_R^e}$ by

$$K(P,Q) = \sum_{n=N}^{+\infty} \frac{2n+1}{4\pi} \kappa_n^2 \left(\frac{R^2}{r_P\, r_Q}\right)^{n+1} P_n(\xi_P \cdot \xi_Q) . \tag{2}$$

Then, making use of the summation theorem (Müller (1966)) as well as the orthogonality of the spherical harmonics in $L^2(\Omega_1)$, it is easy to prove that the function $G : \overline{\Omega_R^e} \times \Omega_R \to \mathcal{R}$

$$G(P,M) = \frac{1}{R} \sum_{n=N}^{+\infty} \frac{2n+1}{4\pi} \kappa_n \left(\frac{R}{r_P}\right)^{n+1} P_n(\xi_P \cdot \xi_M) \tag{3}$$

fulfils

$$K(P,Q) = \int_{\Omega_R} G(P,M) G(Q,M) dw(M) . \tag{4}$$

Because any function K satisfying (4) is at least positive definite on $\overline{\Omega_R^e}$, in order to get positive definite approximations K_α ($\alpha \in [0;\pi]$) of K (2) with local support $[0;2\alpha]$ we opted for

$$K_\alpha(P,Q) = \int_{\Omega_R} G_\alpha(P,M) G_\alpha(Q,M) dw(M) \tag{5}$$

G_α being the most simple approximation of G with support $[0;\alpha]$, i.e.

$$G_\alpha(P,M) = \begin{cases} G(P,M) \text{ for } & \psi_{P,M} \le \alpha, \\ 0 & \text{otherwise} , \end{cases} \tag{6}$$

where $\psi_{P,Q}$ denotes the spherical distance between P and Q.

Thus we obviously have:

$$\begin{array}{ll} G_0 \equiv 0 & K_0 \equiv 0 \\ G_\pi \equiv G & K_\pi \equiv K \end{array} \tag{7}$$

and for all $P \in \overline{\Omega_R^e}$ we get by straightforward computations

$$K(P,P) - K_\alpha(P,P) \ge 0 . \tag{8}$$

Finally, a rough estimation of the error between K and K_α is given in Moreaux et al. (1999b).

3 Legendre's truncation coefficients

Expanding K_α in terms of Legendre polynomials, we get

$$K_\alpha(P,Q) = \sum_{n=0}^{+\infty} \frac{2n+1}{4\pi} \kappa_n(\alpha, \rho_P) \kappa_n(\alpha, \rho_Q) P_n(\xi_P \cdot \xi_Q), \tag{9}$$

where the Legendre coefficients $\kappa_n(\alpha, \rho_P)$ satisfy

$$\kappa_n(\alpha, \rho_P) = \sum_{m=N}^{+\infty} \frac{2m+1}{2} \kappa_m \rho_P^{m+1} I_{m,n}(\alpha) , \tag{10}$$

$\rho_P = R/r_P$, and with

$$I_{m,n}(\alpha) = \int_{\cos\alpha}^{1} P_m(t) P_n(t) dt . \tag{11}$$

A fast and stable pyramidal scheme for the obtention of the integrals $I_{m,n}(\alpha)$ was developed in Moreaux et al. (1999b).

For $\alpha \in]0;\pi[$ and $r \ge R$, using the following majoration (Magnus et al. (1966))

$$|P_n(t)| \le \sqrt{\frac{4}{\pi(2n+1)\sqrt{1-t^2}}} , \tag{12}$$

$n \ge 0$, $|t| < 1$, as well as the relations on $I_{n,m}$, we deduce that $|\kappa_n(\alpha, \rho)| = O(n^{-3/2})$ as $n \to +\infty$ and then

$$\sum_{n=0}^{+\infty} \frac{2n+1}{4\pi} \kappa_n(\alpha, \rho) \kappa_n(\alpha, \rho') < \infty . \tag{13}$$

Because the coefficients $\kappa_n(\alpha, \rho)$ (10) involve infinite series, it was well-founded to investigate the case where such series have closed expressions.

Therefore, by analogy with the different models developed in Tscherning and Rapp (1974), let us consider seqences $\{\kappa_n\}$ of the form

$$\kappa_n = w^{n+1}\frac{(n-u_1)\cdots(n-u_p)}{(n-v_1)\cdots(n-v_q)}, \qquad (14)$$

where $0 < w < 1$, $p < q$, $u_i, v_i \in \mathcal{Z}$ and $v_i \neq v_j$ for $i \neq j$. Then, making use of the relation (5) of Paul (1973) as well as well-known relations on Legendre polynomials we have for $n \geq 1$

$$2\kappa_0(\alpha,\rho) = \kappa_0\rho(1-\cos\alpha) + \check{\kappa}_0(\alpha,\rho) - \hat{\kappa}_0(\alpha,\rho),$$

$$2\kappa_n(\alpha,\rho) = \frac{DP_n(\alpha)}{2n+1}\left[\kappa_0\rho - n(n+1)\bar{\kappa}_n(\alpha,\rho)\right] +$$

$$(2n+1)\kappa_n\,\rho^{n+1}I_{n,n}(\alpha)\,\cdot$$

$$P_n(\cos\alpha)\left[\check{\kappa}_n(\alpha,\rho) - \hat{\kappa}_n(\alpha,\rho)\right], \qquad (15)$$

with

$$DP_n(\alpha) = P_{n-1}(\cos\alpha) - P_{n+1}(\cos\alpha), \qquad (16)$$

and where

$$\check{\kappa}_n(\alpha,\rho) = \sum_{m=N,m\neq 0,n}^{+\infty}\frac{m(m+1)}{(m-n)(n+m+1)}$$

$$\kappa_m\rho^{m+1}P_{m-1}(\cos\alpha) \qquad (17)$$

$$\bar{\kappa}_n(\alpha,\rho) = \sum_{m=N,m\neq 0,n}^{+\infty}\frac{2m+1}{(m-n)(n+m+1)}$$

$$\kappa_m\rho^{m+1}P_m(\cos\alpha) \qquad (18)$$

$$\hat{\kappa}_n(\alpha,\rho) = \sum_{m=N,m\neq 0,n}^{+\infty}\frac{m(m+1)}{(m-n)(n+m+1)}$$

$$\kappa_m\rho^{m+1}P_{m+1}(\cos\alpha). \qquad (19)$$

After decomposition in simple fractions, $\check{\kappa}_n$, $\bar{\kappa}_n$, $\hat{\kappa}_n$ lead to the evaluation of ($\rho \equiv w\rho$)

$$f_n^{\pm}(\alpha,\rho) = \sum_{m=0(m\neq n)}^{+\infty}\frac{\rho^{m+1}}{m\pm n}P_m(\cos\alpha) \qquad (20)$$

$$g_n^{\pm}(\alpha,\rho) = \sum_{m=0(m\neq n)}^{+\infty}\frac{\rho^{m+1}}{(m\pm n)^2}P_m(\cos\alpha) \qquad (21)$$

which satisfy some well-known 3-terms recurrence relations.

Contrary to f_0^-, f_1^- closed expressions of g_0^-, g_1^- could not be obtained, meanwhile, good approximations can be estimated by the Clenshaw technique (Tscherning and Poder (1982)). Moreover, let us remind you that in order to avoid numerical instabilities in the computations of the sequences $\{f_n^{\pm}\}$, $\{g_n^{\pm}\}$, f_n^- and g_n^- have to be estimated in the direction of increasing n when f_n^+ and g_n^+ must be computed in the direction of decreasing n by setting for $M >> 1$

$$\begin{array}{ll} f_{M+1}^+ = 0 & f_M^+ = 0 \\ g_{M+1}^+ = 0 & g_M^+ = 0. \end{array} \qquad (22)$$

4 Numerical examples

All calculations of this section are based on the following reproducing kernel K

$$K(P,Q) = c\sum_{n=361}^{+\infty}\frac{2n+1}{(n+4)^2}\left(\frac{R}{R_E}\right)^{2(n+2)}$$

$$P_n(\cos\psi_{P,Q}) \qquad (23)$$

where $R_E = 6371\ km$ is the mean Earth radius, $R = 6366km$ is the radius of a so-called Bjerhammar sphere and $c = const.$ satisfies $K(P,P) = 100\ mGal^2$. Such a positive definite function can be seen as the covariance function of gravity anomalies at sea-level, i.e. $r_P = r_Q = R_E$, associated with the degree variance model given by

$$\sigma_n = \begin{cases} 0 & i = 0,\ldots,360, \\ \frac{c}{(n-1)^2(n+4)^2} & i = 361,\ldots. \end{cases} \qquad (24)$$

Closed expressions of functions $\check{\kappa}_n$ (17), $\bar{\kappa}_n$ (18), $\hat{\kappa}_n$ (19) corresponding to K are given in Table 1.

In order to get continuous locally supported approximations K_α of K we only investigate the case where α satisfies $G(\alpha) = 0$, i.e.

$$\sum_{n=361}^{2500}\frac{2n+1}{n+4}\left(\frac{R}{R_E}\right)^{n+2}P_n(\cos\psi) = 0, \qquad (25)$$

which leads to $\alpha \in \{0.2116°, 0.6452°, 1.1298°, 2.1222°\}$.

Table 1: Closed expressions of $\check{\kappa}_n(\alpha,\rho)$, $\bar{\kappa}_n(\alpha,\rho)$ and $\hat{\kappa}_n(\alpha,\rho)$ corresponding to $\kappa_n = 2\sqrt{\pi}/(n+4)$ with $0 < \rho < 1$ and $0 \le \alpha \le \pi$.

Quantity	Closed Expression
$\check{\kappa}_0(\alpha,\rho)$	$-\rho f_5^+(\alpha,\rho)$
$\bar{\kappa}_0(\alpha,\rho)$	$-\frac{3}{16}\rho + \frac{1}{4} f_0^-(\alpha,\rho) + \frac{1}{3} f_1^+(\alpha,\rho) - \frac{7}{20} f_4^+(\alpha,\rho)$
$\hat{\kappa}_0(\alpha,\rho)$	$\frac{1}{3} + \frac{1}{4}\rho\cos\alpha - \rho^{-1} f_3^+(\alpha,\rho)$
$\check{\kappa}_3(\alpha,\rho)$	$-\frac{25}{343}\rho^4 P_2(\cos\alpha) + \frac{37}{49}\rho f_5^+(\alpha,\rho) + \frac{12}{49}\rho\left[f_2^-(\alpha,\rho) - 7g_5^+(\alpha,\rho)\right]$
$\bar{\kappa}_3(\alpha,\rho)$	$\frac{1}{48}\rho + g_4^+(\alpha,\rho) + \frac{1}{7}\left[f_3^-(\alpha,\rho) - f_4^+(\alpha,\rho)\right]$
$\hat{\kappa}_3(\alpha,\rho)$	$-\frac{25}{343}\rho^4 P_4(\cos\alpha) + \frac{37}{49}\rho^{-1} f_3^+(\alpha,\rho) + \frac{12}{49}\rho^{-1}\left[f_4^-(\alpha,\rho) - 7g_3^+(\alpha,\rho)\right]$
$\check{\kappa}_n(\alpha,\rho)$	$\frac{1}{n-3}\left[\frac{12}{(n+4)^2} - \frac{n(n+1)}{(2n+1)^2}\right]\rho^{n+1} P_{n-1}(\cos\alpha) - \frac{12}{(n+4)(n-3)}\rho f_5^+(\alpha,\rho)$ $+\frac{n(n+1)}{2n+1}\rho\left[\frac{1}{n-3} f_{n+2}^+(\alpha,\rho) + \frac{1}{n+4} f_{n-1}^-(\alpha,\rho)\right]$
$\bar{\kappa}_n(\alpha,\rho)$	$\frac{1}{4n(n+1)}\rho + \frac{n-3}{(n+4)^2(2n+1)}\rho^{n+1} P_n(\cos\alpha) + \frac{7}{(n+4)(n-3)} f_4^+(\alpha,\rho)$ $-\frac{1}{n-3} f_{n+1}^+(\alpha,\rho) + \frac{1}{n+4} f_n^-(\alpha,\rho)$
$\hat{\kappa}_n(\alpha,\rho)$	$\frac{1}{n-3}\left[\frac{12}{(n+4)^2} - \frac{n(n+1)}{(2n+1)^2}\right]\rho^{n+1} P_{n+1}(\cos\alpha) - \frac{12}{(n+4)(n-3)}\rho^{-1} f_3^+(\alpha,\rho)$ $+\frac{n(n+1)}{2n+1}\rho^{-1}\left[\frac{1}{n-3} f_n^+(\alpha,\rho) + \frac{1}{n+4} f_{n+1}^-(\alpha,\rho)\right]$

In the following, for $i = 1,2,3,4$, we denote by K_i the locally supported function K_α associated with the i-th entry of the last list for which the series (9) have been truncated at order 2500. The figures Fig. 1 and Fig. 2 show the functions K and K_i, $i = 1,2,3,4$, for $0° \le \psi \le 5°$. The statistics of these functions as well as the ones corresponding to the differences $K - K_i$ are summarized in Table 2.

Let us now consider the symmetric positive definite linear systems ($\lambda = 1mGal^2$, $b = (100mGal, \cdots, 100mGal)^T$)

$$\mathbf{K}_\lambda\, x = b\,, \tag{26}$$

where $\mathbf{K}_\lambda(i,j) = K(P_i,P_j) + \lambda\,\delta_{ij}$ is the λ-regularized Gram-matrix of the function K (2) associated with the points $P_i \in \Omega_{R_B}$ belonging to one of the next three regularly gridded data sets

1. $i = 1,\cdots,51^2$
   $80° \le \theta_i \le 100°$ $\quad -10° \le \phi_i \le 10°$
2. $i = 1,\cdots,101^2$
   $80° \le \theta_i \le 100°$ $\quad -10° \le \phi_i \le 10°$
3. $i = 1,\cdots,101^2$
   $70° \le \theta_i \le 110°$ $\quad -20° \le \phi_i \le 20°$.

Table 2: Statistics of the functions K and K_i, $i = 1,2,3,4$. Unit is $mGal^2$

	max	*min*	*mean*	*std*
K	100.00	−15.94	1.94	13.89
K_1	100.00	−0.06	2.33	12.58
K_2	100.00	−14.75	1.96	13.37
K_3	100.00	−15.79	1.92	13.62
K_4	100.00	−16.02	1.91	13.76
$K - K_1$	8.90	−15.89	−0.39	3.91
$K - K_2$	4.58	−6.31	−0.02	2.03
$K - K_3$	3.28	−2.92	0.02	1.38
$K - K_4$	1.64	−1.47	0.03	0.83

Since the functions K_α give good locally supported approximations of the kernel K, its λ-regularized Gram-matrices $\mathbf{K}_{\alpha,\lambda} = K_\alpha(P_i,P_j) + \lambda\,\delta_{ij}$ are multi-banded (when the corresponding Cholesky factors are band matrices) and should give good approximations of the matrix $\mathbf{K}_\lambda$. Therefore we use it as preconditioners of the conjugate gradient (CG) method (Golub and Van Loan (1996)) in the resolution of the linear system (26) as it is described in the following algorithm.

Table 3: Bandwidth of $\mathbf{K}_i$ and convergence rates of the preconditioned conjugate gradient method

Dataset	*Prec.*	*Band.*	k
1	No	–	20
	K_1	52	20
	K_2	155	15
	K_3	258	14
	K_4	514	13
2	No	–	79
	K_1	203	85
	K_2	609	62
	K_3	1115	52
	K_4	2128	38
3	No	–	21
	K_1	102	21
	K_2	305	16
	K_3	607	16
	K_4	1114	14

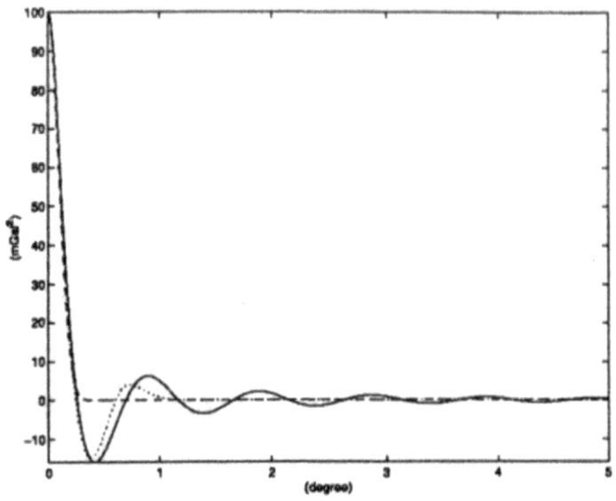

Fig. 1: Functions K (solid), K_1 (long dashes), and K_2 (dashed)

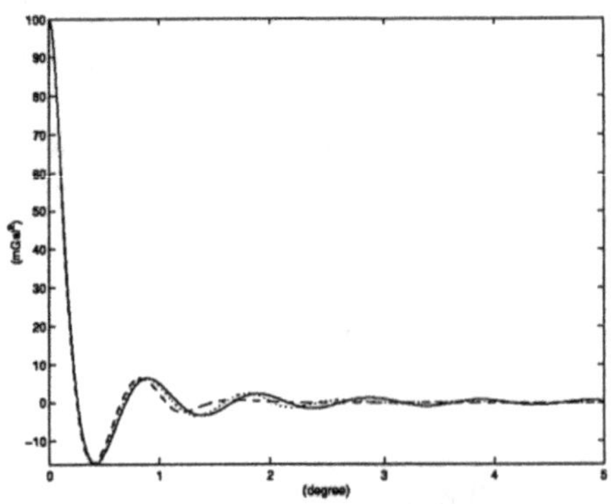

Fig. 2: Functions K (solid), K_3 (long dashes), and K_4 (dashed)

Preconditioned CG Algorihm

Given the initial guess x_0 *satisfying* $\mathbf{K}_{\alpha,\lambda} x_0 = b$, *compute* $r_0 = b - \mathbf{K}_\lambda x_0$ *and solve* $\mathbf{K}_{\alpha,\lambda} z_0 = r_0$. *Set* $p_0 = z_0$. *For* $k = 1, 2, \cdots$

Compute $\mathbf{K}_\lambda p_{k-1}$.

Set $x_k = x_{k-1} + a_{k-1} p_{k-1}$, $a_{k-1} = \frac{\langle r_{k-1}, z_{k-1}\rangle}{\langle p_{k-1}, \mathbf{K}_\lambda p_{k-1}\rangle}$.

Compute $r_k = r_{k-1} - a_{k-1} \mathbf{K}_\lambda p_{k-1}$.

Solve $\mathbf{K}_{\alpha,\lambda} z_k = r_k$.

Set $p_k = z_k + b_{k-1} p_{k-1}$, $b_{k-1} = \frac{\langle r_k, z_k\rangle}{\langle r_{k-1}, z_{k-1}\rangle}$.

In Table 3 are given for the three data sets: the bandwidth of the Cholesky factor of $\mathbf{K}_i$ as well as the iteration step k of the preconditioned conjugate gradient method for which two consecutive iterates x_{k-1} and x_k satisfy

$$\frac{\max_i |x_k(i) - x_{k-1}(i)|}{\min_i |x_{k-1}(i)|} \leq 10^{-10} , \qquad (27)$$

i.e. the conjugate gradient solver will stop when two iterates x_{k-1} and x_k have the same first ten digits.

5 Conclusions

Using the first method developed in Moreaux et al. (1999b) to construct locally supported positive definite approximations of any harmonic reproducing kernel (cf. Section 2) and after having estimated closed expressions of the truncated Legendre coefficients (cf. Section 3) we tested it to the approximation of a kernel K (23) being a covariance function of gravity anomalies. In order to get continuous approximations we opted for cut-off angles α being roots of the function G (25) associated with the kernel K. From the numerical tests (cf. Section 4) we not surprisingly notice that the bigger angle α the smaller are both the error between K and K_α (cf. Table 2) and the convergence rate of the preconditioned conjugate gradient method (cf. Table 3). Moreover because the first α was the first root of G, its associated function K_1 (cf. Fig. 1) is positive when K admits non negligible negative values.

From Table 3 we check that if we multipy α by the integer n we also increase the bandwidth of the matrix $\mathbf{K}_i$ by a factor of n. Noticing that the iteration numbers corresponding to the data sets one and three are roughly the same, we deduce that these numbers depend on the spacing between the data points and not on the size of the area. Moreover, comparing the iteration numbers associated with the second and third data sets, we notice a stagnation of the efficiency of the use of the preconditioner $\mathbf{K}_2$ ($\sim 25\%$) when there is an increase for $\mathbf{K}_3$ (24% to 34%) and $\mathbf{K}_4$ (33% to 52%).

Therefore we believe that if we both increase the size of area and the number of points, the use of our method will be more relevant due also to the fact that more entries of the matrix $\mathbf{K}$ will be negligibly small and then the weight of the near fields will be increased.

Acknowledgements Inspiring discussions with C.C. Tscherning (University of Copenhagen, Denmark), G. Balmino and J.P. Barriot (GRGS-CNES, Toulouse, France) are gratefully acknowledged. This work has been carried out in the framework of the GEOSONAR project and founding was supplied through a grant of the Danish Research Council. The author would also thank P. Fontan for her unconditional support.

References

Arabelos D, Tscherning CC (1996) Collocation with finite covariance functions. Int Geoid Service bulletin 5: 117-135.

Freeden W, Reuter R (1983) Spherical harmonic splines. Meth u Verf d Math Physik 27: 79-103.

Freeden W, Schreiner M, Franke R (1997) A survey on spherical spline approximation. Surv Math Ind 7: 29-85.

Golub GH, Van Loan CF (1996) Matrix computations, third edition. The Johns Hopkins University Press, Baltimore and London.

Magnus W, Oberhettinger F, Soni RP (1966) Formulas and theorems for the special functions of mathematical physics. Springer (Die Grundlehren der mathematischen Wissenschaften in Einzeldarstellungen, Bd 52).

Moreaux G, Barriot J-P, Amodei L (1999a) A harmonic spline model for local estimation of planetary gravity fields from line-of-sight acceleration data. J Geod 73: 130-137.

Moreaux G, Tscherning CC, Sansò F (1999b) Approximation of harmonic covariance functions on the sphere by non harmonic locally supported ones. J Geod. (Accepted for publication)

Müller C (1966) Spherical harmonics. Lecture Notes in Mathematics 17. Springer-Verlag, Berlin.

Paul MK (1973)A method of evaluating the truncation error coefficients for geoidal height. Bull Géod 110: 413-425.

Sansò F, Schuh WD (1987) Finite covariance functions. Bull Géod 61: 331-347.

Tscherning CC, Rapp RH (1974) Closed covariance expressions for gravity anomalies, geoid undulations, and deflections of the vertical implied by anomaly degree-variance models. Reports of the Department of Geodetic Science No 208, The Ohio State University, Columbus, Ohio.

Tscherning CC, Poder K (1982) Some geodetic applications of Clenshaw summation. Boll Geod Sci Aff 4: 349-375.

A wavelet approach to non-stationary collocation

W. Keller
Geodetic Institute
University of Stuttgart, Geschwister-Scholl-Str. 24/D, 70174 Stuttgart, Germany

Abstract. The collocation problem for noisy data with a piecewise constant noise variance is considered. An equivalence to the WIENER-KOLMOGOROV equations in stationary collocation theory is constructed. A numerical solution based on HAAR wavelets is given.

Keywords. Wavelets, instionary noise, collocation, Galerkin's method.

1 Introduction

Collocation is widely used for the treatement of geodetic measurements. The collocation theory is based on the stationarity assumption. In practice this assumption is only piecewise fullfilled, since the variance of the data errors differs in different areas.

A first solution of the instationary collocation problem is given in Sansó and Sideris 1997. There for the instationary case the following extension to the well known WIENER-KOLMOGOROV equation is found:

$$(C_{yy} + w(t)\delta) * h = C_{xy} , \quad \hat{x} = h * y \qquad (1)$$

Here, y is the observed data with the covariance C_{yy} and x is the unknown solution with the cross-covariance C_{xy} . The data y contain white noise with a piecewise varying variance $w(t)$.

Since, in contrast to the classical WIENER-KOLMOGOROV equations w is not constant, the equations cannot be transformed into the frequency-domain and solved by FFT. Therefore, a wavelet solution will be given here, which will exibit the following features

- The piecewise constancy of the noise-variance will be optimally reflected by the underlying HAAR wavelets
- In contrast to the FFT-case the matrix is no longer diagonal but at least sparse.

2 Hilbert space random variables

For the construction of an equivalence to the WIENER-KOLMOGOROV equations a mathematical model for stochastic processes with instationary variance has to be given. In the paper Sanso and Sideris 1997 this model is based on random WIENER measures. Here an alternative description by Hilbert space valued random variables will be given.

Let H be a separable HILBERT space with the scalar product $(\bullet, \bullet)$ and let $[\Omega\, A, P]$ be a probability space.

Definition 1. A mapping $\xi : [\Omega\, A, P] \to H$ is called HILBERT space valued random variable.

For a HILBERT space valued random variable, moments of first and second order have to be defined.

Definition 2. $m \in H$ is called the mean value of a HILBERT space valued random variable ξ, if

$$E(f, \xi) = (f, m) \quad \forall f \in H \qquad (2)$$

holds.

Definition 3. An operator $C : H \to H$ is called the covariance operator of a HILBERT space valued random variable ξ, if

$$E(f, \xi - m)(g, \xi - m) = (f, Cg) \quad \forall f, g \in H \quad (3)$$

holds.

Completely analogously the cross-covariance between two HILBERT space valued random variables ξ and η is defined.

Definition 4. An operator $C_{\xi\eta} : H \to H$ is called the cross-covariance operator of the two random variables ξ and η, if

$$\begin{aligned} &E(f, \xi - m_\xi)(g, \eta - m_\eta) \\ &= (f, C_{\xi\eta} g) \quad \forall f, g \in H \end{aligned} \qquad (4)$$

holds.

International Association of Geodesy Symposia, Vol. 121
Schwarz (ed.), Geodesy Beyond 2000 – The Challenges of the First Decade

3 Prediction problem

Here, the problem to be studied in the sequel will be defined. Let ξ be a random variable with the mean value zero and the covariance operator $C_{\xi\xi}$. To the unknown signal ξ some random noise will be added. This random noise is modeled by a HILBERT space valued random variable n with the mean value zero and the covariance operator

$$(f, C_{nn}g) = (f, w \cdot g) \tag{5}$$

This relation can be interpreted as an uncorrelated random variable n with variance changing timewise according w. The signal ξ itself cannot be observed but the sum $\eta = \xi + n$ of the signal and the noise. The goal is to find an optimal prediction of $z = (f, \xi)$ $f \in H$ from the noisy data η. In this context the concept of an optimal prediction is given by three requirements:

- The prediction has to be linear

$$\hat{z} = (a, \eta) \tag{6}$$

- The prediction has to be unbiased

$$E\hat{z} = z \tag{7}$$

- Among all linear unbiased estimations it must have the minimal error variance

$$E(\hat{z} - z)^2 = \min \tag{8}$$

Theorem 1. The optimal prediction of $z = (f, \xi)$ is

$$\hat{z} = (a, \eta) \tag{9}$$

with

$$(C_{\xi\xi} + C_{nn})a = C_{\eta\xi}f \tag{10}$$

the condition of minimal error variance.

4 Wavelet solution of Wiener-Kolmogorov equations

The equation (10) is an operator equation, which in general can only be solved approximately. One standard technique for the numerical solution of operator equations is GALERKIN method. The basic idea of this method is to look for a solution of (10) not in the whole HILBERT space H but in a finite-dimensional subspace $H_n \subset H$.

Let $e_1, \dots e_n$ be a base of H_n, then the equation for the determination of the approximation a_n for the solution a is

$$\sum_{i=1}^{n}((C_{\xi\xi} + C_{nn})e_i)\alpha_i = C_{\xi\eta}f \tag{11}$$

Of course it cannot be expected that with the elements of a finite-dimensional subspace H_n equation (11) can be fulfilled exactly. There will always be a residual

$$r := C_{\xi\eta}f - \sum_{i=1}^{n}((C_{\xi\xi} + C_{nn})e_i)\alpha_i \tag{12}$$

This residual has to be made as small as possible. One possibility to achieve this is to require that the residual has to be orthogonal to all base elements e_i of H_n.

$$0 = (r, e_i) \qquad i = 1, \dots n \tag{13}$$

This leads to the so-called GALERKIN equations

$$\sum_{j=1}^{n}(e_i, ((C_{\xi\xi} + C_{nn})e_j)\alpha_j = (C_{\xi\eta}f, e_i) \quad i = 1, \dots n \tag{14}$$

or in more detail

$$\sum_{j=1}^{n}[(e_i, C_{\xi\xi}e_j) + (e_i, w(t)e_j)]\alpha_j = (C_{\xi\eta}f, e_i) \quad i = 1, \dots n \tag{15}$$

The question remains how the base functions e_i can be properly chosen. One criterion can be the nature of w. The quantity w is the variance of the data-noise. The data-noise is supposed to have a piecewise constant variance. Hence, w can be optimally represented by HAAR-wavelets

$$w(t) = w_N\varphi_{N,0}(t) + \sum_{n=0}^{N-1}\sum_{m=0}^{2^{N-n}-1} w_{n,m}\psi_{n,m}(t) \tag{16}$$

with

$$\varphi_{N,0}(t) = 2^{-\frac{N}{2}}\varphi(2^{-N}t) \tag{17}$$

$$\psi_{n,m}(t) = 2^{-\frac{n}{2}}\psi(2^{-n}t - m) \tag{18}$$

and

$$\varphi(t) = \begin{cases} 1 & , \quad 0 \le t < 1 \\ 0 & , \quad \text{else} \end{cases} \tag{19}$$

$$\psi(t) = \begin{cases} 1 & , \quad 0 \le t < \frac{1}{2} \\ -1 & , \quad \frac{1}{2} \le t < 1 \\ 0 & , \quad \text{else} \end{cases} \tag{20}$$

This leads to the following form of the GALERKIN equations

$$\sum_{j=1}^{n} [(e_i, C_{\xi\xi} e_j) + w_N(e_i, \varphi_{N,0} e_j) + \sum_{n=0}^{N-1} \sum_{m=0}^{2^{N-n}-1} w_{n,m}(e_i, \psi_{n,m} e_j)] \alpha_j = (e_i, C_{\xi\eta} f) \tag{21}$$

The GALERKIN equations constitute a system of linear equations for the unknown coefficients α_i of the approximate solution a_n. For an efficient solution of these equations it is necessary to find closed expressions for the matrix coefficients

$$a_{ij} = (e_i, C_{\xi\xi} e_j) + w_N(e_i, \varphi_{N,0} e_j) + \sum_{n=0}^{N-1} \sum_{m=0}^{2^{N-n}-1} w_{n,m}(e_i, \psi_{n,m} e_j) \tag{22}$$

of those linear equations. For this purpose the base functions e_i should be as simple as possible. The simplest choice one could think of is the usage of HAAR-wavelets. This choice generates the following form of the GALERKIN equations

$$\begin{aligned} &[(\varphi_{N,0}, C_{\xi\xi}\varphi_{N,0}) + w_N(\varphi_{N,0}, \varphi_{N,0}^2) \\ &+ \sum_{n=0}^{N-1} \sum_{m=0}^{2^{N-n}-1} w_{n,m}(\varphi_{N,0}, \psi_{n,m}\varphi_{N,0})] \alpha_{N,0} \\ &\quad + [\sum_{p=0}^{N-1} \sum_{q=0}^{2^{N-p}-1} [(\varphi_{N,0}, C_{\xi\xi}\psi_{p,q}) \\ &\quad + w_N(\varphi_{N,0}, \varphi_{N,0}\psi_{p,q}) \\ &+ \sum_{n=0}^{N-1} \sum_{m=0}^{2^{N-n}-1} w_{n,m}(\varphi_{N,0}, \psi_{n,m}\psi_{p,q})]] \alpha_{p,q} \\ &\quad = (\varphi_{N,0}, C_{\xi\eta} f) \end{aligned} \tag{23}$$

$$\begin{aligned} &[(\psi_{r,s}, C_{\xi\xi}\varphi_{N,0}) + w_N(\psi_{r,s}, \varphi_{N,0}^2) \\ &+ \sum_{n=0}^{N-1} \sum_{m=0}^{2^{N-n}-1} w_{n,m}(\psi_{r,s}, \psi_{n,m}\varphi_{N,0})] \alpha_{N,0} \\ &\quad + [\sum_{p=0}^{N-1} \sum_{q=0}^{2^{N-p}-1} [(\psi_{r,s}, C_{\xi\xi}\psi_{p,q}) \\ &\quad + w_N(\psi_{r,s}, \varphi_{N,0}\psi_{p,q}) \\ &+ \sum_{n=0}^{N-1} \sum_{m=0}^{2^{N-n}-1} w_{n,m}(\psi_{r,s}, \psi_{n,m}\psi_{p,q})]] \alpha_{p,q} \\ &\quad = (\psi_{r,s}, C_{\xi\eta} f) \end{aligned} \tag{24}$$

$$r = 0, \ldots N-1, \quad s = 0 \ldots 2^{N-r} - 1$$

In matrix notation these equations have the following form

$$\begin{pmatrix} a_{N,0|N,0} & \cdots & a_{N,0|1,2^{N-1}-1} \\ a_{N-1,0|N,0} & \cdots & a_{N-1,0|1,2^{N-1}-1} \\ \vdots & \ddots & \vdots \\ a_{1,2^{N-1}-1|N,0} & \cdots & a_{1,2^{N-1}-1|1,2^{N-1}-1} \end{pmatrix}$$

$$\times \begin{pmatrix} \alpha_{N,0} \\ \alpha_{N-1,0} \\ \vdots \\ \alpha_{1,2^{N-1}-1} \end{pmatrix} = \begin{pmatrix} (\varphi_{N,0}, C_{\xi\eta} f) \\ (\psi_{N-1,0}, C_{\xi\eta} f) \\ \vdots \\ (\psi_{1,2^{N-1}-1}, C_{\xi\eta} f) \end{pmatrix}$$

with

$$a_{N,0|N,0} = (\varphi_{N,0}, C_{\xi\xi}\varphi_{N,0}) + w_N(\varphi_{N,0}, \varphi_{N,0}^2) + \sum_{n=0}^{N-1} \sum_{m=0}^{2^{N-n}-1} w_{n,m}(\varphi_{N,0}, \psi_{n,m}\varphi_{N,0}) \tag{25}$$

$$a_{N,0|p,q} = (\varphi_{N,0}, C_{\xi\xi}\psi_{p,q}) + w_N(\varphi_{N,0}, \varphi_{N,0}\psi_{p,q}) + \sum_{n=0}^{N-1} \sum_{m=0}^{2^{N-n}-1} w_{n,m}(\varphi_{N,0}, \psi_{n,m}\psi_{p,q}) \tag{26}$$

$$a_{r,s|N,0} = (\psi_{r,s}, C_{\xi\xi}\varphi_{N,0}) + w_N(\psi_{r,s}, \varphi_{N,0}^2) + \sum_{n=0}^{N-1} \sum_{m=0}^{2^{N-n}-1} w_{n,m}(\psi_{r,s}, \psi_{n,m}\varphi_{N,0}) \tag{27}$$

$$a_{r,s|p,q} = (\psi_{r,s}, C_{\xi\xi}\psi_{p,q}) + w_N(\psi_{r,s}, \psi_{p,q}\varphi_{N,0}) + \sum_{n=0}^{N-1} \sum_{m=0}^{2^{N-n}-1} w_{n,m}(\psi_{r,s}, \psi_{n,m}\psi_{p,q}) \tag{28}$$

These expressions can be simplified using the relations

$$(\varphi_{N,0}, \varphi_{N,0}^2) = 2^{-\frac{3N}{2}} \int_0^{2^N} dt = 2^{-\frac{N}{2}} \tag{29}$$

$$(\varphi_{N,0}, \psi_{n,m}\varphi_{N,0})$$
$$= 2^{-N}\int_0^{2^N} \psi_{n,m}(t)dt$$
$$= 2^{-N}2^{\frac{n}{2}}\int_0^{2^N} \psi(2^{-n}t-m)dt$$
$$= 2^{-N}2^{\frac{n}{2}}\int_{-m}^{2^{N-n}-m} \psi(z)dz = 0 \quad (30)$$

$$(\varphi_{N,0}, \psi_{n,m}\psi_{p,q})$$
$$= 2^{-\frac{N}{2}}\int_0^{2^N} \psi_{n,m}(t)\psi_{p,q}(t)dt$$
$$= 2^{-\frac{N}{2}}\delta_{np}\delta_{mq} \quad (31)$$

leading to

$$a_{N,0|N,0} = [(\varphi_{N,0}, C_{\xi\xi}\varphi_{N,0}) + w_N 2^{-\frac{N}{2}}] \quad (32)$$
$$a_{N,0|p,q} = [(\varphi_{N,0}, C_{\xi\xi}\psi_{p,q}) + 2^{-\frac{N}{2}}w_{p,q}] \quad (33)$$
$$a_{r,s|N,0} = [(\psi_{r,s}, C_{\xi\xi}\psi_{N,0}) + 2^{-\frac{N}{2}}w_{r,s}] \quad (34)$$

$$a_{r,s|p,q} = [(\psi_{r,s}, C_{\xi\xi}\psi_{p,q}) + w_N 2^{-\frac{N}{2}}\delta_{r,p}\delta_{s,q} + \sum_{n=0}^{N-1}\sum_{m=0}^{2^{N-n}-1} w_{n,m}(\psi_{r,s}, \psi_{n,m}\psi_{p,q})] \quad (35)$$

5 Special case of stationary signal and instationary noise

At the time being no assumptions about stationarity have been made, neither for the signal ξ nor for the noise n. The situation is simplified substantially, if the signal ξ is stationary and only the noise n is instationary. In this case one obtains

$$(\varphi_{N,0}, C_{\xi\xi}\varphi_{N,0})$$
$$= 2^{-N}\int_0^{2^N}\int_0^{2^N} C_{\xi\xi}(s-t)dsdt \quad (36)$$

$$(\varphi_{N,0}, C_{\xi\xi}\psi_{p,q}) = 2^{-\frac{N+p}{2}}$$
$$\times \int_0^{2^N}\int_{-\infty}^{\infty} C_{\xi\xi}(s-t)\psi_{p,q}(2^{-p}t-q)dtds$$
$$= 2^{-\frac{N+p}{2}}\int_0^{2^N}\int_{2^pq}^{2^pq+2^{p-1}} C_{\xi\xi}(s-t)dtds$$
$$- 2^{-\frac{N+p}{2}}\int_0^{2^N}\int_{2^pq+2^{p-1}}^{2^pq+2^p} C_{\xi\xi}(s-t)dtds \quad (37)$$

$$(\psi_{r,s}, C_{\xi\xi}\psi_{p,q}) = 2^{-\frac{r+p}{2}}$$
$$\times \int_{-\infty}^{\infty}\int_{-\infty}^{\infty} C_{\xi\xi}(t'-t)\psi(2^{-r}t'-s)$$
$$\times \psi(2^{-p}t-q)dt'dt = 2^{-\frac{r+p}{2}}$$
$$\times [\int_{2^rs}^{2^rs+2^{r-1}}\int_{2^pq}^{2^pq+2^{p-1}} C_{\xi\xi}(t'-t)dt'dt$$
$$- \int_{2^rs}^{2^rs+2^{r-1}}\int_{2^pq+2^{p-1}}^{2^pq+2^p} C_{\xi\xi}(t'-t)dt'dt$$
$$- \int_{2^rs+2^{r-1}}^{2^rs+2^r}\int_{2^pq}^{2^pq+2^{p-1}} C_{\xi\xi}(t'-t)dt'dt$$
$$+ \int_{2^rs+2^{r-1}}^{2^rs+2^r}\int_{2^pq+2^{p-1}}^{2^pq+2^p} C_{\xi\xi}(t'-t)dt'dt] \quad (38)$$

Introducing the abbreviation

$$I(a,b,c,d) := \int_a^b\int_c^d C_{\xi\xi}(t'-t)dt'dt \quad (39)$$

the following expressions for the scalar products are obtained

$$(\varphi_{N,0}, C_{\xi\xi}\varphi_{N,0}) = 2^{-N}I(0,2^N,0,2^N) \quad (40)$$

$$(\varphi_{N,0}, C_{\xi\xi}\psi_{p,q})$$
$$= 2^{-\frac{N+p}{2}}[I(0,2^N,2^pq,2^pq+2^{p-1})$$
$$- I(0,2^N,2^pq+2^{p-1},2^pq+2^p)] \quad (41)$$

Since for the most common types of covariance functions $C_{\xi\xi}$ closed expressions can be given for (39) the computation of the GALERKIN matrix can be done very efficiently.

6 Numerical example

For a numerical test a stationary signal ξ was chosen having a covariance function

$$C_{\xi\xi}(\tau) = \frac{1}{1+(\alpha\tau)^2} \quad (42)$$

This signal was investigated on the interval $[0,T]$, $T = 512$. For the noise n two types of white noise were used

- on the interval $[0, \frac{T}{2}]$ white noise with the variance $\sigma_1^2 = 1.0$
- on the interval $[\frac{T}{2}, T]$ white noise with the variance $\sigma_2^2 = 4.0$

The first step was the generation of a sample of the staionary signal ξ. For this purpose the following algorithm was applied:

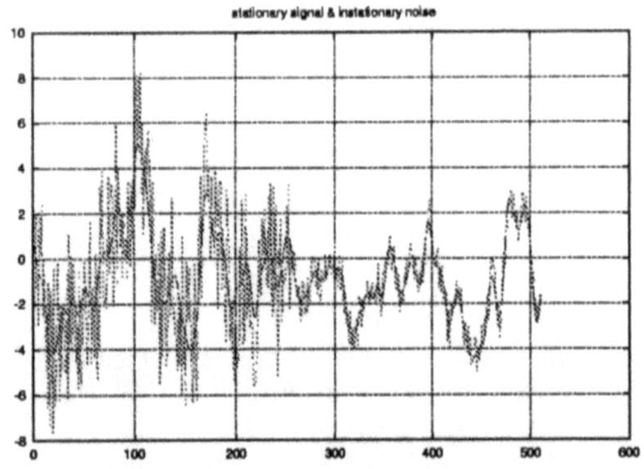

Fig. 1: Noisy signal

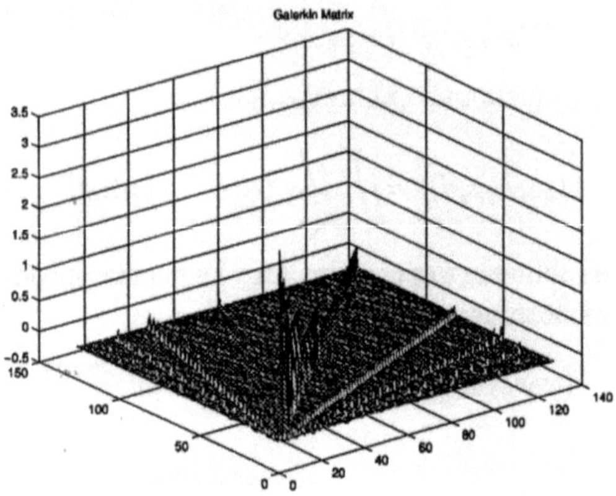

Fig. 2: Galerkin matrix

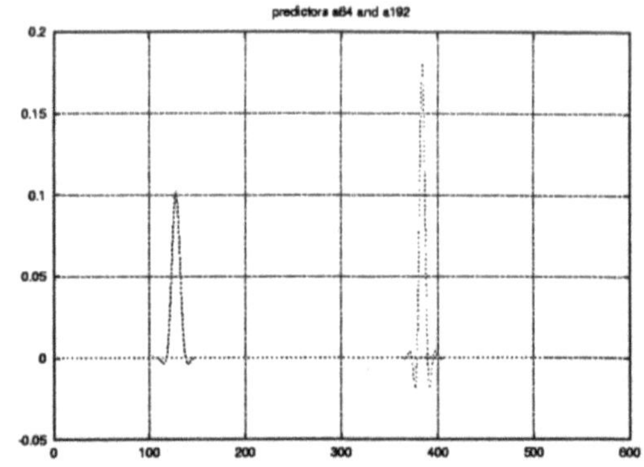

Fig. 3: Predictors

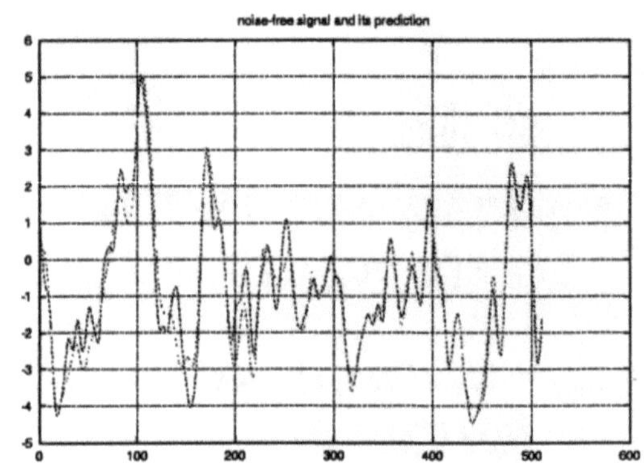

Fig. 4: Instationary prediction

- generation of $N = 512$ independent random variables η_i , $i = 1, \ldots 512$ which are equally distributed on the interval $[-\frac{1}{2}, \frac{1}{2}]$ and which have unit variance.

- computation of the covariance matrix

$$C = (c_{ij}) = C_{\xi\xi}(i-j) \quad \forall i,j = 0,\ldots,511 \tag{43}$$

- CHOLESKY decomposition of the covariance matrix

$$C = LL^\top \tag{44}$$

- construction of the signal

$$\xi_i := \sum_{j=1}^{i} l_{ij}\eta_j \quad i = 1,\ldots 512 \tag{45}$$

With the help of the rules of covariance propagation it can be shown that ξ has the covariance function $C_{\xi\xi}$:

$$\begin{aligned} E\{\xi_i\xi_j\} &= E\{\sum_{k=1}^{i} l_{ik}\eta_k \sum_{l=1}^{j} l_{lj}\eta_l\} \\ &= E\{\eta^\top L^\top L\eta\} = LL^\top = C_{\xi\xi} \end{aligned} \tag{46}$$

Fig. 1 shows a sample of the stationary signal ξ without and with superimposed noise.

For the functional f two instances were chosen

$$\begin{aligned} f_1 &:= \varphi_{0,128} , \\ f_2 &:= \varphi_{0,384} \end{aligned} \tag{47}$$

Since ξ is a rather long-scale featured process, with a reasonable degree of approximation

$$\begin{aligned} \xi(128) &\approx (f_1, \xi) , \\ \xi(384) &\approx (f_2, \xi) , \end{aligned} \tag{48}$$

holds. Due to the fact that ξ and n are uncorrelated $C_{\xi\eta} = C_{\xi\xi}$ is true.

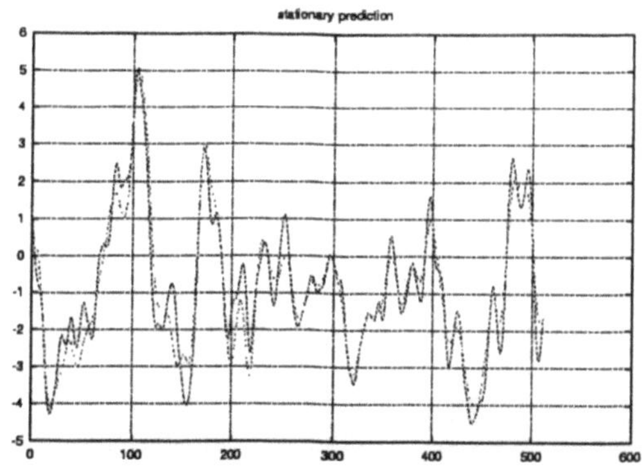

Fig. 5: Conventional prediction

Now, all previously compiled results can be used here

$$
\begin{aligned}
& I(a,b,c,d) \\
&= \int_a^b \int_c^d \frac{1}{1+\alpha^2(t'-t)^2} dt' dt \\
&= \frac{1}{\alpha^2} \int_{\alpha a}^{\alpha b} \int_{\alpha c}^{\alpha d} \frac{1}{1+(x-y)^2} dx dy \\
&= \frac{1}{\alpha^2} [\alpha(b-c) \arctan(\alpha(b-c)) \\
&\quad - \frac{1}{2} \ln(1+\alpha^2(b-c)^2) \\
&\quad - \alpha(b-d) \arctan(\alpha(b-d)) \\
&\quad + \frac{1}{2} \ln(1+\alpha^2(b-d)^2) \\
&\quad - \alpha(a-c) \arctan(\alpha(a-c)) \\
&\quad + \frac{1}{2} \ln(1+\alpha^2(a-c)^2) \\
&\quad + \alpha(a-d) \arctan(\alpha(a-d)) \\
&\quad - \frac{1}{2} \ln(1+\alpha^2(a-d)^2) \qquad (49)
\end{aligned}
$$

With the help of the formulas (23)-(26) and (32) - (37) the GALERKIN matrix can be computed now. Figure 2 shows the structure of the GALERKIN matrix for the instationary noise, given by the wavelet coefficients of its variance

$$
w_N = 2.5 \quad w_N = -1.5 \quad w_{n,m} = 0 \quad \text{else}
$$

The solutions a_i $i \in \{128, 384\}$ are displayed in Fig. 3.

Clearly, the prediction consists in a weighted mean of the data in the neighbourhood of the value which has to be predicted. The stronger the data noise the more smoothing is this weighted mean. In Fig. 4 the noise-free signal and its prediction are compared.

Obviously, the lower the noise is the better the prediction. In Fig. 5 the result of the usual stationary prediction is shown.

A comparison of the results shows that the instationary prediction yields better results, especially in the area of lower noise. The price which has to be paid is that in the instationary case the very efficient tool of FFT cannot be applied, which increases the computational load.

References

Sansó, F. and Sideris, M. (1997). On the similarities and differences between system theory and least-squares collocation in physical geodesy. *Bolletino di Geodesia e Science Affini*, LVI: 173-206.

Wavelets and collocation: An interesting similarity

Christopher Kotsakis
Department of Geomatics Engineering, University of Calgary
2500 University Drive NW, Calgary, Alberta, Canada T2N 1N4

Abstract. The rapid developments in the fields of multiresolution appro-ximation theory and wavelets over the past few years have created an enormous amount of important theoretical knowledge and useful practical tools to be used for various signal processing applications. One of the most attractive properties of multiresolution/wavelet theory is the ability to study the details of a signal locally in various scale levels, according to a *zoom-in/zoom-out* approach. In gravity field modelling, on the other hand, we are used to employing (both theoretically and practically) the concept of collocation in order to approximate and study the behavior of unknown signals based on discrete data. One of the standard formulations of collocation, as a spatio-statistical linear approximation problem in purely deterministic fields, requires the use of a covariance function which is defined via a certain spatial averaging operator over the signal's domain. The "stationary" form of this CV function is often thought to provide strong limitations in the approximation framework, because the actual behavior of the gravity field is "non-stationary". The aim of this paper is to show that there does not really exist any "signal stationarity restriction" problem in statistical collocation, since it can be proven to be equivalent to a wavelet-type expansion. The connection between the two concepts is discussed, and also some recommendations for further work are given.

Keywords. Wavelets, least square collocation, fast Fourier transform.

1 Introduction

The method of least-squares collocation (LSC) represents one of the major foundations in modern physical geodesy. Closely related to Bjerhammar's initial idea on discrete underdetermined boundary value problems, collocation has evolved into a powerful optimal estimation method for either global or local gravity field modelling. Despite the various different interpretations and their associated mathematical concepts upon which collocation has been based (see, e.g., Tscherning 1986; Sanso 1986), a rigorous unified approximation approach that merges both the deterministic (Krarup's formulation) and the stochastic (Wiener's linear prediction theory) viewpoints behind LSC has long been established by Sanso (1980). Such an approach has eliminated, to some degree, most of the "pitfalls" in each individual formulation (e.g., reproducing kernel choice problem, non-stochasticity of the actual gravity field); see also Moritz (1980), Moritz and Sanso (1980). In this way, collocation is usually considered as a rigorous linear *spatio-statistical* method for gravity field approximation, where the term "statistical" is used not to describe some underlying stochastic behavior for the gravity field, but rather to specify the statistical nature of the deterministic norm that is used to quantify the approximation error and to optimize the approximation algorithm.

One of the main characteristics of Sanso's spatio-statistical formulation for the collocation problem is that it leads to the same solution algorithm as the purely deterministic/stochastic approaches. In this case, however, instead of using a reproducing kernel or a covariance (CV) function of a stochastic signal, we only need a *spatial CV function* defined through a certain spatial averaging operator over the unknown deterministic signal. The "stationary" form of this spatial CV function has created the false belief among many geodesists that we still need to model the gravity field of the Earth as a stationary stochastic process, which is furthermore perceived as a strong limitation of the statistical collocation framework since the actual behavior of the gravity field is "non-stationary". However, such a claim is meaningless because no stochastic nature is assigned to the unknown field, and the property of stationarity is not defined at all for deterministic signals; see also the related discussion in Sansò (1980).

International Association of Geodesy Symposia, Vol. 121
Schwarz (ed.), Geodesy Beyond 2000 – The Challenges of the First Decade

In order to eliminate any stationarity concerns about the spatio-statistical collocation framework, and to additionally support the transition towards the use of wavelet/multiresolution approximation techniques in gravity field modelling, the aim of this paper is to show that the optimal signal approximation according to the statistical collocation approach can be expressed in the wavelet-like linear form (e.g., for 1D signals):

$$\hat{g}(x) = \sum_n g(nh)\,\varphi(\frac{x}{h} - n) \tag{1}$$

where $g(x)$ is the unknown field under consideration, h is the resolution level of the discrete data $g(nh)$, and $\varphi(x)$ is a kernel related in a specific way to the "stationary" spatial CV function of $g(x)$. Convolution-based approximation models of the form (1) are a standard tool nowadays in most signal processing applications (Unser and Daubechies 1997; Blu and Unser 1999). Their constantly increasing popularity is due to their close connection with wavelet signal expansions which provide the best available mathematical tool today for localized signal analysis. This important link has actually resulted in the development of the vast field of *multiresolution approximation theory*, originally formulated by Mallat (1989).

2 Statistical collocation and data resolution

In this section, the linear approximation problem for an unknown field $g(x) \in L^2(\Re)$ will be solved in such a way that the immediate connection between the estimated signal $\hat{g}(x)$ and the discrete data resolution will explicitly appear in the solution algorithm. An interesting discussion on the important interplay between data resolution and optimal approximation in gravity field modelling can be found in Sanso (1987). We will assume that the available data represent noiseless point values $g(nh)$ taken on a uniform grid with known resolution level h. The unknown field is considered as 1D for simplicity. The treatment for higher dimensions, i.e. in $L^2(\Re^2)$ or $L^2(\Re^3)$, is just a straightforward extension of the following derivations.

Since we are seeking a linear approximation, the recovered signal will have the general form

$$\hat{g}(x) = \sum_n g(nh)\,\varphi_{n,h}(x) \tag{2}$$

where $\varphi_{n,h}(x)$ is a family of unknown base functions whose dependence (if any) on the data resolution is introduced through the subscript h. If we further impose the condition of *translation-invariance* for the linear approximation with respect to the spatial reference system (in the multi-dimensional case this becomes invariance under more general affine transformations), then the family $\varphi_{n,h}(x)$ should be generated by a single kernel $\varphi_h(x)$ such that $\varphi_{n,h}(x) = \varphi_h(x - nh)$. In this way, eq.(2) is reduced to the simplified form

$$\hat{g}(x) = \sum_n g(nh)\,\varphi_h(x - nh) \tag{3}$$

The above approximation formula can now be illustrated in terms of the linear filtering system shown in Figure 1. Applying the Fourier transform to the "mixed" convolution equation (3), we get

$$\hat{G}(\omega) = \Phi_h(\omega)\,\overline{G}_h(\omega) \tag{4}$$

where $\hat{G}(\omega)$ and $\Phi_h(\omega)$ are the Fourier transforms of the approximated signal and the approximation kernel $\varphi_h(x)$, respectively. The term $\overline{G}_h(\omega)$ corresponds to the aliased periodic Fourier transform:

$$\overline{G}_h(\omega) = \frac{1}{h}\sum_k G(\omega + \frac{2k\pi}{h}) = \sum_n g(nh)\,e^{-i\omega nh} \tag{5}$$

where $G(\omega)$ is the Fourier transform of the original unknown signal (see, Oppenheim and Schafer 1989).

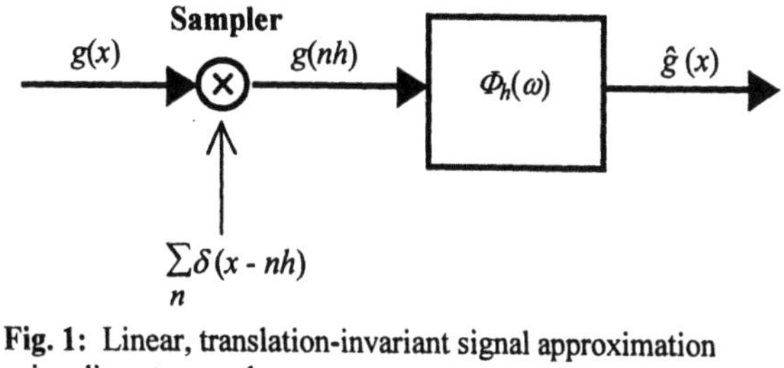

Fig. 1: Linear, translation-invariant signal approximation using discrete samples

Note that the above frequency domain formalism implies that we have sampled our unknown signal over its entire (finite or infinite) support. If the available data grid $g(nh)$ covers only some part of this (finite or infinite) support, then eq.(4) is certainly not valid and a window function should be additionally used. In order to avoid such complications, we will assume that the unknown field we try to approximate covers only the region inside the given grid boundaries. Although such an assumption is unacceptable in applications involving temporal signals (where predictions into the future may be required), it nevertheless provides a very reasonable framework for local approximation studies in spatial fields. It should also be noted that, although $g(x)$ is assumed zero outside the given data grid boundaries, its approximation by eq.(3) may exhibit a non-zero pattern in this region.

2.1 A spatio-statistical optimal principle

The sequence $g(nh)$ is not the only possible information that we could have extracted from the unknown signal at the given resolution level h. If we shift the sampler by an amount x_o, an infinite number of <u>different</u> data sequences can be obtained, which all represent different sampling schemes for the same signal at the same resolution. The situation is illustrated in Figure 2, from which we can see that at a specific resolution value h all the possible sampled sequences of $g(x)$ can be described by the general form $g(nh-x_o)$, where the *sampling phase* parameter x_o varies between the limits $-h/2 \le x_o \le h/2$.

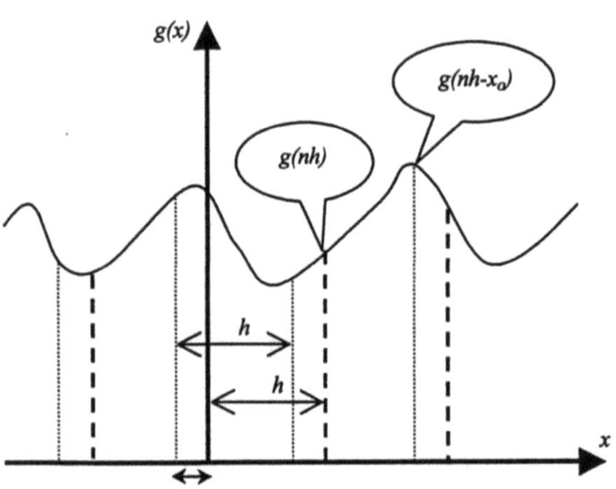

Fig. 2: Different signal sampling configurations at a given resolution level h

In accordance with the translation-invariance condition, we will now have the following general linear equation for the approximated signal from an arbitrary sampled sequence at resolution level h:

$$\hat{g}(x,x_o) = \sum_n g(nh-x_o)\,\varphi_h(x+x_o-nh) \qquad (6)$$

Thus, the approximation error produced by eq.(6) becomes also a function of the sampling phase value x_o associated with the given data set, i.e.

$$e(x,x_o) = g(x)-\hat{g}(x,x_o) \qquad (7)$$

Taking into account eq.(6) and applying the Fourier transform to the last equation (considered as a function of x only), we get

$$E(\omega,x_o) = G(\omega) - \frac{1}{h}\Phi_h(\omega)\sum_k G(\omega+\frac{2k\pi}{h})\,e^{-i\frac{2k\pi}{h}x_o} \qquad (8)$$

The optimal criterion for choosing the best approximation kernel $\varphi_h(x)$ will be

$$P_e(\omega) = \frac{1}{h}\int_{-h/2}^{h/2} \left| E(\omega,x_o)\right|^2 dx_o = \min \qquad (9)$$

Eq. (9) corresponds to a *minimum mean square error* (MMSE) *principle*, expressed in the frequency domain. The quantity $P_e(\omega)$ is nothing else than the *mean error power spectrum*. Note that the term "mean" now has a purely spatio-statistical deterministic meaning, in contrast to Wiener's prediction theory where the mean error is defined in the sense of "experiment repetitions" via an expectation operator. The criterion (9) will minimize the mean error power spectrum over all possible sampling schemes for the given data resolution level h. It is similar to the use of the classic spatial averaging operator $\boldsymbol{M}$ in Moritz's (1980) book and in Sanso's (1980) paper, for the special case of 1D gridded data. It can be easily shown (see, Kotsakis 1999) that

$$P_e(\omega) = C(\omega) - \frac{\Phi_h^*(\omega)C(\omega)}{h} - \frac{\Phi_h(\omega)C(\omega)}{h} + \frac{\Phi_h(\omega)\Phi_h^*(\omega)\overline{C}_h(\omega)}{h} \qquad (10)$$

where the asterisk * denotes complex conjugation, and $C(\omega)$ is just the Fourier transform of the usual spatial CV function $c(\xi)$ of the unknown deterministic signal $g(x)$, i.e.

$$c(\xi) = \int g(x)\,g(x+\xi)\,dx = \Im^{-1}\left\{C(\omega) = G(\omega)G^*(\omega)\right\} \qquad (11)$$

The symbol $\Im^{-1}$ denotes the inverse Fourier transform operator, and the term $\overline{C}_h(\omega)$ has the usual periodized form

$$\overline{C}_h(\omega) = \frac{1}{h}\sum_k C(\omega + \frac{2k\pi}{h}) \tag{12}$$

2.2 Resolution-dependent optimal approximation

Using eqs.(9) and (10), we can easily solve the corresponding variational problem and obtain the optimal approximation kernel (for the analytical procedure, see, e.g., Sideris 1995). The frequency domain form of the optimal kernel will be finally given by the equation

$$\Phi_h(\omega) = \frac{C(\omega)}{\overline{C}_h(\omega)} = \frac{C(\omega)}{\frac{1}{h}\sum_k C(\omega + \frac{2k\pi}{h})} \tag{13}$$

The linear approximation procedure of spatio-statistical collocation, therefore, will be based on the following data filtering formula:

$$\hat{G}(\omega) = \frac{C(\omega)}{\sum_k C(\omega + \frac{2k\pi}{h})} \sum_k G(\omega + \frac{2k\pi}{h}) \tag{14}$$

which is verified by taking into account eqs.(4) and (5). It is seen from the last equation that, as the data resolution increases ($h \to 0$), the approximate will converge to the true unknown signal in the L^2 topology. In order for eq.(14) to correspond to a well defined filtering formula, the power spectrum $C(\omega)$ of the unknown signal has to satisfy some mild conditions that are explained in Kotsakis (1999).

Interestingly enough, the space domain form $\varphi_h(x)$ of the optimal approximation filter $\Phi_h(\omega)$ in eq.(13) can be expressed through the scaling relation

$$\varphi_h(x) = \varphi(\frac{x}{h}) \tag{15}$$

where the generating function $\varphi(x)$ is given by the following inverse Fourier transform

$$\varphi(x) = \Im^{-1}\left\{ \Phi(\omega) = \frac{C(\frac{\omega}{h})}{\sum_k C(\frac{\omega}{h} + \frac{2k\pi}{h})} \right\} \tag{16}$$

The justification of the above fact is trivial and it is based on the fundamental scaling property of the Fourier transform. If we finally substitute eq.(15) into the initial approximation model of eq.(3), we get the wavelet-like expression of eq.(1). It is worth mentioning that the basic optimal kernel $\varphi(x)$ will always be a *symmetric* function, since its Fourier transform $\Phi(\omega)$ given in eq.(16) is always a real-valued function.

2.3 Remarks

Convolution-based linear models of the form of eq.(1) are used in many signal processing applications in the context of classical interpolation, quasi-interpolation, and multi-scale approximation via projections into multiresolution subspaces (see, e.g., Blu and Unser 1999). In such cases, the selection of the approximation model $\varphi(x)$ is usually made a priori and its performance is evaluated according to an assumed behavior for the unknown signal (e.g., bandlimitedness, smoothness, etc.), and/or other theoretical error bounds that depend on the form of the adopted model (i.e. Strang-Fix conditions); for more details, see Unser and Daubechies (1997). In the present paper, on the other hand, the selection of $\varphi(x)$ is adapted to the unknown signal itself through the use of an optimal MMSE principle, and its computation requires some knowledge of the signal's average behavior (i.e. spatial CV function). Note that the basic form of the optimal kernel $\varphi(x)$ depends directly on the data resolution level h, according to eq.(16). As a result, the linear model of eq.(1) in the collocation case will not employ scaled versions $\varphi(x/h)$ of the *same* kernel for each different value of h. This is in contrast to the classic wavelet approximation framework, where a *fixed* scaling kernel $\varphi(x)$ is used for any (dyadic) data resolution level. For more details and discussion, see Kotsakis (1999).

In our derivations we never assumed that the optimally approximated signal should reproduce the available data, i.e. $\hat{g}(nh) = g(nh)$. However, this will always be satisfied since the optimal kernel, defined by eq.(16), is a *cardinal/sampling* function. This simply means that

$$\varphi(n) = \begin{cases} 1 \ , & n=0 \\ 0 \ , & n=\pm 1, \pm 2, \pm 3, \ldots \end{cases} \tag{17}$$

The above property can be easily verified in the frequency domain using the following relationship:

$$\sum_n \Phi(\omega + 2n\pi) = 1 \tag{18}$$

which is of course satisfied by our optimal filter $\Phi(\omega)$ given in eq.(16), for *any* data resolution value h. The validity of the cardinal property (17) can then be ensured through the well known *Poisson summation formula*.

The mean error power spectrum, corresponding to the use of the optimal filter $\Phi_h(\omega)$ given in eq.(13), will be

$$P_e(\omega) = C(\omega) \left(1 - \frac{C(\omega)}{\sum_k C(\omega + \frac{2k\pi}{h})} \right) \tag{19}$$

where $P_e(\omega)$ is the same quantity defined in the MMSE criterion of eq.(9). There is a remarkable similarity between the above error formula and the formula giving the power spectral density (PSD) of the prediction error in noisy stationary random signals according to Wiener's optimal prediction theory (see, e.g., Sideris 1995). The same type of similarity also exists between the optimal approximation filter $\Phi_h(\omega)$ in eq.(13) and the actual Wiener filter.

However, the two formulations correspond to entirely diverse physical situations and they are based on completely different mathematical concepts and assumptions. In the statistical collocation approach, instead of having continuous, noisy and stationary random *input* signals, we deal with purely discrete and deterministic *input* data. Also, in this case the noise takes the form of lost information due to the discretization of the original unknown signal $g(x)$ (see Figure 1). Furthermore, there is no signal stationarity assumption involved in the current formulation. It is actually the translation-invariance condition, imposed for the statistical collocation case, that makes the two approaches algorithmically comparable in terms of *signal-to-noise* ratio (SNR) linear filters which are applied to the input data of each case.

3 Statistical collocation, multiresolution approximation and wavelets

The previous developments can be considered quite general, and they did not involve any special concepts from Mallat's multiresolution approximation theory. It is quite remarkable the fact that the spatio-statistical collocation framework actually leads to a *scale-invariant* approximation scheme (i.e. independent of the scale of the reference system used to describe the position of the gridded data points), similar to the one encountered in wavelet approximation theory. However, there is a significant difference between the optimal collocation model of eq.(1) and the classic wavelet-based approximation methodology, due to the fact that the associated kernel $\varphi(x)$ in the collocation case changes for every different data resolution level h, according to the frequency domain form given in eq.(16).

The most appropriate way to describe the behavior of the signal approximation model of eq.(1), with the associated kernel $\varphi(x)$ given by the optimal frequency domain form in eq.(16), is to characterize it as: (i) *translation-invariant*, (ii) *scale-invariant*, and (iii) *data resolution-dependent*. Regardless of the origin and the scale of the reference system that is used to describe the physical/spatial position of a given set of gridded data points, the approximated field according to the statistical collocation algorithm will always have the same shape. On the other hand, as the *data point density* (h) changes, the linear approximation algorithm of eq.(1) will employ a constantly changing model-kernel $\varphi(x)$, which will be adapted to the average spatial characteristics of the unknown signal and the resolution level h of the currently used data points in a certain optimal fashion, as suggested by eq.(16). For a more detailed treatment, see Kotsakis (1999).

Furthermore, the optimal kernel $\varphi(x)$ in the statistical collocation framework creates a "generalised" type of *multiresolution analysis* (MRA) within the Hilbert space $L^2(\Re)$. Let us denote by $\{V_j\}_{j \in Z}$ an infinite sequence of subspaces in $L^2(\Re)$, each element of which is associated with a specific data resolution level ($h_j > 0$) and it is defined as the closed linear span of the family $\{\varphi(h_j^{-1}x - n) \,|\, n \in Z\}$, where the basic scaling kernel $\varphi(x)$ is given by eq.(16) for $h = h_j$. In this case, it can be shown (Kotsakis 1999) that the multiresolution subspace sequence $\{V_j\}$ will satisfy all the basic properties that define an MRA structure in $L^2(\Re)$ (see, e.g., Mallat 1989), except from the following one:

$$f(x) \in V_j \Leftrightarrow f(2x) \in V_{j+1} \tag{20}$$

Nevertheless, the optimal kernel $\varphi(x)$ in statistical collocation can be viewed as a *scaling function* whose integer translates $\varphi(h_j^{-1}x - n)$, at each resolution level h_j, create a linearly independent and stable system of base functions (i.e. *Riesz basis*) that can span a *nested* sequence of subspaces $\{V_j\}$, which will asymptotically converge to the $L^2(\Re)$ Hilbert space (as $h_j \to 0$), or to the "zero space" (as $h_j \to \infty$). The way that the value of the scaling parameter h_j changes from one subspace (V_j) to the next (V_{j+1}) cannot of course be arbitrary, but it should generally satisfy the following condition:

$$\frac{h_j}{h_{j+1}} = a_j \quad \forall \ j \in Z \tag{21}$$

where a_j is some positive *integer*, different from unity. The actual value of a_j is not needed to remain constant for each subspace pair, and this provides great flexibility in contrast to the classic Mallat's MRA framework where only *dyadic* schemes (i.e. $a_j = 2$) are considered. In a way, the "self-similar" scaling property of eq.(20) between the various subspaces V_j of a classic MRA is now replaced by the freedom to use a much more flexible rule according to which the scaling parameter (data resolution level) h_j changes from one MRA subspace to the next. A couple of mild conditions that the signal power spectrum $C(\omega)$ has to satisfy, in order for the optimal kernel $\varphi(x)$ in eq.(16) to generate such a multiresolution subspace structure in $L^2(\Re)$, are discussed in Kotsakis (1999).

The previous extension of the classic MRA concept suggests that we may be able to achieve a similar extension of the classic wavelet bases associated with Mallat's dyadic MRAs. If such a step becomes successful, we would have essentially generated a "non-stationary" system of base functions in $L^2(\Re)$ that will be explicitly associated with the actual statistical collocation approximation formula (1); i.e. each unknown signal will give rise to a certain type of wavelet-like basis. The potential of such a connection is quite remarkable, in both theoretical and practical terms, and it will presented in future publications.

4 Conclusions and future work

The concept of spatio-statistical collocation, as expressed by the optimal criterion (9) and a translation-invariance condition, leads to signal approximation models commonly encountered in MRA/wavelet theory. It is the opinion of the author that Sanso's formulation for the collocation problem (Sanso 1980) should not be viewed only as a "supplement" to Wiener's stochastic prediction theory for geodetic approximation problems. It actually constitutes a very powerful and autonomous modelling tool, with remarkable connections to multiresolution approximation theory. As far as the "stationarity" issue is concerned, I personally perceive this problem (in the context of optimal estimation in noiseless deterministic fields) as the ability to study locally the approximated signal in a rigorous and consistent manner with the optimal principles. With such an understanding of the problem, MRA/wavelet theory can provide valuable tools without deviating from the widely acceptable collocation spirit (i.e. MMSE principle).

Many theoretical/practical extensions of the issues discussed herein are needed to cover all possible gravity field applications. Some of these topics are: multi-dimensional generalizations (including compact spherical domains), study of the approximation error as a function of the data resolution level h and the used kernel $\varphi(x)$, development of optimal noise filtering methods in multiresolution approximation models, and empirical/numerical determination of the optimal kernel $\varphi(x)$ in eq.(16).

References

Blu T, Unser M (1999) Quantitative Fourier Analysis of Approximation Techniques: Part I (Interpolators and Projectors) and Part II (Wavelets). *IEEE Transactions on Signal Processing*, vol.47, no.10, pp. 2783-2806.

Kotsakis C (1999) The Multiresolution Character of Collocation. *Journal of Geodesy*. (in press)

Mallat SG (1989) Multiresolution Approximations and Wavelet Orthonormal Bases of $L^2(\mathbf{R})$. *Transactions of the American Mathematical Society*, vol.315, no.1, pp. 69-87.

Moritz H (1980) *Advanced Physical Geodesy*. Herbert Wichmann Verlag, Karlsruhe.

Moritz H, Sanso F (1980) A Dialogue on Collocation. *Boll. di Geod. e Sci. Affi.*, vol.39, no.1, pp. 49-51.

Oppenheim AV, Schafer RW (1989) *Discrete-Time Signal Processing*. Prentice-Hall, Inc.

Sanso F (1980) The Minimum Mean Square Estimation Error Principle in Physical Geodesy (Stochastic and

Non-Stochastic Interpretation). *Boll. di Geod. e Sci. Affi.*, vol.39, no.2, pp. 111-129.
Sanso F (1986) Statistical Methods in Physical Geodesy. In: *Sunkel H (ed.) Mathematical and Numerical Techniques in Physical Geodesy*. Springer Verlag, pp. 49-156.
Sanso F (1987) Talk on the Theoretical Foundations of Physical Geodesy. *Contributions to Geodetic Theory and Methodology*, Report No.60006, pp. 5-28, Dept. of Surveying Engineering, University of Calgary, Calgary, Alberta.
Sideris MG (1995) On the use of heterogeneous noisy data in spectral gravity field modelling methods. *Journal of Geodesy*, 70, pp. 470-479.
Tscherning CC (1986) Functional Methods for Gravity Field Approximation. In: *Sunkel H (ed.) Mathematical and Numerical Techniques in Physical Geodesy*. Springer Verlag, pp. 3-48.
Unser M, Daubechies I (1997) On the Approximation Power of Convolution-Based Least Squares versus Interpolation. *IEEE Transactions on Signal Processing*, vol.47, no.7, pp. 1697-1711.

On the wavelet determination of scale exponents in energy spectra and structure functions and their application to CCD camera data

S. Beth, T. Boos, W. Freeden
Geomathematics Group, Department of Mathematics,
University of Kaiserslautern, P.O. Box 3049, 67653 Kaiserlautern, Germany

N. Casott, D. Deussen, B. Witte
Geodetic Institute,
University of Bonn, Nussallee 17, 53115 Bonn, Germany

Abstract. The usage of the wavelet transform as an alternative to the determination of the power law behaviour in energy spectra and structure functions is presented. The relevant aspects of the wavelet theory are summarized and its application on CCD camera data and wind velocity measurements is modelled.

Keywords. Wavelets, structure function, energy spectra, refractive index, fully developed turbulence, wind field modelling, image dancing.

1 Introduction

Compared with the Fourier transform the wavelet transform possesses a more complicated mathematical structure which also suffers at first look from the disadvantage of neither giving an exact frequency localization nor an exact space localization. In general, one obtains a mixture of both under the limitations of Heisenberg's uncertainty principle (e.g. Chui, 1992; Beth, 1999). Thus the application of pure space or frequency orientated methods seems to be more reasonable for the consideration of energy spectra or the power law behaviour of increments in structure functions. However, the experimental situation differs from that ideal conception. In most cases assumptions like stationarity or ergodicity are made to ensure the validity of the underlying theory. Small deviations completely change the results. Therefore an averaging process is installed in order to smooth the data and get adequate results. Thus, one cannot talk of space or frequency localizing methods any more, but a somehow smoothed version of them. This treatment agrees with the way the wavelet transform acts on most types of functions.

In fact, it can be shown that the wavelet transform preserves the power law behaviour of a large class of functions. It goes even further than the structure function approach and defines a complete singularity spectrum.

In this article the relations between wavelet analysis, structure functions and energy spectra are shortly reviewed. Two examples with CCD camera data and wind velocity measurements show the advantages and drawbacks of the wavelet analysis for the representation of these kinds of functions.

2 The wavelet transform

A function $\Psi \in \mathcal{L}^2(\mathcal{R})$ fulfilling the admissibility condition,

$$0 < c_\Psi := 2\pi \int_{-\infty}^{\infty} \frac{|\hat{\Psi}(\omega)|^2}{|\omega|} \, d\omega < \infty, \qquad (1)$$

is denoted as wavelet. If $\Psi \in \mathcal{L}^1(\mathcal{R})$ and continuity of the Fourier transform in 0 is assumed, one gets the more popular definition of a wavelet by its vanishing 0th moment

$$\int_{-\infty}^{\infty} \Psi(t) \, dt = 0 \qquad (2)$$

describing the wave like character of this type of functions. The wavelet transform is then introduced as the convolution of a function $g \in \mathcal{L}^2(\mathcal{R})$ with the wavelet

$$W_\Psi g(b,a) = \frac{1}{\sqrt{a}} \int_{-\infty}^{\infty} g(x) \Psi(\frac{x-b}{a}) \, dx, \qquad (3)$$

where the dilation parameter a controls the size of the region on which the wavelet acts and the translation parameter b gives the centre of the

International Association of Geodesy Symposia, Vol. 121
Schwarz (ed.), Geodesy Beyond 2000 – The Challenges of the First Decade

wavelet transform. In the limit $a \to 0$ the detail information influencing the transform is concentrated around $x = b$ ("zooming-in"), while for a tending to ∞ one has to take care about the whole domain of definition ("zooming-out").

An outstanding property of wavelets is their ability to characterize the local regularity of a function g in terms of its Hölder exponent $h(x_0)$ at a point x_0,

$$|g(x_0 + l) - g(x_0)| \sim Cl^{h(x_0)}. \qquad (4)$$

In particular, negative values for $h(x_0)$ are allowed. (Mallat and Hwang, 1992) showed, that singularities follow a power law behaviour for the scales

$$|W_\Psi g(x_0, a)| \leq Aa^{h(x_0)} \qquad (5)$$

within a so-called cone of influence $|x - x_0| \leq Da$ for wavelets with at least $n > h(x_0)$ vanishing moments, i.e.

$$\int_{-\infty}^{\infty} x^m \Psi(x)\, dx = 0 \qquad (6)$$

for $m = 0, \ldots, n-1$.

Instead of considering the wavelet transform in the whole scale-space domain it suffices to take care only of the modulus maxima of the wavelet transform (WTMM) at each scale. They also satisfy inequality (5) and form lines pointing to the positions of the singularities. The slope of the graph $\log |W_\Psi g(x, a)|$ over $\log(a)$, $(x, a) \in l$ (l maxima line), gives the Hölder exponent of the singularity. Excluded from these examinations are oscillating singularities like $\sin 1/x$, they require a special treatment, cf. (Arneodo et al., 1995).

As experimental data are often available only as discrete values, the introduction of a space discretization is essential. Canonically a uniform subdivision of space and time is used, while the scale is taken in dyadic steps. For numerical purposes wavelets with compact support are of particular interest, because their discretized wavelet transform can be computed by simple, finite summation schemes, cf. e.g. (Maaß und Stark, 1994). Here we only mention that the discrete wavelet transform at scale 2^j and position k may be determined by

$$W_\Psi g(k, 2^j) = \sum_{l=-N_j}^{N_j} c_l^j g(k + l), \qquad (7)$$

where the coefficients satisfy

$$\sum_{l=-N_j}^{N_j} c_l^j = 0 \qquad (8)$$

$j \in \mathbb{Z}$, since the 0th moment of the wavelet is vanishing. Obviously, the behaviour $a \to 0$, equivalently $j \to -\infty$, can only be approximated due to the finite resolution of the given data. Nevertheless, the results of (Mallat and Hwang, 1992) in digital image processing and (Arneodo et al., 1995) in turbulence analysis show the applicability of this approach.

3 Structure functions

A commonly used tool for studying the singularity spectrum of fully developed turbulent flows f are structure functions D_f^p of order p

$$D_f^p(x) = \langle |f(x_0 + x) - f(x_0)|^p \rangle, \qquad (9)$$

$p > 0$. The brackets $\langle\cdot\rangle$ are usually understood as averaging over a time series. In addition, local homogeneity is assumed such that D_f^p is a function of x being independent of x_0. If we moreover require local isotropy, then the function is independent of the orientation of the distance vector. Consequently, D_f^p can be considered as function of a scalar variable $r = |x|$.

The power law scaling behaviour of the structure functions defines the exponents $\eta(p)$

$$D_f^p(r) \sim r^{\eta(p)}, \qquad (10)$$

which are related to the spectrum $D(h)$ of the Hölder exponents by a Legendre transform, cf. (Parisi and Frisch (1984)

$$\eta(p) = 1 + \min_h (ph - D(h)). \qquad (11)$$

The spectrum $D(h)$ is hereby the Hausdorff dimension of the set of points x with $h(x) = h$, i.e.

$$D(h) = d_H(\{x \,|\, h(x) = h\}). \qquad (12)$$

Box counting methods can be used for the approximate calculation of $D(h)$ (for details see Falconer, 1990). However, the structure function approach underlies some drawbacks which can basically be summarized as follows:

- As it is to expect that increments close to zero occur, D_f^p will diverge for negative p.

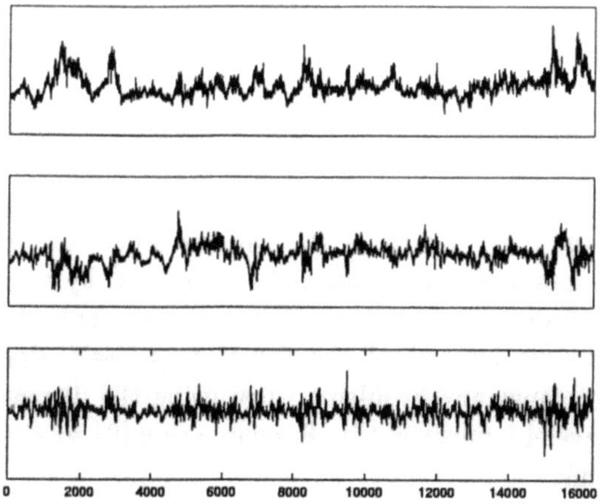

Fig.1: The analyzed experimental data: wind velocity, measurements obtained with an ultrasonic anemometer by the Institute of Geodesy, University of Bonn, Germany. N=16.384 datapoints, sampling frequency 10 Hz, x–, y– and z–component (from the top)

Thus the structure function is only defined for $p > 0$. For $p = 0$ one immediately obtains $\eta(0) = 0$.

- Only Hölder exponents h with $0 < h < 1$ can be detected. Therefore, the method fails to recognize singularities in the derivatives if the function is very smooth (i.e. more than C^1) and the spectrum is trivially $\eta(p) = p$.

These problems can now be avoided using wavelets (Bacry et al. (1991). The difference $f(r_0 + r) - f(r_0)$ may be interpreted as wavelet transform with wavelet $\Psi_\delta(r) = \delta(r-1) - \delta(r)$,

$$f(r_0 + r) - f(r_0) = \frac{1}{r}\int_{-\infty}^{\infty} f(r')\Psi_\delta\Big(\frac{r'-r_0}{r}\Big)\,dr'. \qquad (13)$$

Indeed, a wavelet structure function $\tilde{D}_f^p$ of order p

$$\tilde{D}_f^p(r) = \frac{1}{L}\int_0^L |W_\Psi f(r', r)|^p\,dr' \qquad (14)$$

$L \gg r$, shows the same behaviour like

$$\bar{D}_f^p(r) = \frac{1}{L}\int_0^L |f(r'+r) - f(r')|^p\,dr', \qquad (15)$$

i.e.

$$\bar{D}_f^p(r) \sim r^{\eta(p)} \Leftrightarrow \tilde{D}_f^p(r) \sim r^{\eta(p)}, \qquad (16)$$

if $p > 0$ and $\eta(p) < p$. This property is independent of the choice of the analyzing wavelet.

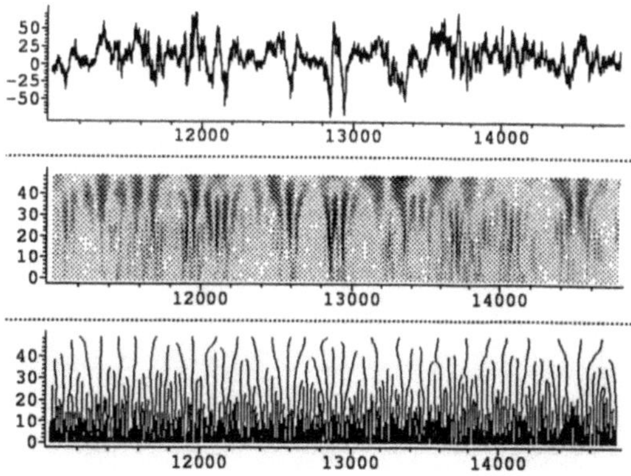

Fig.2: Wavelet analysis of the vertical component of the signal in Figure 1. For better visualization it is shown only a part of the signal. (a) signal data; (b) wavelet transform modulus plotted log(scale) vs. time, fine scales at the bottom; (c) wavelet transform modulus maxima of (b) chained to maxima lines

For the calculation of the wavelet structure function of discrete signals the integral is replaced by a sum

$$\tilde{D}_f^p(r) = \frac{1}{N}\sum_{j=1}^{N} |W_\Psi f(r_j, r')|. \qquad (17)$$

Equation (17) can be simplified by summing over the modulus maxima of the wavelet transform instead,

$$\tilde{\Sigma}_f^p(r) = \sum_{l\in L(r)} \sup_{(x,r')\in l} |W_\Psi f(x, r')|, \qquad (18)$$

where l is a maxima line of the absolut value of the wavelet transform on $[0, r]$ and the supremum is taken over all $(x, r') \in l$ with $r' \le r$. Each influence cone of size r contains maxima lines with the same scaling. Moreover, if the function is everywhere singular, there has to be at least one maxima line in each interval of width r. Thus we take a maximum of each interval of size r leading us to a number of terms proportional to N/r concluding in

$$\begin{aligned}\tilde{D}_f^p(r) &= \frac{1}{N/r}\sum_{l\in L(r)} \sup_{(x,r')\in l} |W_\Psi f(x, r')| \\ &= \frac{1}{N/r}\tilde{\Sigma}_f^p(r). \qquad (19)\end{aligned}$$

We want to discuss two applications in this paper. First, wind velocity measurements were

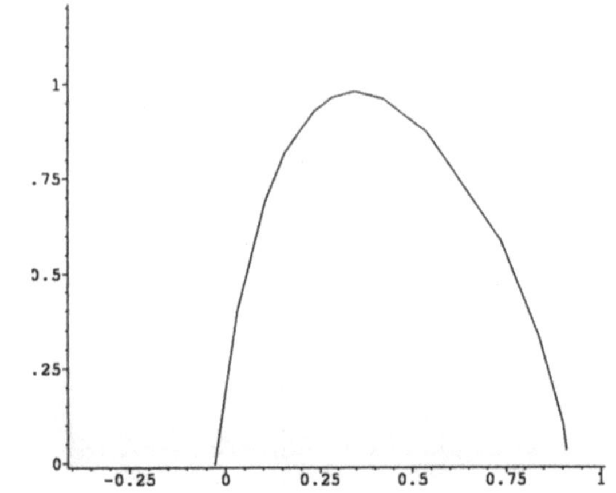

Fig.3: The singularity spectrum D(h) of the analyzed signal calculated from the maxima representation of Figure 2 by the WTMM-method. The maximum value D(h)=0.982 is reached for a Hölder exponent of h=0.343

analysed with the wavelet algorithm. If the components of the wind velocity $c_{w,i}$, $i = 1,2,3$ are known at a fixed point in a time series, the structure functions are calculable under the assumption of stationarity and ergodicity by

$$D^p_{c_{w,i}}(\tau) = \frac{1}{M}\int_0^M |c_{w,i}(t+\tau) - c_{w,i}(t)|^p \, dt. \quad (20)$$

Introducing the mean wind velocity v_0, $D^p_{c_{w,i}}(x)$ can be compared with $D^p_{c_{w,i}}(\tau)$ by $x = \tau v_0$. Figure 1 shows the examined wind velocity signal, while Figure 2 presents the wavelet transform with corresponding maxima lines of a part of the signal. The singularity spectrum in Figure 3 agrees with other experimental results, cf. (Bacry et al., 1991) in the idea that there occur singularities with negative exponent in the signal, too, though they are not as extreme as in the considered data sequences. Furthermore it shows that the most probable value is close to $h = 1/3$ corresponding to $\eta(2) = 2/3$ as predicted by (Kolmogorov, 1941).

As we are not always in the situation to assume stability over a longer time period, it is of particular interest to get in the same way a similar data set in space. Therefore the application of the described method to CCD camera data was examined (within a project supported by the Deutsche Forschungsgemeinschaft (DFG)) and could at least theoretically be solved. The CCD camera takes pictures of a board covered with an equidistant pattern of filled circles. The phase differences $\varphi(m,t) - \varphi(m-1,t)$ between the circles are then calculated by the deviation $z(m,t)$ of the centre of a circle from its mean position

$$\varphi(m,t) - \varphi(m-1,t) = z(m,t)\frac{k_T}{b}, \quad (21)$$

where k_T denotes the carrier wave number and b the focal length of the camera objective. In consequence, one finds $\varphi(l,t) - \varphi(m,t)$ by summation. The fact, that there are given only differences instead of the function φ itself, does not cause any problems when using the discrete wavelet transform (7). Remembering (8) we conclude

$$W_\Psi f(k,2^j) = \sum_{l=-N_j+1}^{N_j} \tilde{c}^j_l(\varphi(k+l,t) - \varphi(k+l-1,t) \quad (22)$$

with

$$\tilde{c}^j_{N_j} = c^j_{N_j}, \tilde{c}^j_l = c^j_{l+1} + \tilde{c}^j_{l+1}, -N_j+1 \le l \le N_j - 1.$$

Therefore the corresponding wavelet structure function to the structure function D^p_φ of the phase given by

$$D^p_\varphi(\varrho, s) = \langle |\varphi(x+\varrho) - \varphi(x)|^p \rangle, \quad (23)$$

$\sqrt{\lambda_T s} \ll \varrho \ll L_0$, λ_T carrier wave length, s distance between camera and target, L_0, could theoretically be computed. But there are two experimental drawbacks to this idea. First, the number of circles is strongly limited by the CCD array. A circle has to possess a certain radius in order to determine its centre exactly, cf. (Beth et al., 1997; Beth and Freeden, 1998). Thus an array of 640x480 pixels does not contain a sufficient number of them contradicting a required signal length $L \gg r$. On the other hand, the assumption $r \gg \sqrt{\lambda_T s}$ is not necessarily assured. These drawbacks are due to the available hardware. Knowing

$$D^2_\varphi(\varrho, s) = 2.92 k_T^2 \varrho^{5/3} C_n^2 s \quad (24)$$

(Deussen and Witte, 1997) used the phenomena of image dancing to compute the structure parameter C_n^2 of the refraction index from the CCD camera data. Their promising results show that the above idea should be kept in mind as the fast progress in the development of hardware may soon overcome the aforementioned problems.

4 Energy spectra

Another interesting function delivering valuable information on turbulence parameters is the spectral energy density S_g,

$$S_g(k) = 2N\Delta t|\hat{g}(k)|^2, \quad (25)$$

$k \geq 0$, for N data points measured with sample frequency $1/\Delta t$. The Fourier transform $\hat{g}$ is now replaced by the wavelet transform in order to define a local spectral wavelet energy density $\tilde{S}_g$

$$\tilde{S}_g(x,k) = N\Delta t \frac{1}{c_\Psi k_T} |W_\Psi g(x, \frac{k_T}{k})|^2, \quad (26)$$

$k \geq 0$. Then the mean wavelet energy density $\tilde{S}_g^m$, defined by

$$\tilde{S}_g^m(k) = \int_0^\infty \tilde{S}_g(x,k)\,dx, \quad (27)$$

follows the same power law behaviour like S_g in case of $S_g(k) \sim k^{-\beta}$ for a wavelet with n vanishing moments, $n > (\beta - 1)/2$, cf. (Perrier et al.). A power law behaviour of the kind $S_g(k) \sim e^{-k^2}$ may not be described in that way by the wavelet transform. The wavelet energy density may be interpreted as an averaged Fourier energy density weighted by the square of the Fourier transform of the wavelet, i.e.

$$\tilde{S}_g^m(k) = \frac{1}{c_\Psi k_T} \int_0^\infty S_g(\omega)|\hat{\Psi}(\frac{k_T \omega}{k})|^2\,d\omega. \quad (28)$$

Although it seems that information gets lost due to the averaging, this is not the case in the experimental application. Figure 4 shows the squared Fourier spectrum of wind velocity measurements from Figure 1. The density should be proportional to $k^{-5/3}$, but it varies around the expected line.

In particular, the deviation becomes stronger with decreasing wave number. Therefore, (Kaimal and Finnigan, 1994) applied a smoothing filter with variable bandwidth which increases for higher frequencies. This technique corresponds to the application of a (global) wavelet based energy density (see Figures 5 and 6).

5 Summary and outlook

The so-called wavelet transform modulus maxima method, cf. e.g. (Muzy et al., 1994), is an alternative to well-known standard techniques like

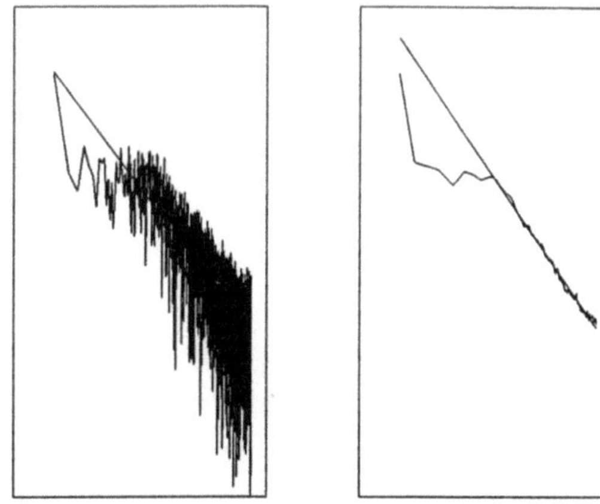

Fig.4: Fourier energy spectrum of the signal of Figure 1(c); by a least square fit we obtain a Kolmogorov exponent β=-1.39, smoothing the spectrum (cf. Kaimal and Finnigan, 1994) leads to β=-1.52

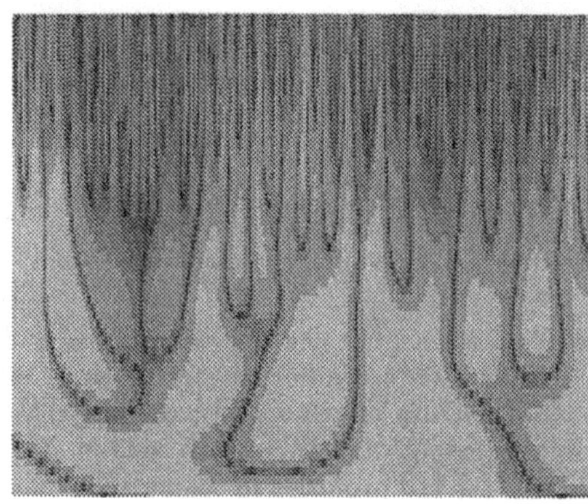

Fig.5: The local wavelet energy spectrum of the same signal shows the local (in time) energy distribution at different scales; analyzing wavelet is the 2nd derivative of a Gaussian (it is shown only a part of the signal)

the structure function approach or in the examination of energy spectra or digital image processing, cf. (Mallat and Hwang, 1992; Beth et al., 1997). The stability of the results is improved by the implicit smoothing. In addition, for the first time the complete singularity spectrum of a function is calculable. Moreover, the wind velocity measurements show the existence of negative scale exponents which can not be given by the structure function approach. In this context it has to be mentioned that not all of the experimental results are mathematically proven.

It also becomes obvious from the examples that the wavelet method requires a sufficient a-

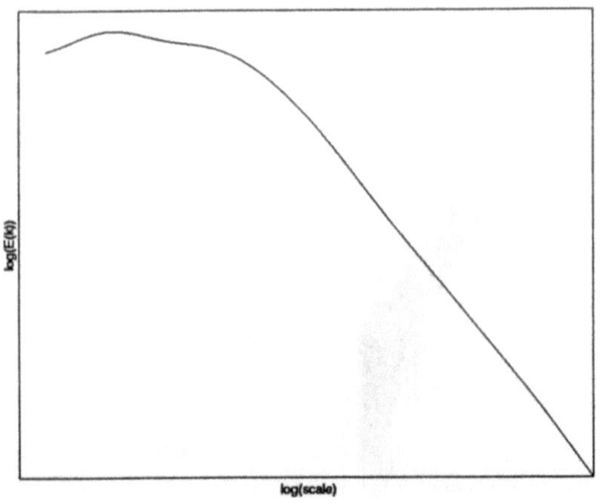

Fig.6: Integration of the above local wavelet energy spectrum over the time gives the global wavelet energy spectrum to be compared with the Fourier spectrum; the slope of the curve in the region of the inertial subrange is β=-1.46, using different wavelets we got values from β=-1.34 to β=-1.56

mount of data. Often this cannot be guaranteed. In a lot of experiments only a few measurements are available, in particular with respect to data spread in space. For the presented analysis of CCD camera data one has to come back to the original structure function approach that allows a simple computation of required values. Nevertheless, the fast evolution in computer sciences will bring up the necessary hardware to overcome the mentioned problems, for example by the development of digital cameras equipped with a larger CCD array. When the time evaluation can be related to the extension of the function in space a variety of problems can be avoided. Then time series deliver data which are easy to analyze by a wavelet transform as we saw in case of the wind velocity measurements.

Besides these aspects the wavelet analysis supplies us with local information of the examined signal, for example, remembering the local wavelet energy density. For the first time we have the chance to consider this local distribution, that is not accessible by Fourier techniques. Although its interpretation is still unclear, this will hopefully help us to better understand the exchange of energy between the different scales or frequencies in turbulent flows.

References

Arneodo A, Argoul F, Bacry E, Elezgaray J, Muzy J-F (1995) Ondelettes, Multifractales et Turbulence. Diderot Editeur, Arts et Sciences

Bacry E (1999) LastWave 1.5 Software, http://www.cmap.polytechnique.fr/~bacry/LastWave

Bacry E, Arneodo A, Frisch U, Gagne Y, Hopfinger E (1991) Wavelet Analysis of Fully Developed Turbulence Data and Measurement Scaling Exponents. Turbulence and Coherent Structures, O. Metais, M. Lesieur (eds.), Selected Papers from 'Turbulence 89: Organized Structures and Turbulence in Fluid Mechanics', Grenoble, 18.-21. September 1989, Kluwer Academics, pp 203–215

Beth S (1999) Spherical Vector Wavelets: Theoretical and Numerical Aspects PhD Thesis, University of Kaiserslautern, Geomathematics Group (in preparation).

Beth S, Casott N, Deussen D, Freeden W, Witte B (1997) Wavelet Methods in Edge Detection and Refraction. Optical 3–D Measurement Techniques IV, Gruen, A, Kahmen, H (eds.), Wichmann Verlag, pp 76–85

Beth S, Freeden W (1998) Waveletmethoden zur Auswertung der geodätischen Refraktion aus CCD–Kamera–Daten. VR, 60, Nr. 7, pp 382–391

Chui CK (1992) An Introduction to Wavelets. Academic Press

Deussen D, Witte B (1997) Detecting 'Image Dancing' for the Determination of Vertical Refraction Using a Digital Camera. Proceedings of the Second Turkish–German Joint Geodetic Days, Berlin, 28.-30. Mai, pp 695–704

Falconer KJ (1990) Fractal Geometry. Mathematical Foundations and Applications. Wiley and Sons Ltd., Chichester

Kaimal JC, Finnigan JJ (1994) Atmospheric Boundary Layer Flows — Their Structure and Measurement. Oxford University Press

Maaß P, Stark H-G (1994) Wavelets and Digital Image Processing. Surveys on Mathematics for Industry, 4, pp 195–235

Mallat S, Hwang WH (1992) Singularity Detection and Processing with Wavelets. IEEE Trans. Inform. Theory, 38, pp 617–643

Parisi G, Frisch U (1984) Turbulence and Predictability in Geophysical Fluid Dynamics and Climate Dynamics, M. Ghil, R. Benzi, G. Parisi, eds., North–Holland, p 84

Muzy J-F, Bacry E, Arneodo A (1994) The Multifractal Formalism Revisited With Wavelets. International Journal of Bifurcation and Chaos, Vol. 4, No. 2, pp 245–302

Perrier V, Philipovitch T, Basdevant C (1995) Wavelet Spectra Compared to Fourier Spectra. J Math Phys, 36, pp 1506–1519

The use of wavelets for the analysis and de-noising of kinematic geodetic measurements

A.M. Bruton and K.-P. Schwarz
Department of Geomatics Engineering, The University of Calgary, Canada.

J. Škaloud
Laboratoire de Topométrie (IGEO/TOPO), Département de Génie Rural (DGR)
École Polytechnique Fédérale de Lausanne (EPFL), Switzerland.

Abstract. The purpose of this paper is to discuss and demonstrate how wavelets can be used to improve the performance of geodetic measurement systems; more specifically to explore the use of wavelets as a tool for analyzing and de-noising the errors that are inherent in GPS/INS systems. Because the wavelet transform provides a time-resolution representation of a signal, it offers the unique ability to analyze the error characteristics of such systems at different frequencies and to localize them in time. This is something not offered by tools that operate in the time or frequency domains. It is seen that these features can be useful for analyzing and removing the errors in kinematic geodetic systems.

Keywords. Wavelet de-noising, wavelet analysis, GPS, INS.

1 Introduction

In this section, background information about the integration of GPS and INS systems is provided and motivation for wavelet-based analysis and de-noising techniques is given.

The goal is to use the measurements provided by these systems to estimate the trajectory of the vehicle in which their sensors are mounted. A standard way to do this is to use a Kalman filter. In a decentralized configuration, there are two filters working independently: an INS filter and a GPS filter. The INS filter is the main one in the configuration and uses the output of the INS mechanization to estimate the state (position, velocity and attitude) along the trajectory. The output of the GPS filter is then used to update the main filter to help in the estimation of the long-term errors of the INS. In order to estimate the state, using a set of measurements, a Kalman filter depends on a measurement and a dynamics model. The quality of the final estimates of the state depends therefore on the quality of both the measurements being made and the models being used. See Schwarz and Wei (1990) for more details. Figure 1a is a conceptual plot of the frequency spectrum of the errors in the measurements made by the inertial sensors. Notice the division into long and short-term errors. Figure 1b shows how each are reduced by the integration process. The long term errors are reduced by updating the filter with the error state vector that comes from the GPS filter (position and velocity). The short term errors are reduced by the smoothing that is done by the numerical integration process of the INS mechanization. It should be noted that the effect of the short term errors can also be reduced by pre-filtering the data as shown in Skaloud (1999). This will be further discussed in Section 4.2. Clearly then, the quality of the final estimates of the state depends on the quality of both the measurements being made and the models being used.

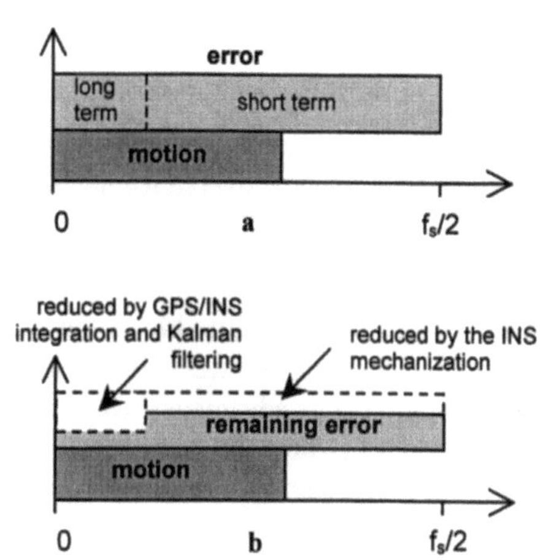

Fig. 1: Conceptual plot of the errors in an INS system: **a** stand-alone, **b** after standard processing (f_s is sampling freq.)

International Association of Geodesy Symposia, Vol. 121
Schwarz (ed.), Geodesy Beyond 2000 – The Challenges of the First Decade

propose and implement techniques for removing some of them. It is shown that through such an analysis, the remaining errors can be separated into deterministic and stochastic errors. Moreover, each error can be characterized as a function of vehicle dynamics. The applicability of such an approach is demonstrated in several ways using real data. Finally, ways of reducing the errors that have been identified are discussed and implemented.

The next section briefly describes the wavelet transform and how it is used for signal analysis, denoising and synthesis.

2 Wavelets as a tool

The continuous wavelet transform (CWT) of a real sequence $x(t)$ is defined as the inner product of the function with a family of functions $\psi_{a,b}(t)$:

$$x(t) \rightarrow X_{a,b} = \langle \psi_{a,b}(t), x(t) \rangle \qquad (1)$$

where in turn that family is defined by continuous dilations (a) and translations (b) of some mother function $\psi(t)$:

$$\psi_{a,b}(t) = \frac{1}{\sqrt{a}} \psi\left(\frac{t-b}{a}\right). \qquad (2)$$

The discrete wavelet transform (DWT) of the same sequence is given by the inner product

$$d_{m,n} = \langle \psi_{n,m}(t), x(t) \rangle \qquad (3)$$

where the basis function $\psi_{n,m}(t)$ is obtained by sampling the continuous parameters a and b. If they are sampled in a dyadic fashion, so that $a = 2^m$ and $b = n2^m$ where m and n are discrete dilation and translation indices, then the basis function is given by

$$\psi_{n,m}(t) = 2^{-m/2} \psi(2^{-m} t - n). \qquad (4)$$

Given the discrete transform coefficients in eq (3), the function $x(t)$ can be recovered using

$$x(t) = \sum_m \sum_n d_{n,m} \psi_{n,m}(t). \qquad (5)$$

Equations (3) and (5) are referred to as the *analysis* and *synthesis* equations. As we will see later, the wavelet transform offers advantages over its Fourier domain counterpart (where the basis function offers only a fixed frequency resolution and no localization in time).

The de-noising procedure consists of 1) performing a wavelet analysis, using the analysis equation, 2) applying a thresholding of the wavelet coefficients, and 3) recovering the de-noised signal using the synthesis equation. Clearly, the choice of threshold in the second step above is crucial to the quality of the de-noising process and should be made carefully. The interested reader should consult a reference such as Donoho and Johnstone (1995) or MATLAB (1998) for details about wavelet domain de-noising and the appropriate selection of a threshold.

3 Time-resolution analyses of real data

3.1 Test description

For these analyses, the data from a controlled kinematic test is used. An airborne survey was conducted where a tactical grade inertial system called the CMigits II was mounted along with an Ashtech Z12 GPS receiver to evaluate the positioning and attitude capabilities of the former. To provide a reference for evaluation, a navigation grade inertial system called the LTN 90-100 was also mounted in the aircraft directly below the CMigits II with a known spatial offset. Although the CMigits II provides a GPS/INS navigation solution in real-time, post-mission processing has been employed so that results can be directly compared to the LTN 90-100 when using the same filter and the same differential GPS code and phase measurements. For the purposes of this paper, it is enough to say that a navigation grade system is of sufficiently better quality than a tactical grade system and that it provides a good independent reference for the sensor errors and the estimated navigation parameters.

The test was approximately 20 minutes in duration and consisted of the takeoff, a short flight line of nearly constant azimuth and velocity (that will be called Line 1), a heading maneuver, a second short flight line of nearly constant azimuth and velocity (that will be called Line 2) and the landing. Since a tactical grade system is not accurate enough to do a static alignment, the initial orientation of the system was supplied by the LTN 90-100.

3.2 Errors in the model assumptions

A conventional Kalman filter is only as smart as you let it be. Its ability to estimate the state and the associated accuracy is no better than the quality of the models used and the quality and type of the observations performed. One of the major problems related to mismodeling occurs when existing dynamics are not modeled. Although it is difficult to diagnose unmodeled state dynamics, a traditional way to do this is to perform an analysis of the innovation sequence of the filter. Intuitively, this sequence can be interpreted as the difference between the incoming measurements and what the filter predicts the measurements to be, based on the estimate of the state prior to an update. If the system is well modeled, the innovation sequence should be uncorrelated meaning that its frequency spectrum will be flat. Any peaks in the frequency spectrum will imply a mismodeled component at that frequency. More details about this type of analysis can be found in Grewal and Andrews, (1993). In the case of GPS/INS integration, the innovation sequence is the difference between the state predicted by INS and the update provided by GPS. There are of course several possible reasons for the non-white nature of such an innovation sequence. These include the facts that 1) the dynamics of the INS may not be fully modeled and 2) the updates coming from the GPS may not have white noise characteristics as we assume they do. In turn, this would imply some unmodeled dynamics in the GPS filter.

Figure 2 shows the frequency domain analysis of the innovation sequence for the z-component of the position state for the CMigits II throughout the survey. This spectrum shows that most of the errors are low frequency in nature (below 0.2 Hz), but it is not possible to quantify when the errors are present in each frequency band. Consider Figure 3 that shows a continuous wavelet decomposition of the same sequence, according to eq. (1), using a Daubechies wavelet. It should be noted that several wavelets were used for this purpose, making no noticeable difference. In this figure, time is along the x-axis and scale is along the y-axis. A darker shade implies that the component of the error at that scale and at that time is larger. Because a scale corresponds to a frequency band, this figure gives a quantitative report of the frequency content of the modeling errors as function of time. Also shown along the left side of the figure is the fact that the scales shown represent frequencies from 0.25 Hz to 0.002 Hz. The power of this type of analysis comes from the fact that the evolution of these errors can be studied as a function of vehicle dynamics, something that is not possible with the frequency domain analysis in Figure 2. Clearly, the errors are larger in the heading maneuver and during the takeoff and landing than during Line 1 and Line 2. This implies that the models being used represent the true dynamics better during the periods of constant velocity and azimuth than during periods where these vary. The low frequency nature of the errors that was seen in the Fourier domain is confirmed here, but there is significantly more information in this figure.

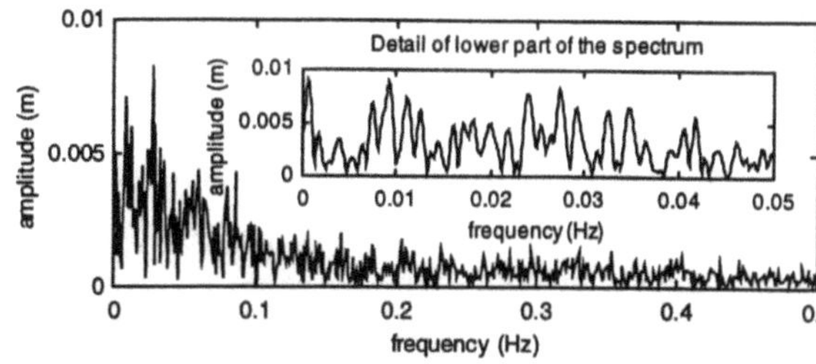

Fig. 2: The amplitude of the frequency spectrum of the z-component of the state vector

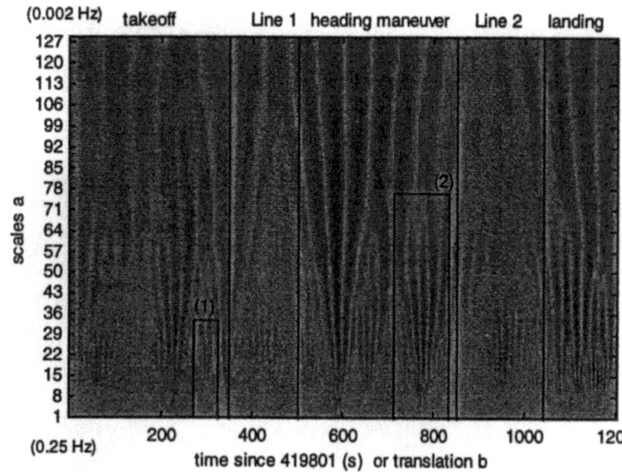

Fig. 3: Innovation of the z-component of the state vector for the C-Migits II in the continuous wavelet domain

For instance, at least three characteristic time-scale features can be identified that clearly repeat throughout the survey and are more pronounced in the ascent, heading maneuver and landing. A sample of the first is shown in the box labeled (1). It occurs at frequencies down to approximately 0.008 Hz which correspondings to periods up to roughly 128 s. The second sample is shown in the box labeled (2), and occurs at frequencies down to approximately 0.003 Hz which corresponds roughly to periods up to 300 s. The last significant feature exists over the whole range of scales down to 0.002Hz and is most pronounced during the heading maneuver.

As mentioned above, these characteristics may come from either the GPS model or the INS model. In order to isolate these from each other, Figure 4 was generated. It shows the same things as Figure 3, except that the LTN 90-100 was used in place of the CMigits II. This means that dynamics that are not modeled in the INS filter are significant when a tactical grade system is used, but not when a navigation grade one is used. In turn, this means that those remaining in Figure 4 are either mismodeling errors in the GPS filter or errors common to both inertial systems. The former could be due to unmodeled dynamics such as multipath or the fact that a constant velocity model was used in the GPS filter. The latter could be due to the fact that any initial alignment error occurs in both cases since the initial orientation of the LTN 90-100 was used in both. Either of these would produce the non-whiteness of the sequence that is observed during the heading maneuver. It can also be hypothesized that features (1) and (2) in Figure 3 are the results of some dynamics that are not modeled in the INS filter that should be modeled when a tactical grade system is used. This will be discussed in more detail in the next section.

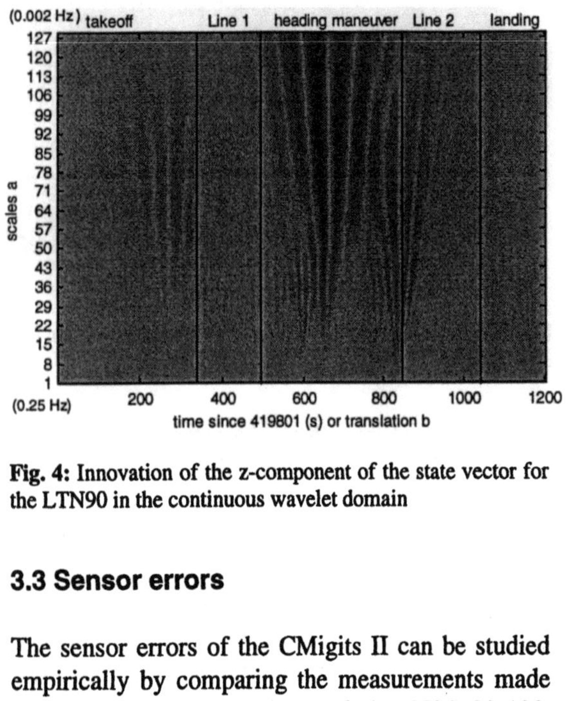

Fig. 4: Innovation of the z-component of the state vector for the LTN90 in the continuous wavelet domain

3.3 Sensor errors

The sensor errors of the CMigits II can be studied empirically by comparing the measurements made by the CMigits II to those of the LTN 90-100. Figure 5 is the result of a ten level discrete wavelet decomposition (again using the Daubechies wavelet) of the errors in the angular rate measurements made by the z-gyro during the warm-up, ascent and Line 1. The top plot in the figure shows the original error signal. The plot below that is the 10th level approximation of it (i.e. it was reconstructed using only the 10th level approximation coefficients) and the ten plots below that are the details going from coarsest to finest. This is the same decomposition that was discussed in Section 2 meaning that Figure 5 is a representation of the error in different frequency bands. The limits of these bands are defined on the right side of the figure given a sampling frequency of 50 Hz.

There are several important observations to be made. First is that much of the error seen in the original signal shows up in the band above 12.5 Hz as seen by comparing the top and bottom plots. Second is that the time evolution of the errors is clearly seen in nearly every band. For example, in the highest frequencies (d_1 and d_2) it is clear that the characteristics of the errors change as the aircraft finishes the ascent and begins Line 1. Also, there are some very obvious changes in the characteristics of the errors in the lowest frequencies (d_{10} and a_{10}, corresponding roughly to periods longer that 20 s). Although it is not shown in this figure, if the same decomposition is plotted over the whole time period, it can be seen that these variations are largest in the heading maneuver and during takeoff and landing where there are larger changes in azimuth and that they are smaller during the periods of constant azimuth and velocity (Line 1 and Line 2). This implies that the error may be a scale factor of the gyro. This is confirmed by considering Figure 6 that shows that the a_9 approximation of the error (i.e. a_{10} + d_{10}) and the actual value of the angular velocity taken from the LTN 90-100. Notice that the LTN 90-100 has a scale factor repeatability of 5 ppm in the gyros while the CMigits II has one of 350 ppm. The correlation between these plots is evident at this resolution while it was certainly not in the original signal. The 15-state error filter we used to process this data in does not model the scale factor and none was applied because it was not known at the time of the survey. This confirms the hypothesis made in the last section that some errors not modeled in the INS filter may become significant when using a tactical grade system. In this case it is clearly a scale factor error, however other errors may show up when other sensors are analyzed. In general, such errors can either be compensated for or can be estimates by augmenting the state vector.

4 Error reduction

To this point, wavelet techniques have been used to identify deterministic errors in GPS and INS Kalman filter models and in the measurements made by the INS sensors. They have been characterized as functions of time, vehicle dynamics and frequency. We have also used wavelet techniques to identify the errors in the low frequency bands of the INS model as scale factor errors. Referring back to Figure 1b, this means that the block called *remaining errors* has been analyzed. In this section, we will discuss ways of reducing those errors.

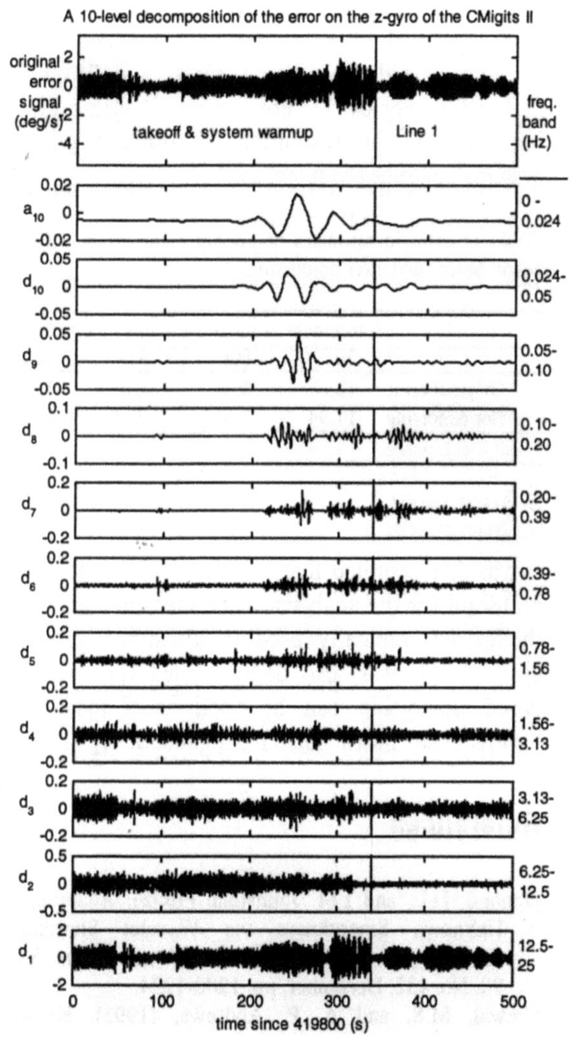

Fig. 5: A 10-level representation of the error on the z-gyro of the CMigits II (created using the db4 wavelet).

4.1 Deterministic errors

The low frequency errors discussed in the last section fall into the broad category of modeling errors. They can be further categorized into INS and GPS errors. To reduce them, they should be either compensated for directly wherever possible, or carefully analyzed as above and as mentioned above appropriately modeled in the error state vector.

4.2 Stochastic errors

High frequency errors are typically stochastic errors. Although they have not been discussed in detail to this point, they can be seen for example in Figure 5 in components d_1 to d_6. (Notice that level 6 is chosen rather approximately here because it is difficult to identify the level at which the separation between stochastic and non-stochastic errors occur.) Such errors obviously exist on any geodetic sensor and will influence the quality of the estimated parameters. As mentioned in the introduction, the reduction of these noise-like errors in the GPS measurements can be done by a Kalman filter, while the noise in inertial sensors is reduced by the data mechanization. Wavelet thresholding techniques can be used to further reduce these errors. We will concentrate on the INS sensor errors although the same techniques could also be applied to other geodetic sensors including GPS receivers.

Two case studies will be reported: the first is de-noising the gyros in the tactical grade system for attitude determination and the second is de-noising the accelerometers in an airborne gravity system for gravity determination.

Because the position and the velocity solutions from both the CMigits II and LTN 90-100 are so heavily weighted by GPS, the attitude solutions will be used for demonstration.

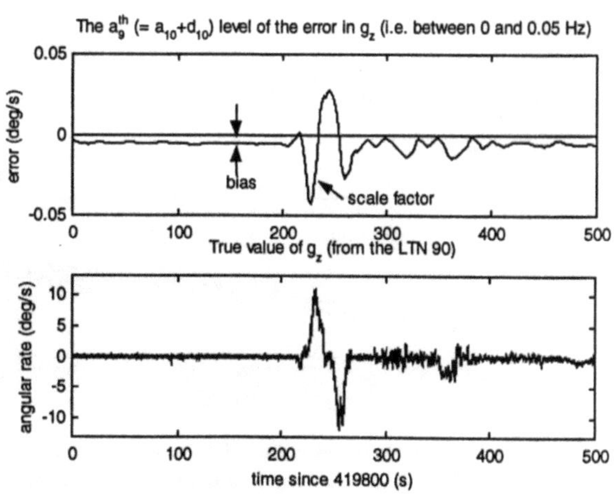

Fig. 6: The coarsest two levels of the decomposition of the error in the z-gyro of the CMigits II and the true angular velocity in the z-gyro.

The general approach is to apply wavelet de-noising to the angular rate measurements made by the CMigits II and use them in our standard filter to derive the attitude parameters. These are then compared to the attitude parameters derived using the LTN 90-100.

The de-noising procedure used was briefly outlined in Section 2. Line 2 was chosen because the modeling errors have been shown to be smaller during that period and because sensor errors make up a larger part of the total error. Figure 7a shows the agreement between the two systems in pitch during this period before de-noising. Figure 7b shows the agreement obtained by means of de-noising using SURE (Donoho and Johnston 1995) soft thresholding and a 5th level decomposition using a Daubechies wavelet of order 20. A large threshold was manually specified for the finest level because it is known that there is no motion in the corresponding frequency band. Although the de-noising has no effect on the long-term component of the error (the sloping bias remains), it clearly reduces the high frequency component. The standard deviation of the error about the same best-fit line is reduced from 18.4" to 13.0", clearly demonstrating the usefulness of this technique. The level of noise reduction is approximately the same in each all three channels. Skaloud (1999) shows that very nearly the same results can be achieved using a low-pass filter, which in fact amounts to the same thing as the manual specification of a high threshold at the finest level.

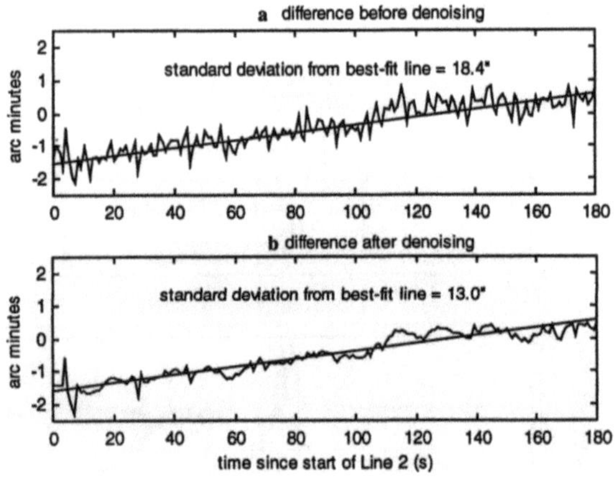

Fig. 7: The difference between the pitch from the CMigits II and the LTN 90-100 during Line 2, **a** before de-noising and, **b** after de-noising.

An airborne gravity system has been developed at the University of Calgary using a strapdown INS/DGPS. In the scalar gravity case, the specific force measurements, after compensation for vehicle motion, are directly compared to the acceleration of the vehicle that is derived from GPS. In such a configuration, the noise level of the acceleration component is a crucial part of the error budget. Further details about this system and its error budget can be found in Wei and Schwarz (1998). Table 1 shows the results for gravity determined along a flight line in the Canadian Rocky Mountains, using an upward continued reference as truth. De-noising was done using a biorthogonal wavelet, a three level decomposition, and SURE soft thresholding. Although the gain is small in this case, the level of noise has been reduced at all filtering periods, becoming more significant at higher frequencies. De-noising may become more important as the resolution of airborne gravity systems continues to increase.

Table 1: Error in estimates of gravity disturbance at flight height before and after de-noising

	Agreement in mGal for a filtering period of			
	30s	60s	90s	120s
Original error	12.65	2.94	2.42	1.91
After denoising	11.34	2.67	2.28	1.88

5 Summary

It was demonstrated that wavelet domain tools can be useful for the analysis of the measurements and the models typically used in kinematic geodetic systems. It was also demonstrated that wavelet based de-noising can be used as a means of removing stochastic noise from geodetic signals.

References

Donoho, D.L. and I.M. Johnstone (1995). Adapting to Unknown Smoothness via Wavelet Shrinkage, *Journal of the American Statistical Association*, Vol. 90, No. 432, December, pp. 1200-1224.

Grewal, M.S. and A. P. Andrews, (1993). Kalman Filtering, Theory and Practice, Prentice Hall, Upper Saddle River, New Jersey.

MATLAB, (1998). Wavelet Toolbox User's Guide, The MathWorks, Inc., Natick, Mass.

Schwarz, K.P. and M. Wei (1990). Testing a Decentralized Filter for GPS/INS Integration, Proceedings of the Poition, Location and Navigation Symposium, 1990, pp. 429-435.

Skaloud, J. (1999). Optimizing Georeferencing of Airborne Survey Systems by INS/DGPS, Department of Geomatics Engineering, University of Calgary, Calgary, Canada, UCGE Report 20126.

Wei, M. and K.P. Schwarz (1998). Flight test results from a Strapdown Airborne Gravity System. *Journal of Geodesy*, 72, pp. 323-332.

A theorem of insensitivity of the collocation solution to variations of the metric of the interpolation space

F. Sansò, G. Venuti
DIIAR, Sez. Rilevamento,
Politecnico di Milano, Piazza Leonardo da Vinci 32, I-20133 Milano, Italy

C.C. Tscherning
Department of Geophysics, University of Copenhagen, Juliane Maries Vej 30,
DK-200 Copenhagen Oe, Denmark

Abstract. The collocation approach to the estimation of a field from observed functionals, is known, by examples and simulations, to display a not very strong dependence from the choice of the specific reproducing kernel-covariance function. In fact the situation is similar to the case of the dependence of least squares parameters on the weight of observations.

The paper, after recalling the basic theory according to its deterministic and stochastic interpretation, shows that the variation of the sought solution is infinitesimal with both, the variation of the metric of the intepolation space going to zero and the quantity of information carried by the observations going to hunderd percent on the specific functional of the field that we want to predict. The combined effect of the two gives an infinitesimal of the second order, namely a theorem of "insensitivity" of the solution to the metric of the interpolation space. Different simulations show the action of this particular effect.

Keywords. Collocation, reproducing kernel, covariance, metric variation.

1 The two versions of collocation and the stability problem

2.1 The deterministic version

This is a simple approximation version where the field $u(t), (t \in T)$ with T any set compatible with the subsequent hypothesis, is assumed to belong to some Hilbert space $\mathcal{H}(T)$ and we have performed on $u(t)$ a finite number (N) of observations which can be represented as bounded linear (or linearized)[1] functionals on $\mathcal{H}(T)$, i.e. via Riesz theorem

$$L_i(u) = \langle h_i, u\rangle_{\mathcal{H}} = Y_i (i = 1, 2, \dots N), \quad h_i, u \in \mathcal{H} . \tag{1}$$

For the sake of simplicity we assume to have "exact observations", i.e. without measurement error. In case we want to include the evaluation functionals

$$\delta_t(u) = u(t) \qquad \forall t \in T \tag{2}$$

in the set of bounded functionals on $\mathcal{H}$ we know that then $\mathcal{H}$ must be a R.K.H.S., i.e. there is a $K(t,t')$ such that (this is useful but not strictly necessary!)

$$\langle K(t,t'), u(t')\rangle \equiv u(t) \ \ \forall t \in T, \ \forall u \in \mathcal{H} . \tag{3}$$

The kernel K is in biunivocal relation with the Hilbert space structure of $\mathcal{H}$, i.e. with the definition of scalar product or, equivalently, with the definition of norm in $\mathcal{H}$.

Here we have a sensible bifurcation of the theory, with two variants:

P1) we want to estimate a field u, with an optimal estimator $\widehat{u}$, satisfying (1) and we decide that optimal here means of minimum norm (maximum smoothing)

$$\|\widehat{u}\|^2 = \mathrm{Min}_{\langle h_i, u\rangle = y_i} \|u\|^2 , \tag{4}$$

P2) we want to estimate some given bounded linear functional of k, i.e.

$$L_0(u) = \langle h_0, u\rangle \tag{5}$$

[1]Note: this is a consistency hypothesis allowing us to say that we do not know which $u \in \mathcal{H}$ is our field (or signal) but we know that in any way we can compute the observational functionals on it.

International Association of Geodesy Symposia, Vol. 121
Schwarz (ed.), Geodesy Beyond 2000 – The Challenges of the First Decade

with an estimator $\widehat{y}_0(u)$ linear in the observations such that the relative error

$$\frac{|\widehat{y}_0(u) - L_0(u)|}{\|u\|} = \mathcal{E}_r(y_0, u) \tag{6}$$

satisfies the minimax principle

$$\mathrm{Min}_{\widehat{y}_0 = \sum \lambda_i y_i} \mathrm{Max}_u \mathcal{E}_r(y_0, u) \,. \tag{7}$$

The solutions are trivial and well known: let us introduce the notations

$$\begin{aligned}
&\underline{Y} = \{Y_i, i = 1 \dots N\} \\
&\underline{h}(t) = \{h_i(t), i = 1 \dots N\} \\
&H = \langle \underline{h}(t), \underline{h}^+(t) \rangle \equiv \langle h_i(t), h_j(t) \rangle \\
&\mathrm{Span}\ (\underline{h}) = \{\underline{\lambda}^+ \underline{h}(t), \underline{\lambda} \in R^N\} \equiv S_N \\
&P = P_N = \text{orthogonal projection on } S_N.
\end{aligned}$$

As we all know the projector P can be explicitely constructed from the Grahamian H as the operator with kernel[2]

$$P(t, t') = \underline{h}^+(t) H^{-1} \underline{h}(t') \,; \tag{8}$$

in fact

$$\langle P(t, t'), v(t') \rangle = 0 \quad \forall v \perp S_N \tag{9}$$

and $\forall w \in S_N$ (i.e. $w(t') = \underline{\lambda}^+ \underline{h}(t') = \underline{h}^+(t') \underline{\lambda}$)

$$\begin{aligned}
&\langle P(t, t'), w(t') \rangle = \\
&\underline{h}^+(t) H^{-1} \underline{h}^+(t) H^{-1} \langle \underline{h}(t'), \underline{h}^+(t') \rangle \underline{\lambda} = \\
&\underline{h}^+(t) H^{-1} H \underline{\lambda} = \underline{h}^+(t) \underline{\lambda} = w(t).
\end{aligned} \tag{10}$$

SOLUTION OF P1:

$$\widehat{u} = Pu = \underline{h}^+(t) H^{-1} \underline{y} \,; \tag{11}$$

this is nothing but observing that Pu is univocally fixed by $\underline{y}$ and then invoking the famous theorem on orthogonal projections in Hilbert spaces.

SOLUTION OF P2:

We put

$$\widehat{y}_0(u) = \underline{\lambda}^+ \underline{y} = \underline{\lambda}^+ \langle \underline{h}, u \rangle = \langle \underline{\lambda}^+ \underline{h}, u \rangle \tag{12}$$

and note that

$$\widehat{y}_0(u) - L_0(u) = \langle \underline{\lambda}^+ \underline{h} - h_0, u \rangle \,. \tag{13}$$

[2]Note: this implicitely assumes that $\{h_i(t)\}$ are linearly independent functionals, what seems very reasonable since observing here without noise we never do redundant observations.

Therefore, by Schwarz inequality,

$$\begin{aligned}
&\mathrm{Max}_u \frac{|Y_0(u) - L_0(u)|}{\|u\|} = \\
&\mathrm{Max}_u \frac{|\langle \underline{\lambda}^+ \underline{h} - h_0, u \rangle|}{\|u\|} = \\
&\|\underline{\lambda}^+ \underline{h} - h_0\| \,.
\end{aligned} \tag{14}$$

Since $\underline{\lambda}^+ \underline{h} \in S_N$, it is indeed

$$\begin{aligned}
&\mathrm{Min}_{\underline{\lambda}} \|\underline{\lambda}^+ \underline{h} - h_0\| \Rightarrow \underline{\lambda}^+ \underline{h} = \widehat{h}_0 = P h_0 = \\
&\underline{\lambda}^+(t) H^{-1} \langle \underline{h}, h_0 \rangle
\end{aligned} \tag{15}$$

Therefore

$$\widehat{y}_0(u) = \left\langle \widehat{h}_0, u \right\rangle = \langle P h_0, u \rangle \,. \tag{16}$$

All that leads to a noteworthy conclusion.

Theorem: the problems P1) and P2) are dual one with respect to the other and their solutions are equivalent in the sense that

$$L_0(\widehat{u}) = \langle h_0, \widehat{u} \rangle = \left\langle \widehat{h}_0, u \right\rangle = \widehat{y}_0(u). \tag{17}$$

□ In fact

$$\langle h_0, Pu \rangle = \langle P h_0, u \rangle$$

because P (as orthogonal projector) is selfadjoint. □

Therefore we might conclude that the deterministic collocation problem can be formulated along two dual variants leading to one and the same solution.

Now our purpose is to study the stability of the solution (11) when we give some change to the metric of the space $\mathcal{H}$; the result will be that it is difficult to say how much $\widehat{u}$ varies in norm, but one can much better study the variation of $\widehat{y}_0 = \langle h_0, \widehat{u} \rangle$, in terms of the relative error (7), showing that this becomes less and less sensitive to a norm change in $\mathcal{H}$, the closer h_0 gets to S_N. This in particular has to happen for N sufficiently large. A motivation for looking into this problem is first of all a logical one; in fact as far as the choice of the particular norm, among the many equivalent, defining the <u>same</u> Hilbert space $\mathcal{H}$ as a set, is arbitrary we would like to be sure that the result does not depend critically on it. Moreover, there are cases (for instance when $\mathcal{H}$ has a reproducing kernel K) in which a relatively small change in $K(t, t')$ can give a much simpler structure to H, simplifying the numerical burden of calculating $H^{-1} \underline{y}$, which could be very large if we have very many data and H is not a sparse matrix.

1.2 The stochastic version

In this case $u = u(t,\omega)$ is interpreted as a generalized random field (GRF) (we assume zero average for simplicity on the Hilbert space $\mathcal{H}$ (Sansò, 1986). This means that $u(t,\omega)$ has realizations which do not belong to $\mathcal{H}$ (Tscherning, 1977) nevertheless we can give a meaning to expressions like

$$L_i(u) = \langle h_i, u\rangle = Y_i \qquad (18)$$

as r.v. on the space $H(u)$ obtained in $\mathcal{L}^2(\omega)$[3] by closing the linear combinations of the type $\sum \lambda_i u(t_i\omega)$.

Already the idea that $u(t,\omega) \in \mathcal{L}^2(\omega)$ implicitely means that $\mathcal{H}$ must have a reproducing kernel; as a matter of fact, recalling the definition of covariance operator (Lauritzen, 1973)

$$E\{\langle\delta_t, u\rangle\,\langle\delta_{t'}, u\rangle\} = \langle\delta_t, C\delta_{t'}\rangle = C(t,t') \qquad (19)$$

which is by definition the covariance function of u. If we choose that the covariance operator C has to be the identity, we see that

$$C(t,t') = \langle\delta_t\delta_{t'}\rangle = K(t,t'), \qquad (20)$$

i.e. the covariance function of the GRF u coincides with the reproducing kernel of $\mathcal{H}$. Otherwise stated: we use as Hilbert space to produce the approximate solutions, the space which has $C(t,t')$ as reproducing kernel. This opens the doors to the empirical estimation of $K(t,t')$, if we assume suitable invariance properties for u.

As we know in this case the problem of approximating $\widehat{u}$ from observations (18) cannot be put in the form of problem P1), but only problem P2) is meaningful: namely we want to approximate the r.v.

$$Y_0 = \langle h_0, u\rangle \qquad (21)$$

from the observation vector (18) $\underline{Y}$. The approximation now is evaluated in terms of $\mathcal{L}^2(\omega)$, i.e. in terms of variance of

$$Y_0 - \underline{\lambda}^+\underline{Y} = \langle h_0 - \underline{\lambda}^+\underline{h}, u\rangle \qquad (22)$$

[3]Note: this is the space of r.v. with finite variance on the same probability space $(\Omega, \mathcal{A}, P)$ on which $u(t,\omega)$ is defined.

In fact the Wiener-Kolmogorov principle is formulated as: find $\underline{\lambda}$ such that

$$E\{[Y_0 - \underline{\lambda}^+\underline{Y}]^2\} = \text{Min.} \qquad (23)$$

But, recallig that $C \equiv I$ in $\mathcal{H}$,

$$E\{[Y_0 - \underline{\lambda}^+\underline{Y}]^2\} = \|h_0 - \underline{\lambda}^+\underline{h}\|^2 \qquad (24)$$

so that we find as before the solutions

$$\widehat{h}_0 = \underline{h}^+(t)H^{-1}\langle\underline{h}, h_0\rangle = Ph_0 \qquad (25)$$

$$\widehat{Y}_0 = \langle\underline{h}_0, h\rangle = \underline{\lambda}^+\underline{Y} = \langle h_0, \underline{h}^+\rangle H^{-1}y\,. \qquad (26)$$

while (25) is exactly the same as (15), (2) is different from (16) in that now $\widehat{Y}_0$ and $\underline{Y}$ are r.v. This is the reason why for the deterministic version we will study the variation of $\widehat{Y}_0$ in terms of its maximum relative variation, while for the stochastic versions, one should consider the variation of $\widehat{Y}_0$ in terms of its variance.

Now if we observe that

$$\langle h_0, \underline{h}\rangle = L_{0t}\underline{L}_{t'}C(t,t') \qquad (27)$$

$$\langle\underline{h}, \underline{h}^+\rangle = L_t\underline{L}_{t'}C(t,t'), \qquad (28)$$

we immediately see that to study the sensitivity of (26) to norm changes is the same as studying its sensitivity to covariance changes. In fact this follows from our choice

$$C = I \qquad C(t,t') = K(t,t') \qquad (29)$$

and the fact that K identifies the metric in $\mathcal{H}$.

This explains the importance of this study because typically our model for the covariance function is not very much close to empirical values and these in any way are themselves not true values.

Then it is very consolating if we can affirm that the final solution is only weakly dependent on the choice of $C(t,t')$, at least when h_0 can be predicted reasonably well from $\underline{\lambda}^+\underline{h}$.

2 The solution stability theorem

The purpose of this paragraph is to prove a theorem of stability of the solutions (16), (26) with respect to first order variations in the definition of norm in $\mathcal{H}$.

This will be done through 7 statements: the first 1), 2), 3) statements establish a general representation of the variation of the norm in $\mathcal{H}$

and the subsequent variations (to the first order) of Riesz representers and of their mutual scalar products; statements 4) and 5) study the stability of $d\widehat{Y}_0$ as a general function of the norm variation in $\mathcal{H}$ and in the particular case that this is given through a reproducing kernel K; statements 6) and 7) apply the former case to the stochastic version, giving a more precise meaning to the concept of a small variation of the covariance/reproducing kernel $dK = dC$.

STATEMENT 1: let $\mathcal{H}$ be given with scalar product $\langle , \rangle$ and assume that to the same set of functions $\mathcal{H}$ is given a slightly different metric with scalar product $\langle , \rangle'$ such a way that the new norm is equivalent to the former; then we have for a suitable bounded selfadjoint operator dQ in $\mathcal{H}$

$$\langle n, v\rangle' = \langle u, (I + dQ)v\rangle \tag{30}$$

with

$$0 < \alpha^2 < (I + dQ) < \beta^2 \, . \tag{31}$$

□ This is a simple consequence of Lax Milgram theorem (Yosida, 1978), implying the existence of a bounded symmetric operator Q such that

$$\langle u, v\rangle' = \langle u, Qv\rangle \, ; \tag{32}$$

it is then enough to put

$$Q = I + dQ \tag{33}$$

and to recall that by hypothesis

$$\alpha\|u\| \leq \|u\|' \leq \beta\|u\| \tag{34}$$

to prove the statement.

Let us notice that

$$(I + dQ) \geq \alpha^2 I \tag{35}$$

implies existence and boundedness of $(I+dQ)^{-1}$. Moreover, since dQ is assumed as "small", e.g. such that $\|dQ\| < 1$, we have as well the relation

$$(I + dQ)^{-1} = I - dQ \tag{36}$$

exact to the first order in dQ. □

STATEMENT 2: when a linear bounded functional $L(u)$ is directly defined on $\mathcal{H}$, we know that, through Riesz theorem, there is a representer h of L such that

$$L(u) \equiv \langle h, u\rangle \; ; \tag{37}$$

now if we change the scalar product in $\mathcal{H}$, also the representer h will change to h' such that

$$L(u) \equiv \langle h', u\rangle' \; ; \tag{38}$$

if we put

$$h' = h + dh \tag{39}$$

then it is, to the first order,

$$dh = -dQh \tag{40}$$

□ Equating (38) and (40) and recalling (30) we get to the first order

$$\begin{aligned}\langle h, u\rangle &= \langle h', u\rangle' = \langle h + dh, (I - dQ)u\rangle = \\ &= \langle h, u\rangle + \langle dh, u\rangle + \langle h, dQu\rangle = \\ &= \langle h, u\rangle + \langle dh + dQh, u\rangle \; , \end{aligned} \tag{41}$$

$\forall u \in \mathcal{H}$, yelding (40). □

STATEMENT 3: let h_1, h_2 be two representer of L_1, L_2 with the scalar product $\langle , \rangle$ and h'_1, h'_2 the representers of the same functionals with the scalar product $\langle , \rangle'$; then we have

$$\begin{aligned}& d\{\langle h_1, h_2\rangle\} = \langle h'_1, h'_2\rangle' - \langle h_1, h_2\rangle = \\ & -\langle dQh_1, h_2\rangle = -\langle h_1 dQh_2\rangle \, . \end{aligned} \tag{42}$$

□ From (40) and (30) we can write to the first order in dQ

$$\begin{aligned}& \langle h'_1, h'_2\rangle' = \langle (I - dQ)h_1, (I + dQ)(I - dQ)h_2\rangle = \\ & \langle (I - dQ)h_1, h_2\rangle = \langle h_1, h_2\rangle - \langle dQh_1, h_2\rangle \; , \end{aligned} \tag{43}$$

because $(I-dQ)(I+dQ) = I-dQ^2$. This proves the first of (42): the second follows because dQ is selfadjoint.

We note that from (42) the formulas

$$d\,\langle h_0, \underline{h}\rangle = -\langle h_0, dQ\underline{h}\rangle \tag{44}$$

$$dH = d\,\langle \underline{h}, \underline{h}^+\rangle = -\langle \underline{h}, dQ\underline{h}^+\rangle \tag{45}$$

follow. □

STATEMENT 4: let $\widehat{y}_0$ be the estimator of y_0

$$\widehat{y}_0 = \langle h_0, \underline{h}^+\rangle \, H^{-1}\underline{y} \; : \tag{46}$$

if we give a variation to the scalar product in $\mathcal{H}$, holding fixed the observations $\underline{y}$, we get a variation $d\widehat{y}_0$ such that

$$\begin{aligned}& \sup_{\|u\|=1} |d\widehat{y}_0| \equiv \|PdQ(I - P)h_0\| \leq \\ & \leq \|dQ\| \cdot \|(I - P)h_0\| \, . \end{aligned} \tag{47}$$

□ Recalling (44), (45) and (8) we find

$$\begin{aligned} d\widehat{y}_0 &= -\langle h_0, dQ\underline{h}^+\rangle H^{-1}\underline{y} \\ &+ \langle h_0, \underline{h}^+\rangle H^{-1}\langle \underline{h}, dQ\underline{h}^+\rangle H^{-1}\underline{y} \\ &= -\langle dQ(I-P)h_0, \underline{h}^+\rangle H^{-1}\langle \underline{h}, u\rangle \\ &= \langle dQ(I-P)h_o, Pu\rangle \\ &= -\langle PdQ(I-P)h_0, u\rangle \ . \end{aligned} \tag{48}$$

Then the first of (47) follows by Schwarz inequality and the second from the fact that for any two bounded operators A, B we have $\|AB\| \leq \|A\|\|B\|$ and $\|P\| = 1$. □

Remark: formula (47) is our fundamental result for the deterministic version of collocation. In fact (47) shows first that the maimum relative error, $\frac{|d\widehat{y}_0|}{\|u\|} = \mathcal{E}_r$, is indeed linear in $\|dQ\|$; but on the same time we have that the sensitivity of $\mathcal{E}_r$ to $\|dQ\|$ is controlled by $\|(I-P)h_0\|$. Therefore, if h_0 is close enough to S_N, or if we assume that N increases with

$$\left[\bigcup_N S_N\right] \equiv \mathcal{H} \ , \tag{49}$$

then this sensitivity is itself small or it goes to 0 when $N \to \infty$ because $Ph_0 \to h_0$ in $\mathcal{H}$.

STATEMENT 5: if $\mathcal{H}$ is endowed with a reproducing kernel $K(t,t')$ with the scalar product $\langle , \rangle$ and with $K'(t,t') = K(t,t') + dK(t,t')$ with the scalar product $\langle , \rangle'$, then the operator dQ corresponds to

$$dQu = -\langle dK(t,t'), u(t')\rangle \ , \quad \forall u \in \mathcal{H} \tag{50}$$

□ In fact, to the first order and $\forall u \in \mathcal{H}$

$$\begin{aligned} u(t) &\equiv \langle K'(t,t'), u(t')\rangle' \\ &\equiv \langle K(t,t') + dK(t,t'), (I+dQ)u\rangle \\ &\equiv \langle K(t,t'), u(t')\rangle + \langle dK(t,t'), u(t')\rangle \\ &+ \langle K(t,t'), (dQu)(t')\rangle \equiv u(t) + (dQu)(t) \\ &+ \langle dK(t,t'), u(t')\rangle \ , \end{aligned} \tag{51}$$

which proves (50).

Let us notice also that from

$$\langle u_1, u_2\rangle = L_{1t}L_{2t'}K(t,t') \tag{52}$$

we derive straightforwardly

$$\begin{aligned} &d\langle u_1, u_2\rangle = -\langle h_1, dQh_2\rangle = \\ &= L_{1t}L_{2t'}dK(t,t') \end{aligned} \tag{53}$$

as one can also directly verify. □

STATEMENT 6: let $\widehat{Y}_0$ be the random estimator

$$\widehat{Y}_0 = \langle h_0, \underline{h}^+\rangle H^{-1}\underline{Y} \ , \tag{54}$$

and dQ the operator of the variation of the metric in $\mathcal{H}$, then

$$\begin{aligned} &E\{d\widehat{Y}_0^2\} = \|PdQ(I-P)h_0\|^2 \leq \\ &\leq \|dQ\|^2\|(I-P)h_0\|^2 \ . \end{aligned} \tag{55}$$

□ In fact, as we saw in Statement 4,

$$d\widehat{Y}_0 = -\langle PdQ(I-P)h_0, u\rangle \ . \tag{56}$$

Now we have only to recall that u is a G.R.F. on $\mathcal{H}$ with covariance operator $C \equiv I$ (i.e. with covariance function $C(t,t') \equiv K(t,t')$) and (55) descends immediately. □

Remark: this is our second fundamental result for the stochastic version. It seems noteworthy that it coincides perfectly with the result (47) for the deterministic version.

STATEMENT 7: when $\mathcal{H}$ is endowed with a reproducing kernel $K(t,t')$, including the stochastic version when $K(t,t') = C(t,t')$, the factor controlling the stability of either solutions $\widehat{y}_0, \widehat{Y}_0$ can be explicitely computed by the formula

$$\left\{\begin{aligned} &\|P\,dQ(I-P)h_0\|^2 = \underline{\Lambda}^+H^{-1}\underline{\Lambda} \\ &\underline{\Lambda} = \underline{L}_t L_{0t'}dK(t,t') \\ &-\left[\underline{L}_t\underline{L}_{t'}^+ dK(t,t')\right]H^{-1}\left[\underline{L}_t L_{0t'}K(t,t')\right] \ . \end{aligned}\right. \tag{57}$$

□ Recall that

$$P^2 = P = \underline{h}^+(t)H^{-1}\underline{h}(t') \tag{58}$$

so that we need only to prove that

$$-\underline{\Lambda} = \langle \underline{h}, dQ(I-P)h_0\rangle \tag{59}$$

coincides with the second of (57). □

To this aim let us adopt a short hand notation like

$$\underline{h}(t) = \underline{L}_{t'}K(t,t') = K(t,\underline{L}) \tag{60}$$

$$h_0(y) = L_{0t'}K(t,t') = K(t,L_0) \tag{61}$$

$$\begin{aligned} &\langle h_0(t), \underline{h}(t)\rangle = L_{0t'}\underline{L}_{t''}\langle K(t,t'), K(t,t'')\rangle = \\ &= L_{0t'}\underline{L}_{t''}K(t',t'') = K(L_0,\underline{L}) \end{aligned} \tag{62}$$

and so forth.

Then from (59)

$$\underline{\Lambda} = -\langle dQ\underline{h}, (I-P)h_0\rangle \qquad (63)$$

$$(-dQ\underline{h})(t) = \langle dK(t,t'), \underline{h}(t')\rangle =$$
$$= \langle dK(t,t'), K(t',\underline{L})\rangle = dK(t,\underline{L}) \qquad (64)$$

$$[(I-P)h_0](t) = h_0(t) +$$
$$-K(t,\underline{L})^{+}H^{-1}\langle \underline{h}(t'), h_0(t')\rangle =$$
$$= K(t,L_0) - K(t,\underline{L}^{+})H^{-1}K(\underline{L},L_0)\ ; \qquad (65)$$

summarizing [4]

$$\begin{aligned}\underline{\Lambda} &= \langle dK(t,\underline{L}), K(t,L_0) \\ &\quad -K(t,\underline{L}^{+})H^{-1}K(\underline{L},L_0)\rangle \\ &= \langle dK(t,\underline{L}), K(t,L_0)\rangle \qquad (66)\\ &\quad -\langle dK(t,\underline{L}), K(t,\underline{L}^{+})\rangle H^{-1}K(\underline{L},L_0) \\ &= dK(\underline{L},L_0) - dK(\underline{L},\underline{L}^{+})H^{-1}K(\underline{L},L_0)\end{aligned}$$

which indeed coincides with (57).

Remark: the meaning of the formula (57) is that it allows the computation of the amplitude of the variations $\frac{d\widehat{y}_0}{||u||}, d\widehat{Y}_0$, through formulas (47), (55), thus understanding to what extent they can be considered as small; (of course in one case one has a deterministic norm, in the other one has a mean square value as measure of this amplitude). To this purpose (57) should be compared with the corresponding (squared) estimation error, namely with

$$||(I-P)h_0||^2 = ||Kh_0||^2 - ||Ph_0||^2 =$$
$$= K(L_0,L_0) - K(L_0,\underline{L}^{+})H^{-1}K(\underline{L},L_0)\ . \qquad (67)$$

Remark: Equation (67) represents the squared <u>relative</u> error for the deterministic case $\left(\frac{|\widehat{y}_0 - y_0|^2}{||u||^2}\right)$ and the <u>absolute</u> mean square error in the stochastic case; this because in the stochastic te information case $K(t,t') = C(t,t')$ includes already information on the amplitude of the signal, while in the deterministic case the $||u||$ is completely unknown.

3 Numerical tests and conclusions

In this paragraph we try to illustrate numerically the theoretical results of §2. Since, at least in part, they are based on inequalities we cannot make a very precise test; nevertheless if we take (47) and (55) in the sense that approximately

$$E\{d\widehat{Y}_0^2\} \cong ||dQ||^2||(S-P)h_0||^1 \qquad (68)$$

where,

$$d\widehat{Y}_0 = \widehat{Y}_{01} - \widehat{Y}_{02} \qquad (69)$$

$\widehat{Y}_{01}$ = estimate with the correct covariance C
$\widehat{Y}_{02}$ = estimate with the modified covariance $C + dC$,

we can find a statement which can be checked numerically.

Namely assume that we have a certain configuration of observables leading to an orthogonal projector P^1; correspondingly we can compute both $\widehat{Y}_{01}^1, \widehat{Y}_{02}^1$ and the mean square variation of the estimator $\widehat{Y}_0$ as

$$(V^1)^2 = E\{d\widehat{Y}_0^{1\,2}\} \qquad (70)$$

as well as the theoetical estimation error

$$(\mathcal{E}^1)^2 \cong ||dQ||^2||(I-P^1)h_0||^2\ . \qquad (71)$$

To be explicit, hereafter we shall use the upper index 1 or 2 to denote two different configurations and a lower index 1 or 2 to denote prediction with C or $C + dC$ covariances.

Formulas (70) and (71) refer to configuration 1 and to certain fixed dC, h_0.

Similarly we can repeat the same computation with another configuration 2, thus deriving the corresponding $(V^2)^2$ and $(\mathcal{E}^2)^2$. Consequently, under our hypothesis, we can claim that

$$\frac{(V^1)^2}{(V^2)^2} \cong \frac{(\mathcal{E}^1)^2}{(\mathcal{E}^2)^2} \qquad (72)$$

and this is of course easy to verify.

Let us examine two examples.

<u>Example 1:</u> here we have use purely simulated data and evaluation functionals only for a process in 1D.

More precisely, assume that $u(t)$ is a stochastic process with zero mean and covariance

$$C(\tau) = 4e^{-0,1|\tau|}\ ; \qquad (73)$$

the wrong (or modified) covariance function for our experiment is

$$C(\tau) + \delta C(\tau) = 0.4\ e^{-0,5|\tau|} \qquad (74)$$

[4]Note: remember that $K(t,t')$ is symmetric but when t,t' are substituted by vectors the order of the product of vectors and their transpose should be kept.

Now we define as configuration 1 of the observation vector $\underline{Y}$ and as functionals to be predicted Y_0 precisely

$$\underline{Y} = \begin{vmatrix} u(-2) \\ u(-1) \\ u(1) \\ u(2) \end{vmatrix} \qquad Y_0 = u(0) \tag{75}$$

Of course form $\underline{Y}$ we can perform the prediction (with 1 upper index)

$$\widehat{u}^1(0) = \widehat{Y}^1_{01} = [C(2)C(1)C(1)C(2)] \cdot \tag{76}$$

$$\begin{bmatrix} C(0) & C(1) & C(3) & C(4) \\ C(1) & C(0) & C(1) & C(3) \\ C(3) & C(1) & C(0) & C(1) \\ C(4) & C(3) & C(1) & C(0) \end{bmatrix}^{-1} \begin{bmatrix} u(-2) \\ u(-1) \\ u(1) \\ u(2) \end{bmatrix}$$

If in (76) we use $C + dC$ instead of C, we derive the modified estimator $\widehat{Y}^1_{02}$ and we can compute the variation $d\widehat{Y}^1_0 = \widehat{Y}^1_{01} - \widehat{Y}^1_{02}$.

Now we have independently sampled our process producing 10 vectors and 10 scalars

$$\left(\underline{Y}^1_i, Y^1_i \qquad i = 1, \ldots 10\right)$$

and then we computed the sample variance

$$(V^1) = E\{(d\widehat{Y}^1_0)\} \cong \frac{1}{10} \sum_{i=1}^{10} \left(\widehat{Y}^1_{01i} - \widehat{Y}^1_{02i}\right)^2 \tag{77}$$

as well as the sample estimation error

$$(\mathcal{E}^1)^2_{emp} \cong \frac{1}{10} \sum_{i=1}^{10} \left(\widehat{Y}^1_{01i} - \widehat{Y}^1_{0i}\right)^2 \tag{78}$$

which in any way has to be compared with the theoretical expression

$$(\mathcal{E}^1)^2 = C(0) - \sum_{i,k} C(\tau_i) C^{(-1)}(\tau_i \tau_k) C(\tau_k) \tag{79}$$

The same procedure adopted for configuration 1 has been repeated for configuration 2 consisting of

$\tau_i = -2,\ -1.5,\ -1,\ -0.5,\ 0.5,\ 1,\ 1.5,\ 2$

$\tau_0 = 0$

$\tau_i =$ measurement points

$\tau_0 =$ prediction point.

the three indexes $(V^2)^2, (\mathcal{E}^2)^2_{emp}$ and $(\mathcal{E}^2)^2$ have also been recomputed obtaining the results shown in Table 1.

Table 1: Results for Example 1

	Conf. 1	Conf. 2	Ratios
$\mathcal{E}$	0.63	0.45	1.40
$\mathcal{E}_{emp}$	0.59	0.54	1.09
V	0.26	0.069	3.77

As we see the experiment provides ratios of the same order of magnitude but not really equal one to the other. It is opinion of the authors that this reflectes the fact that we are approximating an inequality with an equality, which might be too crude a hypothesis.

Example 2: in this case a more realistic example has been worked out from a classical geodetic problem, namely the geoid estimation from gravity anomalies where instead of using many independent samples we have rather computed many sample points $\Delta g(P_{ik})$ and predicted the geoid at the same points $\widehat{N}(P_{ik})$ for one model gravity field only. The model gravity field used here is EGM96 from degree 8 to degree 180. Also the covariances used in the experiment are the true ones

$$\begin{aligned} C(\psi) &= \left(\frac{\mu}{R}\right)^2 \sum_{n=8}^{180} C_n P_n(\cos\psi) && (80) \\ C_n &= \sum_{m=-n}^{m} T^2_{nm} \qquad n = z_1 \ldots 180 \\ T_{nm} &= \text{exact EGM coefficients} \end{aligned}$$

and two modified covariances

$$C(\psi) + dC(\psi) = \left(\frac{\mu}{R}\right)^2 \sum_{n=M}^{180} C_n P_n(\cos\psi) \tag{81}$$

where C_n are as in (80) and M is taken once as $M = 2$ (lower index 2) and once as M=4 (lower index 3). The three covariances are represented in Fig. 1.

Note also here that in (81) the modification of C is obtained by including more degree variances than in (80); in fact without this rather drastic change in the covariance the different prediction $\widehat{N}_1(P_{ik}), \widehat{N}_2(P_{ik})$ were so close to one another that the differences become physically insignificant.

In this case the two configurations correspond to two regular grids on the sphere, namely

Conf. 1 $\to 10^\circ \times 10^\circ$ grid

Conf. 2 $\to 5^\circ \times 5^\circ$ grid.

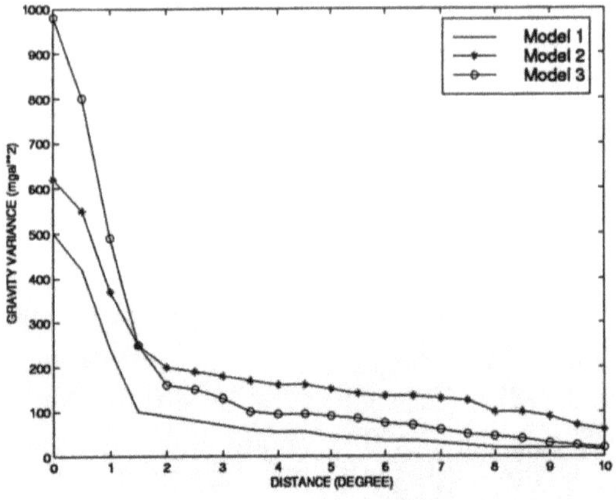

Fig. 1: The three model covariances

For each of the two grids and each of the three covariance functions (80), (81) with $M = 2, 4$, from a synthetic data set $\Delta g(P_{ik})$ the corresponding predictions of geoid undulations have been performed through well-known formulas

$$\gamma \widehat{N}_{ik} = \sum_{\ell,m,j,n} C_{T\Delta g}(P_{ik}, P_{\ell m}) \cdot$$
$$\cdot\, C^{(-1)}_{\Delta g \Delta g}(P_{\ell m}, P_{jn}) \Delta g_{jn} \qquad (82)$$

($\gamma \cong \mu/R^2$) thus obtaining two series of three sets of predicted values $\widehat{N}_1(P_{ik})$, $\widehat{N}_2(P_{ik})$, $\widehat{N}_3(P_{ik})$ (we skip here the upper index 1 or 2).

As it is standard the covariances and cross covariances on (82) are related to $C(\psi)$ by

$$C_{T\Delta g}(\psi) =$$
$$= R\left(\frac{\mu}{R^2}\right)^2 \sum_{n=M}^{180} (n-1) C_n P_n(\cos\psi) \qquad (83)$$

$$C_{\Delta g \Delta g}(\psi) =$$
$$= \left(\frac{\mu}{R^2}^2\right) \sum_{n=M}^{180} (n-1)^2 C_n P_n(\cos\psi) \qquad (84)$$

with $M = 8, 2, 4$ respectively in the three cases.

The results are summarized in Table 2 where in particular the mean square differences

$$(V_{12})^2 = \frac{1}{N} \sum_{(i,k)} \left[\widehat{N}_2(P_{ik}) - \widehat{N}_1(P_{ik})\right]^2 \qquad (85)$$

$$(V_{13})^2 = \frac{1}{N} \sum_{(i,k)} \left[\widehat{N}_3(P_{ik}) - \widehat{N}_1(P_{ik})\right]^2 \qquad (86)$$

Table 2: Results for Example 2

	Conf. 1	Conf. 2	Ratios
$\mathcal{E}$	4.40	3.41	1.29
$\mathcal{E}_{emp}$	4.40	3.16	1.39
V_{12}	6.50	5.50	1.18
V_{13}	4.60	3.50	1.31

are reported in units of m^2,

As we can see the ratios are again in a very reasonable agreement with our hypotesis, the differences being of the same order as the difference between the ratios of empirical values and theoretical values of $\mathcal{E}^2$.

The conclusion is that the analysis performed is confirmed and the dependence of the solution of collocation on the kernel/covariance choice is not very critical particularly where the density of measurement points is high.

References

Lauritzen S. (1973). The probabilistic background of some statistical methods in physical geodesy. Danish Geodetic Inst., no 48, Copenhagen 1973.

Rozanov Y.A. (1982). Markov random fields. Springer Verlag 1982.

Sansò F. (1986). Statistical methods in physical geodesy. Lecture notes in earth sciences, vol 7, "Mathematical and numerical techniques in physical geodesy". H. Suenkel ed, Springer Verlag, 1986

Sansò F., G. Venuti (1998). White noise stochastic BVPs and Cimmino's theory. Hotine-Marussi Symp. Trento, September 1998.

Tscherning C.C. (1977). A note on choice of norm when using collocation for the computation of the approximation to the anomalous potential. Bul. Geod. vol 51, no 2, 1977.

Yosida K. (1978) Functional Analysis. Springer Verlag, 1978.

Biases and accuracy of, and an alternative to discrete nonlinear filters

Peiliang Xu
Disaster Prevention Research Institute, Kyoto University at Uji, Kyoto 611-0011, Japan
email: pxu@rcep.dpri.kyoto-u.ac.jp

Abstract. Dynamical systems encountered in reality are essentially nonlinear. A number of approaches were proposed for nonlinear filter with different accuracy. They are mainly investigated from the Bayesian point of view, and may be classified into three kinds of methods: linearization, statistical approximation and Monte-Carlo simulation (Jazwinski 1970; Tanizaki 1993; Gelb 1994). Very often, linearization of the nonlinear system is done using either a precomputed nominal trajectory or the estimate of the state vector. This second linearization approach is better known as the extended Kalman filter (Jazwinski 1970). Since the solution based on one-step linearization may be poor for a highly nonlinear system, iteration in this case is expected in order to obtain a more accurate estimate. Higher order approaches should probably be taken into account. It should be noted, however, that if the system presents a significant nonlinearity, even the mean and covariance matrix of the nonlinear filter can be misleading, since the means of the estimated state variables may deviate appreciably from their true parameter values. The basic idea of statistical approximation is to replace a nonlinear function of random variables by a series expansion (Gelb 1994) or to approximate the *a posteriori* conditional probability density function of the state vector (Sorenson & Stubberud 1968; Kramer & Sorenson 1988). The Monte Carlo simulation technique may be used to determine the mean and covariance matrix of the nonlinear filter (if properly designed), which requires a large sample to obtain statistically meaningful results (see *e.g.* Brown & Mariano 1989; Carlin et al. 1992; Gelb 1994; Tanizaki 1993).

Under the Bayesian framework, first and second order nonlinear filters have been derived formally by computing the relevant (Bayesian) conditional means and covariances and substituting them into the standard linear Kalman filtering algorithm (Tanizaki 1993; Gelb 1994). Higher order nonlinear filters were proposed, with hope in mind that they might be less biased and more efficient. The Bayesian derivation of the second order nonlinear filter SONF has been critically challenged in this paper by examining the biases and accuracy of the SONF and the corresponding residuals.

We have taken a frequentist standpoint to deal with the problem of nonlinear filtering and derived the biases of the EKF and the SONF. For a nonlinear dynamical and nonlinear measurement system, the frequentist approach and Bayesian method have been shown in this paper to be fundamentally different. Assume that the noises of a nonlinear dynamical and nonlinear measurement system are normally distributed. The interface between Bayesians and frequentists on nonlinear filters has resulted in several new findings in the present paper, which are summarized in the following: (i) the Gaussian or truncated second order nonlinear filter not only cannot guarantee the improvement of the biases of the estimated state vector, but may also exaggerate them; (ii) the variance-covariance measure of the second order nonlinear filter currently given in the literature is correct, only if some extra terms obtained in this paper are included; and (iii) an alternative, almost unbiased second order nonlinear filter is proposed, if the randnomness of some coefficient matrices is neglected. Practically, the new SONF filter in this paper is expected to significantly improve the EKF and the SONF given in the literature, if dynamical and measurement systems are highly nonlinear and if the ratio of signal to noise is not very large. Numerical confirmation requires a large scale of simulations with well designed experiments, which will be left for a future contribution. For other new findings and possible practical pitfalls due to a direct interface between Bayesians and frequentists, the reader is referred to Berger & Robert (1990), Xu (1992) and Xu & Rummel (1994).

International Association of Geodesy Symposia, Vol. 121
Schwarz (ed.), Geodesy Beyond 2000 – The Challenges of the First Decade

Keywords. Nonlinear dynamical and nonlinear measurement system, nonlinear filters, bias and accuracy analysis, almost unbiased second-order nonlinear filter

References

Baheti, R., O'Hallaron, D. & Itzkowitz, H. (1990). Mapping extended Kalman filters onto linear arrays, IEEE Trans. Auto. Contr., **AC-35**, 1310-1319.

Bates, D. & Watts, D.G. (1980). Relative curvature measures of nonlinearity (with discussions), J. R. statist. Soc., **B42**, 1-25.

Bates, D. & Watts, D.G. (1988). Applied nonlinear regression, John Wiley & Sons, New York.

Beale, E.M.L. (1960). Confidence regions in nonlinear estimation (with discussions), J. R. statist. Soc., **B22**, 41-88.

Berger, J.O. & Robert, C. (1990). Subjective hierarchical Bayes estimation of a multivariate normal mean: on the frequentist interface, Ann. Statist., **18**, 617-651.

Box, M.J. (1971). Bias in nonlinear estimation (with discussions), J. R. statist. Soc., **B33**, 171-201.

Brown, B. & Mariano, R. (1989). Measures of deterministic prediction bias in nonlinear models, Int. Economic Rev., **30**, 667-684.

Carlin, B., Polson, N. & Stoffer, D. (1992). A Monte Carlo approach to nonnormal and nonlinear state space modelling, J. Amer. Stat. Assoc., **87**, 493-500.

Clark, G.P.Y. (1980). Moments of the least squares estimators in a nonlinear regression model, J. R. statist, Soc., **B42**, 227-237.

Denham, W. & Pines, S. (1966). Sequential estimation when measurement function nonlinearity is comparable to measurement error, AIAA J., **4**, 1071-1076.

Friedland, B. & Bernstein, I. (1966). Estimation of the state of a nonlinear process in the presence of nongaussian noise and disturbances, J. Franklin Inst., **281**, 455-480.

Gelb, A. (ed) (1994). Applied optimal estimation, The MIT Press, Cambridge.

Jazwinski, A. (1970). Stochastic processes and filtering theory, Academic Press, New York.

Kramer, S. & Sorenson, H. (1988). Recursive Bayesian estimation using piece-wise constant approximations, Automatica, **24**, 789-801.

Kushner, H. (1967a). Dynamical equations for optimal nonlinear filtering, J. diff. Eq., **3**, 179-190.

Kushner, H. (1967b). Approximations to optimal nonlinear filters, IEEE Trans. Auto. Contr., **AC-12**, 546-556.

Mahalanabis, A. & Farooq, M. (1971). A second-order method for state estimation of nonlinear dynamical systems, Int. J. Control, **14**, 631-639.

Schaffrin, B. (1991). Generalized robustified Kalman filters for the integration of GPS and INS, Geodetic Institute, Stuttgart University, Tech. Report No.15, Stuttgart.

Searle, S.R. (1971). Linear models, John Wiley & Sons, New York.

Seber, G. & Wild, C. (1989). Nonlinear Regression, John Wiley & Sons, New York.

Sorenson, H. & Stubberud, A. (1968). Non-linear filtering by approximation of the *a posteriori* density, Int. J. Control, **8**, 33-51.

Tanizaki, H. (1993). Nonlinear filters – Estimation and applications, Springer Verlag, Berlin.

Wishner, R., Tabaczynski, J. & Athans, M. (1969). A comparison of three non-linear filters, Automatica, **5**, 487-496.

Xu, P.L. (1991). Least squares collocation with incorrect prior information, ZfV, **116**, 266-273.

Xu, P.L. (1992). The value of minimum norm estimation of geopotential fields, Geophys. J. Int., **111**, 170-178.

Xu, P.L. & Rummel, R. (1994). A simulation study of smoothness methods in recovery of regional gravity fields, Geophys. J. Int., **127**, 472-486.

The full paper has been published in the Journal of Geodesy, 73, 35-46, 1999.

Are GPS data normally distributed?

C.C.J.M. Tiberius
Department of Mathematical Geodesy and Positioning
Delft University of Technology, Thijsseweg 11, NL-2629 JA Delft

K. Borre
Danish GPS Center
Aalborg University, Niels Jernes Vej 14, DK-9220 Aalborg Ø

Abstract. Knowledge of the probability density function of the observables is not needed to routinely apply a least-squares algorithm and compute estimates for the parameters of interest. For the interpretation of the outcomes, and in particular for statements on the quality of the estimator, the probability density has to be known.

A variety of tools and measures to analyse the distribution of data are reviewed and applied to code and phase observables from a pair of geodetic GPS receivers. As a conclusion the normal probability density function turns out to be a reasonable model for the distribution of GPS code and phase data, but this may not hold under all circumstances.

Keywords. GPS data, probability density function.

1 Introduction

In precise relative GPS positioning, like in many geodetic applications, the data are routinely processed using a least-squares algorithm. The least-squares principle requires the specification of a functional relation between observations and unknown parameters and a weight matrix for the observations. The full probability density function of the observed data is not needed for estimating the parameters. This function however, is needed to interpret the precision of the estimators. Next, a statistical testing procedure to detect and identify outliers and cycle slips in the data, see e.g. Teunissen (1998), does rely directly on the probability distribution. And thirdly, also making probabilistic inferences for resolution of the ambiguities requires knowledge of the probability density function of the data.

We will analyse the probability distribution of GPS data. The purpose is to assess whether (and to what extent) the data distribution fits the (commonly assumed) well-known normal probability density function.

The analysis is based on the least-squares residuals, per channel (satellite), per epoch and per observation type, from a coordinates and ambiguities constrained solution. In this solution, the coordinates and integer ambiguities are treated as a-priori known quantities and hence eliminated from the adjustment. The discussion of the analysis tools and measures in this contribution is illustrated using zero baseline data of a geodetic dual frequency GPS receiver pair. Data were collected at a 1 second interval and we consider a 10 minutes period (thus 600 epochs). The 6 satellites are PRNs 20, 23, 15, 21, 25, and 01. They are sorted by increasing elevation angle; resp. 14°, 17°, 38°, and the latter three are over 45°.

The overview starts with a demonstration of the graphical tools for the analysis, followed by the numerical measures, the statistical tests on the empirical moments and those on the empirical distribution function.

2 Graphical analysis

Two graphical tools for analysing the distribution are the histogram and the normal probability plot. These means are helpful for an impression of the data distribution, but usually it is hard to quantify the fit or misfit with the assumed distribution.

2.1 Histogram

The histogram is probably the oldest and most widely used tool to analyse a data distribution.

For the histogram the range of (experimental) outcomes is divided into classes (bins) and the observed frequencies, the cell counts, are presented in a bar diagram. An example of the histogram is given in Figure 1. The y-axis gives the standardized relative frequency, i.e. the absolute frequency divided by the number of samples and the bin width, see Silverman (1986). The area, the bars together, equals 1, and this enables comparison with histograms of different data sets.

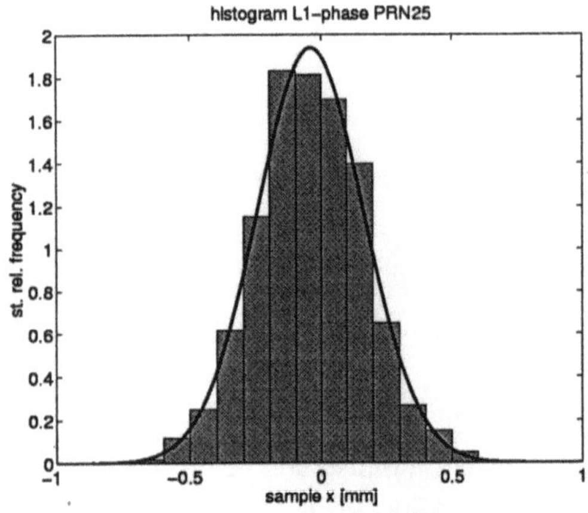

Fig. 1: Histogram of L1 phase residuals, satellite PRN25, 600 samples, 20 classes. Normal density function imposed, μ=-0.04 mm and σ=0.21 mm.

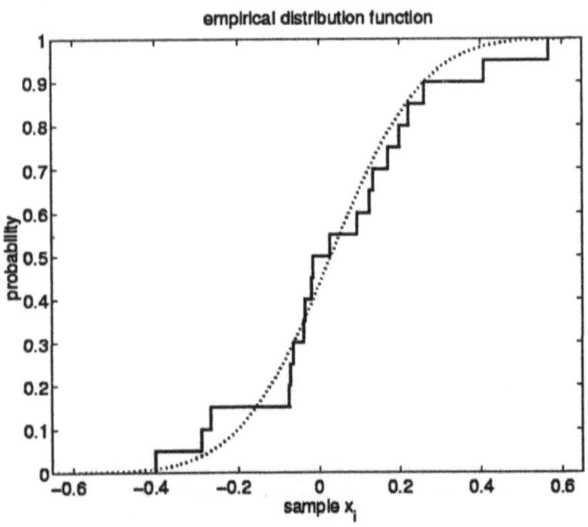

Fig. 2: Empirical distribution function for simulated data; N=20 samples, μ=0.03 and σ=0.18; normally distributed.

The chi-squared test is related. The data are also grouped into classes, and the observed frequencies are compared to the expected ones (computed from the assumed distribution). The weighted sum of squared differences yields the chi-squared test. The reader is referred to Stuart and Ord (1991).

2.2 Normal probability plot

When N samples are independently and identically distributed, each sample represents an equal probability mass, namely $^1/_N$. Sorting the samples ascendingly gives the empirical distribution function, see Figure 2. The dotted line gives the (theoretical) cumulative normal distribution function. Ideally the empirical function and the cumulative one coincide. The curvature of the distribution function may complicate the comparison of the two functions. A solution is to transform the y-axis according to the assumed normal distribution. Using the inverse of the cumulative standard normal distribution function yields the normal probability plot of Figure 3. For interpretation, the y-axis is not labelled by the N(0,1)-quantiles, but by the corresponding cumulative probabilities. The theoretical distribution function (dotted), is now a straight line with intercept $^{-\mu}/_\sigma$ and slope $^1/_\sigma$, see d'Agostino (1986).

Figure 4 gives the normal probability plot for L1 and L2 phase data. The y-axis gives the N(0,1)-quantiles, instead of the probabilities, but the scaling is the same as in Figure 3. One dotted line indicates the mean μ and the other the standard deviation added to the mean $\mu+\sigma$. The empirical distribution is only marked on the edges. At left the fit with the normal distribution is good. At right the samples deviate somewhat from the line, but one should be careful in judging visually as the departures occur at the beginning and end, and represent only a few percent. This distribution is slightly asymmetric, see also Figure 9 later on.

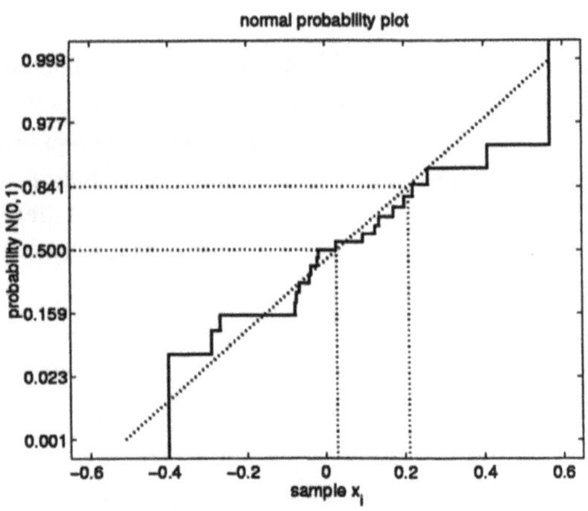

Fig. 3: Normal probability plot for simulated data. The y-axis is labelled by the comulative probability.

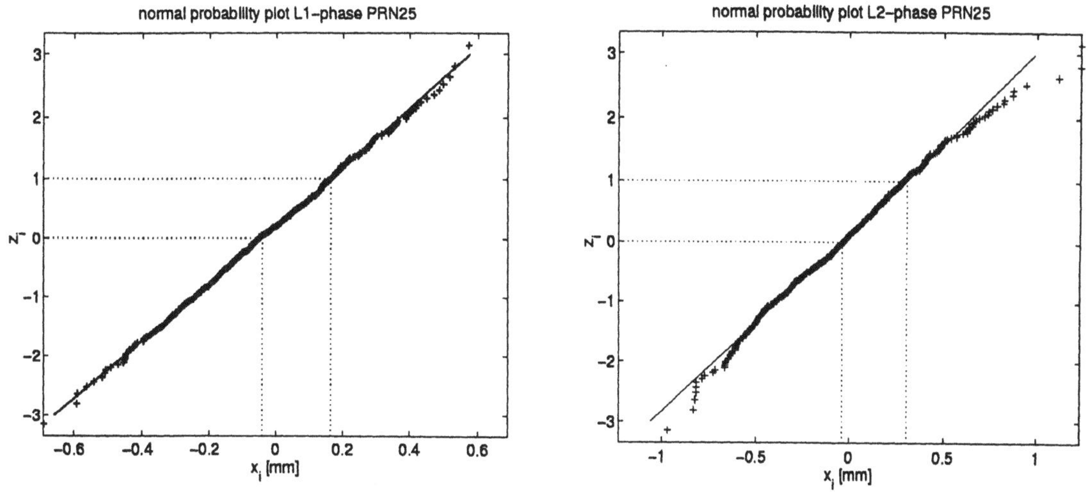

Fig. 4: Normal probability plot for PRN25, (left) L1 phase μ=-0.04mm and σ=0.21mm, (right) L2 phase μ=-0.04mm and σ=0.34mm.

3 Empirical moments

Moments are characteristics of the probability density function. The normal distribution is completely specified by the mean and standard deviation. Figures 7 and 8 give the mean and standard deviation of the GPS observations. The satellites are sorted by increasing elevation, and PRNs 20 and 23 are below 20°. For the standard deviation one can clearly observe the elevation dependence for the observations on the L2 frequency, caused mainly by the reconstruction measurement technique employed to get around Anti-Spoofing.

Besides the first and second order moment, the distribution has a third and fourth moment, namely skewness and kurtosis. These moments are used to analyse normality, see Pearson and Hartley (1966), and Mardia (1988).

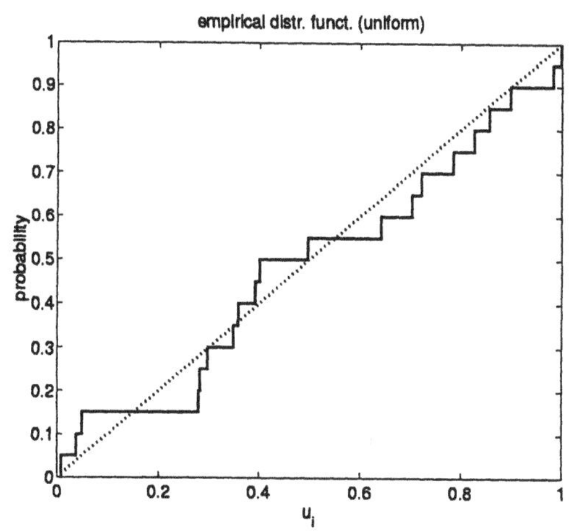

Fig. 5: Empirical distribution function in uniform domain (simulated data). The y-axis is identical to Figure 2.

3.1 Skewness

Skewness measures a-symmetry. The normal distribution is symmetric about its mean. Under the hypothesis of normality, the skewness coefficient is approximately normally distributed, with mean zero. A two-sided critical region about zero is used. Results are given in Figure 9; a 10% level of significance was used (chosen as a common divisor; it is used for all later tests as well). The L1 observations generally pass the skewness test, but a few of the L2 residual series are rejected.

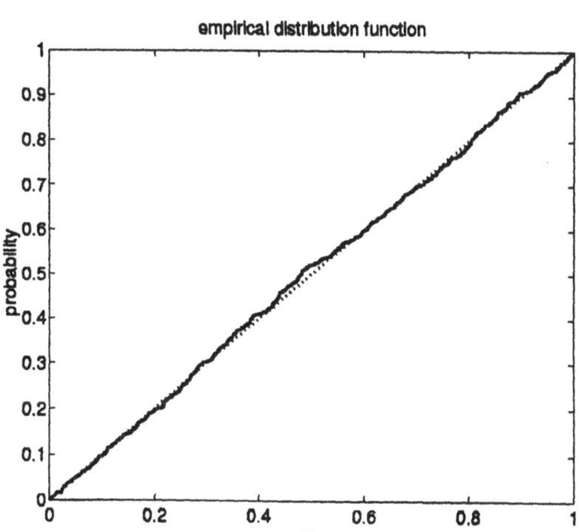

Fig. 6: Empirical distribution function for L1 phase of PRN25.

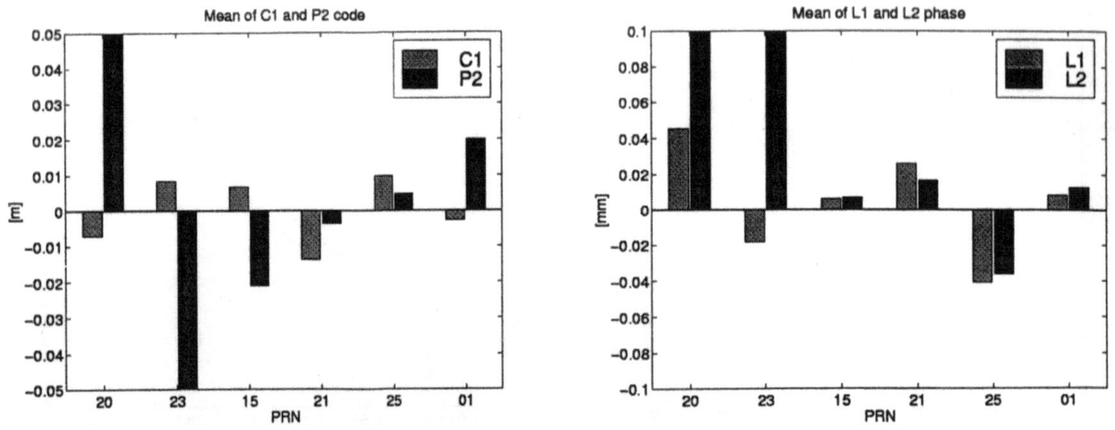

Fig. 7: Mean of C1 and P2 code in [m] (left) and of L1 and L2 phase in [mm] (right).

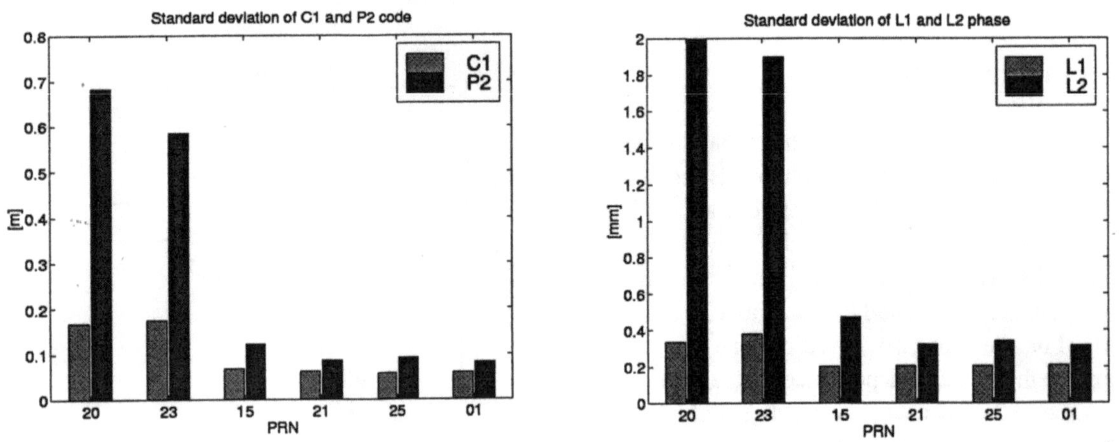

Fig. 8: Standard deviation of C1 and P2 code in [m] (left) and of L1 and L2 phase in [mm] (right).

3.2 Kurtosis

Kurtosis measures peakedness of the distribution. The normal distribution has kurtosis coefficient 3. Similarly the critical region for the empirical kurtosis coefficient is two sided about 3 (but not strictly symmetric). Results are given in Figure 10. There are only a few rejections, at a 10% level of significance.

4 Empirical distribution function

The empirical distribution function was given in Figure 2. Instead of transforming the y-axis, as done for Figure 3, the x-axis is transformed. The sample values are transformed using the assumed cumulative normal distribution function and become samples of a uniform distribution on the interval [0,1]. The (theoretical) cumulative distribution in Figure 5 becomes a straight line. Figure 6 gives an example of the empirical distribution function.

The Kolmogorov-Smirnov test and Anderson-Darling test are two statistical tests that measure the fit of the step function with the dotted line in Figure 5. By their level of significance they indicate and quantify the fit of the data with the assumed distribution, see Stephens (1986). The critical values of these tests, which depend on the sample size N, can be found in tables.

4.1 Kolmogorov-Smirnov test

The Kolmogorov-Smirnov test statistic is the maximum value of the absolute difference between the empirical and theoretical distribution function, the solid and the dotted line in Figure 5.

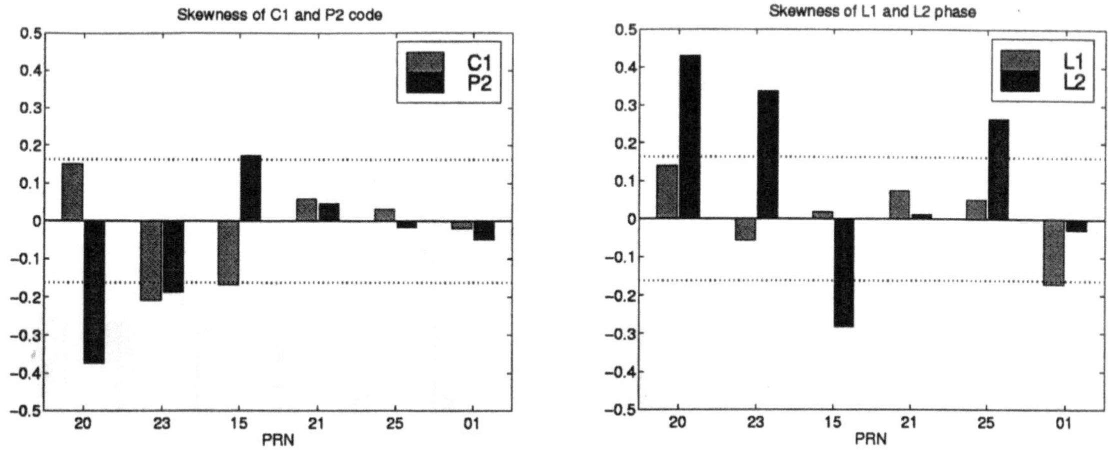

Fig. 9: Skewness coefficient of C1 and P2 code (left) and of L1 and L2 phase (right). For normally distributed data the skewness coefficient equals zero. The two dotted lines mark the critical region (α=0.10, N=600).

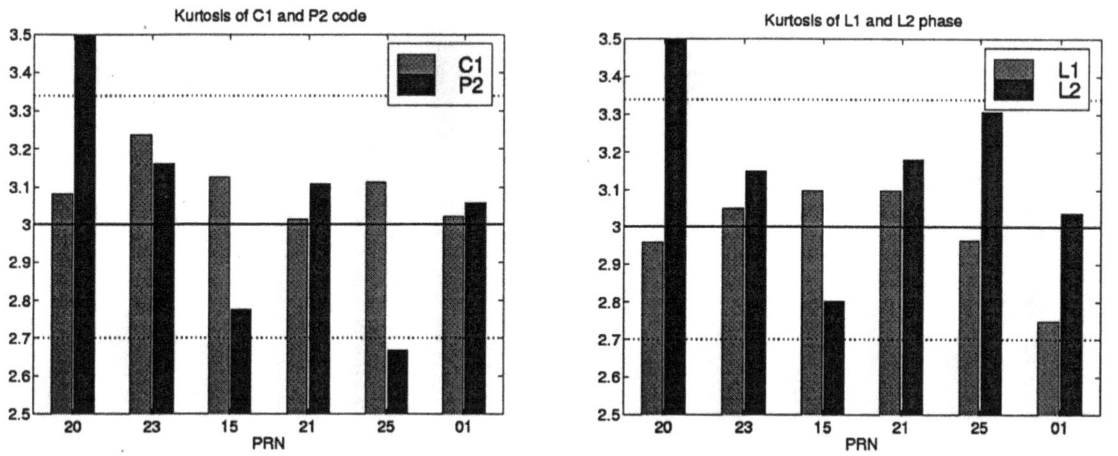

Fig. 10: Kurtosis coefficient of C1 and P2 code (left) and of L1 and L2 phase (right). For normally distributed data the kurtosis coefficient equals 3. The two dotted lines mark the critical region (α=0.10, N=600).

Results of this test are given in Figure 11, again with α=0.10. The phase data of the two low elevation satellites do not pass this test. A bias that is not constant in time, could in these cases obstruct a conclusion on the data probability distribution.

4.2 Anderson-Darling test

Instead of a single measure of distance, the Anderson-Darling test employs the integrated squared difference, weighted over the whole range [0,1], Anderson and Darling (1954). Results of this test, with α=0.10, are given in Figure 12. Again the phase data of the two low elevation satellites do not pass the test.

5 Concluding remarks

Several techniques and tools to analyse the distribution were reviewed, and illustrated by application to GPS data.

Knowledge of the data distribution is of importance for precision description, quality control and ambiguity validation. Our hope for normally distributed GPS code and phase data seems to come true. The results of the graphical tools do point in this direction and the hypothesis of normality was not rejected by the statistical tests in most of the cases at the chosen level of significance (α=0.10). The tests were based on the empirical moments, and on the empirical distribution function.

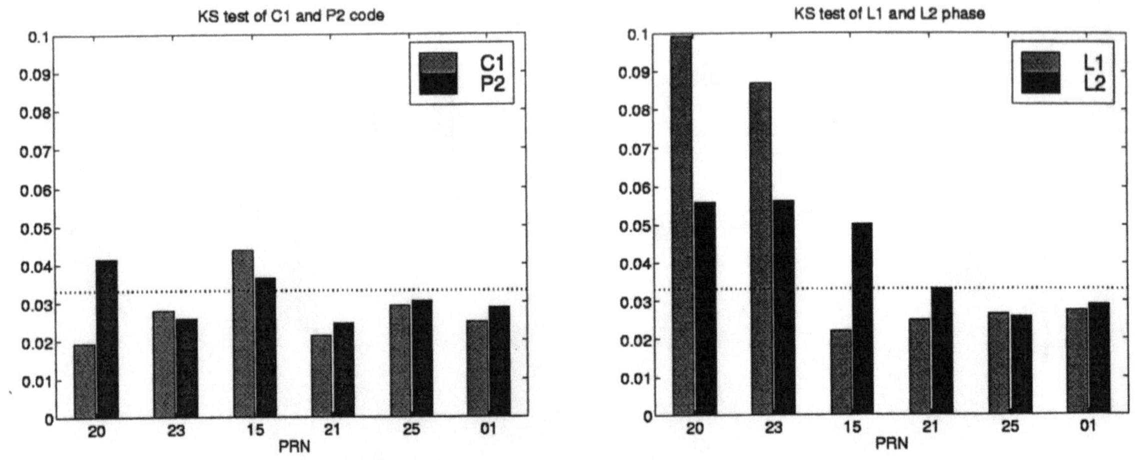

Fig. 11: Kolmogorov-Smirnov test for C1 and P2 code (left) and for L1 and L2 phase (right). The dotted line indicates the critical value (α=0.10, N=600). If the statistic is larger, the KS-test is rejected.

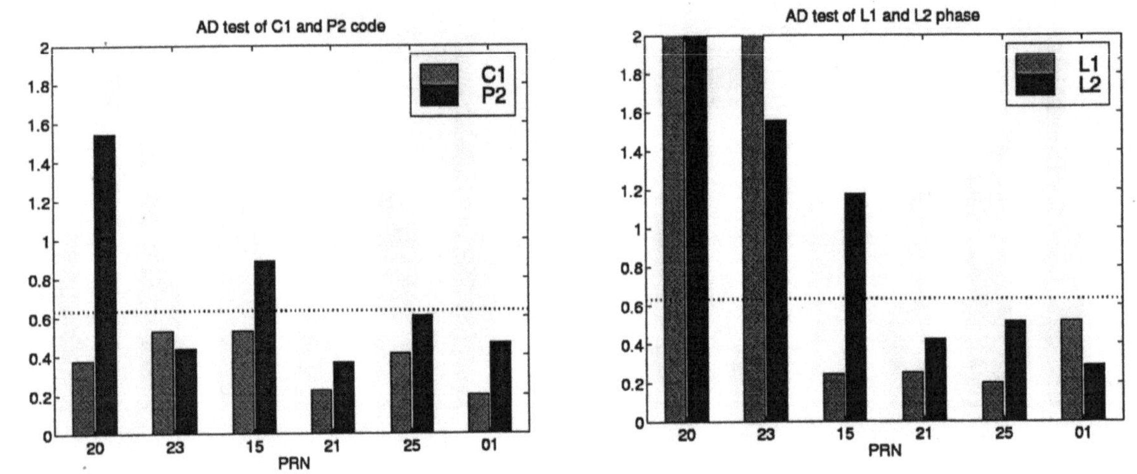

Fig. 12: Anderson-Darling test for C1 and P2 code (left) and for L1 and L2 phase (right). The dotted line indicates the critical value (α=0.10, N=600). If the statistic is larger, the test is rejected.

In this contribution the examples shown originate from a single experiment. Numerical results of various experiments using only the statistical tests can be found in Tiberius and Borre (1999).

References

d'Agostino, R. (1986). Graphical analysis. Chapter 2 in *Goodness-of-fit techniques*, R. d'Agostino and M. Stephens (Eds), Vol. 68 of Statistics: textbooks and monographs. Marcel Dekker Inc., New York.

Anderson, T. and Darling, D. (1954). A test of goodness of fit. *Journal of the American Statistical Association*, Vol. 49, No. 268, pp. 765-769.

Mardia, K. (1988). Tests of univariate and multivariate normality. Chapter 9 in *Analysis of variance*, P. Krishnaiah (Ed), Vol. 1 of Handbook of statistics. Elsevier Science Publishers, Amsterdam, 3rd edition.

Pearson, E. and Hartley, H. (1966). *Biometrika tables for statisticians*, Vol. 1. Cambridge University Press.

Silverman, B. (1986). *Density estimation for statistics and data analysis*. Monographs on statistics and applied probability. Chapman and Hall, London.

Stephens, M. (1986). Tests based on EDF statistics. Chapter 4 in *Goodness-of-fit techniques*, R. d'Agostino and M. Stephens (Eds), Vol. 68 of Statistics: textbooks and monographs. Marcel Dekker Inc., New York.

Stuart, A. and J. Ord (1991). *Classical inference and relationships*. Vol. 2 of Kendall's advanced theory of statistics. Edward Arnold, London, 5th edition.

Teunissen, P.J.G. (1998). Quality control and GPS. Chapter 7 in *GPS for geodesy*, P.J.G. Teunissen and A. Kleusberg (Eds). Springer Verlag, Berlin, 2nd edition.

Tiberius, C.C.J.M. and K. Borre (1999). Probability distribution of GPS code and phase data. *Zeitschrift für Vermessungswesen*, Vol. 124, No. 8, pp. 264-273.

On the precision and reliability of near real-time GPS phase observation ambiguities

H. Kutterer
Geodätisches Institut, Universität Karlsruhe, Englerstraße 7, D-76128 Karlsruhe, Germany

Abstract. During the past decade the integer fixing of the GPS (Global Positioning System) carrier phase ambiguities has been studied in many papers. The available techniques are rather sophisticated regarding the set-up and analysis of the ambiguity search space. Consistently, the number of observation epochs necessary for the unique solution is strongly reduced. As incorrect solutions may occur, it is essential to study their impact on the coordinate estimates and their precision. Thus quality issues concerning the ambiguity resolution techniques are both of practical and theoretical interest.

The paper focuses on the theoretical aspects of the near real-time case with short baseline lengths. Measures of a degree of confidence are introduced for the selection of the ambiguity parameters. An approach is sketched for describing the precision of the ambiguities regarding their integer nature. Reliability aspects are considered taking possibly incorrect ambiguity solutions into account. A proposal concerning the interpretation of the results for practical use concludes the study.

Abstract. Global Positioning System, carrier phase ambiguity.

1 Introduction

In the context of positioning using GPS there are two important observation types: pseudoranges using either C/A- or P-code and carrier phase observations. Differential techniques allow the reduction of external error sources. In the case of code observations sub-meter precision is possible. Nevertheless, results with a precision of a few millimeters can only be achieved by the use of phase observations. To make this observation type useful, it is essential to resolve the phase ambiguity parameters. A formal description shall be given in the next section.

Algorithms dealing with this problem have been studied and published during the last decade. Since it is impossible to review all work on this topic within a short paragraph, a few characteristic approaches shall be mentioned. Nearly all methods use the so-called *float solution* as real-valued estimation of the integer ambiguity parameters. A search space is set up by the float solution and its precision using the covariance matrix of the float parameters.

The Fast Ambiguity Resolution Approach (FARA) by Frei and Beutler (1991) studies the squared sum of observation residuals Ω to identify the *best* solution: that set of integer values is selected that yields the smallest Ω. The ratio between the Ω values of the second-best and of the best integer vector is tested (*ratio test*). The best vector is accepted as solution, if the ratiot is larger than a threshold value. The Lambda strategy was developed by Teunissen and is based on his integer least squares concept (Teunissen, 1996). The integer vector *closest* to the float solution in the ambiguity parameter space extended to real vectors is selected (*minimum distance*). This is equivalent to the *best*-criterion of FARA. The difference lies in the way the search space is set up. A main part of the Lambda strategy consists of the application of a decorrelation procedure to strongly reduce the computational complexity.

Other approaches are based on the ambiguity function (Counselman and Gourevitch, 1981), Bayesian techniques (Betti et al., 1993), and integer optimization (Xu et al., 1995). On-the-fly resolution (OTF, AROF, OTR) was proposed by Abidin (1993) and others applying a variety of test procedures.

The following considerations focus on theoretical aspects of the criteria for the selection of the final integer vector. They aim at the derivation of confidence measures for the integer vectors representing possible choices to resolve the phase ambiguities. If such measures are available, concepts like precision and reliability of discrete quantities can be developed. Teunissen (1998) presented a method for the theoretical derivation of the ambiguity probabilities for several resolution strategies. The procedure described in the following is different since it is based on data-derived quantities. Teunissen (1999) gives the same concepts for ambiguity precision as in this presentation based on the first and second moments of discrete random variables. The reliability measures in Section 5 are defined in analogy to the well-known theory of gross error

International Association of Geodesy Symposia, Vol. 121
Schwarz (ed.), Geodesy Beyond 2000 – The Challenges of the First Decade

detection.

Only the case of simultaneous code and phase differential GPS is studied, assuming systematic errors are handled by parametrization, correction models, differencing techniques, or linear combinations of the observation equations. For general aspects the reader may refer to GPS textbooks like Hofmann-Wellenhof et al. (1997).

2 The model

Teunissen (1997) mathematically describes the characteristics of the float solution for the ambiguity parameters and of its search space, primarily in his tetralogy on the *canonical theory for short GPS baselines*. In the following a basic OTF approach is used which is directed to the immediate evaluation of single epoch solutions.

$$\begin{aligned} E(\Delta\underline{\rho}) &= \mathbf{X}\Delta\xi, & D(\Delta\underline{\rho}) &= \Sigma_{\rho\rho} \\ E(\Delta\underline{\phi}) &= \mathbf{X}\Delta\xi + \mathbf{Z}\Delta\zeta, & D(\Delta\underline{\phi}) &= \Sigma_{\phi\phi} \end{aligned} \quad (2.1)$$

$\Delta\rho$ and $\Delta\phi$ denote the observed-minus-computed values of the code and phase observations, respectively. $\mathbf{X}$ is the design matrix of the coordinate part of the parameters, or more generally spoken, of the real-valued parameters. $\mathbf{Z}$ denotes the design matrix of the ambiguity parameters. $\Delta\xi$ and $\Delta\zeta$ are the symbols for the updates of the approximate values ξ_0 and ζ_0 of the real-valued and integer parameters, respectively. The _-symbol marks a random quantity. Eq. (2.1) represents the result of the linearization of the originally nonlinear observation equations in expectation value notation E(.) as well as the dispersion matrix D(.) of the observables. Previous double differencing and/or linear combining of the observation equations can be assumed as well. However, as this is part of treating external influences, it is not essential for the following.

By means of the code part of Eq. (2.1) a least squares estimate of the position can be computed. Introducing it to the phase part yields

$$E(\Delta\underline{\phi}) = \mathbf{X}\mathbf{X}_L^{-1}\, E(\Delta\underline{\rho}) + \mathbf{Z}\Delta\zeta \quad (2.2)$$

with

$$\mathbf{X}_L^{-1} := (\mathbf{X}^T\mathbf{W}_X\mathbf{X})^{-1}\mathbf{X}^T\mathbf{W}_X$$

a left inverse matrix of $\mathbf{X}$ (Rao and Mitra, 1971, p.20). In addition, the application of

$$\mathbf{Z}_L^{-1} := (\mathbf{Z}^T\mathbf{W}_Z\mathbf{Z})^{-1}\mathbf{Z}^T\mathbf{W}_Z$$

to Eq. (2.2) leads to the final representation for the integer expectation

$$E(\Delta\underline{\hat{\psi}}) = \Delta\zeta \in \mathbf{Z}^s \quad (2.3)$$

of the here defined real-valued least-squares estimator

$$\Delta\underline{\hat{\psi}} := \mathbf{Z}_L^{-1}(\Delta\underline{\phi} - \mathbf{X}\mathbf{X}_L^{-1}\Delta\underline{\rho}) \quad (2.4)$$

approximating the integer-valued ambiguity parameters. Its dispersion matrix reads as

$$\begin{aligned} D(\Delta\underline{\hat{\psi}}) &= \Sigma_{\hat{\psi}\hat{\psi}} \\ &= \mathbf{Z}_L^{-1}(\Sigma_{\phi\phi} + \mathbf{X}\mathbf{X}_L^{-1}\Sigma_{\rho\rho}(\mathbf{X}\mathbf{X}_L^{-1})^T)(\mathbf{Z}_L^{-1})^T \end{aligned} \quad (2.5)$$

J^s denotes the space of s-dimensional integer vectors (s satellites are assumed to be visible).

If normal distribution is assumed for the original observations, a hyperellipsoidal (1-α)-confidence region $K_{1-\alpha}$ for the integer expectation vector $\Delta\zeta$ of $\Delta\underline{\hat{\psi}}$ can be constructed regarding

$$(\Delta\underline{\hat{\psi}} - \Delta\zeta)^T\Sigma_{\hat{\psi}\hat{\psi}}^{-1}(\Delta\underline{\hat{\psi}} - \Delta\zeta) \sim \chi_s^2 \quad (2.6)$$

as

$$K_{1-\alpha} := \ldots \left\{\Delta\hat{\psi} \in \mathbf{R}^s \mid (\Delta\hat{\psi} - \Delta\zeta)^T\Sigma_{\hat{\psi}\hat{\psi}}^{-1}(\Delta\hat{\psi} - \Delta\zeta) \le \chi_{s,1-\alpha}^2\right\} \quad (2.7)$$

with $\mathbf{R}^s$ denoting the s-dimensional Euclidian space and $\Delta\hat{\psi}$ the real-valued numerical approximation of $\Delta\zeta$ using the observed data $\Delta\phi$ and $\Delta\rho$.

3 Ambiguity selection

The critical point of the GPS phase observation processing is the selection of the presumably correct ambiguity parameters. As mentioned above, typical substitutes for the expression *correct* are the equivalent expressions *best* or *closest*. Within this section a proposal for the identification of the integer vector with maximum degree of confidence shall be given (*maximum confidence strategy*).

Eq. (2.7) is the key formula for the following considerations. Let

$$T_{(i)} := (\Delta\hat{\psi} - \Delta\zeta_{(i)})^T\Sigma_{\hat{\psi}\hat{\psi}}^{-1}(\Delta\hat{\psi} - \Delta\zeta_{(i)}) \quad (3.1)$$

denote the real values derived from the quadratic form by inserting all possible s-dimensional integer vectors $\Delta\zeta_{(i)}$, i = 1, 2, ... The values $T_{(i)}$ are assumed to be sorted in ascending order. This is denoted by (.). Thus $T_{(1)}$ points at $\Delta\zeta_{(1)}$, which is the solution to be selected when applying the minimum distance criterion of integer least squares. On the other hand, for each $\Delta\zeta_{(i)}$ a $T_{(i)}$, and thus the corresponding confidence probability $1-\alpha_{(i)}$ can be given. The affiliated confidence region according to Eq. (2.7)

$$K_{1-\alpha_{(i)}} = \left[0, T_{(i)}\right) \quad (3.2)$$

for $\Delta\zeta_{(i)}$ contains all $\Delta\zeta_{(j)}$, with j<i. Each of these confidence probabilities $1-\alpha_{(i)}$ can be understood as degree of distrust in $\Delta\zeta_{(i)}$ being the correct integer vector since up to confidence probability $1-\alpha_{(i)}$, $\Delta\zeta_{(i)}$ is not contained in $K_{1-\alpha_{(i)}}$. Instead, $\alpha_{(i)}$ quantifies a degree of confidence (*conf*) in $\Delta\zeta_{(j)}, j \geq i$.

$$conf\left(E(\Delta\underline{\hat{\psi}}) \in \left\{\Delta\zeta_{(j)}, j \geq i\right\}\right) := \alpha_{(i)} \quad (3.3)$$

This construction results from the centering of the quadratic form connected with T in the float solution. As an example $T_{(1)}$ is studied. If $\Delta\hat{\psi}$ coincides with $\Delta\zeta_{(1)}$ the corresponding $(1-\alpha_{(1)})$-confidence region is empty. Then the degree of confidence expressed for the correct solution coming from the set of all $\Delta\zeta_{(i)}$, $i \geq 1$, is $\alpha_{(1)}=1$. It should be noted that this must not be understood as a measure for the correctness of $\Delta\zeta_{(1)}$ as ambiguity resolving integer vector.

Combining the aspects of distrust and confidence with the exclusiveness of integer ambiguity vectors yields

$$\begin{aligned}
&conf\left(E(\Delta\underline{\hat{\psi}}) \in \left\{\Delta\zeta_{(j)}, j = i\right\}\right) = \\
&= conf\left(E(\Delta\underline{\hat{\psi}}) \in \left(K_{1-\alpha_{(i+1)}} \setminus K_{1-\alpha_{(i)}}\right)\right) \\
&= conf\left(\left(E(\Delta\underline{\hat{\psi}}) \in K_{1-\alpha_{(i+1)}}\right) \wedge \left(E(\Delta\underline{\hat{\psi}}) \notin K_{1-\alpha_{(i)}}\right)\right) \\
&= 1 - \alpha_{(i+1)} + \alpha_{(i)} - 1 = \alpha_{(i)} - \alpha_{(i+1)}
\end{aligned}$$

applying the addition rule for dependent random events. Summarizing this result yields

$$\overline{p}_i := conf\left(E\left(\Delta\underline{\hat{\psi}}\right) = \Delta\zeta_{(i)}\right) = \alpha_{(i)} - \alpha_{(i+1)} \quad (3.4)$$

as data-derived degree of confidence of $\Delta\zeta_{(i)}$ to be the correct integer expectation of $\Delta\hat{\psi}$ and thus the ambiguity vector to be selected. The selection principle reads as: take the unique $\Delta\zeta_{(m)}$ with $\overline{p}_m > \overline{p}_i$ for all natural numbers $i \neq m$ as maximum confidence solution. If there are several solution candidates with equal confidence, the selection fails.

An issue of standardizing has to be mentioned. In general, the probability

$$P\left(E(\Delta\underline{\hat{\psi}}) \in K_{1-\alpha_{(1)}}\right) = 1 - \alpha_{(1)} \neq 0.$$

This is due to the centering of the confidence region in the approximation $\Delta\hat{\psi}$. $K_{1-\alpha_{(1)}}$ does not contain any integer vector. But Eq. (2.3) requires an integer expectation of $\Delta\hat{\psi}$. For this reason the confidence measures $\overline{p}_i$ are conditioned as

$$p_i := \frac{\overline{p}_i}{\alpha_{(1)}} = \frac{\alpha_{(i)} - \alpha_{(i+1)}}{\alpha_{(1)}} \quad (3.5)$$

The resulting confidences p_i add up to 1. Please note that there is no difference in the ambiguity selection procedure and its result for p_i and $\overline{p}_i$. Division by zero is impossible since this would imply complete distrust in the solvability of the ambiguity problem.

The *conf*-measures are not probabilities in the strict sense since they are not derived from the underlying probability distribution. Regarding Eq. (3.4) the presented procedure surely has heuristic aspects. Nevertheless, a multitude of simulation studies shows a high similarity between the confidence measures and the relative frequency of success of the ambiguity resolution. Other numerical results underlining the practicability of the degree-of-confidence measure are given in Section 6.

The presented algorithm yields both integer ambiguity parameters and confidence measures with two essential characteristics. First, if the confidence of a specific integer vector is sufficiently close to one, the selection is nearly certain. Thus for practical data processing a threshold confidence value has to be chosen. If there are more than one integer vectors with positive confidence, then the possibility occurs of selecting an incorrect one.

Second, the degrees of confidence p_i are imprecise. This is due to the fact that they are derived from the observation data. They depend strongly on the actual value of the float approximation $\Delta\hat{\psi}$. This also is a problem of the integer least squares criterion *minimum distance*. The numerical examples in Section 6 shall illustrate the characteristics of the proposed method and shall compare it with the integer least squares method. Apparently there is a mathematical theory for the handling of imprecise probabilities (Walley, 1991). The quantification of the imprecision of the derived confidences and of their impact on the parameter estimation is a necessary task for future studies.

4 Ambiguity precision

In common GPS data processing the postulate of the correctness of the ambiguity parameters is implied when selection was successful. Betti et al. (1993) question the associated significant increase of the coordinate precision from the step of unresolved ambiguities to resolved ones. They introduce a discrete distribution for the ambiguity parameters as prior information. Using a Bayesian approach, the coordinate precision is increasing monotonously epoch by epoch until the ambiguities are resolved. Thus the precision of the ambiguity parameters results from a mixed real and integer density.

Let us now consider the proper discrete density. Is, for example, complete ignorance concerning a certain parameter equivalent to equal probability for all of its possible realizations? This can merely be answered in general, and it is the main objection to the Bayesian approach.

Looking at the results of the previous section, prior information concerning parameter densities doesn't seem necessary. Precision measures for ambiguity parameters can be directly derived from the computed degrees of confidence. This is in analogy to the well-known formulas for discrete random quantities which will be reviewed in the following.

Assuming that $x_1, x_2, \ldots, x_l$ and $y_1, y_2, \ldots, y_l$ are the l possible realizations of the discrete one-dimensional random variables X and Y, respectively, then

$$E(X) = \sum_{i=1}^{l} f_X(x_i)x_i = \sum_{i=1}^{l} p_i x_i =: \bar{x}$$
$$V(X) = E((X-\bar{x})^2) = \sum_{i=1}^{l} p_i (x_i - \bar{x})^2 \qquad (4.1a,b)$$

with

$$p_i := \underbrace{P(X = x_i)}_{\substack{\text{probability}\\ \text{of the event}\\ X=x_i}} = \underbrace{f_X(x_i)}_{\substack{\text{discrete}\\ \text{density}}}$$

(Viertl, 1997). V(.) denotes the variance operator. A similar formula can be derived for the covariance.

$$\mathrm{Cov}(X,Y) = \sum_{i=1}^{l} \underbrace{f_{XY}(x_i, y_i)}_{\substack{\text{joint density}\\ \text{of X and Y}}}(x_i - \bar{x})(y_i - \bar{y})$$
$$= \sum_{i=1}^{l} p_i (x_i - \bar{x})(y_i - \bar{y}) \qquad (4.2)$$

In case of s-dimensional random vectors, $f_X(x_i)$ and $f_{XY}(x_i,y_i)$ represent marginal densities.

In Section 3 a set of l possible ambiguity vectors $\Delta\zeta_{(i)}$ with degrees of confidence p_i for their correctness was derived. Since it is not necessary to sort the values p_i, the (.)-notation for sorted values is dropped for the sake of simplicity.

The vectors $\Delta\zeta_i$ can be seen as events of a discrete random variable $\Delta\underline{\zeta}$. Thus Eqs. (4.1) and (4.2) are suitable for the derivation of quantities which are analogous to the expectation vector and to the variance-covariance matrix (vcm) of the ambiguity parameters.

When discrete random variables are considered, the expectation vector is just the weighted mean of all event vectors $\Delta\zeta_i$

$$\overline{\Delta\zeta} = E(\Delta\underline{\zeta}) = \sum_{i=1}^{l} p_i \Delta\zeta_i \qquad (4.3)$$

in transcription of Eq. (4.1a). The corresponding vcm follows from Eqs. (4.1b) and (4.2) component by component

$$\Sigma_{\zeta\zeta} = D(\Delta\underline{\zeta}) = E((\Delta\underline{\zeta} - \overline{\Delta\zeta})(\Delta\underline{\zeta} - \overline{\Delta\zeta})^T)$$
$$= E((\Delta\underline{\zeta}_j - \overline{\Delta\zeta}_j)(\Delta\underline{\zeta}_k - \overline{\Delta\zeta}_k))^s_{j,k=1} \qquad (4.4)$$
$$= \left(\sum_{i=1}^{l} p_i (\Delta\zeta_{i,j} - \overline{\Delta\zeta}_j)(\Delta\zeta_{i,k} - \overline{\Delta\zeta}_k)\right)^s_{j,k=1}$$

with j, k denoting vector components and i the event.

The Eqs. (4.3) and (4.4) result from basic definitions of probability theory. Introducing the degrees of confidence the two equations represent the calculation of a location and a scale parameter. Obviously, these quantities can be formulated analytically and computed numerically. This is the mathematical part of the considerations. Some interpretation must be added concerning its application. In general, the location parameter, according to Eq. (4.3), describing the selection outcomes is not integer. It is just a convex combination of all possible realizations. But this is not the way the selection works. Successful selection implies ambiguity fixing. If $\Delta\zeta_m$ is selected as an integer vector resolving the phase ambiguities, it defines the center of the dispersion parameter. The associated degree of confidence for the correct decision is p_m. Considering concepts of ambiguity precision, $\overline{\Delta\zeta}$ has to be replaced by $\Delta\zeta_m$ in Eq. (4.4) to describe the situation after fixing. A more general interpretation will be given in Section 7.

5 Reliability aspects

Up to now the selection was based on the comparison of confidence measures. The statistical validity of the selected integer vector has to be checked by a hypothesis test concerning its compatibility with the float approximation.

Let

$$H_0: E(\Delta\underline{\hat{\psi}}) = \Delta\zeta_m, \; H_i: E(\Delta\underline{\hat{\psi}}) = \Delta\zeta_i \neq \Delta\zeta_m$$

denote different hypotheses with the consequence

$$T = (\Delta\underline{\hat{\psi}} - \Delta\zeta_m)^T \Sigma^{-1}_{\hat{\psi}\hat{\psi}} (\Delta\underline{\hat{\psi}} - \Delta\zeta_m) \sim \chi^2_s | H_0$$

The choice of the significance level α (Type I error) leads to a test decision. Only if T is element of the critical region *R*, $\Delta\zeta_m$ is rejected. A measure β in analogy to the probability of a Type II error can be derived from

$$\beta = P(T \in A \mid E(\Delta\underline{\hat{\psi}}) \neq \Delta\zeta)$$

as

$$\beta = \frac{\sum_{i \neq m} P(T \in A \,|\, E(\Delta\underline{\hat{\psi}}) = \Delta\zeta_i) \cdot p_i}{\sum_{i \neq m} p_i} \qquad (5.1)$$

with the region of acceptance A. This measure serves as a separability criterion for the selected integer vector from its alternatives. The smaller β, the higher the reliability of the ambiguity fixing will be.

6 Simulation examples

Within this section the ambiguity selection procedures using the minimum distance and the maximum confidence criterion are studied and compared using simulations. The scenario is a 24 satellites configuration according to Parkinson and Spilker (1996, p.181). Simultaneous code and phase observations are simulated by means of the Matlab© software package using a generator for uncorrelated normally distributed random numbers. 0.3 m and 2 mm are chosen as standard deviations of the simulated code and phase observations, respectively. All other error sources are neglected in studying the core characteristics of the two approaches. Thus for the sake of simplicity double differencing was not applied, and absolute positions were estimated. Several scenaria were evaluated. The number of satellites was controlled by an elevation mask. Different numbers of epochs and different observation times were studied.

Fig. 1 shows the comparison of the minimum distance criterion, the distance ratio between closest and second-closest integer vector, and the confidence probabilities according to Eq. (3.4). The relationship between the two approaches is obvious: the smaller the distance and the higher the separability of integer vectors (indicated by the distance ratio), the higher the degree of confidence.

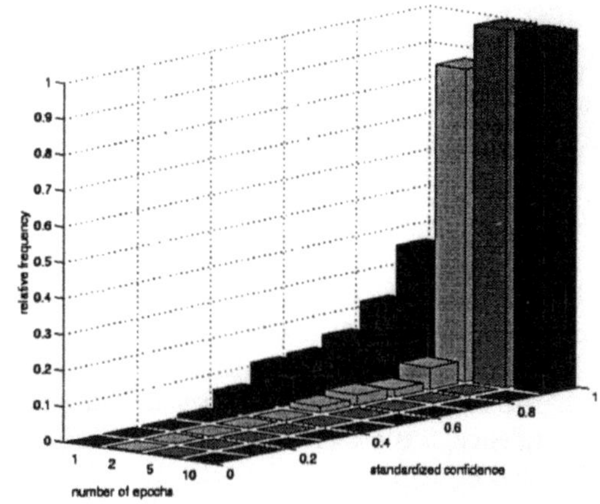

Fig. 2: Relative frequencies of the standardized confidences: 5 satellite configurations, sampling rate 1 sec (columns from left to right: 1,2, 5, and 10 epochs)

Fig. 2 presents a histogram of the standardized confidences for a configuration with five satellites. From left to right a different number of epochs is shown. In the case of 5 and 10 epochs, respectively, nearly all selected integer vectors have confidence one, which means certain selection. In case of one epoch only (leftmost column), the justification of the selection is significantly weaker.

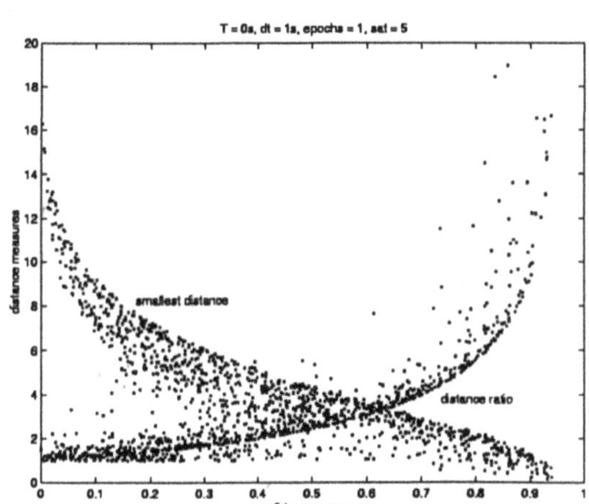

Fig. 1: Dependency plot of the minimum distance and the maximum confidence criteria

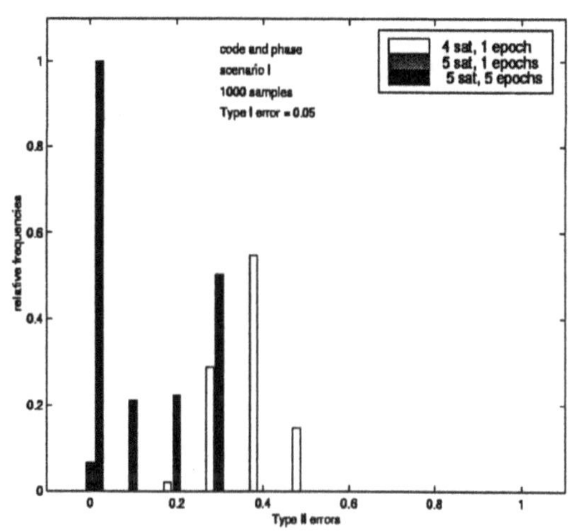

Fig. 3: Relative frequencies of Type II. errors for different satellite constellations

Studying by simulation the selection failures for the two ambiguity selection principles *minimum distance* and *maximum confidence*, similar error rates are found. In most of the successful selections both strategies work well, but there are also correct solutions for the one method when the other fails.

Fig. 3 shows the Type II errors according to Eq. (5.1) for three different elevation masks and tracking rates, respectively. As a result it can be stated that the better the configuration (increasing number of satellites and of epochs) the smaller the type II error will be. If there exists one integer vector with very high confidence, it is very reliable, too. This follows directly from Eq. (5.1).

7 Conclusions

The presented theoretical and data-analytical aspects have to be considered from a practical point of view. If the probability or confidence in the integer candidate is not sufficiently high, it must not be selected as a presumably correct ambiguity vector. When increasing the values of these measures by use of additional epochs or better geometries, the precision approaches zero variance and the reliability the maximum. The natural conclusion to be drawn is obvious. In case of almost certain selection, both precision and reliability measures are more or less blueprints of the underlying probabilities and degrees of confidence, respectively, according to Eqs. (4.4) and (5.1). The practical need for the computation of the moments of the discrete distribution or of the distribution itself, is negligible. Looking back at the beginning of this section, it can be seen that the particular advantage of this knowledge consists in most parts of the probability or degree-of-confidence scale for the judgement of the validity of the solution. Regarding the interpretation such a scale is superior to the separation of possible solutions by means of variance ratio criteria.

References

Abidin, H.Z. (1993): On-the-fly ambiguity resolution: formulation and results. *manuscripta geodaetica* (1993): 18: 380-405.

Betti, B., M. Crespi, and F. Sansò (1993): A geometric illustration of ambiguity resolution in GPS theory and a Bayesian approach. *manuscripta geodaetica* (1993) 18: 317-330.

Counselman, C.C., and Gourevitch, S.A. (1981): Miniature interferometer terminals for earth surveying: ambiguity and multipath with the Global Positioning System. *IEEE Transactions on Geoscience and Remote Sensing*. GE-19(4): 244-252.

Frei, E., and G. Beutler (1991):Rapid static positioning based on the fast ambiguity resolution approach "FARA": theory and first results. *manuscripta geodaetica* (1990) 15:325-356.

Hofmann-Wellenhof, B., H. Lichtenegger, and J. Collins (1997): GPS - Theory and Practice (4th edition). Springer, Wien/New York.

Parkinson, B.W. and J.J. Spilker (1996): Global Positioning System: Theory and Applications Vol. I. Progress in Astronautics and Aeronautics (Vol. 163), American Institute of Aeronautics and Astronautics, Washington DC, U.S.A.

Rao, C.R., and S.K. Mitra (1971): Generalized Inverse of Matrices and its Applications. Wiley, New York.

Teunissen, P.J.G. (1996): GPS carrier phase ambiguity fixing concepts. In: Kleusberg, A., and Teunissen, P.J.G.: *GPS for Geodesy*. Springer, Berlin/Heidelberg.

Teunissen, P.J.G. (1997): A canonical theory for short GPS baselines I-IV. *Journal of Geodesy* (1997) 71: 320-336, 389-401, 486-501, 513-525.

Teunissen, P.J.G. (1998): Success probability of integer GPS ambiguity rounding and bootstrapping. *Journal of Geodesy* (1998) 72: 606-612.

Teunissen, P.J.G. (1999): The probability distribution of the GPS baseline for a class of integer ambiguity estimators. *Journal of Geodesy* (1999) 73: 275-284.

Viertl, R. (1997): Einführung in die Stochastik (2. Aufl.). Springer, Wien/New York.

Walley, P. (1991): Statistical Reasoning with Imprecise Probabilities. Chapman and Hall, London/New York.

Xu, P., E. Cannon, and G. Lachapelle (1995): Mixed integer pro-gramming for the resolution of GPS carrier phase ambiguities. Paper presented at the IUGG95 assembly, July 2-14, 1995, Boulder, Colorado, U.S.A.

PART 4

Geodynamics

Martine Feissel

Degree-one deformations of the Earth

Marianne Greff-Lefftz
Institut de Physique du Globe de Paris, 4 place Jussieu, 75252 Paris 05, France

Hilaire Legros
E.O.S.T., 5 rue R. Descartes, 67084 Strasbourg, France

Abstract. The degree one deformations of the Earth, in a reference frame related to the center of mass of the planet, are computed using a theoretical approach (Love numbers formalism) at short time-scale (from the month up to the century), where the Earth has an elastic behavior. The translations at each interface of the layers of the Earth's model (especially at the surface, at the Core-Mantle boundary (CMB) and at the Inner Core boundary (ICB)) are computed when the excitation source is the atmospheric pressure or a magnetic pressure acting at the CMB and at the ICB. The effects of external and internal tangential tractions are also investigated. The total force, resulting from the excitation sources, in a geographic frame (centered at the center of mass) has to be equal to zero, in order to conserve the center of mass of the Earth. This involves a relation between the different forcing mechanisms; we obtain a Consistency Relation, i.e., a special condition that the degree-one valid solutions have to obey (Farrell, 1972). As geophysical application, we have computed the degree-one static deformations induced by atmospheric loading. To end, at secular and geological time-scales, where the Earth has a viscoelastic behaviour, we have computed the secular and geological variations of the geocenter induced by post-glacial rebound and by mantle density heterogeneities.

Keywords. Deformation, geocenter.

1 Introduction

The degree-one deformations of the Earth (and the induced discrepancy between the center of figure of the outer surface and the center of mass) are computed using a theoretical approach. In a first part, we briefly review the problem of the elasto-gravitational theory for the degree one and then some geophysical applications at different time-scales are presented.

2 Elasto-gravitational theory

We use the classical elasto-gravitational theory (Alterman et al, 1959). This is a theory of perturbation. We start with an initial state where the Earth is assumed to be in hydrostatic equilibrium, radially stratified following the PREM model (Dziewonski and Anderson, 1981). The origin of the reference frame is the center of mass (CM) which is identical to the center of geometry of the sphere at this initial state. At the perturbed state, the Earth is elastic and the static equations are still written in a reference frame related to the center of mass. The center of figure (CF) is defined as the geometric center of the outer deformed sphere with respect to CM. The differential equations governing the spheroidal part of the displacement field, the stresses and the body force potentials are of sixth order. Expanding the displacement vector field $\vec{u}$ and the traction $\vec{T}$ in spherical spheroidal vectors, $\vec{\nabla}_H$ denoting the tangential gradient :

$$\vec{u} = \sum_{n=1}^{\infty}\sum_{m=0}^{n}\Big[y_{1n}(r)Y_n^m(\theta,\varphi)\frac{\vec{r}}{r} + r y_{3n}(r)\vec{\nabla}_H Y_n^m(\theta,\varphi)\Big] \tag{1}$$

$$\vec{T} = \sum_{n=1}^{\infty}\sum_{m=0}^{n}\Big[y_{2n}(r)Y_n^m(\theta,\varphi)\frac{\vec{r}}{r} + r y_{4n}(r)\vec{\nabla}_H Y_n^m(\theta,\varphi)\Big] \tag{2}$$

noting for the potential

$$U = \sum_{n=1}^{\infty}\sum_{m=0}^{n} y_{5n}(r)Y_n^m(\theta,\varphi)$$

and introducing for the radial derivative of the potential a function defined by

$$y_{6n}(r) = \frac{dy_{5n}(r)}{dr} - 4\pi G\rho y_{1n}(r) \tag{3}$$

International Association of Geodesy Symposia, Vol. 121
Schwarz (ed.), Geodesy Beyond 2000 – The Challenges of the First Decade

where Y_n^m are the spherical harmonics, θ the colatitude, φ the longitude, $\rho(r)$ the density and G the gravitational constant, the static elasto-gravitational equations (describing the mass conservation, the equation of Poisson and the Hookean rheological law) can be written (Alterman et al., 1959)

$$\frac{dy_{in}(r)}{dr} = \sum_{j=1}^{6} A_{ij}^n \; y_{jn}(r) \quad \text{for } i = 1..6 \quad (4)$$

where A_{ij}^n is a 6×6 matrix whose elements are a function of the compressibility , the rigidity, the density and the gravity $g(r)$.

When the Earth is submitted to a surface loading potential S_n^e, $y_{in}(r)$ are continuous within the Earth and there are three boundary conditions at the outer surface $r = a$:

- the tangential stress is equal to zero $y_{4n}(a) = 0$
- the radial stress is equal to the pressure effect of the load: $y_{2n}(a) = -\frac{2n+1}{3}\tilde{\rho}S_n^e$
- the discontinuity in the radial derivative of the potential may be written:

$$y_{6n}(a) + \frac{n+1}{a}y_{5n}(a) = \frac{2n+1}{a}S_n^e$$

For the degree $n = 1$ there are two problems: first, the surface boundary conditions are not independent, and second, the center of mass of the Earth is not conserved.

Farrell (1972) has shown that for the degree $n = 1$ there is a Consistency Relation, that is to say a special condition that the degree-one valid solutions have to obey. For a surface load, this relation may be written (in the above section, the subscript $n = 1$ will be suppressed for simplicity of notation):

$$y_2(r) + 2y_4(r) + \frac{g(r)}{4\pi G}\left[y_6(r) + \frac{2}{r}y_5(r)\right] = 0 \quad (5)$$

this is a relationship between the stresses and the potential within the Earth (Saito, 1974). Because of this relationship, the boundary conditions are not independent: only two of the three surface boundary conditions are needed and the Consistency Relation ensures that the third boundary condition is met automatically. To solve the elasto-gravitational differential system (4) we have to add a new boundary condition: the conservation of the center of mass of the Earth. From the MacCullagh theorem (Munk and MacDonald, 1960), we show that the conservation of the center of mass will simply require that the degree-one surface potential is equal to zero:

$$y_5(a) = 0 \quad (6)$$

The Consistency Relation may be interpreted as implying that, in the static case, there cannot be net force on any portion of the Earth (Okubo and Endo, 1986). As a matter of fact, if we rewrite this Relation when the Earth is submitted to external and internal pressure and tangential traction (noted respectively P^e and T^e at the surface, P^c and T^c at the CMB (r=b) and P^{ic} and T^{ic} at the ICB (r=c)), we will obtain (Greff-Lefftz and Legros, 1997):

$$\frac{d}{dr}\left[r^2\left(y_2(r) + 2y_4(r) + \frac{g(r)}{4\pi G}\left[y_6(r) + \frac{2}{r}y_5(r)\right]\right)\right] = 0 \quad (7)$$

The boundary conditions within the Earth are not independent and imply that:

$$a^2(-P^e + 2T^e) + b^2(P^c + 2T^c) + c^2(-P^{ic} + 2T^{ic}) = 0 \quad (8)$$

The total degree one force acting at the various fluid-solid interfaces and at the surface of the planet has to be equal to zero. There may be a static equilibrium between stresses at ICB and CMB (which vary at decadal time-scale) and a static equilibrium between pressure and tangential traction at the Earth's surface (which vary at annual time-scale).

Solving the elasto-gravitational static system for the degree one and using the new boundary condition described above, we obtain the displacement field within the Earth, when the Earth is submitted to a surface load. Expanding the surface loading potential in non-normalized degree-one spherical harmonics:

$$S_1^e = S_1^{0e}\cos\theta + [\cos\varphi S_1^{1e} + \sin\varphi \tilde{S}_1^{1e}]\sin\theta \quad (9)$$

the displacement at the Earth's surface may be written using a Love number formalism:

$$y_1(a) = h_1\frac{S_1^e}{g_o}; \quad y_3(a) = l_1\frac{S_1^e}{g_o} \quad (10)$$

where g_o is the surface gravity. For the PREM model, we find: $h_1 = -1.2858$ and $l_1 = -0.8958$.

Note that these values differ from the values obtained by Farrell (1972) by a factor +1: this is because we work in a reference frame related to the center of mass of the Earth.

We are now able to compute the geocenter, that is to say the displacement of CF with respect to CM. This last one may be simply computed from the averaged displacement over the outer surface:

$$\frac{1}{4\pi} \int_{\text{surface}} \vec{u} \, \sin\theta d\theta d\varphi \tag{11}$$

and consequently, the coordinates of the geocenter (X along the Greenwich meridian, Y along the 90°E meridian and Z perpendicular to the equator) may be written using loading Love numbers:

$$\left[X, Y, Z\right] = \frac{h_1 + 2l_1}{3} \left[\frac{S_1^{1e}}{g_o}, \frac{\tilde{S}_1^{1e}}{g_o}, \frac{S_1^{0e}}{g_o}\right] \tag{12}$$

3 Geophysical application

3.1 Atmospheric continental loading

In this part we want to investigate the effects of the atmospheric pressure on the geocenter. To compute the degree-one deformations, we use a model proposed by Gegout (1995) for the atmospheric loading. The data are the 11-years pressure field record provided by ECMWF. The atmospheric load which has dominant annual and semi-annual components, is separated in two parts: a part over the continents and a part over the oceans. On the one hand, the load over the oceans does not deform the Earth, because oceans react as Inverted Barometer at this time-scale. On the other hand, the load over the continent deforms the whole Earth, i.e the continental surface but also the bottom of the oceans. As a consequence, we have to take into account the variations of the water thickness induced by the deformation of the bottom of oceans. To do that, Gegout (1995) has introduced a coefficient β_n which is a combination of Love numbers (h_n in radial displacement and $1 + k_n$ in surface potential)

$$\beta_n = 1 - \frac{\rho^w}{\tilde{\rho}}(1 + k_n - h_n) \tag{13}$$

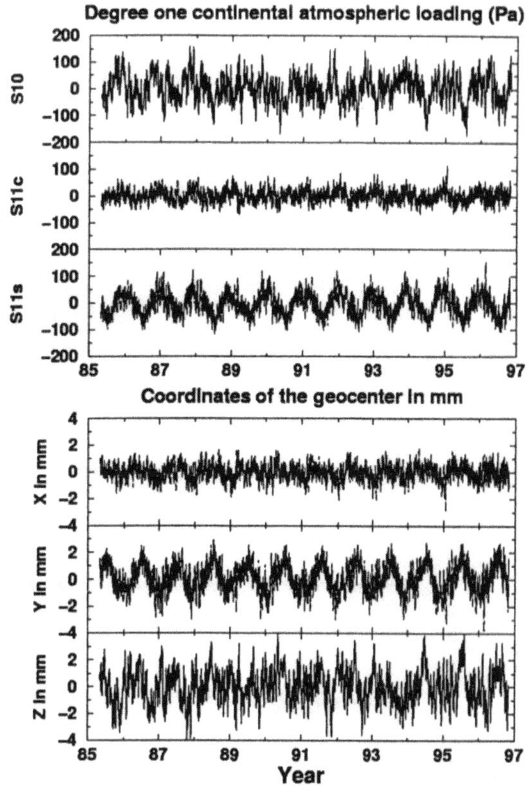

Fig. 1: (top): Degree one non-normalized spherical harmonics coefficients for the continental atmospheric loading computed by Gegout (1995). (bottom): Coordinates of the geocenter derived from the above load.

where ρ^w is the water density and $\tilde{\rho}$ the mean density of the Earth. For the degree-one, because of the conservation of the center of mass, β_1 may be simply written: $\beta_1 = 1 + \frac{\rho^w}{\tilde{\rho}} h_1$. The surface displacement is then written using classical Love numbers and this β_1 coefficient:

$$y_1(a) = \frac{h_1}{\beta_1} \frac{S_1^e}{g_o}; \quad y_3(a) = \frac{l_1}{\beta_1} \frac{S_1^e}{g_o}; \tag{14}$$

where S_1^e is the degree-one spherical harmonics coefficients of the continental part of the atmospheric loading. The effect of the variation of the water thickness is an increase of the amplitude of the degree-one deformations by about 30%.

We have plotted, on Figure 1, the degree-one non-normalized coefficients of the continental atmospheric loading S_1^{0e}, S_1^{1e}, $\tilde{S}_1^{1e}$ and the associated variation of the geocenter coordinates from 1985 up to 1997.

The geocenter variation is about a few millimeters and has essentially annual and semi-

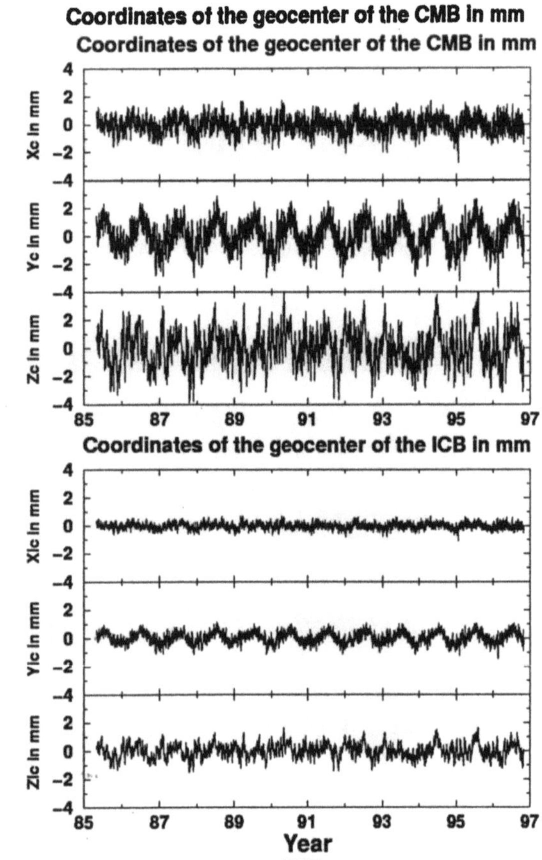

Fig. 2: (top): Temporal variation, in millimeter, of the coordinates of the figure center of the CMB. (bottom) : Same than the top for the ICB

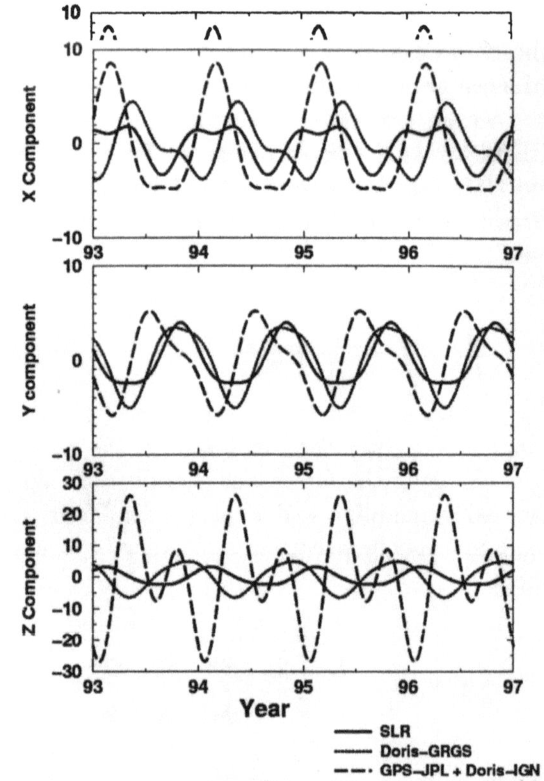

Fig. 3: Observed variation of the geocenter's coordinates since 1993, from Sillard, 1998

annual components. Z has the largest amplitude but has noise. Y has a major annual component. These results are similar to those obtained by Dong et al. (1997) in phase but the amplitude is 30 % larger because of the amplification factor β_1.

We have also computed the variations of the figure center of the CMB and of the ICB with respect to the center of mass (in Figure 2). These are about one millimeter for the CMB and 0.1 mm for the ICB.

Some recent space geodesy measurements (DORIS, GPS and SLR) give the variations of CF with respect to a fixed reference frame. In Figure 3 we present a compilation of these observations done by Sillard (1998) for the X, Y and Z annual and semi-annual components of the geocenter since 1993.

Note that these different kinds of data are neither consistent in phase nor in amplitude. The amplitude of the observations seems to be about 10 mm, that is to say larger than the geocenter variations induced by atmospheric loading: this source cannot alone explain the annual and semi-annual variations of the geocenter. Dong et al. (1997) have shown that the ground water storage may create an annual variation of the geocenter with the same order of magnitude than the observations.

3.2 Secular and geological variation of the geocenter

In this part, we present an application of the same theory at longer time-scale, where the effect of the viscosity of the mantle has to be taken into account: this is the degree-one Visco-Elasto-Gravitational theory. We assume a Maxwell model of rheology for the mantle (for more details, see Greff-Lefftz and Legros, 1997).

At secular time-scale, we have computed the geocenter variation induced by the Pleistocenic deglaciation, for the present-time (almost 7000 years after the end of the deglaciation). We find a surface translation of about one meter in the direction ($\theta_o = 20.7^o$ and $\lambda = -46.6^oE$), which tends to be relaxed in the future, because of the surface isostatic compensation. The present rate of the geocenter's variation is strongly de-

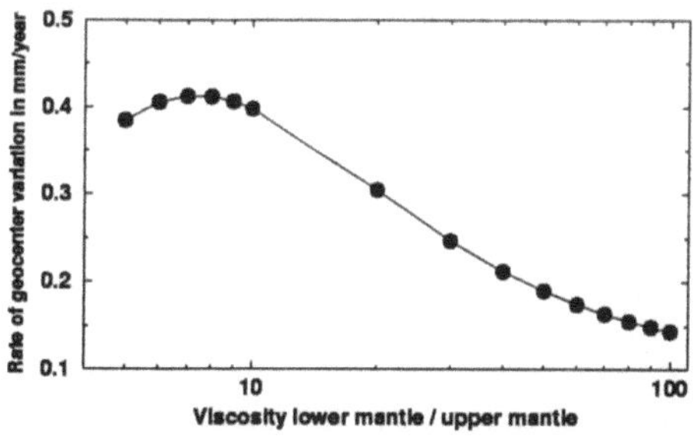

Fig. 4: Rate of the geocenter temporal variation in the antipodal direction of the load ($\theta_o = 20.7^o$ and $\lambda = -46.6^o E$)

pendent on the viscosity profile within the mantle. Assuming a viscosity of the upper mantle of about 10^{21} Pa.s, we plot in Figure 4 the rate $\sqrt{\dot{X}^2 + \dot{Y}^2 + \dot{Z}^2}$ (the dot denotes here the time-derivative) obtained for viscosity increases in the lower mantle varying from 5 up to 100.

The value of this rate varies from 0.2 up to 0.4 mm/years and consequently, in 10 years, one can expect a variation of the geocenter induced by the post-glacial rebound comparable to the one due to atmospheric loading.

At geological time-scale, we have computed the degree one deformation induced by the present-day mantle density heterogeneity derived by Ricard et al. (1993). This model uses plate motion reconstructions under the assumption that subducted slabs sink vertically into the mantle. It assumes, at 670 km depth, no density jump and a viscosity contrast of about 40. We find that the difference between CF and CM, i.e. the coordinates of the geocenter, is of about some hundred meters. At the time-scale of the observations (i.e. few years) it could be interesting to see how this kind of constant translation may be observed in geodetical measurements.

4 Conclusion

In this paper we have computed the degree one elastic and viscoelastic deformations of the Earth using a theoretical approach. We have noted that for the degree-one deformations, a Consistency Relation exists implying that the total force acting on the Earth has to be equal to zero.

As geophysical application, we have computed the degree one elastic deformations induced by atmospheric continental loading: the geocenter's variation has semi-annual and annual components and is about a few millimeters.

At secular time scale, we find that the post-glacial rebound involves a present rate of the geocenter variation of about 0.2 - 0.4 mm/year. At geological time-scale, we obtain a geocenter about a few hundred meters, varying on the time-scale of the convection, i.e some millions years.

Acknowledgments. We thank Pascal Gegout for giving us the data of the continental loading and for discussions.

References

Alterman, Z., Jarosch, H. & Pekeris, C.H., 1959. Oscillation of the Earth. Proc. R. Soc. London, **A252**, 80-95.

Dong, D., Dickey, J.O., Chao, Y. & Cheng, M.K., 1997. Geocenter variations caused by atmosphere, ocean and surface ground water. *Geophys. Res. Let.*, **24**, No 15, 1867-1870.

Dziewonski, A.M., & Anderson, D.L., 1981. Preliminary Reference Earth Model PREM, *Phys. Earth Planet. Int.*, **25**, 297-356.

Farrell, W.E., 1972. Deformation of the Earth by Surface Loads. Reviews of Geophysics and Space Physics, **10**,n^o3, 761-797.

Gegout, P., 1995. De la variabilité de la rotation de la Terre et du champ de gravité, conséquente aux dynamiques de l'Atmosphère et des Océans. Thesis, Strasbourg, France.

Greff-Lefftz, M. & Legros, H., 1997. Some remarks about the degree-one deformation of the Earth.*Geophys. J. Int.*, **131**, 699-723.

Munk, W.H. & MacDonald, G.J.F. 1960. The rotation of the Earth, Cambridge University Press, 323 pp.

Okubo, S. & Endo, T., 1986. Static spheroidal deformation of degree I. Consistency relation, stress solution and partials, *Geophys. J.R. astr. Soc.*, **86**, 91-102.

Ricard, Y., Richards, M., Lithgow-Bertelloni,C. & Le Stunff Y., 1993. A geodynamical model of mantle density heterogeneity. J. Geophys. Res., **98**, 21895-21909.

Saito, M., 1974. Some problems of static deformation of the earth. *J. Phys. Earth*, **22**, 123-140.

Sillard, P., 1998. Les variations du geocentre determinees par DORIS et les autres techniques de geodesie spatiale. Journees DORIS 1998, Toulouse, 28 Avril 1998.

Geodynamics from the analysis of the mean orbital motion of geodetic satellites

P. Exertier, S. Bruinsma, G. Métris, Y. Boudon and F. Barlier
Observatoire de la Côte D'Azur, départ. CERGA, av. Copernic, F-06130 Grasse

Abstract. Secular and long periodic perturbing forces acting on the orbit of an artificial satellite are modelled and included in a semi-analytical theory of the mean orbital motion. Gravitational and non-gravitational forces are averaged using analytical transformations and numerical quadratures, respectively. As non-gravitational modelling is the first source of problems arising in the determination of geodynamical parameters, special attention is given to this question. Resulting averaged equations of motion are integrated numerically allowing the visualization and decorrelation of long periodic signals in the classical mean orbital elements. Observed mean orbital elements of LAGEOS (18 years) and Starlette (14 years) have been computed from Satellite Laser Ranging (SLR) measurements which had first been reduced by classical orbit fits on short arcs. The results of these long arc solutions exhibit variations of geodynamical coefficients at different periods ranging from semi-annual and annual to 9.3 and 18.6 years. Values of coefficients are compared to other results obtained recently in this field.

Keywords. Geodynamics, mean orbital motion, geodetic satellites.

1 Introduction

The Earth is a dynamic system: it has a continually changing distribution of ice, snow, and groundwater, a fluid core undergoing hydromagnetic motion, a mantle undergoing both thermal convection and rebound from glacial loading of the last ice age, and moving tectonic plates. With only a few exceptions on the Earth's surface, the history and spatial pattern of such mass transports are often not amenable to direct observations. Space geodetic techniques, however, have the capability of monitoring certain direct and global consequences of the mass transport, including Earth's variations in rotation, gravitational field, and in its geocenter motion.

Since more than 20 years Satellite Laser Ranging (SLR) to the geodetic satellites LAGEOS (launched in 1976, altitude of 6000 km) and Starlette (launched in 1975, altitude of 800 km) provides a unique and truly global measure of natural changes in the atmosphere, oceans, and interior of the Earth on a variety of time-scales from days to decades. The principle of global measurements of the Earth's gravity field, which have to be made with appropriate spatial and temporal sampling, consists in analyzing satellite orbit perturbations from a nominal model thanks to a set of tracking observations (e.g., Kaula 1966). Today, this technique of analysis is based on the numerical integration of the satellite equations of motion, which includes the integration of the variational equations. Highly precise determinations of the low-degree zonal harmonic coefficients of the Earth's gravity field have been provided by this technique on time-scales from month to years. This has permitted a better understanding of the processes of mass motions within the Earth, as well as on and above its surface (e.g., Gegout and Cazenave, 1993; Nerem et al., 1993; Cazenave et al., 1996; Nerem and Klosko, 1996).

The very small orbital signatures of mass redistribution within the Earth are, however, the result of a long integration process that gradually changes the variables characterizing the satellite motion. The need for longer arc lengths (to assure stronger and unambiguous orbital signals) matched the difficulties in establishing a long-arc analysis. This last technique is in fact the most effective for determining and decorrelating secular and long-period variations in the Earth's gravitational field (e.g., Cheng et al., 1997), because orbital signals that exhibit linear, quadratic and long-period signatures, are integrated continuously (without truncations). At present, a single dynamically consistent orbit for LAGEOS, over a 20-year period, can be determined using a purely numerical approach (Tapley et al., 1993). However, owing to the coupling between errors in the dynamic force and measurement models and the numerical integration process, it is difficult to converge a single orbit over a time span longer than 1-3 years for low-altitude satellites such as Starlette

International Association of Geodesy Symposia, Vol. 121
Schwarz (ed.), Geodesy Beyond 2000 – The Challenges of the First Decade

(Cheng et al., 1997). However, since they are at a lower altitude, such satellites are more sensitive to the zonal gravity variations than other satellites such as LAGEOS or Etalon.

In order to be able to carry out long-arc analysis with satellites at both low and high altitudes, we have re-developed the "old" idea, which consists in using "mean orbital elements" (to be defined rigorously) for analyzing the secular and long-period perturbations due to the Earth's gravity field (static long-wavelength part, and temporal variations). The most important point to emphasize is that we have derived the explicit equations of the mean orbital motion of an artificial satellite (Métris and Exertier, 1995). Indeed, this allows to compute very long arcs, from 1 to 20-25 years, whith all geophysical phenomena, including drag, taken into account together within a dedicated numerical integration technique (Exertier et al., 1996). Because short periods have been removed, the integration step can be several hours which guarantees a great numerical stability and is low on cpu-time consumption. This principle is applied here to the determination of geodynamical parameters, but it can be applied also to the extrapolation of orbits in view of orbit design for future space missions (Bruinsma et al., 1997a).

In this paper we analyze SLR observations of the geodetic satellites Starlette and LAGEOS over 15 and 18 years, respectively. Precise mean orbital elements are computed by filtering of short orbital arcs (of 20-25 days) that are previously adjusted on these tracking data; they form the basis for observing long periodic signals from semi-annual to 18.6 years. Through these long-period perturbations, temporal variations in the low-degree zonal harmonic coefficients of the Earth's external potential are computed, and values are discussed and compared to recent results.

2 Concept of a mean orbital motion

2.1 Principle

The leading idea of the averaging method consists in filtering the short-period perturbations (orbital and sidereal periods, mainly) appearing in the orbit of a satellite. This field of orbital dynamics has been explored essentially in the 70's using analytical and/or semi-analytical methods of order one or two, thus with a relative precision of roughly 10^{-3} or 10^{-6}, respectively (Douglas et al., 1973). The temporal variations of the geopotential produce, however, orbital signals that are of the order of several mas/yr (milli-arc-second per year) on the node and eccentricity notably (e.g., Cheng et al., 1997), that is to say 10^{-8} to 10^{-9}/yr depending on the satellite altitude. In order to reach this level of precision and a stability which is at least at the same level in the orbital variables, we have developed an averaging method based on a semi-analytical theory of the mean orbital motion (Métris and Exertier, 1995). Gravitational forces, which derive from a potential, have been analytically averaged up to order 4 (relative precision of 10^{-12}, theoretically), whereas dissipative forces, like drag and shadowing effects, are averaged step by step by precise numerical quadratures along with the numerical integration process. This process uses a step size of several hours (from 3 to 12 hours, depending on the orbital altitude). It provides consistent dynamical long-arcs of several years which include all the modelled secular and long-period orbital effects and their coupling.

The averaging method used here is very different, in a dynamical sense, from other methods, which are based on a purely numerical approach. In order to use SLR data to study the temporal variations of the Earth's gravity field, some methods use the orbital information contained in successive short arcs, as for example in Cazenave et al. (1996), while other methods use the information contained in the differences between short arcs and a nominal long arc, as for example in Eanes and Bettadpur (1996). This last procedure does not permit to comput a real differential correction of the long arc, although the latter differences are based on a single long-arc (best fit to the SLR data over a duration depending on the satellite being analyzed), which are then converted to mean orbital elements.

Finally, comparing filtered short arcs to the result of a numerical integration of transformed equations has been very helpful in understanding what the mean orbital elements of an artificial satellite actually are. Indeed, to get mean elements one has not only to remove short-period variations but also to keep all the long-period variations (the mean elements must be "centered" on the osculating elements). This is exactly what a filtering does, but this is not necessarily what is produced by the analytical transformed system. More precisely, these mean, centered elements can not be directly generated by a Hamiltonian system. The simplest way to compute centered mean elements is to add purely analytical long-periodic terms to the mean elements resulting from the integration of the averaged differential system (Exertier et al., 1994).

2.2 Method

Thanks to the averaging of the perturbing potentials and forces, new mean models can have simpler expressions. For example, as explained by Kaula (1966), the mean geopotential is reduced to zonal terms, plus some tesseral coefficients, which sometimes give non-negligible secular and long period perturbations of order 4 due to coupling phenomena (Métris et al., 1993). On the other hand, in case of ocean tides, tesseral terms (prograde) of order zero, one, and two have been retained because they produce long period perturbations in the orbit. Concerning drag and radiation pressures, which are non-conservative surface forces, numerical quadratures have been developed, dividing an orbital revolution in up to 250 segments. In these cases only purely numerical averaging is done step by step during the numerical integration of the long arc. With a step size of 3 hours at low altitude (Starlette) and of 12 hours for satellites such as LAGEOS, uncertainties can be due to brutal changes in the force amplitude. For instance, shadowing effects due to the Moon or erratic variations in the geomagnetic activity can appear between two steps. Nevertheless, these effects can be strongly reduced by automatically adjusting a local empirical coefficient for the averaged force, here the radiation pressures or drag.

2.3 Tests on the consistency of the method

A test procedure has been developed in order to validate the whole process explained above. It uses classical numerical integration to produce orbital arcs of 10 years, which serve as reference. The integrator used to generate these arcs has been carefully verified (Balmino et al., 1990). Tests have been implemented to check all the orbit-averaged perturbing forces, separately and together. Schematically, simulations (the validation of the method) show that mean gravitational perturbing potentials give relative uncertainties of around 5.10^{-9} to some 10^{-8} in the mean variables, depending on the altitude of the satellite being tested. On the other hand, residual effects coming from small numerical approximations when averaging of atmospheric drag may corrupt notably the mean semi-major axis and mean eccentricity at the level of some 10^{-8} in case of a low altitude and a high solar activity. The mean eccentricity can be locally affected by an artificial secular trend (of about 3.10^{-8}/yr), which can be empirically modelled, however, thanks to these tests. But, considering this modelling is not perfect, slight erroneous quadratic or long-periodic signals could appear, for example in the mean ascending node, due to the coupling effects between the metric and the angular variables during the computation of a very long arc. Because the determination of geodynamical parameters is based notably on the motion of the node, it suggests for these parameters an error budget to be established as exhaustively as possible.

Today, the semi-analytical theory has been implemented as an analysis tool, which is based on the differential correction procedure of mean orbital motion using a least-squares approach. From a set of observed mean elements, it is thus possible to adjust the initial conditions of the motion, and estimate geodynamical parameters and empirical coefficients (global or local) within some iterations (two to three, generally).

3 Data, analysis and results

3.1 Data

SLR data that have been acquired on Starlette have the form of normal points (NP) during the period from 1990 to the end of 1998. From 1983 to 1989, we have used full rate data, which are noisier and less precise than NP-data. Short orbital arcs fitted to these older data are certainly less precise than those computed recently, the precision of which is at the level of about 10 cm (e.g., Tapley et al., 1996; Schwintzer et al., 1997). The same scheme exists with the SLR data of LAGEOS since 1980. In fact, tracking data are available since the launch of LAGEOS (1976), but due to their low precision (a few decimeters) we preferred to start our analysis in 1980.

Consecutive 24 day short arcs have been fitted to these data, for both Starlette and LAGEOS. Thanks to the filtering process applied to these short arcs, the precision with which mean orbital elements are computed is quasi-constant with time. The computation being realized exactly in the middle of each arc, this avoids a great part of the orbit errors (including the well-known butterfly effect appearing when using the least-squares to fit an orbit) to be propagated into the mean observed elements. On the other hand, small biases may affect the short arcs, in the orbital inclination and/or ascending node for example, in the middle of the arc. This effect has been estimated at the level of a few mas locally.

The 24-day arc length has been chosen in order to have the greatest efficiency in the filter at periods below 3 days. This permits the removal of notably the m-daily and weak resonance effects due to the tesseral harmonics, the periods of which range from 1 day to 2.7-2.9 days both for Starlette and

LAGEOS. All other contributions of tesseral harmonics are removed analytically (Exertier, 1990). The overall precision with which the six mean orbital elements are computed for each short arc has been estimated at 5-7 mas, which corresponds to a relative precision of 5.10^{-9} to some 10^{-8}. Now, we have at our disposal a set of precise mean elements per satellite. Regarding the data processing and the large amount of computing time involved, the fact that these new data sets have been computed once for all is very efficient to establish an iterative long-arc analysis.

3.2 Nominal model

In this study, the 18.6- and 9.3-year ocean tides were modelled at equilibrium with amplitudes of 1.2 cm according to Trupin and Wahr (1990) and 0.02 cm (Cartwright and Edden, 1973), respectively. The nominal ocean tide model uses the coefficients from Schwiderski (1980), whereas the coefficients of the annual and semi-annual ocean tides (Sa and Ssa, respectively) have been set to zero. The reason for that is twofold: (i) the air mass redistribution, which is the strongest annual contributor, undergoes important interannual variations, (ii) the ocean tide amplitudes at those long periods are still not reliable.

The contribution of the atmospheric mass redistribution has been accounted for using the ECMWF (European Centre for Medium-range Weather Forecasts) meteorological analysis data over the period 1985-1998, assuming the inverted-barometer (IB) effect of the ocean; only 3 days-averaged zonal coefficients, from degree 2 to 8, have been considered. As an example, if we compute the theoretical variations of the gravity field based upon these ECMWF-data, variations of some 10^{-7} can be found in the mean eccentricity of Starlette at periods ranging from 68 to 264 days. These periods are the result of the combination of semi-annual and annual periods with the period of the orbit's perigee.

On the other hand, the solid Earth tide model includes all classical tides using a value of k_2=0.30, and takes into account the major frequency dependencies of the Earth's tidal response, assuming no tidal dissipation (Wahr, 1981). Finally, the adopted static gravity field model is the EGM-96 model, which has been used up to degree 70 (Lemoine et al., 1998). A thermospheric model is used to take into account the air drag effects for Starlette. The nominal model is DTM (Berger et al., 1998), although some tests have been performed using MSIS (Hedin, 1987). For LAGEOS, the effects of thermal forces have been taken into account in the nominal solution (Scharroo et al., 1991; Métris et al., 1997); the motion of the spin axis of the satellite being provided by the model of Farinella et al. (1996). Indeed, these forces have been averaged (analytically) following the requirements given in section 2 for removing short periods.

From this nominal model, low-degree zonal coefficients have been estimated: (i) of degree 2 and 3 for semi-annual and annual periods, (ii) of degree 2 for periods at 18.6 and 9.3 years. These "lumped" coefficients represent in fact the sum of the effects of several geophysical phenomena. At the semi-annual and annual periods, these effects can be due to: (i) oceans (tidal and non tidal parts), (ii) hydrology as reported by Chao and O'Connor (1988), and (iii) a residual atmosphere taking into account the fact that the IB assumption is not valid at all frequencies. At the 18.6- and 9.3-year periods, in addition to the ocean tide, effects can be due to the inelastic tidal response of the Earth, which becomes more distinct with decreasing frequency, especially at the 18.6-year period. Other effects due to atmospheric and hydrospheric mass redistributions can also be added at these long periods.

3.3 Analysis

An analysis of the secular and long-period perturbations of an artificial satellite orbit is based on the following principles, which come from satellite dynamics (Kaula 1966). Schematically, the odd degree zonal harmonics generate long periods especially in the orbital eccentricity and perigee, whereas the even zonal terms generate secular effects in the angular variables. The ascending node, which is modeled more accurately is particularly studied. As a consequence, a secular variation of an odd coefficient will change gradually the amplitude of the long-period effects appearing in the eccentricity without change of their frequency. On the other hand, a given long-period tidal effect of odd degree (annual period, for example) will introduce long-period effects in the eccentricity due to the mixing of both frequencies (satellite perigee and annual tide). Similarly, the secular variation of an even coefficient will introduce a quadratic effect in the node, while a given long-period tidal effect of even degree will introduce its own frequency in this last angular variable. As an example, it is easy to understand that a quadratic effect (e.g., $J_2^{sec.}$) and a long period (e.g., 18.6-yr tide) are orbital signals, which are very correlated in the node on a duration of less than 14 years.

As we show below, the analysis of the mean orbital motion of the geodetic satellites Starlette and LAGEOS permits to derive a very high sensitivity of the effects of geophysical phenomena in the orbital variables. As an exemple of these sensitivities, Table 1 shows the maximum amplitude of periodic signals, which can be observed in the mean eccentricity (units in 10^{-9}) and ascending node (units in mas) for a given annual periodic variation of 10^{-10} in a zonal gravity harmonic J_n for degrees $n = 2$ to 6. For instance, the long periodic signal appearing in the node of Starlette, which would correspond to a 9.3-yr tide of 0.1 cm (around 0.12 10^{-10}, and of degree 2) would have a total amplitude of 86 mas (0.12x9.3x76.1).

Table 1: Sensitivity of LAGEOS and Starlette in the orbital node (in mas) and eccentricity (in 10^{-9}), to a periodic variation of 10^{-10} in the zonal harmonic coefficients of degrees 2 to 6

$\delta J_n = 10^{-10}$		n = 2	n = 3	n = 4	n = 5	n = 6
Nœud asc. (mas)	LAGEOS-1	6.5		2.4		0.5
	Starlette	76.1		3.0		-42.2
Excent. (10^{-9})	LAGEOS-1		9.3		-2.6	
	Starlette		63.8		66.0	

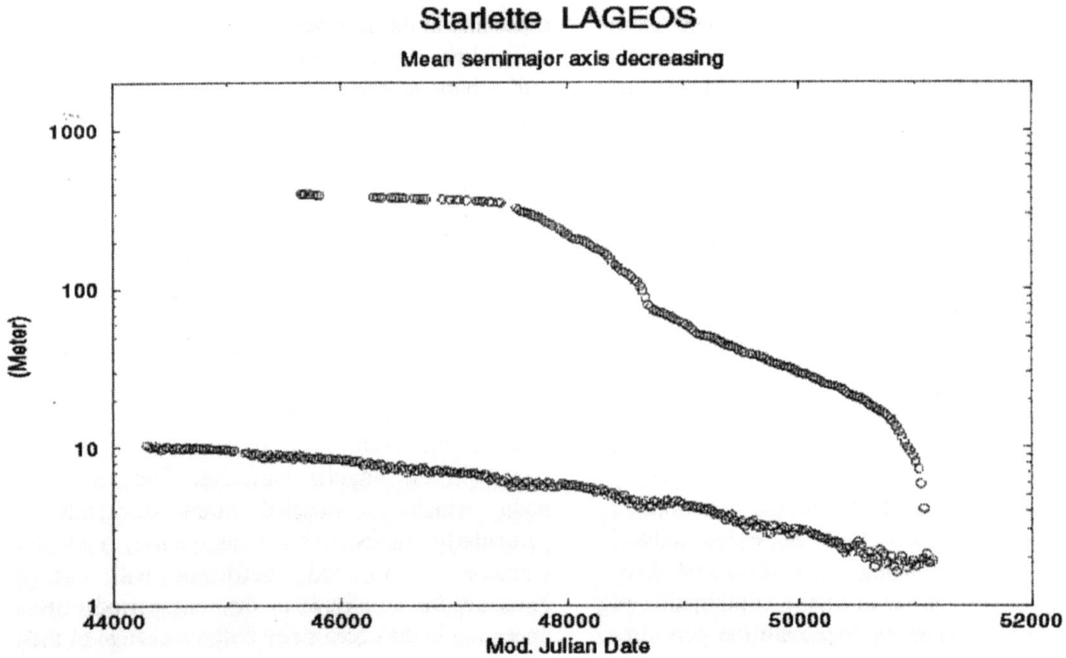

Fig. 1: Decreasing of the mean semi-major axis of Starlette and LAGEOS geodetic satellites from 1980 to 1998

Following these rules, several analyses of long-term SLR tracking data of geodetic satellites have been realized in the past to provide precise determinations of low-degree zonal harmonic variations (see e.g. Exertier 1995, for a review). In this field, the objective of any improvement is notably to assess the validity of the equilibrium assumptions for long-period tides, and to better separate the non-tidal mass redistributions in atmosphere, ocean, and continental water storage from purely tidal effects. The number of geophysical effects, considering their spatial (several wavelengths) and temporal (several periods) coverage, is great however with respect to the relatively small information content available per orbit analysis (the ascending node and eccentricity, essentially). Thus, multi-arc solutions as well as dynamically independent information coming from the available global geophysical data have been used more recently (e.g., Chao and Eanes, 1995; Dong et al., 1996; Cheng et al., 1999). Moreover, the non-gravitational effects acting on LAGEOS have been studied more accurately (Métris et al., 1997; Slabinski 1996), in order to avoid well-known correlations in the eccentricity residual variations with the effects of odd degree zonal variations (e.g., Nerem et al., 1993).

Starlette
The mean motion of Starlette has been integrated over 15 years as a single consistent arc, from 1983 to 1998, or approximately 5500 days. The semi-major axis variations are shown in Figure 1 (logarithmic scale). About 90% of the total decrease (about 270 meters) occurs in three years between 1989-1992 when the solar activity is maximal. In a first step, only the initial mean orbital elements as well as empirical drag coefficients are determined (values range from 0.7 to 1.2, with an average of 1.03). In comparing the integrated mean semi-major axis variations with individual mean semi-major axis computed (observed) every 24 days, we find an rms difference of 26 cm with the DTM model and 32 cm with MSIS model. The mean semi-major axis of Starlette can be adjusted with a std deviation of 5 cm over the two periods 1983-1988 and 1991-1998, but this precision decreases to 40 cm during the period 1988-1991, which corresponds to the maximum of the last solar cycle (period of about 11 years). The main differences between DTM and MSIS happen in seasonal terms regulated by the annual period and can generate pseudo seasonal variations in the mean eccentricity with amplitudes of about $1\text{-}2.10^{-7}$. As a matter of fact, this is the main limit in estimating the semi-annual and annual variations in the odd zonal harmonics through the variations of the mean eccentricity, although empirical drag coefficients have been fitted.

Finally, after fitting the mean orbital motion, taking into account the different known phenomena (see the nominal model), a residual signal can be found with a period of 110 days and an amplitude of 4.10^{-8} in the mean eccentricity and of 3.10^{-6} radians in the mean perigee. This can be interpreted a priori in terms of corrections to the odd zonal harmonics. It will be used in the elaboration of the new gravity field model GRIM5 (Schwintzer et al., 1998). In the mean ascending node variations, a residual signal can be clearly evidenced with a period of 9.3 years. A priori, it could be interpreted in terms of long period tide. However, its determination is questionable because its amplitude and phase appear to be dependent on the considered period, for example 1983-1992 or 1990-1998. Its determination depends also on the quality of the fit of the mean semi-major axis. Indeed, there are some coupling effects in the satellite equations of motion between the node evolution and the variations of the semi-major axis. Therefore, the determination of the 9.3-year signal as a "tide" is difficult and thus poorly estimated.

LAGEOS
The mean orbital motion of LAGEOS has been integrated as a single consistent arc from 1980 to 1998, or approximately 6700 days. The decrease of the mean semi-major axis is shown in Figure 1. As for Starlette, the quality of the fit depends on the considered period and clearly the dispersion of individual mean semi-major axis values is greater at the end (1997-1998). This is due to the fact that there is a coupling between non-gravitational forces and the direction of the spin axis of the satellite (Farinella et al., 1996), and that the rotation of LAGEOS is more and more difficult to predict (the rotation speed has strongly decreased since 1976). Finally after the fitting, the residuals in the mean semi-major axis are of about 2.4 cm rms, and they are of $7.6\ 10^{-8}$ rms in the mean eccentricity, when 1997-1998 are excluded. Regarding the lesser sensitivity of LAGEOS, with respect to Starlette, to variations in the Earth's gravity field, the latter residuals seem to be great compared to those obtained for Starlette. As a result, geodynamical signals coming from odd zonal harmonics can not be precisely extracted from LAGEOS. On the other hand, the node of LAGEOS exhibits again clear signatures at 18.6 and 9.3 years, after having taken into account the nominal effects of tides at these periods.

3.4 Results and discussion

Our analysis is focused on periodic variations ranging from 6 months to 18.6 years. All the known phenomena have been taken into account before the analysis. In particular, the secular variations of J_2 and J_4 have been taken from Bruinsma et al. (1997b), who computed the following values from a combined analysis involving the mean orbital motion of LAGEOS, LAGEOS-2, and Starlette (un-normalized):

$$J_2^{sec.} = -2.7\ 10^{-11}\ /\ yr \pm 0.2$$
$$J_4^{sec.} = -1.1\ 10^{-11}\ /\ yr \pm 0.4$$

Concerning the long period effects, we have determined: (i) the tide at 18.6 yr with an amplitude and a phase of: 1.66 cm ± 0.2, and -15 deg ± 5, from the LAGEOS mean node, and (ii) a "tide" at 9.3 yr with an amplitude and a phase of: 0.3 cm ± 0.2, and -56 deg ± 30, from the LAGEOS and Starlette mean nodes. In both cases, the effects of the inelasticity of the Earth reported e.g. by Zhu et al. (1996) and of the ocean tides (at equilibrium) seem not quite sufficient to totally explain what it has been observed. Atmospheric mass redistribution effects of about 0.1 cm to 0.15 cm could better fit our results. Apart from this last possible effect, values are in good agreement with the ones published by Eanes and Bettadpur (1996) or Cheng et al. (1997).

Finally, annual and semi-annual variations have been found in the lumped degrees 2 and 3. They are reported in Table 2.

Table 2: Annual and semi-annual tidal variations in the lumped zonal gravity coefficients of degree 2 and 3 (un-normalized, amplitudes are in 10^{-11} and phases in degree), deduced from the analysis of the LAGEOS and Starlette mean orbital motions

		Deg. 2 Ampl. / Phas.	Deg. 3 Ampl. / Phas.
Semi-annual	LAGEOS	21. ± 2 /175 ± 5	
	Starlette	22. ± 2 /195 ± 25	6. ± 4 /315 ± 45
	Both	21.5 ± 2 /185 ± 25	
Annual	LAGEOS	10. ± 2 / 65 ± 15	22. ± 10 / 20 ± 20
	Starlette	20. ± 6 / 55 ± 15	9. ± 4 /300 ± 20
	both	15. ± 6 / 60 ± 15	14. ± 10 /340 ± 20

As a result of using atmospheric mass redistribution effects given by ECMWF-data in the nominal model, the semi-annual and annual variations in Table 2 should be originated from the ocean and the effects of hydrospheric mass redistributions. In fact, the semi-annual variations are dominated by ocean tide effects, while annual variations are dominated by hydrological effects.

The effects of atmospheric changes, under assumptions given by the IB effect of the ocean, are certainly not ideally modelled and ECMWF-data could also have possible discontinuities as reported by Chao and Eanes (1995). These slight uncertainties, in addition to those coming from the non-gravitational models have been included in the error budgets (Table 2). The annual value of degree 2 obtained with Starlette and the one of degree 3 obtained with LAGEOS seem to be contaminated by non-gravitational uncertainties, which explains the relatively large amplitudes.

Taking into account these remarks, the values are in good agreement with results found by e.g. Dong et al. (1996) or Cheng et al. (1999).

4 Conclusion

A new analysis of the secular and long-period perturbations arising in a satellite orbit has been performed on very long periods of time, of about 15-20 years, thanks to the concept of mean orbital motion. Single consistent dynamical long arcs of Starlette and LAGEOS have been computed using a semi-analytical approach and SLR tracking observations available since 1980. It is of importance to emphasize that the effects of the gravity field variations, which have to be determined, are directly transformed into effects of changes in the mean orbital elements. This technique has allowed us to better determine and decorrelate the secular, quadratic and long-period orbital signals coming from mass redistributions within the Earth system.

We have provided accurate measurements of the long-period tidal variations at periods from 6 months to 18.6 years. Regarding the effects with periods of 18.6 and 9.3 years, we think that atmospheric mass redistributions could play a role, at the level of 0.1 cm in terms of water height, but have been neglected so far. Due to larger dispersive effects in the estimate of the 9.3-year tide, the 18.6-year tide solution actually provides the most effective constraint on mantle rheology. Non-tidal effects (hydrospheric) influence the determination of the ocean tide coefficients of degree 2 at the annual period, whereas non-tidal signals are smaller and ocean tide clearly less influenced at the semi-annual period. On the other hand, it is still difficult to accurately constrain tidal terms of odd degree with LAGEOS, whereas Starlette permits to give a better information.

Error analysis can be quickly made because the cpu-time is comparatively short and many runs can be carried out to estimate the role of different effects including the mis-modelling of a given force. So, progress can be expected by an improvement of the non-gravitational forces, especially concerning the determination of odd zonal coefficients in the gravity field.

References

Balmino, G., J.P. Barriot, N. Valès (1990) Numerical integration techniques revisited, *J. of Geodesy*, 15, 1-10

Berger, Ch., R. Biancale, M. Ill, and F. Barlier (1998) Improvement of the empirical thermospheric model DTM: DTM94 - a comparative review of various temporal variations and prospects in space geodesy applications, *J. of Geodesy*, 72, 161-178

Bruinsma, S.L., P. Exertier, G. Métris, J. Bardina (1997a) Semi-Analytical Theory of Mean Orbital Motion : A New Tool for computing Ephemerides, *Proceedings of the 12th Intern. Symp. on Space Flight Dynamics,* ESOC, Darmstadt, Germany, 2-6 June, pp. 289-294

Bruinsma, S.L. and P. Exertier (1997b) Geodynamic Parameter Estimation using a Mean Motion Theory, American Geophysical Union - Fall Meeting, Poster Session, G31A-18, San Francisco, Dec 8-12

Cartwright, D.E., A.C. Edden (1973) Corrected tables of tidal harmonics, *Geophys. J. R. Astr. Soc.*, 33, 253-264

Cazenave, A., P. Gegout, G. Ferhat and R. Biancale (1996) Temporal variations of the gravity field from LAGEOS 1 and LAGEOS 2 observations, in "Global Gravity Field and its Temporal Variations", Proceed. of IAG Symp. G3, Vol. 116, pp. 141-151, Springer-Verlag, New York

Chao, B.F. and R.J. Eanes (1995) Global gravitational changes due to atmospheric mass redistribution as observed by the Lageos nodal residual, *J. Geophys. Int.*, 122, 755-764

Chao, B.F. and W.P. O'Connor (1988) Global surface-water-induced seasonal variations in the Earth's rotation and gravity field, *Geophys. J.*, 94, 263-270

Cheng, M.K., B.D. Tapley (1999) Seasonal variations in low degree zonal harmonics of the Earth's gravity field from satellite laser ranging observations, *J. Geophys. Res.*, 104 (B2), 2667-2681

Cheng, M.K., Shum C.K., and B.D. Tapley (1997) Determination of long-term changes in the Earth's gravity field from satellite laser ranging observations, *J. Geophys. Res.*, 102 (B10), 22377-22390

Dong, D., R.S. Gross, J.O. Dickey (1996) Seasonal variations of the Earth's gravitational field: an analysis of atmospheric pressure, ocean tidal, and surface water excitation, *Geophys. Res. Lett.*, 23(7), 725-728

Douglas, B.C., J.G. Marsh, N.E. Mullins (1973) Mean elements of Geos-1 and Geos-2, *Celest. Mechan.*, 7, 195-204

Eanes, R.J. and S.V. Bettadpur (1996) Temporal variability of Earth's gravitational field from satellite laser ranging observations, in *"Global Gravity Field and its Temporal Variations", Proceed. of IAG Symp G3*, Vol. 116, pp. 30-41, Springer-Verlag, New York

Exertier, P., G. Métris, F. Barlier (1996) Mean Orbital Motion of Geodetic Satellites and its Applications, *Proceedings of IAU Colloquium 165, Dynamics and Astrometry of Natural and Artificial Celestial Bodies*, Poznan, Poland, July 1-5, pp. 333-340

Exertier, P. (1995) Temporal Variations of the Geoid, *Quadriennal Report of the CNFGG 1991-1994 / UGGI-1995*, Munschy Sauter and Schlich Eds., pp. 43-48

Exertier P., G. Métris, Y. Boudon, F. Barlier (1994) Long Term Evolution of Mean Orbital Elements of Artificial Satellites,*Geophysical Monograph - IUGG*, 82, Vol. 17, 103-108

Exertier, P. (1990) Precise determination of mean orbital elements from osculating elements by semi-analytical filtering, *Manus. Geod.*, 15, 115-123

Farinella, P. , D. Vokrouhlicky, and F. Barlier (1996) The rotation of lageos and its long-term semimajor axis decay: A self-consistent solution, *J. Geophys. Res.*, 101, 17861-17872

Gegout, P. and A. Cazenave (1993) Temporal variations of the Earth gravity field for 1985-1989 derived from Lageos, *Geophys. J. Int.*, 114, 347-359

Hedin, A.E. (1987) MSIS-86 Thermospheric model, *J. Geophys. Res.*, 92(A5), 4649-4662

Kaula, W.M. (1966) Theory of Satellite Geodesy, Blaisdell Pub. Co., Walthan, New-York

Lemoine, F.G., S.C. Kenyon, J.K. Factor, R.G. Trimmer, N.K. Pavlis, D.S. Chinn, C.M. Cox, S.M. Klosko, S.B. Luthcke, M.H. Torrence, Y.M. Wang, R.G. Williamson, E.C. Pavlis, R.H. Rapp,T.R. Olson (1998a) The Development of the Joint NASA GSFC and the National Imagery and Mapping Agency (NIMA) Geopotential Model EGM96, NASA/TP--1998-206861, GSFC, Greenbelt, Maryland, July 1998

Métris, G., D. Vokroulicky, J.C. Ries, R.J. Eanes (1997) Non gravitational effects and the LAGEOS eccentricity excitations, *J. Geophys. Res.*, 102(B2), 2711-2729

Métris, G. and P. Exertier (1995) Semi-analytical Theory of the Mean Orbital Motion, *Astron. Astrophys.*, 294, 278-286

Métris, G., P. Exertier, Y. Boudon, F. Barlier (1993) Long period variations of the motion of a satellite due to non-resonant tesseral harmonics of a gravity potential, *Cel. Mechanics*, 57, 175-188

Nerem, R.S., B.F. Chao, A.Y. Au, J.C. Chan, S.M. Klosko, N.K. Pavlis, R.G. Williamson (1993) Temporal variations of the Earth's gravitational field from satellite laser ranging to LAGEOS, *Geophys. Res. Lett.*, 20(7), 595-598

Nerem, R.S. and S.M. Klosko (1996) Secular variations of the zonal harmonics and polar motion as geophysical constraints, Global Gravity Field and its Temporal Variations, Int. Assoc. of Geod. Symp., Springer-Verlag, New York, 116, pp. 152-163

Scharroo, R., K.F. Wakker, B.A.C. Ambrosius, R. Noomen (1991) On the along-track acceleration of the LAGEOS satellite, *J. of Geophys. Res.*, 96, 729-740

Schwiderski, E.W. (1980) Ocean tides, Part 1 : Global ocean tidal equationbs, *Marine Geodesy*, 3, 161-207

Schwintzer, P. et al. (1998) A New Global Earth Gravity Field Model from Satellite Orbit Perturbations for Support of Geodetic/Geophysical and Oceanographic Satellite Missions, GFZ Sci. Tech. Rep., STR98/18, 34p.

Schwintzer, P., et al. (1997) Long-wavelength global gravity field models : GRIM4-S4, GRIM4-C4, *J. of Geodesy*, 71, 189-208

Slabinski, V.J. (1996) A numerical solution for LAGEOS thermal thrust: The rapid-spin case, *Celest. Mech.*, 66, 131-179

Tapley, B.D., M.M. Watkins, J.C. Ries, G.W. Davis, R.J. Eanes, S.R. Poole, H.J. Rim, B.E. Schutz and C.K. Shum (1996) The joint Gravity Model 3, *J. Geophys. Res.*, 101(B12), 28029-28049

Tapley, B.D., B.E. Schutz, R.J. Eanes, J.C. Ries, M.M. Watkins (1993) Lageos Laser Ranging Contributions to Geodynamics, Geodesy, and Orbital Dynamics, Geodynamics Series --- Contribution of Space Geodesy to Geodynamics: Earth Dynamics, 24, pp. 147-174

Trupin, A. and J. Wahr (1990) Spectroscopic analysis of global tide gauge sea level data, *Geophys. J. Int.*, 100, 441-453

Wahr, J.M. (1981) Body tides on an elliptical, rotating, elastic and oceanless Earth, Geophys. J. R. Astr. Soc., 64, 677-703

Zhu, Y., C.K. Shum, M.K. Cheng, B.D. Tapley, B.F. Chao (1996) Long-period variations in gravity field caused by mantle anelasticity, *J. Geophys. Res.*, 101(B5), 11243-11248

Geodynamics of S.E. Asia: First results of the Sulawesi 1998 GPS campaign

W.J.F. Simons, D. van Loon, A. Walpersdorf[1] and B.A.C. Ambrosius
Delft Institute for Earth-Oriented Space Research (DEOS), Kluyverweg 1, 2629 HS Delft, The Netherlands

J. Kahar, H.Z. Abidin, D.A. Sarsito
Institut Teknologi Bandung (ITB), Jalan Ganesha 10, 40132 Bandung, Indonesia

C. Vigny
École Normale Supérieure (ENS), Laboratoire de Géologie, 24 rue Lhomond, 75005 Paris, France

S. Haji Abu
Department of Survey and Mapping Malaysia (DSMM), Jalan Semarak, Kuala Lumpur, Malaysia

P. Morgan
University of Canberra (UC), School of Computing, ACT 2600 Canberra, Australia

Abstract. In November 1998, concurrently with the third GPS campaign of the international GEODYSSEA (GEODYnamics of South and South-East Asia) project, a dense geodetic network in Sulawesi, Indonesia was re-measured with GPS. The network is situated in a tectonic complex region near the triple junction of the Eurasian, the Philippine and the Indo-Australian tectonic plates. The GPS campaign included second and third repeat measurements for densification sites of the GEODYSSEA network in this region, and also a re-measurement of a Palu-Koro fault transect in Sulawesi. The acquired data, together with data from IGS stations in the same region, have been processed at DEOS, using the JPL GIPSY-II software, while the DEOS 3D-Motion software was used for the network deformation analysis. The station coordinates were analyzed together with those of the previous Sulawesi campaigns and the GEODYSSEA campaigns of 1994 and 1996. Our kinematic model for the present motions was further validated and improved, and contains both steady state velocity and non-linear displacements. The tectonic block and fault motions are compared with previous results and are geophysically interpreted, and provide detailed new information for this region.

[1]Now at Laboratoire de Meteorologie Dynamique (LMD), Ecole Polytechnique, 91128 Palaiseau Cedex, France.

Keywords. GPS, positioning, geodynamics, Sulawesi, S.E. Asia.

1 Introduction

Over the last decade, the Global Positioning System (GPS) has become an important space-geodetic tool for geodynamic studies. Besides for extensive permanent global and regional GPS networks, GPS is also widely used to measure relative position changes within dense networks, through repeated short-term observation campaigns. The results of these GPS campaigns can also be used to improve the models of the geophysical processes taking place in the regions of interest.

At present, the GPS research activities at DEOS include a study of the geodynamics of South and South-East Asia (participation in the GEODYSSEA project (Wilson et al., 1998)), with an emphasis on the triple junction of the Eurasian, Australian and Phillipine plates located east of Sulawesi, Indonesia. Because the present GPS research activities in this region of South-East Asia would strongly benefit from the addition of GPS data for regions near the triple plate junction, a GPS field campaign was organized in 1997, followed by a repetition in 1998, on the island of Sulawesi in Indonesia. Both campaigns were organized in coopera-

International Association of Geodesy Symposia, Vol. 121
Schwarz (ed.), Geodesy Beyond 2000 – The Challenges of the First Decade

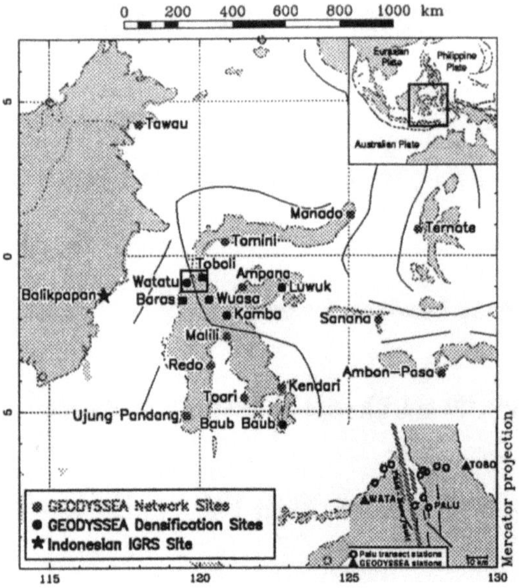

Fig. 1: Observation sites and major tectonic boundaries around Sulawesi

tion with the Institut Teknologi Bandung (ITB), Bandung, Indonesia, thereby receiving additional support (equipment and/or personnel) from DSMM, Kuala Lumpur, Malaysia (1997 and 1998), the Indonesian National Coordination Agency for Surveys and Mapping (BAKOSURTANAL), Cibinong, Indonesia (1997), the École Normale Supérieure (ENS) (1997), Scripps Institute for Oceanography (1997) and the University of Canberra, Canberra, Australia (1997 and 1998). This paper discusses the geodetic analysis of the 1998 GPS campaign data, and gives a preliminary geophysical interpretation of the results, in conjunction with previous campaign results.

2 Region of interest

South-East Asia is a region with complex tectonic kinematics (Rangin et al., 1990), especially in and around the locus of convergence of three major plates (Eurasian, Philippine and Indo-Australian plate), which collide here at velocities up to 10 cm a year. Although the movements of the major plates are relatively well known, the regional motions and deformation in tectonic areas and plate boundaries could not be measured until recent years.

During the GEODYSSEA project (Simons et al., 1999), defined to deduce plate motions and crustal deformations with GPS in South-East Asia, a geodetic network was established which includes 42 observation points and covers an area of about 4000 by 4000 km. Although the selected points provide a good coverage of all the major tectonic blocks in South and South-East Asia, near the triple junction of the 3 tectonic plates, the geodynamic processes are even more complex, and a densified network is required here to study the region in more detail. Therefore additional points were already installed by the École Normale Supérieure (ENS) in the course of the GEODYSSEA project (Fig.), but they are not part of the official GEODYSSEA network, and therefore not measured by the German institutes (the GeoForschungsZentrum Potsdam (GFZ) and the Bundesamt für Kartographie und Geodäsie (BKG, formerly known as IfAG)), who organized and conducted the GPS field campaigns of the GEODYSSEA project in 1994, 1996 and 1998. Also there is an interesting fault system in Sulawesi, namely the Palu-Koro fault, which is responsible for the main part of the tectonic processes in Central Sulawesi, by being a major transformation zone in the region of convergence of the Australian, Philippine and Eurasian plates. Along a transect across this fault, stretching from the city of Palu to Watatu in the west and to Toboli in the east, 13 observation points were installed in 1992 and measured yearly until 1994 by RPI (Rensselaer Polytechnic Institute), SIO and BAKOSURTANAL teams. In 1995 and 1996 the transect was measured by ENS and BAKOSURTANAL, after which it has been measured in 1997 and 1998, along with the densified GEODYSSEA network during the Sulawesi GPS campaigns. In 1997, also the main GEODYSSEA points in and around Sulawesi were observed as part of the Sulawesi campaign.

3 GPS field campaign

The 1998 Sulawesi GPS campaign took place from 19 to 28 November 1998. The field campaign was split into two parts, a 5 day and a 4 day period which clearly can be distinguished in the campaign observation summary shown in Fig. (2). The first part was carried out concurrently with the GFZ/BKG GEODYSSEA-98 GPS campaign. Eight teams were sent into the field, each equipped with one DEOS/UC Trimble 4000SSi/SSE GPS receiver which was operated by 2 surveyors provided by ITB, DEOS and UC. In both parts of the campaign, the field

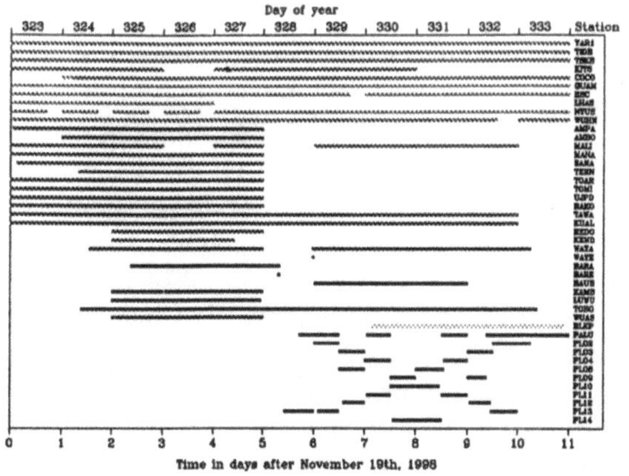

Fig. 2: Sulawesi 1998 GPS campaign observation summary and additional GEODYSSEA-98 and IGS GPS data

teams daily collected 24 hour data at in total 7 sites of the GEODYSSEA densification network (WUAS, TOBO, LUWU, KAMB, BAUB, BARA, and WATA), 3 sites of the main GEODYSSEA network (MALI (only in the second part), REDO, and KEND) and 1 IGRS site (BLKP) as shown in Fig. (2). All the GEODYSSEA points are monumented (mostly in bedrock) with specially designed markers, yielding a re-centering accuracy of 0.2 mm (Reinking et al., 1995). The 11 Palu-Koro transect points (PALU and PL02 to PL14, all with pin type markers), in between Watatu and Toboli, were each occupied at least twice for 12 hours in the second part of the campaign, while the end points at Watatu and Toboli were observed continuously throughout the entire campaign, see Fig. (2). In total 73 observation days were collected at 21 sites.

The GEODYSSEA-98 measurements at the two main GEODYSSEA sites in Malaysia (KUAL and TAWA) were extended by DSMM so that data from these sites was also available for the entire duration of the 1998 Sulawesi GPS campaign. Figure (2) also shows the observations taken during the GFZ/BKG GEODYSSEA-98 GPS campaign for 12 of the main GEODYSSEA sites (AMPA, AMBO, MALI, MANA, SANA, TERN, TOAR, TOMI, UJPD, BAKO, TAWA, and KUAL) in and around Sulawesi. This data was generously provided by BKG/GFZ to DEOS on an exchange basis for GEODYSSEA data which was collected during the 1997 Sulawesi GPS campaign. With the addition of this data, a geodetic network identical to the one of the Sulawesi 1997 GPS campaign could be analyzed. To facilitate the positioning of the network in a global reference frame, data from 11 stations (YAR1, TIDB, TSKB, KIT3, COCO, KARR, GUAM, IISC, LHAS, NTUS, and WUHN in Fig. (2)) of the International GPS Service for Geodynamics (IGS) tracking network (Beutler et al., 1994) in the region was included.

4 GPS data analysis

In this section the results of the DEOS GPS data analysis with the GIPSY-OASIS II v2.5 (Blewittl et al., 1988) and the 3D-Motion software (Noomen et al., 1993) software are discussed. Other groups (ENS, ITB, and UC) are also computing solutions with (part) of the available Sulawesi 1998 campaign data using their favored software and analysis strategy, and when these become available, all results will be evaluated and combined with into a final solution.

For the complete data set of 1998 (and 1997), daily fiducial-free network solutions were computed with the GIPSY precise point positioning (PPP) strategy. With this strategy, each station can be solved for individually using precise (JPL) satellite orbits, clocks and corresponding earth rotation parameters. Because the orbits and clocks are held fixed in this technique, all correlations among the different stations are removed. Therefore a network can be processed much more computationally efficient, as opposed to the traditional technique where the complete network is processed simultaneously.

The individual PPP station coordinate solutions were computed using the ionosphere-free linear combination of GPS phase and pseudorange data at 5 minute intervals with an elevation cut-off angle of 15 degrees. To account for tropospheric effects, the zenith path delay was estimated every 5 minutes for each station. Because different antennae (Trimble 4000ST L1/L2 GEOD, Trimble GEOD L1/L2 GP Compact and Dorne Margolin T) were used in the network, the IGS antenna phase center corrections (Rothacher et al., 1996b) were also applied.

In a final step, the solutions of each station can be simply merged into one covariance matrix for the complete network. It is then possible to directly map the daily PPP network coordinate solution into (at present) the Terrestrial Reference Frame solution of 1996 (ITRF-96), with Helmert transformation parameters from JPL

Table 1: Daily coordinate repeatability Sulawesi campaigns

Stations	Station Solutions	RMS Residuals (mm) North	East	Up
Sulawesi 1998				
Sulawesi	102 (-15)	1.5	3.7	7.2
Transect	29 (- 2)	3.3	11.3	11.1
IGS	99 (- 7)	1.5	3.6	7.3
Total	230 (-24)	2.1	5.7	8.6
Sulawesi 1997				
Sulawesi	132 (-21)	1.8	5.4	7.7
Transect	22 (- 1)	4.5	15.6	20.1
IGS	100 (-13)	1.7	4.3	8.3
Total	254 (-35)	2.2	6.6	9.7

which are compatible with the JPL precise orbits and clocks. Also other techniques can be used to map the results in the ITRF, of which one will be explained further down in this paper.

4.1 Multi-day averaged campaign coordinate solution

The daily PPP coordinate solutions were combined into a multi-day averaged campaign coordinate solution with the 3D-Motion software, which eliminated any systematic differences between the various daily network solutions by computing optimized 7-parameter Helmert transformations using a least squares adjustment. In this process, station solutions identified as outliers with respect to the averaged solution were removed.

The daily station coordinate repeatabilities, together with the number of final and removed station coordinate solutions, are given in Table (1) for both Sulawesi campaigns. Besides the Root-Mean-Square (RMS) of the daily station coordinate residuals with respect to the multi-day averaged campaign solution for the complete network, the RMS statistics are also shown for only the Sulawesi, transect and IGS stations. The repeatabilities for the Sulawesi and the IGS stations, ranging in both campaigns from 1.5 to 5 mm for the horizontal components and averaging at about 8 mm in the height, show that the internal precision of the Sulawesi network, with exception of the Palu-Koro transect points, is very high.

The transect points have significantly poorer coordinate repeatabilities in 1998 and 1997, with 3 to 11 mm and 5 to 16 mm respectively for the horizontal position and 11 mm respectively 20 mm for the vertical position. This indicates that while the GIPSY PPP strategy gives very good results for 24 hour session data, it does not perform as well for site observation times of 6 (in 1997) and 12 (in 1998) hours. However, also the traditional strategy, where the whole network is processed at once, did not give significantly better results for the transect points. Because the repeatabilities improve as the observation time becomes larger, this suggests that there might be some effects that are averaged in the processing of 24 hour session data, but are not or not sufficiently accurately modeled for the processing of shorter sessions. Also most outlying station coordinate solutions, see Table (1), were those computed with less than 24 hour data available. Therefore the data for the transect points needs to be further analyzed in a different way, on a session per session basis, thereby probably also taking into account the effects of ocean loading. The latter because the transect is located relatively close to the coast. These additional computations still need to be completed, and therefore this paper will only present and discuss the results for the two end points of the Palu-Koro fault transect (Watatu and Toboli, both with daily observation times of 24 hour in 1997 and 1998).

4.2 Mapping solutions in ITRF-97 and displacement rates estimation

The Sulawesi 1997 and 1998 campaign coordinate solutions were both mapped in the latest ITRF solution of 1997 (ITRF-97) (Boucher et al., 1999), using a regional IGS site approach. The potential of this technique, when applied to local network solutions in S.E. Asia, has significantly increased with the arrival of ITRF-97, as this reference frame now also includes accurate station coordinates and velocity estimates of relatively new IGS sites in the region (e.g., COCO, GUAM, IISC, KARR, LHAS, NTUS, WUHN, and XIAN. The Sulawesi campaign solutions each include the coordinates of 11, of which 10 identical, IGS sites. The IGS station at Singapore (NTUS) was not used in the final network transformations, because the solved station coordinates are still not reliable enough due to unresolved problems with the amount and quality of the GPS data collected by this site. Also, the daily data stream from NTUS was frequently interrupted during the Sulawesi 1998 GPS campaign, see Fig. (2).

Table 2: Residuals ITRF-97 transformation of used IGS sites

	1998 Campaign			1997 Campaign		
	Coordinate Residuals (mm)					
IGS Station	North	East	Up	North	East	Up
Kitab	1.4	-3.3	-8.0	-1.7	-4.5	-1.0
Tidbinballa	-0.3	1.5	-2.2	5.4	3.0	-2.4
Tsukuba	2.0	-0.8	3.7	-0.9	0.2	-0.5
Yarragadee	-0.5	-3.4	-7.0	-1.2	0.7	-4.7
Cocos Island	3.4	3.1	11.4	2.4	0.4	6.0
Guam	-1.2	4.1	9.7	-1.6	7.3	-2.1
Bangelore	0.4	-9.0	-11.2	2.7	-7.1	7.1
Lhasa	-4.1	0.1	-12.9	-1.7	0.5	-2.4
Wuhan	-2.0	0.7	6.2	-1.7	-0.8	4.6
RMS	2.1	3.8	8.7	2.5	3.9	4.0

The mapping in ITRF-97 was done, by projecting each campaign solution with 3D-Motion onto the ITRF-97 coordinate set, that contains the positions of the same 9 IGS stations at the time of each campaign. The coordinate residuals of the IGS sites for the 1997 and 1998 epoch transformations are shown in Table (2).

The coordinate residuals of all IGS stations are consistent and, with an RMS value of only 2 to 4 mm for the horizontal and 4 to 9 mm for the vertical position, small for both campaign solutions. The vertical position residuals are higher in the 1998 solution, which may be due to the propagation in time of inaccuracies in the ITRF-97 vertical velocity estimates (given at epoch 01/01/1997) for some of the new IGS sites and/or more noise in the 1998 IGS data.

The results of Table (2) together with the daily coordinate repeatabilities of the Sulawesi sites, indicates that the global accuracy with respect to ITRF-97 of both campaign solutions is better than 1 cm for the horizontal position, and around 1 cm for the height. This means that, with only a time interval of 13 months, an accurate estimate of the (relatively high) horizontal site motions in and around Sulawesi could be obtained from only 2 campaigns. This is of great importance as frequent earthquake activity regularly results in co-seismic motions at some sites. These can be more easily detected and quantified if the time interval between subsequent observations can become smaller. Of course also permanent GPS sites play a key role here, but at present it is not feasible to install a large number of them in Sulawesi. Finally the displacement rates were computed by subtracting the ITRF-97 mapped 1997 and 1998 campaign solutions. They are shown in Fig. (3), together with their formal (1-σ) uncertainties of typically

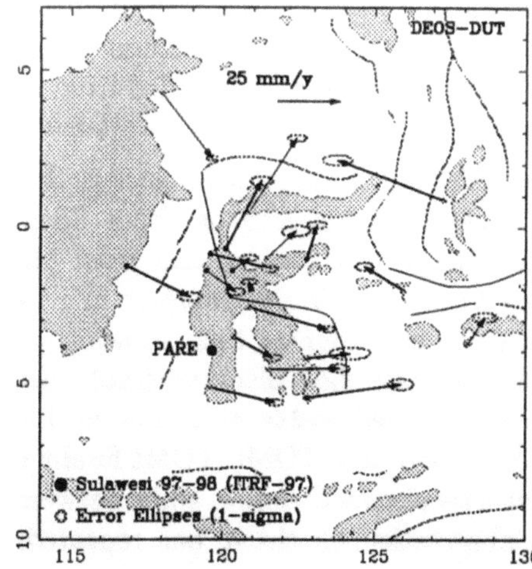

Fig. 3: Sulawesi 97-98 horizontal station velocities in ITRF-97

2 to 5 mm/yr in the horizontal direction.

5 Results and discussion

To get a better view on the site motions in Sulawesi, the motion of the Eurasian Sundaland block was subtracted from the velocity field. This was done, by first estimating the Sundaland motion with the 4 available sites (UJPD, BAKO, KUAL, and BLKP) which are located on this block. The velocity estimates in this new reference frame are given in Fig. (4), where also the previous results from the combination of the 1997 Sulawesi and the 1996 main GFZ/BKG and ENS densification GEODYSSEA campaigns are included.

The velocity field estimates, see Fig. (4) in the southwest of Sulawesi indicate that this area moves as part of the Eurasian Sundaland block, with some compression taking place in the Macassar Strait. The first repetition measurements now available for Balikpapan (BLKP) and Baras (BARA) seem to confirm this, although the observed closure rate of 15 mm/yr might be too high. Inclusion of the data from the permanent station at Pare Pare (PARE in Fig. 3) operated by BAKOSURTANAL will provide more information. The motion of the sites in the southeast of Sulawesi and on Ambon, all located on micro-blocks, seem to be influenced by the Australian plate northward motion. The double subduction zone in the Molucca sea, resulting in a previously observed closure rate of about 7.5 cm/yr (measured between MANA and

TERN), can not yet be verified with the 97-98 results, since the MANA measurement took place on a still unidentified secondary marker and therefore MANA was excluded from the velocity field.

The existence of a rigid Sula block in Sulawesi with a clockwise rotation of about 3deg/Myr and a rotation pole located east from Manado was confirmed by (Walpersdorf et al., 1998a). Our results also confirm this, showing a good fit between their rigid block modeled velocities and the observed stable velocities at LUWU, AMPA, TOBO, and TOMI. TOMI finally seems to have a secular motion now, while between previous observation epochs it was repeatedly affected by earthquake events. The increased westward velocity components at WUAS, KAMB, and BARA require further investigation. If correct, the result at BARA, together with the relative motions of REDO, MALI, TOAR, KEND, BAUB would suggest that the southern part of Sulawesi is also rotating counterclockwise around Ujung Pandang (UJPD). However, an explanation for the observed differences could also be that one of the 2 transformations into the Sundaland reference frame was not optimally computed. This is plausible as only 4 sites representing this block are available in both campaign solutions. The relative velocity between Tomini (TOMI) and AMPA (Ampana) indicates an opening of the Gulf of Tomini, which suggests an N-S extension in the Sula Block as it is pulled towards the subduction zone in the North Sulawesi Trench.

The first velocity estimate for Baras, together with the result from Wuasa (WUAS) provide a new Palu-Koro transect measurement about 50 km south of the existing fault transect in Palu. The observed motion along the fault is smaller here, which might be explained by a higher degree of deformation in the southwest of the Sula block (at WUAS and TOBO) and/or co-seismic motion in the upper transect due to the magnitude 6.6 earthquake that occurred on the Palu-Koro fault north of Palu shortly before the 1998 campaign. Both transect results are being analyzed further. Finally, Fig. (4) confirms a motion for the Malaysian site at Tawau (TAWA), which seems to be influenced by intra-plate deformation in the east of Borneo. Data from the permanent Malaysian GPS network, which is now operational for more than one year, would certainly contribute to a better understanding of

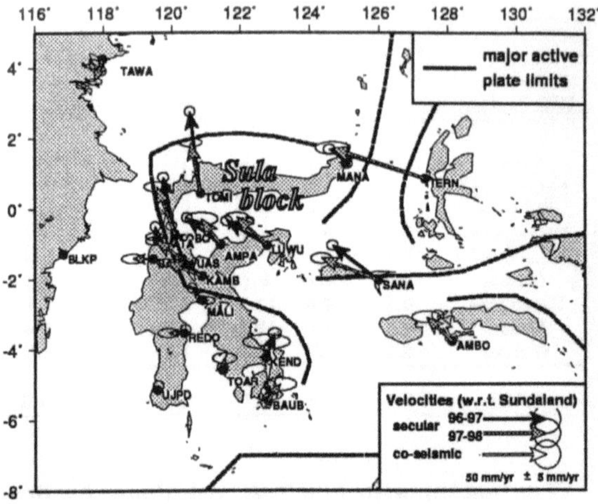

Fig. 4: Horizontal velocities 96-97 and 97-98 in Sulawesi w.r.t Sundaland

the underlying geodynamic processes.

Finally the latest result in the time evolution of the WATA-TOBO baseline over the Palu-Koro fault was compared with (Walpersdorf et al., 1998b). The observed (left lateral) strike-slip rate (3.3 cm/yr) has increased over the last year, probably due to the October 1998 earthquake (M 6.6) on the fault. This was also observed after the January 1996 earthquake (M 7.9) 80 km up north in the Minahassa trench (Walpersdorf et al., 1998b). As the second earthquake can be interpreted as a response to the increase in strain generated by the first one, this clearly shows strain transfer from the Minahassa trench to the Palu Fault. This is being modeled by ENS, and preliminary results indicate that the Minahassa earthquake may have forced a small opening on the fault, followed by a fluid migration which filled up the created gap, thereby finally forcing the fault lips to open and cause the recent earthquake on the fault. The interpretation of the velocities of the intermediate points is beyond the scope of this paper, and will be published elsewhere. If the results for the intermediate transect points can confirm this theory, it might become possible to use (permanent) fault monitoring for the detection of earthquake hazards in this region.

Future Sulawesi GPS campaigns already have been scheduled in 1999 and 2000, and DEOS, ENS, ITB and BAKOSURTANAL also intend to have a total of 4 permanent GPS stations in West-Sulawesi to monitor the Palu-Koro fault.

Acknowledgments. This work is a continuation of the DEOS and ENS research activities in S.E. Asia, and was initiated by their participation in the GFZ/BKG GEODYSSEA project. Thanks and appreciation are extended to GFZ and BKG, for making the Sulawesi part of the GEODYSSEA-98 data available, and to DSMM for extending the 1998 measurements at the Malaysian GEODYSSEA sites. Special recognition is given to all other people at DEOS, ITB and UC who contributed to and enabled the organization and execution of the 1998 Sulawesi GPS field campaign.

References

Beutler, G., I. I. Mueller, and R. E. Neilan (1994), The International GPS Service for Geodynamics (IGS): Development and start of official service on January 1, 1994, *Bull. Géod.*, *68*, 39–70.

Blewitt, G., *et al.* (1988), GPS geodesy with centimeter accuracy, in *Lecture Notes in Earth Sciences*, edited by E. Groten and R. Strauss, Springer-Verlag, New York.

Boucher, C., Z. Altamimi, and P. Sillard (1999), The 1997 international terrestrial reference frame ITRF97, *IERS technical note No. 27*, Observatoire de Paris, France.

Noomen, R., B. A. C. Ambrosius, and K. F. Wakker (1993), Crustal motions in the Mediterranean region determined from laser ranging to LAGEOS, *Contributions of Space Geodesy to Geodynamics*, *23*, 331–346.

Rangin, C., L. Jolivet, and M. Pubellier (1990), A simple model for the tectonic evolution of South-East Asia and the Indonesian region for the past 43 my., *Bull. Soc. Geol. Fr.*, *6*, 889–905.

Reinking, J., D. Angermann, and J. Klotz (1995), *Zur Anlage und Beobachtung grossraeumiger GPS-Netze fuer geodynamische Untersuchungen. Allgemeine Vermessungsnachrichten*, H. Wichmann Verlag.

Rothacher, M., and G. Mader (1996), *Combination of Antenna Phase Center Offsets and Variations: Antenna Calibration set IGS_01*, IGS Central Bureau / University of Berne, Switzerland.

Simons, W. J. F., B. A. C. Ambrosius, R. Noomen, D. Angermann, P. Wilson, M. Becker, E. Reinhart, A. Walpersdorf, and C. Vigny (1999), Observing plate motions in s.e. asia: Geodetic results of the GEODYSSEA project, *Geophys. Res. Lett.*, *26*(14), 2081–2084.

Walpersdorf, A., C. Vigny, P. Manurung, C. Subarya, and S. Sutisna (1998*a*), Determining the Sula block kinematics in the triple junction area in Indonesia by GPS, *Geophys. J. Int.*, *135*, 351–361.

Walpersdorf, A., C. Vigny, C. Subarya, and P. Manurung (1998*b*), Monitoring of the Palu-Koro fault by GPS, *Geophys. Res. Lett.*, *25*(13), 2313–2316.

Wilson, P., J. Rais, C. Reigber, E. Reinhart, B. A. C. Ambrosius, X. Le Pichon, M. Kasser, P. Suharto, A. Majid, P. Awang, R. Almeda, and C. Boonphakdee (1998), Study provides data on active plate tectonics in Southeast Asia region, *Eos Trans. AGU*, *79*(45), 545–549.

Four-dimensional geodesy: time dependent inversion for earthquake and volcanic sources

Paul Segall
Geophysics Department, Stanford University, Stanford CA, 94305

Abstract. The past five years has witnessed a tremendous expansion in the number of permanent Global Positioning System (GPS) receivers. There are now GPS networks in Japan, California, and Hawaii that provide crustal deformation data that are dense in both space and time. These data, combined with tilt and bore hole strain measurements, can be used to invert for spatial and temporal variations in fault slip and magma chamber dilation. Previously, space-time inversions had been hampered by poor signal to noise in the data, contaminating non-tectonic motions near the instrument (benchmark wobble), and our lack of knowledge of the temporal character of aseismic motions. The recently introduced Network Inversion Filter (Segall and Matthews., 1997) yields estimates of quasi-static fault slip as a function of space and time using data from dense continuous geodetic networks. The NIF employs time domain, Kalman filtering, and allows for non-parametric descriptions of slip velocity, local benchmark motion, and measurement error. A state-space model for the full geodetic network is adopted, so that all data from a given epoch are analyzed together. This allows the filter to distinguish between non-steady fault slip and local surficial effects like benchmark wobble.

We have used the Network Inversion Filter to analyze the time-dependence of post-seismic slip following the 1989 Loma Prieta earthquake, (Segall and others, 1999) to image rift-zone deformation opening and aseismic fault slip following the January 30, 1997 eruption at Kilauea volcano, Hawaii, and to image a propagating dike off the Izu Peninsula in Japan (Aoki and others, 1999). In situations where the geometry of the causative geologic structures is unknown it is advantageous to use filtering techniques to image the strain-rate fields in space and time.

Keywords. Global Positioning System, network inversion filter, seismic slips.

References

Aoki, Y., Paul Segall, Teruyuki Kato, Peter Cervelli, Seiichi Shimada, Imaging magma transport from he 1997 seismic swarm off the Izu Peninsula, Japan, *Science*, in press, 1999.

Segall, P., R. Bürgmann, and M. Matthews, Time-dependent triggered afterslip following the 1989 Loma-Prieta earthquake, *Journal of Geophysical Research*, in press, 1999.

Segall, P. and M. Matthews, Time dependent inversion of geodetic data, *Journal of Geophysical Research.*, v. 102, p. 22,391-409, 1997.

International Association of Geodesy Symposia, Vol. 121
Schwarz (ed.), Geodesy Beyond 2000 – The Challenges of the First Decade

An interdisciplinary approach to studying seismic hazard throughout Greece

P.R. Cruddace, P.A. Cross
Department of Geomatic Engineering, University College London, London, WC1E 6BT, U.K.

G. Veis, H. Billiris, D. Paradissis, J. Galanis
Dionysos Satellite Observatory,
National Technical University of Athens, 9 Heroon Polytechnou Str, Zographos, GR-15780 Athens, Greece.

H. Lyon-Caen, P. Briole
Départment de Sismologie,
Institut de Physique du Globe de Paris, 4 Place Jussieu, F-75252 Paris Cedex 05, France.

B.A.C. Ambrosius, W.J.F. Simons, E. Roegies
Delft Institute for Earth-Oriented Space Research,
Delft University of Technology, Kluyverweg 1, PO Box 5058, 2600 GB Delft, The Netherlands.

B. Parsons, P. England
Department of Earth Sciences, Oxford University, Parks Road, Oxford, OX1 3PR, U.K.

H.-G. Kahle, M. Cocard, P. Yannick
Geodesy and Geodynamics Laboratory,
Institute of Geodesy and Photogrammetry, ETH Hönggerberg, CH-8093 Zürich, Switzerland.

G. Stavrakakis
Institute of Geodynamics,
National Observatory of Athens, Lofos Nymfon, PO Box 20048, GR-11810 Athens, Greece.

P. Clarke
Department of Geomatics, University of Newcastle, Newcastle upon Tyne, NE1 7RU, U.K.

M. Lilje
National Land Survey, 80182 Gavle, Sweden.

Abstract. This paper describes and reviews the progress of a three year (11/97-11/00) European Commission FP4 (Climate and Natural Hazards) funded project entitled GPS Seismic Hazard in Greece (SING), A major international interdisciplinary consortium is investigating and comparing strain derived using both geodetic and seismic methods.

The specific objectives of SING are to assess strain accumulation throughout Greece, to identify areas of high seismic hazard, to develop new and more efficient operational and computational methods for GPS, and to improve our understanding of the relationships between geodetic strain, seismic catalogues and geological data. New GPS networks have been installed in regions of significant hazard and initial computations have been carried out. To date, a primary result of SING is the integration of 33 historical geodetic data sets to provide a national strain map, giving the first full picture of geodetic strain in Greece and providing the basis for the setting up of the new geodetic networks.

This paper presents an overview of the project's goals, the methodologies employed and initial results.

Keywords. GPS, seismology, geodynamics, Greece, plate tectonics, geology, seismic hazard, strain accumulation.

1 Introduction

Greece lies within a region of intense intra-plate deformation. Three major plates (African, Eurasian and Anatolian) surround the region and induce both a highly complex and varied series of geophysical phenomena, described in figure 1. There are on

International Association of Geodesy Symposia, Vol. 121
Schwarz (ed.), Geodesy Beyond 2000 – The Challenges of the First Decade

three year GPS Seismic Hazard average eight >M5 earthquakes annually in Greece Ambraseys and

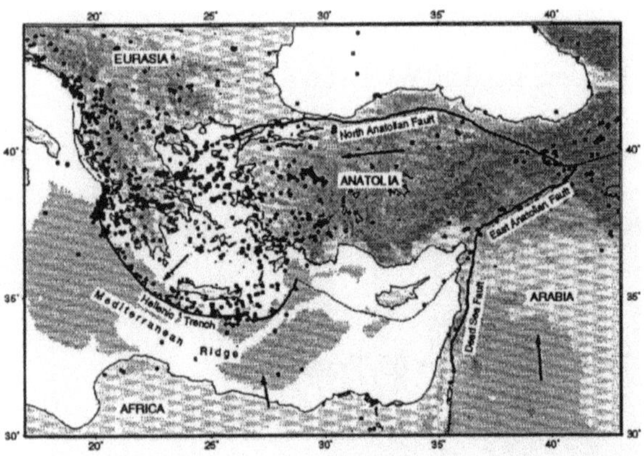

Fig. 1: The tectonics of the Eastern Mediterranean region. Earthquakes shown are $M_b > 4.5$ (<40km depth) between 1964 and 1993 from Clarke et al. (1998)

Jackson (1990), making the area both a natural scientific laboratory and region of extreme seismic hazard. The aim of the in Greece (SING) project is to identify areas of high seismic hazard in this highly tectonically active region. This is achieved by utilising a multidisciplinary approach whereby ground strain accumulation from precision Global Positioning System (GPS) positions is compared to geological, geophysical and seismic activity records. The region studied through SING extends from longitude East 20.5° to 28° and from latitude North 35° to 41.5°.

There have been a number of geodetic measurement campaigns in the region to study this phenomena in the last ten years, including, for example, Billiris et al. (1991), Kahle et al. (1996), and Clarke et al. (1998). This project aims to build on the successes of the previous projects through the combination of results, the extending and combining of geodetic networks and the identification and study of specific areas of seismic interest.

The overall goal from this GPS network is to obtain an integrated 'kinematic model' that will accurately describe (spatially and quantitatively) the geodetic strain distribution throughout Greece. This will then be integrated with the seismic catalogue and other geologic data to serve as a basic tool for seismic hazard assessment.

The project utilises a novel approach to data collection whereby a minimal number of GPS receivers occupy a large number of stations in an accurate and cost effective manner. Geodetic conclusions are expected from the approach with respect to the optimum observation strategies to obtain the necessary accuracy.

The task of seismic hazard assessment within zones of intra-plate deformation is entirely different in character and more complicated from that at plate boundaries where the main task is to monitor usually well defined fault(s) and to accumulate data that relate to the risk of future earthquakes on that fault(s). Where the deformation is spread out over a wider area, as it is in Greece, the primary task is to determine where, within the area, the tectonic strain accumulates most rapidly. Only then can the task of assessing hazards associated with slip on individual faults within the region begin.

One of the primary initial objectives of SING is to search and integrate all available strain data. To achieve this, a data-base of previously obtained geodetic data is being developed. Coordinates, velocities and strain rates determined by the groups involved in this proposal, and abstracted from publications of other groups will be integrated. Existing seismic hazards maps of Greece, computed by utilising different methodologies, are to be collected in order to delineate areas of high seismic hazard based on pure seismological data.

The work carried out in this project is of significant importance for more reliable seismic hazard assessment, for land use planning, GPS data capture and processing methods and to help towards earthquake surveillance. The application of the techniques used and developed in this project is of extreme importance in other areas of the world, especially in areas where money is at a premium.

To summarise, the four primary objectives of SING are:

- To identify areas of high seismic hazard in Greece.
- To assess strain accumulation throughout Greece, and in specifically targeted areas.
- To develop new and more efficient operational and computational procedures for the use of GPS in the delivery of high quality positional data within regional and global control networks.
- To improve the understanding of the relationship between strain accumulation and seismic hazard by integrating geodetic derived strain information with existing seismic catalogues and other geological data.

2 Methodology

The approach undertaken uses GPS geodetic techniques to provide the spatial and temporal accumulation of strain across Greece and compares this to the regional seismic and geological data to ameliorate understanding of the mechanisms of earthquake hazards.

The target is to observe 300 stations annually using geodetic quality GPS receivers. The siting and repeat occupation intervals of stations follows a three-tiered approach; small station spacing in areas of identified high seismic risk, large equidistant spacing country-wide (to fill in gaps from previous networks) and a number of permanently or semi permanently occupied sites. A forced antenna centring mechanism has been developed for the project, allowing the antenna to be screwed directly on to the station mark. GPS data from both permanent International GPS Service (IGS) sites and semi-permanent stations occupied for relatively short time periods (three days plus) is also collected. This allows the stations in Greece to be linked to the International Terrestrial Reference Frame (ITRF) as well as producing an observation strategy whereby local stations can be observed with respect to 'local' permanent (during the observations) stations, reducing baseline lengths and observation periods. New GPS networks have been installed in regions of significant hazard,

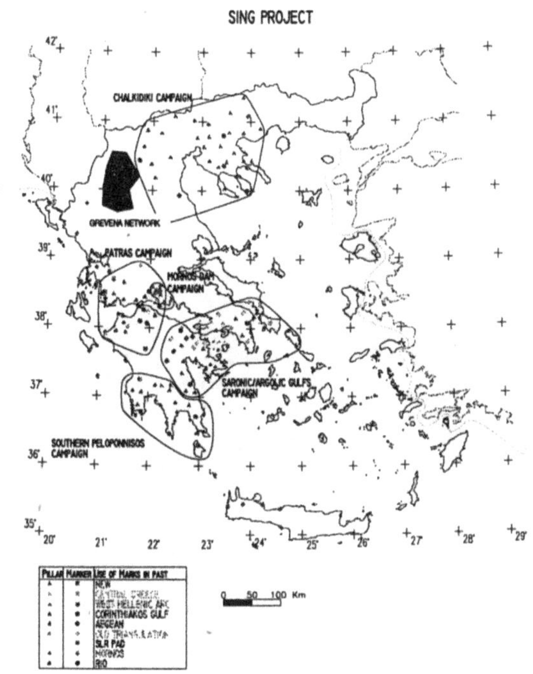

Fig. 2: Observed stations in the SING project during 1998

figure 2; Chalkidiki, Patras, Saronic/Argolic Gulfs and Southern Peloponnisos. Further measurements were made during 1998 at the existing networks of Grevena, Mornos Dam and the Western Hellenic Arc.

The efficient monitoring of the SING network requires the consideration of alternative strategies to the more commonly adopted permanent or campaign style of GPS data collection. As part of this work, the concept of a quasi-continuous approach has been developed. This approach is one in which small numbers of receivers are moved from station to station as and when they (and the logistical support) are available. In this way large numbers of stations can be monitored in a flexible way without the need for either a dense array of permanently located receivers or campaigns involving large groups of people and the associated complex logistics. The receivers can be quickly mobilised to regions of seismic interest if necessary. The practical implementation of a quasi-continuous procedure involves the establishment of temporary base stations whilst other receivers are moved in the local regions. The efficiency of the system is then a function of the time spent on each station and the distance between the stations, and one goal of the project is to investigate these issues. The particular observation system that is used consists of two permanent observing receivers at ITRF linked sites within Greece, a limited number of semi-permanent stations (observing for a few days) and a small number of mobile GPS receivers moving from point to point in a quasi-continuous manner. The kinematic GPS observing technique will be used to measure road profiles in the Korinthos and Evia regions.

Existing seismic hazards maps of Greece, computed using different methodologies, have been collected in order to delineate areas of high seismic hazard based on pure seismological data. These maps can be categorised into two groups depending upon the approaches used to compute them. The first includes seismic hazard maps computed on the basis of seismological data without considering geological or other geophysical data. The second category includes hazard maps computed on the basis not only of pure seismological data but also of seismotectonic data and other geophysical information. An estimation of the rates of crustal deformation in some specific seismotectonic zones of Greece, from earthquake moment tensor mechanisms, has been made. Owing to different fault planes solutions for the same earthquakes being available, the rates of crustal deformation cannot be uniquely determined. However, focal parameters are considered as being highly reliable, in the cases where they are obtained by computing synthetic seismograms. The analysis was made using Kostrov's formulation according to the hypothesis that the average strain rate of the seismic deformation in a region is a function of the sum of symmetric moment tensors, the seismogenic volume of the deforming zone, the rigidity and of the time considered in the analysis. Emphasis has been given on estimating seismic hazard based on the physical parameters of the seismogenic fault, such as:

- fault area
- slip rate
- seismic moment.

All these parameters were considered for computing the prior estimates of the seismicity together with the mean rate of earthquake occurrence.

3 GPS processing strategy

GPS data processing has been carried out with the Bernese, Rothacher and Mervart (1996), and GIPSY-OASIS II (GIPSY), Webb and Zumberge (1997) scientific software packages. IGS raw data, precise orbit and earth orientation products have been used in the processing along with NOAA antenna phase centre information. Investigations into baseline repeatability's have been carried out to aid the scaling of the formal errors output by the GPS processing packages allowing more realistic estimates of attainable precision to be calculated.

GPS processing methods follow two strategies:

1) Calculate an average position for the permanent station CG54 (in the ITRF) over the period of observations of the local region being measured. This is calculated either through the Precise Point Positioning (PPP) technique using GIPSY, Zumberge et al. (1997), or through baselines to a sub-set of 'A' class IGS sites using Bernese. The coordinate is transformed to ITRF 96 at the mid point of the observations and the weighted average coordinate taken. The PPP strategy solves for each station individually using a minimum of 24 hours of raw GPS data, fixed precise orbits, satellite clock correction files, satellite shadow event files and precise earth orientation parameters.

The two methods used different processing software and philosophies (only similarities being the orbits, earth rotation parameters and antenna phase centre offsets used) and were compared. Eight daily solutions were combined for the two approaches and gave very encouraging solution differences of:

East (m)	North (m)	Vertical (m)
-0.0011	0.0082	0.0006

CG54 was then fixed and the semi-permanent site in the region of observations is calculated from all days of data in a network approach and the weighted average value taken.

2) A daily network solution containing the fixed local semi-permanent station and the local stations is calculated. The parameterisation and modelling of the atmosphere and GPS satellite and receiver locations, the motion of the earth and the receiver and satellite clocks is key to the highest quality processing. For example, with GIPSY;

Tropospheric modelling uses an elevation dependent model for the hydrostatic component and a random walk function for the water vapour content.

Transmitter and receiver clocks are estimated as white noise functions. Stations with high quality clocks (for example the Hydrogen Maser at Wettzell) have greater weight in the solution.

Formal errors are scaled using a combination of baseline repeatability measures and the χ^2 value.

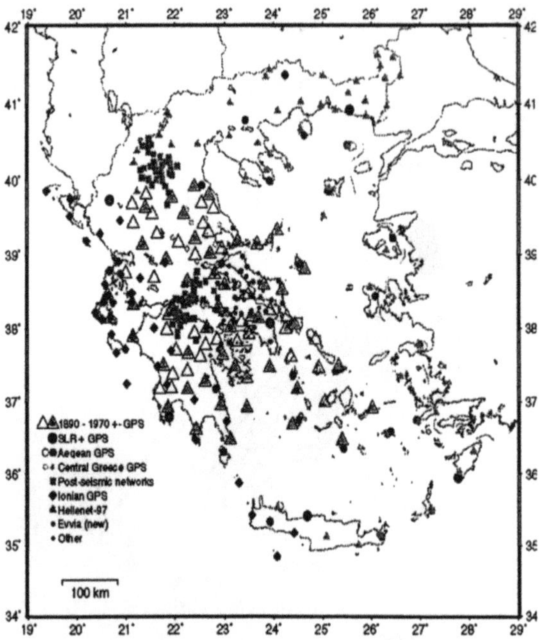

Fig. 3: Stations used in the combination of historical geodetic coordinate sets

4 Results

To date, a primary result of SING is the integration of 33 historical geodetic data sets, as seen in figure 3, from the past ten years to provide a velocity field, figure 4 and national strain map, figure 5. Coordinates and covariance matrices from the datasets were combined into the same reference frame and an overall database, velocity field and strain map produced. The velocity field shows actual annual station velocities relative to a stable European reference frame. Open arrows show the motion of the Anatolian and African plates derived from the Nuvel 1A NNR model. The strain field is based on a spline interpolation of the whole velocity field. The strain field provides an accurate estimate across the Aegean Sea and mainland Greece, but does however wrongly determine strain at the southern extents, owing to the current lack of geodetic data in that region. The combined field gives the first full picture of geodetic strain in Greece and provides the basis for the siting of the new geodetic networks. This data-set will continuously evolve over the time span of the project and will be combined with new geodetic data from the region. The geodetic strain map is being used to compare with actual seismic ground deformation to gauge potentially hazardous regions throughout Greece.

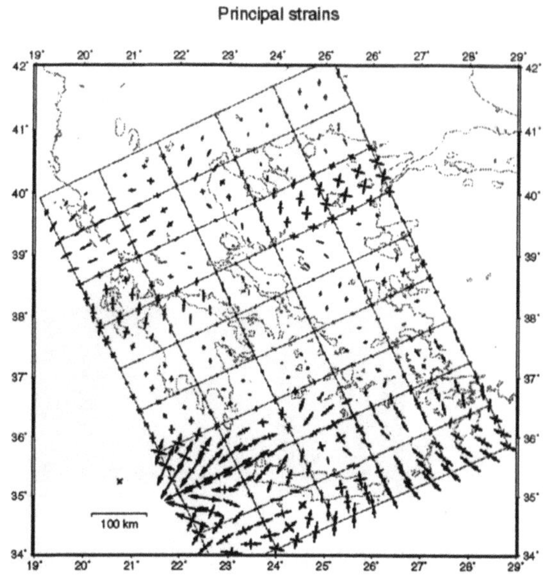

Fig. 5: Strain field based on a spline interpolation of the velocity field in figure 4. The scale bar equals 0.1ppm/year

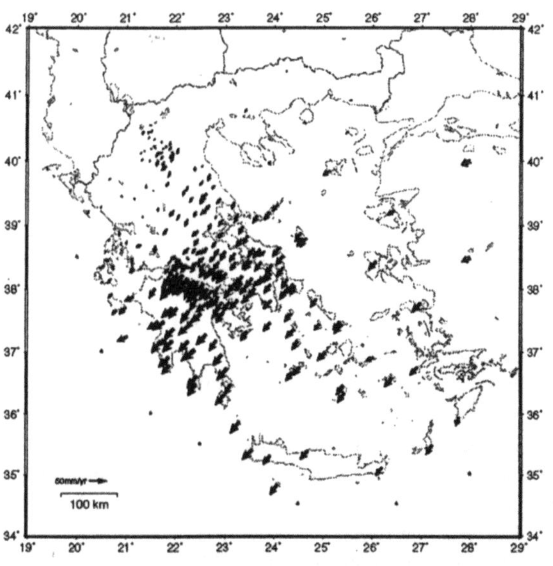

Fig. 4: Velocity field built up from the combination of 33 geodetic data sets relative to a stable European reference frame. The scale bar equals 50mm/year

The investigation into optimal observing methods has led to an optimised observation strategy for the remaining project fieldwork/processing. The method involves looking at the relationship between processing strategy, baseline length, occupation time and precision. A subset of data collected during the 1998 SING fieldwork has been processed using the GIPSY software. Results indicate minimum observation periods necessary to obtain the required precision.

Approximately fifty subsets of SING data have been processed. Different length baselines between stations with greater than four hours of common data were used. The baseline was then processed four times; with all available data and with four, three and two hour sub-sets. The processing run with all data was assumed to be 'truth' and the three other runs compared to this The standard deviations of all differences between 'truth' and the computed sub-set values are shown in table 1. Figure 6 shows a plot of the differences from truth for the derived two-dimensional component.

Table 1: Standard deviation of component differences from the defined 'truth', (m)

σ	4 hours	3 hours	2 hours
East	0.002	0.003	0.005
North	0.003	0.004	0.010
Vertical	0.011	0.016	0.020
2D	0.003	0.003	0.010

This study has shown that for baselines of up to 100km, there is no significant difference between the 'truth' and coordinates derived from either three or four hours of data. However for observation periods of two hours, the difference from truth changes dramatically and the solution becomes significantly less accurate. This conclusion has helped in observation planning for the 1999 fieldwork.

The results from this work have immediate implications for groups with limited resources

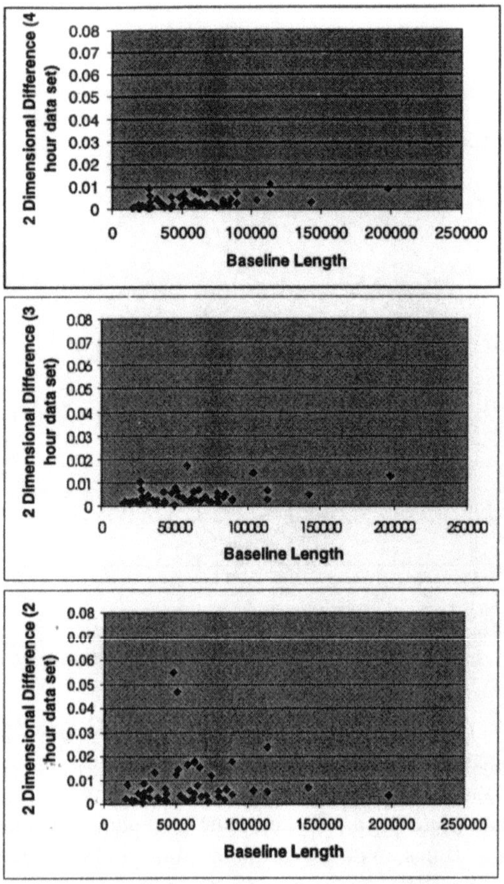

Fig. 6: Differences from truth for the derived two-dimensional component

(manpower, money or equipment) who wish to observe large GPS networks in a quasi-continuous manner, as long as similar careful processing strategies are employed.

Results from initial investigations in to historical seismicity indicate that for northern Greece, seismic hazard seems to be overestimated using solely historical data and underestimated using only seismotectonic data. The next step is to incorporate information from different sources to reliably assess the seismic hazard within a seismogenic zone.

Of special interest is the situation that has been identified around the Gulf of Corinth where geodetically measured strain is as much as four times higher than that implied through the study of the seismicity and geology. Although there has been significant debate, for example Clarke (1998) and Roberts (1996), in the scientific literature on this topic, a full explanation of this difference has not yet been found. SING aims to add to the understanding of this anomaly.

Acknowledgements. The SING project (ENV4-519) is funded by the European Commission, Directorate General XII, for Science, Research and Development, under the Environment and Climate Programme (1994-98).

References

Ambraseys, N. N. and Jackson, J. A., (1990). Seismicity and associated strain of central Greece between 1890 and 1988. Geophys. J. Int., 101, 663-708.

Billiris, H., Paradissis D., Veis, G., England, P., Featherstone, W., Parsons, B., Cross, P., Rands, P., Rayson, M., Sellers, P., Ashkenazi, V., Davison, M., Jackson, J., and Ambraseys, N., (1991). Geodetic determination of tectonic deformation in central Greece from 1900 to 1988. Nature, 350, 124-129.

Clarke, P.J., Davies, R.R., England, P.C., Parsons, B.E., Billiris, H., Paradissis, D., Veis G., Cross, P.A., Denys, P.H., Ashkenazi, V., Bingley, R., Kahle, H.-G., Muller, M.-V. and Briole, P., (1998). Crustal strain in central Greece from repeated GPS measurements in the interval 1989-1997. Geophys. J. Int., 135, 195-214.

Davies, R.R., England, P.C., Parsons, B.E., Billiris, H., Paradissis, D. and Veis, G., (1997). Geodetic strain of Greece in the interval 1892-1992, J. geophys. Res., 102(B11), 24 571-24 588.

Kahle, H.-G., Muller, M.-V. and Veis, G., (1996). Trajectories of crustal deformation of western Greece from GPS observations 1989-1994. Geophys. Res. Lett., 23, 677-680.

Roberts., G.P., (1996) Noncharacteristic normal faulting surface ruptures from the Gulf of Corinth, Greece. Journal of Geophysical Research, 101, 25,255-25,267.

Rothacher, M and Mervart L (1996). Bernese GPS software version 4.0, Astronomical Institute, University of Berne.

Webb, F H and Zumberge, J F (1997). An introduction to GIPSY-OASIS II. Jet Propulsion Laboratory document JPL D-11088, California Institute of Technology, USA.

Zumberge, J F., Heflin, M.B,. Jefferson, D.C., Watkins, M.M., and Webb, F.H., (1997). Precise point positioning for the efficient and robust analysis of GPS data from large networks. J. Geophys Res., 102, B3, 5005-5017.

Crustal deformation monitoring of volcanoes in Japan using L-band SAR interferometry

Makoto Murakami, Satoshi Fujiwara, Takuya Nishimura, Mikio Tobita, Hiroyuki Nakagawa, Shinzaburo Ozawa, Masaki Murakami

Geographical Survey Institute, Kitasato-1, Tsukuba-shi, Ibaraki-ken, 305 Japan
Tel : +81-298-64-6925
Fax: +81-298-64-2955
E-mail: mccopy@gsi-mc.go.jp

Abstract. Since 1994 the Geographical Survey Institute (GSI) has been conducting studies applying interferometric SAR (InSAR) for detection of crustal deformations mainly using JERS-1 data. With this technology we obtained interferograms depicting crustal deformations associated with large earthquakes. We also detected deformations related to volcanic activities. Subsidence of caldera in Izu-Oshima, deformation associated with earthquake swarm of volcanic origin in Izu-peninsula and deformation around Mt. Iwate. In some applications the InSAR data played a key role to construct models for crustal deformations in above studies.

We carried out a repeatability study over Izu-peninsula where coverage of JERS-1 data is fine and crustal deformations are persistent. We processed many scenes and compared them. We found to model water vapor distribution is difficult without additional information. Water vapor effects sometimes show a correlation with topography but it is not always consistent. A pragmatic way to eliminate this effect is to perform temporal averaging for multiple interferograms obtained at different epochs. After this averaging the precision of measurement of crustal deformation using JERS-1 SAR is estimated better than 1 cm.

The GSI operates a nationwide continuous GPS array that consists of 890 permanent GPS stations all over the country; we started a study to improve the precision of SAR interferometry using information about water vapor distribution derived from this GPS array. Although JERS-1 ceased to operate on October 12, 1998, this study is essential for JERS-1 InSAR using enormous amount of existing data and for the preparation for next generation L-band SAR sensors such as one onboard ALOS to be launched in 2002 and airborne SAR.

Keywords. Synthetic Aperture Radar (SAR), Crustal Deformation, GPS.

1 Introduction

The Geographical Survey Institute (GSI) has been conducting studies on driving mechanisms of crustal deformations. Since 1994 we use SAR (Synthetic Aperture Radar) interferometry as a tool to detect changes of the surface of the earth. We are using SAR data acquired mainly by JERS-1 satellite developed and launched by the National Space Development Agency (NASDA) and the Ministry of International Trade and Industry (MITI). SAR interferometry is a 2-dimensional measurement of the surface deformations of the earth and is very powerful to construct a model of driving mechanism of crustal deformations. GPS, on the other hand, measures 3-dimensional displacement of sites, but the distribution of the site is usually sparse; in the even extreme case the nearest spacing of those sites is a few kilometers. When GPS and InSAR data are available we use both to construct a model. In case only InSAR data is available it is possible to come up with a reasonable model.

The power of InSAR when applied to studies of crustal deformation has been fully demonstrated for the crustal deformation associated with the 1994 Landers, California, earthquake by Massonnet *et al.*(1993) and Zebker *et al.*(1994) using ERS-1 data. Since 1994, we have been using JERS-1 SAR data and detected crustal deformations associated with the 1994 Northridge earthquake (Murakami *et al.*, 1996); the 1995 Hyogo-ken Nanbu earthquake (Murakami *et al.*, 1995; Ozawa *et al.*, 1997); the 1995 Northern Sakhalin (Neftegorsk) earthquake (Tobita *et al.*, 1997a); persistent earthquake swarm in Izu-hanto (Fujiwara *et al.*, 1997); and the March

International Association of Geodesy Symposia, Vol. 121
Schwarz (ed.), Geodesy Beyond 2000 – The Challenges of the First Decade

1997 Kagoshima-ken Hokuseibu earthquake (Tobita *et al.*, 1997b) as well as surface deformations at the volcanic caldera in Izu-Oshima and glacier migration in Antarctica (Ozawa *et al.*, 1997).

2 Deformation associated with earthquake swarm in Izu-Peninsula

Fujiwara *et al.*(1997) carried out a comprehensive study to identify crustal deformation in the eastern part of Izu-Peninsula due to persistent earthquake swarm, which is of a volcanic origin, as well as to identify the effects of ground surface and atmospheric characteristics on SAR interferometry. JERS-1/SAR data acquired from 1993 to 1994 were used comprehensively and eleven independent interferograms were obtained. An uplift as large as 5 cm had been observed from repeated precise leveling surveys. In the interferograms, we can see the fringes indicating shortening of range to the sensor. Our interpretation is that this fringe corresponds to uplift of the area. However, there exist other fringes where the ground truth data indicate no crustal deformations. Shapes and intensities of those fringes are not the same in different interferograms. Some of them have correlation with topography. One possible explanation is meteorological effects such as rotor clouds or mountain waves on the leeward side of mountains; other features can be related to cloud cover. We assume that most of those fringes are caused by a heterogeneous distribution of water vapor in the troposphere. The similar effect of water vapor exists in positioning using GPS, which also uses L-band microwave signal. This effect is called tropospheric "wet" delay of microwave. We have eliminated fringes due to water vapor in SAR interferograms by averaging two or more interferograms obtained for different seasons. After averaging of multiple interferograms we obtained a picture depicting only surface deformation as large as 5 cm.

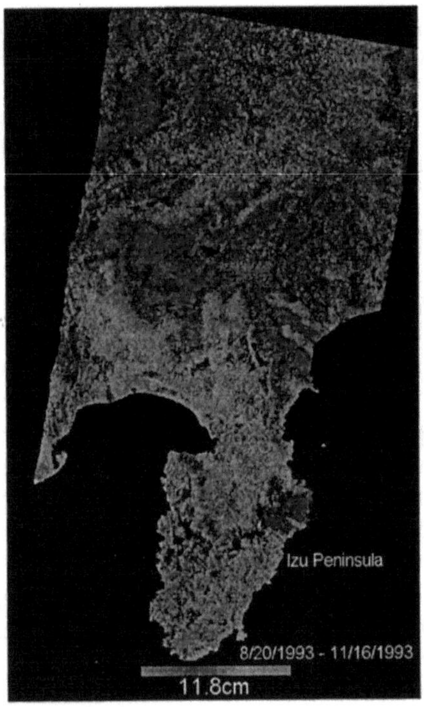

Fig. 1: SAR interferogram in Izu peninsula showing the effect of water vapor.

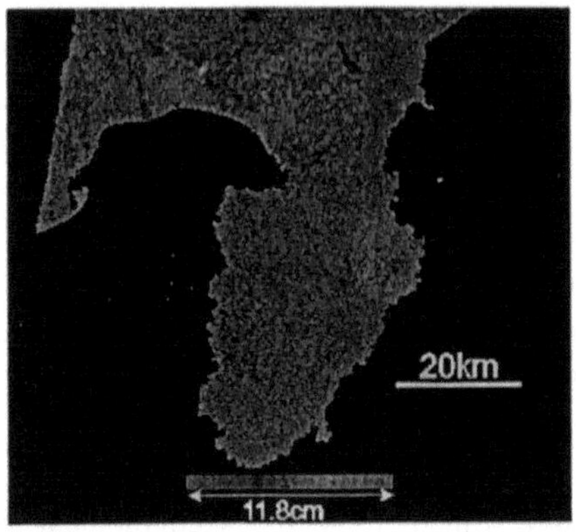

Fig. 2: An averaged image of two interferograms to reduce the effect of water vapor. Uplift associated with earthquake swarm is found in the north-eastern part of the peninsula.

3 Deformation after volcanic eruption in Izu-Oshima

We made an interferogram for a volcanic island, Izu-Oshima. JERS-1/SAR imageries we used were taken on October 15, 1992 and March 14, 1995 (Figure 3). We found an inflation of the island and subduction of a caldera of Mt. Mihara. This volcano erupted in November 1986. The subsidence indicated by the interferogram was not associated with any volcanic activities because the volcano has been quiet during the period of SAR observations. One possible explanation is that the mass loading of the lava has caused the subsidence.

The interferogram shows good coherency in and around the caldera despite a long time interval of 29 months between the two observations. However, coherency in other areas is worse. Because the areas with low coherency are mainly steep slopes, the loss of coherence were caused by the ruggedness of the terrain.

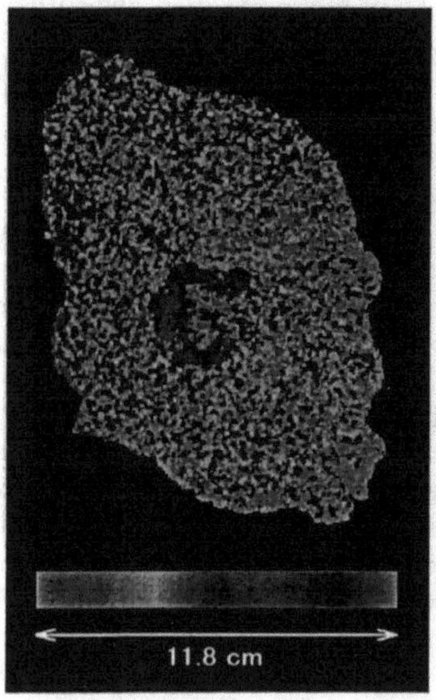

Fig. 3: SAR interferogram showing subsidence in the caldera of Izu-Oshima, a volcanic island.

4 Deformation detection of Mt. Iwate volcano in 1998

Mt. Iwate became active in 1998. The activity was firstly noticed by increase of micro-seismisity and lateral expansion of crust around the volcano beginning February 1998. Those phenomenon continued and in September an earthquake of M6.1 occurred at the southwestern edge of the mountain. By JERS-1 SAR interferometry we detected dome-like uplift of 10 cm at the center about 10 km to the west of the volcano. We also found fringes associated with crustal deformation caused by the earthquake. We constructed a model to explain the surface fringe pattern and concluded. An inflation source (Mogi model) is located in the western part of the volcano at the depth of 8 km. The total volmetric increase was 0.033 km^3. The calculation of Coulomb failure function showed that ΔCFF increased in the vicinity of epicenter of the earthquake by 0.2MPa which corresponds to 4% of coseismic stress drop. It strongly suggests that the inflation triggered the earthquake.

5 Future challenges

5.1 Combination with GPS

In this country a continuous nationwide GPS array is operating. To have better understanding of a crustal deformation it is useful to combine both InSAR and GPS data. Both data are complimentary. GPS is continuous in terms of time and capable of measuring 3-dimensional components of the deformation, but its spatial distribution is sparse. An average spacing between the sites of GSI's network is 25 km even in one of the most dense network. SAR interferometry is, on the other hand, one dimensional measurement, but it provides better spatial resolution of about 100m or less. Table 1 summarizes comparison of the advantages and disadvantages of SAR and GPS.

It is much more powerful to combine those data to construct a source model of any earthquake or volcanic activity. An example of such an approach is analysis of the March 1997 Kagoshima-ken Hokuseibu earthquake is one of successful examples using both SAR and GPS data.

Table 1: Comparison of the two geodetic techniques: SAR and GPS

....	Continuous GPS array	SAR interferometry
Observables	Three dimensional (horizontal and vertical)	one dimensional (along the line of sight)
Temporal distribution	Dense (every 30 sec)	Sparse (JERS-1: 44 days recurrence)
Spatial distribution	Sparse (separation~25km)	Dense (pixel size<100m)
Siting	Receivers needed	No facility on the ground

5.2 Airborne InSAR

Another topic of future challenge is airborne SAR interferometry. We just started a study to apply repeat path airborne SAR interferometry in X-band using a commercial equipment by a Japanese manufacturer. A joint study by universities and other institutes has demonstrated a potensiality of X-band repeat path airborne SAR interferometry. They obtained very clear coherence between 2 observations collected from different flight paths. The key idea of the experiment is to use corner-cube reflectors as ground control points and careful navigation of the aircraft using differential GPS guidance system. We will participate in the joint study to make this technology more robust and effective.

Acknowledgments. Most of this work is based on the collaborative works with the National Space Development Agency (NASDA), National Institute of Polar Research (NIPR), and the Jet Propulsion Laboratory (JPL). In the study we used JERS-1 SAR data provided by NASDA as a part of the joint study of NASDA and GSI. The ownership of original SAR data is retained by NASDA and the Ministry of International Trade and Industry (MITI).

References

Fujiwara, S., P. A. Rosen, M. Tobita and M. Murakami, Crustal Deformation Measurements Using Repeat-pass JERS 1 SAR Interferometry near the Izu Peninsula, Japan, submitted to J. Geophys. Res., 1997.

Goldstein, R. M. , H. A. Zebker, and C. L. Werner, Satellite radar interferometry: Two -dimensional phase unwrapping, Radio Sci., 23, 713-720, 1988.

Hiramatsu, A., The shortest discharge tree problem for 2D-image, Memo of Hiramatsu on January 10, 1992.

Massonnet, D., M. Rossi, C. Carmona, F. Adragna, G. Peltzer, K. Fiegl, and T. Rabaute, The displacement field of the Landers earthquake mapped by radar interferometry, Nature, 364, 138-142, 1993.

Miyazaki, S., T. Saito, M. Sasaki, Y. Hatanaka and Y. Iimura: Expansion of GSI's Nationwide GPS Array, Bull. Geogr. Surv. Inst., 43, 23-34, 1997.

Murakami, Mak., M. Tobita, S. Fujiwara, T. Saito and H. Masaharu, Coseismic crustal deformations of 1994 Northridge, California, earthquake detected by interferometric JERS 1 synthetic aperture radar, J. Geophys. Res., 101, 8605-8614, 1996.

Murakami, Mak., S. Fujiwara, and T. Saito, Detection of Crustal Deformations Associated with the 1995 Hyogoken-Nanbu Earthquake by Interferometric SAR, Journal of the Geographical Survey Institute, 83, 24-27, 1995 (in Japanese).

Murakami, Mas., S. Fujiwara, M. Tobita, K. Nitta, H. Nakagawa, S. Ozawa, and H. Yarai, Progress in SAR interferometry for the detection of crustal deformation by the Geographical Survey Institute, Journal of the Geographical Survey Institute, 88, 1-9, 1997 (in Japanese).

Ozawa, S., M. Murakami, S. Fujiwara, and M. Tobita, Synthetic aperture radar interferogram of the 1995 Kobe earthquake and its geodetic inversion, Geophys. Res. Letter, 24, 2327-2330, 1997.

Ozawa, T., K. Doi, K. Shibuya, H. Nakagawa, S. Fujiwara, and M. Murakami, Detection of Ice Sheet Motion by JERS-1 SAR Interferometry, presented at the Second Workshop on SAR Interferometry, SAR WORKSHOP '97 TSUKUBA, at Tsukuba, Japan on November 19, 1997 (in this proceedings).

Tobita, M., S. Fujiwara, S. Ozawa, P. Rosen, E. J. Fielding, C. L. Warner, Mas. Murakami, H. Nakagawa, K. Nitta, and Mak. Murakami, Deformation of the 1995 North Sakhalin Earthquake Detected by JERS-1/SAR Interferometry, submitted to Journal of Earth Physics, 1997a.

Tobita, M. et al., Crustal Deformation Associated with the March 1997 Kagoshima-ken Hokusei-bu Earthquake Detected by InSAR and GPS, presented at the Second Workshop on SAR Interferometry, SAR WORKSHOP '97 TSUKUBA, at Tsukuba, Japan on November 19, 1997b .

Tsuji, Hatanaka and Miyazaki, Tremors! Monitoring Crustal Deformation in Japan, GPS World, pp18-30, April 1996.

Werner, C. L. and R. M. Goldstein, 1997.

Zebker, H. A., P. A. Rosen and S. Hensley, Atmospheric effects in interferometric synthetic aperture radar surface deformation and topographic maps, J. Geophys. Res., 102, 7547-7563, 1997.

Zebker, H. A., P. A. Rosen, R. M. Goldstein, A. Gabriel, and C. L. Werner, On the derivation of coseismic displacement fields using differential radar interferometry; The Landers earthquake, J. Geophys. Res., 99, 19617-19634, 1994.

The sea surface of the Baltic – a result from the Baltic Sea Level Project (IAG SSC 8.1)

Markku Poutanen and Juhani Kakkuri
Finnish Geodetic Institute, Geodeetinrinne 2, FIN-02430 Masala, Finland
e-mail: Markku.Poutanen@fgi.fi, Juhani.Kakkuri@fgi.fi

Abstract. We describe the final results of the Baltic Sea Level (BSL) 1997 GPS campaign and the combination of three successive BSL campaigns. Because the last campaign was made simultaneously with the EUVN (European Vertical Reference Network) GPS campaign, we were able to obtain a well-defined connection between the two networks. Combining tide gauge data and a gravimetric precise geoid model, we recomputed the Sea Surface Topography (SST) of the Baltic Sea with a better accuracy than before. We also used satellite altimetry data to obtain detailed spatial and temporal information of the SST.

Keywords. GPS, height determination, sea surface topography (SST), satellite altimetry.

1 Introduction

The Baltic Sea Level (BSL) project was initiated as an *ad hoc* working group in the General Meeting of the IAG in Edinburgh in 1989. After the IUGG General Assembly in Vienna 1991, it received the status of a Special Study Group (IAG SSG 5.147) and after the IUGG General Assembly in Boulder in 1995, status of Subcommission 1 of IAG Commission 8. All the countries around the Baltic Sea participated in the project.

One of the goals of the Baltic Sea Level Project is the unification of the vertical datums in the countries around the Baltic Sea. The other goals, set in the first BSL meeting in Helsinki, December 1991, were to contribute to the determination of the gravity field and the geoid in the Baltic Sea region, to determine the sea level and sea surface topography of the Baltic Sea, to monitor postglacial rebound, especially in the sea area, and to remeasure the Baltic Ring for horizontal crustal deformation studies.

Series of repeated GPS measurements have been performed, namely in 1990 (BSL I), 1993 (BSL II) and 1997 (BSL III). The observing geometry is shown in Fig. 1 and the stations of BSL III in Fig. 2.

In the BSL I a total of 30 tide gauges and nine fiducial stations were observed, whereas in the BSL II 35 tide gauges and 12 fiducial stations were observed, the total number of GPS receivers being more than 50. The BSL III was arranged simultaneously with the EUVN (European Vertical GPS Reference Network) GPS campaign. The total number of stations was about 60.

Also, a great number of gravimetric data have been collected from Russia, the Baltic countries, Poland and the Nordic countries for the geoid computation in the Baltic Sea region.

2 Computation of the BSL GPS campaigns

During the seven years from 1990 to 1997, there has been a vast development on receiver technology, precision of satellite ephemeris and data processing. Also, the Solar activity reached its maximum around the year 1990, whereas in 1997 the activity was still near its minimum. As one might expect, the increase in accuracy has been as dramatical as the changes in technology and environmental conditions.

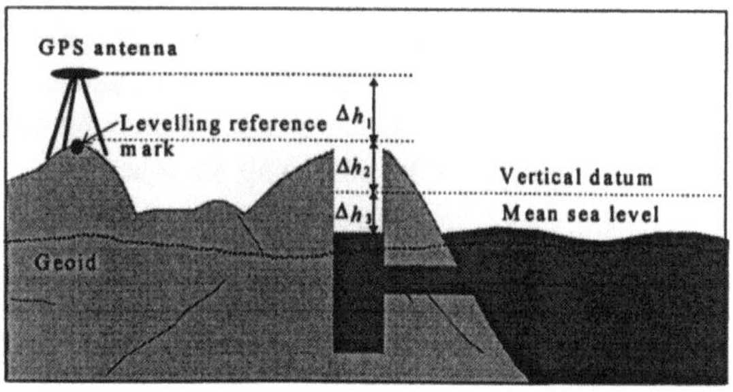

Fig. 1: Observing geometry at a tide gauge in the BSL campaigns. The height difference between the geoid and the mean sea level is the sea surface topography (SST). The level of a national vertical datum does not necessarily coincide with the geoid. □h_1 is the GPS antenna height which is eliminated already during the data processing. □h_2 is the orthometric height of the GPS marker in the national height system and □h_3 is the height of the mean sea level at the epoch of the GPS observations.

International Association of Geodesy Symposia, Vol. 121
Schwarz (ed.), Geodesy Beyond 2000 – The Challenges of the First Decade

All three campaigns have been computed in different reference frames: 1990 in ITRF89 (epoch 1988), 1993 in ITRF1993 (epoch 1993.45), and 1997 in ITRF96 (epoch 1997.4). Results of the BSL campaigns have been published in the Reports series of the Finnish Geodetic Institute (Nos. 94:2, 95:2, and 99:4).

In the combined solution of the BSL III campaign (Poutanen *et al.*, 1999) we fixed the EUVN stations to the coordinate values given in Ineichen *et al.* (1998). This ensures a good tie between these two networks. A detailed description of the observing and data processing procedures is found in Poutanen (1998) and Poutanen *et al.* (1999).

The repeatability of the BSL III is about 1/10th of that of the 1990 campaign, namely 1 mm in the horizontal and 3 mm in the vertical component. However, systematic errors can be considerably larger than might be expected from the repeatability as was seen in the BSL II campaign (Poutanen 1995).

2 Sea surface topography from the 1997 BSL GPS campaign

The benchmarks observed were connected to the national precise levelling network and to the readings of the tide gauges. Long series of tide gauge observations give us a connection to the mean sea surface. From this, we can compute the height of the mean sea surface at the epoch of the observations. On the other hand, if the orthometric height of the GPS marker near a tide gauge is known, we can compute the height difference between the GPS marker and the mean sea level (Fig. 1).

From a geoid model we get the height of the geoid above the ellipsoid, N, and from GPS observations the height of the point above the ellipsoid, h. Using the symbols in Fig. 1, we can compute the sea surface topography at the observing point as

$$SST = h - N - \Delta h_2 - \Delta h_3 - \Delta H. \qquad (1)$$

ΔH is to convert heights above the non-tidal geoid to the heights above the mean geoid (Ekman, 1989)

$$\Delta H = 0.296\, \gamma(\sin^2 \varphi_N - \sin^2 \varphi_S), \qquad (2)$$

where γ is the Love number, here we used value 0.8, and φ_N and φ_S are latitude of the northern and southern station, respectively. For the southern station we used $\varphi_S = 54.8°$ which is the latitude of the Danish-German border (see Ekman, 1989, Ekman and Mäkinen, 1996).

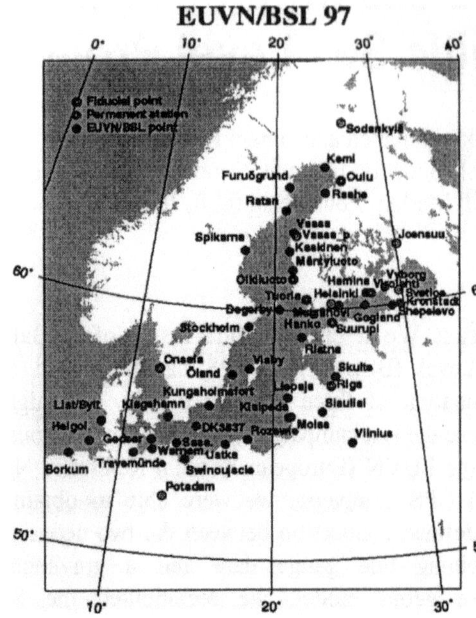

Fig. 2: The BSL 1997 GPS network

In Kakkuri and Poutanen (1997) we computed the Sea Surface Topography (SST) of the Baltic Sea from the BSL 1993 GPS campaign using the geoid model for the Baltic sea computed by Vermeer (1995).

Here we recompute the SST using the final coordinates of the BSL/EUVN 1997 GPS campaign (Poutanen *et al.*, 1999) and the NKG96 geoid model (Forsberg, 1997). The results are shown in Fig. 3. It can be seen that some anomalous features of Kakkuri and Poutanen (1997), Fig. 6 have disappeared. The resulting SST seems to fit very well to the other determinations. It seems that the NKG96 geoid model has less deformations than the earlier used geoid models, especially in the southern part of the area. Also, the improvement of the GPS observations may play a small role on this.

3 Sea surface topography from the satellite altimetry

The Baltic Sea is surrounded by a dense network of tide gauges, most of them connected to the national precise levelling networks. At many tide gauges there have been GPS measurements, either during the BSL/EUVN campaigns or on some other occasions. Above we used GPS observations to obtain the SST at tide gauges. However, the method can be used at shorelines and islands only and its time resolution is poor, i.e. it gives no contribution in the open sea. Satellite altimetry can be used to increase the spatial and temporal resolution.

There are some earlier works on satellite altimetry in the Baltic Sea using SEASAT data (e.g. Vermeer, 1983) but there exists no extensive temporal coverage. We used here a set of CERSAT, the French processing and archiving facility (CERSAT, 1994) processed ERS-1 and ERS-2 altimetry data from about 7 years (1991 – 1998).

Cross-over adjustment is a traditional way to fix the altimetry errors. However, it requires both ascending and descending arcs, and only cross-over points of arcs can be used which cause some additional problems in such narrow areas like the Gulf of Bothnia and the Gulf of Finland. Instead, we decided to use the good coverage of tide gauges around the sea area and the GPS observations to give the necessary information for data error corrections.

The first test was made with the gradient method, i.e. computing the height difference of two successive altimetry values instead of the height itself. Most of the height errors due e.g. to the atmosphere are removed in the gradient method. We need several fixed points to fix the level (and, in practice, also tilt) of the computed surface. For this, the tide gauge/GPS points can be used.

From two successive ellipsoidal heights of the instantaneous sea surface, h_i, we get the deflection along the track

$$\varepsilon_i = (h_i - h_{i+1}) / s \tag{3}$$

separately for ascending and descending tracks where s is the distance between two successive points. Analogically to the astrogeodetic levelling, we can write for the sea surface at the point of observation

$$\varepsilon = \xi \cos A + \eta \sin A \tag{4}$$

where ξ and η are here the north and east deviations of the sea surface normal from the normal of the reference ellipsoid, and A is the azimuth of the track at the point of computation.

Writing Eq. (4) separately for ascending and descending arcs, ε_A, ε_D, we can solve the components of the surface normal:

$$\begin{cases} \eta = \dfrac{\varepsilon_A \cos A_D - \varepsilon_D \cos A_A}{\sin A_A \cos A_D - \sin A_D \cos A_A} \\ \xi = \dfrac{\varepsilon_D - \eta \sin A_D}{\cos A_D} \end{cases} \tag{5}$$

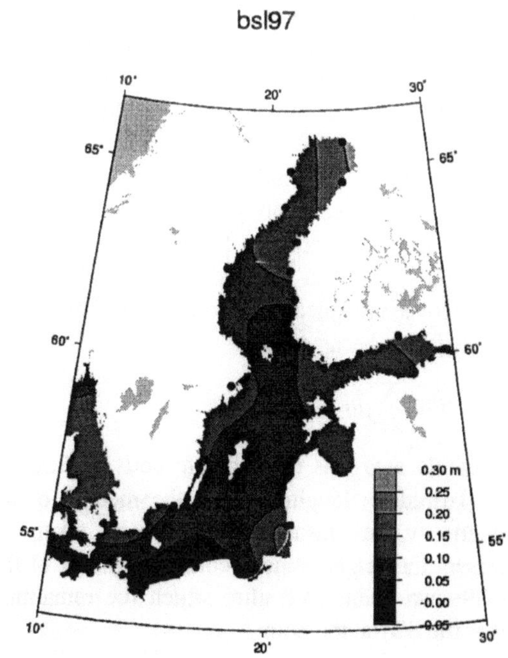

Fig. 3: Sea Surface Topography, computed from the BSL97 GPS campaign according to Eq. 1

For solving the unknowns ξ and η in (5) the following procedure was used: The whole sea area was divided into 0.25° × 0.5° (latitude × longitude) cells. Inside each box, medians of ascending and descending arc gradients were taken to represent the gradient values ε_A, ε_D in that cell. These are inserted into (5) after which the deflection of the vertical from the geoid model was subtracted.

The final computation of the SST was made with a program originally written some 15 years ago for astrogeodetic geoid determination by Dr. *M. Ollikainen* (private communication, 1999). A slight modification of input routines and data arrays allowed us to apply it also for this work. In the adjustment, we fixed the tide gauge/GPS points.

During processing we discovered some problems of the gradient method. The idea itself turned out to be applicable but the narrowness of the Gulf of Bothnia and the Gulf of Finland and also the Åland archipelago caused that only a few points were accepted in these areas. Noisy data caused more rejections.

The gradient method could be used in more open sea areas, provided that there exist enough GPS control points to fix the computed surface. It is an appropriate way to avoid most of the satellite-borne errors but the tilt and level of the surface is totally dependent on the control points.

In the final computation of the altimetric SST we first fitted a second degree polynomial surface to the SST obtained with GPS at tide gauges. The purpose of this surface is just to serve as a preliminary reference surface to fix only the bias of the altimetry observations.

Next, each arc of altimetry observations were fitted to this surface. Only the level of the arc was shifted but no correction to the tilt was made. We shifted the arc so that the sum of the height differences between the measured points h_i and the reference surface s_i became zero:

$$dh = \sum h_i - s_i = 0 \qquad (6)$$

The whole area was divided into cells of 0.25° × 0.5° (latitude × longitude) and the median of all altimetry values inside the box was taken to represent the height value. Geoidal heights N of the NKG96 was subtracted after which the remaining part is the SST at the point.

The overall agreement with the GPS based SST is good. One has to remember that although we used the tide gauge GPS points to define the reference surface for altimetry orbital error correction, we did not affect the tilt of the arcs.

In Fig. 5 we have a gray scale plot of ERS2 altimetry from the year 1997 and in Fig. 6 we show the residual between the SSTs based on GPS and satellite altimetry. If one ignores the details coming from the satellite altimetry, there are no global biases or tilts visible bigger than normal yearly sea level variation. The agreement is in general in the range of ±5 cm.

One stable feature is SW–W of island Saaremaa where the sea surface is 10-20 cm higher than in the surrounding area. The flow of the river Daugava into the Gulf of Riga probably does not alone explain this because no such features can be seen near other big rivers, nor is the surface of the Gulf of Riga exceptionally high. The effect of the sea bottom shape or an error in the geoid model cannot be ruled out before an oceanographic study has been made.

Another exceptional feature can be seen near Kattegat. In the autumn of the year 1997, the sea level appears to be even 20 cm higher than during the first half. It is possible that noisy data near the coastline can cause this but it alone cannot explain the systematics.

The yearly mean fits nicely to the GPS determination at Klagshamn but deviates significantly from other determinations, e.g. from that of Ekman and Mäkinen (1996), hereafter EM96. The effect of wind can be quite large in the narrow strait.

The Kronstadt area, at the east end of the Gulf of Finland shows a pronounced high water feature at spring time; this can be explained by the increased spring flow of the River Neva. In this respect, the tide gauge situates on the most unfavourable place. The Kronstadt water level is critical if one wants to compare directly the height system zero points, e.g. Kronstadt vs. Helsinki.

There are several other determinations of the sea surface topography of the Baltic Sea. One should mention oceanographic determinations of Lisitzin (1974) and Carlsson (1998). EM96 have determined the SST at the tide gauges using levelling connection to the national precise levelling networks.

The model of Lisitzin gives too steep slope of the sea surface as discussed in Carlsson (1998) and EM96. Determinations of Carlsson and EM96 give similar results, mostly within 2 cm. The 20–25 cm S–N slope of the SST is mainly due to the change of salinity; at the end of the Gulfs the salinity is practically zero.

In their paper EM96 makes a readjustment of the levelling loop around the Gulf of Bothnia, using the geostrophic connection between Åland and Sweden to close the loop. The original reason was that sea level appeared to be 5 cm higher at the Åland islands than at the Swedish coast which is unrealistic from the oceanographic point of view.

In our determination the SST is lower at Åland than at the Swedish coast. This is in agreement with the oceanographic models. The sea surface height difference is due to the southward going current between Sweden and Åland.

The heights are given in NH60 height system, defined in EM96. We converted the altimetric and GPS heights to the approximate NH60 subtracting the eustatic rise of the sea level, 1 mm/year, or the total amount 0.037 m from 1997 to 1960.

In Fig. 6 we plot the SST from various determinations. Although the overall agreement is reasonably good, there are some anomalous values, most notably at Ratan, Furuögrund, Mäntyluoto and Degerby. Because both the altimetric and GPS SST are solely dependent on the same geoid model, we cannot rule out an error in the model.

Another problematic place is Klagshamn (and sites at Kattegat) where GPS and altimetric SST agree but both are, for some unknown reason, 10–15 cm higher than what they should be. Other parts of the Baltic Sea south coast is on the correct level.

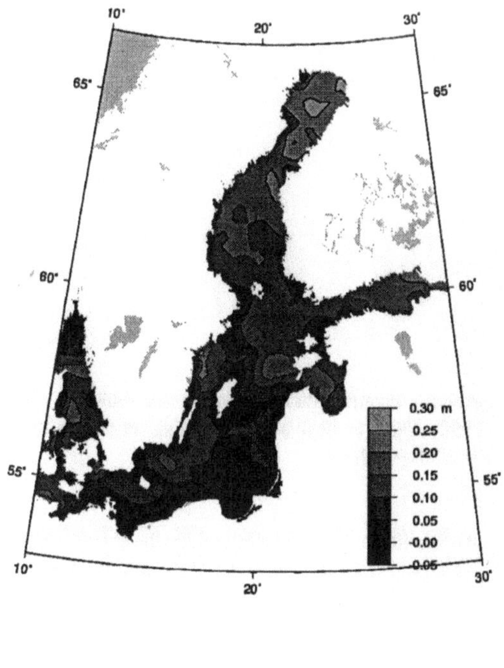

Fig. 4: Sea Surface Topography computed from ERS-2 altimetry data of the year 1997

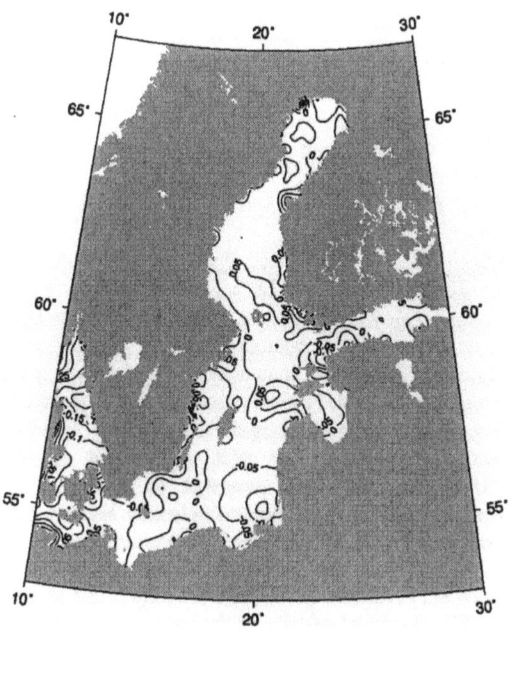

Fig. 5: Difference between the altimetric and GPS/tide gauge SST. The contour interval is 5 cm

The accuracy of the NKG96 is in the order of ±5 cm which allows errors of this size. On the other hand, GPS and altimetric SST deviate mostly just on these sites which may indicate another reason.

5 Conclusions

The current state-of-the-art of the GPS determination gives a centimetre accuracy in the height above the ellipsoid. It is interesting to compare different independent GPS determinations. In Table I we compare heights on some tide gauges from the EUVN official solution (Ineichen *et al.*, 1998), BSL solution of Poutanen (1998) and EUREF densification in Finland (Ollikainen *et al.*, 1999).

Differences between solutions are greater than one may expect according to the repeatability, and in spite of the same reference frame there is also a small bias. A part of this is explained by differences in coordinates of fixed points. Another part is coming from different geometry of networks and processing strategy.

Nowadays there exists a dense network of permanent GPS stations in the Baltic Sea region. Therefore, one is tempted to ask if large GPS campaigns are needed anymore. There could be some special needs for simultaneous observations but in most cases regional GPS measurements with a good tie to the existing reference frame will be sufficient. In international campaigns, the problem of the schedule and availability of receivers is minimised because everyone can make plans according to their own schedule. Also, performing of the whole project becomes more flexible when each country can plan its part more independently and fit observations to its field work plans.

One should not forget the hierarchy of geodetic reference frames; ephemeral campaigns like EUVN or BSL cannot be as high in hierarchy as the permanent networks which may have a data history of several years. If this is not accepted, there will be a lot of confusion with different coordinates, all of which may have an equal "official" status.

If one want to expand the SST determination methods discussed here to other seas, a precise geoid model of the area will become necessary. Accuracy of the GPS height determination is already better than the accuracy of the best geoid models which, however, are needed in computation of orthometric or normal heights. Using the tide gauge/GPS points as a control, errors of the satellite altimetry can be diminshed. The general slope of the SST can be fixed with GPS and details in the open sea are obtained with the satellite altimetry.

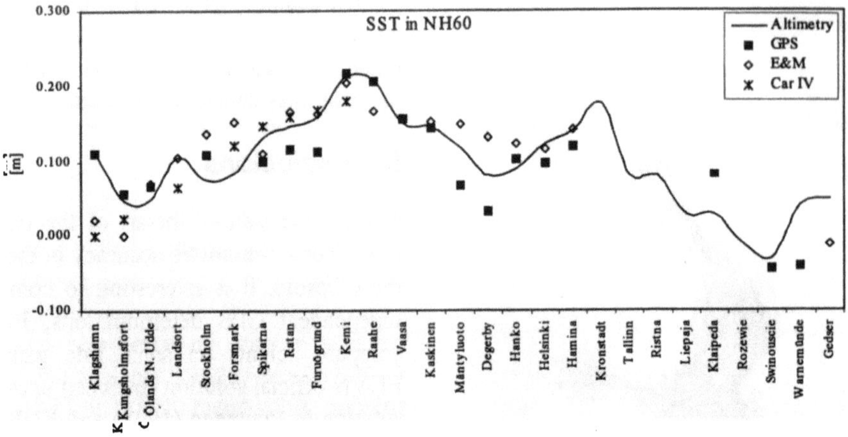

Fig. 6: Comparison of SST from satellite altimetry, GPS/tide gauge, geodetic determination of Ekman and Mäkinen (E&M, 1996) and oceanographic determination of Carlsson, model IV (Car IV, 1998). The altimetric and GPS values have been brought to an approximate NH60 height by subtracting 0.037 m for the eustatic rise of the sea level

Table 1: Differences between independent solutions on six tide gauges. All but Raahe has a concrete pillar for antenna. EUREF solution is from a totally different campaign than the BSL and EUVN which are from the same campaign but an independent solution and a slightly different set of stations

Station	EUREF [m]	BSL [m]	EUVN [m]	EUREF–BSL	EUREF–EUVN	BSL–EUVN
Degerby	21.658	21.657	21.663	0.001	-0.005	-0.006
Helsinki	24.173	24.186	24.193	-0.013	-0.020	-0.007
Hanko	24.858	24.868	24.873	-0.010	-0.015	-0.005
Kaskinen	25.290	25.284	25.292	0.006	-0.002	-0.008
Raahe	21.290	21.305	21.313	-0.015	-0.023	-0.008
Kemi	26.132	26.130	26.136	0.002	-0.004	-0.006

References

Carlsson M. (1998): Mean sea-level topography in the Baltic Sea determined by oceanographic methods. *Marine Geodesy*, **21**, 203–217.

CERSAT (1994): Altimeter products. User manual. C1-EX-MUT-A21-01-CN, issue 2.6.

Ekman M. (1989): Impacts of geodynamic phenomena on systems for height and gravity. *Bulletin Géodésique* **63**, 281–296.

Ekman M., and J. Mäkinen (1996): Mean sea surface topography in the Baltic Sea and its transition area to the North Sea: A geodetic solution and comparisons with oceanographic models. *J. Geoph. Res.* **101**, C5, 11993–11999.

Forsberg R. (1997): Geoid information and GPS – A review and Nordic status. In *Geodetic Applications of GPS* (Ed. B. Jonsson). LMV Rapport 1997:16, Gävle, Sweden. 235–255.

Ineichen D., W. Gurtner, T. Springer, G. Engelhart, J. Luthard, J. Ihde (1998): EUVN combined solution. In *Report of the results of the European vertical reference network GPS campaign 1997*. Euref symposium, Bad Neuenahr – Ahrweiler, Germany, June 10–12, 1998, p. 19–36.

Kakkuri J and Poutanen M. (1997): Geodetic determination of the surface topography of the Baltic Sea. *Marine Geodesy* **20**, 307–316.

Lisitzin E. (1974): *Sea level changes*. Elsevier oceanography series, vol. 8. Amsterdam.

Ollikainen M., Koivula H., and Poutanen M. (1999): The densification of the EUREF network in Finland. *Publications of the Finnish Geodetic Institute.* In preparation.

Poutanen M., (1995): A combined solution of the Second Baltic Sea Level GPS campaign. In *Final results of the Baltic Sea Level 1993 GPS campaign* (Ed. J. Kakkuri). Reports of the Finnish Geodetic Institute, **95:2**, 115–123.

Poutanen M. (1998): The FGI computation of the EUVN/BSL 1997 GPS campaign. In *Report of the results of the European vertical reference network GPS campaign 97 (EUVN'97)*. Euref symposium, June 10–12, 1998. Bad-Neuenahr – Ahrweiler, Germany. 151–160.

Poutanen M., Z. Malkin, A. Voinov, G. Liebsch, and M. Pan, (1999). Combined solution of the Baltic Sea Level 1997 GPS campaign. In *Final results of the Baltic Sea Level 1997 GPS campaign. Research works of the SSC 8.1 of the International Association of Geodesy* (Ed. M. Poutanen and J. Kakkuri). *Rep. Finn. Geod. Inst.* **99:4**, 9–40.

Vermeer M. (1983): A new SEASAT altimetric geoid for the Baltic. *Rep. Finn. Geod. Inst.* **83:4**.

Vermeer M. (1995): Two new geoids determined at the FGI. *Rep. Finn. Geod. Inst.* **95:5**.

Preliminary study of block rotation model in North China area using GPS measurements

Caijun Xu, Jingnan Liu, Dingbo Chao, Chuang Shi, Ting Chen
School of Geoscience and Surveying Engineering, Wuhan Technical University of Surveying and Mapping, 129 Luoyu Road, Wuhan, 430079, P.R.China. E-mail: cjxu@hpb1.wtusm.edu.cn

Yanxing Li
First Crustal Deformation Monitoring Center, State Seismological Bureau, Tianjin, 300180, P.R.China

Abstract. Repeated GPS surveys in North China provide a direct measurement of current crustal motions. GPS surveys have been carried out in this region in 1992, 1995,and 1996 respectively. The horizontal crustal deformation and tectonic activity have been studied using GPS measurements .The GPS velocity fields are estimated and they are used to estimate the North China sub-plate and its intra-block rotation motions. The sub-plate and blocks present some rigid features. Sub-plate and blocks rotations are discussed in details by the study of GPS measurements, geology structural, and seismicity and paleomagnetic measurement. The Euler model using GPS data agrees with NNR-NUVEL-1A model. The current rotation pattern of Eerduosi block, Taihang block, Jiluxi block and Jiaoliao block all rotate in counterclockwise in ITRF96 frame. Yinyan block relative to Eerduosi block, Yinyan block relative to Taihang Jiluxi block, and Eerduosi block relative to Taihang Jiluxi block all have extension trend. Finally, the dynamics of rotation is preliminarily studied .It is associated with the collision of India, Pacific with Eurasia.

Keywords. Block rotation model, GPS, North China.

1 Introduction

North China as a lithospheric dynamics sub-plate is located in the middle part of eastern China. It is characterized by more intense activities of lithospheric dynamics in comparison with Northeast China and South China. According to the seismic tectonic setting geological structure and the concept of activity tectonic block, North China sub-plate may also be divided into two first-order activity blocks: Eerduosi activity block and THJLX activity block (Taihang block+Jiluxi block), and three secondary activity blocks: Taihang –Baoding block (Taihang block), Huanghua-Luxi block (Jiluxi block) and Jiaodong-Subei block (Jiaoliao block), in addition to the Yinyan block in northern North China (Fig.1, Xu, 1996). Since 1992, about 87 GPS stations have been established for the purpose of monitoring crustal movement and implied earthquake precursor in Capital Circle area and the northern part of North China by the First Crustal Deformation Monitoring Center, State Seismological Bureau, China(FCDMC-SSB) in Figure 2. Geologists and geodesists carefully explore all the locations of GPS stations, which are all situated in important structures. There are 42 GPS stations in North China GPS monitoring network, and three observation campaigns carried out in 1992,1995,1996 respectively. There are 57 GPS stations in Capital Circle area (including 20 GPS stations of North China GPS monitoring network), and two observation campaigns carried out in 1995 and 1996. The GPS velocity fields are estimated and they are used to estimate those North China sub-plate and intra-block rotation motions in this paper.

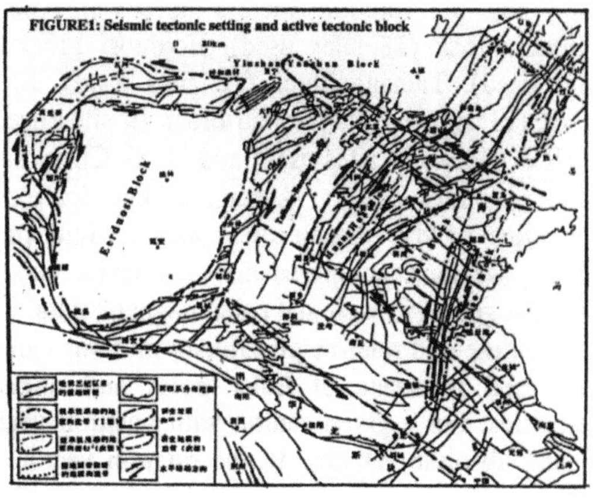

Fig. 1: Seismic tectonic setting and active tectonic block of North China (Xujie, 1996)

International Association of Geodesy Symposia, Vol. 121
Schwarz (ed.), Geodesy Beyond 2000 – The Challenges of the First Decade

2 Tectonic block rotation model

2.1 GPS velocity field

From March 13th, 1992 to April 26th, 1992, 42 stations had been observed by Ashtech MDX II dual-frequency GPS receivers, at each station observation was made four sessions for one day: 19:00 to 23:00, 23:00 to 3:30,3:30 to 8:00, 8:00 to 12:00, the sampling rate was 1/20sec.

Ashtech-Z12 dual-frequency GPS receivers were used in both measurements of 1995 (from August 10 to November 22) and 1996 (from September 10 to December 20). At each station observation was made continuously for at least 72 hours in each campaign, and at a few stations, the observation was made even for 96 hours successively, the sampling rate was 1/30 sec.

We analyzed the GPS data using the GAMIT software [King and Bock, 1995] and PowerAdj Ver. 3.0 [Liu and Shi, 1999]. We included BANGALORE (IISC), SHANGHAI (SHAO), USUDA (USU3), KITAB (KIT3), IRKOUTSK (IRKT), TAEJON (TAEJ), TAIPEI (TAIW), TSUKUBA (TSKB), LHASA (LHSA), POLIGAN/BISHKEK (POL2) etc. available global tracking data from IGS, and orbital data from the global tracking network using a geodetic reference frame defined by ITRF96 In our analyses. We show our velocity estimates computed with respect to IRKT in the Siberia block in Fig. 2. The mean formal uncertainty in northward velocities relative to IRKT in Siberia block for 37 sites common to the 1992 and 1995 surveys is ± 7.2 mm/yr., and ± 6.1mm/yr for northward and eastward components respectively. The mean formal uncertainty in northward velocities relative to IRKT in Siberia block for 87 sites common to the 1995 and 1996 surveys is ± 4.3 mm/yr. and ± 3.2mm/yr for northward and eastward components respectively. We mainly discuss the results of GPS campaigns 95 and 96 in the paper.

Table 1: The datum precept for integrated adjustment of North China GPS network

Data	IGS stations	Frame	Epoch
GPS92		ITRF96	1992.22
GPS95	IISC,SHAO,USU3,KIT3, TAIW	ITRF96	1995.62
GPS96	IRKT,KIT3,SHAO,TAEJ, TAIW,TSKB,USU3,LHSA, POL2	ITRF96	1996.83

2.2 Block rotation model

From the GPS velocity solutions, we have computed the rotation vectors of seven blocks: Yinyan block (18 sites used), Eerduosi block (5 sites used), Taihang block (12 sites used), Jiluxi block (21 sites used), Jiaoliao block (9 sites used), THJLX block (33 sites used) and North China sub-plate (77 sites used). Assuming that horizontal velocities of the chosen sites obey a rigid plate model, we have compared the GPS Euler vectors with the geology result and paleomagnetic data.

Absolute and relative plate/block motions can be expressed in terms of Euler poles and angular velocities. For a given station with position vector r and absolute velocity v, the rotation vector ω relating the motion of the plate/block in a given reference system is

$$v(r) = \omega_i \times r \tag{1}$$

Relations between the plate angular velocity coordinates (ω_x, ω_y *and* ω_z), pole coordinates (Φ, Λ), angular velocity magnitude (ω), and the station velocities (v^n north velocity, v^e east velocity) located on this plate are

$$\begin{cases} v_i^n = R\sin\lambda_i\omega_x - R\cos\lambda_i\omega_y \\ v_i^e = -R\sin\phi_i\cos\lambda_i\omega_x - R\sin\phi_i\sin\lambda_i\omega_y \\ \quad + R\cos\phi_i\omega_z \end{cases} \tag{2}$$

where R is mean radius of the earth, $\omega_x, \omega_y, \omega_z$ are components of angular velocity,

ϕ_i , λ_i are the geodetic latitude and longitude.

We adjusted by least square fit a purely rigid block motion to the stable nodal velocity vectors of such blocks. The angular velocity (rotation rate and pole) of block can be written as

$$\begin{cases} \omega = sqrt(\omega_x^2 + \omega_y^2 + \omega_z^2) \\ \Phi = \arcsin({}^{\omega_x}/_{\omega}) \\ \Lambda = arctg({}^{\omega_y}/_{\omega_x}) \end{cases} \quad (3)$$

If we know angular velocities ω_a and ω_b of two blocks sharing a boundary in the same reference frame, we may get the relative angular velocity for blocks sharing the same boundary

$$\begin{cases} \omega' = sqrt(\Delta\omega_x^2 + \Delta\omega_y^2 + \Delta\omega_z^2) \\ \Phi' = \arcsin({}^{\Delta\omega_x}/_{\omega'}) \\ \Lambda' = arctg({}^{\Delta\omega_y}/_{\Delta\omega_x}) \end{cases} \quad (4)$$

where $\Delta\omega_x = \omega_{ax} - \omega_{bx}$, $\Delta\omega_y = \omega_{ay} - \omega_{by}$, $\Delta\omega_z = \omega_{az} - \omega_{bz}$

2.3 Rotation vectors and horizontal motions

In this section we discuss the results of rotation model in the ITRF96 frame and relative to IRKT (Siberia block) in ITRF96 frame. The main features of Euler model are shown on tables2, 3. North China sub-plate rotates counterclockwise about a pole 71.1°± 1.4°N, 199.1°± 3.2°E, at a rate of 0.389°± 0.074°/my in ITRF96 frame .Its rotation ratio is 0.258°± 0.041°/my relative to Siberia block in ITRF96 frame, and its Euler pole is 53.6°± 0.6°N, 128.5°±1.7°E. The Yinyan block, Eerduosi block, Taihang block, Jiluxi block and Jiaoliao block all rotate counterclockwise in ITRF96 frame. We have compared the motions measured by GPS with the motions described by angular velocities, and with the predictions of NNR-NUVEL-1A [DeMets, C et al., 1994]. This comparison allows us to assess whether the geodetic estimate is representative of tectonic motions. Because NNR-NUVEL-1A is based on rigid plate assumption, only sites far from deformation zones are considered. They are showed in Figures 3 and 4.

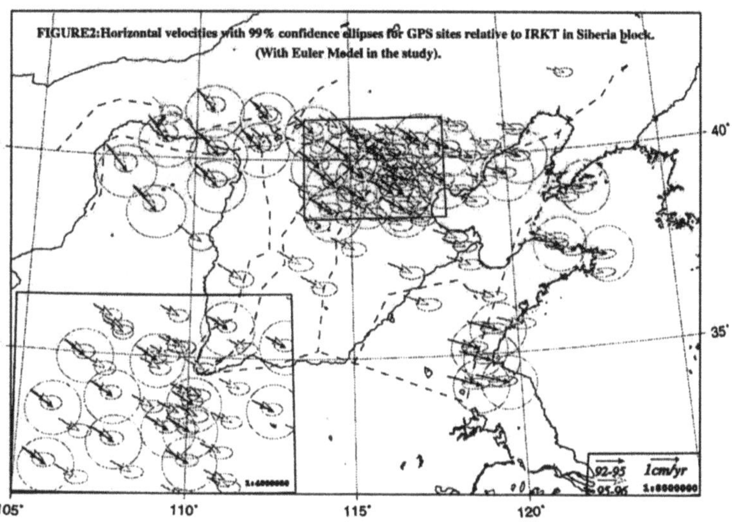

Fig. 2: Horizontal velocities with 99% confidence ellipses for GPS relative to IRKT in Siberia block (with Euler model)

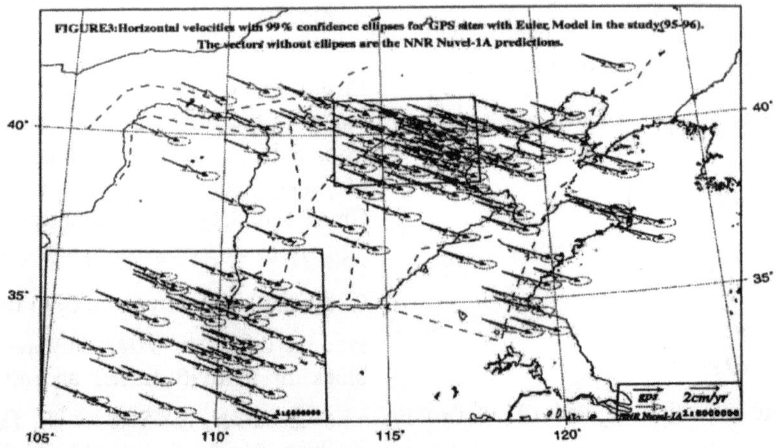

Fig. 3: Horizontal velocities with 99% confidence ellipses for GPS sites with Euler Model and NNR-NUVEL-1A predictions

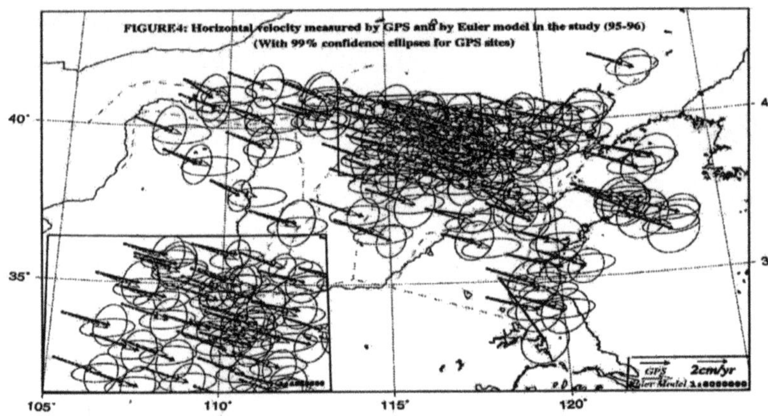

Fig. 4: Horizontal velocity measured by GPS and by Eular model in the study (95-96)

Table 2: Angular velocity describing block motion in ITRF 96(95-96)

Block Name	Angular Velocity			Pole Error Ellipse			
	Φ	Λ	ω	σ_{min}	σ_{max}	ψ	σ_{ω}
	deg	deg	deg/my	deg	deg	deg	deg/my
North China	71.9	199.1	.389	1.4	3.2	89.3	.074
Yinyan	71.1	220.6	.335	1.5	6.1	89.7	.050
Eerduosi	63.4	140.3	.466	2.9	22.2	-89.9	.086
Jiaolao	38.2	267.2	.328	1.9	17.3	89.9	.026
THJLX□	71.9	180.0	.453	2.2	9.5	89.7	.095

Table 2a: Angular velocity describing block motion in ITRF 96(92-95)

Block Name	Angular Velocity			Pole Error Ellipse			
	Φ deg	Λ deg	ω deg/my	σ_{min} deg	σ_{max} deg	ψ deg	σ_{ω} deg/my
North China	69.0	188.4	.563	0.3	13.3	-89.6	.011
Yinyan	-0.1	275.3	.516	1.4	17.1	-89.6	.016
Eerduosi	57.5	125.9	1.045	1.0	16.5	-90.0	.023
Jiaolao	55.1	138.1	1.274	2.1	44.3	89.9	.062
THJLX$^{\square}$	56.2	239.7	.435	4.4	26.5	90.0	.085

Table 3: Angular velocity describing block motion relative to IRKT in ITRF 96 (95-96)

Block Name	Angular Velocity			Pole Error Ellipse			
	Φ deg	Λ deg	ω deg/my	σ_{min} deg	σ_{max} deg	ψ deg	σ_{ω} deg/my
North China	53.6	128.5	.258	0.6	1.7	87.0	.041
Yingyan	54.2	124.4	.198	0.7	3.5	89.5	.012
Eerduosi	37.6	115.8	.412	0.6	2.1	89.6	.028
Jiaolao	18.1	275.6	.115	1.7	9.6	90.0	.011
THJLX$^{\square}$	53.6	127.8	.331	2.1	9.2	-89.9	.067

Table 3a: Angular velocity describing block motion relative to IRKT in ITRF 96 (92-95)

Block Name	Angular Velocity			Pole Error Ellipse			
	Φ deg	Λ deg	ω deg/my	σ_{min} deg	σ_{max} deg	ψ deg	σ_{ω} deg/my
North China	49.1	125.5	.405	0.1	6.2	-90.0	.0001
Yingyan	-37.5	289.3	.361	1.3	4.5	-89.3	.051
Eerduosi	41.0	112.1	.978	0.2	3.3	89.6	.011
Jiaolao	41.4	123.5	1.712	1.4	31.3	89.6	.069
THJLX$^{\square}$	59.3	158.9	.161	1.4	11.7	89.9	.012

These figures and tables indicate that a. North China sub-plate is rigid block as a whole. GPS velocity field is basically agreement with NNR-NUVEL-1A predictions b. The inner deformation for every tectonic unit has its character.

2.4 Relative motion for block pairs

We can get the relative angular velocities for block pair share a boundary from Eq. (4). These results are showed in table 4,the relative motions for block pairs are showed in Fig.5a, b, c, and d.

Table 4: Relative angular velocities for block pairs sharing a boundary in ITRF96(95-96)

Block Pairs	Φ'(deg)	Λ (deg)	ω(deg/my)
Yinyan/Eerduosi	-24.3±25.4	292.2±17.0	.237±.073
Jiluxi/Jiaolao	38.6±18.4	114.7±44.5	.134±.038
Yinyan/ THJLX□	-50.7±75.1	312.7±20.7	.145±.082
Eerduosi/ THJLX□	-5.9±50.9	98.4±29.9	.132±.078

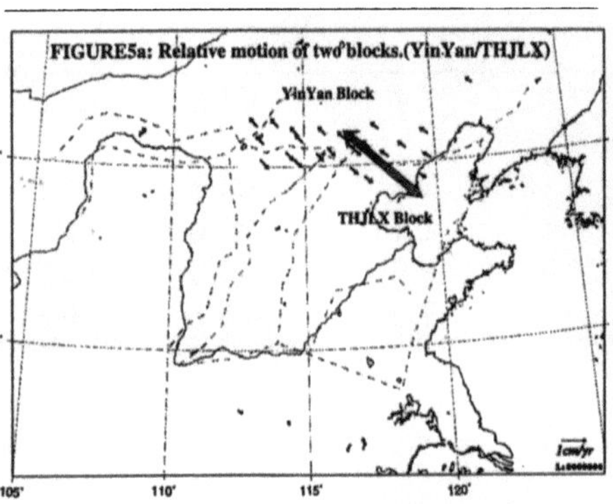

Fig. 5a: Relative motion of two blocks, YinYan/THJLX

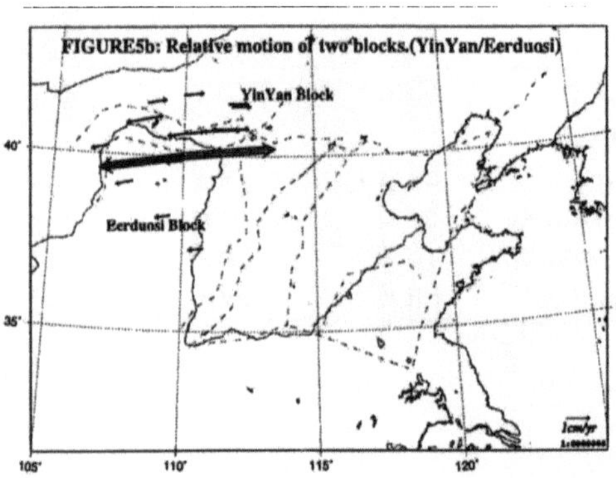

Fig. 5b: Relative motion of two blocks, YinYan/Eeduosi

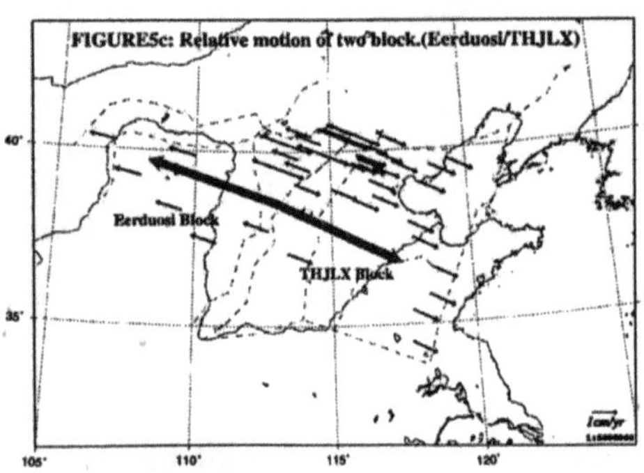

Fig. 5c: Relative motion of two blocks, Eerduosi/THJLX

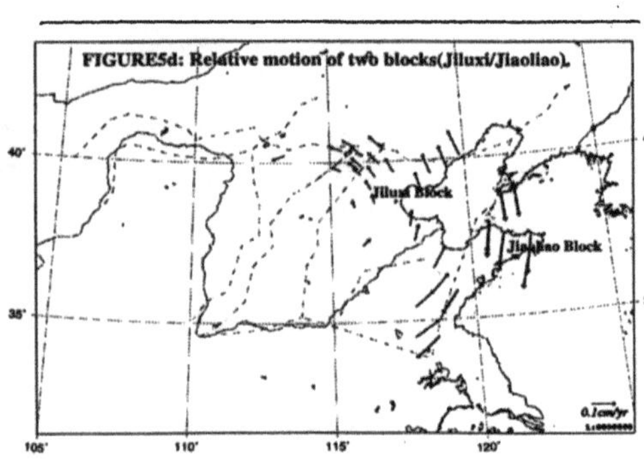

Fig. 5d: Relative motion of two blocks, Jiluxi/Jiaoliao

The Table 4 and Fig. 5a, b, c, d indicate that THJLX and Eerduosi block appear as extensional motion with slight left lateral component relative to Yinyan block in their north, and THJLX block appears as stronger extensional motion with slight right lateral component relative to Eerduosi block in its west. There are of compressional motion in the underside of the Tan-Lu fault zone,and there are of stronger extensional motion with right rotation in the upside of the Tan-Lu fault zone.

3 Geological evidences

Since Neocene (about 12 million years ago), North China plain block, Taihang Mst. Block, and Eerduosi block have been rotating counter-clockwisely.

In the idealized case of rigid blocks and straight parallel faults with a variable spacing that terminate on a straight reference boundary fault displacements are proportional to the widths of blocks, i.e., to fault spacing. Therefore the ratio k=d/w, where d is fault displacement and w is fault spacing, is constant throughout the domain. If the block rotate the angle δ, and original angle between fault to reference boundary is α, then k, δ and α have the relation as following [Garfunkel and Ron, 1985,Xu Xiwei, 1994]:

$$k = d/w = \sin\delta / \sin(\alpha+\delta) \quad (5)$$

Since Pliocene in Cenozoic afternoon, there are 9.83km right lateral displacement in Taihang Mts. Block, whose width is 252km,then the k=0.039. However, the angle of NNE-orientated Shanxi rift valley system to E-W direction boundary is that the Taihang Mts. Block rotate at ratio 0.18°/my (=.δ/12my) In addition, another evidence of counterclockwise for every block in North China subregion is that there are interphase extending and companding distributions, which fit the counterclockwise rotation, in southern end and northern end near main boundaries. It is estimated that the rotational magnitude is about 2.2° to 3.4° for blocks in North China subregion [Xu Xiwei, 1994]. 85° ($\alpha+\delta$), thus the Taihang Mts. Block counterclockwise about δ=2.2° since Pliocene in Cenozoic afternoon [Xu Xiwei, 1994]. That is to say

Thus we can estimate that the North China block rotate at average ratio is about 0.18°/my to 0.28°/my, which fits the GPS result.

4 Cenozoic paleomagnetic evidences

It is determined that rotational magnitude for different stratum block may be obtained using primary magnetic declination D° and magnetic dip I° in different period comparing with the pre-magnetic declination D°_{exp} and pre-magnetic dip I°_{exp}. The rotational magnitude to vertical axis is

$$R = D_{exp} - D \quad (6)$$

The Cenozoic paleomagnetic data for the North China is listed in Table 5.

It shows that Taihang Mts. Block rotate in counterclockwise 1.3° since Miocene, and North China plain block rotate in counterclockwise 3.7° to 6.3° since Miocene-Pliocene. Their average rotational ratio (0.302°/my) fit to the rotational ratios from GPS measurements.

5 Discussion and conclusion

Eerduosi block, Taihang block, Juluxi block all rotate counterclockwise. The relative motion of blocks for GPS data show that activities of blocks are agree with the activities of seismic and geological structure as a whole.

Block Euler model using GPS data show concord with that of using active fault data. The velocity field of the sub-plate of North China with GPS measurements is agree with that of the NNR-NUVEL-1A model, which show that North China sub-plate is of rigid characteristic as a whole.

Chinese continent is bounded on the Eurasian plate, Indian plate and the Pacific plate, its tectonic motion and deformation are the result of these interactive plates. Indian plate moves about 50mm/yr to Eurasian plate in the southwest of Chinese continent, which results in an intensive push and press force northward in the western part of China. The Pacific plate subducted under the Eurasian plate, which caused a push and press force in SWW direction in the eastern part of China. From east to west in the Chinese continent, the principal compressive stress axis of tectonic stress field was deflected into NEE direction with nearly NS direction. Most of areas in the eastern part of China rotated clockwise from the beginning of Oligocene. However, the Qinghai-Tibet plateau uplifted quickly and extended outside with the future colliding of the Eurasian plate and Indian plate from Neogene, which resulted in an intensive push and press action in the NEE direction in the southern part of western boundary of the North China sub-plate. Thus there were a set of push and press forces acted on the western boundary of the Eerduosi block, these forces's magnitude decrease progressively from south to north, and their direction deflected into NEE direction with nearly EW direction from West to East.

Table 5: Cenozoic paleomagnetic data and rotational ratio for the North China

No.	Name	Time	Lat.	Long.	D°	I°	D°_{exp}	R°	ω $(°/my)$	Ref.
1	XYT	N_1	37.6	113.8	11.7	58.4				Xu[1994]
2	XYE	N_1	37.6	113.8	12.6	59.3				Xu[1994]
3	XYD	N_1	37.6	113.8	6.4	50.5				Xu[1994]
4	HBX	N_1	38.0	114.0	3.5	59.5				Xu[1994]
	Mean.				7.5	58.4	8.8	-1.3	-.072	
5	SDY	N_{1-2}	35.8	118.6	2.0	55.5				Xu[1994]
6	SDL	N_{1-2}	36.5	118.5	0.1	55.8				Xu[1994]
7	SDW	N_{1-2}	36.6	119.1	187.6	-56.4				Xu[1994]
8	SDL	N_{1-2}	36.6	118.5	15.9	61.3				Lin[1987]
9	SDL	N_{1-2}	36.2	118.5	3.2	53.4				Lin[1987]
10	SDL	N_{1-2}	36.4	118.8	181.7	-53.3				Lin[1989]
	Mean.				5.3	56.2	9.0	-3.7	-.308	
11	NJP	N_{1-2}	32.3	118.3	18.6	42.9				Xu[1994]
12	NJL	N_{1-2}	32.5	118.3	354.7	42.2				Liu[1976]
13	NJF	N_{1-2}	31.9	118.8	195.5	-54.2				Liu[1976]
14	NJX	N_{1-2}	32.5	118.8	180.4	-43.0				Shao[1989]
	Mean.				2.3	48.4	8.6	-6.3	-.525	
Average									-.302	

N : Neocene N_1 : Miocene N_2 : Pliocene N_{1-2} : Miocene-Pliocene XYT : Xiyangtaoyipo XYE : Xiyangerlanggo XYD : Xiyangdongjiedu HBX : Hebeixihuashan SDY : Shangdongxishui SDL : Shangdongliangq SDW : Shangdongweifan NJP : Nanjingpinshan NJL : Nanjinglinyanshan NJF: Nanjingfangshan NJX : Nanjingxiaopanshan

Considering the rigid characteristic of Eerduosi block and its adjacent blocks constrains, these forces resulted in counterclock rotating of Eerduosi block, Taihang block and Juluxi block.

Acknowledgment. Mr. Zhizhao Liu and Mr. Xingkang Hu accomplish the GPS data preprocessing. Mr. Lixiang Dong with GMT plots the figures in the paper. This work has been supported by National Nature Science Foundation of Chinese (No. 49725411).

References

Ding G.Y. Lu Y.Z. (1986). A preliminary analyzes of present-day motion in the intraplate of China, Chinese Science Bulletin, 18, pp.1412-1415.

Editorial Board for Lithospheric Dynamics Atlas of China (1990). Lithospheric Dynamics Atlas of China, China Cartographic Publish House, Beijing, China, pp.142-153

Garfunkel Z. (1985). Block rotation and deformation by strike slip faults, 2, the properties of a type of Macroscopic discontinuous deformation. Journal of Geophysical Research, Vol.90, No.B10, pp.8589-8599.

Hong Hanjing Wang Yipeng Shen Jun Li Chuanyou (1998).Mean Velocity Field and its Dynamic Implication of Continental Crustal-Block's Motion in China. Research on Active Fault,6,pp.17-29

Huang Liren Guo Liangqian Ma Qing (1998). Preliminary Analysis of Re-measurement Data from GPS Monitoring Network in North China. Resarch on Active Fault, 6, pp.31-41

Jean-Francois Cretaux, (1998), Present- day tectonic plate motions and crustal deformations from the DORIS space system. Journal of Geophysical Research, Vol.103, No.B12, pp.30167-30181.

Li Yanxing Hu Xinkang Zhao Chengkun Wang Min Guo Liangqian Xu Jusheng (1998). Establishment of the GPS Monitoring Network in North China and Relationship between the crustal horizontal motion and the stress field and seismicity, Earthquake Research in China, Vol.14,No.2,pp.116-125

Lin Jinglu (1987). Paleomagnetic data for China (1). Geology Science, 2,pp.183-187

Lin Jinglu (1989). Paleomagnetic data for China (2). Geology Science, 4, pp. 400-404

Liu Chun Cheng Guoliang (1976). Journal of Geophysics, 2, pp.125-137, Shao Jiaji (1989). Geology Review, 35(2). pp.97-106

Xu Caijun Chao Dingbo Jing Biaoren Xu Jusheng (1997). Block Rotation Model and its dynamics in North China area. In: Proceedings of International Symposium on Current Crustal Movement and hazard reduction in east Asia and southeast Asia, Wuhan, P.R.China, Nov.4-7,pp.235-245

Xu Xiwei Cheng Guoliang Ma Xingyuan Sun Yuihang Han Zhujun (1994). Rotation Model and Dynamics of Blocks in North China and its Adjacent Areas. Earth Science—Journal of China University of Geosciences, Vol.19, No.2, pp.129- 138

Realization of a terrestrial reference frame for large-scale GPS networks

D. Angermann*, J. Klotz, C. Reigber

GeoForschungsZentrum Potsdam, Division 1, D-14473 Potsdam, Germany

Abstract. In cooperation with various partners in Europe, North America and in the host countries, GFZ Potsdam has established and repeatedly observed three major GPS networks: in South America the SAGA network, in Central Asia the CATS- and in South-East Asia the GEODYSSEA-network.

This paper focusses on data processing and datum definition issues for these networks. In general GPS data for these networks were processed simultaneously with IGS network data to tie the fiducial-free network solutions with the ITRF. Approximately 20 globally distributed IGS stations were included when computing global solutions, a few IGS stations in or adjacent to the respective region, when computing regional solutions. Mean station position residuals between our global network solutions and ITRF97 coordinates are 1cm horizontally and 1-2 cm vertically. For the regional solutions the station position residuals range from 2-5 mm for the horizontal to 10 mm for the vertical component.

Our GPS-derived results are interpreted in terms of regional tectonics and relative plate motions. For this the datum of the station velocities is defined in a regional reference frame. Datum definition results are presented for the CATS and SAGA networks.

Keywords. Terrestrial reference frame, GPS.

1 Introduction

In the last years, space geodetic observations, such as Very Long Baseline Interferometry (VLBI), Satellite Laser Ranging (SLR), Global Positioning System (GPS) and Doppler Orbit Determination and Radio Positioning Integrated by Satellite (DORIS) have become a key tool in plate tectonic studies.

*) since July 1, 1999: Deutsches Geodätisches Forschungsinstitut, Marstallplatz 8, D-80539 München, Germany

The data of these systems provide an excellent basis for many geodynamic applications, such as the observation of global plate motions and intraplate deformations. To contribute to the understanding of the geodynamic processes at plate margins, the GFZ Potsdam established, in cooperation with various organizations in the host countries, large-scale GPS networks in Chile and Argentina (SAGA), in the Pamir-Tienshan region (CATS) and in South-East Asia (GEODYSSEA), see Figure 1.

The observation of global plate motions and intraplate deformations require an appropriate definition of the reference frame. The latest and most precise realization of the International Terrestrial Reference System (ITRS) is the ITRF97, Boucher et al. (1999), Sillard et al. (1998), which was realized by the International Earth Rotation Service (IERS). The ITRF97 solution is produced from a combination of sets of station coordinates and velocities determined by means of the techniques mentioned previously. It consists of a set of station coordinates, a velocity field and a full variance-covariance matrix for about 300 globally distributed stations (see Figure 1).

This paper deals with the data processing of the regional GPS networks using the GFZ Software EPOS and the datum definition of these networks in the ITRF97. We concentrate on the realization of a reference frame for the CATS and SAGA network. Since the main focus of the geodynamic studies in Central Asia and South America is on regional deformations, we define the CATS network with respect to »stable« Eurasia and the SAGA network with respect to the stable part of the South American continent.

2 Data analysis

The GPS data of these large-scale networks were processed simultaneously with data of the IGS station network (International GPS Service for Geodynamics, 1998) to tie the network solutions to the ITRF. The processing is done with the GFZ Software EPOS, Angermann et al. (1997a). This software package makes use of undifferenced phase

International Association of Geodesy Symposia, Vol. 121
Schwarz (ed.), Geodesy Beyond 2000 – The Challenges of the First Decade

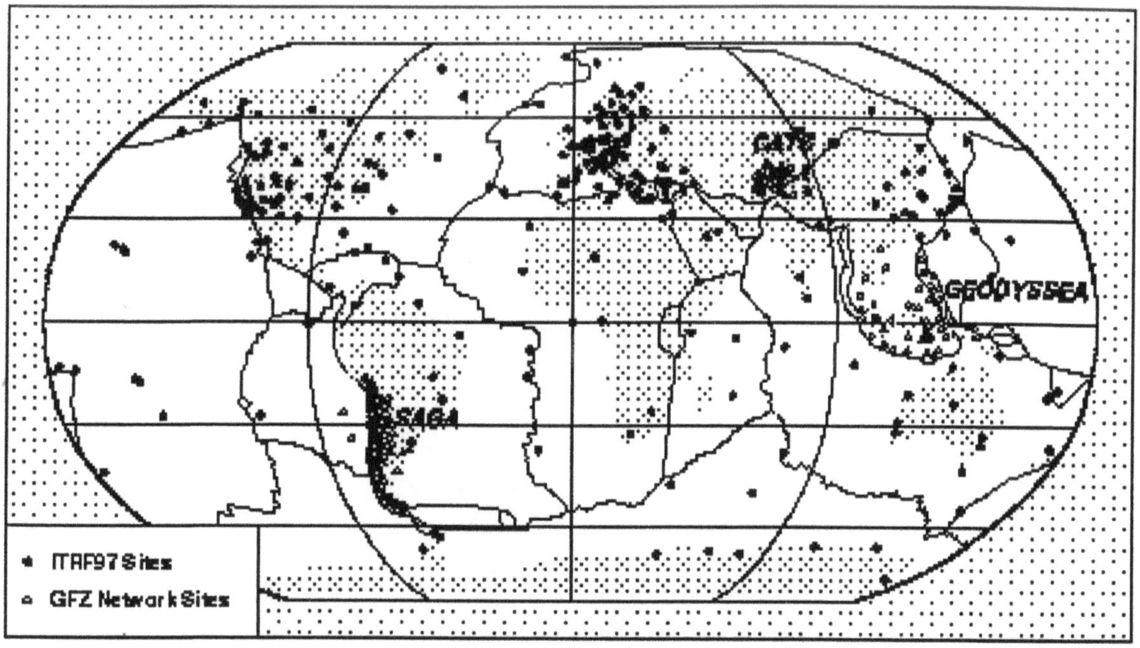

Fig. 1: ITRF97 Sites and large-scale GPS Networks of GFZ Potsdam; CATS=Central Asian Tectonic Science, SAGA=South American Geodynamic Activities, GEODYSSEA=Geodynamics of South and South East Asia.

and pseudorange observations, and applies a rigorous least squares adjustment for the parameter coordinates, ambiguities, receiver and satellite clock parameters for each observation epoch and tropospheric parameters for each station. Optional is the solution for orbit and earth orientation parameters (EOP). The network computation is done in the form of daily solutions with ionosphere-free linear combinations of phases and pseudoranges. We used an elevation cut-off angle of 15 degree and a sampling rate of 4 minutes. The combined IGS orbits and the corresponding EOP's were fixed and all station coordinates were solved without any constraints (fiducial-free solution). The ambiguities resulting from the ionosphere-free linear combination of phases were estimated as real values (float solution). The tropospheric effects were modelled with the CfA mapping function, Davis et al. (1985) using meteorological data from a standard atmosphere. The tropospheric zenith path delay was estimated for each station at 4 hour intervals. To account for the antenna phase center variations, the IGS correction tables were applied, Rothacher and Mader (1996).

3 Reference frame realization

The reference frame of the fiducial-free network solutions is defined by the combined IGS orbits (accuracy 5-10cm) and the corresponding EOP's. In order to derive more precise station coordinates we tie the network solutions to the International Terrestrial Reference Frame (ITRF) by estimating a seven parameter Helmert-transformation. We use regional and global IGS station constellations for this transformation. Results for the datum definition of the SAGA network show, that the residuals for the IGS stations are significantly smaller in the case of a regional transformation, when compared with those derived from a global transformation, Angermann et al. (1999a). In that study we obtained mean residuals for the IGS stations between the global network solutions and the ITRF97 coordinates in the order of 1 cm horizontally and 1-2 cm for the height. With respect to a South American reference frame the mean station coordinates residuals are 2-5 mm horizontally and below 1 cm for the height. In case of a global transformation a significant amount of coherent noise may affect station coordinates in specific regions. This is due to the fact, that both the network computations as well as the realization of the ITRF may be influenced by errors in the underlying physical and geometrical models (e.g. plate motion models, solid earth tides, ocean and atmospheric loading, satellite orbits, tropospheric and ionospheric models), which may affect the station coordinates systematically. Due to this, the mean residuals of the IGS stations are reduced significantly in the case of the regional transformation.

Site velocities can be related to either a global or

a regional reference frame. The relative velocities in a regional frame are more precise, as a result of spatial correlations of the estimated parameters. For the datum definition of the SAGA network and the CATS network we prefer a regional frame with respect to the South American plate and the Eurasian plate, respectively. The site velocites of the GEODYSSEA network are related to a global reference frame, Angermann et al. (1998), Simons et al. (1999), Wilson et al. (1998).

3.1 CATS network

In the scope of the CATS (Central Asian Tectonic Science) project, the GFZ Potsdam installed, in cooperation with different geodetic and geologic institutions in China, Kyrgyztan, Kazakhstan, Uzbekistan and Tadjikistan, a regional 1200 * 1800 km GPS network that covers a major part of the Tien Shan, the Northern Pamir and the Tarim. The CATS network has been periodically re-observed since 1992.

As the focus of the geodynamic studies in Central Asia is on regional deformations, the CATS network has been tied to the ITRF by using regional distributed IGS stations. The mean residuals of a Helmert-transformation between the CATS98 solution and the ITRF97 station coordinates for 9 IGS stations in Eurasia are 2-4 mm horizontally and 5 mm for the height (see Table 1). These values represent the regional network accuracy.

In order to realize a regional reference frame for the station velocities, we derive in a first step a best fitting Euler vector using a least squares adjustment approach and ITRF97 velocities of 35 IGS stations distributed over the entire »stable« Eurasian plate. The ITRF97-based Euler vector is located at 59.4 degree N, 98.5 degree East with a rate of 0.26 degree/Myr, which in close agreement with the estimations of Larson et al. (1997) and the estimations of the Actual Plate KInematic and Deformation Model (APKIM 8.80), Drewes (1997). The space-geodetic derived Euler vector differs significantly from that of the geophysical-geological global plate motion model NNR-NUVEL-1A, Argus et al. (1991), DeMets et al.(1990), DeMets et al. (1994), (see Figure 2). The mean residuals of the ITRF97 velocities with respect to our estimated Euler vector are in the range between 1-2 mm/yr, indicating that the 35 stations selected represent stable Eurasia (see Figure 2). Furthermore this result proves the high precision of the realization of the Eurasian reference frame.

Table 1: Residuals of Helmert-Transformation between CATS98 solution and ITRF97

IGS stations	Residuals [mm]		
	Latitude	Longitude	Height
Zwenigorod, Russia	-5	-4,4	-4,1
Tsukuba, Japan	0,4	-4,8	-2,1
Shanghai, China	1,3	-0,6	4,1
Xian, China	0	6	2,7
Kitab, Uzbekistan	2,3	-3	-0,6
Bishkek, Kyrgiztan	4,3	0,9	3,2
Krasnoyarsk, Russia	-4,5	0,7	9,7
Wuhan, China	-1,6	1,2	-10,4
Irkutsk, Russia	-1,9	3,5	-2,6
Mean	2,9	3,4	5,4

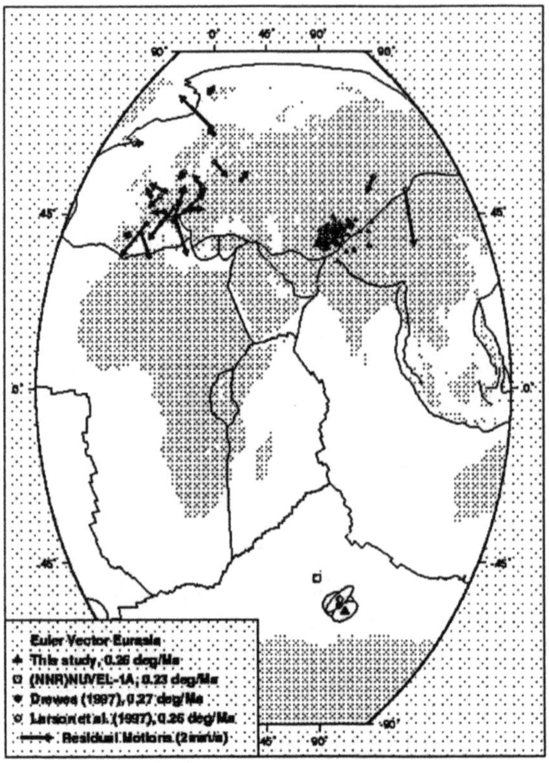

Fig. 2: ITRF97-derived Euler Vector for Eurasian plate along with the 95% confidence ellipse is shown in comparison to (NNR)NUVEL-1A model and the results of earlier space geodetic studies. The residual motions are shown for 35 sites on the Eurasian plate.

In a second step, we use our ITRF97-based Euler vector to transform the CATS network solutions to the Eurasian reference frame. As a result, we get relative velocities for the CATS network stations with respect to the Eurasian plate, which have been used for the geodynamic interpretation of the CATS results, which will be presented in seperate papers.

3.2 SAGA Network

Within the SAGA (South American Geodynamic Activities) project, the GFZ Potsdam established in cooperation with many partner institutions of the host countries in 1993/94 a network of 215 GPS stations covering the whole territory of Chile and the western part of Argentina. Results of the periodically re-observed SAGA stations are discussed in seperate papers, Klotz et al. (1996), Klotz et al. (1999).

The datum of the SAGA network is defined with respect to the South American plate. The SAGA network has been tied to the ITRF97 by using regional distributed IGS stations. The mean residuals of a Helmert-Transformation between the SAGA solutions and the ITRF97 station coordinates are in the order of 2-5 mm, Angermann et al. (1999a).

In order to derive a regional South American reference frame for the station velocities we assume that the IGS stations Fortaleza (FORT), Kourou (KOUR), Brazilia (BRAZ), La Plata (LPGS) and Ascension Island (ASC1) represent the stable body of South America. With the ITRF97 velocities of these sites we estimate an Euler pole position of -21.9 ± 3.9°N, 204.6 ± 7.9°E with a rotation rate of 0.10 ± 0.01°/Myr. Our result is in good agreement with the estimations of the Actual Plate KInematic Model (APKIM8.80), Drewes (1997) of -19.4 ± 5.1°N, 210.1 ± 12.3°E and 0.127 ± 0.010°/Myr. The corresponding values of the global plate motion model (NNR) NUVEL-1A are -25.4°N, 235.6°E and 0.12°/My. The rate and the latitude of the pole agree with the space geodetic estimations, however for the longitude there is a difference of more than 25°.

The internal consistency of the reference frame applied here is best assessed by examining the residuals of all stations located within the stable part of South America. The residuals of the ITRF97 velocities with respect to the ITRF97-based Euler vector are shown in Table 2. The horizontal velocity residuals for the five stations of the stable part of South America (upper part of Table 2) are in the range of 1-2 mm/yr and thevelocity residuals in vertical directions are below 1 cm/yr. The residual motions of these stations are below their three sigma standard deviations, expect for the north component of Fortaleza and the vertical component of Brazilia.

Table 2: Site velocities with respect to the stable part of South America along with their one sigma standard deviations. The five stations in the upper part of the table belong to the South American plate. The velocities of the stations Robinson Crusoe Island (IRCR) and San Felix Island (IRCR) are estimations from the SAGA94/96 campaigns; the other stations show ITRF97 velocities. The ITRF97 velocity of station GALA is replaced by the CODE solution, CODE (1999).

Stations	North [mm/yr]	East [mm/yr]	Height [mm/yr]
KOUR	1.1 ± 0.4	0.4 ± 0.8	1.5 ± 0.9
FORT	1.6 ± 0.4	0.0 ± 0.6	1.8 ± 0.6
BRAZ	-1.3 ± 0.7	-1.2 ± 1.0	-8.9 ± 1.0
LPGS	-1.2 ± 2.2	0.1 ± 2.1	-3.1 ± 2.4
ASC1	-0.4 ± 1.6	0.7 ± 3.5	-8.1 ± 5.9
BOGT	3.6 ± 0.9	10.6 ± 2.0	2.0 ± 1.1
AREQ	1.9 ± 0.4	14.3 ± 0,5	2.7 ± 0.5
SANT	4.1 ± 0.6	20.8 ± 0.6	7.9 ± 0.7
EISL	-16.1 ± 0.7	68.7 ± 0.8	4.0 ± 0.9
IRCR	11.9 ± 1.8	67.2 ± 1.9	-7.2 ± 3.7
ISFE	10.7 ± 2.2	62.2 ± 2.5	3.8 ± 3.8
GALA	4.7 ± 1.4	48.4 ± 2.6	21.6 ± 2.2

These results indicate that the five stations selected represent stable South America. Using the ITRF97-based Euler vector the SAGA velocities, including the GPS derived velocities of the IGS sites derived at GFZ and the CODE solution for Galapagos, were transformed to the datum of stable South America, making sure that the different solutions are related to the same reference frame. Fig. 3 shows the horizontal station velocities for the 8 IGS stations in South America and the 4 stations of the Nazca Plate. The SAGA 94/96 velocities and the corresponding ITRF97 velocities agree within 1-3 mm/yr, the differences being insignificant. The stations Bogota, Arequipa and a Santiago, which are located within the plate boundary zone, show significant motions of 1-2 cm/yr with respect to the South American reference frame. Kendrick et al. (1999) realized in a recent paper a reference frame that represents the rigid core of the South American continent by minimising the motions of those stations located within the stable part of South America. The station velocites given by Kendrick et al. (1999) agree

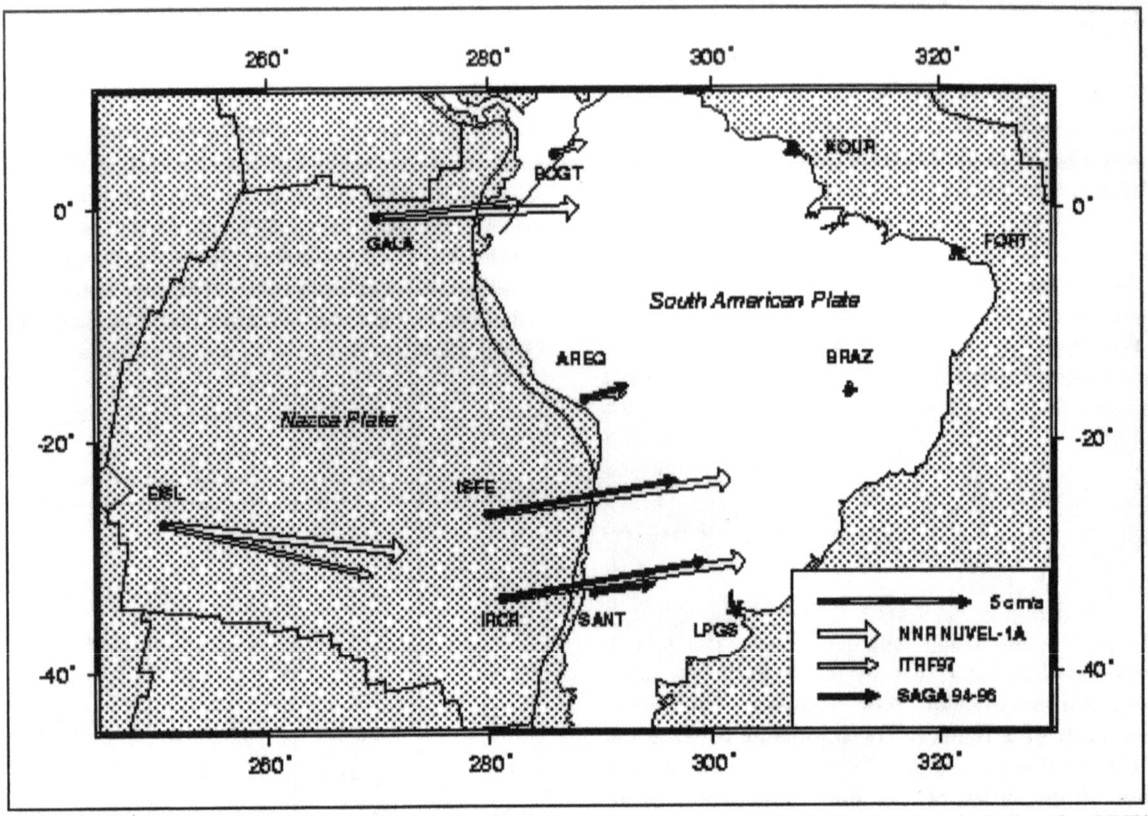

Fig. 3: ITRF97 and SAGA-derived horizontal displacements with respect to stable South America, including the (NNR) NUVEL-1A velocities for the stations located on the Nazca plate. The ITRF97 velocity of station GALA is poorly estimated, therefore the CODE solution is used, CODE (1999)

within their standard deviations with the results derived here. GPS data from four sites in the Nazca Plate (Easter Island, Galapagos, Robinson Crusoe and San Felix Islands) and from five sites in the stable core of the South American Plate enabled us to estimate the Euler vector of the Nazca Plate with respect to South America. The observed velocities of Easter Island (EISL), Galapagos (GALA), Robinson Crusoe (IRCR) and San Felix (ISFE) are significantly slower than the global plate model (NNR) NUVEL-1A velocites for those four sites. (see Fig. 3). With the velocities of these four stations we derived an Euler vector for the Nazca plate with respect to the South American plate located at 42.5 ± 3.4°N, 267.0 ± 1.5°E with a rate of 0.63 ± 0.02°/Myr. This result agrees with the previously estimated Euler vector of 48.8 ± 3.4°N, 268.3 ± 1.6°E with a rate of 0.59 ± 0.02°/Myr in the ITRF96 frame using the same set of stations, Angermann et al. (1999b). The (NNR)NUVEL-1A and earlier space-geodetic studies of, Larson et al. (1997) and Norabuena et al. (1998), give rotation rates that are significantly faster.

4 Conclusions

In the last years the scope and accuracy of space geodetic techniques expanded greatly. This leads to a continously improvement for the realization of the International Terrestrial Reference System (ITRS). The latest realization, the ITRF97 provides an excellent reference frame for the positioning of regional network solutions. In this study we defined the datum of the CATS and SAGA network with respect to a regional frame. The mean residuals for regional distributed IGS stations are 2-5 mm, which is assumed to be a realistic value for the regional network accuracy. In case of a global Helmert-transformation between the network solutions and the ITRF97 station coordinates the mean residuals are in the order of 1 cm. The reason for the significant smaller residuals in the case of a regional transformation is that a significant amount of spatial correlations may affect the estimated parameters.

From the SAGA network data we derived an Euler vector for the Nazca plate with respect to the South American plate, which differs significantly from NUVEL-1A and the results of earlier geodetic studies (e.g. Larson et al. (1997), Norabuena et al. (1998). Our result derived from four sites on the Nazca plate suggests a significantly smaller convergence rate than that predicted by NUVEL-1A.

In the near future further improvements for the

definition and realization of the Terrestrial Reference System are expected due to the ongoing extension and densification of the IGS station network and due to international projects aiming at the improvement of reference frames in specific regions, such as the SIRGAS (Sistema de Referencia Geocentrico para a America do Sul) Project, SIRGAS (1997) and the APSG (Asia Pacific Space Geodynamics) Project, APSG Report (1998).

References

Angermann, D., G, Baustert, R. Galas, S.Y. Zhu (1997a). EPOS.P.V3 (Earth Parameter and Orbit System): Software User Manual for GPS Data Processing.*Scientific Technical Report STR97/14*, GeoForschungsZentrum Potsdam, 52pp.

Angermann, D., J. Klotz, G. Michel, C. Reigber, J. Reinking (1997b). Großräumige GPS Netze zur Bestimmung der Krustenkinematik in Zentralasien und Südamerika. *DVW Schriftenreihe, 41. Fortbildungseminar: GPS-Anwendungen und Ergebnisse '96*, Heft 28, 79-93.

Angermann, D. (1998). Datum Definition of the GEODYSSEA Network. In: GEODYSSEA Final Report, P. Wilson, G. Michel (eds), *Scientific Technical Report STR98/14*, GeoForschungsZentrum Potsdam.

Angermann, D., J. Klotz, C. Reigber (1999a). Geodetic Datum Definition of the SAGA Network. *Proceedings of IAG Section II Symposium*, Munich, in press.

Angermann, D., J. Klotz, C. Reigber (1999b). Space geodetic Estimation of the Nazca-South America Euler Vector. *Earth and Planetary Science Letters*, 171, 329-334.

APSG (1998). Report on the second international meeting of the Asia-Pacific Space Geodynamics Program, M. Pearlman, A. Bonneville (eds), Tahiti, French Polynesia.

Argus, D. F., R. G. Gordon (1991), No-net-rotation model of current plate velocities incorporating plate motion model NUVEL-1, *Geophys. Res. Lett.*, 18, 2038-2042.

Boucher, C., Z. Altamimi, P. Sillard (1999). The International Terrestrial Reference Frame (ITRF97), IERS Tech.Note, 27, Observatoire de Paris.

CODE (1999): CODE Solution »SSC/SSV, CODE, P01 of 1999« submitted to IERS for the contribution to the ITRF98, University of Berne, Berne, Switzerland.

Davis J. L., T. A. Herring, I. I. Shapiro, A. E. Rogers (1985). Geodesy by radio interferometry: Effects of atmosheric modeling on estimates of baseline length, *Radio Sci.* 20, 1593-1607.

DeMets, C., R. G. Gordon, D. F. Argus, and S. Stein (1990). Current plate motions, *Geophys. J. Int.*, 101, 425-478.

DeMets, C., R. G. Gordon, D. F. Argus, and S. Stein (1994). Effect of recent revisions on to the geomagnetic reversal time scale on estimates of current plate motions, *Geophys. Res. Lett.*, 21, 2191-2194.

Drewes, H.: (1997). Combination of VLBI, SLR and GPS determined station velocities for actual plate kinematic and crustal deformation models. *IAG Symp. 119*, R. Forsberg, M. Feissel, R. Dietrich (eds), 377-382.

International GPS Service for Geodynamics. 1997 Technical Report, eds. I. Mueller, K. Gowey, R. Neilan, IGS Central Bureau, Jet Propulsion Laboratory, California Institute ofTechnology, Pasadena, California, U.S.A.

Kendrick, E., M. Bevis, R. Smalley, O. Cifuentes, F. Galban (1999). Current Rates of Convergence across the Central Andes: Estimates from Continuous GPS Observations, *Geophys. Res. Lett.*, 26, 541-544.

Klotz, J., J. Reinking, D. Angermann (1996). Die Vermessung der Deformation der Erdoberfläche, *Geowissenschaften 14*, Heft 10, 389-394.

Klotz, J., D. Angermann, G.W. Michel, R. Porth, C. Reigber, J. Reinking, J. Viramonte, R. Perdomo, V.H. Rios, S. Barientos, R. Barriga, O. Cifuentes (1999). GPS-derived Deformation of the Central Andes Including the 1995 Antofagasta Mw=8.0 Earthquake, *Pure and Appl. Geophys.*, Vol. 154, 709-730.

Larson, K., J. T. Freymueller, S. Philipsen (1997). Global plate velocities from the Global Positioning System, *J. Geophys. Res.*, 102, 9961-9981.

Norabuena, E., L. Leffler-Griffin, A. Mao, T. Dixon, S. Stein, I. S. Sacks, L. Ocola, M. Ellis (1998). Space Geodetic Observation of Nazca-South America Convergence Across Central Andes, *Science*, 279, 358-362

Reigber, C., Xia, Y., Michel, G., Klotz, J., Angermann, D. (1997). The Antofagasta 1995 Earthquake: Crustal Deformation Pattern as observed by GPS and D-INSAR. *Proceedings of 3. ERS Symposium*, Florence, 507-513.

Rothacher, M. and G. Mader (1996). Combination of Antenna Phase Center Offsets and Variations: Antenna *Calibration set IGS 01, IGS Central Bureau / University of Berne*, Berne, Switzerland.

Sillard, P., Altamimi, Z., Boucher, C. (1998). The ITRF96 realization and its associated velocity field, *Geophys. Res. Lett.*, 25, 3223-3226.

Simons, W.J.F., B.A.C. Ambrosius, R. Noomen, D. Angermann, P. Wilson, M. Becker, E. Reinhart, A. Walpersdorf, C. Vigny (1999). Observing Plate Motions in S.E. Asia: Geodetic results of the GEODYSSEA project, *Geophys. Res. Lett.*, 26, 2081-2084.

SIRGAS (1997): SIRGAS Final Report, Working Groups I and II. Edited by Instituto Brasileiro de Geografia e Estatistica (IBGE), Rio de Janeiroimons.

Wilson, P., J. Rais, C. Reigber, E. Reinhart, B.A.C. Ambrosius, X. Le Pichon, M. Kasser, P. Suharto, A. Maijid, P. Awang, R. Almeda,C. Boonphakdee (1998). Study provides data on active plate tectonics in Southeast Asia region, *Eos Trans. AGU*, 79(45), 545-549.

PART 5

Kinematic Systems and Precision Engineering

Fritz Brunner
Paul De Jonge
Heribert Kahmen
Chris Rizos

Building structures as kinematic systems - dynamic monitoring and system analysis

R. Flesch
Österreichisches Forschungs- und Prüfzentrum, Arsenal GmbH, Geotechnisches Institut
A-1030 Wien, Farradaygasse 3

H. Kahmen
Technische Universität Wien, Institut für Geodäsie und Geophysik, Abteilung für Angewandte Geodäsie und Ingenieurgeodäsie, A-1040 Wien, Gusshausstrasse 27 - 29/E1283

Abstract. With different damage szenarios on bridges could be stated that dynamic monitoring of building structures is an interesting contribution to the assessment of the actual structural behaviour and localisation of damages. As up to now mainly the high frequencies of the spectrum of the vibrations are used for system identification, the results are still limited. A survey of the sensors which can be used for the monitoring processes shows that the detection of the total spectrum of the vibrations is possible. It is demonstrated that for instance fluid pressure tiltmeters provide an interesting completion of the system.

Keywords. Forced Vibration Testing, Ambient Vibration Testing, bridge monitoring.

Introduction

Owners of structures and authorities responsible for structural safety realize more and more that important existing structures should be monitored. The goals are assessment of the structural behaviour (safety inspection), improvement of maintenance (optimization of repair, early detection of damages). The basis for monitoring is a very good „understanding" of the actual structural behaviour.

In the case of „dynamic monitoring" the dynamic properties are identified from measured structural response due to a vibration input. The input can be forced- or ambient vibrations. "Forced Vibration Testing (FVT)" is based on controlled exitation of the structure at one point and measurement of the structural response at a large number of points on the structure. Calculating the Frequency Response Functions for every response spectrum and the input force spectrum yields the information necessary for system identification. In contrast to FVT "Ambient Vibration Testing AVT" does not require any controlled excitation. Instead the response of the structure to "ambient" excitation sources like temperature, wind, erection loading, traffic on the structure and micro tremors is recorded. As with FVT the measured time signals are then processed in the frequency domain. Because the forcing function is unknown Frequency Response Functions cannot be calculated. Instead the natural frequencies can e. g. be identified from the peaks of the Power Spectral Densities, computed for the individual time series.

With the natural frequencies gained from FVT or AVT the individual vibration modes can be identified. The natural frequencies and mode shapes can be used to calibrate and verify the Finite Element models (FE). The result is a FE of the structure which is as close to reality as possible. This can be used to analytically determine

- the system`s response to any static or dynamic load within the limits of linearity
- the consequences of structural changes like strengthening, cracking, etc.

1 Transducers, measurement techniques, and analysis

1.1 A survey of the sensors which can be used for bridge monitoring

Normally the instrumentation is provided to monitor wind data, displacements, accelerations, velocities and strains in characteristic parts of the structure during the construction and after it is opened for traffic. Generally the instrumentation consists of a selection of the following sensors:

(a) *DGPS and measurement robots* (automatically pointing and motorized tacheometers [7]) are suitable for monitoring of 3D-relative and absolute displacements. The structure to be monitored can

International Association of Geodesy Symposia, Vol. 121
Schwarz (ed.), Geodesy Beyond 2000 – The Challenges of the First Decade

have an extension of 1 km and more. However, if high accuracy are required, the sampling rate of the measurements should not exceed frequencies of about 0,1 Hz.

(b) *Hydrostatic tiltmeters* are used for the detection of vertical displacements. Depending on the working principles the instruments are characterized by different dynamical properties. While with "fluid level tiltmeters" vibrations can be monitored if the vibrations do not exceed 10^{-3} Hz, with "fluid pressure gauges" vibrations up to about 1 Hz can be detected [10].

(c) *Strain gauges* measure the structural strain. There are electronic resistance, vibrating wire type and electrooptical strain gauges. All the instruments can be used for short term and long term monitoring. Especially electrical resistance gauges are sensitive against temperature drifting.

(d) *Laser alignement systems* consisting of a transmitter (a laser source) and a receiver unit (e. g. a CCD device) which measures the position of a laser beam in relation to an optical target. Vibration and refraction effects in the surrounding of the light source are limiting the accuracy. The systems are suitable for measuring short term and long term vibrations.

(e) *Linear accelerometers and velocity transducers* especially are used to determine dynamic displacements of higher frequencies. The instruments are gravity referenced devices, that means they are sensitive to rotations as well as to linear accelerations. Especially on large structures the displacements are normally forced by a superposition of angular and linear accelerations. Experiences show that the error caused by neglecting the change in g-components can be significant for low frequencies but they diminish rapidly with increasing frequencies. Besides, because of instrumental drifting these instruments are not suitable for long term monitoring.

(f) *Servo inclinometers* are used to measure the rate of inclinations of parts of a structure. Like accelerometers the gauges are also gravity referenced and sensitive to both rotation and linear accelerations. However, errors caused by neglecting the change in g-components diminish in that case with decreasing frequencies. Fig. 1.1 shows the frequency spectrum of the displacements, which can in general be monitored on bridges. The peak with a period of 1 year is mainly caused by temperatures, that with a period of 7 days by traffic load, that with a period of 1 day by temperature and traffic load, that with a period of about 0,5 days by tidal loads and the frequency spectrum with frequencies between 0,1 and 20 Hz meanly by wind and traffic loads. As we can see from fig. 1.1 some sensors are only suitable to detect the lower frequencies, some can only monitor the higher frequencies and some can be used to capture data for the computation of the total spectrum.

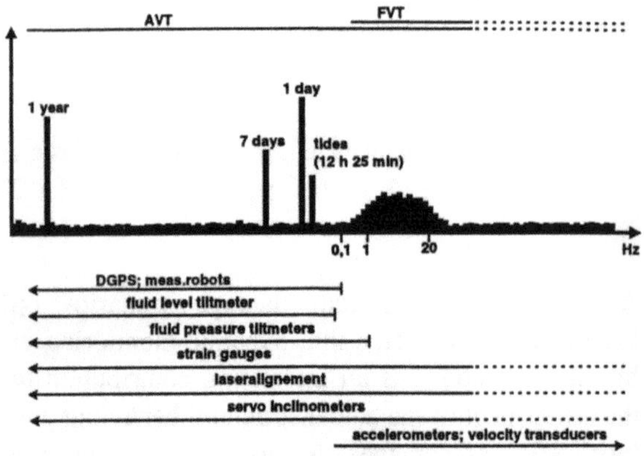

Fig. 1.1: Frequency spectrum of dynamical displacements of bridges and the sensors which serve for monitoring of the displacements

In the past mainly FVT was applied for dynamic bridge monitoring and linear accelerometers or velocity transducers were used to detect the vibrations.. As Fig. 1.1 shows, then only a small part of the spectrum is used for system identification. In that case, however, there is one advantage, the system can be a mobile system and only a few sensors are needed, which can be used step by step, to get the measurement data of a greater number of profiles. On the other hand, as only a small part of the information of the spectrum is used, we have to study by further investigations

- how far is system identification and damage location limited then and
- which sensors could be an interesting completion of the monitoring system to get a more extended spectrum.

1.2 Dynamic insitu test equipment of Vienna arsenal [1]

Within the last years tests of approximately 10 bridges were carried out in order to get a „baseline" for the long term behaviour [3,4,5]. The latest tests were carried out in April 99 at *Talübergang Warth.* The bridge is located on motorway A2, 63 km south of Vienna. One of the non coupled twin bridges (direction to Graz) was tested. The bridge has 7 spans and a total length of 459 m (Fig. 1.2).

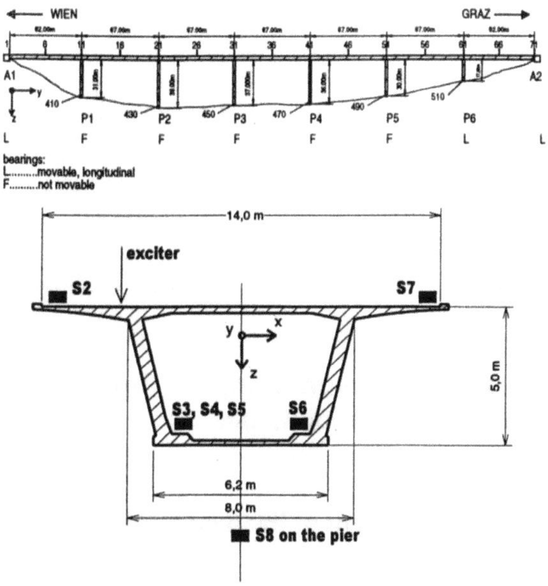

Fig. 1.2: Bridge WARTH, axial and cross section and sensors S2, S3, S4, S5, S6 and S8

It was one goal of this investigation to demonstrate the high capability of measurements using swept sinusoidal force excitation. Arsenal research has developed a very specialized test equipment. The experiments can be carried out under traffic, only one lane has to be closed during the tests.

The bridge was excited by a reaction mass exciter, which is driven by a hydraulic actuator. The maximum excitation force is 25 kN, which can be kept constant from 3,4 Hz upwards. In the case of a maximum force 10 – 15 kN this is between 1,5 and 2 Hz. The maximum reaction mass is 1260 kg (removable steel plates). The total mass of the exciter is about 3000 kg. The maximum stroke of the actuator is 0,1 m and the maximum velocity is 0,938 m/s. The hydraulic pump needs an electrical supply of 380 V / 100 A (diesel- electric power generator). As the exciter could not be connected to the bridge by dowels, the maximum force of excitation was 10 kN in horizontal- and 15 kN in vertical direction. The frequency range 0,5 –11 Hz was swept within 10 minutes.

The frequency sweeps were repeated many times in order to measure the response of the deck in profiles at distances between 5 and 6,7 m. After each sweep the transducers were moved to the next profile. At the piers the response was measured at points close to the foundation. Due to the fact that the frequency sweep is repeated many times, the vibrations induced by traffic, which are (heavy) distortions of the response, can be eliminated quite well.

8 velocity transducers Hottinger SMU 30A with an accuracy of $0,5.10^{-3}$ mm/s were used to measure the response. In the most sensitive measurement range they give 10 V per 0,1 mm/s. Consequently for each measurement point transfer functions and tape recorded time histories are available.

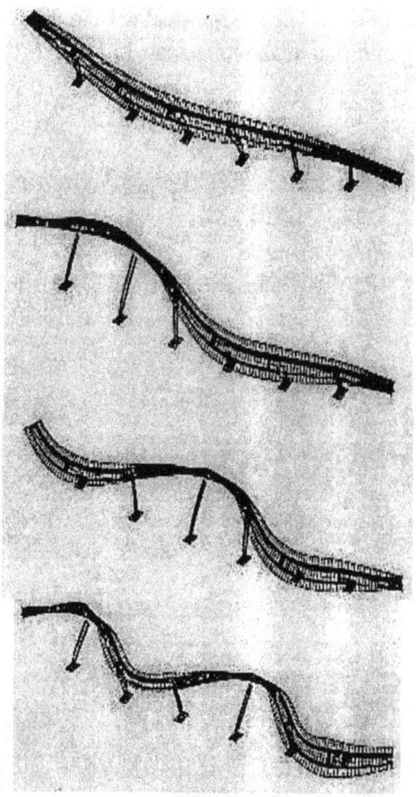

Fig. 1.3: First four calculated horizontal modeshapes, bridge Warth [6].

The bridge was excited in horizontal and vertical direction and the response was obtained for each of the 71 profiles. 5 additional (special) sweeps with the transducers in the same profile but with different force levels and different duration of the sweeps were carried out. In addition, in some of the profiles also the traffic- and ambient induced vibrations were recorded for several minutes (about 2min 40s). In general, small differences in the identified modeshapes will be found when exciting structures from different driving points. Hence for structural monitoring and damage detection always measurements from the same driving point must be compared.

13 vertical and 6 horizontal modes could be identified with driving point E2 for vertical excitation and 9 horizontal modes could be identified with driving point E1 for horizontal excitation. Fig. 1.3 shows the first four calculated horizontal modeshapes [6].

1.3 Dynamic insitu test equipment of T.U. Vienna/example "Rose Bridge" Tulln

The rose bridge is an asymetrical two-span cable-stayed bridge. The bridge deck of the short span is 103 m and that of the long span 177 m.

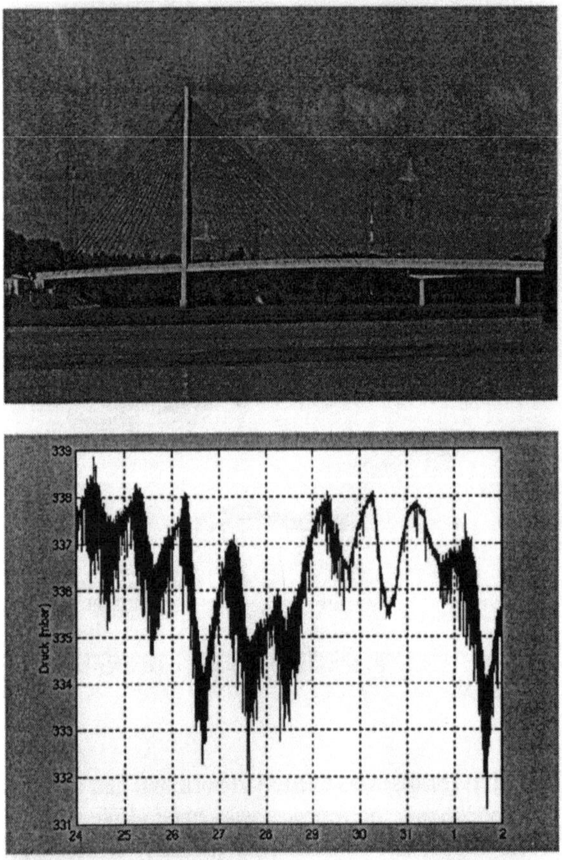

Fig. 1.4: Vertical displacements in the middle of the long span of the "Rose Bridge" [10]

During the last years a research project war organized to investigate the capability of fluid pressure tiltmeters for bridge monitoring. [10]. Fig. 1.4 shows some results of the experiments: vertical deflections registered in the middle of the long span during the Easter days in 1997. Traffic load was largest in the middle of the week and in the middle of the days. The load decreased after Saturday afternoon before Easter and encreased again after Monday evening after Easter. Further comprehensive tests convinced of the high effectivity of the fluid pressure tiltmeters, as not only the low frequency bands but also those up to about 1 Hz could be monitored. Other investigations are on the way, to state, that frequencies up to 10 Hz can be detected if some parts of the sensor are modified. Then this measurement system will be an interesting completion of the method, described in chapter 1.2.

2 Dynamic monitoring of bridges for localization of damages

2.1 Development of the dynamic method

The idea is the localization and quantification of damages using measured changes of modal properties. The approach is a combination of dynamic insitu testing and FE-modelling. The measured data are used to update the FE model, which should be as close as possible to reality. The first experiments in Austria were carried out on bridge Raach in 1981 [9]. The framed bridge had 3 fields and a total length of 234 m. Before the bridge was replaced by a new construction, damages could be applied to the structure. By cutting tendons the bending stiffness at midspan and in one sidespan was reduced to 5/8 and 1/8 respectively. Before and after the damaging dynamic insitu tests were carried out with an eccentric mass exciter. The resulting decrease of modal frequencies was only around 0,015 Hz but could be identified very well.

In 1983 tests were carried out at Gänstor bridge before and after repair work [3,5]. The dynamic tests were carried out after the following phases:

- Installation of four additional tendons
- Tensioning of the tendons
- Injection of a main crack in the 1/3 point of the outer girder

From the increase of the modal frequencies (around 0,03 – 0,04 Hz) the increase of stiffness was elaborated. In the case of the main crack this was 24%.

2.2 The SIMCES project

The Brite Euram Project BE96-3157 System Identification to Monitor Civil Engineering Structures (SIMCES) was started in January 1997 and was finished in April 1999 [1,2].

Bridge Z24 in Switzerland was first extensively instrumented to set up a long term test (one year) for quantifying the degree of variance due to environmental influences and also due to differences induced by the parameter choice of the selected system identification methods. Then progressive damage tests of the Swiss bridge were carried out in order to prove that changes in dynamic properties be linked to damage in a particular part of the structure. The applied damage scenarios are listed in Table 2.1.

No	Date	Scenario
1	04.08.98	1. Reference measurements
2	09.08.98	2. Reference measurements
3	10.08.98	Settlement of pier, 20 mm
4	12.08.98	Settlement of pier, 40 mm
5	17.08.98	Settlement of pier, 80 mm
6	18.08.98	Settlement of pier, 95 mm
7	19.08.98	Tilt of foundation
8	20.08.98	3. Reference measurements
9	25.08.98	Spalling of concrete 12 m²
10	26.08.98	Spalling of concrete 24 m²
11	27.08.98	Landslide
12	31.08.98	Failure of concrete hinges at abutment colu.
13	02.09.98	Failure of anchor heads #1
14	03.09.98	Failure of anchor heads #2
15	07.09.98	Rupture of tendons #1
16	08.09.98	Rupture of tendons #2
17	09.09.98	Rupture of tendons #3

Table 2.1: Progressive Damage Test Scenarios for bridge Z24

The changes of the modal frequencies due to the applied damage scenarios are shown in Fig. 2.1.

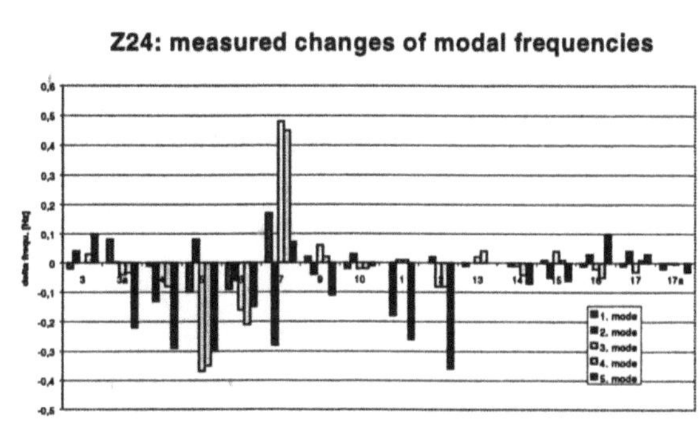

Fig. 2.1: Measured changes of the modal frequencies 1 –5 due to the scenarios 3 – 17 given in Table 2.1 The changes presented under no. 3a and 17a were calculated with the ABAQUS shell model.

2.3 Latest concept for safety inspection of bridges

It is difficult to give general apriori statements, to which bridge types dynamic monitoring could be applied successfully. In each case a preliminary analysis with the following steps seems to be necessary:

- Consideration of the most probable damage scenarios which are safety - relevant and/or cause expensive repair work. It is the main goal of bridge monitoring to detect these damages at an early stage.
- Detailed elaboration of the dynamic parameters of the bridge (measurements plus calculations). The influence of environmental conditions on dynamic parameters must be estimated at least roughly.
- A damage which reduces stiffness at a certain location can be localized and quantified, if the damage location coincides with areas of large modal curvatures. In this case modal parameters will change. A local damage will always have a different influence on different modes (e.g. in a point of contraflexure).
- If areas of probable damages really coincide with areas of large modal curvatures of certain modes, then the method is applicable and these modes could be continuously monitored by a „tailored monitoring system".
- Then, „warning levels" and safety relevant limits for changes of certain modal parameters can be found from further sensitivity studies using the updated FE model.

In general the changes of modal frequencies will be small because stiffness is always „integrated" over the whole structure. Hence, it has to be checked if the environmental influences on the modal frequencies can be separated from the damage - related influences. It is suggested to give modal differences always in absolute as well as relative numbers.

It is emphasized that changes of boundary conditions have strong influences on the modal parameters of certain modes.

For the localization and quantification of damages the sensitivity matrix has to be obtained (sensitivity of modal parameters to certain local changes, e.g. local stiffness changes). If one looks only after frequencies the results of updating procedures, which should show the position of a damage, are not necessarily unique, because damages at different locations could have the same effects especially for symmetric structures. In trial

& error approaches one should consider, that the location of considerable changes is not necessarily identical with the location of the stiffness changes. It is an advantage that bridges normally are not exactly symmetric.

Concerning prestressed structures with bond between steel and concrete, it is emphasized that there is no „normal force" in the structure (no buckling can occur). Hence no changes of geometric stiffness will result from changes of the prestressing force. Modal parameters will only change, if cracks occur or boundary conditions or the span length changes due to a decrease of the prestressing force.

Acknowledgements. A great part of the reported research work was carried out in the framework of the BRITE-EURAM Programme CT96 0277, SIMCES with a financial contribution by the Commission. Partners in the project were:

- K.U.Leuven (Department Civil Engineering, Afdeling Bouwmechanica)
- Aalborg University (Insitut for Bygningsteknik),
- EMPA (Swiss Federal Laboratories for Materials Testing and Research, Section Concrete Structures)
- LMS (Leuven Measurements and Systems International N.V., Engineering and Modelling)
- WS Atkins Consultants Ltd. (Science and Technology)
- Sineco SpA (Ufficio Promozione e Sviluppo)
- Technische Universität Graz (Structural Concrete Institute) + ÖFPZ Arsenal Ges.m.b.H.

References

Dynamicinsitu test of bridge WARTH/Austria. TU Graz report no. TUG TA 99/0314.

Brite/EuRam BE96-3157, SIMCES (1999): Technicalreport Task C TU Graz report no. TUG TC 99/0401.

Flesch, R., Kernbichler, K. (1987): Bridge inspectionby dynamic tests and calculations - dynamic investigations of Lavant Bridge. Int. Workshop on structural safety evaluation based on system identification approaches, Lambrecht/FRG.

Flesch, . G., Kernbichler, K. (1990): A dynamic method for the safety inspection of large prestressed bridges. NATO US-European Workshop on Bridges. Baltimore/USA.

Flesch, R.G. (1992):Dynamic testing and system identification for bridge inspection.Proc. of 3rd Int.Workshop on Bridge Rehabilitation , pp. 383 Ernst & Sohn.

Flesch, R.G., Stebernjak, B., Freytag, B. (1999) System identification of bridge Warth / Austria. EURODYN'99, Prague.

Kahmen, H. (1997): Vermessungskunde. Walter de Gruyter, Berlin, New York.

Kahmen, H., Mentes, G. (1998): Contribution of the dynamics of hydrostatic tiltmeters. Proc. of the Intern. Symp. Geodesy for Geotechnical and Structural Engineering, Eisenstadt.

Kernbichler K., Flesch, R. (1984): Static and dynamic tests, their qualification for bridge inspection and long-term observations of bridge structures.RILEM-Symposium,Long-term observation of concrete structures, sess.III, Budapest.

Kuhn, M. (1998): Deformationsmessung an vertikal schwingenden Bauwerken mit Differenzdruckgebern. Dissertation, Techn. University Vienna.

Mobile multi-sensor systems: The new trend in mapping and GIS applications

Naser El-Sheimy
Department of Geomatics Engineering,
The University of Calgary, 2500 University Drive N.W., Calgary, Alberta, Canada, T2N 1N4
Tel: (403) 220 7587, Fax: (403) 284 1980
E-mail: naser@ensu.ucalgary.ca

Abstract. Major progress has been made in Mobile Multi-sensor Systems (MMS) over the last few years in terms of sensor resolution, data rate, and operational flexibility. Thus, the use of such sensors in mapping and Geographic Information Systems (GIS) applications has become very attractive. Examples of such systems can be found in airborne remote sensing, airborne laser scanning and mobile mapping from vans and trains.

The paper will cover both, the concept of multi-sensor integration and implementation aspects. Based on experience with a number of different systems, features common to most systems will be identified and a unified model for georeferencing of multi-sensor data will be formulated. Emphasis will be on systems that integrate navigation and imaging sensors. All major features will be illustrated by examples. Finally, results of van and portable system surveys used to illustrate the salient features of the MMS.

Keywords. Mobile mapping, multi-sensor systems, GPS/INS, georeferencing.

1 Introduction

Mobile Multi-sensor Systems (MMS) have become an emerging trend in mapping applications because they allow a task-oriented implementation of geodetic concepts at the measurement level (El-Sheimy, 1996). Examples of such systems (see Table 1) can be found in airborne remote sensing (Cramer et al, 1997), airborne gravimetry (Wei and Schwarz (1995)), airborne laser scanning (Wagner (1995)), and mobile mapping vans, trains and 3D robotics systems (El-Sheimy et. al. (1995), Kahmen H. and G. Retscher (1999), and Blaho and Toth (1995)). All of these systems have a common feature in that the sensors necessary to solve a specific problem are mounted on a common platform. By accurately synchronizing the data streams, the solution of the mapping problem is possible by using data from one integrated measurement process only. The post-mission integration of results from a number of disjoint measurements processes and the unavoidable errors inherent in such a process are avoided.

Table 1: Examples of existing MMS and their hardware structure

System	Positioning			Imaging		Other
Name / Developers[1]	GPS	INS	DR	CCD	VHS	Sensors
Land Systems (Van, Railway, etc.)						
GPS Van/ OSU	✓		✓	2	2	
KISS/UBW	✓	✓		2	1	
ARAN/ Roadware	✓		✓	2	1	Ultrasonic
CDSS/AUT	✓		✓	2		
TruckMAP -JECA	✓			1	1	Laser
VISAT-UofC/VII	✓	✓	✓	8		
3-D Robot/ TUV				✓		Motorized theodolits
Airborne Systems						
AIMS-OSU	✓	✓		1		
Dual Camera/ UofC	✓	✓		2		
3-linear scanner/ LH Sys. And GAC	✓	✓		3*		* line scanner
Toposys / Toposys GmbH	✓	✓		1**		laser+ **RGB line scanner

[1] **OSU:** Ohio State Univ., **UBW:** Univ. der Bundeswehr Munchen, **AUT:** Aachen Univ. of Technology, **JECA:** John E. & Chance, **UofC:** The Univ. of Calgary, **TUV:** Technischen Universität Wien, **GAC:** German Aerospace Center

International Association of Geodesy Symposia, Vol. 121
Schwarz (ed.), Geodesy Beyond 2000 – The Challenges of the First Decade

The trend towards MMS in mapping and GIS applications is fuelled by the demand for fast and cost-effective data acquisition techniques. Two developments are especially important in this context: Digital imaging sensors and precise navigation systems. Digital imaging sensors, such as digital cameras and lasers, considerably reduce the data processing effort by eliminating the digitizing step. In the form of digital frame cameras, they are inexpensive enough to make redundancy a major design tool. In the form of multi-spectral pushbroom scanners, they provide additional layers of information, not available from optical cameras. Precise navigation has developed to a point where it can provide the solution of the exterior orientation problem without the use of GCPs or block adjustment procedures.

Combining these two developments, the concept of direct georeferenced of imaging sensor data emerges. This means that the imaging sensor data is stamped with its georeferencing parameters, namely three positions and three orientations, and can be combined with any other georeferenced data of the same area by using geometric constraints, such as epipolar geometry or object-space matching. This is a qualitatively new step because the georeferencing parameters are obtained in a direct way by independent measurement. This is conceptually different from the notion that a block of connected measurements and sufficient ground control is needed to solve for the georeferencing parameters. It is especially intriguing to consider its use for mapping applications which use either digital frame cameras, pushbroom scanners, or laser scanners as imaging components, for details of the principle, see Schwarz (1995) for mapping applications and Kahmen (1997) for robotics applications.

2 Components of mobile multi-sensor systems

In the following, common features in the design and analysis of MMS will be discussed. System design and analysis comprises the following steps as a minimum; for more details see El-Sheimy (1996):

1. Data acquisition
2. Kinematic Modeling
3. Synchronization and Calibration
5. Georeferencing
6. Integration and data fusion
7. Quality control and data flow optimization.

The conceptual layout of multi-sensor systems for mapping applications is shown in Figure 1. The selection of sensors for MMS obviously depends on system requirements, such as accuracy, reliability, operational flexibility, and range of applications. The data acquisition module has, therefore, to be designed keeping both the carrier vehicle and the intended applications in mind. The data acquisition module contains navigation sensors and imaging sensors. Navigation sensors are used to solve the georeferencing problem. Although a number of different systems are used in general navigation, the rather stringent requirements in terms of accuracy and environment make the integration of an INS with GPS receivers the core of any sensor combination for an accurate MMS applications. This combination also offers considerable redundancy and makes the use of additional sensors for reliability purposes usually unnecessary. However, the addition of an odometer type device, such as the ABS, for van applications may be useful for operational reasons, as for instance keeping a fixed distance between camera exposures.

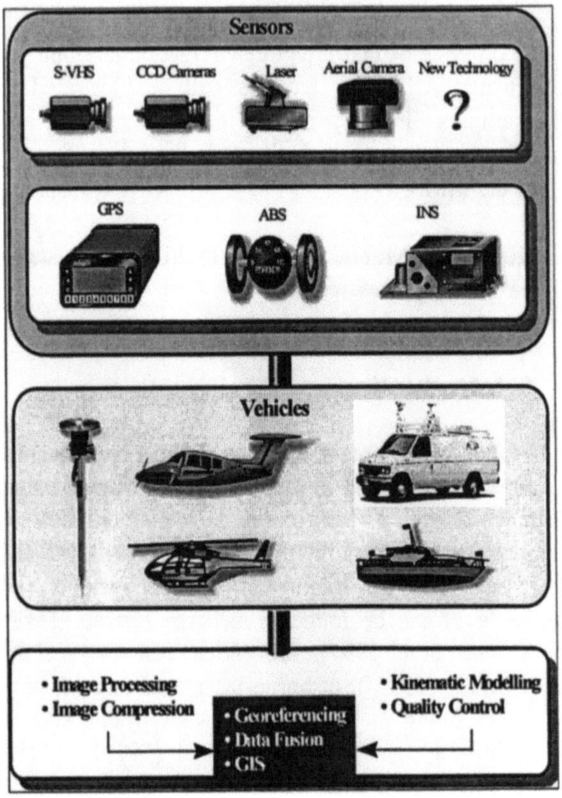

Fig. 1: Components of mobile multi-sensor systems

The imaging/navigation sensors can be subdivided by the way they contribute to the information about the object space. They may provide descriptive information, as for instance grey scales, or geometric information, as for instance direction or ranges from the camera to the object. Table 2 summarizes the contribution of sensors typically used in mapping applications.

Table 2: Summary of close-range photogrammetry related sensors

Type of sensor	Type of information	Characteristics
CCD Cameras	Descriptive /Geometric	Image, geometric accuracy depends on sensor resolution
Imaging Laser	Descriptive /Geometric	Image + Distance between the object and the sensor
Laser Profiles, Laser Scanners	Geometric	Distance between the object and the sensor, scanning angle
Impulse Radar	Descriptive /Geometric	Thickness of objects (mainly used for pavement structure voids)
Ultra-sonic sensors	Geometric	Distance between the object and the sensor Road rutting measurements, and cross section profiles measurement
GPS positioning and attitude determination	Geometric	High accuracy position and medium accuracy attitude, global reference
Inertial sensors	Geometric	Low-to-high accuracy relative position and attitude.
Odometers	Geometric	Distances.

The selected sensor configuration requires a certain data processing sequence. Part of the processing will have to be performed in real time, such as data compression for the imaging data and initial quality control processing for the navigation data. Most of the data, however, will immediately be stored for post-mission use. In post-mission, the data processing hierarchy is determined by the fact that all mapping data have to be georeferenced first before they can be used in the integration process. The first step is, therefore, the georeferencing of all mapping data and their storage in a multimedia data base.

To determine 3-D coordinates of objects from imaging sensors such as digital images, the following information is needed for a pair of cameras:

- Position of the camera perspective center at exposure time (3 parameters per image),
- Camera orientation at exposure time (3 parameters per image),
- Interior geometry of the camera sensor, and
- Lens distortion parameters.

The first two sets of parameters are known as georeferencing, also exterior orientation, parameters, while the other two sets are known as interior orientation parameters. The georeferencing parameters are determined from the output of the navigation sensors, while the interior orientation parameters are determined by field or laboratory calibration. This means that georeferencing is tied to a real-time measurement process and its parameters change quickly. In contrast, interior orientation is obtained by using a static field calibration procedure and can be considered as more or less constant for a period of time. Thus, it can be done before or after the survey mission and is not generally affected by the data acquisition process.

Implied in the georeferencing process is the synchronization of the different data streams. The accuracy of georeferencing is dependent on the accuracy with which this can be achieved. The synchronization accuracy needed is dependent on the required system performance and on the speed with which the survey vehicle moves. It is, therefore, much more critical for airborne applications than for marine and land vehicle applications. Fortunately, GPS provides a well-defined time signal to which the other sensors can be slaved. Still, the implementation of sensor synchronization is not a trivial process and should be done with care.

Model integration and data fusion comprises all steps necessary to extract the desired result from the georeferenced data. If the objective is to extract 3-D coordinates of objects from digital images, then the application of geometric constraints, the handling of redundant images of the same object, and the fusion of data of different type and quality are important considerations. Figures 2 and 3 show typical examples of data fusion in mapping application. Figure 2 shows an example of an airborne mapping system which uses digital full-frame images and laser profile data for the imaging component, and GPS and INS data for georeferencing of the

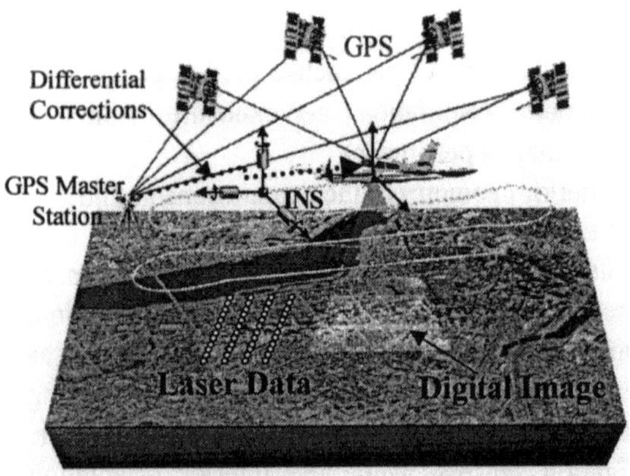

Fig. 2: An example of data fusion in airborne applications

imaging component data. The objective may, however, be much wider than object coordinates as for example the production of digital orthophotos or digital terrain models. Figure 3 shows an example of a MMS which uses digital images and laser profile data for the imaging component and GPS, INS, and ABS data for georeferencing of the imaging component data. Data fusion generally means that data from various sources and different nature are merged together to provide a versatile resource for mapping applications, a typical example could be a combination of the two systems shown in Figures 2 and 3.

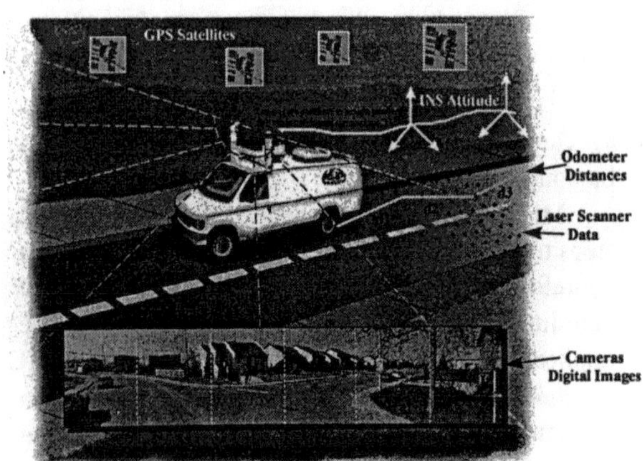

Fig. 3: An example of data fusion in close-range applications

3 Georeferencing of MMS data

The integration of MMS data requires a unified model for georeferencing such data. Unified in this context means that the model can be applied to most, if not all, sensor data without the need to account for a different set of parameters for each sensor. Georeferencing multi-sensor data can be defined as the problem of transforming the 3-D coordinate vector r^S of the imaging sensor frame (S-frame) to the 3-D coordinate vector r^m of the mapping frame (m-frame) in which the results are required. The m-frame can be any earth-fixed coordinate system, such as a system of curvilinear geodetic coordinates (latitude, longitude, height), UTM or 3TM coordinates. The georeferencing process can be described by the following formula, Table 3 outlines how different quantities are obtained:

$$r_i^m = r_{nav}^m(t) + R_b^m(t)[S^i \bullet R_S^b \bullet r^S + a^b] \quad (1)$$

Table 3: Elements of the georeferencing formula

Variable	Obtained from
r_i^m	is the coordinate vector of point (i) in the mapping frame (m-frame) **Unknown** (3)
$r_{nav}^m(t)$	is the interpolated coordinate vector of the navigation sensors (INS/GPS) in the m-frame
S^i	is a scale factor, determined by stereo techniques, laser scanners or DTM
$R_b^m(t)$	is the interpolated rotation matrix between the navigation sensor body frame (b-frame) and the m-frame
(t)	is the time of exposure, i.e. the time of capturing the images, determined by synchronization
R_S^b	is the differential rotation between the S-frame and the b-frame, determined by calibration
r^S	is the coordinate vector of the point in the S-frame (i.e. image coordinate),
a^b	is the offset between the b-frame and the S-frame, determined by calibration

The georeferencing formula implies that it is necessary to calibrate the entire system before estimating position and attitude as functions of time for the georeferencing process. As part of the calibration, the imaging sensor (e.g. camera) geometry, the relative orientation between the sensor and the b-frame R_S^b, the offset between the sensor a^b, must be determined. The calibration is accomplished by using a bundle adjustment and a test field of control points; for more details on calibration of MMS, see El-Sheimy et. al. (1995).

Equation (1) can be used to evaluate the georeferencing requirements for photographic systems, scanning systems, linear array systems, and radargrammetric systems. The overall accuracy will depend on the resolution of the imaging sensor and the accuracy with which the parameters in equation (1) can be determined. The important parameters are the accuracy of the position and attitude determination on the one hand and the stability of the sensor configuration on the other. It should be noted, however, that for radargrammetric systems, velocity is an additional parameter which has to be determined with high accuracy. It is required for motion compensation which strictly speaking is not part of the georeferencing process but part of the remote sensing process and therefore has to be accomplished in real-time from the navigation sensors output. For more details on georeferencing of remotely sensed images, see Schwarz (1995).

4 Examples of MMS

4.1 Airborne-based MMS

An airborne MMS for digital image capture and georeferencing has been developed on a joint research project between The University of Calgary and The University of California at Berkeley. Its major application is in remote sensing of various vegetation species in areas where ground control is neither available nor needed. In this system, a medium class INS, two low-cost GPS receivers, and two digital cameras (vertical and oblique) are integrated. The digital cameras capture strips of overlapping images. Camera exposure stations and INS digital records are time-tagged in real time by the GPS pulse through a real-time logging software. Preliminary system testing indicates that positioning accuracies of 0.9 m for horizontal coordinates and 1.8 m for height can be achieved. For more details on the system hardware configuration, see Mostafa et al (1998). Other fully digital airborne MMS comprising GPS, INS, and digital cameras have been developed by the Ohio State Center for Mapping (Toth and Brzezinska, 1998) and University of Stuttgart (Cramer et al, 1997).

4.2 Land-based MMS

To illustrate the concept of land-based MMS, the development of the VISAT system will be taken as an example. This system has been jointly developed by the University of Calgary and Geofit Inc. to satisfy the growing demands for a precise and up-to-date urban GIS data acquisition system. The survey van is equipped with an integrated system consisting of a GPS receiver, a strapdown inertial system, and a cluster of 8 CCD cameras. All components are precisely time-tagged to minimize errors from insufficient sensor synchronization. The system has been called VISAT to indicate the integration of video, inertial, and satellite technology.

Results achieved with this system in numerous tests show that the design accuracy has been considerably surpassed. Instead of an expected standard deviation of 0.3 m for the GIS data acquisition, horizontal and vertical coordinate errors are currently within an envelope of 0.2 m with a standard deviation of about 0.1 m (see Figure 4). Figure 4 shows the difference between the GCP coordinates and the coordinates obtained from the VISAT system. They are obtained by deriving 3-D coordinates from the VISAT system and transforming them to 3TM coordinates. These coordinates were then compared to the completely independent GCP coordinates. The GCP were about 10-30 m away from the van. With some minor system modifications, the system accuracy could be brought below 0.05m, which would be sufficient for a number of precise mapping applications. For a more detailed on the VISAT system, see El-Sheimy (1996). Similar land-based MMS have been developed by the Ohio State Center for Mapping (Blaho and Toth, 1995), Aachen University of Technology (Benning and Aussems, 1998), and the University of Federal Armed Forces (Sternberg et al, 1998).

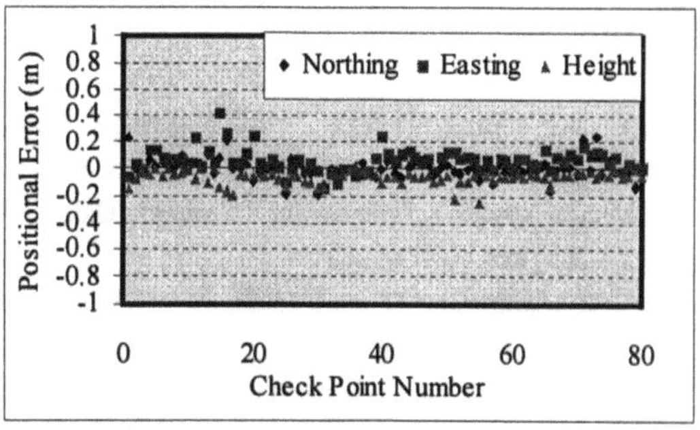

Fig. 4: The absolute accuracy of the VISAT system

4.3 Portable MMS

A portable MMS, see Figure 5 for schematic diagram, integrating a small and low-cost Inertial Measurement Unit (IMU) with a GPS receiver and a digital camera is being developed at the University of Calgary. The system differs from the previous ones because the carrier in this case is a survey stick carried by the system's operator. The system overcomes the drawbacks of current mobile mapping systems – namely their high cost, large size, and complexity – which have restricted their widespread adoption in the mapping industry. The development of such a system will satisfy the demand for MMS that can compete both cost-wise and in user friendliness with current GPS and conventional terrestrial survey systems.

Possible applications for such a system include pipeline right-of-way mapping, urban GIS data acquisition, accident investigations, precise industrial alignment, and highway inventory applications. A prototype system will be tested the winter of 2000 and the operational system will be ready by September 2000.

Fig. 5: A portable mobile multi-sensor system

References

Blaho, G. and Toth, C. (1995), "Field Experience With a Fully Digital Mobile Stereo Image Acquisition System", The Mobile Mapping Symposium, Columbus, OH, USA, May 24-26, 1995, pp. 97-104.

Cramer, M, Stallmann, D. and Halla, N. (1997), "High Precision Georeferencing Using GPS/INS and Image Matching", KIS97 Proceedings, Banff, Canada, pp. 453-462, June 3-6, 1997.

El-Sheimy, N., Schwarz K.P., and Gravel, M. (1995), "Mobile 3-D Positioning Using GPS/INS/Video Cameras, The Mobile Mapping Symposium", Columbus, OH, USA, pp. 236-249, May 24-26.

El-Sheimy, N. (1996), "A Mobile Multi-Sensor System For GIS Applications In Urban Centers", ISPRS 1996, Commission II, Working Group 1, Vienna, Austria, July 9-19, 1996.

Kahmen, H. (1997), "A New Kind of Measurement Robot Systems for Surveying of Non Signalized Targes", 4th International Symposium on "Optical 3-D Measurement Techniques". Zürich, Septmber 29th – October 2nd, 1997.

Kahmen H., G. Retscher (1999): "Precise 3-D Navigation of Construction Machine Platforms", in: Papers presented at the Workshop on Mobile Mapping Technology, April 21-23, 1999, Bangkok, Thailand, pg. 5A-2.1-5A-2.5.

Mostafa, M.M.R., K-P Schwarz, and M.A. Chapman (1998), "Development and Testing of an Airborne Remote Sensing Multi-Sensor System", ISPRS-Commission II, Symposium on Data Integration: Systems and Techniques, Cambridge, pp. 217-222, UK, July 13-17, 1998.

Schwarz, K.P (1995), "Integrated Airborne Navigation Systems for Photogrammetry" The Wichmann-Verlag, ISBN 3-87907-277-9, Photogrammetric week'95, Stuttgart, Oberkochen, Germany, pp. 139-153.

Sternberg, H., Caspary, W., and Heister, H. (1998), "Determination of the Trajectory Surveyed by the Mobile Surveying System KiSS", Proceedings of The International Symposium on Kinematic Systems In Geodesy, Geomatics and Navigation, Eisenstadt , Austria, pp. 361-366, April 20-23, 1998.

Toth, C. and Brzezinska, D. (1998), "Performance Analysis of the Airborne Integrated Mapping System", ISPRS-Commission II, Symposium on Data Integration: Systems and Techniques, Cambridge, 320-326, July 13-17, 1998.

Wagner, M. J. (1995), "Seeing in 3-D Without the Glasses", Earth Observation Magazine (EOM), July 1995, pp. 51-53.

Wei, M. and Schwarz, K.P. (1995), "Analysis of GPS-derived Acceleration From Airborne Tests", IAG Symposium G4, IUGG XXI General Assembly, Boulder, Colorado, July 2-14, 1995, pp. 175-188. Published by Department of Geomatics Engineering, UofC, Report Number 60010.

Benning, W. and Aussems, W. (1998), "Mobile Mapping by a Car-Driven Survey System (CDSS), Proceedings of The International Symposium on Kinematic Systems In Geodesy, Geomatics and Navigation, Eisenstadt, Austria, pp. 367-374, April 20-23, 1998.

Skaloud, J, Cosandier, D., Schwarz, K.P., Chapman, M.A. (1994), "GPS/INS Orientation Accuracy Derived From A Medium Scale Photogrammetry Test", Proceedings of the International Symposium on Kinematic Systems in Geodesy, Geomatics and Navigation, KIS94, Banff, Canada, pp. 341-348, August 30-September 2.

Adaptive Kalman filtering for integration of GPS with GLONASS and INS

J. Wang, M. P. Stewart and M. Tsakiri
School of Spatial Sciences
Curtin University of Technology
GPO Box U 1987, Perth, WA 6845, Australia

Abstract. In an integrated kinematic system, the Kalman filter is commonly used to integrate the data from different sensors (such as GPS/GLONASS and INS) for precise positioning. Reliable Kalman filtering results rely heavily on the correct definition of both the mathematical and stochastic models used in the filtering process. Whilst the mathematical models for various positioning measurements are (sufficiently) known and well documented in the current literature, stochastic modelling is not trivial, in particular for real-time applications.

In this paper, a newly developed adaptive Kalman filter algorithm is introduced to directly estimate the variance and covariance components for the measurements. Example applications of the proposed algorithm in GPS/GLONASS kinematic positioning and GPS/INS integration are discussed using test data sets. Test results show that the proposed algorithm can improve the performance of the filtering process.

Keywords. Kalman filtering, stochastic modelling, GPS, GLONASS, INS.

1 Introduction

Kinematic positioning as a geodetic tool has been widely used in a variety of areas, such as real-time mapping and precise navigation. During the last decade, GPS has been the major technology for kinematic positioning techniques. A recent development in kinematic GPS positioning is the technique of using pseudo-range and carrier phase measurements to resolve the carrier phase ambiguities almost instantaneously. Kinematic techniques improve productivity in surveying applications, and also apply well to navigation. Reliable kinematic positioning results depend on the availability of a large number of GPS satellites. GPS kinematic positioning techniques, therefore, may not be feasible for some situations, such as the positioning in built up areas, where the number of visible satellites is limited. One possible scheme to increase the availability of satellites is to combine GPS and GLONASS, which is the Russian counterpart to GPS.

Satellite-based kinematic positioning techniques can offer consistent precision during the period of positioning. However, such systems require line-of-sight between the orbiting satellites and receiver antennas. In some situations, in highway tunnels and under trees or bridges, GPS/GLONASS receivers cannot track satellite ranging signals, and thus will fail to provide positioning results. Integration of GPS with a self-contained inertial navigation system (INS) can overcome this disadvantage of the satellite-based systems. On the other hand, precise GPS positioning results are ideally suited for continuous INS calibration. GPS/INS integration has been successfully used in modern real-time mapping systems (e.g. Grejner-Brezinska, 1997; Schwarz, 1998).

In an integrated kinematic positioning system, the Kalman filter is commonly used as a data fusion tool because a kinematic system will include some time-dependent unknown states, such as coordinates of a moving platform. The Kalman filter, as a set of mathematical equations, can provide an efficient computational (recursive) solution of the least-squares estimation problem. Therefore, the Kalman filter can produce optimal estimation results in real-time if the mathematical and stochastic models used in the filtering process are both correctly defined. Whilst the mathematical models for major positioning systems (such as GPS, GLONASS and INS) are sufficiently known and well documented, the stochastic modelling for kinematic positioning is a critical issue to be investigated.

For example, in the commonly used procedures for constructing the covariance matrices of the differenced GPS and GLONASS measurements, all the one-way code or carrier phase measurements are assumed to be independent and to have the same accuracy. In fact, these assumptions do not fit

International Association of Geodesy Symposia, Vol. 121
Schwarz (ed.), Geodesy Beyond 2000 – The Challenges of the First Decade

reality (e.g. Gianniou and Groten, 1996; Wang *et al*, 1997). First, the code or carrier phase observations are correlated (the single- and double-difference methods are based explicitly on this fact). Secondly, measurements obtained from different satellites cannot have the same variance because of various noise sources. This is particularly true when the measurements come from different systems in combined GPS and GLONASS positioning. Any deficiency in the stochastic models for the differenced GPS and GLONASS measurements in data processing will inevitably result in unreliable statistics for ambiguity resolution and biased positioning results (e.g. Cannon and Lachapelle, 1995; Wang *et al*, 1997).

In the Kalman filtering process, online stochastic modelling can be implemented using the so-called adaptive filtering techniques (Chin, 1979; Mehra, 1972). Applications of adaptive Kalman filtering techniques in real-time GPS kinematic positioning (Wang *et al*, 1997) and GPS/INS integration (Mohamed and Schwarz, 1999) have shown encouraging results. This paper will introduce a simplified explanation for the algorithms used in the adaptive Kalman filtering procedure. Latest test results in GPS/GLONASS and GPS/INS integration are discussed in detail.

2 Adaptive Kalman filter procedure

Since the true values of the model errors (measurement noise or process noise) are unknown, stochastic modelling has to be based on the filtering residuals of the measurements and state corrections, which are generated in the process of parameter estimation. A problem here is that the parameter estimation process itself relies on the estimated measurement covariance matrix R and process covariance matrix Q. To tackle this problem, an *adaptive* procedure can be used.

The basic idea of the adaptive filtering procedure is that the residuals collected from the previous segment of positioning results are used to estimate the covariance matrices of the measurement noise and process noise for the current epoch. (Preset default covariance matrices are needed to seed the adaptive estimation process.) Within a segment, the covariance matrices for each epoch is assumed to be the same. Therefore, the formulation of an adaptive stochastic modelling method includes two critical steps, namely: (a) to derive suitable formulae for use in estimating the covariance matrices, and (b) to determine the optimal segment (or window size), which is application-dependent and will be discussed in Section 3. The formulae for constructing R and Q can be derived using the so-called 'covariance matching' method (Mehra, 1972).

2.1 Estimating the measurement covariance matrix

Suppose the measurement filtering residuals are:

$$v_{z_k} = z_k - H_k \hat{x}_k \tag{1}$$

where z_k is the measurement vector and H_k is the measurement design matrix. Equation (1), obviously, is the optimal estimator of the measurement noise level because the estimated values $\hat{x}_k$ (not the predicted values $\bar{x}_k$) of the state parameters are used in their computations. In order to obtain the covariance matrix of the measurement filtering residuals, equation (1) is modified

$$v_{z_k} = z_k - H_k(\bar{x}_k + G_k d_k) = (E - H_k G_k) d_k \tag{2}$$

where G_k is the gain matrix and d_k is the innovation vector. By applying the error propagation law to equation (2), after extensive computations, one obtains

$$Q_{v_{z_k}} = R_k - H_k Q_{\hat{x}_k} H_k^T \tag{3}$$

In equation (3), if the covariance matrix $Q_{v_{z_k}}$ is computed using the measurement filtering residuals from the previous m epochs, the covariance matrix R_k can be estimated as (Wang, 1998)

$$\begin{aligned} \hat{R}_k &= \hat{Q}_{v_{z_k}} + H_k Q_{\hat{x}_k} H_k^T \\ &= \frac{1}{m} \sum_{i=0}^{m-1} v_{z_{k-i}} v_{z_{k-i}}^T + H_k Q_{\hat{x}_k} H_k^T \end{aligned} \tag{4}$$

which can be used in the computation of epoch $k+1$. In equation (4), m is called *the width of moving windows*. It is noted that the covariance matrix $\hat{R}_k$ estimated with equation (4) is always positive definite because it is the sum of the two positive definite matrices. Equation (4) requires some extra computations for both v_{z_k} and $H_k Q_{\hat{x}_k} H_k^T$, which are not generated by the standard Kalman filtering process. Fortunately, the amount of these additional computations is small and leads to no significant time delay in data processing.

2.2 Estimating the process covariance matrix

Similar to the measurement covariance matrix, the process noise covariance matrix can be derived using the state corrections as follows

$$v_{x_k} = \hat{x}_k - \bar{x}_k = G_k d_k \tag{5}$$

By applying the error propagation law to equation (5), one obtains

$$\begin{aligned} Q_{v_{x_k}} &= G_k Q_{d_k} G_k^T \\ &= Q_{\bar{x}_k} - Q_{\hat{x}_k} \\ &= \Phi_{k,k-1} Q_{\hat{x}_{k-1}} Q_{k,k-1}^T + Q_k - Q_{\hat{x}_k} \end{aligned} \tag{6}$$

In equation (6), if the covariance matrix $Q_{v_{x_k}}$ is computed using the state corrections from the previous m epochs, the covariance matrix Q_k can be estimated as

$$\begin{aligned} \hat{Q}_k &= \hat{Q}_{v_{x_k}} + Q_{\hat{x}_k} - \Phi_{k,k-1} Q_{\hat{x}_{k-1}} \Phi_{k,k-1}^T \\ &= \frac{1}{m} \sum_{i=0}^{m-1} v_{x_{k-i}} v_{x_{k-i}}^T \\ &\quad + Q_{\hat{x}_k} - \Phi_{k,k-1} Q_{\hat{x}_{k-1}} \Phi_{k,k-1}^T \end{aligned} \tag{7}$$

which can be used in the computation of epoch $k+1$. Unlike equation (4), however, with equation (7), it cannot be guaranteed that the estimated process noise matrix will be positive definite.

In many physical situations, therefore, it is very difficult to model the dynamic behaviour of a moving platform with the accuracy that GPS/GLONASS measurements can give. In order to avoid the effect of errors in the process noise covariance matrix on the state estimation, the filter can be set up to operate only on the measurement noise. The basic concept regarding this filter operating mode has been discussed extensively in the literature (e.g. Hwang and Brown, 1990; Lachapelle *et al*, 1992). In the following experiment, the process covariance matrix will be constructed using default values.

It is noted that the formulae derived above, equations (4) and (7), can also be derived using a maximum likelihood criterion (Mohamed and Schwarz, 1999).

3 Experiments for GPS/GLONASS and GPS/INS integration

3.1 GPS/GLONASS positioning

The test data set was collected on a 1.2km (static) baseline, on February 16, 1998, in Perth, Australia, using two Ashtech GG24 GPS/GLONASS receivers. The data span was 25 minutes with a data interval of 1 second. During the whole session of observation, 7 GPS and 5 GLONASS satellites were tracked.

The realistic estimation of measurement covariance matrices provides reliable statistics for ambiguity resolution. To demonstrate this more clearly, both the solutions with the preset and estimated covariance matrix were generated. The *ambiguity dilution of precision* (ADOP) (Teunissen and Odijk, 1997) is illustrated in Figure 1. It indicates that, with the estimated covariance matrix, the accuracy of the estimated ambiguities was greatly improved. As a consequence of this, the ambiguity search volume was significantly reduced and faster search process could be expected (Landau and Euler, 1992; Teunissen and Odijk, 1997).

The improved float ambiguity estimates are of great importance for the ambiguity validation test, which is one of the critical steps in ambiguity resolution. The ambiguity validation results, using the ambiguity validation test statistic Ws, see description in Wang *et al* (1998), are graphed in Figure 2. In all cases, the ambiguity validation test statistics with the estimated measurement covariance matrices are much larger than those with the preset measurement covariance matrices. The averaged confidence levels of the ambiguity resolution for the preset and estimated stochastic models are 0.83 and 1.0, respectively. This improvement in confidence level is statistically significant.

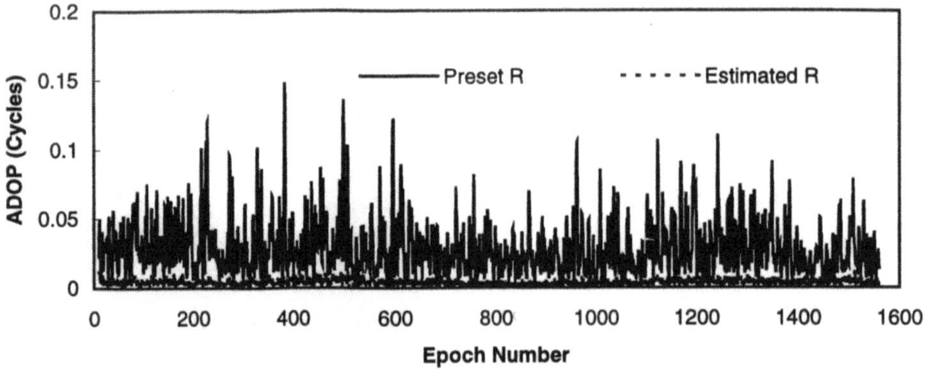

Fig. 1: Ambiguity dilution of precision (ADOP, after Teunissen and Odijk, 1997)

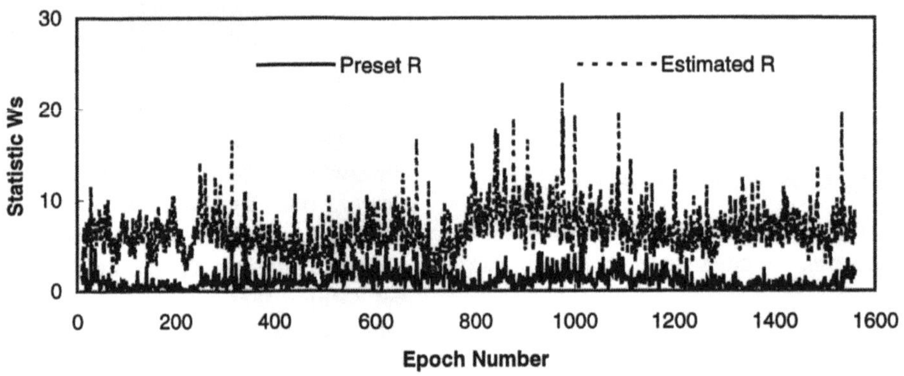

Fig. 2: Ambiguity discrimination test statistic Ws (After Wang et al,1998)

Table 1: Impacts of the width of moving windows on ambiguity resolution

Width	6	7	8	9	10	30	120
Averaged F-ratio	3.5	5.2	13.3	12.0	11.1	6.0	4.1
Averaged Ws	2.4	3.5	6.6	6.3	6.2	4.3	3.3
Success Rate (%)	62.7	96.9	100	100	100	99.5	94.2

A realistic stochastic model can also improve the accuracy of the ambiguity fixed solutions. For instance, at epoch 8 when the filtering process was using the preset measurement covariance matrix, the standard deviations of (x,y,z) coordinated were 0.008m, 0.016m and 0.012m, respectively. At epoch 9 when the filtering process began to utilise the estimated measurement covariance matrix, the standard deviations of the coordinates were down to 0.006m, 0.013m and 0.009m, respectively. The differences in the coordinates between the two solutions are around 0.005 m.

To determine the optimal width of the moving window, some further tests for various widths have been conducted. The averaged ambiguity validation test statistics and success rate of ambiguity resolution are listed in Table 1, which show that for the tested GPS/GLONASS data set, the optimal value for the width of the moving window is 8.

3.2 GPS/INS integration

In a combined GPS/INS positioning system, GPS measurements can be used to calibrate INS instrument and alignment errors. Alignment is the process whereby the orientation of the axes of an inertial navigation system are determined. Whilst the initial position and velocity can be directly obtained using GPS measurements, the orientation parameters may be determined using the so-called (INS and GPS) velocity matching method (e.g. Titterton and Weston, 1997). This method may be implemented using a Kalman filter procedure, in which the measurements are differences of the GPS and INS velocity.

A GPS/INS data set was used to test the performance of the adaptive Kalman filtering procedure in the GPS-INS velocity match alignment. The data set was collected using a Litton IMU and Trimble 4000SSE GPS receivers.

In the filter, the orientation parameters (pitch, roll and yaw) are of direct interest. Therefore, the major concern is the precision of the estimated orientation parameters. The test results show that the precision of the yaw parameter can be significantly reduced (see Figure 3). However, only a slight improvement in the precision of the pitch and roll parameters was observed. The possible reason for this is that the pitch and roll precision may rely heavily on the geometry of the system instead of the measurement covariance matrix. This, however, needs further investigation.

For this data set, the different window sizes (ranging from 2 to 10) did not influence significantly the filtering results.

4 Concluding remarks

In an integrated GPS/GLONASS kinematic positioning system, ambiguity resolution and positioning results rely on the correctness of the measurement noise covariance matrix. The common practice of assuming that the one-way (code or carrier phase) measurements are statistically independent, and have the same accuracy, is certainly not realistic, and thus inevitably leads to an unsuitable measurement covariance matrix.

With the proposed adaptive Kalman filtering procedure, the measurement noise covariance matrix can be applied to real-time data processing. By using an estimated covariance matrix, the reliability of ambiguity resolution and the accuracy of kinematic positioning can be significantly improved.

The proposed adaptive Kalman filtering procedure can also be used in other integrated kinematic positioning systems. Test results from an integrated GPS/INS data set indicate that by using the estimated measurement covariance matrix, the precision of the orientation parameters can be improved, particularly that of the estimated yaw parameter in the velocity matching alignment process.

It is noted that the optimal window size used in the adaptive filtering procedure may depend on the geometry of the solution. In practical applications, the optimal window size may need to be determined in real-time. On the other hand, the adaptive estimation of the filtering process noise covariance matrix needs further investigation.

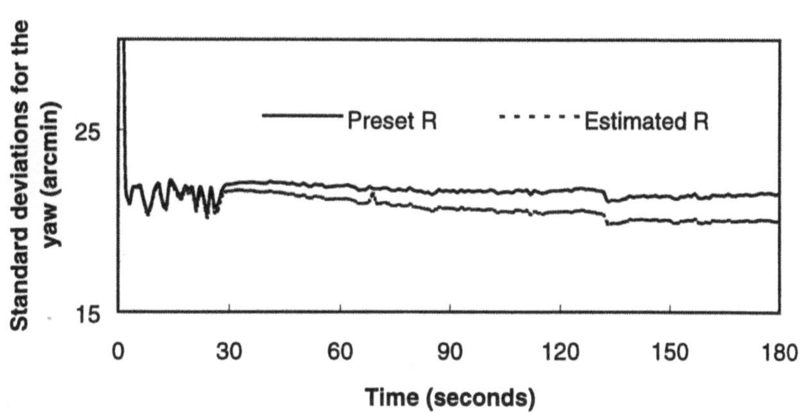

Fig. 3: Standard deviation for the yaw parameter

Acknowledgment. The authors thank Dr. Dorota A. Grejner-Brzezinska (The Ohio State University Center for Mapping) for kindly providing us GPS/INS data.

References

Cannon, M.E. and G. Lachapelle (1995) Kinematic GPS Trends — Equipment, Methodologies and Applications. In: Beutler, Hein, Melbourne and Seeber (Ed.): *GPS Trends in Precise Terrestrial, Airborne, and Spaceborne Applications, IAG Symposium No. 115*, Boulder, USA, July 3-4, 161-169.

Chin, L. (1979) Advances in Adaptive Filtering. In: Leondes C.T. (ed.): *Advances in Control Systems Theory and Applications*, Academic Press, No. 15, 277-356.

Gianniou, M. and E. Groten (1996) An Advanced Real-Time Algorithm for Code and Phase DGPS. Paper presented at the DSNS'96 Conference, St. Petersburg, Russia, May 20-24.

Grejner-Brezinska, D. A. (1997) High-accuracy Airborne Integrated Mapping System. In Brunner K. F. (Ed.) *Advances in positioning and reference frames*, IAG scientific Assembly, Rio de Janeiro, Brazil, Sept. 2-9, Springer-Verlag, 337-342.

Hwang, P.Y.C. and R. G. Brown (1990) GPS Navigation: Combining Pseudo-range with Continuous Carrier Phase Using a Kalman Filter. *Navigation*, Vol. 37, No. 2, 181-196.

Lachapelle, G., P. Kielland and M. Casey (1992) GPS for Marine Navigation and Hydrography. *International Hydrographic Review*, LXIX(1), 43-69.

Landau, H. and H.J. Euler (1992) On-the-fly Ambiguity Resolution for Precision Differential Positioning. *Proceedings of 5th International Technical Meeting of the Division of the Institute of Navigation, ION GPS-92*, Albuquerque, USA, September22-24, 607-613.

Mehra, R.K. (1972) Approaches to Adaptive Filtering. *IEEE Transactions on Automatic Control*, AC-17, 693-698.

Mohamed, A.H. and K.-P. Schwarz K.P. (1999) Adaptive Kalman Filtering for INS/GPS. J. of Geodesy, 73, 193-203.

Schwarz, K.-P. (1998) Sensor Integration and Image Georeferencing. Invited Lecture, Duane C. Brown International Summer School in Geomatics, The Ohio State University, July 9-11, 34 pp.

Titterton, D.H. and J.L. Weston (1997) Strapdown Inertial Navigation Technology. Stevenage, U.K., Peregrinus, 455 pp.

Teunissen, P.J.G. and D. Odijk (1997) Ambiguity Dilution of Precision: Definition, Properties and Application. *Proceedings of ION GPS-97*, September 16-19, Kansas City, USA, 891-900.

Wang, J. (1998) Combined GPS and GLONASS Kinematic Positioning: Modelling Aspects. *Proceedings of 39th Australian Surveyors Congress*, Launceston, Tas, November 8-13, 227-235.

Wang, J., M. Stewart and M. Tsakiri (1997) Kinematic GPS Positioning with Adaptive Kalman Filtering Techniques. . In Brunner K. F. (Ed.) *Advances in positioning and reference frames*, IAG scientific Assembly, Rio de Janeiro, Brazil, Sept. 2-9, Springer-Verlag, 389-394.

Wang, J., M. Stewart and M. Tsakiri (1998) A Discrimination Test Procedure for Ambiguity Resolution On-the-Fly. *Journal of Geodesy*, Vol. 72, No. 11, 644-653.

A GPS/INS/Imaging system for kinematic mapping in fully digital mode

Mohamed M.R. Mostafa and Klaus-Peter Schwarz
Department of Geomatics Engineering, The University of Calgary, 2500 University Drive N.W., Calgary, Alberta, Canada T2N 1N4, Tel: (403) 220-8794, Fax : (403) 284-1980, e-mail: mrashad@ensu.ucalgary.ca

Abstract. This paper describes the development and testing of a multi-sensor system for kinematic mapping in fully digital mode. The system consists of a navigation-grade strapdown Inertial Navigation System, two GPS receivers, and two high resolution digital cameras. The two digital cameras capture strips of overlapping nadir and oblique images. The INS/GPS-derived trajectory describes the full translational and rotational motion of the carrier aircraft. During postprocessing, image exterior orientation information is extracted from the trajectory. This approach eliminates the need for ground control to provide 3D position of objects that appear in the field of view of the system imaging component. Test flights were conducted over the campus of The University of Calgary. In this paper, the multi-sensor system configuration and calibration is briefly reviewed and results of the test flights are discussed in some detail. First results indicate that major applications of such a system in the future are the mapping of utility lines, roads, pipelines (at mapping scales of 1:1000 and smaller), and the generation of digital elevation models for engineering applications.

Keywords. GPS/INS, georeferencing, digital mapping, multi-sensor systems.

1 Mapping the Earth's surface - a kinematic approach

Helmert (1880) defined geodesy as the science of measuring and mapping the Earth's surface. Interpretations of this definition cover the whole spectrum of geomatics. The geodetic positioning specialist considers the measurement of the Earth's surface as a point positioning problem with reference frame implications. The accurate determination and monumentation of points on the surface of the Earth is therefore seen as the major task. In order to express these points in a consistent coordinate system over larger parts of the Earth's surface, networks are established and the datum problem must be solved. Once this has been done, the network points are used for point densification in local areas. The concept implied by this approach is that the higher the point accuracy, the better the mapping. This is true for pointwise mapping, but obviously not for surface mapping. Simple interpolation between network points will, for instance, create large errors in a topographic map. Thus, the accuracy of the surface representation will not be uniform. In addition, although networks may stretch over a large part of the Earth's surface, they are globally disconnected when established by conventional procedures. This means that the datum problem cannot be solved without extraterrestrial measurements.

The photogrammetrist, on the other hand, considers the measurement of the Earth's surface as an imaging problem. It is solved by deriving a model of the surface from digital or photographic images. In this case, patches of the Earth's surface are actually measured and mapped in accordance with Helmert's definition. The concept behind this method is that the surface of the Earth can be presented by pixels measured in projected images. The smaller the pixel size and the better the geometry, the better the mapping. In this case, the accuracy is more or less uniform across the image and interpolation of specific image features is possible with high accuracy, once the image has been properly georeferenced. This is done by solving the datum problem using geodetic ground control in the survey area. Comparing the view of the positioning specialist with that of the photogrammetrist shows that they are essentially complementary. One provides highly accurate point positions in an adopted reference system which can be used by the other to georeference measurements and solve the datum problem for the precise local maps derived from images.

Because of this complementarity, the problem of measuring and mapping the surface of the Earth can be solved by combining geodetic point positioning with digital imaging. To do this efficiently, mobile mapping

International Association of Geodesy Symposia, Vol. 121
Schwarz (ed.), Geodesy Beyond 2000 – The Challenges of the First Decade

methods should be employed and the use of ground control should be avoided. This is possible by using differential GPS together with an inertial measuring unit (IMU). Currently, GPS is widely seen as a highly accurate relative positioning method.Interstation vectors are the output of differential GPS methods and, in this sense, GPS is viewed as a sophisticated tool for network densification. What is lost in this view of GPS positioning is the fact that the receiver output is directly connected to the global reference frame by way of satellites. Thus, ground control can be replaced by sky control. This means that it is possible to determine globally referenced positions without direct access to networks or dense ground control. Instead of tying into monumented control points one links into satellites which, in their orbital positions, carry accurate reference system information with them.

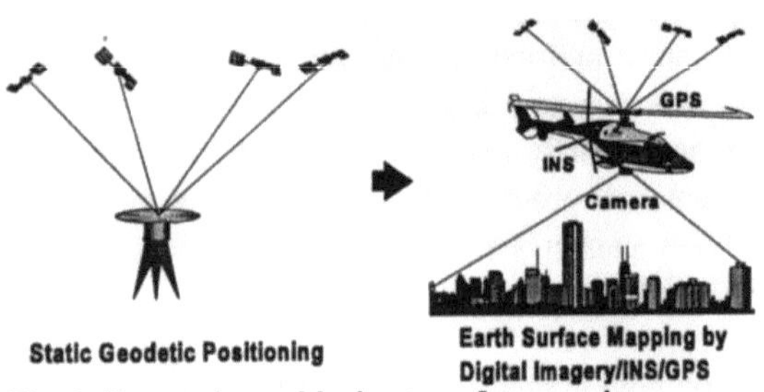

Fig. 1: From point positioning to surface mapping

By making use of this aspect of GPS, it is possible to design kinematic systems that integrate the georeferencing and imaging aspects of the problem discussed here. Such systems will provide a consistent and uniform representation of the Earth's surface worldwide. This is possible because a highly accurate global reference frame now exists which can be accessed everywhere by using GPS receivers as measurement tools. Since GPS receivers work in kinematic mode, there is no reason to separate the positioning process from the imaging process. By determining the perspective centre of the imaging sensor lens by DGPS at the moment of exposure, the first three parameters of exterior orientation are obtained in an accurate global reference frame, such as the WGS84. The other three parameters describing the orientation of the camera at the moment of exposure can be obtained by integrating an IMU with DGPS and the camera. This has the advantage that each individual image will now get its full set of exterior orientation parameters. Thus, any two images with overlapping image content can be directly used for mapping part of the Earth's surface in a consistent global coordinate frame, see Figure 1. The implementation of such a system and its current accuracy will be described in the following sections.

2 Kinematic motion determination by geodetic tools

The 3D motion of a rigid body can be described by the sum of two vectors. One models the position vector to a reference point (o) of the rigid body with respect to an Earth-fixed coordinate system (e-frame), while the other accounts for the orientation of the rigid body in 3D space. This can be written as:

$$\mathbf{r}_i^e(t) = \mathbf{r}_o^e(t) + \mathbf{R}_b^e(t)\,\Delta\mathbf{r}^b \qquad (1)$$

where,

- $\mathbf{r}_i^e(t)$ is the position vector of an arbitrary point i of the rigid body in the e-frame at time t;
- $\mathbf{r}_o^e(t)$ is the position vector of point o of the rigid body in the e-frame at time t;
- $\Delta\mathbf{r}^b$ is a constant offset vector between point i and o in the rigid body b-frame; and
- $\mathbf{R}_b^e(t)$ is the rotation matrix between the b-frame and the e-frame at time t.

If an IMU is located at point o and a GPS antenna at point i, then their common carrier platform motion can be fully described by Equation 1 and estimated from inertial and satellite measurements. For a detailed discussion on GPS/INS integration modelling estimation, and applications, see Schwarz (1998).

3 Photogrammetric reconstruction of a 3D point position

Although mapping patches of the Earth's surface is the final product of imaging applications, the basic analytical form, needed to relate an image frame (c) and a mapping frame (M), is still expressed pointwise. Therefore, the process of reconstructing a 3D object position from 2D image data follows the collinearity concept. As shown in Figure 2, the object point, its image on the acquired photo, and the perspective centre of the lens in use have to be collinear. This can be expressed as:

$$\mathbf{r}_G^M = \mathbf{r}_E^M(t) + s_g\,\mathbf{R}_c^M(t)\,\mathbf{r}_g^c(t) \qquad (2)$$

where,

- $\mathbf{r}_G^M$ are the 3D object point coordinates of point G in the M-frame;
- $\mathbf{r}_E^M(t)$ are the 3D coordinates of the exposure station E at the instant of exposure t, in the M-frame;
- $\mathbf{r}_g^c(t)$ are the 3D coordinates of the image point g in the c-frame at the instant of exposure t (2D image coordinates plus lens focal length);
- s_g is an image point scale factor implicitly derived during the 3D reconstruction of objects using

image stereopairs; and

$R_c^M(t)$ is a rotation matrix rotating the c-frame into the M-frame at time t, utilizing the three c-frame orientation angles $\omega(t)$, $\varphi(t)$, and $\kappa(t)$, and the primary, secondary, and tertiary elementary rotation matrices $\mathbf{R}_1$, $\mathbf{R}_2$, and $\mathbf{R}_3$, respectively. This rotation matrix can therefore be expressed by:

$$\mathbf{R}_c^M(t) = \mathbf{R}_3(\kappa(t))\,\mathbf{R}_2(\varphi(t))\,\mathbf{R}_1(\omega(t)) \quad (3)$$

In other words, once a coordinate system has been attached to the imaging sensor, the mapping coordinates of any point that appears in an image stereopair can be determined using Equation 2, if the position and orientation parameters of the image at the instant of exposure are known.

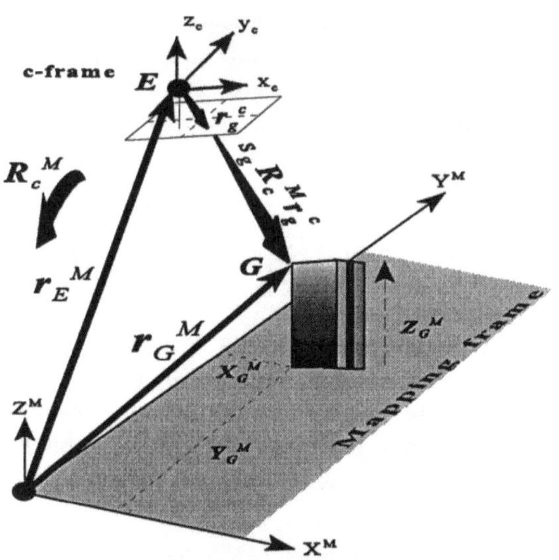

Fig. 2: Image-to-object space relationship

4 Integration of kinematic geodesy and aerial imaging for surface mapping

The measurable quantities in Equation 2 vary according to the sensors available onboard. In standard aerial photography applications, known ground control points are used to relate the camera frame to the mapping frame. In this case, $\mathbf{r}_G^M$ and $\mathbf{r}_g^c(t)$ are the observables used to recover $\mathbf{r}_E^M(t)$ and $\mathbf{R}_c^M(t)$ of each image. They are then subsequently (or simultaneously) used to determine $\mathbf{r}_G^M$ of other points of interest. This process is indirect from the standpoint that camera position and orientation, needed to determine positions of points on the ground, are determined using other ground points with known coordinates. This process is called aerotriangulation.

With the development of GPS kinematic techniques, the combination of kinematic positioning with aerial triangulation was proposed in Schwarz et al (1984). First test results were published in the late 1980s and since then, GPS-assisted photogrammetry has become an operational procedure. Yet, GPS only provides the position component of the exterior orientation parameters, and thus, blocks of images are always needed to recover the orientation matrix for each image. To orient individual images which are not part of a block structure, an IMU is needed in the data acquisition system to fully describe the platform motion, as briefly discussed in section 2. GPS/INS integration in support of photogrammetric mapping has, therefore, been an active area of research over the past decade. The mathematical model and proposed applications were demonstrated by Schwarz et al (1993). Lechner and Lahmann (1995) and Škaloud et al (1996) showed the success of this method to georeference aerial photos by GPS/INS. On the other hand, Mostafa et al (1997), Cramer et al (1997), and Toth and Grejner-Brzezinska (1998) reported on systems that integrated GPS/INS with digital cameras. Economically, such systems have a great potential to replace the currently used aerial mapping systems due to their lower cost, shorter turn-around-time, and the rapid development in the digital camera industry. Figures 3 and 4 show a comparison between current aerial mapping systems and the integrated navigation/imaging systems discussed here. For details, see Lyon et al (1995) and Congalton et al (1998).

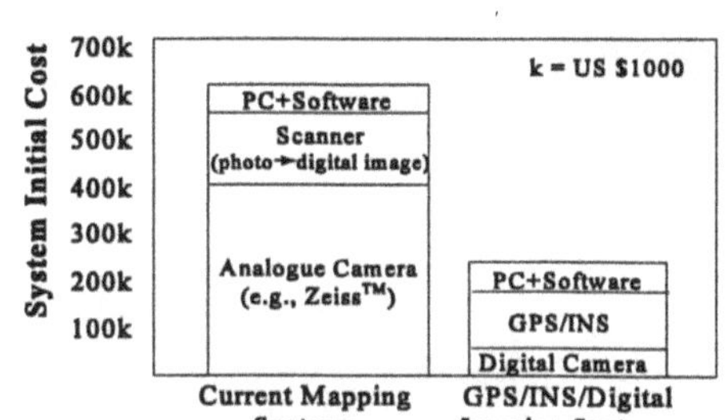

Fig. 3: Economics of mapping systems

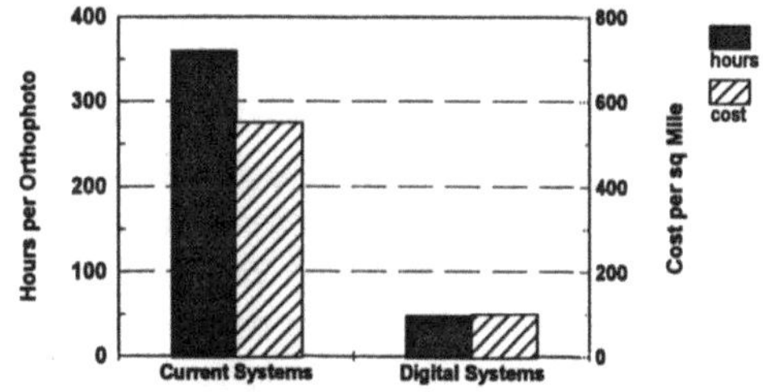

Fig. 4: Operating cost of mapping systems

5 System configuration

The two major components of such a system are the navigation component and the imaging component. The navigation component comprises two GPS receivers to allow for ground-to-aircraft DGPS positioning, and a navigation-grade inertial navigation system. The imaging component includes two digital cameras. One is mounted vertically, while the other is oblique. They were configured in such a way as to reduce the geometric limitations that arise when using a single camera; for details, see Mostafa et al (1998b). Accommodated onboard a CESSNA 310 twin engine aeroplane, two dual-frequency GPS receivers, two Kodak DCS 420 digital cameras, and a Honeywell LRFIII INS, were used for the test. The GPS receivers, the INS, and the cameras were interfaced to three laptops, which control the different tasks required for data acquisition, as shown in Figure 5.

Fig. 5: The multi-sensor system installed

6 System calibration

An overall system calibration is required to relate GPS-derived positions, INS-derived attitude parameters, and image-derived object point coordinates. In addition, the digital cameras have to be calibrated to establish their interior geometry and to determine their lens distortion.

Both digital cameras were calibrated at the University of Calgary using a self-calibrating software (Lichti and Chapman, 1997). The precision of calibration of each camera was at the level of 5-7 cm on the ground, for a flight altitude of 450 m. For details, see Mostafa and Schwarz (1999). Sensor position offsets were precisely surveyed to mm accuracy. A critical part of the calibration process is the computation of the rotation offset between the INS and each of the two camera coordinate systems. Since both have invisible coordinate system axes, their orientation offset has to be computed indirectly. To do this, either in-flight calibration or static calibration can be used. In in-flight calibration, known ground control is used to determine the camera orientation angles. They are then compared to the INS-derived ones to determine the orientation offset

$$\mathbf{R}_{c_i}^{M}(t) = \mathbf{R}_{INS}^{M}(t)\, \mathbf{R}_{c_i}^{INS} \tag{4}$$

where, $\mathbf{R}_{c_i}^{INS}$ is a rotation matrix rotating the c-frame of each camera i into the INS-frame. The second approach is similar to the former one, except that it is implemented in static mode on the ground using a close range target field. Both approaches have their inherent advantages and drawbacks because they differ in geometry, resolution and motion. For details, see Mostafa and Schwarz (1999). Substituting Equation 4 into Equation 2 and accounting for a 3D position offset vector from the INS to each camera's coordinate system origin $\mathbf{a}_i^{INS}$, the model represented by Equation 2 can be written as:

$$\mathbf{r}_G^M = \mathbf{r}_{E_i}^M(t) + \mathbf{R}_{INS}^M(t)\left(\mathbf{a}_i^{INS} + s_{g_i}\mathbf{R}_{c_i}^{INS}\mathbf{r}_g^{c_i}(t)\right) \tag{5}$$

The right-hand side of Equation 5 contains all the measurable/processed quantities except for $\mathbf{R}_{c_i}^{INS}$ and $\mathbf{a}_i^{INS}$, which are determined by calibration. The data flow of the entire georeferencing process is briefly shown in Figure 6; for details, see Mostafa et al (1998a).

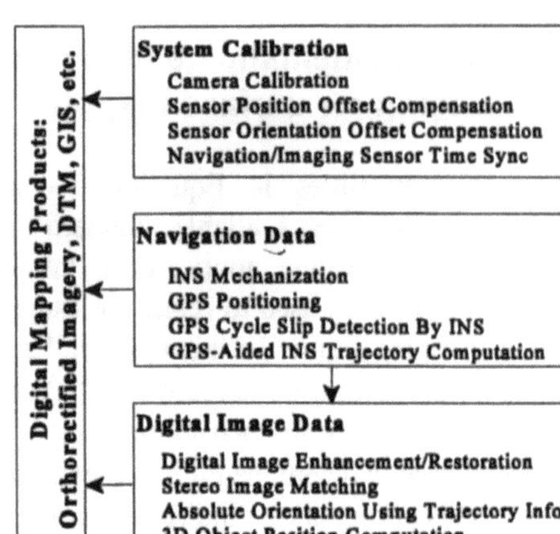

Fig. 6: Multi-sensor system data flow

7 In-flight system testing

A test flight was conducted over the campus of the University of Calgary. The university campus was chosen because of its detailed urban character as well as the ease with which ground control points could be placed on roads, in parking lots, and on building rooftops. Almost a hundred ground control points were established by GPS all over the campus. They were distributed in such a way that an optimal geometry for in-flight calibration of the digital cameras and the entire system could be achieved, see Figure 8.

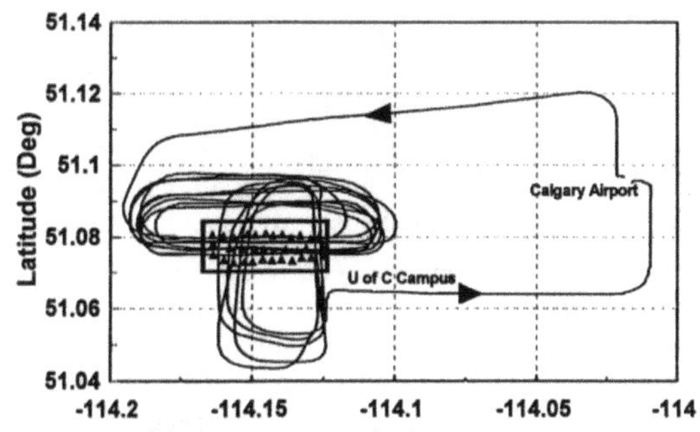

Fig. 7: GPS/INS-derived trajectory

Using the characteristics of the digital cameras (9mm x 13mm image size and f = 28 and 52mm, respectively, and their data rate, typically, 4 seconds per image, the flight pattern was designed to cover the entire campus by a block of standard 60% endlap and 40% sidelap, using repeated flight lines. The flight pattern is shown in Figure 7. Figure 8 shows a nadir image over the Calgary campus.

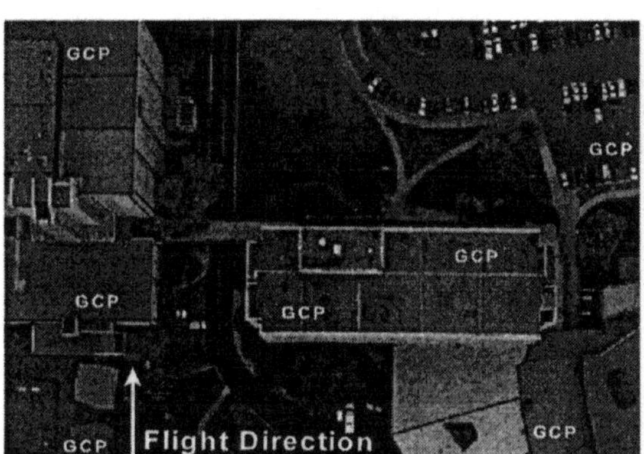

Fig. 8: A digital image over U of C campus

8 System performance

The GPS/INS raw data were processed using the KINGSPAD software. Figure 9 shows the Kalman filter-derived estimates of the positioning accuracy for the three components as well as the Position Dilution of Precision (PDOP) during the entire test flight.

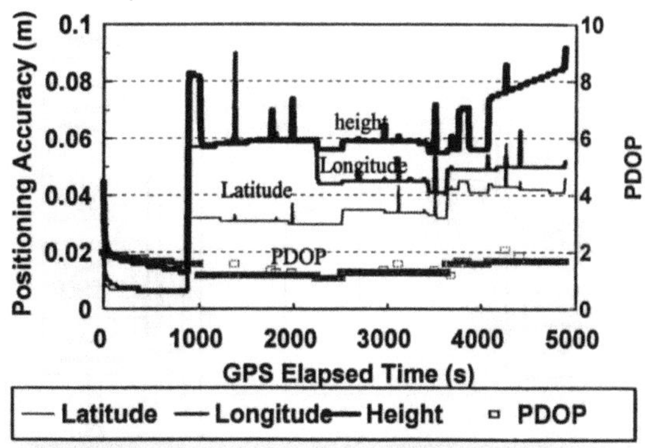

Fig. 9: GPS/INS positioning accuracy

The attitude at the interpolated camera exposures was accurate to about twenty arcseconds in roll and pitch and about one arcminute in azimuth. Using those parameters, together with the digital imagery which was processed in stereopairs, strips, and blocks, using the PCI OrthoEngine software, the ground control point coordinate values were independently computed and compared with their reference values obtained from the GPS ground survey. The statistics of the differences are shown in Table1. These accuracies are consistent with those achieved by simulations. The composition of the total error budget is given in Figure 10, for details see Mostafa et al (1998b).

Table 1: Accuracy of georeferencing by GPS/INS

Processing Mode	East (m)	North (m)	Height (m)
Stereo*	0.45	0.51	0.75
Strip # 1*	0.34	0.42	0.65
Strip # 2*	0.39	0.47	0.59
Strip # 3*	0.37	0.46	0.57
3 x 3 Block**	0.22	0.24	0.34

*using nadir and oblique images
**using nadir and oblique images and cross strips in two directions

9 Conclusions and future work

Results achieved with a prototype system show that the concept of using georeferenced digital images to map patches of the Earth's surface in kinematic mode is sound and that accurate 3D positions can be obtained without the use of ground control. It thus reduces turnaround-time for airborne mapping very considerably. Accuracies achieved are at the level 20 cm in horizontal and 30 cm in height. Higher accuracy is possible if larger format digital cameras are used For highest accuracy, especially in the height component, it is recommended that a laser scanner be added to the system.

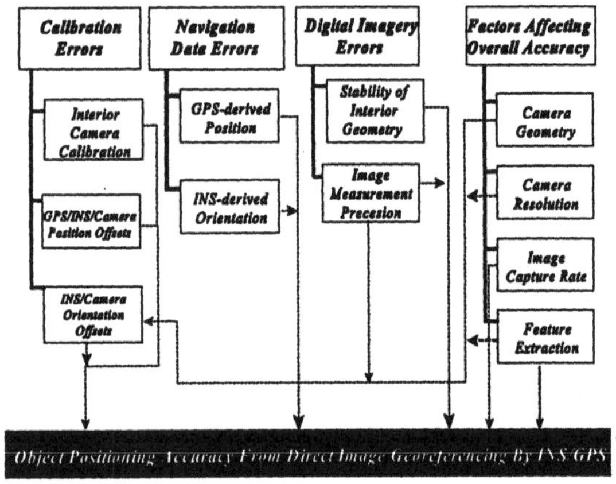

Fig. 10: Factors affecting system accuracy

Acknowledgements. Financial support for this research was obtained through an NSERC operating grant of the second author, an Egyptian scholarship of the first author as well as Graduate Research Scholarships and Special Awards from The University of Calgary. Messrs. T. Ludwig, J. Yom, A. Bruton, and J. Škaloud are gratefully acknowledged for their cooperation during the testing period.

References

Cramer, M., D. Stallmann, and N. Halla, (1997). High Precision Georeferencing Using GPS/INS and Image Matching. Proceedings of the International Symposium on Kinematic Systems in Geodesy, Geomatics and Navigation, Banff, Canada, 453-462.

Congalton, R.G., M. Balogh, C. Bell, K. Green, J.A. Milliken, and R. Ottman, (1998). Mapping and Monitoring Agricultural Crops and other Land Cover in the Lower Colorado River Basin. PE&RS, 64(11):1107-1113.

Helmert, F.R., (1880). Die Mathematischen und Physikalischen Theorien der Höheren Geodäsie. Leipzig, 1880 (1): 3.

Lichti, D. and M. A. Chapman, (1997). Constrained FEM Self-Calibration. PE&RS, 63(9): 1111-1119.

Lechner, W. and P. Lahmann, (1995). Airborne Photogrammetry Based on Integrated DGPS/INS Navigation. Proceedings of The 3rd International Workshop on High Precision Navigation (K.Linkwitz and U. Hangleiter, editors), 303-310.

Lyon, J.G, E. Falkner, and W. Bergen, (1995). Estimating Cost for Photogrammetric Mapping and Aerial Photography. Journal of Surveying Engineering, 121(13):63-86.

Mostafa, M.M.R., K.P. Schwarz, and P. Gong, (1997). A Fully Digital System for Airborne Mapping. Proceedings of the International Symposium on Kinematic Systems in Geodesy, Geomatics and Navigation, Banff, Canada, 463-471.

Mostafa, M.M.R., K.P. Schwarz, and P. Gong (1998a). A GPS/INS Integrated Navigation System In Support of Digital Image Georeferencing. Proceedings of the ION 54th Annual Meeting: 435-444.

Mostafa, M.M.R., K.P. Schwarz, and M.A. Chapman, (1998b). Development and Testing of an Airborne Remote Sensing Multi-Sensor System. International Archives of Photogrammetry and Remote Sensing, 32 (2): 217-222.

Mostafa, M.M.R. and K.P. Schwarz (1999). An Autonomous Multi-Sensor System for Airborne Digital Image Capture and Georeferencing. Proceedings of the ASPRS Annual Convention, Portland, Oregon, May 17-21, 976 - 987.

Schwarz, K.P., C.S. Fraser, and P.C. Gustafson, (1984). Aerotriangulation without ground control. International Archives of Photogrammetry and Remote Sensing, 25 (A1): 237:250.

Schwarz, K.P., M.A. Chapman, M.E. Cannon and P. Gong, (1993). An Integrated INS/GPS Approach to The Georeferencing of Remotely Sensed Data. PE&RS, 59(11): 1167-1674.

Schwarz, K.P., (1998). Mobile Multi-Sensor Systems: Modelling and Estimation. Proceedings, International Association of Geodesy Special Commission 4, Eisenstadt, Austria, 347-360.

Škaloud, J., M. Cramer, and K.P. Schwarz, (1996). Exterior Orientation by Direct Measurement of Position and Attitude. International Archives of Photogrammetry and Remote Sensing, 31 (B3):125-130.

Toth, C. and D.A. Grejner-Brzezinska, (1998). Performance Analysis of The Airborne Integrated Mapping System (AIMS™). International Archives of Photogrammetry and Remote Sensing, 32 (2):320-326.

GPS-based attitude determination for airborne remote sensing

K.F. Sheridan, P.A. Cross, and M.R. Mahmud
Department of Geomatic Engineering,
University College London, Gower Street, London, WC1E 6BT, United Kingdom.

Abstract. The design and testing of a GPS based attitude determination system for the direct georeferencing of airborne imagery is discussed. By combining dual-frequency, high data rate GPS receivers with a relatively simple gyroscopic attitude and heading reference system, position and orientation data can be collected at a rate which matches that of a typical airborne scanner, with an accuracy adequate for georeferencing imaged features to a few metres.

Data collected in two flight trials in October 1998 have been processed using software developed at University College London and the University of Newcastle upon Tyne, to determine a best estimate of the aircraft trajectory. The software employs single epoch ambiguity resolution, a modified double-differencing algorithm for direct determination of attitude parameters, and a Kalman filter for data integration. The method presented here uses a robust and efficient routine for ambiguity resolution, and exploits the additional redundancy of the modified double-differencing algorithm to detect and remove any poor observations from the final solution. Details of the computation methods are presented, together with an analysis of the results.

Keywords. GPS, attitude, direct georeferencing, outlier detection, ambiguity resolution.

---

1 Introduction

The provision of on-board navigation instruments to measure the position and orientation of airborne sensors enables ground coordinates to be determined for imagery with minimal, or even no, ground control. The potential reductions in time and cost and the increased operational flexibility that so-called *direct georeferencing* offers, have made it an increasingly active research area over the past few years. For details of the mathematical model and examples of applications see Schwarz et al (1993). The majority of systems that have been developed to measure both the position and attitude of airborne platforms have integrated some form of Inertial Navigation System (INS) with GPS. Cramer et al (1997), Toth and Grejner-Brzezinska (1998), and Skaloud and Schwarz (1998) discuss the design and performance of a number of such systems. In general, a high quality INS comprising accelerometers and gyroscopes is used to derive all six exterior orientation parameters of the imaging device. GPS position and velocity data allow the time dependent errors of the inertial positioning to be constrained. The resultant system combines accurately derived attitude at a high measurement rate (typically 50-100 Hz) from the INS, with the long term positional stability of GPS. The high data rate characteristic is particularly important when each individual scan line in an image requires orientation parameters. A typical scanner may record 20 to 100 lines per second.

An alternative approach is to determine attitude from GPS carrier phase signals measured by an array of antennas. A number of commercially available systems have been developed (e.g. the Ashtech ADU2 and Trimble TANS Vector), that consist of either three or four GPS antennas rigidly fixed together and linked to a special receiver that 'measures' the phase differences between the carrier signals received by different antennas. Essentially, kinematic phase GPS processing takes place between these antennas, and the coordinate differences obtained are converted to attitude components – usually pitch, roll and heading.

Attitude determination can also be achieved from independent multiple receivers in a similar way. GPS phase data are collected, vectors between antennas are computed, and the orientation parameters are estimated (using least squares if more than the minimum of three antennas have been deployed). In principle, any of the currently available methods can be used for the critical part of the processing – the estimation of the baseline vectors. Here a technique based on independent ambiguity resolution at every epoch is used.

International Association of Geodesy Symposia, Vol. 121
Schwarz (ed.), Geodesy Beyond 2000 – The Challenges of the First Decade

The technique used to resolve integer ambiguities using only data from a single epoch is based on the Ambiguity Function Method, first considered by Counselman and Gourevitch (1981). The algorithms have been implemented in a software package known as GPS Ambiguity Searching Programme (GASP), developed by Corbett (1994). By computing positions on a single epoch basis this method is immune to cycle slips or loss of lock. Full details of this method and successful results from comparisons of GASP with other commercial post-processing software, may be found in Corbett and Cross (1995) and Al-Haifi et al (1998).

The system presented in this paper uses high data rate, dual-frequency independent GPS receivers (two Ashtech Z-Surveyors, one Z-Sensor, all recording at 10Hz, and a Z-XII, recording at 5Hz) to determine both position and attitude. In addition, a low cost Attitude and Heading Reference System (AHRS), the Litef LCR-92H (Litef, 1997) which uses fibre-optic gyroscopes (FOGs) to derive angular rates and attitude parameters at 64Hz, was used. This solution provides reliable position and attitude data at a frequency and accuracy adequate for a range of commonly used airborne scanning devices.

In the next section details of an approach to directly estimate attitude parameters from GPS phase data based on modified double-differenced observation equations are given. A method to detect and reject outliers in the observations by exploiting the increased redundancy of the approach is then presented. Details of static and airborne testing of the system, including results, are given and potential improvements to the original method are discussed. Finally some conclusions are drawn and a number of possible applications are suggested. In the following text, *robust* is used to refer to a system's ability to successfully resolve integer ambiguities, whereas *reliability* describes the ability to detect and remove data that do not fit the model due to the presence of larger than expected errors.

2 Attitude determination from independent GPS receivers

The steps involved in attitude determination (on an epoch by epoch basis) may be summarised as follows.

1. Compute the coordinates of one of the antennas on the dynamic platform from a static base station.
2. The coordinates of this antenna on the platform are now held fixed, and treated as known values. Compute the vectors (and hence coordinates) to any additional antennas on the platform in the WGS84 reference frame. Providing successful ambiguity resolution has been obtained, the relative positions between the antennas on the platform will be very precise.
3. Using *a priori* knowledge of the relative positions of the antennas in a fixed body reference frame, and the relative positions of the vectors in WGS84 (or any local level reference frame), the orientation of the platform may be computed.

To identify outliers amongst observations it is essential for the system to contain significant redundancy – something that does not exist in kinematic GPS baseline estimation. Once ambiguities have been fixed (either correctly or incorrectly) each baseline only provides a redundancy of $f(p-1)-3$, (where p is the number of satellites in view, and f is the number of frequencies used) and there would clearly be advantages if the redundancy could be increased. Since the relative positions of the antennas in the fixed body frame system would always in practice be known, one way to increase redundancy is to combine all GPS phase data collected at a single epoch and estimate only the three attitude parameters, leading to a redundancy of $3[f(p-1)]-3$ in the case of four antennas. For example, with five satellites in view, there would be a redundancy of twenty-one. To do this, it is necessary to develop equations that relate the GPS double-differenced phase observables directly to the three attitude parameters.

The attitude of a rigid body platform is defined as the orientation of the specific body frame with respect to a local coordinate system or reference frame (Kleusberg, 1995). In GPS attitude determination, the two systems are related through the antenna locations. A minimum of three antennas are necessary to solve for pitch, roll and yaw, any additional antennas will of course increase redundancy.

The mathematical relationship between the body frame coordinate system and the local level coordinate system can be expressed as follows:

$$X^{LLS} = R_{\psi\theta\gamma} X^{BFS} \tag{1}$$

where

X^{LLS} is the matrix of antenna locations in the local level coordinate system,

X^{BFS} is the corresponding matrix in the body frame coordinate system, and

$R_{\psi\theta\gamma}$ is the rotation transformation matrix.

The rotation matrix $R_{\psi\theta\gamma}$ is the product of three matrices describing the rotations of yaw γ followed by roll θ and finally pitch ψ.

$$R_{\psi\theta\gamma}=\begin{bmatrix}\cos\theta\cos\gamma & \cos\theta\sin\gamma & -\sin\theta \\ \sin\psi\sin\theta\cos\gamma-\cos\psi\sin\gamma & \sin\psi\sin\theta\sin\gamma+\cos\psi\cos\gamma & \sin\psi\cos\theta \\ \cos\psi\sin\theta\cos\gamma+\sin\psi\sin\gamma & \cos\psi\sin\theta\sin\gamma-\sin\psi\cos\gamma & \cos\psi\cos\theta\end{bmatrix} \quad (2)$$

The method described in this paper estimates the attitude parameters directly from the GPS phase observations. The *a priori* knowledge of the antenna body frame coordinates is incorporated into the measurement models, and then all measurements are used to estimate the three attitude parameters. To achieve this, the standard double-differenced phase observation equation is modified to include the attitude parameters. The attitude parameters appear explicitly in the modified equation using the elements of the composite rotation matrix, $R_{\psi\theta\gamma}$. For ease of use these terms are abbreviated to m_{11} to m_{33}, as follows,

$$R_{\psi\theta\gamma}=\begin{bmatrix}m_{11} & m_{12} & m_{13} \\ m_{21} & m_{22} & m_{23} \\ m_{31} & m_{32} & m_{33}\end{bmatrix}; \qquad (3)$$

so that the double-differenced observation equation can be expressed directly in terms of attitude as,

$$\left\{\left[x_i^{LLS}-\left(m_{11}x_A^{BFS}+m_{12}y_A^{BFS}+m_{13}z_A^{BFS}\right)\right]^2+\left[y_i^{LLS}-\left(m_{21}x_A^{BFS}+m_{22}y_A^{BFS}+m_{23}z_A^{BFS}\right)\right]^2+\left[z_i^{LLS}-\left(m_{31}x_A^{BFS}+m_{32}y_A^{BFS}+m_{33}z_A^{BFS}\right)\right]^2\right\}^{1/2} -$$

$$\left\{\left[x_i^{LLS}-\left(m_{11}x_B^{BFS}+m_{12}y_B^{BFS}+m_{13}z_B^{BFS}\right)\right]^2+\left[y_i^{LLS}-\left(m_{21}x_B^{BFS}+m_{22}y_B^{BFS}+m_{23}z_B^{BFS}\right)\right]^2+\left[z_i^{LLS}-\left(m_{31}x_B^{BFS}+m_{32}y_B^{BFS}+m_{33}z_B^{BFS}\right)\right]^2\right\}^{1/2} -$$

$$\left\{\left[x_j^{LLS}-\left(m_{11}x_A^{BFS}+m_{12}y_A^{BFS}+m_{13}z_A^{BFS}\right)\right]^2+\left[y_j^{LLS}-\left(m_{21}x_A^{BFS}+m_{22}y_A^{BFS}+m_{23}z_A^{BFS}\right)\right]^2+\left[z_j^{LLS}-\left(m_{31}x_A^{BFS}+m_{32}y_A^{BFS}+m_{33}z_A^{BFS}\right)\right]^2\right\}^{1/2} -$$

$$\left\{\left[x_j^{LLS}-\left(m_{11}x_B^{BFS}+m_{12}y_B^{BFS}+m_{13}z_B^{BFS}\right)\right]^2+\left[y_j^{LLS}-\left(m_{21}x_B^{BFS}+m_{22}y_B^{BFS}+m_{23}z_B^{BFS}\right)\right]^2+\left[z_j^{LLS}-\left(m_{31}x_B^{BFS}+m_{32}y_B^{BFS}+m_{33}z_B^{BFS}\right)\right]^2\right\}^{1/2} -$$

$$+\ \lambda N_{AB}^{ij}$$

$$=\lambda\left(\Phi_A^i-\Phi_B^i-\Phi_A^j+\Phi_B^j\right) \qquad (4)$$

where

x^{LLS}, y^{LLS}, z^{LLS} are the coordinates of satellites i and j in the local level system;

x^{BFS}, y^{BFS}, z^{BFS} are the coordinates of antennas A and B in the body frame system;

N_{AB}^{ij} is the double-differenced integer ambiguity;

λ is the wavelength; and

Φ_A^i, Φ_B^i, Φ_A^j, Φ_B^j are the phase observables measured at receivers A and B from satellites i and j.

In practice there will be p antennas ($A,B,\ldots$) and q satellites ($i,j,\ldots$) leading to $(p-1)(q-1)$ double-differenced equations. These are linearised to

$$\frac{\partial\rho_{AB}^{ij}}{\partial\psi}d\psi+\frac{\partial\rho_{AB}^{ij}}{\partial\theta}d\theta+\frac{\partial\rho_{AB}^{ij}}{\partial\gamma}d\gamma+\frac{\partial\rho_{AB}^{ij}}{\partial N}dN=\left[\Phi_{AB}^{ij}(observed)-\Phi_{AB}^{ij}(computed)\right]+v \quad (5)$$

where

$\frac{\partial\rho_{AB}^{ij}}{\partial\psi}, \frac{\partial\rho_{AB}^{ij}}{\partial\theta}, \frac{\partial\rho_{AB}^{ij}}{\partial\gamma}, \frac{\partial\rho_{AB}^{ij}}{\partial N_{AB}^{ij}}$ are the partial derivatives relating double-differenced geometric ranges to attitude parameters and integer ambiguities;

$d\psi, d\theta, d\gamma$ are the corrections to the provisional values of the attitude parameters;

$\Phi_{AB}^{ij}(observed)$ is the observed double-differenced phase for each pair of satellites;

and

$\Phi_{AB}^{ij}(computed)$ is the computed double-differenced phase for each pair of satellites using the provisional values;

and then assembled to the well-known matrix form

$$Ax = b + v \qquad (6)$$

The full derivation of the modified double difference-phased observation equation, including formulae for computing the partial derivatives needed to construct the design matrix, and a more thorough description of the general mathematics of attitude determination are given in Cross et al (1997), and Mahmud (1999).

These algorithms have been implemented in a software package called GPS Routine for Attitude Parameter Estimation, or GRAPE (Mahmud, 1999).

3 Outlier detection and rejection

Static trials have been carried out to test the reliability and accuracy of the proposed method. Data were collected over a two hour period in which there were between five and nine satellites in view, above the 15° mask angle. Four antennas were arranged with baseline lengths between 20 and 40m. The results have demonstrated a 100% success rate in ambiguity resolution, and accuracies (95%) of the order of 1.5 arc-minutes in pitch and roll, and around 20 arc-seconds in heading, equivalent to positioning baselines to 1cm and 2mm respectively (Mahmud, 1999). The heading component is more accurate as it is not affected by the height component of GPS, the least well determined.

The increased redundancy produced by directly determining attitude parameters, rather then using a baseline-by-baseline approach, should improve the reliability and robustness of the GPS attitude determination

Outlier detection and rejection has been investigated with this static dataset, using a procedure outlined in Mahmud (1999), based on the tau statistic, as described in Pope (1976). Initially an error of 0.005m was introduced into the L1 phase observation on SV24. The rest of the observations remained the same. The data were then reprocessed using the direct attitude system and a single baseline system separately. Next the error was increased by 0.005m and the whole data set was processed again. This process was repeated until, for all epochs, either the direct attitude system or the single baseline system managed to reject the observations in which the constant error was included. This process was then repeated for SV25 to see if the results differed significantly. Table 1 shows the number of epochs at which observations containing outliers were successfully detected and rejected by the two approaches, as the introduced error was increased. Figures in brackets express the number of epochs rejected as a percentage of the total number of epochs tested.

Table 1: Outlier rejection comparison

	Direct Attitude System		Single Baseline System	
Error (m)	SV24	SV25	SV24	SV25
0.005	4802 (66%)	6718 (92%)	4327 (59%)	5044 (69%)
0.010	6824 (93%)	7267 (99.8%)	6425 (88%)	6290 (86%)
0.015	7209 (98%)	7281 (100%)	7002 (96%)	6837 (94%)
0.020	7318 (99.96%)	-	7235 (99%)	-
0.025	7321 (100%)	-	7167 (98%)	-

The direct attitude system is consistently able to detect outliers at more epochs than the single baseline system, due to the additional redundancy, which leads to more accurate and reliable attitude solutions.

4 Flight trial

In October 1998 two trial flights were carried out to gather data to test the proposed system in an airborne, operational environment. The aircraft used was the Natural Environment Research Council's Piper Chieftain, which is equipped with a number of imaging and navigational instruments to perform aerial surveys.

The data from four dual-frequency high data receivers have been processed using the GRAPE software package, and the results have been compared against the AHRS solution. Figure 1 shows the heading from the two systems during a ninety second flight line. In this example an offset of 1° has been deliberately introduced so that the two lines can be differentiated. Figure 2 is a ten second period from the same flight line with the GRAPE solution at each epoch represented by a large diamond and the AHRS solution by smaller point symbols, which because of the high frequency (64Hz) of the solution appear as a single line. The GRAPE solution for this period is not a complete 10Hz data set due to logging problems experienced during the flights.

These plots illustrate some important points. Firstly, ambiguity resolution has been successful throughout the flight line, any incorrect ambiguities would result in a significant difference in the two solutions. Secondly, despite the much higher frequency of the AHRS solution, it does not appear to detect any aircraft motions that the GRAPE solution fails to measure. Although this may be

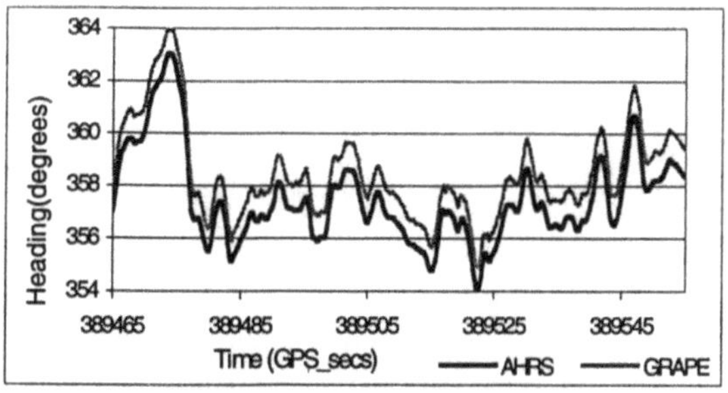

Fig. 1: GRAPE and AHRS heading (90 seconds)

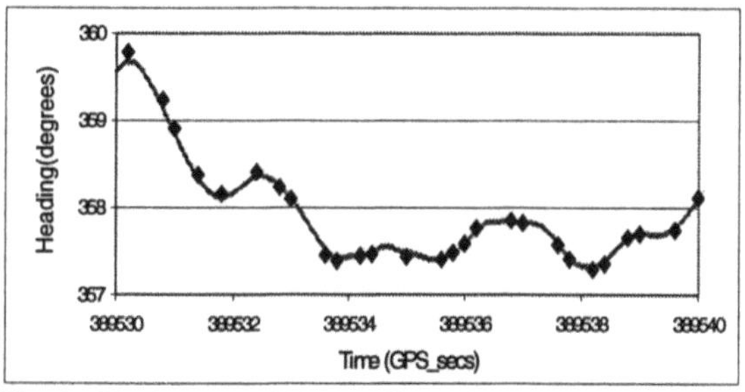

Fig. 2: GRAPE and AHRS heading (10 seconds)

partly due to a filter applied to the raw gyro measurements before they are recorded, it is reasonable to assume that the GPS solution is adequately measuring any high frequency attitude variations. Prior to the flight it was the intention to combine AHRS measurements with the GPS solution (using a Kalman filter), so that the higher frequency data would act as an 'interpolator' between GPS epochs, but in the light of these results this approach does not appear to offer any advantages over a GPS only solution. It is possible, however, to make use of the AHRS values, by using them as starting estimates to improve the robustness and efficiency of the single epoch ambiguity resolution method, as described in the following section.

Tests were also carried out to compare the success of GRAPE ambiguity resolution, with that of a single epoch, single baseline system, GASP. Essentially, if two baselines between three antennas on the aircraft can be computed successfully using the method outlined in section 2, then it is possible to derive the attitude components. It is assumed that ambiguity resolution has been successful if the baseline lengths agree with the *a priori* values to within a few cms. Table 2 shows the percentage of epochs at which sufficient vectors to derive attitude can be determined using the single baseline system. On all the flight-lines tested, the direct attitude solution correctly resolved ambiguities (based on AHRS comparison) for all epochs. In periods of fairly good observation conditions, the baseline-by-baseline approach performs relatively well although it is less successful than the direct attitude system. However, when data quality deteriorates, the direct attitude system proves to be far more robust. During lines L001 and L003, one of the winglet antennas was not logging data, and the other was recording data with significant noise levels, almost certainly due to multipathing effects.

Table 2: Ambiguity resolution from single baselines

Flight Line	% Success
L001	21
L003	16
L006	67
L013	91

5 Modified ambiguity resolution

The ambiguity resolution method used in this research computes the set of double-differenced ambiguities for each trial attitude within a search volume centred on a starting approximation. To ensure the search volume contains a trial attitude leading to the correct ambiguities, its size must be related to the accuracy of the starting approximation. With no good starting values available, a search volume can be constructed around some arbitrary value (eg. zero in all components) that is sufficiently large to include any possible attitude values that an aircraft could experience. This involves computations for a huge number of trial attitudes and is therefore very inefficient. In periods when sufficient good GPS observations are available for a successful solution, final values from a previous epoch can be used as starting approximations, or values can be predicted based on two or more previous epochs. This approach has proved successful in periods of relatively stable attitude, as experienced during flight lines. If, however, there is an increased change in attitude between two epochs, due to greater aircraft dynamics or a longer time interval between measurements (which may occur when the aircraft turns and some antennas are unable to receive satellite signals for a time), this approach may not be adequate, or may only be successful with a very large search volume.

One possible solution is to use values from the AHRS as starting approximations to resolve ambiguities. It has been found, that provided any axis misalignments and time differences between the two systems are taken into account, ambiguities can be resolved using a single attitude value, i.e. no searching is necessary. Float ambiguities are determined for the starting attitude, these are then rounded to integer values, and a least squares process solves for the corrections to the initial values. Applying this method for periods when ambiguities can also be resolved with a GPS only solution, shows that the ambiguities and final attitude values are identical. Table 3 shows how the processing time of the GRAPE solution is related to the number of trial attitudes that must be considered in the ambiguity search.

Table 3: Search volume and processing time

Start Value	Range (radians)	Trial attitudes	Processing time per epoch
GRAPE	+/- 0.1	9261	100.0%
GRAPE	+/- 0.05	1331	7.5%
GRAPE	+/- 0.02	125	3.6%
AHRS	0	1	1.5%

Note. The searching increment is fixed at 0.01 radians.

The actual time will depend on the processing speed of a specific computer, so the time taken for the largest search volume has been considered as a benchmark. The figures clearly show how the processing time increases with the search volume. Using AHRS data there is only a single trial attitude to consider, making the processing significantly more efficient. This method does not rely on any values from previous epochs, so is a truly independent single epoch method for attitude determination.

6 Conclusions

A method has been developed that determines attitude from a single epoch of dual-frequency GPS phase data collected by three or more antennas. Modified double-differencing equations that estimate attitude parameters directly from phase measurements have been derived. The increased redundancy resulting from this method allows outliers to be identified and rejected from the computation, leading to a more accurate and reliable final solution based on less, but better, data. A robust and efficient method for resolving ambiguities using attitude measurements from a relatively simple gyroscopic instrument has also been developed. A system combining both these ideas should be able to work with very noisy data (that may result in incorrect or no ambiguity resolution with other approaches) because the noisy GPS observations are not used for ambiguity resolution, and can then be rejected in the subsequent outlier detection phase in the least squares processing

Results from static testing and test flights prove the potential of the system, and demonstrate the accuracy, robustness and reliability that such an approach offers.

Acknowledgements. The authors would like to acknowledge the support received during the course of this research from the NERC Airborne Remote Sensing Facility who carried out the test flights, and Colin Beatty International Ltd., and Ashtech Europe Ltd., who supplied the GPS equipment necessary for the independent receiver solution. Kevin Sheridan is supported by a NERC PhD studentship.

References

Al-Haifi, Y. M., S. J. Corbett, and P. A. Cross (1998). Performance Evaluation of GPS Single Epoch on the fly Ambiguity Resolution. *Navigation*, 44 (4): 479-487.

Corbett, S. J., (1994). GPS Single Epoch Ambiguity Resolution for Airborne Positioning and Orientation. PhD Thesis, University of Newcastle upon Tyne, UK.

Corbett, S. J., and P. A. Cross (1995) GPS Single Epoch Ambiguity Resolution. *Survey Review*, 33 (257): 149-160.

Counselman, C. C., and S. A. Gourevitch (1981). Miniature Interferometer Terminals for Earth Surveying: Ambiguity and Multipath in Global Positioning System. *IEEE Transactions on Geoscience and Remote Sensing*. GE-19(No.4): 244-252.

Cramer, M., D. Stallmann, and N. Halla (1997). High Precision Georeferencing Using GPS/INS and Image Matching. *Proc. of International Symposium on Kinematic Systems in Geodesy, Geomatics and Navigation (KIS97)*, Banff, Canada, June 3-6, 453-462.

Cross, P. A., S. J. Corbett, and M.R. Mahmud (1997). Benchmarking of Commercial Offshore GPS-Based Attitude Determination Systems. *Proc. of International Symposium on Kinematic Systems in Geodesy, Geomatics and Navigation*, Banff, Canada, June 3-6, 379-388.

Kleusberg, A.(1995). Mathematics of Attitude Determination with GPS. *GPS World*, Vol. 6, No. 9: 72-78.

Litef (1997). LCR-92 uAHRS®: Attitude and Heading Reference System. Technical data supplied by Litef GmbH Freiburg, Germany. February 1997.

Mahmud, M.R., (1999). Precise Three-Dimensional Attitude Estimation from Independent GPS Arrays. PhD Thesis, University of London, UK.

Pope, A.J, (1976). The Statistics of Residuals and the Detection of Outliers. *NOAA Technical Report* NOS 65 NGS 1, National Oceanic and Atmospheric Administration, Rockville, Maryland, USA, 133pp.

Schwarz, K.P., M.A. Chapman, M.W. Cannon, and P. Gong (1993). An Integrated INS/GPS Approach to the Georeferencing of Remotely Sensed Data. *Photogrammetric Engineering and Remote Sensing*, Vol. 59, No. 11, November 1993: 1667-1674.

Skaloud, J., and K.P. Schwarz (1998). Accurate Orientation for Airborne Mapping Systems. *Proc. of ISPRS Commission II Symposium*, Cambridge, UK, July 13-17, 283-290.

Toth, C.K., and D.A. Grejner-Brzezinska (1988). Performance Analysis of the Airborne Integrated Mapping System (AIMS™). *Proc. of ISPRS Commission II Symposium*, Cambridge, UK, July 13-17, 320-326.

Absolute kinematic GPS positioning using satellite clock estimation every 1 second

Jay Hyoun Kwon, Christopher Jekeli, Shin-Chan Han
Department of Civil and Environmental Engineering and Geodetic Science,
The Ohio State University, 2070 Neil Avenue, Columbus, OH 43210-1275, USA

Abstract. An algorithm for absolute positioning through satellite clock estimation has been developed. Using IGS precise orbits and measurements, the GPS clock errors were estimated at 30-second intervals and these estimates were compared to values determined by JPL. The agreement was at the level of about 0.1 nsec (3 cm). The clock error estimates were then used in an application of a single-differenced (between satellite) positioning algorithm in static and kinematic mode. For the static case, an IGS station was selected and the coordinates were estimated. The estimated absolute position coordinates and the known values had a mean difference of up to 18 cm with standard deviation less than 2 cm. For the kinematic case, data (every second) obtained from a GPS buoy were tested and the result from the absolute positioning was compared to a DGPS solution. The mean difference between two algorithms is less than 45 cm and the standard deviation is less than 30 cm. It was proved that a higher rate of satellite clock determination is necessary to do absolute kinematic positioning at better than 10 cm precision.

Keywords. Global Positioning System (GPS), satellite clock error, absolute kinematic positioning.

1 Introduction

In order to conduct precise absolute (stand-alone) point positioning using GPS, the satellite clock error as well as the precise orbit should be determined (Lachapelle et al, 1996). Because of Selective Availability (SA) the precision of the signal degrades up to 60 meters. Usually, differential techniques are used to circumvent this error. However, there are some situations when differential positioning is not possible. Then, the determination of the satellite clock error will be the most important factor in absolute positioning.

Currently the International GPS Service (IGS) provides a GPS satellite orbit and clock error estimations every 900 seconds with about 5 and 10 centimeters accuracy, respectively (IGS, 1999). It should be noted that a higher rate of orbit and clock error determinations are required in some applications related to precise positioning of moving-base platforms. The higher-rate orbits, e.g. with 30-second resolution can be obtained by an interpolation with centimeter level accuracy (Remondi, 1989, 1991). The interpolation, however, is no longer feasible to obtain the satellite clock errors because of the SA effect on the clock errors causing high frequency variation.

In this paper, a post-processing method with interpolated orbits is introduced to estimate the satellite clock error every 30 seconds using the observations from the globally distributed IGS control stations. In addition, results from a kinematic as well as an absolute static positioning that uses the estimates of the satellite clock error are presented.

2 Data processing

2.1 The observation equations

The observation equations of the GPS phase measurements are given as follows (Goad and Yang, 1995):

$$\Phi^k_{r,1}(t) = \rho^k_r(t) + c(dt_r(t) - dt^k(t)) + T^k_r(t) - \frac{I^k_r(t)}{f_1^2} + \lambda_1 N^k_{r,1} + \lambda_1[\varphi_r(t_0) - \varphi^k(t_0)]_1 + \varepsilon^k_{r,1}, \quad (1)$$

$$\Phi^k_{r,2}(t) = \rho^k_r(t) + c(dt_r(t) - dt^k(t)) + T^k_r(t) - \frac{I^k_r(t)}{f_2^2} + \lambda_2 N^k_{r,2} + \lambda_2[\varphi_r(t_0) - \varphi^k(t_0)]_2 + b^k_{r,1}(t) + \varepsilon^k_{r,2}. \quad (2)$$

The subscript r indicates the index of the receiver and superscript k that of the satellite. c is the speed

International Association of Geodesy Symposia, Vol. 121
Schwarz (ed.), Geodesy Beyond 2000 – The Challenges of the First Decade

of light; f_1, f_2 and λ_1, λ_2 are the L1 and L2 carrier frequencies and wavelengths; $\Phi^k_{r,1}(t)$, $\Phi^k_{r,2}(t)$ are the phase range measurements from the satellite k and at the receiver r; $\rho^k_r(t)$ is the geometric distance between the satellite's antenna at the signal transmission time and the receiver's antenna position at the signal reception time; $T^k_r(t)$ is the troposphere delay; $I^k_r(t)/f^2_{1or2}$ is the frequency-dependent ionospheric refraction; and $N^k_{r,1}$ and $N^k_{r,2}$ are the ambiguities of the L1 and L2 phase measurements. The one-way phase observables also include a fixed nonzero initial fractional phase term $\lambda[\varphi_r(t_0)-\varphi^k(t_0)]$ that is contained in the receiver- and satellite-generated phase signals. The remaining term $b^k_{r,1}(t)$ is the relative interchannel bias between $\Phi^k_{r,1}(t)$ and $\Phi^k_{r,2}(t)$, and results from the fact that the L1 and L2 signals travel through different hardware paths inside the receiver as well as the satellite transmitter (Coco, 1991).

From the above measurement equations, some combinations are possible for eliminating the nuisance parameters such as the ionosphere refraction, the receiver clock errors, and the ambiguities.

2.2 Ion-free wide lane combination

Using the dual-frequency signal it is possible to eliminate the first order ionospheric effect by a combination of phase measurements (Hofmann-Wellenhof et al, 1992). Because the maximum contributions of the 2nd and 3rd order terms are about 3 cm and less than 1 cm, respectively (Seeber, 1993), eliminating the first-order effect would be enough for most applications. The so-called ion-free wide lane signal (86 cm wavelength) can be obtained by first multiplying equation (1) and (2) by the combination coefficients $f_1^2/\{(f_1+f_2)\cdot c\}\approx 2.95$ and $f_2^2/\{(f_1+f_2)\cdot c\}\approx 1.79$ and then taking the differences between L1 and L2 measurements:

$$\begin{aligned}&\varphi^k_{r,ion-free}(t)\\&=\frac{f_1-f_2}{c}\rho^{*k}_r(t)+(f_1-f_2)\{dt_r(t)-dt^k(t)\}\qquad(3)\\&+\left(\frac{f_1}{f_1+f_2}N^{*k}_{r,1}-\frac{f_1}{f_1+f_2}N^{*k}_{r,2+}\right)+b^k_{r,phase}(t)+\varepsilon^k_r.\end{aligned}$$

Note that $\rho^{*k}_r(t)$ includes the geometric range $\rho^k_r(t)$ and troposphere delay $T^k_r(t)$. Furthermore, N^{*k}_r is no longer an integer and consists of the integer ambiguity N^k_r and the fractional phase offset $\lambda_1[\varphi_r(t_0)-\varphi^k(t_0)]$.

2.3 The time and satellite-differenced measurements

If the measurements have no cycle slips or loss of lock, the ambiguity can be eliminated when two independent measurements are differenced with respect to time. Similarly, the interchannel bias could be eliminated if it is assumed to be a constant. Furthermore, the receiver clock error will be eliminated if the measurements from the two satellites k and l are differenced.

Applying the time and satellite differencing to the ion-free phase combination successively;

$$\begin{aligned}\varphi^{k,l}_{r,ion-free}(t_{i,j})&=\varphi^k_{r,ion-free}(t_{i,j})-\varphi^l_{r,ion-free}(t_{i,j})\\&=\frac{f_1-f_2}{c}\rho^{*k,l}_r(t_{i,j})-(f_1-f_2)dt^{k,l}(t_{i,j})+\varepsilon\ ,\end{aligned}\qquad(4)$$

where

$\rho^{*k,l}_r(t_{i,j})\equiv(\rho^{*k}_r(t_i)-\rho^{*k}_r(t_j))-(\rho^{*l}_r(t_i)-\rho^{*l}_r(t_j))$,

$dt^{k,l}(t_{i,j})\equiv(dt^k(t_i)-dt^k(t_j))-(dt^l(t_i)-dt^l(t_j))$.

Now, two nuisance parameters, namely the receiver clock error and the ambiguity, no longer exist. Using IGS globally distributed station coordinates, precise orbits and a tropospheric model, $\rho^{*k}_r(t_{i,j})$ can be obtained, leaving the measurement $\varphi^k_{r,ion-free}(t_{i,j})$. Thus, the only unknown parameter is the relative satellite single-differenced GPS clock error.

2.4 The satellite clock error and absolute positioning

Rearranging equation (4) in terms of the unknown quantity, $dt^{k,l}(t_{i,j})$, equation (5) follows:

$$dt^{k,l}(t_{i,j})=\frac{1}{c}\rho^{*k,l}_r(t_{i,j})-\frac{1}{(f_1-f_2)}\varphi^{k,l}_{r,ion-free}(t_{i,j})+\varepsilon.\ (5)$$

The satellite- and time-differenced, ion-free, phase combination produces the relative variations of the single-differenced satellite clock error with respect to the initial epoch. Thus, the satellite clock error at the nth epoch can be expressed as equation (6) using the initial clock error $dt^{k,l}(t_1)$:

$$dt^{k,l}(t_n) = dt^{k,l}(t_1) + \sum_{i=2}^{n} dt^{k,l}(t_{i,i-1}). \qquad (6)$$

Therefore, if the satellite clock error at an initial or an arbitrary epoch is available, the satellite clock errors of all epochs are calculated according to equation (6). Once the satellite clock errors have been estimated, the absolute positioning can be conducted as described in the following paragraphs.

Let's assume that the relative satellite clock error was estimated and measurements were obtained at an unknown site, whose coordinates are to be determined. After taking the ion-free, wide-lane combination, performing satellite single-differencing, and substituting the estimated satellite clock errors, the measurements are described as follows:

$$\varphi^{k,l}_{r,ion-free}(t_1) - \frac{f_1 - f_2}{c} T^{k,l}_r(t_1) = \frac{f_1 - f_2}{c} \rho^{k,l}_r(t_1) + \tilde{N}^{*k,l}_w + \varepsilon, \qquad (7)$$

$$\varphi^{k,l}_{r,ion-free}(t_2) + (f_1 - f_2)dt^{k,l}(t_{2,1}) - \frac{f_1 - f_2}{c} T^{k,l}_r(t_2) = \frac{f_1 - f_2}{c} \rho^{k,l}_r(t_2) + \tilde{N}^{*k,l}_w + \varepsilon, \qquad (8)$$

$$\ldots$$

$$\varphi^{k,l}_{r,ion-free}(t_n) + (f_1 - f_2)\sum_{i=2}^{n} dt^{k,l}(t_{i,i-1}) - \frac{f_1 - f_2}{c} T^{k,l}_r(t_n) = \frac{f_1 - f_2}{c} \rho^{k,l}_r(t_n) + \tilde{N}^{*k,l}_w + \varepsilon, \qquad (9)$$

where
$\tilde{N}^{*k,l}_w \equiv N^{*k,l}_w + b^{k,l}_{phase} - (f_1 - f_2)dt^{k,l}(t_1)$.

Note that the new variable, $\tilde{N}^{*k,l}_w$, is defined by merging several constant terms. The unknowns are the newly defined ambiguity term $\tilde{N}^{*k,l}_w$ and the position coordinates of the (moving) receiver $x_r(t)$, $y_r(t)$, and $z_r(t)$ contained in $\rho^{k,l}_r(t)$. With the measurement $\varphi^{k,l}_{r,ion-free}(t)$, estimated clock error $dt^{k,l}(t_{i,i-1})$, and modeled tropospheric effect $T^{k,l}_r(t)$, these unknowns can be determined.

The overall procedure of satellite clock estimation is depicted in Figure 1. After calculating the periodic relativistic effect (Gibson, 1983), the satellite position at signal emission time is determined using the IGS precise orbit. Next, time-differencing of all observation data at the receiver station is performed to obtain time-differenced clock errors (both satellite and receiver clock errors still remain). The satellite position is calculated with a 9th-order polynomial interpolator, and the tropospheric refraction model is the modified Hopfield model (Goad and Goodman, 1974). Finally, with the satellite differencing the receiver clock error is eliminated. The results of this process are the relative satellite-differenced GPS clock errors with respect to the initial clock error. Because the same satellite clock error can be calculated from other stations as well, a final estimate is an equally weighted average. The slight differences in transmission time due to different locations of the stations are disregarded.

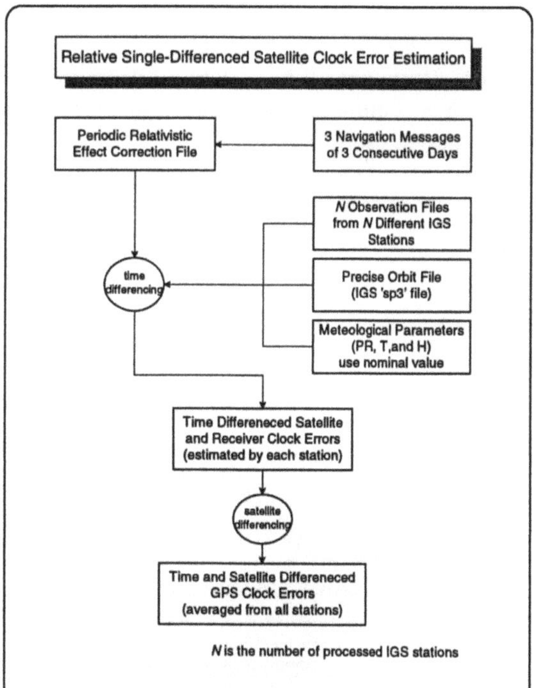

Fig. 1: Satellite clock error estimation procedure

3 Test for the absolute positioning

3.1 Satellite clock error estimates

Figure 2 shows the distribution of five stations used for satellite clock estimation. The station 'USNA' is treated as an unknown site. The baseline lengths between the 'USNA' and other stations are in the range of 600-1300 km. Observations from the five

IGS stations are processed independently and the satellite clock errors, estimated from each station, are averaged with same weight.

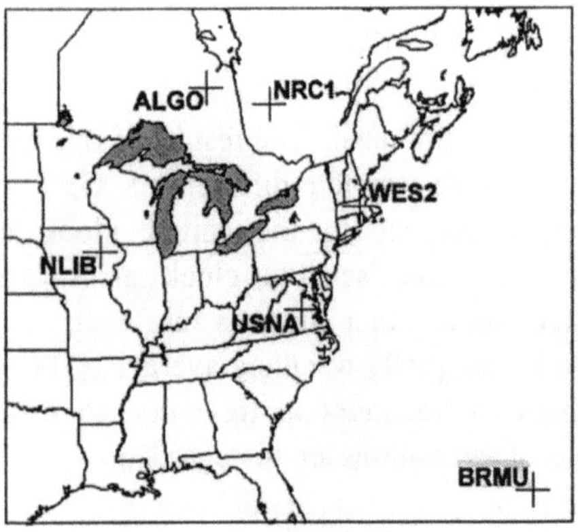

Fig. 2: Location of 6 IGS stations for static positioning test

Figure 3 represents the satellite-differenced GPS clock error estimates and their differences with JPL's estimates for PRN4 & PRN7.

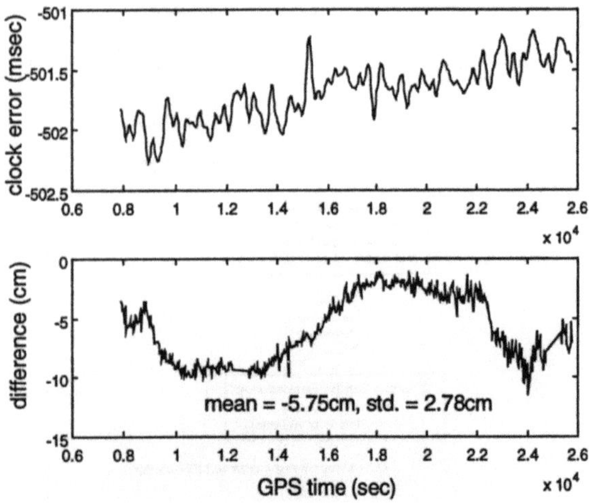

Fig. 3: Satellite-differenced GPS clock error estimates (top), and differenced with JPL estimates (bottom) for PRN4 & PRN7

The estimates are calculated by summing the time-differenced estimates and using the initial clock error provided by IGS. One can see a linear trend as well as high frequency fluctuations. The linear trend can be well determined using the clock error information in the navigation message. The high frequency error due to SA, however, cannot be corrected using the navigation message and the magnitude of the fluctuation is about 200 nanoseconds (60 meters). The difference with the JPL solution (JPL, 1999) can be explained as follows: 1) different orbits: the orbit affects the range directly, resulting in the difference in clock error estimation; 2) unmodeled tropospheric delay: in this method, the tropospheric delay is not estimated but is modeled, while JPL estimates the zenith tropospheric delays and uses them for estimating the clock error.

3.2 Absolute static positioning

Using the GPS clock error estimates, the coordinates of the station 'USNA' were estimated. This was done in a two-step process, whereby the float ambiguities, $\tilde{N}_w^{*k,l}$, were first determined using the ambiguity function method (Leick, 1995). With these fixed, the absolute position for subsequent times was estimated on the basis of the single difference equations (7)-(9).

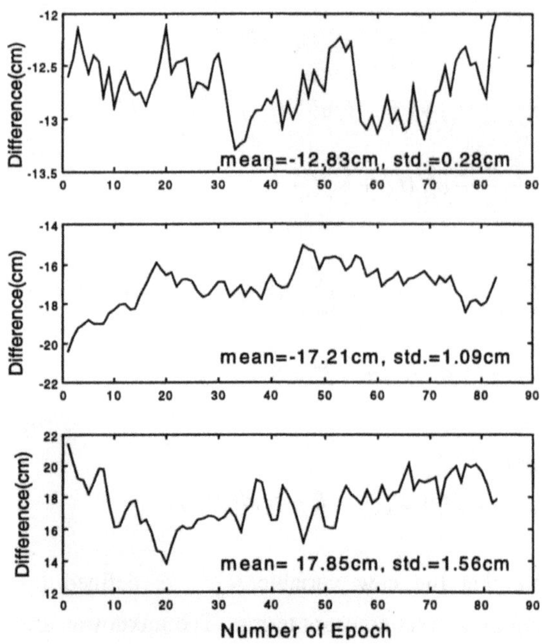

Fig. 4: Differences between estimated coordinates of USNA and its known coordinates: x (top), y (middle), z (bottom)

Figure 4 shows the differences between the known coordinates and the estimated coordinates of the station 'USNA'. The standard deviations are less than 2 cm for a span of 40 minutes of data; for longer periods the standard deviations would increase as more systematic effects enter (such as unmodeled tropospheric delays not already accounted for in equation 9). In addition, there are relatively large mean differences of up to 18 cm. These may be the result of imperfect float ambiguity determination.

3.3 Absolute kinematic positioning

For the kinematic application the data from a GPS buoy survey on Lake Michigan were processed and compared to the result of the DGPS solution. Figure 5 shows the location of a base station (7031D, NOAA benchmark) with three IGS stations (NLIB, ALGO, GODE).

The GPS buoy was deployed about 350 meters away from the base station and the baseline lengths to NLIB, ALGO, and GODE are 450km, 740km, and 890km, respectively. The data were collected every second at both base station and buoy.

Again, the satellite clock errors were estimated using the measurements from the IGS stations at every 30 seconds and 1 second clock errors were calculated using a third-degree Lagrange interpolator. After estimating the ambiguities, the buoy's position coordinates were estimated. The only difference in data processing with the static case is that the interpolation was applied to obtain 1-second clock estimates. This interpolation generates about 8.5 cm RMS interpolation error for the estimates of the satellite clock error originally estimated at 30 seconds resolution (Zumberge et al, 1998).

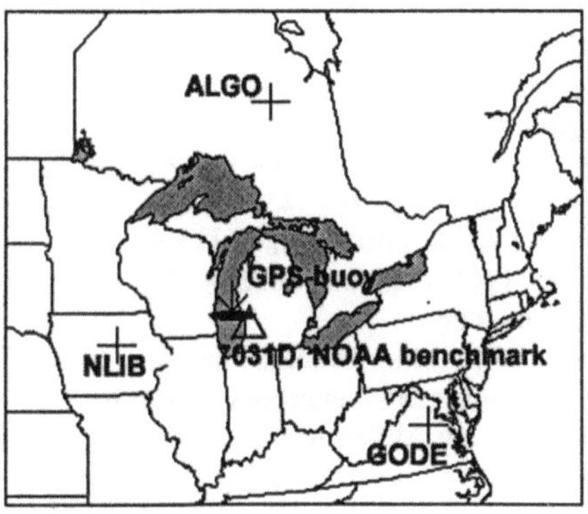

Fig. 5: Kinematic GPS-buoy test area

Observations collected at 1-second intervals from the receiver established on the NOAA benchmark were used to estimate the satellite-differenced clock error, also at 1-second intervals. Figure 6 shows the consistency of these estimates by plotting the differences (crosses) between every 30-th 1-second estimate and the 30-second IGS-based clock error estimates. Also shown (solid line) are the differences between the NOAA benchmark clock error estimates and the interpolated values from the IGS 30-second series at 1-second intervals.

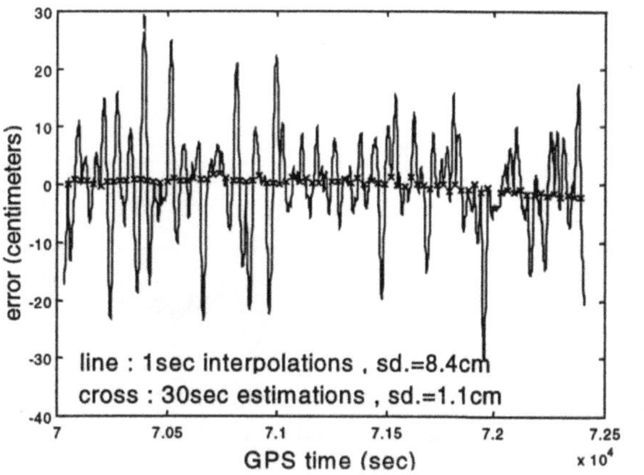

Fig. 6: GPS clock error interpolation effect (1sec.)

The 30-second GPS clock error estimates from the remote stations are well matched to those from the NOAA station with 1.1 cm standard deviation. The 1-second interpolation errors, however, show short-term fluctuations of relatively large magnitude. The maximum interpolation errors occur in the middle of 30-second spacing and the standard deviation is about 8.4 cm. This short-term variation is due to SA, specifically the •-process (dithering the fundamental frequency of the satellite clock). Therefore, one cannot avoid less than 8-9 cm (s.d.) range error for every satellite or satellite pair when GPS clock errors are interpolated from 30-second estimates.

Figure 7 shows a comparison of the kinematic positions estimated by two different methods. The dash lines represent the positions calculated by the KARS software developed by G. Mader at NGS (National Geodetic Survey) using double-differenced dual-frequency carrier phase data. The solid lines represent the positions calculated by the proposed absolute positioning method. One can see that the general trends are well matched; the differences in each coordinate include biases and high frequency fluctuations. The biases in the X, Y, and Z coordinates are 14 cm, 43 cm, and 33 cm, and the standard deviations are 6 cm, 23 cm, and 14cm, respectively. The major error source for high frequency fluctuation is the interpolation error, as explained above. This conclusion is consistent with static positioning analyses done for the NOAA station. Differences between the known coordinates of station 7031D and absolute-positioned coordinates using 30-second clock error estimates have standard deviations of (0.5 cm, 2.0 cm, 2.0 cm), while corresponding standard deviations with 1-second interpolated clock error estimates are significantly larger: (5 cm, 21 cm, 15 cm).

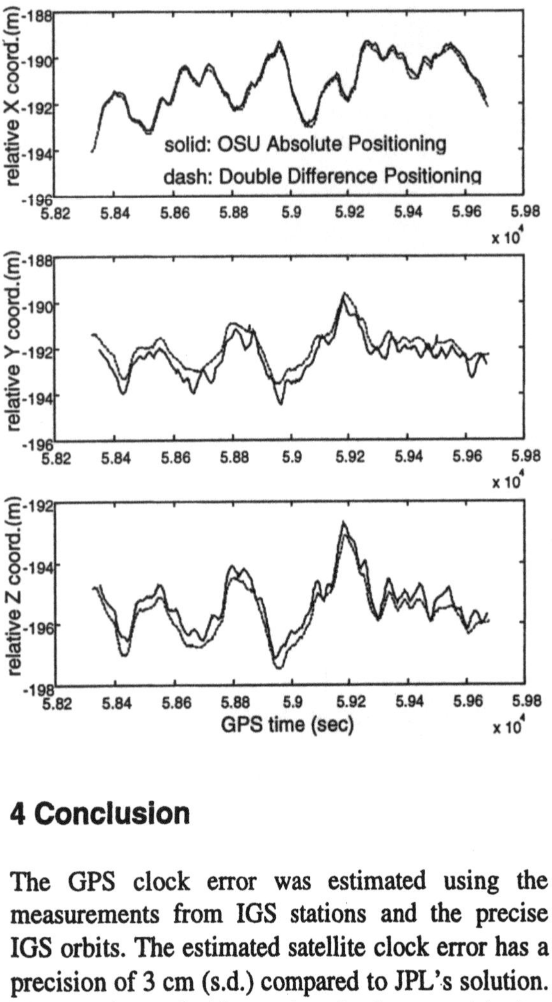

4 Conclusion

The GPS clock error was estimated using the measurements from IGS stations and the precise IGS orbits. The estimated satellite clock error has a precision of 3 cm (s.d.) compared to JPL's solution. The static and kinematic absolute positioning conducted using the estimated clock errors indicate standard deviations of 2 cm and 30 cm, respectively. The mean differences were less than 20 cm and 45 cm. The reason for the worse precision in the kinematic positioning is due to the clock interpolation error. Thus, higher rate clock error estimates will be necessary to perform precise absolute positioning.

Acknowledgments. This work was supported by the National Imagery and Mapping Agency under Air Force Phillips Laboratory contracts F19628-95-K-0020 and F19628-96-C-0169. The authors thank the reviewers for their comments.

References

Coco, R. (1991). GPS-satellite of Opportunity for Ionospheric Monitoring. *GPS World*, 2(9), 47-50

Gibson, R. (1983). A Derivation of Relativistic Effects in Satellite Tracking. *Technical report*, TR 83-55, Naval Surface Weapons Center, Dahlgren, Virginia.

Goad, C.C., and L. Goodman (1974). A Modified Hopfield Model Tropospheric Refraction Correction Model. Presented at the Fall Meeting of the American Geophysical Union, San Francisco, December, 1974.

Goad, C.C., and M. Yang (1995). A New Approach to Precision Airborn GPS Positioning for Photogrammetry. *Photogrammetric Engineering and Remote Sensing*, 63(9), 1067-1077.

Hofmann-Wellenhof, B., H. Lichtenegger, and J. Collins (1992). GPS Theory and Practice. 4th revised edition, Springer, New York.

IGS 900-second orbit and clock (1999), ftp://cddisa.gsfc.nasa.gov/pub/gps/products.

JPL 30-second orbit and clock (1999), ftp://sideshow.jpl.nasa.gov/pub/jpligsac/hirate.

Lachapelle, G., M.E. Cannon, W. Qiu, and C. Varner (1996). Precise Aircraft Single-point Positioning Using GPS Post-mission Orbits and Satellite Clock Corrections. *Journal of Geodesy*, Vol. 70, 562-571.

Leick, A. (1995). GPS Satellite Surveying. 2nd edition, Wiley-Interscience.

Remondi, B.W. (1989). Extending the National Geodetic Survey Standard GPS Orbit Formats. National Information Center, Rockville, Maryland, *NOAA Technical Report* NOS 133, NGS 46.

Remondi, B.W. (1991). NGS Second Generation ASCII and Orbit Formats and Associated Interpolation Studies. *Paper presented at the XX General Assembly of the IUGG at Vienna*, Austria, August 11-24.

Seeber, G. (1993). Satellite Geodesy. Walter de Gruyter, Berlin New York.

Zumberge, J.F., M.M. Watkins, and F.H. Webb (1998). Characteristics and Applications of Precise GPS Clock Solutions Every 30 Seconds. *Journal of The Institute of Navigation*, Vol. 44, No. 4, Winter 1997-1998.

GNSS long baseline ambiguity resolution: Impact of a third navigation frequency

N.F. Jonkman, P.J.G Teunissen, P. Joosten, D. Odijk
Department of Mathematical Geodesy and Positioning,
Delft University of Technology, Thijsseweg 11, 2629 JA Delft, the Netherlands

Abstract. Both GPS and the European second generation satellite navigation system GNSS-2 or "Galileo", are intended to transmit signals on three navigation frequencies in order to enhance ambiguity resolution performance for real-time high-precision position determination. In this paper the performance of GPS and GNSS-2 three frequency ambiguity resolution will be analyzed and compared with GPS two frequency ambiguity resolution for long baselines, i.e. baselines for which atmospheric distortions in the observations can not be neglected. The analyses and comparisons will be based on ambiguity resolution success-rates, being the probability of estimating ambiguities at their correct integer values. The success-rates are evaluated for the simple geometry-free observation model using the integer least-squares estimator.

Keywords. GNSS, GPS, ambiguity resolution.

1 Introduction

On January 25th 1999, Vice President Gore of the United States made an announcement on a new Global Positioning System modernization initiative. The announcement was the latest in a series, ending years of deliberations on the design of the future generation of GPS satellites, the Block IIFs. Key feature of Block IIF design is the transmission of additional unencrypted signals for civil use.

GPS satellites currently transmit signals on two frequencies in order to account for the dispersive part of atmo2sphere induced distortions, the so-called ionospheric delays. Civil access to the signal on the second of the two GPS frequencies however is restricted by a United States Department of Defense imposed encryption of the signal. Although this encryption can be circumvented through specially adapted receiver design, a second non-encrypted signal has long been advocated for civilian users of the GPS. In response to this lobby, the announcement of Vice President Gore contained the promise to implement a second civil signal on the GPS block IIF satellites. In addition to this second civil signal however, the announcement also indicated the implementation of an unencrypted signal on a third frequency. The current two and future three GPS frequencies are listed in table (1). The reason

Table 1: GPS two and three frequencies (GPS2 and GPS3) and proposed frequencies for GNSS-2

	F1 (MHz)	F2 (MHz)	F3 (MHz)
GPS2	1575.420	1227.600	-
GPS3	1575.420	1227.600	1176.450
GNSS-2	1589.742	1561.098	1256.244

for incorporating two rather than one additional civil signal in the Block IIF satellite design was hinted at in the announcement of the Vice President as "the new signals will enable unprecedented real-time determination of highly accurate positions anywhere on earth". The third signal appears therefore to be specifically aimed at enhancing the possibilities for the most precise form of GPS positioning, relative carrier phase positioning by means of ambiguity resolution. The signal is apparently expected to shorten the time span necessary for the successful estimation at the correct integer values of these ambiguities or to stretch the baseline length for which a successful resolution is possible in real-time. In this paper the latter type of improvement will be investigated in more detail.

In order to judge the third civil signal on its merit for long baseline ambiguity resolution, the probability of correct integer ambiguity estimation, or the success-rate, will be considered in this paper. Success-rates will be presented and compared for the current GPS two frequency and

International Association of Geodesy Symposia, Vol. 121
Schwarz (ed.), Geodesy Beyond 2000 – The Challenges of the First Decade

the future three frequency set-up. In addition, the effect on the success-rate of varying the spacing between the three frequencies will also be considered. Although the actual GPS frequencies have already been established, this is not yet the case for the proposed European satellite navigation system GNSS-2 or "Galileo". Hence, the success-rates obtained for different frequency spacings may give some indication for an optimal choice for the GNSS-2 frequencies.

2 The success of geometry-free integer least-squares ambiguity estimation

The probability of estimating GPS or GNSS-2 carrier phase ambiguities at their correct integer values depends on the mathematical model used to adjust the observations and on the method employed to incorporate the integer constraint on the ambiguities in the adjustment. In this paper we will consider the success-rates for the geometry-free observation model adjusted according to the integer least-squares criterion. The geometry-free model will be discussed in the first subsection, the integer least-squares adjustment in the second. In the third subsection, the computation of the ambiguity resolution success-rates for this approach will briefly be explained.

2.1 The geometry-free model

The geometry-free model is the simplest possible mathematical model for the adjustment of GPS or GNSS-2 observations that still allows the estimation of carrier phase ambiguities, see (Euler and Goad, 1990), (Hatch, 1982), (Melbourne, 1985), and (Teunissen, 1996). In its most basic form, the model consists of the double differenced pseudo range and carrier phase observations of two receivers to two satellites, parametrized in terms of an unknown double differenced satellite-receiver range, unknown ambiguities and an unknown ionospheric delay

$$\begin{aligned} p_i &= \rho + \left(\tfrac{\lambda_i}{\lambda_1}\right)^2 I \\ \phi_i &= \rho - \left(\tfrac{\lambda_i}{\lambda_1}\right)^2 I + \lambda_i a_i \end{aligned} \tag{1}$$

where p and ϕ indicate the pseudo range and carrier phase observations in meters, ρ, a and I denote the unknown range, phase ambiguity and ionospheric delay and λ denotes the wavelength of the carrier. The lower index i indicates the dependence of the observations and unknowns on the observation frequency.

For the case of three frequency observations, the single epoch geometry-free model consists of three of these pairs of pseudo range and carrier phase observation equations: one pair for each frequency. For any additional epochs of data, the model has to be extended with additional unknowns for the range and ionospheric delay. The ambiguities however remain constant if no loss-of-lock occurs.

The above-mentioned equations contain unknowns for the ionospheric delays in the observations. The delays are however negligible for short baselines, i.e. baselines up to 10 to 20 kilometers. Hence, for such short baselines the ambiguities could be estimated from a model without ionospheric unknowns. To distinguish this model from the model with ionospheric unknowns, it is usually indicated as the ionosphere-fixed model, whereas the model with ionospheric unknowns is indicated as the ionosphere-float model.

Contrary to the ionosphere-float model however, the ionosphere-fixed model already yields very high success-rates if observations on only two frequencies are available. With pseudo range and carrier phase observations of moderate precision, the single epoch two frequency success-rate will be of the order of 0.9 if the ionospheric delays can be neglected, while it is less than 0.1 if the delays have to be included in the model. The impact of the third observation frequency on the ionosphere-float success-rates is therefore by far the more important and the analyses in this paper will be limited to this long baseline case.

2.2 Integer estimation

One of the first to realize the potential contribution of a third frequency to ambiguity resolution was Ron Hatch. In his 1996 GPS World article "The Promise of a Third Frequency", he analyzed this contribution for GPS by considering the performance of the geometry-free model in combination with a technique called "wide-laning", (Hatch, 1996). Briefly summarized, wide-laning aims to resolve phase ambiguities by rounding the ambiguities of between-frequency phase differences to their nearest integer values. In various guises this technique is also used in other feasibility studies for GPS and GNSS-2 three frequency ambiguity resolution, see (Ericson, 1999), (Forsell et al., 1997), and (Vollath et al., 1998).

Rather than Hatches wide-laning technique however, in this paper we will consider the geometry-free observation model in combination with an integer least-squares adjustment. Ambiguity estimation by integer least-squares was first introduced in (Teunissen, 1993), and is made operational in the Least-Squares Ambiguity Decorrelation Adjustment or LAMBDA-method. Although mathematically more complex than the simple rounding inherent to wide-laning, integer least-squares has a superior performance in the sense that it maximizes the probability of estimating ambiguities at their correct integer values, (Teunissen, 1999). In other words, of all possible integer estimation techniques, integer least-squares will yield the highest possible ambiguity resolution success-rates. It is this optimality property of integer least-squares estimation that allows a proper evaluation of the added value for ambiguity resolution of a third navigation frequency.

Starting point for the integer least-squares ambiguity estimation with the LAMBDA-method is an ordinary least-squares adjustment of the observations according to the geometry-free model. The constraining of the resulting real-valued ambiguity estimates to integers can be phrased in terms of the following minimization problem

$$\begin{array}{l} \min ||\hat{a} - a||^2_{Q_{\hat{a}}} \;\; ; \quad a \in Z^n \\ \text{with } ||(.)||^2 = (.)^T Q_{\hat{a}}^{-1}(.) \\ \text{and } Z^n \; n\text{-dimensional integer space} \end{array} \qquad (2)$$

where $\hat{a}$ indicates the vector of real-valued ambiguity estimates, the so-called float solution, and $Q_{\hat{a}}$ denotes its variance-covariance matrix. The solution to this problem, indicated as the fixed solution, is the integer vector nearest to the float solution, where nearness is measured in the metric of the variance-covariance matrix of the float solution. If this matrix were to be a diagonal matrix and the least-squares ambiguity estimates thus uncorrelated, the fixed solution could be determined by a simple rounding of the least-squares estimates. For GPS observations however, the estimates are usually highly correlated, and the fixed solution has to be identified by a discrete search in a subspace of Z^n, the integer ambiguity search space.

The search for the solution of the integer least-squares minimization problem is hampered by the strong correlation and poor precision of the least-squares ambiguity estimates. In geometrical terms, the ambiguity search space is highly elongated due the correlation of the estimates and stretches over a considerable range of wavelengths or cycles as a result of their low precision. To improve the computational efficiency of the discrete search, the LAMBDA-method therefore employs a decorrelating Z-transformation. This Z-transformation manages to retain the integer character of the minimization problem whilst at the same time lowers the correlation and improves the precision of the least-squares ambiguity estimates. The transformed search space allows a relatively easy identification of the fixed solution, as it has a more sphere-like shape. For a detailed description of both the search and the construction of the Z-transformation, the reader is referred to (Teunissen et al., 1997).

2.3 Ambiguity success-rates

For the analysis described in this paper two approaches have been used to determine integer least-squares success-rates. The first approach consisted of generating large numbers of samples from a multi-variate normal distribution with the stochastic characteristics of the geometry-free least-squares ambiguity estimates. By counting the number of samples that is mapped with the LAMBDA-method to the integer vector at which the distribution of the samples is centered, the rate of successful ambiguity resolution could very accurately be determined.

The second, computationally less demanding approach consisted of determining success-rates for the bootstrapped ambiguity estimator, (Blewitt, 1989) and (Dong and Bock, 1989). Bootstrapped ambiguity estimates are obtained by means of a sequential rounding scheme. Although this scheme does not fully account for the correlation between the ambiguities and the bootstrapped estimator is consequently suboptimal from the point of view of ambiguity resolution, the estimator does have the desirable property that its success-rates can be expressed analytically in terms of the cumulative probability density function of the normal distribution, (Teunissen, 1998). A relatively easy numerical integration of the normal probability density function therefore suffices for the computation of bootstrapped success-rates.

As the bootstrapped success-rates are known

to be smaller than the integer-least squares success-rates, they can serve as a lower bound for the success of integer least-squares estimation. This lower bound is particularly tight if it is computed for the Z-transformed ambiguity estimates rather than for the original estimates. In fact, a comparison with the success-rates obtained with the first, simulation-based, approach showed that for our present study the differences are negligible.

3 GPS three frequency vs. two frequency ambiguity resolution

In this section the success-rates for GPS three frequency observations will be confronted with those for GPS two frequency observations. The two and three frequency success-rates for the ionosphere-float geometry-free observation model with a single epoch of data are summarized in table (2). As the success of ambiguity resolution with the geometry-free model tends to depend strongly on the precision of the pseudo range observations, the success-rates are provided for different assumptions with regard to the standard deviation of the undifferenced pseudo range observations, ranging from 10 to 30 centimeters. The standard deviation of the undifferenced phase observations was in all cases kept fixed to 3 millimeters.

Table 2: Single epoch ionosphere-float success-rates for GPS2 and GPS3. Success-rates refer to undifferenced pseudo range standard deviations of 10, 15 and 30 centimeters. The undifferenced phase standard deviation was assumed equal to 3 millimeters

σ_p	GPS2	GPS3
10 [cm]	0.0773	0.1245
15 [cm]	0.0491	0.0941
30 [cm]	0.0178	0.0565

Judging by table (2), instantaneous ambiguity resolution is highly unlikely with the ionosphere-float geometry-free model, even with the observations from the additional third frequency. The three frequency success-rates in table (2) indicate that even under favorable conditions, less than 15% of the ambiguities is estimated at their correct integer values. Although this is an improvement with respect to the likelihood of estimating ambiguities at their correct integer values with two frequency observations, the reliability of the integer ambiguity estimation is of course still far too poor.

The third GPS frequency however does manage to achieve a considerable shortening of the observation time span necessary for a reliable ambiguity estimation at long ranges. The number of observation epochs needed to achieve certain predefined success-rates for ambiguity resolution with the ionosphere-float model are summarized in table (3). It can be seen from the table, that in order to achieve a high probability of correct integer ambiguity estimation with two frequency data, up to several thousands of epochs of data may have to be included in the observation model. With three frequency observations available however, such high probabilities can be achieved with less than 800 epochs of data even with a relatively large pseudo range standard deviation of 30cm.

Table 3: Number of epochs required to achieve predefined levels of success, 0.90, 0.95, 0.99, with the ionosphere-float geometry-free model for GPS2 and GPS3. Number of epochs refers to undifferenced pseudo range standard deviations of 10 centimeters (left) and 30 centimeters (right)

Success-	$\sigma_{code} = 10[cm]$		$\sigma_{code} = 30[cm]$	
rate	GPS2	GPS3	GPS2	GPS3
0.90	300	120	2700	310
0.95	420	160	3800	440
0.99	730	280	6500	760

To fully appreciate the shortening of the observation time span for ambiguity resolution, the conversion of observation epochs into time has to be given carefully consideration. The future three frequency GPS signal structure will not only mean that we will have more observations available, but the observations will also be of better quality as they can be derived directly from unencrypted signals. At present, due to the encryption of signal on the second GPS frequency, dual frequency observations can only be obtained with civilian GPS receivers by means of reconstruction techniques. The quality of the reconstructed observations is however comparatively poor, particularly if short observation intervals are used. Hence, with the future unencrypted signals the required observation epochs could be collected at a very high rate, while with the present encrypted signal a lower rate has to be used or additional observation epochs need to be collected. Real-time carrier phase relative positioning over long distances may thus not seem possible with three observation frequencies and the geometry-free model, a significant improvement is certainly achieved.

4 Success-rate vs. location of third frequency

The analysis in the previous section indicate that the performance of GPS three frequency ambiguity resolution falls short of the mark for long baselines. In this section we will therefore consider the question whether a better performance could have been achieved if the location of the third GPS frequency had been chosen differently. This seems to be an academic question for GPS as this location has already been established, but the answer is certainly of interest for the European satellite navigation system GNSS-2 that is still in its design phase. Like the GPS Block IIF satellites, GNSS-2 satellites are intended to transmit on three frequencies, and again specifically with the aim to enable real-time carrier phase positioning.

Figure (1) shows the success-rate for the single epoch ionosphere-float geometry-free observation model as a function of the location of the third GPS frequency. It is immediately clear from the figure that the choice of the third frequency has a considerable impact on the ionosphere-float success-rate: the success-rate varies with the third frequency between approximately 0.03 and 0.35. Judging by the figure, optimal ambiguity resolution performance is achieved if the third frequency is located as far away as possible from the two other frequencies. Moreover, it appears that a large frequency separation is particularly beneficial if the third frequency is located near the lower end of the spectrum. The observation that frequency separation plays such a predominant role in ambiguity resolution for long baselines has an interesting consequence for the design of GNSS-2. During the development of GNSS-2 quite a number of frequency distribution schemes have been proposed. The frequencies of one of the more likely schemes are presented in table (1). This scheme has served as a reference in an European Space Agency (ESA) funded study of three frequency ambiguity resolution (Vollath et al., 1998) and with a minor variation also features in an earlier ESA study on ambiguity resolution, (Forsell et al., 1997). This

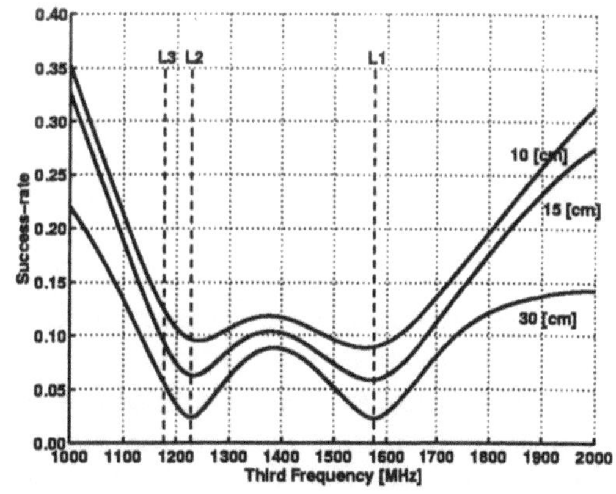

Fig. 1: Single epoch ionosphere-float success-rates vs. location of the third GPS frequency. Success-rate curves apply to pseudo range observations with a standard deviation of 10, 15 and 30 centimeters. The dashed vertical lines indicate the current L1- and L2 frequency, as well as the chosen third frequency.

scheme is therefore taken as the point of departure for the following analysis.

In the GNSS-2 frequency distribution scheme, one of the frequencies is located near the GPS L2 frequency, while the two other frequencies are both located very closely to the GPS L1 frequency. The rationale for considering two frequencies with little separation was indicated in (Forsell et al., 1997) as that the very large wavelength of the corresponding difference frequency would enable a relatively simple resolution of its ambiguity with the wide-laning technique. From our analysis of the success-rates however, it would appear that this choice of the frequencies will give sub-optimal results for long baseline ambiguity resolution.

Table 4: Number of epochs required to achieve predefined levels of success, 0.90, 0.95, 0.99, with the ionosphere-float geometry-free model for GPS3 and GNSS-2. Number of epochs refers to undifferenced pseudo range standard deviations of 10 centimeters (left) and 30 centimeters (right).

Success-rate	$\sigma_{code} = 10[cm]$		$\sigma_{code} = 30[cm]$	
	GNSS-2	GPS3	GNSS-2	GPS3
0.90	260	120	1700	310
0.95	380	160	2400	440
0.99	620	280	4000	760

To illustrate the relatively poor ambiguity resolution performance of GNSS-2 for long baselines with the proposed frequencies, the numbers of observation epochs required to attain certain predefined success-rates with the ionosphere-float model are summarized in table (4) together with the numbers of epochs required for the three frequency GPS set-up. It can be seen from the table that GNSS-2 with the proposed frequency distribution does indeed require many more epochs to achieve a reliable ambiguity resolution than GPS with three frequencies. Under unfavorable conditions the initialization time for the proposed GNSS-2 set-up may be more than five times longer than for GPS, thereby seriously hampering its use for fast long range precise phase positioning. From the point of view of long baseline ambiguity resolution this particular GNSS-2 frequency distribution scheme seems therefore less well suited.

5 Conclusion

In this paper the added value for real-time long range carrier phase ambiguity resolution of a third navigational frequency was evaluated. The evaluation was based on ambiguity resolution success-rates obtained with the geometry-free model and integer least-squares estimation. The success-rates, representing the probability of estimating ambiguities at their correct integer values, are maximized by the integer least-squares estimation, allowing an undistorted judgment of the merit of the additional frequency.

The success-rates indicate that the third GPS frequency will not allow a reliable instantaneous ambiguity resolution for long baselines, at least not with the geometry-free model. Compared with the present two frequency GPS set-up however, the third frequency will achieve a very substantial reduction of the observation time span in which a reliable ambiguity estimate can be obtained.

The analyses of the success-rates do in addition show that the interfrequency spacing plays an important role in the success of long range ambiguity resolution. For the part of the frequency spectrum considered in the analysis, i.e. from 1000 to 2000 MHz, best ambiguity resolution performance is achieved if the three frequencies are located as far apart as possible. One of the proposed frequency distribution schemes for the European GNSS-2 or "Galileo" satellite navigation system does not quite conform to this condition, and its performance was shown to be sub-optimal.

References

Blewitt, G. (1989), Carrier-phase ambiguity resolution for the Global Positioning System applied to geodetic baselines up to 2000 km, Journal of Geophysical Research 94: 10187-10203.

Dong, D. and Y. Bock (1989), Global Positioning System network analysis with phase ambiguity resolution applied to crustal deformation studies in California. Journal of Geophysical Research 94: 3949-3966.

Ericson, S. (1999), A study of Linear Phase Combinations in Considering Future Civil GPS Frequencies. In: Proceedings ION GPS NTM-99, pp. 677-686.

Euler, H.J. and C.C. Goad (1990), On optimal filtering of GPS dual-frequency observations without using orbit information. Bulletin Geodesique 65: 130-143.

Forsell, B., M. Martin-Neira, R.A. Harris (1997), Carrier Phase Ambiguity Resolution in GNSS-2. In: proceedings ION GPS-97, pp. 1727-1736.

Hatch, R. (1982), The synergism of GPS code and carrier-phase measurements. In: proceedings 3rd International Geodetic Symposium on Satellite Positioning 2: 1213-1231.

Hatch, R. (1996), The Promise of a Third Frequency. GPS World, May 1996: 55-58.

Melbourne, W.G. (1985), The case for ranging in GPS-based geodetic systems. In: proceedings First International Symposium on Precise Positioning with the Global Positioning System, Rockville, MD: National Geodetic Survey: 373-386.

Teunissen, P.J.G. (1993), Least-squares estimation of the integer GPS ambiguities. In: LGR-series 6, Delft Geodetic Computing Center, Delft.

Teunissen, P.J.G. (1996), An analytical study of ambiguity resolution using dual frequency code and carrier phase. Journal of Geodesy 70: 518-528.

Teunissen, P.J.G. (1998), Success probability of integer GPS ambiguity rounding and bootstrapping. Journal of Geodesy 72: 606-612.

Teunissen, P.J.G. (1999), An optimality property of the integer least-squares estimator. Accepted for publication in the Journal of Geodesy.

Teunissen, P.J.G., P.J. de Jonge and C.C.J.M. Tiberius (1997), Performance of the LAMBDA Method for Fast GPS Ambiguity Resolution. NAVIGATION: Journal of the Institute of Navigation 44 (3): 373-377.

Vollath, U., S. Birnbach, H. Landau (1998), Analysis of Three-Carrier Ambiguity Resolution (TCAR) Technique for Precise Relative Positioning in GNSS-2. In: proceedings ION GPS-98: 417-426.

Monitoring the height deflections of the Humber Bridge by GPS, GLONASS, and finite element modelling

G. W. Roberts, A. H. Dodson
Institute of Engineering Surveying and Space Geodesy (IESSG), The University of Nottingham, UK. NG7 2RD, email: gethin.roberts@nottingham.ac.uk

C.J. Brown, R. Karuna
Department of Mechanical Engineering, Brunel University, UK. UB8 3PH

R. A. Evans
The Humber Bridge Board, Ferriby Road, Hessle, East Yorkshire. HU13 0JG

Abstract. The following paper details trials recently conducted by the IESSG whereby kinematic GPS and the Russian equivalent to GPS, Global'naya Navigatsionnaya Sputnikovaya Sistema (GLONASS), were used to monitor the deflection of the Humber Bridge under a loading of 160 tons, moved across the bridge by five articulated lorries. The trial consisted of placing four dual frequency GPS receivers and a GPS/GLONASS receiver on the bridge deck, a dual frequency GPS receiver upon one of the lorries and reference receivers placed on land. The data, gathered at 5 Hz, was then processed in an On the Fly manner resulting in a precision of the order of a few millimetres. Finite element modelling of such structures can take advantage of similar trials and results. The results were compared to a finite element model by the Department of Mechanical Engineering at Brunel University, and showed remarkable comparisons.

Keywords. Kinematic GPS, finite element modelling, deformation monitoring.

1 Introduction

Research is being conducted at the Institute of Engineering Surveying and Space Geodesy (IESSG) to investigate the use of kinematic GPS to monitor and measure the deflections of large structures. This work began early in 1996, and was first published in May 1996 by Ashkenazi et al. (1996). Since these initial trials, many more have been conducted and analysed at the IESSG by Ashkenazi et al. (1997) and Ashkenazi and Roberts (1997), leading to the most recent series of trials detailed in this paper.

The trials discussed in this paper were conducted using kinematic GPS to simultaneously position multiple points upon the bridge, all relative to reference GPS and GPS/GLONASS receivers. Similar trials have been conducted on tall buildings by Lovse et al. (1995). Ultimately, this could lead to the development of a real-time structural failure alarm system as well as the computation of long term cumulative deterioration factors, which are essential to the maintenance and integrity of the structure. Some famous structural disasters may have benefited from early warning systems that could detect the initial characteristics of dangerous movements, such as the Tacoma Narrows suspension bridge failure. Bridges and other large structures are designed to withstand calculated and predicted forces due to, for example, wind loading, traffic loading and tidal loading. The life expectancy and characteristics of the structure are partly governed by these variables. In reality, however, these forces may or most probably may not be as those predicted. Therefore, a GPS based monitoring system could be used to assess whether the structure is deflecting as it was initially designed to do so. In addition, this could benefit future bridge design techniques and the development of traffic management schemes.

The kinematic GPS deflection data can be integrated with a Finite Element Model (FEM). Once the model has been calibrated with the GPS data, the predicted bridge deflections and characteristics will be more reliable and could also be updated with a continuous stream of real time data.

The trial consisted of measuring the deflection of the Humber bridge at 5 strategic locations, whilst a load with a measured weight of approximately 160 tons travelled over. The lorries carrying the load conducted a series of manoeuvres whilst the

International Association of Geodesy Symposia, Vol. 121
Schwarz (ed.), Geodesy Beyond 2000 – The Challenges of the First Decade

deflection were measured using the Ashtech receivers.

Work conducted in parallel at the Brunel University has resulted in a FEM of the Humber Bridge by Karuna et al. (1997a). This work has been combined with the monitoring work conducted at the IESSG, resulting in a system that combines the FEM with the GPS results by Brown et al. (1999). Some of the results of this work are mentioned in the paper.

2 Bridge description

The Humber suspension bridge is located across the Humber estuary on the east coast of England. The bridge lies in a generally north-south direction, and consists of three spans supported by two towers 155.5 m in height. The main span is 1, 410 m long, and was the longest span suspension bridge in the world since its construction in 1981 until Japan's Akashi Kaiko bridge opened in April 1998 with a 1, 990.8 m suspended span. The two side spans on the Humber Bridge lie 530 m (Barton span on the south side of the bridge) and 280 m long (Hessle span on the north side of the bridge).

The Humber bridge is a large suspension bridge, which exhibits deflections of the order of tens of centimetres. The streamlined low drag box-girder steel decks exhibit "negative lift" in wind conditions. The bridge accommodates dual two lane carriageways, with footpath tracks on both sides; each wide enough for maintenance and emergency vehicle access. The total width is 28.5 m and the box depth is 4.5 m.

3 Trial description

The bridge trial was conducted in order to assess the use of satellite positioning and using an FEM to evaluate the bridge's deflections. Five lorries, weighing in total 159.742 tons were contracted by the Humber Bridge Board to travel across the bridge.

The trial was conducted on the night of the 16 February 1998 at around 2-am local time. The bridge was equipped with 5 GPS receivers, strategically located to measure the deflections. These receivers consisted of 4 Ashtech ZXII dual frequency receivers and an Ashtech GG24 single frequency GPS/GLONASS receiver. Additionally, an Ashtech ZXII receiver was located upon one of the lorries, with reference receivers placed upon the bridge's control tower, situated on land adjacent to the bridge. The control tower is situated at the northern end of the bridge, making the baseline lengths of the order of up to 1, 970 m. The data was post-processed using in-house and commercial Ashtech PNAV software. Specially developed clamps were used to allow the antennas to be attached directly to the bridge, thus reducing unwanted vibrations experiences in past trials. Two ZXII receivers were located at the mid span of the main span, one on the eastern side of the bridge (REC1), and one on the western side (REC2). A third ZXII receiver was placed upon the main span, this time located at the quarter span location on the eastern side of the bridge (REC3). In addition a ZXII was situated at the mid span of the Barton side span on the western side of the bridge (REC4) and a GG24 receiver placed at the mid span of the main span on the eastern side of the bridge (REC5).

The weather conditions were quite calm when the equipment was being set up, but by the time the trials started, the temperature had dropped to 9 °C, and the wind had increased to approximately 14.8 m/s from a southwesterly direction. Peak values of 19.6 m/s were recorded on the approach spans.

During the trial, southbound traffic was stopped from crossing the bridge, although it was not possible to stop northbound traffic. However, traffic was light at this time of night.

The GPS receivers' locations, Figure 1, were chosen so that the movement of the main span could be assessed, as well as that of the Barton side span. The configuration demonstrated that the bridge did in fact displace vertically, but also that the bridge deck twisted, and the weight on one of the spans would cause a cantilever effect, pulling up the other span.

The trials consisted of the lorries being driven over the bridge in three formations. The first comprised of all three lorries travelling together southbound on the eastern carriageway, at a constant speed of approximately 30 km/h. They were configured in a group so that there were three lorries in the outer (most easterly) running carriageway, and two in the other southbound lane, see Figure 2. A compact group was formed which occupied no more than a total of five lorry lengths (approximately 50 m). The second run consisted of all five lorries travelling together northbound on the western carriageway, again at a constant speed of approximately 30 km/h. For the third run, lorries A and B were started at the Barton end of the bridge, whilst lorries C and D were started from the Hessle end. The lorries were then run from the ends to the mid span of the main span, and remained stationary there for a while, so that a static, approximately symmetric load was placed at the mid span.

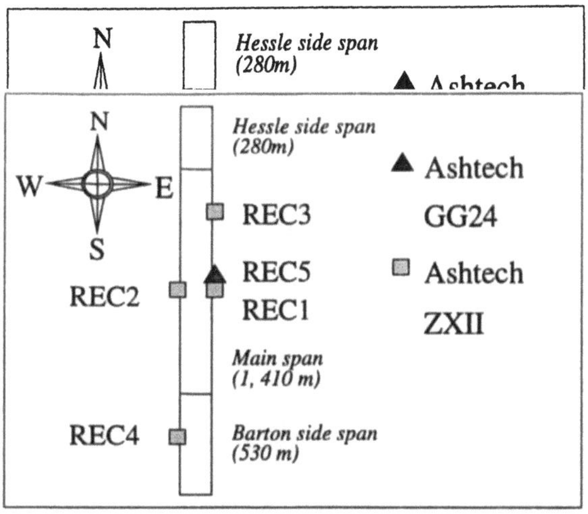

Fig. 1: Plan of the Humber Bridge showing the receivers' locations.

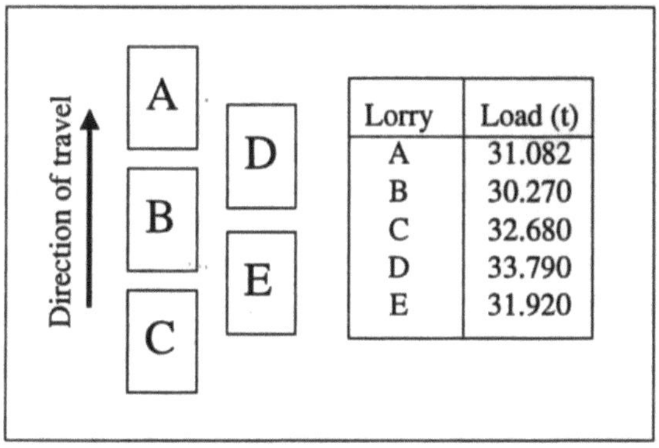

Fig. 2: Configuration of the lorries during the first two formations

4 Height deflection results

The results obtained from the trial are detailed in the following sections. The graphs illustrate the integer fixed solutions resulting from an On The Fly search. The 3D deflection of each epoch has been converted into distances along the bridge (longitudinal), distances across the bridge (lateral) and height. Only the height component will be discussed. All the transformations from WGS84 were conducted by the University of Nottingham's CODA software.

The WGS84 ellipsoidal heights over time for the stations REC1 to 5 are shown in Figures 3, 4, 5, 6 and 7 respectively. It is evident from these figures that the deflection experienced by the bridge is in the decimetre range. The three receivers located at the middle of the main span, REC 1, 2 and 5, all exhibit similar magnitudes of deflection at the same time.

At approximately 1:33 am all five lorries travelled along the bridge from north to south on the eastern carriageway. It is evident from the figures that REC1 deflects by approximately 600 mm, whilst REC2 deflected by approximately 500 mm. The maximum displacement of 530 mm corresponds very well to the value predicted by the FE model of 519 mm for the pedestrian walkway. The displacement is slightly less than would be expected if no wind loading was present due to the bridge's negative lift characteristics.

On the second run, at approximately 1:52 am, the lorries travelled south to north on the western carriageway, this time causing REC2 to deflect by approximately 600 mm and REC1 and REC5 by approximately 500 mm. The fact that the deflections on the opposite sides of the carriageway are mirror images of each other suggest that the bridge deck not only deflects in height, but also twists due to the loading being biased towards one side.

Similarly, it can be seen that the height deflection of REC4, Figure 6, is seen to occur at an offset time to those at the middle of the midspan. This is as expected due to the lorries travelling over these points at different times.

Another characteristic of the bridge evident in the results is that of a continuity effect. As the traffic load is on the midspan, pushing it down, the weight causes the cables to pull on the support tower and hence pull the other span up. The deck does not provide structural continuity, but the tension in the main cable and the consequent movement of the towers gives rise to uplift in the main and Barton spans. This is evident in Figures 4 and 6. Here, Figure 4 illustrates the height displacement results of REC2 located at the centre of the main mid span and Figure 6 illustrates the results acquired from REC4 located at the centre of the southern (Barton) span. As the lorries travel from north to south, the main mid span, Figure 4, drops in height at approximately 1:34 am, whilst the south span raises, Figure 6. Similarly as the traffic continues to travel, reaching the south span, this then reduces in height whilst the midspan raises, at approximately 1:36 am.

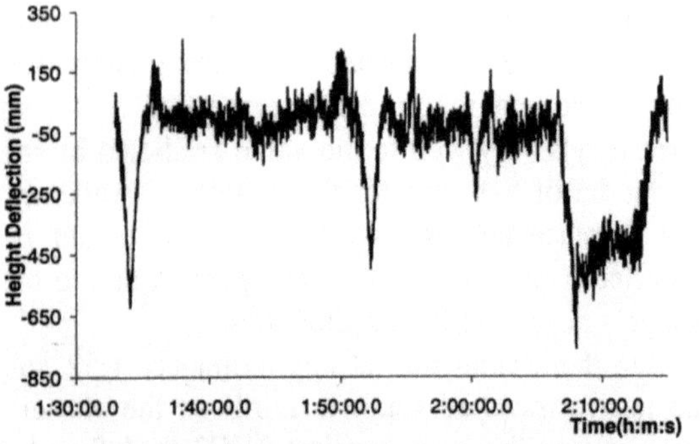

Fig. 3: Height deflection of REC1.

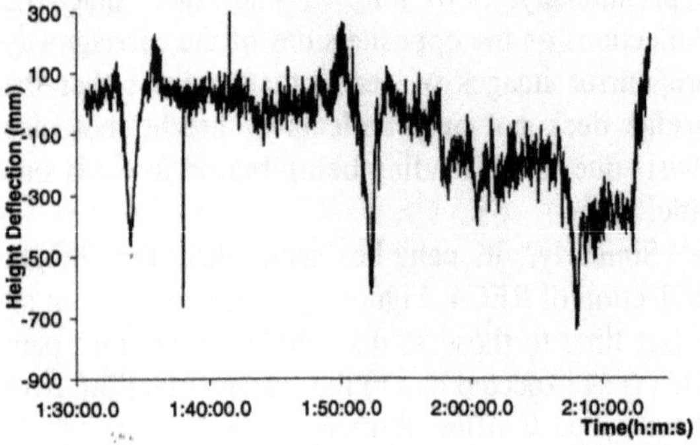

Fig. 4: Height deflection of REC2.

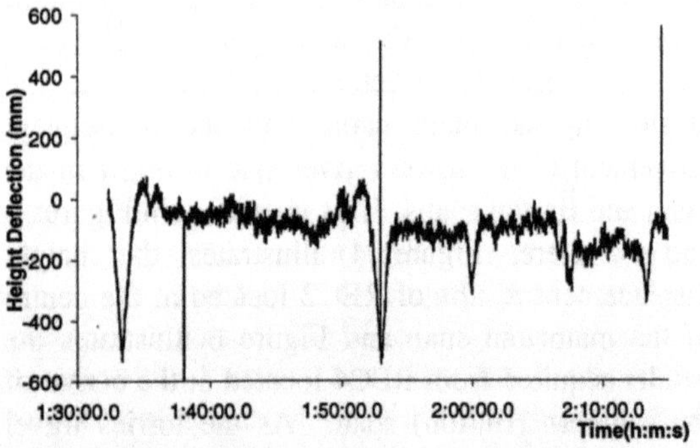

Fig. 5: Height deflection of REC3.

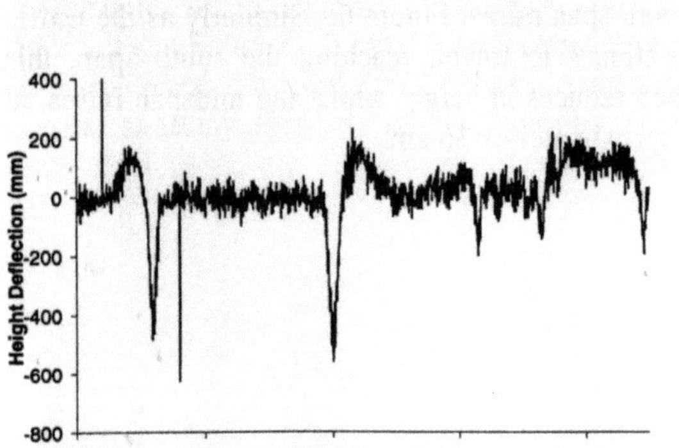

Fig. 6: Height deflection of REC4.

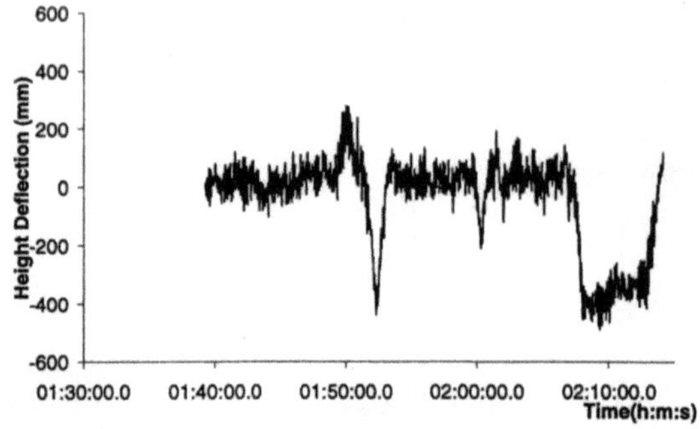

Fig. 7: Height deflection of REC5.

representation of the deck. They are run as geometrically non-linear analyses while the major design forces change very little from an elastic analysis; the predicted deformations can be significantly changed.

The full model has 68, 924 degrees of freedom, and uses link, plate and beam elements. The cable is modelled as a series of link elements, while the hanger elements have "birth and death" capability that ensures that they carry tensile only. The towers are modelled as a series of continuous beam elements, while the deck consists of inter-connected plate elements. The deck cables and hangers are constructed in steel, while the towers are in-situ reinforced concrete. The plate model simplifies the deck to a two-dimensional flat plate system with equivalent flexural properties. Simplification reduces run time, due to the reduced total number of degrees of freedom at 22, 760, with virtually no change in predicted hanger or cable forces. Fuller explanations have been presented elsewhere by Karuna et al. (1997a), and Karuna et al. (1997b). A key feature of the development of a FEM is its validation, to make sure that the FEM is a true representation of the structure itself.

Previous trials have been conducted by Brownjohn et al. (1994) in order to validate the natural vibration frequencies and associated modes of the bridge through conducting ambient vibration testing at various locations. Natural frequencies of the deck section had also been determined for design purposes from wind tunnel tests carried out at the National Physical Laboratory (unpublished).

The use of accelerometers is an important method for modal analysis. At a fundamental level, it is evident that mode shapes with deformation occurring at midspan are likely to be detected by accelerometers placed there. As well as modal analysis, Brownjohn et al. (1994), and Zasso et al. (1993) conducted additional work on the Humebr Bridge to determine the vertical displacements through visual techniques. Data from tests involving a single lorry estimated to weigh 172 t from the haulier's records was used.

Due to the uncertainties associated with the data previously available for displacements under known loading conditions, additional data was required to validate the FEM. The use of GPS located upon the bridge was seen to be an ideal method, allowing 3D movements to be gathered.

6 Comparison of the GPS results with FEM

The data was further analysed by Brown et al. (1999), whereby the response of the structure under the action of wind loading was examined. The advantage of GPS over other deflection monitoring systems such as strain gauges, is that it is a 3-dimensional displacement transducer. In addition, the subsequent analysis enables the determination of velocities and accelerations in any direction (from the displacement/time information). In this case, the two velocities orthogonal to the axis of the bridge, as the one vertically are taken. The vertical displacements were monitored at different positions along the main span, and the results suggest that the deflected dynamic form of the bridge is consistent with the predicted mode shape for the lowest natural frequency. The lateral mode shapes are consistent with simultaneous lateral motion of the deck. The value of the frequency at which oscillation occurs can be determined using readily available computer-based numerical analysis. By use of a fast-Fourier transform based power spectral density analysis, it can be shown that for the vertical mode the frequency of oscillation is 0.116 Hz, while for the lateral movement the frequency of oscillation is 0.052 Hz. These values agree extremely well with those calculated from previous trials by Brownjohn et al. (1994), Brownjohn et al. (1986) at 0.116 Hz and 0.056 Hz respectively. Those predicted by the plate model at 0.107 Hz and 0.052 Hz, and the full model at 0.108 Hz and 0.054 Hz also agree well.

7 Conclusions

The results obtained by these latest trials have been discussed briefly in this paper. Further work is underway to analyse the results.

Due to the lorries travelling over the bridge being of a known weight, the results will prove very useful data set to build and validate a FEM of the bridge. Obviously, this type of monitoring work is not limited to large suspension bridges. Further trials will be conducted on other types of structures, enabling their characteristics to be determined.

The data gathered was at a rate of 5 Hz. However, it is now possible to gather dual and single frequency GPS carrier phase data at a rate of 10 Hz, thus allowing the vibration of such structures to be detailed and quantified.

The instantaneous multipath or cycle slip errors exhibited in these results require further investigation. The work described in this paper used choke ring and ground plane antennas. Future work will use less expensive antennas, which will be more prone to multipath. This issue clearly requires addressing, especially when the GPS receivers are located upon such a complex structure.

The work detailed here is now being extended at the IESSG through a joint project with Leica Geosystems Ltd. This will involve the use of dual frequency GPS receivers gathering data at a rate of 10 Hz or more.

Acknowledgements. The initial field trials, which led to this paper, were carried out as part of a much wider research, sponsored by the EPSRC on "Real Time Positioning Using GPS for Civil Engineering". The authors would like to acknowledge Mr Roger Evans, the Humber Bridge Master and Engineer for the assistance given and the enthusiasm shown during the trials and for his help and contribution towards the organisation of the trials. The trials conducted leading to this paper were partly funded by the University of Nottingham's Quick Response Research scheme. The authors would like to acknowledge Mr C Young for the work conducted towards his MSc project. Additionally, the authors would like to acknowledge Mr Ken Gibson Mr Andy Evans for their help with the field trials.

References

Ashkenazi, V., A. H. Dodson, T. Moore, G. W. Roberts (1996). Real Time Monitoring of the Humber Bridge. *Surveying World* May/June 1996. Volume 4 issue 4. ISSN 0927-7900.

Ashkenazi V., A. H. Dodson, T. Moore, G. W. Roberts (1997). Monitoring the Humber Suspension Bridge by GPS. *Proc. GNSS'97, First European Symposium on Global Navigation Satellite Systems.* 21 - 25 April 1997, Munich, Germany.

Ashkenazi V., and G. W. Roberts (1997) Monitoring the Humber Bridge with GPS. *Proc. Institution of Civil Engineers*, Civil Engineering, Nov 1997, vol 120, Issue 4., pp. 177-182. ISSN 0965 089 X.

Brown C. J., R. Karuna, V. Ashkenazi, G. W. Roberts, and R. A. Evans (1999) Monitoring of Structures using the Global positioning System. *Proc. Instn Civ Engrs* Structs & Bldgs, 1999, vol 134, Feb 1999, pp 97 – 105.

Brownjohn J. M. W., A. A. Dumanoglu, R. T. Severn, and C. A. Taylor (1986) Ambient Vibration Survey of the Humber Suspension Bridge. Civil Engineering Department, Bristol University, 1986, Research Report UBCE-EE-86-2.

Brownjohn J. M. W., M. Bocciolone, A. Curami, M. Falco and A. Zasso (1994) Humber Bridge full-scale measurement campaigns 1990-1991. *Journal of Wind Engineering and Industrial Aerodynamics*, 1994, 52, 185-218.

Karuna R., M. S Yao., C. J. Brown , and R. A. Evans (1997a) Modelling and Analysis of the Humber Bridge" *IASS International Colloquium on Computation of Shell and Spatial Structures (ICCSS'97)*, Taiwan, November 1997.

Karuna R., M. S. Yao, C. J. Brown, and R. A. Evans (1997b) Behaviour of the Humber Suspension Bridge. *Seventh International Conference on Computing in Civil and Building Engineering* (ICCCBE-VII), Seoul, Korea, 1997.

Lovse J. W., W. F. Teskey, G. Lachapelle, and M. E. Cannon (1995) Dynamic deformation monitoring of tall buildings using GPS technology. *ASCE Journal of Surveying Engineering*, 1995, 121, Mo. 1. pp 35 – 40.

Sember B. A. (1993) Mathematical model for suspension bridge vibration. *Mathematical Computing and modelling*, 1993, 18, No. 11, pp 17 – 28.

Zasso A., M. Vergani, M. Bocciolone, and R. A. Evans (1993) Use of a newly designed optimetric instrument for long-term, long distance monitoring of structures, with an example of its application on the Humber Bridge. *Second International Conference on Bridge Management*, University of Surrey, 1993.

Continuously operating GPS-based volcano deformation monitoring in Indonesia: the technical and logistical challenges

Chris Rizos, Shaowei Han, Craig Roberts & Xiujiao Han
School of Geomatic Engineering, The University of New South Wales, Sydney, AUSTRALIA

Hasanuddin Z. Abidin
Department of Geodetic Engineering, Institute of Technology Bandung, Bandung, INDONESIA

Ony K. Suganda, A. Djumarma Wirakusumah
Volcanological Survey of Indonesia, Bandung, INDONESIA

Abstract. In the past decade or so there has been increasing interest in the use of permanent, continuously-operating GPS networks, with a small number of continuous networks having been deployed in the USA, Japan, Canada and Europe for large scale crustal motion studies. However, only in Japan has a country-wide continuous GPS network to support seismic research been established by the Geographical Survey Institute (GSI). Even with such a dense network of dual-frequency GPS receivers the station separation is of the order of 20km or more. There are, however, applications of GPS-based deformation systems which require receiver densities of the order of just a few kilometres. Furthermore, the high cost of geodetic GPS receivers means that many countries cannot afford to establish such networks. Applications of *dense* permanent GPS arrays include monitoring of volcano flanks, micro-faults, ground subsidence due to underground mining or fluid extraction, slope stability, and even engineering structures such as dams, bridges, etc. This paper describes a low-cost design of an automatic GPS-based volcano deformation system that has recently been deployed by the authors on the Papandayan volcano in Indonesia. The critical problems that had to be overcome will be described, and early results presented.

Keywords. deformation, monitoring, low-cost, continuous, GPS.

1 Introduction

The continuously-operating, GPS-based, ground deformation monitoring system described in this paper is designed to "fill the gap" by providing a technology appropriate for small-scale monitoring systems such as that required for an active volcano. The system design requirements dictate the use of low-cost components so that a station setup will cost less than US$3000 (including monuments). The objective of this deformation monitoring system is to provide centimetre accuracy, three-dimensional coordinates of stationary points. Re-measuring these same points on a continuous basis (up to many times a day) will produce a time series of coordinates which can be further analysed in order to extract possible deformational signals of the volcano.

The motivation for this project reflects the trend toward continuously operating, GPS networks for geoscientific research. Examples of such networks include the SCIGN network in Southern California (Bock et al., 1997), the GSI network in Japan now consisting of almost 1000 permanent GPS stations (Sagiya et al., 1997), and the WCDA network in Canada (Dragert et al., 1994). All of these arrays, however, utilise dual-frequency GPS receivers, resulting in very high capital costs for each station. These networks are designed for large-scale geodynamic studies, and typically the station separations are in the range of many tens to several hundred kilometres. The location of receiver stations is further restricted by the need for mains power and a telephone line connection for the daily download of the recorded data.

International Association of Geodesy Symposia, Vol. 121
Schwarz (ed.), Geodesy Beyond 2000 – The Challenges of the First Decade

The system described in this paper is intended for small-scale networks (~15 x 15km, or smaller) with on-site battery power (augmented with solar generated power), radio modems and a base station located in a stable location away from the zone of deformation. The subject volcano, Gunung Papandayan, is located approximately 60km SSE of Bandung in West Java, Indonesia (Figure 1). Papandayan erupted violently on 11 & 12 August 1772, ejecting the summit portion of the mountain over 250 square kms surrounding the area. However, although it is now a tourist destination, it remains an eruptive threat with heightened activity observed in the 1920s, with the last explosion of mud occurring in 1993 (VSI, 1998). The Vulcanological Survey of Indonesia monitors this volcano with a continuous seismic station, periodic temperature and gas observations, as well as by EDM and levelling surveys.

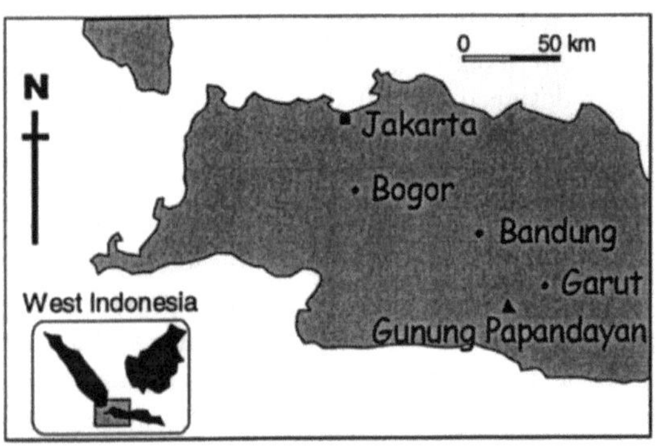

Fig. 1: Map of West Java showing location of Gunung Papandayan.

2 Collaboration & management

The School of Geomatic Engineering, The University of New South Wales (UNSW) is collaborating with the Department of Geodetic Engineering, Institute of Technology, Bandung, Indonesia (ITB) and the Vulcanological Survey of Indonesia (VSI) on this project. In August 1998, one of the authors (Craig Roberts) travelled to Indonesia to meet with ITB and VSI colleagues, and to determine whether Gunung Papandayan would be a suitable volcano to make first tests of the GPS-based deformation monitoring system. This volcano was chosen principally because of its proximity to the city of Bandung and the easy accessibility of the receiver stations to be emplaced on the volcano.

It was anticipated that UNSW staff would visit Indonesia on frequent short-term visits to ensure that any technical problems could be solved with a minimum of effort. However, the key to the ongoing operation (and maintenance) of this system is the involvement of in-country (ITB and VSI) collaborators. Furthermore, the local knowledge of the ITB and VSI experts is vital in such a project, for concerns ranging from local weather patterns, human and animal vandalism, harsh volcanic environments, to cultural and financial issues. Station visits can be carried out on a regular basis by either VSI staff resident at the observatory/base station at Papandayan, or by interested 4th year geodesy students from the ITB. ITB students will be encouraged to use the data for their own research purposes. All these measures are intended to strengthen the collaboration between the respective institutions, and in so doing ensure the test of a robust and sustainable monitoring system in Indonesia.

3 Equipment & logistics

UNSW has coordinated the design and assembly of the deformation monitoring system. The array will initially consist of one base station situated at the Papandayan observatory and three "slave" stations located on the volcano. More stations will be added at a later date. Each station consists of five components: (1) GPS/PC module, (2) radio modem sub-system, (3) monument, (4) GPS antenna, and (5) power supply.

3.1 GPS/PC module

The GPS/PC module was custom built by GPSat Systems, Melbourne. It comprises a Canadian Marconi Co. Allstar, 12 channel, single-frequency GPS receiver. The receiver board is installed within a small aluminium box, and connected to a JED PC541 microcomputer board (in the case of the slave stations) mounted in the same box (see Figure 2). These computers have no screen, no keyboard and no harddisk, and because they are only required to run rudimentary software low power, low-cost 8086 PCs, with 240Kb of RAM, have been chosen. On the other hand, the base station GPS/PC module will collect data from all slave stations (and its own receiver), and run the baseline processing software. Therefore a 150MHz Pentium computer board with 7Mb of RAM is utilised. The GPS/PC modules will run on a 12V DC power supply and the slave stations will use approximately 2.5 Watts, while the base station module with the Pentium PC board will use 6 Watts of power.

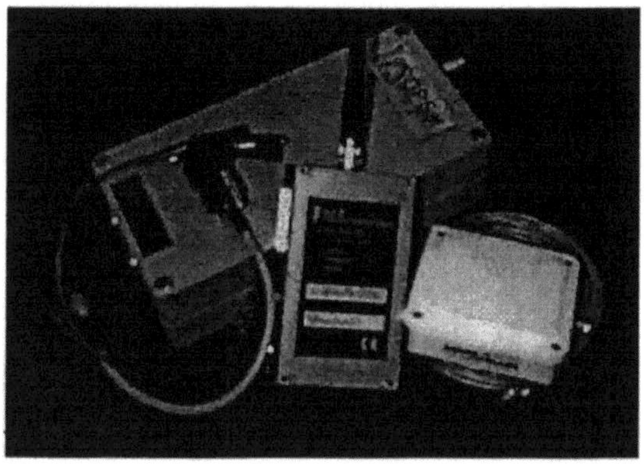

Fig. 2: GPS/PC module, radio modem, GPS antenna and cables

3.2 Radio modem and antenna

Radio modems are used to transfer GPS data collected at the slave stations to the base station at the VSI observatory. Almost all of the 60 active volcanoes monitored by the VSI in Indonesia have an observatory based up to 15km away from the zone of deformation, with a clear line-of-sight to the summit. VHF radios from GLB Radio Data Systems are being used (with Yagi antennas at all slave stations and a dipole Omni at the base). Monuments have been emplaced to ensure line-of-sight communications is achieved.

3.3 Monuments

A VSI network of Electronic Distance Meter (EDM) stations already exists on and in the crater of Papandayan. However, because the VSI experts believe that the volcano is "dying", and expect to measure a deflation of the crater, the EDM network was established to detect this deflation. It was therefore suggested that the continuous GPS network should be placed near the EDM network as a "rough yet independent" check of any deformation that may be detected. Therefore the three GPS slave station monuments were placed near existing EDM marks during the first visit in August 1998. The fourth mark (for the base station receiver) was established at the VSI observatory.

3D control marks from Berntsen International Inc. were used for the GPS monuments. These marks comprise screw-in sections of 0.6m and 0.9m lengths which are hammered into the ground until refusal. A 1m x 0.4m x 0.4m reinforced concrete block also surrounds the upper 1m section of the mark. A screw thread at the surface of the mark is connected to a fibreglass pole which elevates the GPS antenna 1.5m above the ground surface.

Fig. 3: Continuous slave station at Kawah showing radio antenna & solar panels mounted on PVC struts. The GPS antenna can be seen behind the solar panels mounted on a fibreglass pole. The tripod is temporarily in place for an EDM measurement

3.4 GPS antenna

MicroPulse Lightweight Survey L1 GPS antennas are used at all stations. These antennas are not the cheapest on the market, costing approximately US$300 each. The United States National Geodetic Survey estimated the phase centre variation of the antenna, and the offset model will be incorporated into the GPS baseline processing software.

Johnson et al. (1995) recommends mounting an antenna about 1.5m above the ground surface in order to reduce multipath disturbance. Corrosive gases are also continuously emitted from the main crater hence a fibreglass pole is used to elevate the lightweight GPS antenna above the ground surface. (Fibreglass will not be as readily affected by the acidic atmosphere and is therefore expected to last longer.) A protective fence has been erected around, but separate from, the fibreglass pole, to offer some protection against vandalism (Figure 3).

3.5 Power supply, solar panels and installation

Lead acid batteries are used on-site at the remote slave stations to supply continuous power to the GPS/PC module, GPS antenna and radio modem. Solar panels recharge the batteries during daylight hours. Two solar panels, each capable of generating 75 Watts, are used at each slave station (Figure 3).

The longevity of the system will be determined by its robustness and minimal maintenance. The slave station components are located on a concrete pad. The GPS/PC module, radio modem, battery and solar regulator are covered by a metal drum which is locked to the concrete pad to protect against vandalism. Solar panels are mounted on PVC struts in which bolts are concreted in position. They are probably the most vulnerable piece of equipment to vandalism.

The internal temperature of the system is a major concern. Although all components are rated to an operating temperature of +55°C, it is not inconceivable that such temperatures could be exceeded in tropical Indonesia, inside a poorly ventilated box. The ground surface is also warm from the volcanic activity underfoot. It was decided to ventilate the box with holes that allow air to pass through but to avoid water entering these holes during times of heavy rain. Caging over the ventilation holes protects against animals. PVC tubing covering the GPS, radio antenna and solar power cables also offers some protection against animals. Attempts have been made to minimise erosion around both the box and the monument.

3.6 Software sub-system

All software for the deformation monitoring system has been written at UNSW. The GPS receiver at the slave station logs data continuously. The PC on board the GPS/PC module creates a new data file every hour which is stored on the RAM at the slave station. The data file is stored in the receiver's binary proprietary format. The base station controls the data download via the radio modems by polling each slave station in sequence, downloading the latest file from its RAM and storing it on a harddisk located in the observatory. The combined session of four files will then be automatically processed using the UNSW *Baseline* program. *Baseline* scans the four data files and then processes the data using the standard double-differencing algorithm. The results will be transferred via a wire connection to a desktop PC also located in the observatory. At this stage the raw data and the baseline results will be saved on the PC's harddisk and FTP'ed to UNSW for further analysis. A display system will be developed to allow observers in the observatory to monitor the GPS-derived deformations.

Fig. 4: Shows the positions of the three slave stations installed on Gunung Papandayan.

4 Error mitigation strategies

The most important component for ground deformation studies of a volcano is the vertical coordinate. Given that the vertical derived by GPS is about three times less accurate than the horizontal components, the challenge is therefore to achieve centimetre vertical accuracy using low-cost, single-frequency GPS receivers. Some of the major error sources are briefly discussed below.

4.1 Antenna phase centre variation

For almost all GPS antennas the electrical phase centre, or the point to which all measurements are made, and the physical centre of the antenna do not coincide. This discrepancy can be split into two components: an offset component and a zenith dependent variation. This error was only discovered in the early 1990s, and is particularly problematic when mixing antennas of a different make and model (Rothacher et al., 1995). Many geodetic, dual-frequency antennas have since had their APCVs modelled. However this is only the case for a few single-frequency GPS antennas, as they are not frequently used for high precision applications. While it is true that at this stage only MicroPulse antennas will be used on Gunung Papandayan (ie no mixing of antenna models), it is planned to integrate this inner volcano array with an outer fiducial array, which will be equipped with high precision, choke-ring antennas (see "Future Proposals & Concluding Remarks"). Therefore it is essential that the APCV model be known. The National Geodetic Survey in Virginia, USA, determined the APCV models for two identical

MicroPulse GPS antennas. The rms between the APCV models for the two antennas was found to be very low, indicating that the repeatability (or quality) of this brand of antenna was good enough for this application.

4.2 Multipath

Multipath is the effect where direct and reflected indirect signals superimpose at the antenna to corrupt the measured signal. Multipath is unique at all station locations as it depends on the material composition and geometry of the surroundings. In this application the antennas are mounted 1.5m above the ground surface to minimise multipath effects. Code ranges are more highly affected by multipath than carrier phases, however carrier phases are used for precise baseline determinations. Carrier phase multipath may reach several cm's.

The static points in the deformation monitoring array, and the fact that carrier phase multipath is periodic, can be exploited using digital signal processing techniques. Because the GPS antennas are stationary, the carrier phase multipath will repeat itself every day at the same time minus about 236 seconds (the time difference between a solar day and two periods of the GPS satellites). The multipath signal from the previous day's observations can be identified using a discrete Fourier transform and subtracted from the present day's data using an appropriately designed finite impulse response filter. Han & Rizos (1997) describe how such an approach was successfully implemented on GPS data collected in an opencut mine environment.

4.3 Troposphere

The effect of tropospheric delay on GPS measurements is most pronounced in the vertical component. It is usually assumed that for short baselines of up to 15km in length, no tropospheric modelling is necessary as double-differencing largely eliminates this bias. However, in volcano environments the difference in height between the base station and slave stations is large enough to cause a different delay at the different stations. Existing tropospheric models cannot accurately model this discrepancy. Scientific GPS software generally requires that a tropospheric bias parameter be estimated at each station. Abidin et al. (1998) compare the results obtained using commercial software (no troposphere parameter estimation) with those obtained using scientific software. With respect to the vertical component, it is clearly shown that the stability of the parameter estimation is enhanced in the troposphere parameter estimation approach.

4.4 Ionosphere

The assumption that short baselines can be determined using single-frequency data because the residual (double-differenced) ionospheric bias is negligible is potentially a dangerous one as the solar sunspot cycle maximum (associated with the greatest ionosphere disturbances) is due to peak in the period mid-1999 to 2001. Furthermore, this will be particularly a problem in equatorial regions. However, the first generation system will not be able to overcome the effect of residual ionospheric biases. Subsequent systems will have the capability to account for residual ionospheric delay (see "Future Proposals & Concluding Remarks").

5 Early results

The first continuous *low-cost* GPS deformation monitoring system in Indonesia was setup by the third author on Gunung Papandayan in July 1999. Although many hardware problems still need to be addressed, it has been demonstrated that the system functions as designed. Preliminary data is yet to be satisfactorily processed and will not be presented in this paper. Software upgrades will be incorporated to automatically and reliably process baselines at the base station on site.

6 Future proposals & concluding remarks

The continuously operating, low-cost, GPS-based, deformation monitoring system described in this paper is a first step towards a more sophisticated system which will be capable of operating over larger areal extents. An outer "fiducial" network of dual-frequency receivers will be used to construct empirical data bias corrections which can be applied to the single-frequency receiver network in order to improve the accuracy of these results by mitigating the orbit bias, multipath effects, and the unmodelled residual ionospheric and tropospheric delays (Han & Rizos, 1996).

The GSI network in Japan is a good example of such a fiducial GPS network that could be "densified" using a low-cost array such as that described in this paper. Tests in Japan of the UNSW system will commence in July 1999 to investigate how such a low-cost system could be integrated into an existing dual-frequency GPS receiver network, and used to monitor volcano deformation, ground subsidence due to fluid extraction or underground mining, slope stability, or other engineering deformation applications, at a fraction of the cost of existing GPS monitoring systems.

The main objective of this project has been to demonstrate the feasibility of developing and deploying a low-cost, GPS-based, volcano deformation monitoring system, for an extended period of time, capable of reliably determining centimetre level accuracy baselines. The challenges in developing and deploying such a system are both technical and non-technical in nature. Resolving the technical issues such as choosing the appropriate hardware, building the monuments and installing the system on a volcano in Indonesia is important. However, without careful planning, good management and close cooperation between the partners, a project such as this is prone to failure.

The in-house data processing software provides the authors with the ability to improve the system based on experience gained on Papandayan. This is a significant advantage over using a commercial off-the-shelf system. An innovative methodology has been proposed for the next generation of GPS-based, deformation monitoring systems, whereby existing permanent GPS networks are "densified" using low-cost GPS arrays based on the robust single-frequency design described in this paper.

References

Abidin, H.Z., Meilano, I., Suganda, O.K., Kusuma, M.A., Muhardi, D., Yolanda, O., Setyadji, B., Sukhyar, R., Kahar, J. & Tanaka, T., 1998. Monitoring the deformation of Guntur Volcano using repeated GPS survey method, *Pres. XXI FIG Congress*, Brighton, U.K., 18-25 July.

Bock, Y., Wdowinski, S., Fang, P., Zhang, J., Williams, S., Johnson, H., Behr, J., Genrich, J., Dean, J., Van Domselaar, M., Agnew, D., Wyatt, F., Stark, K., Oral, B., Hudnut, K., King, R., Herring, T., Dinardo, S., Young, W., Jackson, D. & Gurtner, W., 1997. Southern California Permanent GPS Geodetic Array: continuous measurements of regional crustal deformation between the 1992 Landers and 1994 Northridge earthquakes, *J. Geophys Res*, 102(B8), 18013-18033.

Dragert, H., Schmidt, M. & Chen, X., 1994. The use of continuous tracking for deformation studies in southwestern British Columbia, *Proc. 7th Int. Tech. Meeting of the Satellite Division of the U.S. Inst. of Navigation*, Salt Lake City, Utah, 20-23 September, 97-103.

Han, S. & Rizos, C., 1996. GPS network design and error mitigation for real-time continuous array monitoring systems, *Proc. 9th Int. Tech. Meeting of the Satellite Division of the U.S. Inst. of Navigation,* Kansas City, Missouri, 17-20 September, 1827-1836.

Han, S. & Rizos, C., 1997. Multipath effects on GPS in mine environments, *Proc. Xth International Congress on the International Society for Mine Surveying*, Fremantle, Australia, 2-6 November, 447-457.

Johnson, J., Braun, J., Rocken, C. & VanHove, T., 1995. The role of multipath in antenna height tests at Table Mountain, UNAVCO website, July 1995, http://www.unavco.ucar.edu/science_tech/technology/publications/tblmtn/
tblmtn_2.html.

Rothacher, M., Schaer, S., Mervart, L. & Beutler, G., 1995. Determination of antenna phase center variations using GPS data, *Pres. 1995 IGS Workshop*, Potsdam, Germany, 15-17 May.

Sagiya, T., Yoshimura, A., Iwata, E., Abe, K., Kimura, I., Uemura, K. & Tada, T., 1997. Establishment of permanent GPS observation network and crustal deformation monitoring in the Southern Kanto and Tokai Areas, *GSI Website*, http://www.gsi-mc.go.jp/ENGLISH/RESEARCH/BULLETIN/vol-41/gps.html.

Vulcanological Survey Institute (VSI) of Indonesia website, http://www.vsi.dpe.go.id/papandayan

A national network of continuously operating GPS receivers for the UK

A. H. Dodson, R. M. Bingley, N. T. Penna and M. H. O. Aquino
Institute of Engineering Surveying and Space Geodesy (IESSG),
University of Nottingham, University Park, Nottingham NG7 2RD, UK

Abstract. A national network of continuously operating GPS receivers (COGRs) can provide a temporal and spatial density of continuous GPS data that can greatly expand the potential uses of IGS data. For example, flood defence monitoring on a national scale can be related to the mean sea level monitoring carried out on a global scale. Similarly, the use of data from a national network, supplemented with data from the IGS global network, will permit detailed studies of tropospheric and ionospheric activity. The potential of COGRs has already been recognised by four government organisations in the UK, namely the Environment Agency, the Meteorological Office, the Ministry of Agriculture Fisheries and Food and the Proudman Oceanographic Laboratory. By the end of 1998, a national network of 12 COGRs was effectively in place in the UK. In the near future, these 12 COGRs will be supplemented by data from the Ordnance Survey of Great Britain, who have recently announced the establishment of a number of COGRs as active survey reference stations. The IESSG at the University of Nottingham is in the process of establishing a national centre responsible for the transfer, archiving, processing and analysis of data from this network of COGRs in the UK, which will include about 30 stations by the end of 1999. This paper discusses how the national data archive will be structured and presents preliminary results from a number of IESSG research projects that are using the COGR data.

Keywords. National Network, Continuously Operating GPS Receiver (COGR).

1 Introduction

GPS technology has now been available for well over a decade. While the system was primarily designed for, and is operated by, the US military, it is being used for applications in many diverse fields of civilian activity.

With the continuous development and refinement of techniques associated with GPS, it has become essential to the geodetic community. Central to these developments has been the establishment of the International GPS Service (IGS). The IGS operates and maintains a global GPS network, which is currently based on approximately 200 COGRs distributed around the world, and an extensive data archiving and communication infrastructure. It has proved invaluable as a source of GPS tracking data from precisely located stations, not only for the routine production of very precise GPS satellite orbits, but also as a reference for regional geophysical and environmental monitoring projects. However, many regional and local studies require more densely distributed data, which the IGS cannot realistically provide. Conversely, the regional detail cannot be reliably obtained without reference to the global information from the IGS. Investigations on national or more regional scales can therefore only be carried out if both levels of network are present.

This fact has already been recognised in other countries, notably the networks in Norway, Sweden and Finland (Johansson et al, 1997), and in other regions, eg Southern California (Bock et al, 1997).

The aim of the national network of COGRs in the UK is to provide a temporal and spatial density of continuous GPS data that can greatly expand the potential uses of the IGS.

The relevant data will be made available to the scientific community via an Internet site, which is currently under development. As part of the project, data processing and analysis procedures are also being developed, through the interaction with other IESSG research projects focusing on geophysical monitoring and meteorology. An additional goal of the project is to develop products, such as the coordinates of the COGRs in the European Terrestrial Reference System (ETRS 89), to provide the wider environmental research community with an active reference network, for example for use in GIS and mapping applications.

Section 2 describes the current status and the near future of the national network of COGRs in the

International Association of Geodesy Symposia, Vol. 121
Schwarz (ed.), Geodesy Beyond 2000 – The Challenges of the First Decade

UK. Section 3 presents the details on how the national data archive will be structured. Section 4 discusses the results of several IESSG research projects that are using the COGR data and conclusions are presented in section 5.

2 Current status and the near future

The initial national network comprised 12 COGRs (Figure 1), which were operational by the end of 1998. These COGRs were established through several IESSG research projects, funded by the Environment Agency (EA), for flood defence monitoring in the Thames Region; the Meteorological Office (UKMO), for investigating the use of GPS for measuring atmospheric water vapour; and the Ministry of Agriculture Fisheries and Food (MAFF), for the monitoring of vertical land movement at UK tide gauge sites.

The 12 COGRs currently in operation are either Ashtech Z-XII or Trimble 4000 SSI dual frequency geodetic receivers, equipped with choke ring antennas. The 'raw' data are recorded at a 30 seconds epoch rate and downloaded on a daily basis. The data are then archived in RINEX format.

The network will be expanded in the near future to include 4 more COGRs being established by the UKMO and a further 3 existing COGRs, from the Institute of Terrestrial Ecology (ITE), the Military Survey (MS) and the National Physical Laboratory (NPL). This will form a national network of 19 COGRs established purely for scientific purposes (see Figure 1).

Although funded by various government organisations, every attempt has been made to ensure that these 19 COGRs are useable for all scientific applications. For example, the COGRs installed for meteorological applications have, where possible, been sited on concrete foundations connected to 'solid' rock so that they may also be used for geophysical monitoring.

In parallel with the establishment of a national network of COGRs for purely scientific purposes, the Ordnance Survey of Great Britain (OS) are in the process of upgrading their network of GPS control stations. Part of this upgrading is the planned establishment of about 30 active survey reference stations, deployed across Great Britain such that all locations are within 100 km of one (Davies et al, 1999).

A number of these active survey reference stations are operated by the General Lighthouse Authorities (GLAs) as part of a marine differential GPS service. At 9 sites the GPS receivers have been upgraded to log data files which include carrier phase observations, so that they can be used as COGRs as well as DGPS reference stations. The other active reference stations will be operated directly by the OS. At present the OS have installed 8 COGRs at OS offices mainly located along the 'central spine' of UK (see Figure 2).

All of the OS and GLA COGRs are equipped with at least one geodetic quality receiver, most with choke-ring antennas. However, the stations are mostly sited on building roofs. This means that they are not ideal for geophysical monitoring, but should still be useful for meteorology and for GIS and mapping applications.

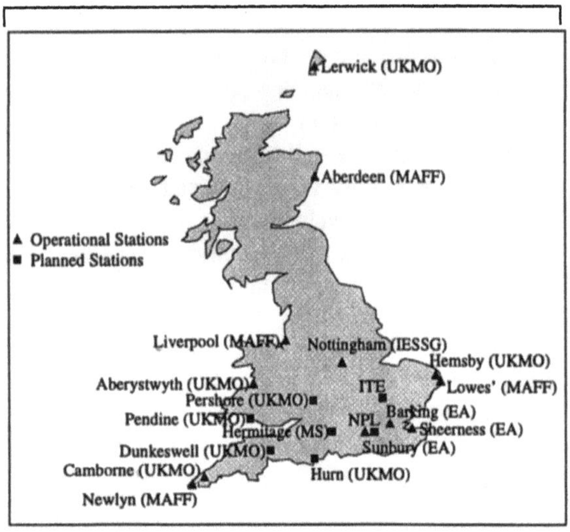

Fig. 1: Current Status of the National Network of COGRs

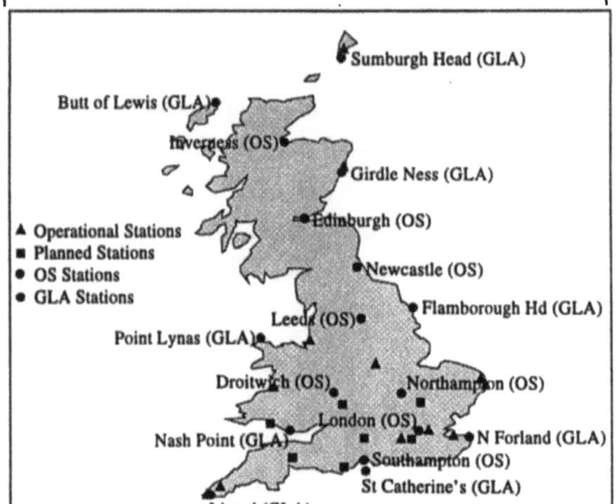

Fig. 2: Near Future of the National Network of COGRs

The OS plan to establish an Internet server with ISDN links to all of its active survey reference stations. The short term aim is to provide 24 hour access to station information and data for use by the OS and for commercial users, such as surveyors. In the longer term this may develop into a real-time positioning service at 10 to 50 cm accuracy levels. However, through discussions between the IESSG and the OS, arrangements are being made for the data from the OS and GLA COGRs to be included in the national data archive.

Combining the COGRs installed for purely scientific purposes, with those of the OS and the GLAs, will lead to a national network that will include about 50 stations available for non-commercial, scientific purposes.

3 National data archive

The national data archive will be maintained on a dedicated UNIX workstation with associated data archiving hardware, housed at the IESSG.

It has already been stated that the COGRs which form the national network are funded by a number of government organisations. This creates various levels of input to the national data archive. The IESSG is currently responsible for the management of the operational stations shown in Figure 1. The 'raw' data from these COGRs is downloaded from the receiver at the remote site to a PC at the IESSG via modems. It is then converted to RINEX format before being transferred to the UNIX workstation. For any COGRs that are not managed by the IESSG, the data is downloaded and reformatted by the responsible organisation before it is transferred to the UNIX workstation at the IESSG using ftp.

The management of the COGRs that are currently contributing to the national data archive is such that the receivers are downloaded and archived on a daily basis. However, it is anticipated that most remote sites will change to hourly downloading in the near future, and some sites may be upgraded to ISDN lines to facilitate faster recording intervals.

The management of the COGRs operated by the OS and the GLAs will be carried out by the OS. This will be done in a way that makes the data available to the OS and the commercial users for near real-time post-processing. It is proposed that the COGR data will remain on the OS Internet server for a period of 30 days. This is considered to be the 'sell by date' of active survey reference station data, after which there will be very little demand from commercial users. After 30 days, the RINEX format data will then be transferred from the OS to the IESSG, where it will become part of the national data archive and will be available for non-commercial, scientific purposes.

Information on the status of the network and on data availability will be provided via an Internet site, which is currently under development. Through this site there will be a series of forms enabling enquiries from organisations wanting to contribute new data to the network and enquiries from organisations wanting to access the national data archive.

4 Results from COGRs in the UK

As stated previously, the 12 COGRs in the initial national network were established through several IESSG research projects. Of these 12 COGRs, 4 have been operational since March/April 1997 and 4 of them since April 1998. This section presents some of the results from these COGRs, and highlights their specific application in the UK.

All of these results have been obtained using the IESSG GPS Analysis Software (GAS), with a series of UNIX Shell scripts and programs that have been developed to enable automated processing.

In this processing, the final IGS precise ephemerides have been used along with corrections for solid Earth tides, ocean tide loading and antenna phase centre variations. The tropospheric delay has been mitigated by solving for time-variant zenith delay parameters, and using the Neill mapping function. The height time series have been computed in the ITRF94 reference frame, through the inclusion of data from IGS stations in Europe during the processing.

4.1 Monitoring of mean sea level at tide gauge sites

A major application of high precision GPS is the study of mean sea level, in which GPS is used for the monitoring of vertical ground movements at tide gauge sites. These ground movements corrupt the changes in mean sea level recorded by tide gauges. Hence, to properly monitor changes in mean sea level, the rates of any vertical ground movements at the tide gauge sites must be determined. However, the rates of change in mean sea level are of the order of 1 or 2 millimetres per year, and the rates of vertical ground movements are typically of a similar order of magnitude.

The application of high precision GPS for monitoring vertical ground movements at selected sites of the UK National Tide Gauge Network has been on-going at the IESSG and the Proudman

Oceanographic Laboratory since 1990. Initially, a series of nine episodic GPS campaigns were carried out at annual and sub-annual intervals, in the period from late 1991 to late 1996 (Ashkenazi et al, 1997).

The results from these projects highlighted the improvements in high precision GPS over this period, but also served to demonstrate the limitations in using episodic GPS campaigns for monitoring vertical ground movements on a national or continental scale. These limitations mean that for a height precision of 10 mm, a series of annual episodic GPS campaigns would need to be carried out over a 20 year period in order to measure vertical station velocities to a precision of 0.5 mm/yr.

The use of continuous GPS data could potentially reduce this monitoring period to 5 to 10 years. Hence, in March 1997, the first COGR to be installed at a UK tide gauge site became operational at Sheerness. Since then, four more COGRs have been installed at the tide gauge sites of Newlyn and Aberdeen in September 1998, and Liverpool and Lowestoft in February 1999.

The preliminary results for Sheerness are given in Figure 3, which is a two year time series of weekly height solutions.

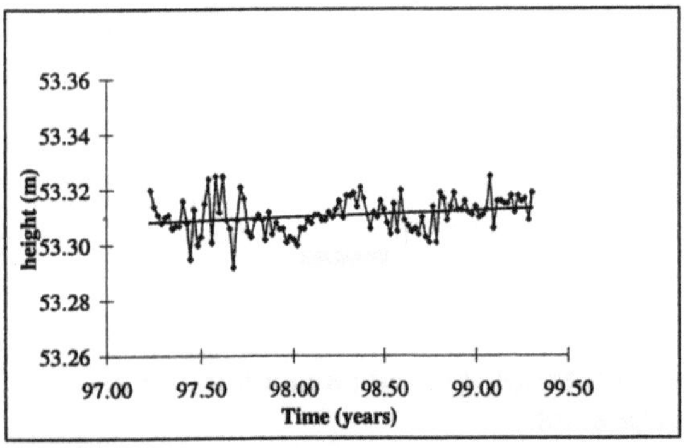

Fig. 3: Sheerness Tide Gauge COGR Height Time Series

This two year height time series for Sheerness illustrates the potential of continuous GPS data in monitoring vertical ground movements at tide gauge sites. The preliminary vertical velocity estimate is + 2.6 ± 3.0 mm/yr. On closer inspection, there is also some evidence for an annual cycle in the height time series, which would obviously not have been apparent in any time series based on campaign-type episodic GPS data.

The establishment of the national network will enable studies of vertical ground movements on a national scale, at both tide gauge sites and at other stations considered to be connected to 'solid' rock. This is particularly important when analysing the ground movements at tide gauge sites, which may be due to a combination of short-term engineering processes and short and long-term geological processes. In this context, the location of a second COGR on 'solid' rock, but relatively close to the COGR at the tide gauge site, gives the ability to monitor these types of movements. Such pairs of COGRs already exist at Newlyn-Camborne and Lowestoft-Hemsby. These sites will be part of a European network of primary tide gauges, which is currently being planned through the European Commission COST Action 40.

4.2 Monitoring changes in regional ground level

GPS can be used to monitor changes in ground level that occur on a regional scale, ie over tens of kilometres, due to various geological processes. Examples include the consolidation of deposits in low lying river estuaries and coastal regions, ground movements due to aquifer drawdown/recharge and the swell-shrink of clay.

In 1995, the IESSG and the British Geological Survey started work on a project for monitoring changes in regional ground level using high precision GPS, and providing an interpretation of such changes in terms of local and regional geology. The monitoring is based on a network of survey stations observed using GPS in continuous and episodic modes. The proposed strategy uses a small number of COGRs, acting as reference stations for a dense network of monitoring stations, observed via episodic GPS measurements using a single 'roving' GPS receiver. The network has been designed such that the reference stations are about 50 km apart and the monitoring stations are within 50 km of the nearest reference station. The sites of the monitoring stations were chosen specifically to facilitate the geological interpretation of the short and long term changes in local and regional ground level.

The Thames Estuary and Greater London was selected as the pilot region on which to test the monitoring strategy. The Thames tidal defences, which include the Thames Barrier, were constructed in the late 1970s to upgrade the protection of London and low lying areas of Essex and Kent from flooding. Historical evidence suggests an apparent rise in the water level of the Thames Estuary of 800 mm over the last century. This may be due partly to a rise in global sea level and partly to ground subsidence. The aim of the project is to develop a method for assessing the precise contribution of each of these two factors (Ashkenazi et al, 1998).

In March 1997, 3 COGRs were installed at specific sites along the Thames Estuary, namely Sheerness, Barking and Sunbury. The preliminary results for Sunbury are given in Figure 4, which is a two year time series of weekly height solutions

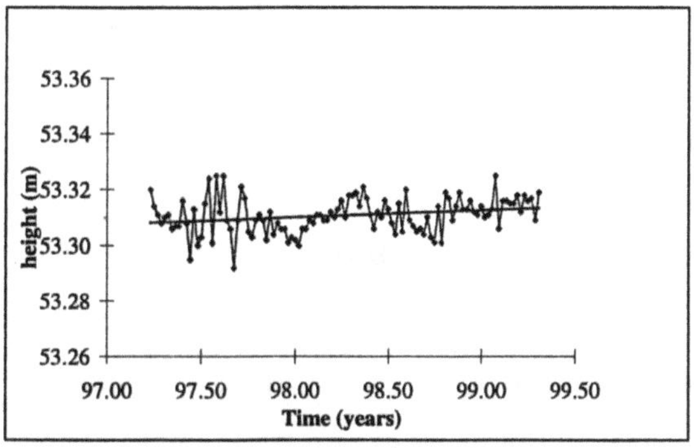

Fig. 3: Sheerness Tide Gauge COGR Height Time Series

In this project, the COGRs are being used to provide a small number of reference stations in the region, which have well defined vertical velocity estimates. This two year height time series for Sunbury has a preliminary vertical velocity of 0.0 ± 3.0 mm/yr. With an extended time series, these velocity estimates should approach the current level of precision of ITRF stations. This will effectively mean that for this region, the IGS will have been densified to a level where all of the monitoring stations are within 50 km of an 'ITRF quality' station.

Through this project a network of stations has been put in place for the long term monitoring of changes in ground level in the Thames region well into the next century.

4.3 Meteorological applications

Recently there has been considerable worldwide interest in the capability of GPS to provide continuous, all weather measurement of integrated tropospheric water vapour (IWV). This is accomplished by exploiting the fact that the GPS signals transmitted by the different satellites are delayed on passing through the Earth's troposphere. This propagation delay can be used, together with information on surface pressure and temperature, to obtain a time series of estimates of the water vapour content vertically above a COGR site.

Previous studies to date have mainly relied on single station comparisons with radiosondes and radiometers. However, the meteorological community is actively investigating the applicability of such observations to numerical forecasting, as well as the use of time series of water vapour data for long-term climate studies.

The IESSG are currently involved in the WAVEFRONT (GPS / WAter Vapour Experiment For Regional Operational Network Trials) project (Baker et al, 1998). Related to this project, in March 1998, the UK Meteorological Office installed four COGRs at the radiosonde sites of Camborne, Aberystwyth, Hemsby and Lerwick.

In Figure 4 the GPS zenith IWV estimates are compared with radiosonde IWV estimates at Hemsby. These results are typical of those obtained at all four sites, with an RMS agreement between GPS and radiosonde estimates of 1 to 2 kg/m^2. These time series also show interesting differences between the two techniques, with the IWV time series formed from 15 minute GPS estimates exhibiting variations which are not seen by the radiosondes, due to their lower temporal resolution of typically 6 hours.

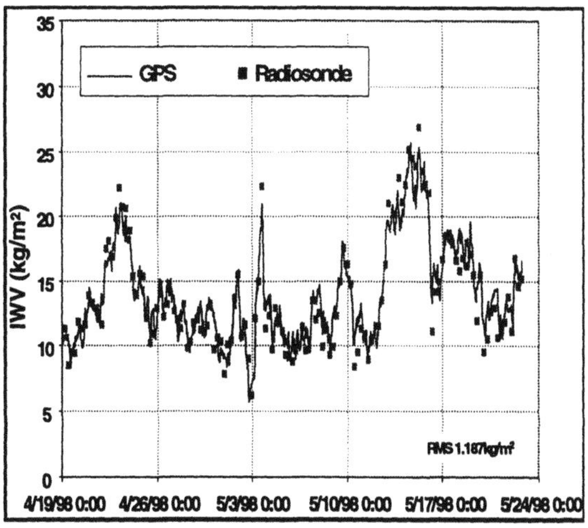

Fig. 5: Water Vapour Estimates from GPS and Radiosondes at Hemsby

The next stage in this research is to evaluate the impact of GPS water vapour estimates on numerical weather prediction models. This will require a much denser network than the four radiosonde sites. Hence, the UK Meteorological Office are about to install four more COGRs. These are being installed in South Wales and the South-West of England in order to densify this part of the national network. From Figure 1, it can be seen that the addition of these 4 COGRs will provide a 100 km station separation in the Southern part of the national network, which will provide the data for impact assessments on the quality of the mesoscale forecasts.

5 Conclusions

This paper illustrates how a national network of COGRs is being established in the UK through a combination of stations that were originally installed for a number of different reasons. These include DGPS reference stations, active survey reference stations and stations installed for different scientific objectives.

It is anticipated that the national network could include about 50 stations in the near future. For lower order scientific applications all of the COGRs will be available, to provide an active reference network. This will be suitable for GIS and mapping applications.

A sub-set of the 50 stations will also be useable for higher order scientific applications, such as geophysical monitoring and meteorology, as illustrated in Section 4.

Ultimately, it is anticipated that some of the stations in the UK national network will also be offered for inclusion in the EUREF Permanent GPS Network.

Acknowledgements. The establishment of the national data archive is funded by the Natural Environment Research Council Earth Observation Science Initiative (NERC Contract F14/G6/44). The IESSG would like to acknowledge the Environment Agency, the General Lighthouse Authorities, the Institute of Terrestrial Ecology, the Ministry of Agriculture, Fisheries and Food, the Meteorological Office, the Military Survey, the National Physical Laboratory, the Ordnance Survey of Great Britain and the Proudman Oceanographic Laboratory, for their co-operation.

The UKGAUGE project is funded by the Ministry of Agriculture, Fisheries and Food through the Proudman Oceanographic Laboratory (MAFF Contract FD305). The 'Thames Region' project is funded by the Environment Agency and the Natural Environment Research Council (NERC Contract GR3/C0003). The WAVEFRONT project is funded by the European Commission Environment and Climate Programme (EC Contract ENV4-CT96-0301).

For further information on the network, readers can visit the Internet site at: http://ukcogr.iessg.nottingham.ac.uk/~eosi or contact Dr Marcio Aquino E-mail: marcio.aquino@nottingham.ac.uk

References

Ashkenazi, V, Bingley, R M, Booth, S J, Greenaway, R G, Nursey, K, Bedlington, D, Ellison, R A, and Arthurton, R S (1998). Monitoring Long Term Vertical Land Movements in the Thames Estuary and Greater London. Proceedings of Commission 5 of the FIG '98 Congress, Brighton, UK, July 1998, pp 176-188.

Ashkenazi, V, Bingley, R M, Dodson, A H, Penna, N T, and Baker, T F (1997). Monitoring Vertical Land Movements at Tide Gauges in the UK. Proceedings of the IGS/PSMSL Workshop on Methods for Monitoring Sea Level: GPS and Tide Gauge Benchmark Monitoring, GPS Altimeter Calibration, Jet Propulsion Laboratory, Pasadena, CA, USA, 17-18 March 1997, JPL Publication 97-17 3/98, pp 97-106.

Baker, H C, Dodson, A H, Jerrett, D, and Offiler, D (1998). Ground-Based GPS Water Vapour Estimation for Meteorological Forecasting. Proceedings of the American Meteorological Society 9th Conference on Satellite Meteorology and Oceanography, 25-29 May 1998.

Bock, Y, Wdowinski, S, Fang, P, Zhang, J, Williams, S, Johnson, H, Behr, J, Genrich, J, Dean, J, van Domselaar, M, Agnew, D, Wyatt, F, Stark, K, Oral, B, Hudnut, K, King, R, Herring, T, Dinardo, S, Young, W, Jackson, D, and Gurtner, W (1997). Southern California Permanent GPS Geodetic Array: Continuous Measurements of Regional Crustal Deformation between the 1992 Landers and 1994 Northridge Earthquakes. Journal of Geophysical Research, Vol 102, No B8, pp 18,013-18033.

Davies, P, Greaves, M, Dodson, A H, and Bingley, R M (1999). National Report of Great Britain. Report on the Symposium of the IAG Subcommission for the European Reference Frame (EUREF) held in Prague, 1-4 June 1999.

Johansson, J M, Scherneck, H-G, Vermeer, M, Koivula, H, Poutanen, M, Davis, J L, and Mitrovica, J X (1997). BIFROST Project: Three Years of Continuous GPS Observations. Proceedings of the IGS/PSMSL Workshop on Methods for Monitoring Sea Level: GPS and Tide Gauge Benchmark Monitoring, GPS Altimeter Calibration, Jet Propulsion Laboratory, Pasadena, CA, USA, 17-18 March 1997, JPL Publication 97-17 3/98, pp 125-140.

The impact of the atmosphere and other systematic errors on permanent GPS networks

S. Schaer, G. Beutler, M. Rothacher
Astronomical Institute, University of Berne, Sidlerstrasse 5, CH-3012 Bern, Switzerland

E. Brockmann, A. Wiget, U. Wild
Swiss Federal Office of Topography, Seftigenstrasse 264, CH-3084 Wabern, Switzerland

Abstract. The Swiss Federal Office of Topography is currently building up the Automated GPS Network Switzerland, called AGNES. This GPS network will serve various purposes, like national surveying, engineering surveying, navigation, meteorology, and research.

For wide-area differential GPS (WADGPS), in particular precise real-time positioning using the GPS phase data of one or more reference stations of the AGNES network, one may expect that atmospheric refraction plays the crucial role, since the reference stations may be rather widely separated in Switzerland, typically by 50–100 km. This assumes that orbits and "absolute" tropospheric refraction for at least one of the sites are known with sufficient precision.

In the time period of the next solar maximum, the ionospheric refraction will probably be the most crucial error source, in particular for resolving the initial phase ambiguities, which is mandatory when aiming at a coordinate accuracy on the centimeter level. Nowadays, global and regional total electron content (TEC) maps, like those produced by the Center for Orbit Determination in Europe (CODE), are available and may be used to model the mean ionospheric refraction. However, on time scales of few minutes, ambiguity resolution usually suffers from short-term ionospheric fluctuations which are not accounted for by such TEC maps.

Significant height differences with respect to the nearest reference stations may also occur, especially in mountainous areas like Switzerland. This implies that not perfectly modeled tropospheric refraction may cause biases too.

We study the impact of tropospheric as well as ionospheric biases on WADGPS results and investigate whether it is possible to extract atmospheric information from networks of the AGNES-type for correcting data of mobile GPS receivers in (near) real time.

Keywords. Permanent GPS networks, WADGPS, atmosphere modeling.

1 Global, regional, and local permanent GPS networks

Many permanent GPS networks were deployed for various purposes in recent years. We have to distinguish global, regional, and local networks.

1.1 The IGS and its products

The IGS (International GPS Service) network is probably the best known global GPS network. It is documented, e. g., in (Beutler et al., 1996) and in a series of annual reports, e. g., (Zumberge et al., 1997). The IGS products are of importance for all other permanent GPS arrays (at least for those where the reference frame is an important issue). Let us mention in particular

- coordinates and velocities for all IGS sites,
- precise orbits (final, rapid, predicted),
- daily values for the earth rotation parameters x, y and LOD (length of day),
- troposphere estimates for IGS sites, and
- global ionosphere maps.

It is our understanding that these products must be used (to the extent possible) by the operators of regional or local networks in order to assure consistency on the sub-centimeter level of the reference frames with (the latest IGS realization of) the ITRF.

1.2 Permanent regional and local networks

Permanent regional or local arrays have been deployed and are operated for various purposes, e. g., for precise relative navigation, crustal deformation monitoring (e. g., SCIGN network, the

International Association of Geodesy Symposia, Vol. 121
Schwarz (ed.), Geodesy Beyond 2000 – The Challenges of the First Decade

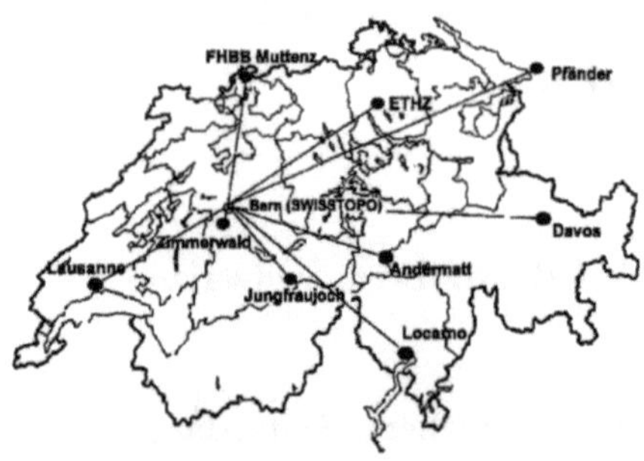

Fig. 1: The reference stations of the AGNES network 1999

Table 1: Ellipsoidal station heights in the AGNES network

Station ID	Height (m)	Dist. wrt. ZIMM (km)
FHBB	378	74
EPFL	459	79
ETHZ	595	99
SIER	602	65
ZIMM	956	0
PFAN	1090	190
JUJO	3635	54

•

Japanese 1000-receiver network), or for precise (near) real-time relative positioning using short data pieces, often addressed as *wide-area differential GPS (WADGPS)*.

The Canadian Active Control System (CACS) is a good example for networks of the latter type. Other examples may be found in Germany, see, e. g., (Wanninger, 1999) or in the Netherlands, see (van der Marel, 1999). The accuracy aimed at may vary between the 1-m and the 1-cm level.

2 The AGNES: A local GPS array

AGNES stands for *Automated GPS Network Switzerland*. The network serves several purposes, in particular monitoring and maintenance of the Swiss first-order control network, navigation, and research. The AGNES is being deployed and operated by the Swiss Federal Office of Topography.

The AGNES sites are shown in Figure 1. The distances of the AGNES sites with respect to the Zimmerwald IGS site as well as the station heights are given in Table 1.

Table 1 indicates a particular difficulty in the AGNES, namely a height difference of more than 3000 m between the lowest and the highest site.

2.1 AGNES data analyzed

Only a small amount of the AGNES data were analyzed in our study, namely one week of data in 1998 and two days in 1999. The two data sets may be characterized as follows:

- 1998 data set: 16–22 August, 30-second data from 7 stations (EPFL, ETHZ, FHBB, JUJO, PFAN, SIER, ZIMM);
- 1999 data set: 8–9 April, 5-second data from 9 stations (DACH, DAVO, EPFL, ETHZ, FHBB, JUJO, THUN, WICH, ZIMM).

The non-AGNES sites DACH, THUN, and WICH were observed only from noon to noon.

The following information was taken over from the CODE/IGS data processing or from other sources:

- position of the Zimmerwald observatory for the two periods,
- troposphere estimates (one-hour tropospheric zenith path delay parameters) for Zimmerwald from the CODE/EUREF data processing,
- broadcast, IGS precise, or IGS (2-day) predicted orbits,
- IGS earth rotation parameters, and
- CODE global ionosphere maps.

3 Remarks concerning theory

When analyzing data from a permanent network like AGNES we must ask the question: *What do we have to do in order to let the AGNES user take maximum profit out of the network and its data?* The question must be answered differently for post-processing and for (near) real-time applications.

In the *post-processing mode*, the user would download the required data from the entire AGNES, use the best possible information available from the IGS (orbits, ERPs, troposphere parameters for the Zimmerwald site, if available troposphere parameters for all other AGNES sites from the AGNES operator), and produce a correct network solution using a scientific software package. From the quality point of view, there is no better solution. We will not pursue the post-processing issue further.

In the *real-time mode*, other solutions have to be found. Let us mention a minimum and a maximum solution:

- As a *minimum solution*, the data from the closest AGNES site are downloaded and processed. The AGNES user only has to be given the coordinates of the AGNES site and maybe the current values for the troposphere corrections for the site requested. It is then entirely up to the user to process the data in the best possible way, e.g., by tabbing the best possible orbit files, etc.

- As a *maximum solution*, the AGNES operator might perform the following steps:
  (1) produce in (near) real time a correct network solution (based on the ionosphere-free linear combination) of the entire AGNES with ambiguities fixed on the double-difference level,
  (2) reconstruct zero-difference observations and relocate them to one site, e.g., the Zimmerwald site,
  (3) combine the phase and code observations for all AGNES receivers with the goal to reduce the observation noise and site-specific biases,
  (4) produce satellite- and epoch-specific linear corrections of ionospheric refraction with latitude β and longitude λ as independent arguments (epoch-specific may be replaced by "specific for a few epochs"),
  (5) produce for time intervals of, let us say, 30 minutes linear models of tropospheric zenith corrections as a function of latitude β, longitude λ, and height h, and
  (6) make available the sum of the above mentioned corrections for each satellite and epoch for each user position. In addition, one would add the antenna phase center corrections due to the user antenna in this step. Alternatively, the correction model parameters might be made available.

The proposed maximum solution is closely related to proposals made by (Wanninger, 1999) or (van der Marel, 1999). Our approach differs in the following respects:

- We try to avoid epoch-by-epoch corrections wherever it is possible, because they have the tendency to increase the noise.
- We do *not* make the attempt to model tropospheric refraction for each satellite and each epoch. Experience tells that it is sufficient to introduce one zenith path delay parameter per hour and station.
- Estimation of orbit and troposphere parameters is routinely performed in the above step (1). This step is not time critical. One solution every hour would be sufficient.
- If IGS predicted orbits of sufficient quality are available the orbit improvement may be skipped in step (1). If not, one (maybe two) orbit parameters for each satellite are sufficient to remove all problems.
- An epoch-by-epoch correction is only considered for modeling ionospheric refraction.
- We strive for full consistency of all phase and code observations in the above combination step (2), a non-trivial issue, which we will not address here.

In Section 4 we address issues related to processing step (1); in Section 5 we will address issues related to a user of AGNES products.

4 Processing data of permanent GPS networks

For the AGNES operator it must be the goal to generate a phase solution based on the ionosphere-free linear combination (LC). To achieve this goal he has to resolve the ambiguities using, at one point or another, the basic L1 and L2 observations.

The network operators are in a good position because they may assume their GPS receiver/antenna locations to be known in the ITRF with centimeter accuracy. If code observations with an rms error smaller than about 0.3 m rms were available, wide-lane ambiguity resolution would be trivial using the Melbourne-Wübbena LC of phase and code measurements, see (Beutler et al., 1990).

In the AGNES, at least under the AS regime, this is not the case. Consequently, we were obliged to apply ambiguity resolution techniques *without* using code observations. It was decided to resolve the wide-lane ambiguities first and to resolve the L1 ambiguities afterwards. Other techniques, like the QIF technique, see (Rothacher and Mervart, 1996), might have been used as well. The method used here gives more insight into the dependency on different bias types. We were processing one-day data pieces.

4.1 Wide-lane ambiguity resolution and ionospheric refraction

Figure 2 gives an impression of the wide-lane ambiguity resolution. We see a histogram of the

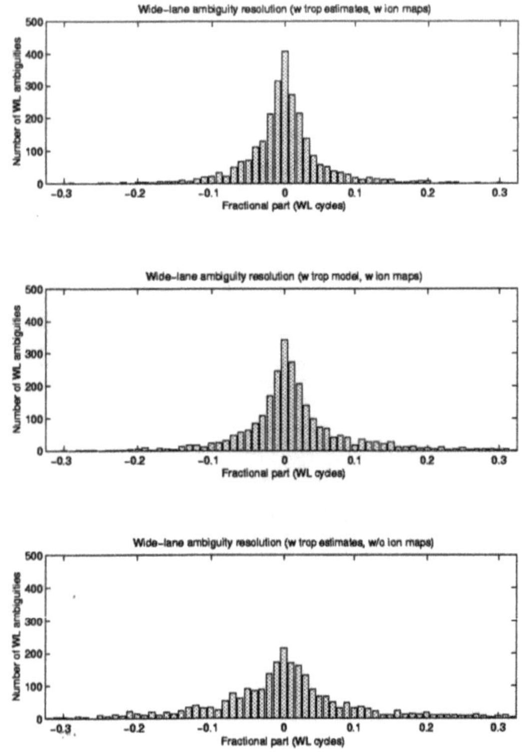

Fig. 2: Wide-lane (WL) ambiguity resolution

fractional parts of the wide-lane ambiguities at the moment they were fixed to integer numbers. The best strategy (taking over the troposphere estimates from an ambiguity-free network solution and applying the corrections from our global ionosphere maps) is reflected in the top subfigure, the second strategy (using an a-priori troposphere model) is only marginally worse, the third strategy (not taking into account the ionosphere maps), given in the bottom subfigure, is clearly inferior. Precise orbits were used in all cases, but this was not a crucial issue—even broadcast orbits would have been sufficient to resolve the wide-lane ambiguities.

In the first case 98.2%, in the second 97.4%, in the third 95.2% of the wide-lane ambiguities could be resolved using an elevation cut-off angle of 10°. The ambiguities that could not be resolved were always associated with low elevations.

In summary we may state that wide-lane ambiguity resolution does not pose a serious problem even if the orbit information is not of the best quality. Double-difference wide-lane ambiguity resolution is in practice and in essence limited by our ability to model the ionosphere—let us say on the 0.2 wide-lane wavelength level. Under usual conditions this should not pose a serious problem.

We used in essence a post-processing scheme. May it be transferred to real time? One might argue that this is not the case, because in practice we will always have the problem of rising satellites over an area, when we will have to make the attempt to resolve ambiguities based on short time spans. Whereas this is true, one has to point out that the problem is limited to exactly one satellite—all other ambiguities (in addition to all receiver positions, orbits, etc.) are known with sufficient accuracy to allow for a rapid resolution of the wide-lane ambiguity. The practical consequences, when ambiguity resolution is not possible instantaneously, are limited, as well: good differential atmospheric corrections are not available immediately after a satellite rises over the area.

4.2 Narrow-lane ambiguity resolution, tropospheric refraction, and orbit quality

After having successfully resolved the wide-lane ambiguities using the wide-lane LC, we may resolve the L1 ambiguities using the ionosphere-free LC of L1 and L2.

In this step we have the advantage that the ionosphere is completely eliminated and cannot bias our solution. The effective wavelength in the ionosphere-free LC, however, is now about 11 cm, which means that other systematic errors become relevant. We have to consider orbit-induced errors and unmodeled tropospheric effects.

Figure 3 shows a histogram of the fractional parts of the narrow-lane ambiguities at the moment of fixing them to integers. The top subfigure shows our best solution: precise orbits are taken over from the CODE/IGS data processing. In the second subfigure tropospheric refraction is not estimated, but computed using a standard atmosphere model (Saastamoinen in connection with a standard atmosphere). In the bottom subfigure of Figure 3, broadcast instead of CODE precise orbits were used.

The standard deviations for the fractional parts of the narrow-lane ambiguities are 0.092 cycles in the first case (top subfigure), 0.112 cycles in the second, 0.133 cycles in the third case, where broadcast orbits are used. In the first case 95.2%, in the second 92.4%, in the third 91.0%

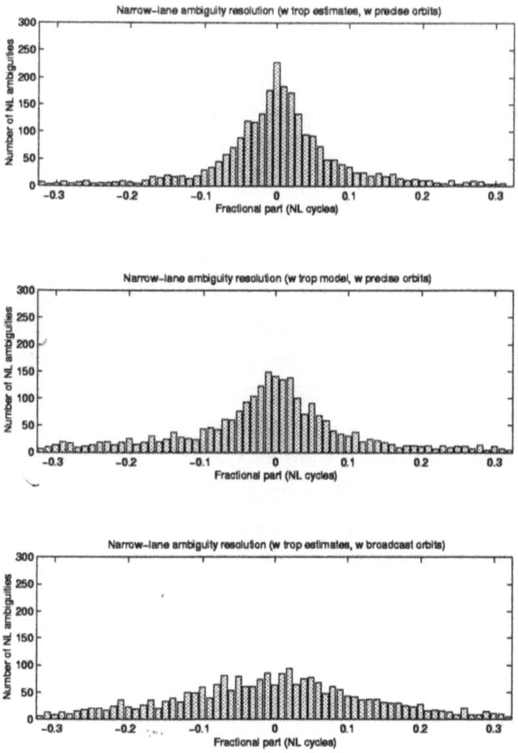

Fig. 3: Narrow-lane (NL) ambiguity resolution

of the narrow-lane ambiguities could be resolved.

Figure 3 tells us that both orbits of excellent quality and accurate troposphere parameters are mandatory for a network of the size of AGNES.

It is obvious that much better than broadcast orbits are required. Wanninger (1999) and other authors propose to reduce the orbit and tropospheric effects by a linear model over the area. We have proposed an alternative method above, and test runs were made using one day of data: By using the IGS predicted orbits as a-priori orbits and introducing only the argument of latitude u_0 as a free orbit parameter for each satellite we obtained the same observation rms error for the ambiguity-free solution, and we were able to resolve the same number of ambiguities as in the best case when using (and not altering) the IGS final orbits as a-priori orbits. The results when improving the broadcast instead of the predicted orbits were not of comparable quality. This is explained by the fact that the reference frame is slightly different for broadcast orbits.

Let us get one aspect straight. Orbits, as established by CODE or other IGS analysis centers, based on a world-wide network may be extrapolated for, let us say, six hours in such a way that the quality is sufficient to provide results of the type of the top picture in Figure 3. The problem is at present that the IGS predicted orbits are based on data with a latency of one to two days instead of few hours.

What about tropospheric refraction? Figure 3 gives an answer as well: For a network of the AGNES-type it is necessary to solve for tropospheric zenith delays for each site with a relatively high time resolution (say 30 minutes to one hour). The aspect is somewhat less dramatic than the orbit aspect, but the impact is clearly visible as well (compare middle subfigure with the top subfigure in Figure 3).

5 AGNES user aspects

Let us first address the question what kind of biases an AGNES user has to cope with, assuming that he successfully removed all phase ambiguities. The following experiments were all performed with the data of 19 August 1998. We assumed 5-minute data pieces. We were considering the Zimmerwald site as an unknown user receiver in the rest of the AGNES, when processing the data in the network mode; we were processing the baseline `FHBB-ZIMM` assuming `FHBB` known, when performing baseline-specific solutions. 288 5-minute solutions could be made using the data material of 24 hours. Our experiments may be summarized as follows:

- With few exceptions the ionosphere-free LC was processed. In view of the network size this is in general the only acceptable solution. Exceptions might be considered for short distances to the closest AGNES site.
- With few exceptions the rms repeatability is of the order of 5–10 mm for the horizontal coordinates and of the order of 15–30 mm for the height. These values would probably be acceptable in most cases in practice.
- The gain in quality, when processing the data in the network-, as opposed to the baseline-mode was about 1.3. This may be expected because one may easily show that the rms error σ_{new} of the user position may be computed in the following way as a function of the rms of observation σ_{obs}:

$$\sigma_{new} \sim \sigma_{obs} \cdot \sqrt{\frac{n+1}{n}} \qquad (1)$$

The case $n = 1$ corresponds to a baseline-specific processing, $n \to \infty$ corresponds to the best possible case. This explains the gain of 1.3 mentioned above (with $n = 6$).

- IGS precise orbits gave repeatabilities better by a factor of about 1.5–2 when compared to broadcast orbits. IGS predicted orbits did not give better results than broadcast orbits.
- 5-minute data pieces are not sufficient to solve for tropospheric parameters *and* station coordinates.

For a common user of AGNES it will be absolutely essential to resolve the ambiguities with short data pieces (one epoch to fifteen minutes). There are many degrees of freedom "in this game," however. Let us mention a few of them:

- One parameter is the baseline length. We performed tests using the data of the 1999 campaign. The baselines Zimmerwald-Wabern (6 km), Zimmerwald-Thun (17 km), and Zimmerwald-FHBB (74 km) were analyzed using 5-minute data pieces for the two short and 15-minute pieces for the long baseline.
- We may use different LCs and phase/code combinations. We used phases only and processed L1 and L2 simultaneously (FARA method).
- We may or may not introduce epoch- and satellite-specific stochastic ionosphere parameters; if we do so, we may vary the a-priori variance.

Already this (very much reduced) list shows that a thorough analysis would contain an exorbitant number of test runs. Our results may be summarized briefly as follows:

5.1 6-km baseline Zimmerwald-Wabern

Using 5-minute data pieces and a cut-off angle of 15°, most tests lead to a safe resolution of L1 and L2 ambiguity parameters. It was not necessary to introduce stochastic ionosphere parameters. Even broadcast orbits were sufficient to resolve the ambiguities. From 132 possible 5-minute intervals 131 could be successfully resolved, one was rejected. The result is as expected. If a user is operating at a distance of less than 10 km to a reference site, he is best advised to use the data of this station without any modification.

5.2 17-km baseline Zimmerwald-Thun

Tests without setting up stochastic ionosphere parameters were still quite successful: from 143 possible 5-minute intervals only between 5 and 20 could not be resolved. 20 failures were reported, when the global ionosphere maps were not used. This clearly demonstrates that a good ionosphere model is a prerequisite for successfully resolving ambiguities on baselines longer than 10 km.

All ambiguities could be safely resolved when epoch- and satellite-specific stochastic ionosphere parameters were set up. 7 (correct) solutions were rejected when the a-priori ionosphere maps were not used. Even the least accurate orbits (broadcast) and the most trivial troposphere modeling technique (a-priori model based on standard atmosphere) proved to be sufficient for the job.

5.3 74-km baseline Zimmerwald-Muttenz

After a few attempts, only tests with 15-minute data pieces were made. Even then, ambiguity resolution without introducing stochastic parameters was very problematic: failure rates up to 50% were common. This situation might be improved by absorbing part of the ionosphere signal by a linear, satellite-specific model.

The situation improved when stochastic ionosphere parameters were introduced. Success rates of well over 90%—even when the a-priori model for tropospheric refraction was used—let us believe that this technique is very promising in the context of AGNES.

It is interesting to note that (Wanninger, 1999) concludes that procedures including the estimation of (what we call) stochastic ionosphere parameters (and what he calls ambiguity resolution strategies for long baselines) have to be recommended even when using ionosphere-corrected RINEX data (corresponding to a virtual reference station). We come to similar conclusions. For ambiguity resolution on baselines longer than about 20 km, this option seems to be almost mandatory.

6 Modeling the atmosphere over a ground network

Which fraction of the *tropospheric* zenith correction and of the satellite-specific *ionospheric* delay can be absorbed by linear models in latitude β,

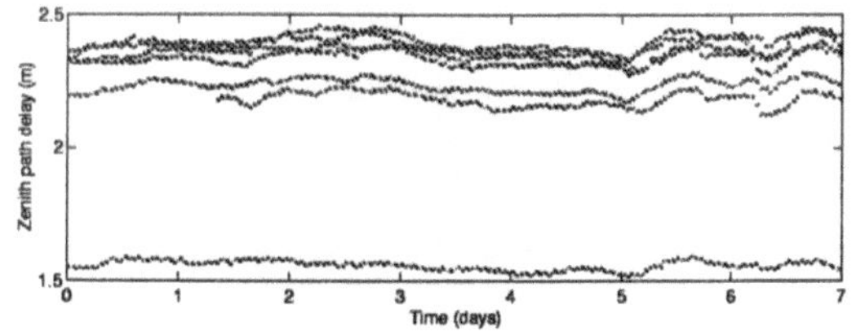

Fig. 4: Hourly tropospheric zenith path delay estimates

longitude λ, and height h? Let us consider the two effects separately.

6.1 Modeling troposphere parameters

We analyze the station-specific troposphere parameters for the 1998 data set as they were generated by a network solution of type (1).

Figure 4 shows estimated zenith path delays for all stations and for every hour.

The individual curves in Figure 4 correspond to the AGNES sites. The offsets are related to the station heights, the lowest curve corresponds to the highest, the Jungfraujoch site, the highest curve to the lowest site (Muttenz).

We have to address two issues:

- Question 1: What are the independent parameters? Obviously, we should use latitude β, longitude λ, *and* height h. We are sure, however, that we do not have enough stations to model the height dependency in an adequate way.
- Question 2: What quantity do we model as a function of our independent arguments? We decided to analyze

$$q = \frac{\Delta\rho_{trop}}{\Delta\rho_{trop,0}(h)} \tag{2}$$

where $\Delta\rho_{trop,0}(h)$ are a-priori values stemming from an atmosphere model.

Our model reads as:

$$q = c_0 + c_\beta \cdot \Delta\beta + c_\lambda \cdot \Delta\lambda + c_h \cdot \Delta h + c_{h^2} \cdot \Delta h^2 \tag{3}$$

Table 2 characterizes the results of our tests. n_{par} is the number of parameters, f the degree of freedom (of an individual determination). The standard deviation σ_q of the quantity q is dimensionless. When multiplying this value with an average value for $\Delta\rho_0 \approx 2.2$ m we get an impression of the rms of the fit of the original zenith path delays.

From Table 2 we conclude that in the present realization phase of AGNES either a model of

Table 2: Solution types fitting the zenith path delay estimates

n_{par}	f	σ_q (1)/(mm)	Parameters
1	6	0.0085/17.8	c_0
2	5	0.0083/17.3	c_0, c_h
3	4	0.0076/15.9	c_0, c_β, c_λ
4	3	0.0056/11.7	c_0, c_β, c_λ, c_h
5	2	0.0041/ 8.5	c_0, c_β, c_λ, c_h, c_{h^2}

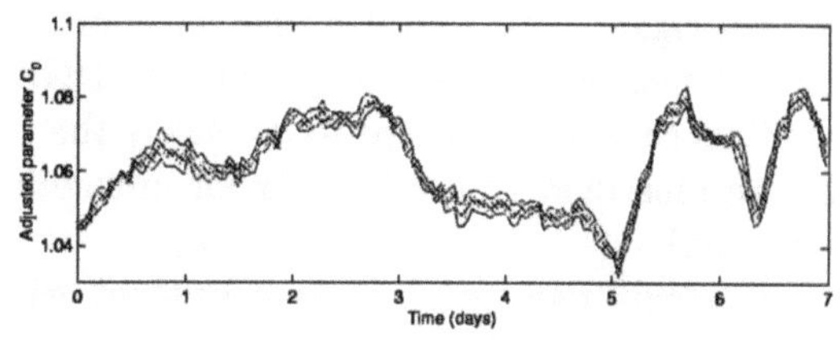

Fig. 5: Estimated constant term c_0

3 or 4 parameters seems to be an optimum and that a fit on the 1-cm level, leading to height RMS errors of about 3 cm are achievable. Figures 5 and 6 show the estimated terms c_0 and c_β (together with the associated rms band). They underline that a significant fraction of the tropospheric zenith delays may be absorbed by our models.

6.2 Modeling ionospheric refraction

After having reduced the ionospheric delay using global ionosphere maps (based on procedures developed by (Schaer, 1999) the residual ionospheric (slant) delay must be analyzed separately for each satellite and for each epoch. This prevents us from modeling the entire ionosphere over the area of the network and allows us to consider at each epoch only a small portion of the ionosphere above the ground network. Thinking in terms of a single layer, we only have to consider an area of the size of the network in the ionospheric layer.

We made tests with the following linear model:

$$\Delta\rho_{ion} = c_0 + c_\beta \cdot \Delta\beta + c_\lambda \cdot \Delta\lambda \tag{4}$$

where $\Delta\rho_{ion}$ is the ionospheric observable and c_0, c_β, and c_λ are the model parameters established separately for each epoch and each satellite.

First attempts estimating ionospheric gradient parameters c_β and c_λ showed that these parameters may absorb a significant part of the satellite- and epoch-specific biases. Nevertheless,

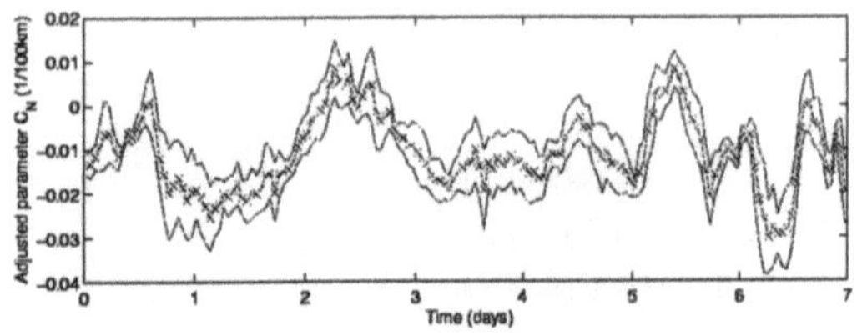

Fig. 6: Estimated (North-South) gradient c_β

our investigation also revealed that the approach does not remove the entire signal.The WADGPS user thus cannot be advised to use solely the L1 observations (and to use L2 only for ambiguity resolution).

The result reported here is consistent with those given by other authors, e.g., (Wanninger, 1999).

7 Summary and outlook

Two small portions of the observations stemming from the AGNES were analyzed. We studied the impact of atmospheric and orbit biases (a) on network solutions to be produced by the AGNES operator and (b) on the results a WADGPS user may expect from the AGNES.

We have seen that, despite the disappointing fact that the quality of code observations did not allow for an ambiguity resolution using the Melbourne-Wübbena LC, the wide-lane ambiguities can be safely resolved provided ionosphere maps are considered.

For narrow-lane ambiguity resolution, the orbit quality is a critical issue. Neither the broadcast nor the IGS predicted orbits are at present of sufficient quality for that purpose. We have argued that this problem will be resolved as soon as IGS orbits based on a sub-daily turnaround cycle will become available. We have also shown that simple orbit improvement schemes applied are well suited to cure the orbit problem.

For the AGNES user, tropospheric refraction is the ultimate accuracy limiting factor. Our study indicates that under normal conditions it should be possible, even in a mountainous environment, to model the tropospheric zenith path delay on the 1-cm level in real time. This would imply (sub-)cm accuracy in horizontal baseline components and 3–5-cm accuracy for the height estimates.

For ambiguity resolution using short data pieces, ionospheric refraction is the crucial element. The approach using pseudo-stochastic parameters is very promising. It remains to be seen by how much the ambiguity resolution performance is improved, if the information derived from the satellite- and epoch-specific models is used in the ambiguity resolution procedure.

Acknowledgements. This article summarizes one part of the work performed in the context of a contract between the Swiss Federal Office of Topography and the Astronomical Institute of the University of Berne. The contributions by the Federal Office of Topography are gratefully acknowledged.

References

Beutler, G., W. Gurtner, M. Rothacher, U. Wild, and E. Frei (1990). Relative Static Positioning with the Global Positioning System: Basic Technical Considerations. In: *Global Positioning System: An Overview*, IAG Symposia, No. 102, pp. 1–23.

Beutler, G., I. I. Mueller, R. E. Neilan (1996). The International GPS Service for Geodynamics (IGS): The Story. In: *International Association of Geodesy Symposium, No. 115, GPS Trends in Precise Terrestrial, Airborne, and Spaceborne Applications*, pp. 3–13, Springer-Verlag, ISBN 3-540-60872-6.

Kouba, J. and J. Popelar (1994). Modern Geodetic Reference Frames for Precise Satellite Positioning and Navigation. In: *Proceedings of the KIS 94 International Symposium on Kinematic Systems in Geodesy.*

Van der Marel, H. and C. D. de Jong (1999). Active GPS Reference System for the Netherlands. In: *Physics and Chemistry of the Earth* (in preparation).

Rothacher, M. and L. Mervart (1996). Bernese GPS Software, Version 4.0, *Astronomical Institute, University of Berne.*

Schaer, S. (1999). Mapping and Predicting the Earth's Ionosphere Using the Global Positioning System. Ph.D. Theses Series, *Astronomical Institute, University of Berne.*

Wanninger, L. (1999). Der Einfluss ionosphärischer Störungen auf die präzise GPS-Positionierung mit Hilfe virtueller Referenzstationen. In: *Zeitschrift für Vermessungswesen* (in press).

Zumberge, J. F., D. E. Fulton, and R. E. Neilan (1997). International GPS Service for Geodynamics, 1996 Annual Report. *IGS Central Bureau, Jet Propulsion Laboratory.*

Re-weighting of GPS baselines for vertical deformation analysis

Henk de Heus[†], Marcel Martens, Hans van der Marel
Delft University of Technology, Department of Mathematical Geodesy and Positioning,
Thijsseweg 11, 2629 JA Delft, The Netherlands,
E-mail: H.vanderMarel@geo.tudelft.nl

Abstract. In the Netherlands, GPS is used to monitor subsidence caused by extraction of natural gas. GPS campaigns are organized on a yearly basis. Each campaign consists of GPS observations on 30 to 60 points, using typically 5 dual frequency GPS receivers. The duration of a session is between one and two hours, but as every point is observed several times, up to 56 sessions have been observed for each campaign. The area is about 75 by 60 km, resulting in an average baseline length of 10 km.

The GPS data is processed using an iterative stepwise approach. First the GPS data is processed session by session. Then the sessions are combined, using their full covariance matrix in a network adjustment. The network adjustment is used to estimate the proper variance of the GPS observations. This involves removal of observation errors using statistical procedures, estimation of variance components, re-scaling and reprocessing of the GPS sessions. The best results are obtained using L1 and L2 phase data with an ionospheric constraint determined from the network adjustment, and using distinct weighting factors for the horizontal and vertical components.

The precision of the heights after the network adjustment is on the average 6 mm for individual campaigns. This has been confirmed by independent levellings.

Keywords. GPS, ionosphere, subsidence monitoring, variance component estimation.

1 Introduction

In 1959 natural gas was discovered in the province of Groningen in the Northeast of the Netherlands. The Groningen gasfield, which stretches over an area of 900 square kilometers with initial gas reserves of some 2900 billion cubic meters, is one of the largest and most compact gasfields in the world. The Nederlandse Aardolie Maatschappij (NAM) took the field in production in 1964. This resulted in reservoir compaction and consequently surface subsidence. To date, a maximum of 23 centimeters of subsidence has been measured since the start of the production, see Figure 1. Prognosis indicates that at the end of the field's life, in 2050, subsidence will amount to 33-43 cm maximum.

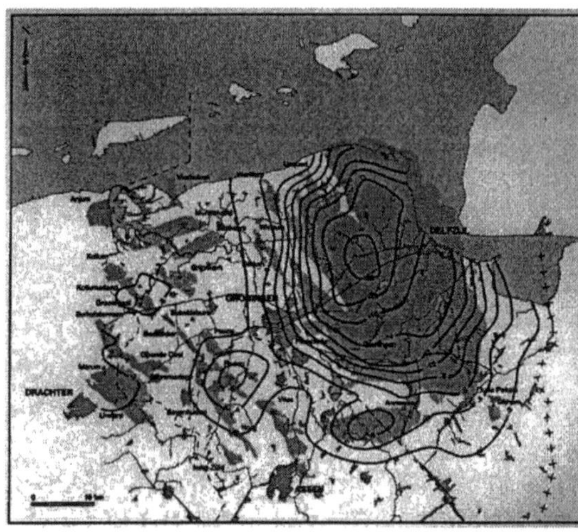

Fig. 1: Groningen gas field with measured subsidence (1995)

Since large parts of the Netherlands, including Groningen, are situated below sea level surface, subsidence has serious consequences for the water management systems in this area. The bowl shaped subsidence will affect the drainage of the area, hampering agricultural or ecological functions, and making additional investments in sluices and pumping stations necessary. Furthermore, the safety of embankments and the headway at bridges is reduced. The NAM, by agreement with the province of Groningen, has to compensate for the costs of the water management works that are a result from surface subsidence induced by gas production. Therefore, since the start of the production, subsidence is monitored with great care.

From the beginning levelling networks have been measured to monitor subsidence. In time the area of subsidence has grown wider, and newer, adjacent fields, were taken into production. The area of subsidence now extends beyond the shore. As a

International Association of Geodesy Symposia, Vol. 121
Schwarz (ed.), Geodesy Beyond 2000 – The Challenges of the First Decade

consequence the size of the networks changed accordingly. Nowadays, a combination of levelling and GPS is used. GPS is not only faster and more cost effective than levelling, it can also bridge more easily large distances and thus make up for lost tie points. In this paper we will report about the strategies we have used to monitor subsidence by GPS.

2 Subsidence monitoring in Groningen

The first levelling survey took place in 1964. At first these surveys were repeated every three years. Since 1980, a coarse grid of about 750 km of levelling lines is surveyed on a yearly basis, and every 6 years, a more extensive network of 1600 km of levelling lines is measured. After some initial experimenting, in 1994 the first supplementary GPS campaign was organized. The campaign was repeated in 1995, 1996 and 1997.

The networks are connected to a number of underground benchmarks. The underground benchmarks consist of a capped steel rod founded on Pleistocene sand layers 20-40 meter below the surface. Initially, most of these benchmarks were outside the area of subsidence, and it was rather straightforward to determine land subsidence by comparing heights from the individual levelling networks once they were connected to stable underground benchmarks. However, in time the area of subsidence has grown, making it necessary to reevaluate the stability of the underground benchmarks.

In December 1990, the Delft University of Technology, in co-operation with the NAM and Survey Department of Rijkswaterstaat (MD), started a project with the following objectives:

- develop procedures to estimate a kinematic deformation model from repeated levelling data, to control the quality of the data and to analyse the stability of the benchmarks,
- implement the procedure in software,
- test the procedure on levelling data of over 20 epochs of the Groningen gas field.

The procedure consisted of three steps:

1. analysis of single epochs (free network)
2. evaluation of benchmark stability
3. multi-epoch analysis

In every step first a consistent mathematical model must be established. This is achieved by hypothesis testing for the functional part and variance component estimation for the stochastical part. Hypothesis testing is used for the detection of outliers and to identify possible extensions of the functional model, such as the polynomial order and inclusion of breakpoints in the deformation model. The deformation model was at first limited to functions of time for the points in the network, consisting of a stable part (no subsidence), one or more breakpoints, and a polynomial part of varying degree. The degree of the polynomial part and the time of the breakpoint are estimated in the multi-epoch analysis. Statistical tests are used to identify points that do not follow this model. Also, spatial deformation models using a 3D polynomial, and more recently a bowl shaped function, were investigated. The results of this project have been published by De Heus et al. (1994a, 1994b, 1995) and Verhoef and De Heus (1995).

3 Subsidence monitoring with GPS

3.1 Levelling and GPS

Although, repeated levelling campaigns have proofed to be very adequate for subsidence monitoring, the NAM became in 1992 very interested in the application of GPS for subsidence monitoring. Subsidence monitoring in Groningen could benefit from GPS in a number of ways:

- it is a fast and relatively inexpensive method when compared to levelling,
- it can bridge long distances easily and can be used over bodies of water,
- it would be a second, independent method, making it possible to assess systematic effects in the levelling

As the area of subsidence extends beyond the shore there are few stable underground benchmarks along the northern edge of the field. In addition, new fields have been taken into production, making existing stable underground markers unusable in the future. GPS can be used to provide ties to new reference points in non-subsiding areas around the area of subsidence.

The main concern was whether GPS could provide the necessary sub-centimeter accuracy for the deformation problem at hand, especially because GPS derived heights are known to be less accurate than the horizontal components. Another problem is that GPS produces ellipsoidal heights, whereas a levelling survey yields orthometric or normal heights. However, if one is only interested in deformation, it is possible to combine height changes derived from GPS with levelling. In other words, relative geoid separation values between the network points are treated as nuisance parameters in a multi-epoch solution. Verhoef and De Heus (1995) have worked out the procedural and computational aspects of this approach. Theoretical studies by Strang van Hees (1978) have shown that the geoid separation is changed by the exploitation of the field, with a maximum of 9 mm at the center of the field, in 2035, at the end of the exploitation period. At present, for the 4 years of GPS campaigns, the effect is very small and disappears in the observation noise.

Another problem of GPS is that a different type of benchmarks had to be used. The levelling benchmarks, mostly bronze bolts cemented into walls of buildings, could not be used for GPS directly. Therefore separate GPS markers were designed, consisting of a 25x25x100 centimeter concrete pillar, fully dug into the ground. The pillars are nearby a levelling marker, not only to allow checks on the possible settlement of the GPS marker, but also to connect to the levelling network.

3.2 Design of the GPS network

As has been said the main concern was the accuracy of the GPS network. The NAM has taken great care in the design of the network. Using a network design tool, the precision and reliability of various network configurations were analysed. The design criteria for the networks, as discussed by Krijnen and De Heus (1995), were

- maximum distance between stations of 15 km
- each station has to be occupied at least twice by different receivers
- baselines along the perimeter of the network have to be measured at least twice
- adequate siting in terms of obstructions, multipath, stability, accessibility and safety

From a logistics analysis it was concluded that the optimum number of receivers for simultaneous measurements would be five. The length of the sessions was one hour.

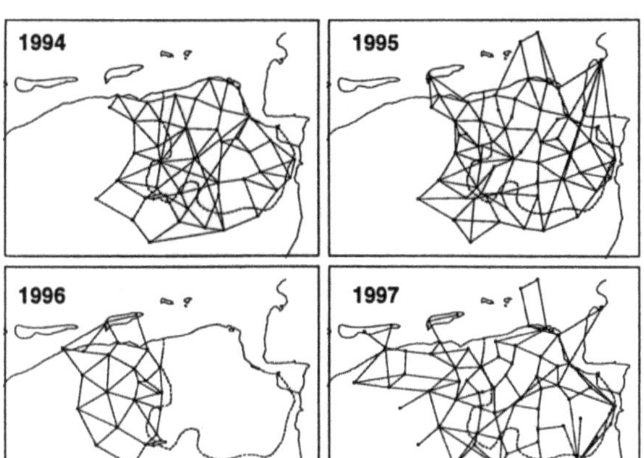

Fig. 2: The GPS networks.

GPS campaigns were organized in May/June 1994, May 1995, June 1996 and July/August 1997. All networks have been measured using five Trimble SSE, and later SSI, receivers per session. The networks are shown in Figure 2. The baselines shown in the figure are baselines that have been used in the preprocessing. In the actual adjustment the data is processed session by session. Over the years the size of the networks has grown, except for the 1996 campaign, which is much smaller. Because of the large travelling time needed to reach the islands in the North several sessions have been observed with fewer than five receivers. In table 1 the number of observed points and sessions are given. Stations have been occupied several times, more than strictly needed. The 2nd column on the right gives the total number of occupancies. The number of redundant occupancies is given in the last column of table 1.

Table 1: Characteristics of the GPS networks.

Year	points	Sessions	station visits	redundancy
1994	44	33	165	89
1995	62	56	234	117
1996	26	22	99	52
1997	66	47	195	83

For the GPS networks in Groningen an approach has been chosen involving a large number of short baselines. The main reason for this approach is that subsidence must be monitored over the whole of the area, involving many points. The length of the baselines is short enough to make use of the fact that the atmosphere delays above the stations are related, ambiguities are easy to find and observation times permit several occupancies during normal working hours.

Alternative approaches to monitor the height exist. For instance, in the European Vertical Network (EUVN) campaign, stations separated by several hundreds of kilometers were occupied continuously for up to 8 days. However, this approach is not very practical if a small and dense network is needed. Another alternative would be to use permanent stations. This approach would not work for the same reasons, although it is used in coastal areas in the Netherlands to measure the subsidence of platforms, see Flouzat et al. (1995).

4 GPS processing strategy

The Delft University of Technology processed the GPS data of the Groningen campaigns with the aim to fully assess the accuracy of the data, to establish the most adequate processing method and to compute the final solution, see Beckers et al. (1995, 1996, 1997, 1998). The GPS data was processed session by session using the Bernese GPS software. The sessions were then combined in a network adjustment using the SCAN-3 software.

4.1 Bernese software

The GPS data was first processed using the Bernese software, see Rothacher et al. (1996). The main characteristics of the processing were:

- sessions are processed independently
- approximate values for the receiver clocks are obtained from a pseudo range solution,
- a-priori ionosphere model (polynomial function of the geographic latitude and the hour angle of the Sun) is estimated for every session,
- cycle slip detection using triple differences,
- free network solution for every session, using precise IGS orbits, with one of the coordinates fixed to a-priori values in the reference frame of the orbits, using different scenarios,
- the final a-priori coordinates for the fixed stations are computed with SCAN-3 using an earlier iteration with the Bernese software,
- the model of Saastamoinen is used in combination with a standard atmosphere to correct for tropospheric delays,
- double difference ambiguities are fixed (in most cases, all of the ambiguities could be fixed).

The processing with the Bernese software results for every session in a set of coordinates with their full covariance matrix.

A standard troposphere model is used to correct for tropospheric delays. No zenith delays are estimated, since this was not considered useful over 15 km baselines in view of the duration of the sessions (one hour). Further experimentation will be needed to verify this assumption.

Several scenarios for combining the observations have been tested:

- L1&L2 with a-priori ionosphere model
- L3 (ionosphere free solution)
- L1&L2 with differential ionosphere parameters (DIP) and a-priori ionosphere model

Furthermore, the effect of elevation cut-off has been evaluated. Before we will discuss the results, we have first to set up a method to compare the different scenarios. This is done, amongst other things, using the SCAN-3 network adjustment software.

4.2 Network adjustment with SCAN-3

The results of the Bernese processing of sessions are subsequently subjected to a free network adjustment using the SCAN-3 software. The networks were designed to contain a considerable amount of redundancy with the objective to analyse precision and to eliminate poor data. The input for SCAN-3 is the set of coordinates for each session with their covariance matrix. Because each session was arbitrarily fixed to one station in the Bernese step, three translation parameters per session are added to the functional model in SCAN-3. Rotation and scale parameters can be neglected, because the reference system provided by the precise orbits is so well defined that the rotation and scale are actually the same for every session. Only a free network adjustment is performed using SCAN-3. The resulting network is tied to stable underground benchmarks in the so-called multi-epoch analysis, after integration with the levelling data.

A combination of hypothesis tests and variance component estimation is used in an iterative way to find the model that best fits the data. The procedure is very similar to the procedure used on levelling data, which is described in De Heus et al. (1994). In each step the most likely adaptation is indicated by the largest test ratio, the test quantity divided by its critical value, when the critical values are linked according to the B-method of testing. For the theoretical background see Verhoef and De Heus (1995, 1996). The following tests are executed:

a. conventional alternative hypotheses: one (coordinate) observation is in error
b. three dimensional tests of a coordinate set of one point: one point occupancy is in error
c. two-dimensional tests of the horizontal components of a point: centering is in error
d. one-dimensional test of the vertical component of the points: antenna height is in error
e. multi-dimensional tests of an entire observation session,
f. overall model test

The conventional alternative hypotheses are less useful in this context. It does not make much sense to remove just one coordinate component of a point. The three dimensional tests b) are very useful for spotting problems with a particular station in an observation session. The test c) and d) provide the possibility to trace specific errors such as centering and antenna height errors. The process of adaptation is repeated until the global test f) has the largest test-ratio.

4.3 Variance component estimation

Once the overall model test has the largest test ratio the covariance matrices can be scaled by the a posteriori variance factor (the overall model test quantity). However, to obtain more meaningful estimates of precision the covariance matrices are scaled by different factors for the horizontal and vertical components. The scaling factors are determined by comparing the histograms of the one- and two dimensional test quantities with the theoretical density function of the test quantities. After the covariance matrices have been re-scaled the process of finding the most likely adaptations must be iterated. Most of the time, re-scaling of individual components of the covariance matrix resulted in different adaptations at the end of the process. Application of this procedure to the GPS campaigns of 1994-1997 resulted in a very significant overall improvement of up to 30% of the standard deviation of the height component. The scaling factors for the final solution (DIP;2.5, 20° cutoff angle) are given in table 2.

Table 2: Scale factors to the covariance matrix.

Year	one scale	scale height & planimetry	
1994	9.9	11.1	9.1
1995	12.9	8.8	15.1
1996	12.2	7.4	13.9
1997	n/a	8.5	9.7

As can be observed from table 2 the precision of the height is generally improved when horizontal and vertical components are scaled differently. Usually, with GPS it is the other way round, but remember the precision in height is, without independent scaling, a factor three worse than the horizontal components. So, even with this improvement the heights are still not as good than the horizontal components. It is not yet understood why the scale factor for the height is smaller, but it is possible that this is related to the fact that all observations have been given equal weights, regardless of their elevation or signal to noise ratio.

5 Optimization of the GPS processing

Using the procedure of the previous section, it is now possible to compare different GPS processing scenarios. Although the Bernese software gives some elementary indications of precision, such as the estimated variance factor (sigma of single difference observation) and the covariance matrix scaled by the estimated variance factor, these cannot be used for comparison purposes because the various scenarios have different degrees of freedom. It is only after the scaling factors, estimated by SCAN-3, have been applied that different covariance matrices can be compared.

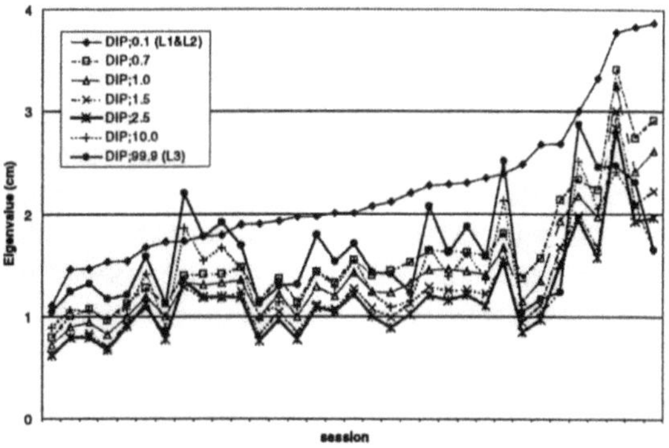

Fig. 3: Largest eigenvalues for GPS sessions, after scaling with SCAN-3, using various DIP parameter settings.

Before the start of the processing it was thought, that for a network with 15 km baselines, a full ionosphere free solution (L3) would not be necessary. However, we found that the relative precision of the heights after scaling was much better for the L3 solution than for the L1&L2 solution (Figure 3), despite the fact that the formal standard deviation for the heights in the L3 solution is a factor 4.2 larger. Similarly it was found that a 20 degrees elevation cutoff angle gave slightly better results than 15 degrees, see De Heus (1998).

Although the L3 solution did give better results than the L1&L2 solution, the feeling remained that it was a rather rigorous approach to deal with the ionospheric delays. Something in between was needed: differential ionosphere parameters (DIP) were the answer. With differential ionosphere parameters the mathematical model of single difference observations is extended with an additional parameters in each epoch to model remaining ionospheric effects. The parameter is assumed to be zero, but with a standard deviation σ_{DIP}. With $\sigma_{DIP}=0$ the original L1&L2 solution is obtained, with $\sigma_{DIP}=\infty$ the original L3 solution is simulated. Several values for σ_{DIP} were evaluated. In Figure 3 the largest eigenvalue of the covariance matrix of the coordinates per session, after scaling with the SCAN-3 results, is shown for the 1994 campaign. The value $\sigma_{DIP}=2.5\sigma_o$, with σ_o the a-priori standard deviation of the observations, gave the best results. Also, clearly visible, is that the L3 solution is better than the combined L1 and L2 solution. These results are confirmed by the other campaigns.

6 Concluding remarks

The relative precision in height is shown in Figure 4, as function of the distance between the points, for the four GPS networks and second order levelling. The relative precision of the GPS heights is well below the one-centimeter level. This is the internal precision of the networks, and this does not automatically guarantee a good external precision. When GPS heights (after a correction for geoid separations) are compared with levelling heights systematic effects show up. These effects disappear when the networks are connected to stable points, and the remaining differences are acceptable in relation to the internal precision of the networks.

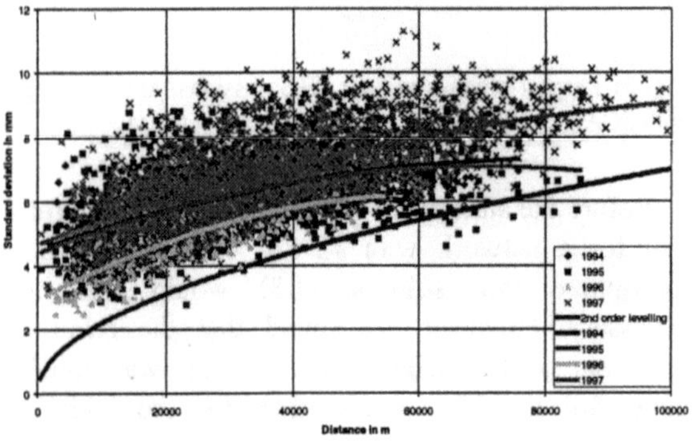

Fig. 4: Relative precision of the height as function of the distance between points for GPS and levelling

It is for our application more interesting to compare subsidence from levelling and GPS. After the networks were connected to common stable points, the difference in subsidence, over a period of 3 years, between GPS and levelling was computed. The RMS difference for the DIP;2.5 solution is about 5 mm/year, and for L3 some 8 mm. These results confirm that the estimates of internal precision are indeed correct, as well as the better results with the DIP;2.5 solution compared to L3 and L1&L2.

It must be realized that the precision of the GPS heights, despite the recent progress, is still less than the precision of a conventional second order levelling survey. Therefore, GPS height measurements will rather supplement than replace levelling for monitoring subsidence in Groningen.

References

Beckers, G.W.J., H.M. de Heus, H. van der Marel, M.H.F. Martens and H.M.E. Verhoef (1995). GPS-onderzoek bodemdalingsmetingen Groningen, In: *Publications of the Delft Geodetic Computing Centre,* No. 10, Delft (in Dutch).

Beckers, G.W.J., H.M. de Heus, H. van der Marel, M.H.F. Martens and H.M.E. Verhoef (1996). Gecombineerde GPS- en waterpasdata t.b.v. de bodemdalingsmetingen in Groningen, In: *Publications of the Delft Geodetic Computing Centre,* No. 15, Delft (in Dutch).

Beckers, G.W.J., H.M. de Heus, H. van der Marel and M.H.F. Martens (1997). Verwerking NAM GPS-campagne 1996, *Technical report department of Mathematical Geodesy and Positioning,* Delft, 1997 (in Dutch).

Beckers, G.W.J., H.M. de Heus, H. van der Marel and M.H.F. Martens (1998). Onderzoek bodemdalings-metingen Groningen '95-'96-'97, In: *Publications of the Delft Geodetic Computing Centre,* No. 18, Delft (in Dutch).

Flouzat, M., D. Fourmaintraux, R. Camphuysen (1995). Advanced continuous monitoring of subsidence above gas fields using spatial geodetic measurements. In: *Proc. Fifth Int. Symp. on Land Subsidence, The Hague,* pp. 269-279.

Heus, H.M. de, P. Joosten, M.H.F. Martens and H.M.E. Verhoef (1994a). Geodetische Deformatie Analyse. In: *Publications of the Delft Geodetic Computing Centre,* No. 5, Delft (in Dutch).

Heus, H.M. de, P. Joosten, M.H.F. Martens and H.M.E. Verhoef (1994b). Stability analysis as a part of the strategy for the analysis of the Groningen gas field levellings. In: *Proc. of the Perelmutter Workshop on Dynamic Deformation Models,* Haifa, Israel, pp. 259-272.

Heus, H.M. de, P. Joosten, M.H.F. Martens and H.M.E. Verhoef (1995). Strategy for the analysis of the Groningen gasfield levellings. In: *Proc. Fifth Int. Symp. on Land Subsidence, The Hague,* pp. 301-312.

Heus, H.M. de (1998). Subsidence monitoring with GPS in the Netherlands. In: *Proc. Int. Symposium on Current Crustal Movement and Hazard Reduction,* IUGG IAG, Wuhan, China, Seismological Press, pp. 184-195.

Krijnen, H. and H.M. de Heus (1995). Application of GPS with sub-centimeter accuracy for subsidence monitoring. In: *Proc. Fifth Int. Symp. on Land Subsidence, The Hague,* pp. 333-344.

Strang van Hees, G.L. (1978). Hoogte- en zwaartekrachts-veranderingen in het gaswinningsgebied Oost-Groningen. Internal report (in dutch).

Rothacher, M. et al. (1996). Bernese GPS Software Version 4.0. *Astronomical Institute University of Berne.*

Verhoef, H.M.E., and H.M. de Heus (1995). On the estimation of polynomial breakpoints in the subsidence of the Groningen gasfield. In: *Survey Review,* Vol.33, no. 255 (January 1995), pp. 17-30.

Verhoef, H.M.E., and H.M. de Heus (1996). Subsidence analysis from integrated GPS and precise levelling data. In: *Proc. EGS XXI General Assembly, Session G7, The Hague,* 7p..

† Henk de Heus died unexpectedly on January 21st 1999 at the age of only 50. He has been the driving force behind the project right from the start in 1990, and was responsible for many of the results reported in this paper.

Stochastic modelling of the ionosphere for fast GPS ambiguity resolution

Dennis Odijk
Department of Mathematical Geodesy and Positioning,
Delft University of Technology, P.O. Box 5030, 2600 AA Delft, The Netherlands
e-mail: D.Odijk@geo.tudelft.nl

Abstract. Integer carrier phase ambiguity resolution is the key to precise GPS positioning. For baselines over 10 km the errors due to the ionosphere limit the ability to resolve the integer ambiguities within short observation time spans. Of course, one can estimate the ionospheric delays, but the additional unknowns in the model limit the use of very short time spans. On the other hand, very short time spans can be successfully used by not estimating the delays at all, but only if the ionospheric delays in the data are within certain bounds.

In this contribution the effect of a priori weighting of the ionosphere is investigated on integer least-squares estimation with the LAMBDA method. Stochastic modelling of the ionosphere can mathematically be regarded as the addition of pseudo-observations to the model of GPS observation equations, together with a variance-covariance matrix (vc-matrix) in which the uncertainty of the ionospheric delays is accounted for. Note, on the one hand, if "infinite" weights are used for these ionospheric pseudo-observables, the model degenerates to the one in which no ionospheric delays are estimated at all. On the other hand, if these observables are not weighted at all, the model in which the ionospheric delays are estimated is obtained. Stochastic modelling in a way interpolates between these two "extreme" models. For a medium-length baseline it can be shown that with this technique it is possible to determine the correct integer ambiguities within a shorter time span than with the model in which the ionospheric delays are considered as completely unknown parameters.

Keywords. GPS integer ambiguity resolution, ionosphere, stochastic modelling.

1 Introduction

For precise (mm-cm) GPS positioning which requires very short time spans, integer phase ambiguity resolution is a prerequisite. Reliable resolution of these (double-difference) ambiguities for long baselines is mainly hampered by errors due to the ionospheric delay in the GPS data, as the model in which unknown parameters for the DD ionospheric delays are estimated (the ionosphere-float model) becomes too weak. Another possibility to handle these ionospheric biases is to assume them completely known and priori correct the data for them (the ionosphere-fixed model), but this requires precise corrections.

In this paper the effect of stochastic modelling of the ionosphere on integer ambiguity resolution is investigated: by adding pseudo-observations to the GPS model in which the ionospheric uncertainty is modelled with a vc-matrix. The advantage of this so-called ionosphere-weighted model is that the benefits of the ionosphere-fixed model (relatively short observation time spans) are realized, as well as those of the ionosphere-float model (relatively long baseline lengths). In section 2 this concept is explained.

In the third section the generation of these ionospheric pseudo-observations (sample values plus stochastic model) is discussed. Several possibilities have been reported in the literature and they are briefly reviewed.

In the fourth section a 100 km baseline in the Netherlands is investigated, and the manner in which the ionospheric weights should be chosen in order to minimize the observation time to obtain the correct integer ambiguities is discussed.

2 Ionosphere-weighted ambiguity precision

In this section the influence of ionosphere modelling on ambiguity resolution is investigated. Three types of models are considered: the ionosphere-float, -fixed, and -weighted model.

2.1 The ionosphere-float model

International Association of Geodesy Symposia, Vol. 121
Schwarz (ed.), Geodesy Beyond 2000 – The Challenges of the First Decade

In the case that unknown parameters for the ionospheric delays are modelled, the linearized functional model for dual-frequency phase and code (or pseudo-range) observation equations for a single GPS baseline involving m satellites, which are commonly tracked during k observation epochs (in double-difference or DD mode), is

$$E\left\{\begin{pmatrix}\varphi\\ p\end{pmatrix}\right\} = \left(\begin{pmatrix}e_2\\ e_2\end{pmatrix}\otimes G \quad \begin{pmatrix}\Lambda\\ 0\end{pmatrix}\otimes F \quad \begin{pmatrix}-\mu\\ \mu\end{pmatrix}\otimes J\right)\begin{pmatrix}b\\ a\\ I\end{pmatrix} \qquad (1)$$

where $E\{\,.\,\}$ denotes the expectation operator, and $\otimes$ the matrix Kronecker-product (Rao, 1973).

Model (1) contains the following types of observations: φ, the $2(m\text{-}1)k$ observed-minus-computed DD phase observations on L1 and L2, and p the $2(m\text{-}1)k$ observed-minus-computed DD code observations on L1 and L2. Unknown parameters are: (1) b, the unknown baseline parameters, with G their coefficient matrix; (2) a, the $2(m\text{-}1)$ unknown but time-constant DD phase ambiguities on L1 and L2, in units of cycles, where their coefficient matrix is formed by a 2x2 diagonal matrix $\Lambda = diag(\lambda_1, \lambda_2)$, containing the known wavelengths λ_1 and λ_2, plus the matrix product $F = e_k \otimes I_{m-1}$; (3) I, the $(m\text{-}1)k$ unknown DD ionospheric delays on L1, with $\mu = (\mu_1, \mu_2)^T$, and, in order to estimate delays on L1, $\mu_j = \lambda_j^2/\lambda_1^2$ (j=1,2), and $J = I_k \otimes I_{m-1}$.

Furtermore, note that e_j is a jx1 vector consisting of ones and I_j a jxj unit matrix.

This model, in which the DD ionospheric delays are explicitly solved for, is called the *ionosphere-float* model (Teunissen, 1997).

The stochastic model for these DD phase and code observables is

$$D\left\{\begin{pmatrix}\varphi\\ p\end{pmatrix}\right\} = \begin{pmatrix}C_\varphi & 0\\ 0 & C_p\end{pmatrix}\otimes T \qquad (2)$$

where $D\{\,.\,\}$ denotes the dispersion operator, and $T = I_k \otimes 2D^TD$.

In this model C_φ and C_p denote the 2x2 undifferenced vc-matrices for phase and code respectively. The matrix product D^TD represents the correlations due to the double differencing of the observables (D is the differencing operator).

In order to estimate the parameters of interest (i.e. the baseline coordinates) with a high precision, it is necessary to fix the *correct* integer ambiguities. The success of this integer ambiguity estimation (the *validation*) heavily depends on the precision of the *float* least-squares DD ambiguities from model (1), which is described by its vc-matrix $Q_{\hat{a}}$. Since - especially for short time spans- the covariances between the ambiguities can be very large, use is made of the so-called Ambiguity Dilution Of Precision (ADOP) measure, to gain an insight into whether a successful integer estimation can be expected. This ADOP is defined as $|Q_{\hat{a}}|^{1/(2n)}$ (see Teunissen and Odijk, 1997), where n is the order of the ambiguity vc-matrix (n=2(m-1) for dual-frequency observations).

As an example, consider one observation epoch and 6 satellites, an ADOP of 0.94 cycle is obtained with the ionosphere-float model. From experience it has been determined however that the ADOP value should be smaller than about 0.2 cycle to obtain the correct integers with sufficient probability (see Teunissen et al., 1997a), hence this ADOP is way too large. Of course, the ADOP will decrease with the change of receiver-satellite geometry in time.

2.2 The ionosphere-fixed model

A much faster ambiguity resolution than with the previous model is expected if a priori corrections for the DD ionospheric delays were available and unknowns for the ionosphere did not need to be estimated. The model of dual-frequency phase and code becomes:

$$E\left\{\begin{pmatrix}\varphi+\mu\otimes I\\ p-\mu\otimes I\end{pmatrix}\right\} = \left(\begin{pmatrix}e_2\\ e\end{pmatrix}\otimes G \quad \begin{pmatrix}\Lambda\\ 0\end{pmatrix}\otimes F\right)\begin{pmatrix}b\\ a\end{pmatrix} \qquad (3)$$

with vc-matrix:

$$D\left\{\begin{pmatrix}\varphi+\mu\otimes I\\ p-\mu\otimes I\end{pmatrix}\right\} = \begin{pmatrix}C_\varphi & 0\\ 0 & C_p\end{pmatrix}\otimes T \qquad (4)$$

This model is the so-called *ionosphere-fixed* model. Note that it is possible to apply 'zero'-corrections for the DD ionospheric delays (I=0). This is what is often done for sufficiently short baselines, for which the relative delays are so small that they do not influence the ambiguity resolution.

For the example in the previous section the ADOP when the ionosphere-fixed model is used can be computed, giving an *ADOP* of 0.094 cycle, which is a factor of 10 smaller than for the ionosphere-float model. It is obvious that with this model successful resolution may be expected, even with only one observation epoch, as has already been extensively reported in recent literature (for example, Hwang, 1991, and Teunissen et al., 1997b).

2.3 The ionosphere-weighted model

If it is assumed that the ionospheric corrections applied in the ionosphere-fixed model are *stochastic* quantities, rather than deterministic, the same functional model (3) as the ionosphere-fixed model is valid, but with a vc-matrix in which the uncertainty of the ionospheric corrections is adequately propagated:

$$D\{\begin{pmatrix}\varphi+\mu\otimes I\\ p-\mu\otimes I\end{pmatrix}\} = \begin{pmatrix}(C_\varphi+\sigma_I^2\mu\mu^T) & -\sigma_I^2\mu\mu^T\\ -\sigma_I^2\mu\mu^T & (C_p+\sigma_I^2\mu\mu^T)\end{pmatrix}\otimes T \qquad (5)$$

with σ_I the standard deviation of the undifferenced ionospheric corrections.

In the model above the vc-matrix is no longer block-diagonal. However, note that the mathematical model, which is defined by (3) and (5), is completely equivalent with

$$E\{\begin{pmatrix}\varphi\\ p\\ I\end{pmatrix}\} = \left(\begin{pmatrix}e_2\\ e_2\\ 0\end{pmatrix}\otimes G \;\; \begin{pmatrix}\Lambda\\ 0\\ 0\end{pmatrix}\otimes F \;\; \begin{pmatrix}-\mu\\ \mu\\ 1\end{pmatrix}\otimes J\right)\begin{pmatrix}b\\ a\\ I\end{pmatrix} \qquad (6)$$

and a block-diagonal vc-matrix

$$D\{\begin{pmatrix}\varphi\\ p\\ I\end{pmatrix}\} = \begin{pmatrix}C_\varphi & 0 & 0\\ 0 & C_p & 0\\ 0 & 0 & \sigma_I^2\end{pmatrix}\otimes T \qquad (7)$$

This model, formed by (6) and (7), can be transformed to the model (3) and (5) without loss of information.

The stochastic ionospheric corrections can be regarded as *pseudo-observations,* which have an appropriate vc-matrix to model the ionospheric weights. This so-called *ionosphere-weighted model* was introduced by Bock et al. (1986).

The ionosphere-float and ionosphere-fixed models, as discussed in sections 2.1 and 2.2, are in fact the two 'extreme' versions of the ionosphere-weighted mode. On the one hand, the ionosphere-float model can be regarded as the model in which the a priori information does not contribute to the solution, i.e. $\sigma_I = \infty$. On the other hand, the ionosphere-fixed model is the model in which the a priori information is regarded as being deterministic, i.e. $\sigma_I = 0$.

How does the *ambiguity precision* behave as a function of the a priori ionospheric uncertainty? To answer this question, in figure 1 the ADOP of the

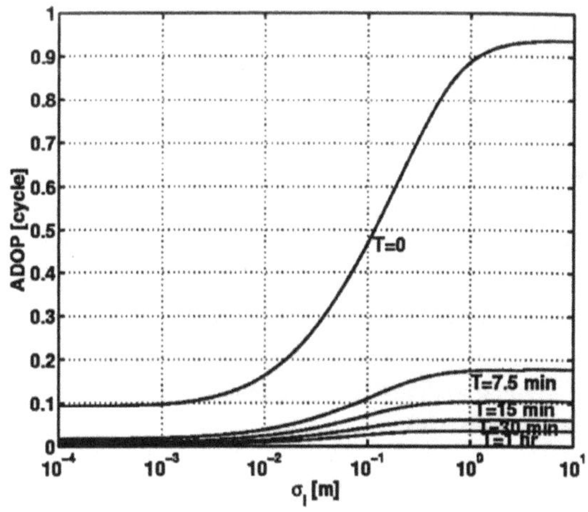

Fig. 1: ADOP as a function of the a priori ionospheric standard deviation for 6 satellites and dual-frequency phase and code data under different time spans (denoted with "T")

ionosphere-weighted model is plotted as a function of σ_I for 6 satellites continuously tracked. The a priori standard deviation of the undifferenced phase is assumed σ_φ = 3mm, and that of the undifferenced code as σ_p = 30cm. In addition, no correlation is assumed between both frequencies. The ADOPs are plotted for 5 observation time spans with different lengths (1 hour, 30 min., 15 min., 7.5 min., and instantaneous), which are all taken from a baseline in the Netherlands measured between 12:01:30-13:01:00 UTC, with a sampling-interval of 30 seconds.

In figure 1 it can be seen that the ADOP exhibits an S-curve behaviour: before $(\sigma_I \rightarrow 0)$ or after $(\sigma_I \rightarrow \infty)$ a certain value for the ionospheric standard deviation the ADOP is insensitive to changes in the value of σ_I. The longer the time span, the more insensitive the ambiguity precision becomes against the a priori ionospheric precision. In the case of 1 hour the ADOP only changes a few hundredths of a cycle over the whole range of σ_I. This insensitivity of the ADOP if σ_I is close to infinity stems from the use of the *code data,* which strengthen the model: their influence increases when the precision of the ionospheric observables decreases. If the code is not used, the ADOPs will grow to infinity if the ionospheric standard deviation increases towards infinity. With the code data included, it is remarkable how strong, concerning ambiguity validation, the ionosphere-float model already is after 15 minutes.

Like the ambiguity precision, also the *baseline precision* interpolates for the ionosphere-weighted model between the ionosphere-fixed and -float models. It can be proved that if $\sigma_I > 0.10$ m the baseline precision after ambiguity fixing is about the same as with the ionosphere-float model, and if $\sigma_I < 0.001$ m approximately the same as with the ionosphere-fixed model (see Teunissen, 1998).

When the ionosphere is weighted, the choice of the a priori ionospheric standard deviation should correspond to the expected statistical behaviour of the noise of the ionospheric delays.

3 Choice of the ionospheric pseudo-observations

The Bernese software (version 4.0) is able to process ionospheric pseudo-observables as explained in the previous section, see Rothacher and Mervart (1996), Schaer (1994) and De Heus et al. (1999). However, for this paper the GPSveQ software of the Delft University of Technology is used (De Jonge, 1998), as it uses the LAMBDA method for fast integer ambiguity estimation.

The GPSveQ software has been extended with ionospheric pseudo-observations according to the model in (6) and (7), so the ionospheric uncertainty in the data can be accounted for by tuning σ_I. These ionospheric pseudo-observations are assumed to be uncorrelated in time: it is not possible yet to implement correlations due to the temporal behavior of the ionosphere. Furthermore, with GPSveQ it is not possible yet to implement different ionospheric variances for different baselines if a network is processed. It is known that the spatial correlation of the ionosphere depends on the length of the baseline, and one might want to implement the ionospheric variances in a way as, for example, is suggested in Schaffrin and Bock (1988). Another feature that has not been studied yet is an elevation-dependent weighting of the ionospheric observables, though this elevation dependence seems to be present.

With respect to the *sample values* of the ionospheric pseudo-observables, possibilities are to use values obtained from certain ionosphere models or from (global) ionosphere maps. In this paper only *zero sample* values are used, as many existing models hardly influence ambiguity resolution for short time spans.

In the next section some results are presented for ambiguity resolution performance of the ionosphere-weighted approach with respect to the ionosphere-float approach. As an example, data from a single baseline with a length of approximately 100km are used.

4 Test results for a medium-length baseline

An 1 hour data set for a 100km baseline is used as a reference for the subsequent computations. In a first step, the ambiguities are estimated as float values with the ionosphere-float model (i.e. no a priori ionospheric assumptions), and due to the 1 hour time span the ADOP was so small that in a second step the integers could be estimated and be assigned as the correct ones with a sufficient amount of probability.

The data set used is summarized as follows:

- 2 permanent stations DELF (reference) and KOSA in the Netherlands (Turbo-Rogues);
- dual-band observations under A/S: L1 L2 C1 P2;
- observation time span: 12:01:30-13:01:00 UTC at 1 January 1999 with sampling interval 30sec;
- 6 satellites continuously tracked by both receivers;
- no cycle-slips and outliers in the data;
- a priori tropospheric corrections with the Saastamoinen-model;
- precise IGS orbits used for satellite positions;
- ionospheric activity: absolute effects were maximally 9m on L1 according to CODE's Global Ionosphere Map.

A skyplot of the 6 satellites during the observation period is given in figure 2.

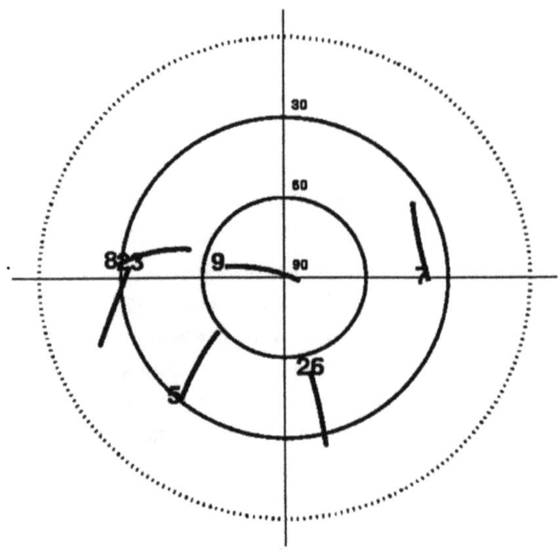

Fig. 2: Skyplot (azimuth vs elevation) for the station DELFT on January 1, 1999 12:01:30 - 13:01:30 UTC

In figure 3 a time-series for the DD ionospheric delays for the 1 hour period is given. These ionospheric delays are estimated from the ionosphere-float model with the integer ambiguities held fixed. As a consequence, their precision is at the mm-cm level. In the graph all time-series have been plotted with respect to the reference satellite PRN 5. For all satellites in the graph the time-series of the DD ionospheric delays are quite smooth when two consecutive epochs are considered, but over a

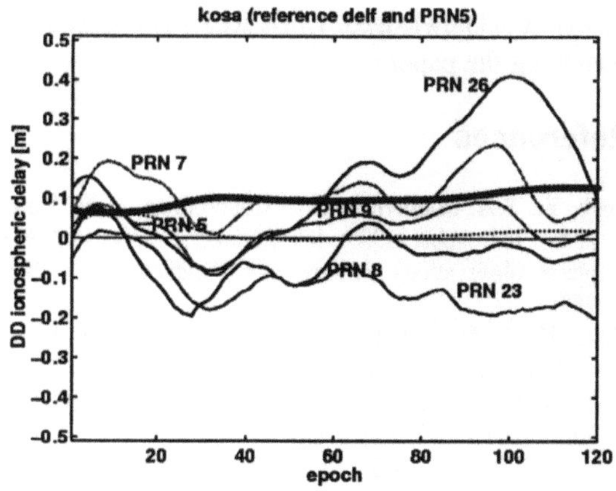

Fig. 3: Ambiguity-fixed DD ionospheric delay estimates for the station KOSA on Jan 1, 1999 12:01:30 - 13:01:30 UTC

time span of say 5 to 10 minutes they fluctuate. Such behaviour could be caused by irregular ionospheric effects such as 'travelling ionospheric disturbances' (TIDs, see Wanninger, 1995).

In fact, the time-series of the estimated delays can provide insight to the statistics of the ionosphere to use in the ionosphere-weighted model. So in figure 3 two other time-series are also plotted: the averaged cumulative DD ionospheric delay over all satellites evaluated at every epoch (the dotted curve), and its standard deviation (the bold curve). So for the complete time span of 1 hour the time-averaged mean DD ionospheric delay m_I is 0.03m and its standard deviation is 0.12m. This standard deviation is double-differenced; undifferenced it yields $s_I = 0.12/2 = 0.06\,m$.

For the baseline involved the values of σ_I for the ionosphere-weighted model are investigated so as to reduce the observation time span, while ensuring successful estimation of the integer ambiguities. Starting with the complete time span of 1 hour, and then systematically dividing the span into two halves and performing the processing, and so on.

In table 1 the results for the processing of the complete time span are given. Although with the ionosphere-float model the correct integers are obtained, if $0.06\,m \le \sigma_I < \infty$ also the ionosphere-weighted model yields these values (the baseline precision will only significantly improve with the ionosphere-weighted model if $\sigma_I < 0.10\,m$).

Table 1: Allowed range of σ_I which lead to correct ambiguity resolution for a time span of *1 hr*

Time span [hr:min:sec]	Empirical mean+stdev [m]	A priori stdev ionosphere [m]
12:01:30-13:01:00	m_I=0.03;s_I=0.06	$0.06 \le \sigma_I \le \infty$

Note that the empirical standard deviation for this time span, s_I, coincides very well with the tolerated lower bound σ_I for the a priori standard deviation of the ionosphere-weighted model. If $\sigma_I < 0.06\,m$ the wrong integers are obtained; obviously the real behaviour of the ionospheric delays in the data does not fit such a priori assumed behaviour.

This simple investigation of the relation between the a priori weights of the ionospheric delays and the used time span is also carried out for the two halves of the time span which are each 30 minutes. In table 2 the results are summarized.

Table 2: Allowed range of σ_I which lead to correct ambiguity resolution for a time span of *30 min*

Time span [hr:min:sec]	Empirical mean+stdev [m]	A priori stdev ionosphere [m]
12:01:30-12:31:00	m_I=-0.01;s_I=0.05	$0.09 \le \sigma_I \le \infty$
12:31:30-13:01:00	m_I=0.04;s_I=0.08	$0.13 \le \sigma_I \le \infty$

For a time span of half an hour it is still possible to obtain the correct integer ambiguities with the ionosphere-float model. Applying the ionosphere-weighted model with $0.13 \le \sigma_I < \infty$ ambiguity resolution is also successful.

Note in table 2 that the lowerbounds of σ_I do not coincide with the empirical values s_I. The empirical standard deviations are systematically smaller. A possible reason for this could be the negligence of time-correlation in the ionospheric pseudo-observations, but more research is needed to verify this. In a third step, the observation time span is divided into 15 minutes; see table 3 for the results. From table 3 it becomes clear that a time span of 15 mins is still not too short for successful

Table 3: Allowed range of σ_I which lead to correct ambiguity resolution for a time span of *15 min*

Time span [hr:min:sec]	Empirical mean+stdev [m]	A priori stdev ionosphere [m]
12:01:30-12:16:00	m_I=0.01;s_I=0.05	$0.15 \leq \sigma_I \leq \infty$
12:16:30-12:31:00	m_I=-0.03;s_I=0.04	$0.12 \leq \sigma_I \leq \infty$
12:31:30-12:46:00	m_I=0.04;s_I=0.06	$0.15 \leq \sigma_I \leq \infty$
12:46:30-13:01:00	m_I=0.05;s_I=0.09	$0.20 \leq \sigma_I \leq \infty$

ambiguity fixing with the ionosphere-float model. The lower bounds for σ_I to obtain the correct integers with the ionosphere-weighted model have increased somewhat, compared to table 2.

In a last step, the time span is divided into 8 parts of each 7.5 minutes, see table 4.

Table 4: Allowed range of σ_I which lead to correct ambiguity resolution for a time span of *7.5 min*

Time span [hr:min:sec]	Empirical mean+stdev [m]	A priori stdev ionosphere [m]
12:01:30-12:08:30	m_I=0.06;s_I=0.03	$0.17 \leq \sigma_I \leq \infty$
12:09:00-12:16:00	m_I=-0.04;s_I=0.05	$0.18 \leq \sigma_I \leq 2.4$
12:16:30-12:23:30	m_I=-0.05;s_I=0.03	$0.11 \leq \sigma_I \leq 1.1$
12:24:00-12:31:00	m_I=0.00;s_I=0.04	$0.20 \leq \sigma_I \leq 1.2$
12:31:30-12:38:30	m_I=0.04;s_I=0.05	$0.15 \leq \sigma_I \leq \infty$
12:39:00-12:46:00	m_I=0.04;s_I=0.07	$0.26 \leq \sigma_I \leq \infty$
12:46:30-12:53:30	m_I=0.07;s_I=0.10	$0.62 \leq \sigma_I \leq \infty$
12:54:00-13:01:00	m_I=0.02;s_I=0.07	$0.28 \leq \sigma_I \leq 2.7$

Now finally with these short observation time spans, the ionosphere-weighted model seems to be superior to the ionosphere-float model. With the last model, integer estimation was only successful for 50% of the cases, while with the ionosphere-weighted model still in all cases the correct integers could be resolved.

5 Concluding remarks

For GPS baselines in which the ionospheric delays may not be neglected it is necessary to estimate unknowns for these delays together with the other parameters, or to a priori correct the data. However, already very short time spans are possible if a priori ionospheric stochastic information is added to the model. By simply tuning a single ionospheric standard deviation, in all investigated cases it was possible to resolve the ambiguities for a 100km baseline within 7.5 minutes, without introducing an external ionosphere model or corrections. Trying to resolve these ambiguities with the model in which the ionospheric delays are considered as completely unknown resulted in a success rate of only 50%.

Acknowledgements. This contribution was performed under a contract with the Triangulation Department of the Dutch Cadastre. Furthermore, the help of Frank Kleijer and Peter Joosten of the Department of MGP with the preparation of the figures is acknowledged. Chris Rizos is thanked for reviewing the paper.

References

Bock, Y., S.A. Gourevitch, C.C. Counselman, R.W. King, and R.I. Abbot (1986). Interferometric analysis of GPS phase observations. *Manuscripta Geodaetica,* 11, 282-288.

Heus, H.M. de, M.H.F. Martens, and H. van der Marel (1999). *Re-weighting of GPS baselines for vertical deformation analysis,* Paper presented at IUGG99, Birmingham, UK, July 18-30.

Hwang, P.Y.C. (1991). Kinematic GPS for differential positioning: Resolving integer ambiguities on the fly. *Navigation: Journal of the Institute of Navigation,* Vol. 38, No. 1, 1-15.

Jonge, P.J. de (1998). A processing strategy for the application of the GPS in networks. *Netherlands Geodetic Commission, Publications on Geodesy,* no. 46, 225pp.

Rao, C.R. (1973). *Linear Statistical Inference and Its Applications.* 2nd edition, Wiley, New York.

Rothacher, M., and L. Mervart (eds.) (1996). *Bernese GPS Software, Version 4.0.* Astronomical Institute, University of Berne.

Schaer, S. (1994). *Stochastische Ionosphärenmodellierung beim "Rapid Static Positioning" mit GPS.* Astronomisches Institut, Universität Bern, 87pp.

Schaffrin, B., and Y. Bock (1988). A unified scheme for processing GPS dual-band phase observations. *Bulletin Géodésique,* 62, 142-160.

Teunissen, P.J.G. (1997). The geometry-free GPS ambiguity search space with a weighted ionosphere. *Journal of Geodesy,* 71, 370-383.

Teunissen, P.J.G., and D. Odijk (1997). Ambiguity Dilution Of Precision: definition, properties and application. *Proc. of ION GPS-97,* Kansas City, USA, September 16-19, 891-899.

Teunissen, P.J.G., P.J. de Jonge, D. Odijk, and C.C.J.M. Tiberius (1997a). Fast ambiguity resolution in network mode. *Proc. IAG Scientific Assembly,* Rio de Janeiro, Brasil, September 3-9, 313-318.

Teunissen, P.J.G., P.J. de Jonge, and C.C.J.M. Tiberius (1997b). Performance of the LAMBDA method for fast GPS ambiguity resolution. *Navigation: Journal of the Institute of Navigation,* Vol. 44, No. 3, 373-383.

Teunissen, P.J.G. (1998). The ionosphere-weighted GPS baseline precision in canonical form. *Journal of Geodesy,* 72, 107-117.

Wanninger, L. (1995). Improved ambiguity resolution by regional modelling of the ionosphere. *Proc. of ION GPS-95,* Palm Springs, USA, September 12-15, 55-62.

Mitigating multipath errors using semi-parametric models for high precision static positioning

M. Jia, M. Tsakiri, M. Stewart
School of Spatial Sciences,
Curtin University of Technology, GPO Box U 1987 Perth WA 6845, Australia

Abstract. Carrier phase multipath is currently the main error source for high precision and short distance Global Positioning System (GPS) applications, such as deformation monitoring of open pit walls and dams. Although important progress has been made to mitigate the impact of multipath errors on positioning results, more practicable methods need to be developed.

In this paper, the semi-parametric model and penalised least squares method have been developed for mitigation of multipath errors. In the semi-parametric model, multipath is described by a complicated, but smoothly varying, function with time. The functions, parameters, such as the coordinates of sites and ambiguities, and observation noise are decomposed using the penalised least squares method. The potential of the semi-parametric model to reduce multipath to the level of receiver noise is demonstrated on a static GPS baseline. The improvement of the precision and reliability of ambiguity resolution and positioning results using the semi-parametric model over the parametric model. (standard GPS data processing techniques) is also discussed.

Keywords. Global Positioning System (GPS) multipath, semi-parametric model, penalised least squares method.

1 Introduction

GPS has been used in a wide variety of high precision applications, such as deformation monitoring of open pit walls, dams and bridges. However, the influence of multipath on carrier phase measurements is the main limitation to achieving high accuracy positioning required by these applications. Firstly, multipath can cause the ambiguities not to be fixed. Secondly, multipath can lead to incorrect ambiguity resolution. Moreover, multipath can still degrade positioning accuracy even though correct ambiguities may be fixed.

Multipath is the corruption of the direct GPS signal by one or more signals reflected from the environment surrounding the antenna. There is also diffraction of the direct GPS signal due to obstructions between the GPS satellite and the antenna (Brunner *et al.*, 1999). Unlike other GPS errors (e.g., satellite and receiver clock errors), multipath cannot be eliminated or even reduced by double differencing. Especially for short occupation, multipath errors cannot be eliminated simply through averaging even in static applications (van Nee, 1991). Fig. 1 shows a plot of double difference residuals of satellite pair PRN 2-7, which exhibits significant multipath errors.

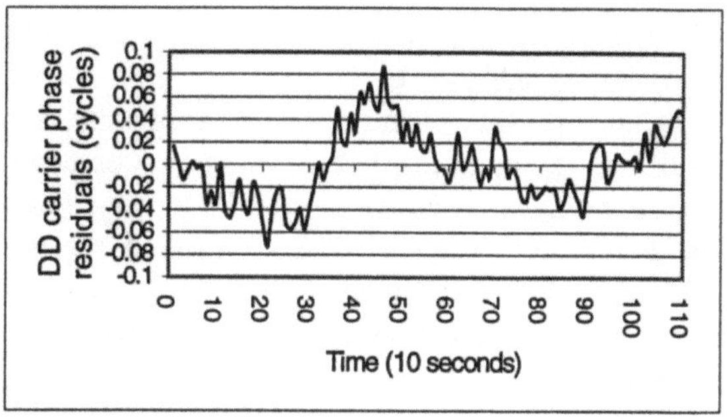

Fig. 1: Double difference residuals of satellite pair PRN 2-7 from 215.929 m baseline (data were collected on June 7, 1999 with two Ashtech Z12 receivers)

Significant progress on multipath mitigation has been achieved by a variety of techniques. Generally, these techniques can be classified as hardware and software techniques. Special antenna designs, like choke ring antenna and antenna with ground planes, and receiver architecture, such as narrow correlation architecture (Van Dierendonck *et al.*, 1992), are some examples amongst hardware techniques currently used to prevent multipath signals entering a receiver.

Mitigation using software techniques includes parametric approaches and digital filtering techniques that deal with the observations during the data processing stage (Genrich and Bock, 1992). Strategies to mitigate multipath errors using parametric models are that multipath delay is first expressed by some relative parameters, and then the parameters and multipath delay are estimated and subtracted from observations. Georgiadou and Kleusberg (1988) derived a parameterised mathematical model for carrier phase multipath errors based on the analysis of the strength of the reflected signal and the satellite-reflector-antenna geometry. To calculate multipath delay using this approach, dual frequency combinations are required. According to Hardwick and Liu (1996), multipath errors are functions of satellite elevation and azimuth. Based on this assumption, spherical harmonics were employed to model double difference residuals. This technique was also employed by Cohen (1992) in GPS-based attitude determination. A relatively stable antenna environment, however, is compulsory for the implementation of this technique. Axelrad *et al.* (1996), and Comp and Axelrad (1998) have developed a signal-to-noise (SNR) based carrier phase multipath mitigation technique. This method, however, requires knowledge of the antenna gain pattern. A stochastic model estimation technique has been developed by Brunner *et al.* (1999) based on the evaluation of SNR. Walker *et al.* (1996) and Hannah *et al.* (1998) have developed a parabolic equation technique to model the GPS signal propagation. This technique can be applied to data collected in complicated observation environments. However, *a priori* environmental knowledge (such as a Digital Terrain Model) is required for this technique. Ray and Cannon (1998) have developed a mitigation approach using multiple closely-spaced antennae.

This paper proposes a new carrier phase multipath mitigation technique, namely, a semi-parametric model and the penalised least squares method technique. The important facts, on which the semi-parametric model and penalised least squares is used to mitigate multipath, are: 1) without the presence of multipath, GPS carrier phase double difference observation errors should be random; 2) the parametric model and least squares method (standard GPS data processing techniques) cannot fit the data with the presence of multipath; 3) an ideal model should fit the data very well even with the presence of multipath; in other words, the residuals should be random.

In the remainder of this paper, the semi-parametric model and penalised least squares principle are introduced, followed by solution methods of the semi-parametric model using the penalised least squares method. Then, a field experiment is described followed by analysis of the influences of multipath on ambiguity resolution and baseline accuracy. Finally, concluding remarks and future work are provided.

2 Semi-parametric model and penalised least squares method

Since multipath errors result from the mixture of direct and indirect signals, it is very difficult to model these errors using existing parametric models. The semi-parametric models can be an ideal tool to model and mitigate multipath errors. The semi-parametric model is a new statistical method developed in the last two decades (e.g., Green, 1987; Green and Silverman, 1994). In the semi-parametric model, multipath errors can be expressed as complicated functions that varying smoothly with time. Using the penalised least squares principle, the functions can be decomposed from parametric components and observation error components.

The semi-parametric model can be expressed as

$$y_i = A_i x + g(t_i) + \varepsilon_i, \quad i = 1, \ldots n \tag{1}$$

$$\varepsilon_i \sim N(0, \Sigma_i), \quad E(\varepsilon_i \varepsilon_j) = 0, \quad i \neq j \tag{2}$$

where $g(t_i), \varepsilon_i, y_i \in \Re^m$ are the systematic error functions, the random errors and the observations, respectively, at ith epoch; t_i is the time index; $x \in \Re^p$, $A_i \in \Re^{m \times p}$ are the estimated parameters and the observation matrix, respectively; Σ_i denotes the error variance matrix.

Equation (1) contains $m \times n$ observations, but $m \times n + p$ unknowns. Therefore, it cannot be resolved using the least squares principle. An alternative is to use the penalised least square principle, which can be expressed as

$$\sum_{i=1}^{n} \left(y_i - A_i x - g(t_i)\right)^T \Sigma_i^{-1} (y_i - A_i x - g(t_i)) + \sum_{j=1}^{m} \lambda_j \int (\ddot{g}_j(t))^2 \, dt = \min \tag{3}$$

where λ_j is the smoothing parameter and $\ddot{g}_j(t)$ is the second-order derivative of the jth function with respect to time.

Equation (3) defines a penalised sum of squares. The first part of equation (3) is the least squares residual sum of squares and the second part is the *roughness* penalty term. The parameter vector x and the function vector $g(t_i)$ are estimated not only by their goodness-of-fit to the data as quantified by the residual sum of squares

$$\sum_{i=1}^{n}(y_i - A_i x - g(t_i))^T \Sigma_i^{-1}(y_i - A_i x - g(t_i)),$$

but also by the roughness of functions

$$\sum_{j=1}^{m}\lambda_j \int (\ddot{g}_j(t))^2 dt.$$

The smoothing parameters λ_j control the tradeoff between fit to the data and the smoothness of the functions and can be determined automatically by the following *cross validation* equations:

$$CV(\lambda) = \frac{1}{n}\sum_{i=1}^{n}(y_i - A_i\hat{x}_{-k} - \hat{g}_{\lambda,-k}(t_i))^T \Sigma_i^{-1} (y_i - A_i\hat{x}_{-k} - \hat{g}_{\lambda,-k}(t)) = \min \tag{4}$$

$$\sum_{i=1, i\neq k}^{n}(y_i - A_i x - g(t_i))^T \Sigma_i^{-1}(y_i - A_i x - g(t_i)) + \sum_{j=1}^{m}\lambda_j \int (\ddot{g}_j(t))^2 dt = \min \tag{5}$$

where $CV(\lambda)$ is the cross validation to automatically select the smoothing parameters λ_j, and $\hat{x}_{-k}, \hat{g}_{\lambda,-k}(t_i)$ are the estimates of the parameters and the functions, respectively, without the *k*th epoch observations.

3 Solution methods of the semi-parametric model

If the observation is a scalar, the solution $\hat{g}$ of the function $g(t_i)$ is a cubic spline (Reinsch, 1967). For vector measurements, such as in GPS positioning applications, the solution $\hat{g}$ of the function vector $g(t_i)$ is a vector cubic spline (Fessler, 1991). Therefore, the roughness penalty term can be expressed as

$$\sum_{j=1}^{m}\lambda_j \int (\ddot{g}_j(t))^2 dt = ((R^{-1}\otimes I_m)(Q^T \otimes I_m)g)^T (R\otimes D(\lambda))((R^{-1}\otimes I_m)(Q^T\otimes I_m))g \tag{6}$$

where R and Q are penalised constraint matrices which are only relative to the time index t_i, $\otimes$ denotes Kronecker product (Rao, 1973), and

$$D(\lambda) = diag(\lambda_1, \lambda_2, \ldots \lambda_m).$$

Substituting equation (6) into equation (3) and assuming the variance matrix for each epoch is invariant, the following equations are obtained.

$$A^T(\Sigma^{-1}\otimes I_n)Ax + A^T(\Sigma^{-1}\otimes I_n)g = A^T(\Sigma^{-1}\otimes I_n)y \tag{7}$$

$$(\Sigma^{-1}\otimes I_n)Ax + ((\Sigma^{-1}\otimes I_n) + (Q\otimes D(\lambda))(R^{-1}\otimes I_m)(Q^T\otimes I_m))g = (\Sigma^{-1}\otimes I_n)y \tag{8}$$

where

$$A = (A_1^T, A_2^T, \ldots A_n^T)^T,$$
$$\Sigma = \Sigma_i,$$
$$y = (y_1^T, y_2^T, \ldots y_n^T)^T$$

$$g = (g_1^1, g_1^2, \ldots g_1^m, g_2^1, g_2^2, \ldots g_2^m, \ldots g_n^1, g_n^2, \ldots g_n^m)^T$$

Equations (7) and (8) can be solved by the back-fitting method (Buja *et al.*, 1989; Yee, 1998) or the direct method (Green and Silverman, 1994; Yee and Wild, 1996)

4 Experiment and analysis

In order to test the semi-parametric method, an experiment was conducted in Perth, Western Australia on June 7, 1999. Data were collected with two Ashtech Z12 receivers at a sampling interval of 1 second and a cut off elevation angle of 15°. The two receivers were mounted on two pillars that belong to a first-order terrestrial survey network. The network was established by the Western Australia Department of Land Administration (DOLA). The known length of the baseline is 215.929 ± 0.001 m.

The data were first processed using standard GPS data processing software with an interval of 10 seconds and a session length of 19.5 minutes. The time series of double difference residuals in Fig. 2 shows that there exist significant multipath errors in several satellite pairs PRN 2-7, 2-19, 2-26. Then, the data are processed by the semi-parametric model and penalised least squares method with the same interval and session length. The double difference residuals are plotted in Fig. 3, which shows that no significant error patterns in that series and the residuals are very close to double difference observation errors. As an example, Fig. 4 shows the extracted function and double difference residual series of satellite pair PRN 2-7.

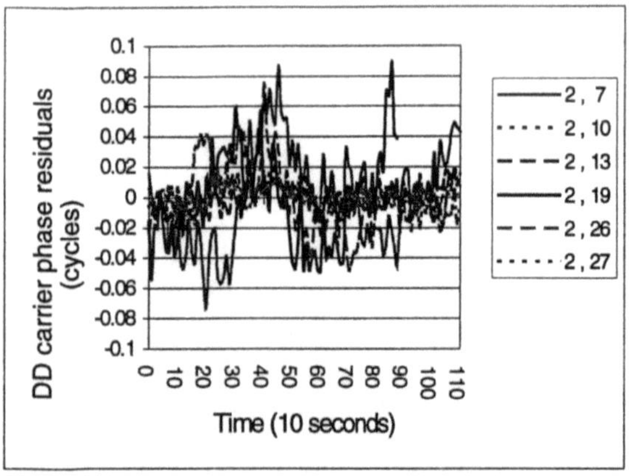

Fig. 2: Double difference carrier phase residuals from the parametric model

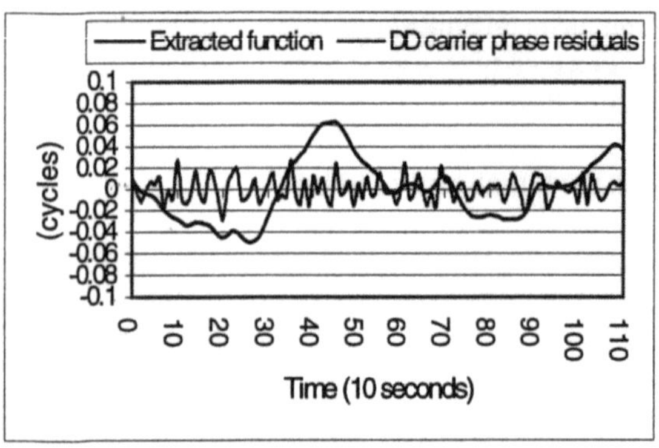

Fig. 3: Double difference carrier phase residuals from the semi-parametric model

In order to analyse the influence of multipath errors on the ambiguity resolution and baseline accuracy, as well as to demonstrate the potential of the semi-parametric model and penalised least squares method over the parametric model and least squares method, the data are processed with the same interval, but session lengths were increased sequentially by one minute. The F-ratios from the two models are shown in Fig. 5 and Table 1, and the difference of baseline length between the known (215.929m) and the estimated by using two models are shown in Fig. 6 and Table 1. Fig. 5, and columns 6 and 11 in Table 1, show that the reliability of the ambiguity resolution has been improved using the semi-parametric model and penalised least squares method over the parametric model and least squares method. Fig. 6, and columns 7 and 12 in Table 1 show, that because of the presence of multipath errors, the ambiguities cannot be fixed with the session lengths of 3.5, 4.5, 8.5, 9.5 and 13.5 minutes. Moreover, incorrect ambiguities have been fixed with the session lengths of 12.5, 14.5-18.5 minutes using the parametric model and least squares method. On the contrary, correct ambiguities can be fixed over all of session lengths after 3.5 minutes using the semi-parametric model and penalised least squares method. Furthermore, in the cases where the correct ambiguities have been fixed using two models (with the session lengths of 5.5, 6.5, 7.5, 10.5 and 11.5 minutes) an improvement of 2-4mm in baseline length is obtained by using the semi-parametric model and penalised least squares method over the parametric model and least squares method.

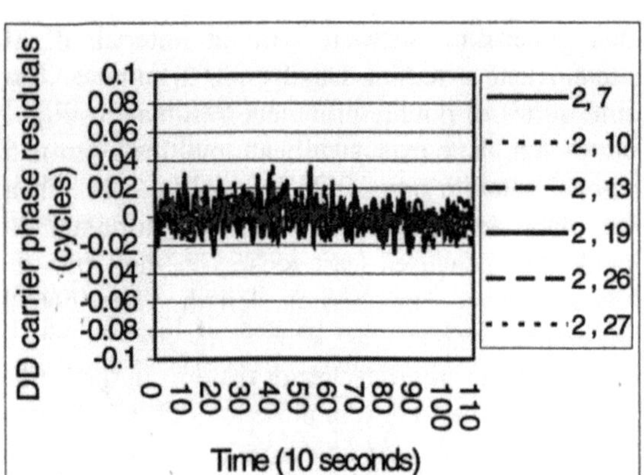

Fig. 4: The extracted function and double difference residuals of satellite pair PRN 2-7 from the semi-parametric model

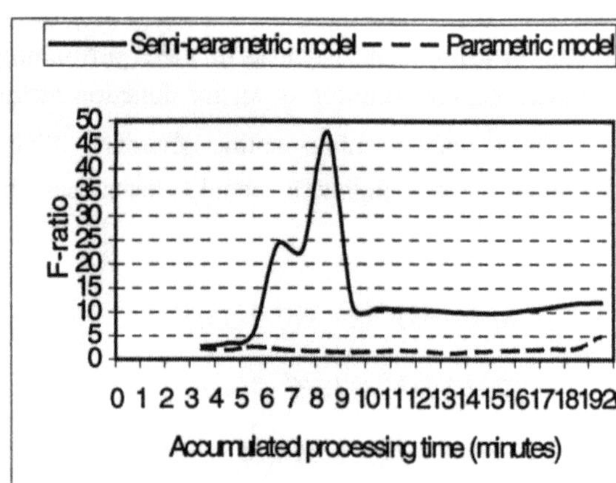

Fig. 5: Comparison of F-ratios from the parametric and semi-parametric models

Table 1: Comparisons between baseline lengths and between F-ratios

Start time	End time	Semi-parametric model & penalised least squares					Parametric model & least squares				
		Pass		Fail	F-ratio	Length of baseline	Pass		Fail	F-ratio	Length of baseline
		True	False				True	False			
12:40:00	43:30	*			2.71	215.924			*	2.30	
	44:30	*			3.33	215.924			*	1.96	
	45:30	*			3.34	215.925	*			2.66	215.921
	46:30	*			23.99	215.925	*			2.07	215.923
	47:30	*			22.89	215.926	*			1.75	215.924
	48:30	*			47.65	215.926			*	1.54	
	49:30	*			11.08	215.926			*	1.51	
	50:30	*			10.70	215.927	*			1.61	215.927
	51:30	*			10.52	215.927	*			1.81	215.927
	52:30	*			10.31	215.927		*		1.60	215.812
	53:30	*			10.00	215.928			*	1.28	
	54:30	*			9.79	215.928		*		1.67	216.073
	55:30	*			9.83	215.928		*		1.85	216.073
	56:30	*			10.43	215.928		*		2.00	216.072
	57:30	*			11.01	215.928		*		2.17	216.072
	58:30	*			11.80	215.928		*		2.29	216.071
	59:30	*			12.03	215.928	*			4.90	215.928

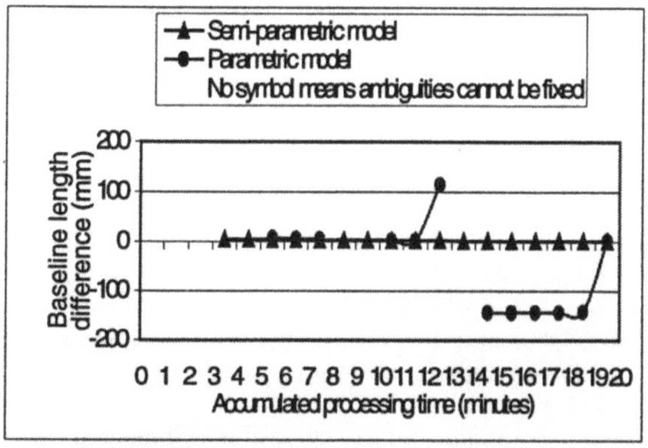

Fig. 6: Differences between the estimated and real baseline length

5 Concluding remarks and future work

The semi-parametric model and penalised least squares method have been developed to mitigate multipath errors for short distance and high precision GPS applications. The following important conclusions may be extracted from this study

- Increasing the session length does not mean improvement of the reliability of the ambiguity resolution using standard GPS data processing methods under the presence of multipath errors (e.g., unfixed ambiguities with data of 8.5, 9.5 and 13.5 minutes, and incorrect fixed ambiguities with data of 12.5 and 14.5-18.5 minutes).
- Using the semi-parametric model, ambiguities can be successfully fixed in cases where they cannot be fixed using the parametric model (e.g., with data of 8.5, 9.5 and 13.5 minutes).
- The semi-parametric model can fix the correct ambiguity sets, in cases where the parametric model cannot (e.g., with data of 12.5, 14.5-18.5 minutes).
- Baseline length accuracy can be improved using the semi-parametric model over the parametric model when ambiguities can be resolved successfully for both of the models (e.g., with data of 5.5-7.5, 10.5 and 11.5 minutes).
- The semi-parametric model and penalised least squares method is an alternative of the parametric model and least squares in some data processing cases.

In addition, the proposed method does not need any *a priori* antenna environmental knowledge, hence, it can be easily transplanted into existing GPS data processing software.

Future investigations include:

- Develop and further test the theories of the semi-parametric model and penalised least squares method;
- Apply the semi-parametric model and penalised least squares method to mitigating other GPS/GLONASS systematic errors (e.g., residual ionospheric errors).

Acknowledgments. The first author would like to thank Associate Professor Will Featherstone for his encouragement during this research. Thanks also to Mr. T. Forward and Dr. J. Wang for their assistance in data collection. Discussions with Dr. T. Yee (Department of Statistics, University of Auckland) and Professor J. A. Fessler (University of Michigan), and their subroutines have been very helpful. The first author is partially supported by Overseas Postgraduate Research Scholarship and Australian Research Council funding.

References

Axelrad, P., Comp, C.J. and Macdoran, P.F. (1996) SNR-Based Multipath Error Correction for GPS Differential Phase, *IEEE Transactions on Aerospace and Electronic Systems*, Vol. 32, pp. 650-660.

Brunner, F.K., Hartinger, H. and Troyer, L. (1999) GPS signal diffraction modelling: the stochastic SIGMA-Δ model, *Journal of Geodesy*, Vol. 73, pp. 259-267.

Buja, A., Hastie, T.J. and Tibshirani, R.J. (1989) Linear Smoothers and Additive Models, *Annual of Statistics*, Vol. 17, No. 2, pp. 453-555.

Comp, C.J. and Axelrad, P. (1998) Adaptive SNR-Based Carrier Phase Multipath Mitigation Technique, *IEEE Transactions on Aerospace and Electronic Systems*, Vol. 34, pp. 264-276.

Cohen, C.E. (1992) Attitude Determination Using GPS, *PhD Dissertation*, Department of Aeronautics and Astronautics, Stanford University, Stanford, CA.

van Dierendonck, A.J., Fenton, P. and Ford, T. (1992) Theory and Performance of Narrow Correlator Spacing in a GPS Receiver, *Navigation*, Vol. 39, pp. 265-283.

Fessler, J.A. (1991) Nonparametric Fixed-Interval Smoothing with Vector Splines, *IEEE Transactions on Signal Processing*, Vol. 39, No. 4, pp. 852-859.

Genrich, J. and Bock, Y. (1992) Rapid Resolution of Crustal Motion at Short Ranges with the Global Positioning System, *Journal of Geophysical Research*, Vol. 97, No. B3, pp. 3261-3269.

Georgiadou, Y. and Kleusberg, A. (1988) On the Carrier Signal Multipath Effects in Relative GPS Positioning, *Manuscripta Geodaetica*, Vol. 13, pp. 172-179.

Green, P.J. (1987) Penalised Likelihood for General Semi-parametric Regression Model, *International Statistical Review*, Vol. 55, pp. 245-259.

Green, P.J. and Silverman, B.W. (1994) *Nonparametric Regression and Generalized Linear Models*, Chapman & Hall, London, 182 pp.

Hardwick, C.D. and Liu, J. (1996) Characterization of Phase and Multipath Errors for an Aircraft GPS Antenna, *Navigation*, Journal of the ION, Vol. 43, pp. 41-54.

Hannah, M.B., Walker, R.A. and Kubik, K. (1998) Toward a Complete Virtual Multipath Analysis Tool, *Proceedings of ION GPS-98*, Nashville, Tennessee, USA, pp. 1055-1063.

van Nee, R. (1991) Multipath Effects on GPS Code Phase Measurement, *Proceedings of ION GPS-91*, Albuquerque, Washington, DC, USA, pp. 915-924.

Rao, C.R. (1973) *Linear Statistical Inference and its Applications*, 2nd ed, Wiley, New York.

Ray, J.K., Cannon, M.E. and Fenton, P. (1998) Mitigation of Static Carrier Phase Multipath Effects Using Multiple Closely-Spaced Antennas, *Proceedings of ION GPS-98*, Nashville, Tennessee, USA, pp. 1025-1034.

Reinsch, C.H. (1967) Smoothing by Spline Function, *Numer. Math.*, Vol. 10, pp. 177-183.

Walker, R.A. (1996) Numerical Modelling of GPS Signal Propagation, *Proceedings of ION GPS-96*, Kansas City, USA, pp. 709-717.

Yee, T.W. and Wild, C.J. (1996) Vector Generalised Additive Models, *J. R. Statist. Soc. B*, 58, No. 3, pp. 481-493.

Yee, T.W. (1998) On an Alternative Solution to the Vector Spline Problem, *J. R. Statist. Soc. B*, 60, No. 1, pp. 183-188.

Geotechnical exploration - wider fields of activities for geodesists and geophysicists

E. Brückl
Institute of Geodesy and Geophysics
Vienna University of Technology, A-1040 Wien, Gusshausstrasse 27-29, Austria

Abstract. The geodetic community has a traditional position within geotechnical exploration, however, as it turns out geodesists are rarely included in the planning and decision making team and new fields of activities are not occupied by geodesists considering their knowledge and expertise.

Based on an operational model of a multidisciplinary exploration team, a strategy is developed to improve the situation described above by utilizing the synergy between geodesy and geophysics. Essential impulses could come from the universities by combining an education in geodesy with a good geophysical training. This can diversify the methods to be offered and improve the ability to develop geoscientific models. The construction and maintenance of databases including visualization and data exchange, as well as the global optimization of complex systems is fundamental to geodetic and geophysical work and additional geoscientific applications of these techniques should be emphasized. Furthermore, a basic understanding of geology and related geosciences is important to promote the integration into a geotechnical exploration team. As an example of a joint geodetic-geophysical activity a field named "Applied Geodynamics" is presented.

Keywords. Geotechnique, geodesy, geophysics, exploration, applied geodynamics.

1 Introduction

Geodesy has a strong position within it's traditional market and is progressing to new activities in many scientific and technical fields. Similarily, due to the increasing demands for optimum use and exploitation of the earth's resources, geotechnical exploration is becoming more and more important. However, it is the author's personal experience as a geophysicist having been involved in many geotechnical projects, geodesy is not integrated in this field of activities as much as it could be. This impression was supported by papers presented by geotechnical experts at the Symposium on Geodesy for Geotechnical and Structural Engineering, Eisenstadt, Austria, April 20-22, 1998.

In this paper the characteristics of geotechnical exploration will be presented and the traditional position of geodesy within this field will be discussed. On the basis of this analysis a strategy to open new fields of activities for geodesits will be developed, based on the synergy of geodesy and geophysics.

2 Geotechnical exploration

The demands of energy supply, traffic, transportation, and storage initiate major geotechnical projects like the construction of dams, tunnels, slope cuts, deep foundations, and underground caverns. To be save and cost effective these projects need a thorough geotechnical exploration. According to Kleberger (1998) a geotechnical exploration has the following intentions:

- optimum location for the project
- avoid natural hazards
- minimize a negative impact on the environ-ment
- minimize technical and financial efforts
- create a geotechnical model.

Conventionally a geotechnical exploration is synonymous to a geotechnical site investigation. By this investigation the necessary information is supplied about the geological structure, the physical and chemical properties of the geological bodies and the stress field. Additionally, relevant geological processes are considered like weathering, and groundwater flow. Beyond the classical tasks of a geotechnical site investigation we will have a wider view on the activities necessary for a comprehensive study of a geotechnical site. One additional task is the exploration of construction materials like clay, sand gravel and stones which might be used for sealing, dam construction, production of concrete, and other construction purposes. A further issue in every large

International Association of Geodesy Symposia, Vol. 121
Schwarz (ed.), Geodesy Beyond 2000 – The Challenges of the First Decade

geotechnical project is environmental protection. Care must be taken not to reduce the quality of live

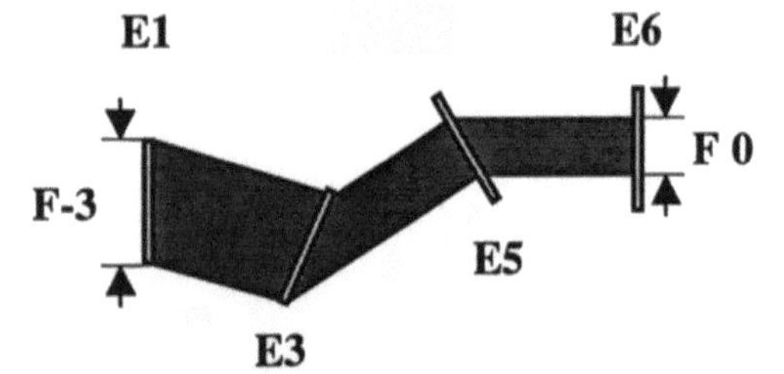

Fig. 1: Pipe-line model

in the area of influence of the geotechnical project. That means, e.g., vibrations and acoustic noise caused by the construction work and by the later operation must be reduced to an acceptable level. The groundwater level must be kept within limits not to influence agriculture. Also the temporal or permanent storage of excavation material or waste cause problems. natural hazards must be taken into consideration in order to make geotechnical constructions save. Geodynamic and geomorphological processes impose hazards like earthquakes, volcanic eruptions, landslides and debris flows, subsidence and other deformations caused by neotectonic activities.

Geotechnical exploration is a multistage process with a continuing improvement of the geotechnical model. As an example, the sequence of phases according Swiss legislation (Schneider, 1998) is given in the following:

- feasibility study
- preliminary design
- documentation for public approval
- basic design
- detailed design
- tender documents and call for tender
- construction period.

Another striking feature of geotechnical exploration is it's multidisciplinarity. According to Kleberger (1998) the following remarkable list of individual disciplines might be involved in the exploration work:

- engineering geology and geophysics,
- mineralogy, chemistry, hydrology,
- seismology, geodesy, remote sensing,
- civil engineering, soil and rock mechanics,
- palaeontology, archaeology,
- public relations, administration, and law.

3 Present position of geodesy and geophysics within geotechnical exploration

Geodesy has a strong position for classical geodetic tasks like the production of maps, the determination of coordinates of drillholes, and the control of the direction of the headings in tunnels. The geodetic work is highly specialized, it is carried out and evaluated with little feedback from other geoscientific disciplines. The deliverables are generally uninterpreted data. High-tech monitoring systems were developed by geodesists, however, frequently not operated and interpreted by geodesists. Geodesists are rarely integrated in the team establishing the geotechnical model. Also the expertise of geodesists in data-management and GIS is not used very frequently. Generally, geodesists are not included into the decision making process.

Besides the fact of being the key technology for the exploration of oil and gas (Granser, 1998) geophysics is still in an introductory stage concerning geotechnical exploration. The budget for geophysics rarely exceeds 10% of the expenses for drillholes. On the other hand, by the strong interpretive component of their work, geophysicists are generally included in the team developing the geotechnical model and making planning decisions.

4 Operational models

Köhler (1998) pointed out that interdisciplinary scientific cooperation is the key to obtain results of a high standard, but it requires a high degree of coordination between the various working groups. Furthermore, clear organisational structures are necessary for an effective collaboration and data processing and communication within the project groups is one of the keys to rapid progress of interdisciplinary work. In the following two operational models for geotechnical exploration are presented and discussed.

4.1 Pipe-line model

Figure 1 shows an organizational structure where the information necessary to construct the geotechnical model is collected by experts of the individual geoscientific disciplines and handed over to another expert combining these data and his own findings to a higher level model. At the end a "master" expert will collect all results into one report describing the final geotechnical model. So we find several hierachical levels of the work of the

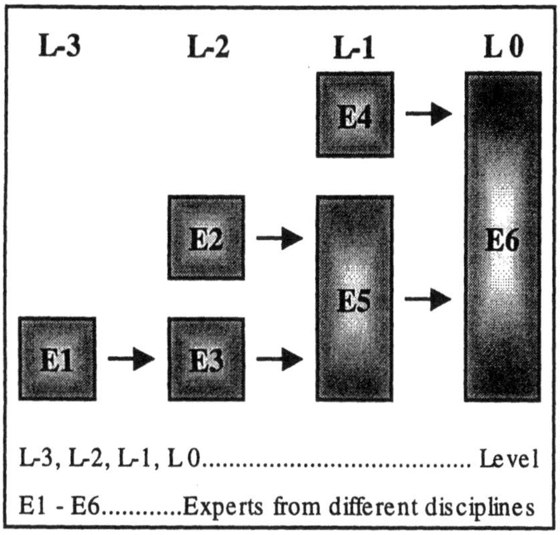

Fig. 2: Loss of information inherent to the pipe-line model

experts from the different scientific disciplines. Geodesists are frequently found at the lowest hierachical level, geophysicists mostly at an intermediate level According to the sequential character of this model, it will be called the pipe-line model.

The advantages of the pipe-line model are its simple organisation, the strict schedule, and, in case of very experienced master experts, consistent results. Disadvantages are poor feedback to the experts at the lower hierachical levels. Furthermore, to the author's personal experience an essential loss of information may occur during the transfer of information from one level or expert to the next. The mechanism of this loss of information is illustrated in figure 2. According to this figure the information from an expert at a lower level discipline is projected to the expert of the next higher level. However, as the parameter spaces of the individual scientific disciplines differ, only correlating information is accepted and orthogonal information is discarded.

4.2 Round table model

To overcome the disadvantage of loss of information and lack of feedback inherent in the pipe line model, a so called round table model is presented. All experts from the different disciplines meet at the round table (figure 3) and exchange their information. There is no hierachical order between experts of the different scientific disciplines. Several times the author has participated in expert teams working according to this operational model. Also Köhler (1998) addresses this operational model when speaking about interdisciplinary work. The advantages are the preservation of the full information from all disciplines, a good feedback, and an increase of experience for all participants.

As hierarchical levels are no more existing the development of the geotechnical model is an issue. Figure 4 illustrates how the individual experts deliver their data and models to different layers. Each layer may be regarded as a projection of the geotechnical reality to the parameter spaces of the models of the individual scientific disciplines. The establishment of the final geotechnical model should be a reconstruction from these projections.

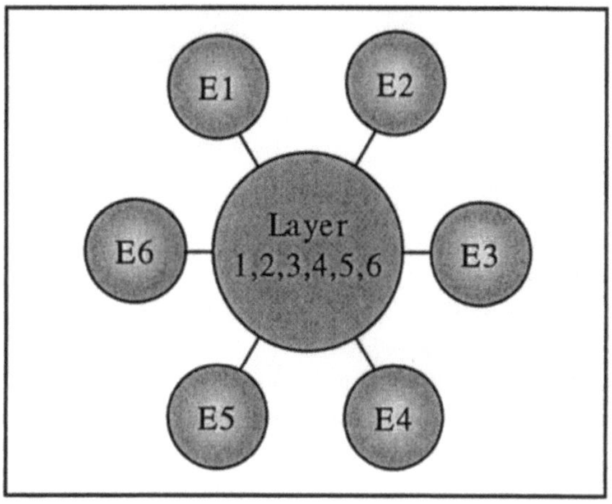

Fig. 3: Round table model, plan view

We may view this process as interdisciplinary tomographic imaging. Obviously, there are more difficulties in realizing the round table model rather than the pipe-line model. However, one can accept these problems also as a chance for future activities.

5 Strategy to open up wider fields of activities for geodesy and geophysics

A strategy to open up wider fields of activities for geodesy and geophysics is based on the assumption that the round table model will prevail on the pipe-line model.

5.1 Synergy of geodesy and geophysics

For a paper presented at IUGG it is not necessary to stress the common scientific background of geodesy and geophysics. However, in practice and even in education at the universities there is sometimes a poor awareness of the common roots in matematics and physics. But there is a broad common scientific background which should be highlighted.

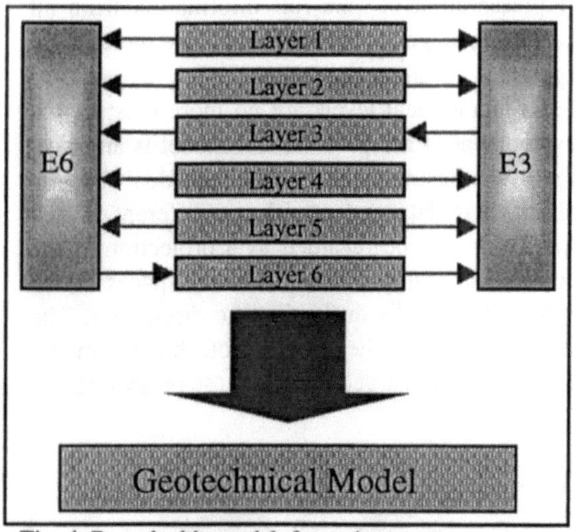

Fig. 4: Round table model, front view

They comprise of:
- gravity field and figure of the earth
- reference systems, height
- structure of the earth
- continuation of potential and wave fields
- time / space-series analysis
- interpolation and extrapolation
- modelling and inversion of physical fields
- optimization and parameter estimation

And in practice there are many tasks using the same or similar techniques and skills like:
- field data acquisition
- surveying, navigation
- quality control, logistics
- airborne and satellite remote sensing
- data management, GIS, visualization.

5.2 First step

The discussion of the round table model showed that the individual scientific disciplines deliver their informations to different layers. However, geodesy and geophysics must not give different projections of the real world. The data from both disciplines may be combined and delivered as one model to the round table. So, the first step of a strategy to open wider fields of activities would be joining geodesy and geophysics and to get promoted from a subcontractor to a partner of the round table.

5.3 Second step

As pointed out in the discussion of the round table model the organization of data upgrade, exchange of information, and the final reconstruction of the geotechnical model from the individual projections is an issue. As indicated by Mattanovich (1998) and Otepka (1998), geodesy and geophysics may have some expertise to manage such an operation. After having done the first step as described above, some highly qualified expert of geodesy and geophysics may advance from a partner of the round table to become the manager of the round table. This process may concern only a few experts, however, such a development will have a positive impact on the increased integration of geodesy and geophysics to the geotechnical exploration.

5.4 Educational measures

Traditionally geodesy and geophysics are different studies at the univerities. As an exception, at the Vienna University of Technology there is a study named geodesy and geophysics. However, the geophysical courses within this study have more educational aspects and it is not really intended to make geodesists ready for applying geophysical techniques in their later professional work. On the other hand many students did their geophysical courses quite well and to the author's opinion there is no great gap to get them prepared for successful practical work.

The minimum geophysical courses to reach this goal comprise of lectures on global and applied geophysics, field practice on geophysical data acquisition, and practice of geophysical processing and interpretation. Now these courses are distributed over four semesters and have a total of 16 hours per week and semester. Based on a solid knowledge of mathematics and physics, geodesy and geophysics may be joined in one study. Additional educational requirements for successfully joining the round table in geotechnical exploration are a good understanding of the fundamentals of geology and an enzyclopedic knowledge of givil engineering, soil- and rock mechanics, hydrology, and environmental protection.

6 Example of a geodetic-geophysical approach for geotechnical exploration

Traditionally, geodynamics observes and interpretes continuous (e.g. sea floor spreading) and discontinuous (e.g. earthquakes) deformations of the earth crust on a global scale. Here geodesy and geophysics are already cooperating in teams. Downscaling the techniques developed within global geodynamics and by combining it with techniques from engineering geodesy and geophysics a new field of practical activiy may be defined and named applied geodynamics.

Applied geodynamics will be based on the measurements of displacements of the earth's crust at all relevant spatial scales and frequency ranges, comprising both geodetic and seismic measurement techniques. Remote sensing methods as well as geotechnical multi-sensor monitoring may complete these observations. Furthermore the consideration of the stress field and gravitational forces is an integrated factor of applied geodynamics. Information about the geological structure and physical properties of the geological bodies are supplied by geophysical methods. Modelling, inversion, and interpretation of all data should be based on advanced numerical methods and a comprehensive understanding of the relevant geological processes like plate tectonics, the effect of gravitational and inertial forces, or the cycle of disintegration, erosion, transport, deposition, and compaction. The application of GIS and visualisation techniques will be important for data management and communication. The work of an expert in applied geodynamics will be in a close relation to geology in all spatial ranges and to geomechanics and geohydrology at least in small to medium spatial ranges.

Geotechnical problems to be solved by experts in applied geodynamics may be the following:

- characterisation of geotechnical sites with respect to their structure, physical properties and especially the fault system,
- stress field and tectonics,
- long term stability of selected sites (e.g. for nuclear waste deposits),
- geological hazards like earthquakes or landslides,
- reaction of geological bodies to technical measures like slope cuttings or damming up of water reservoirs,
- 4D-monitoring and modelling (4D is variation in space and time).

Definite examples supporting the idea of applied geodynamics were given by Altan et al. (1998) and Scharler and Stolitzka (1998) for the fields of seismic risk and monitoring work during tunnel construction.

7 Summary

The present image of geodesy within the geotechnical exploration community is to deliver valuable and precise data without much interpretation. In most geotechnical projects it is not perceived as necessary to include experts from geodesy in the team establishing the geotechnical model. This situation restricts the activities of geodesists to traditional or highly specialized tasks and does not take advantage of the full scientific and technological potential they would have to offer. In this paper an idea has been presented to improve this situation based on the following strategy:

1) For the organization of the interdisciplinary work in an expert team for geotechnical exploration the round table model should be supported and established.

2) Experts in geodesy and geophysics should become partners or even managers at the round table. This steps should be achieved by using the synergy between both disciplines and integrating geodesy and geophysics. This integration will bring more interpretational aspects to the work of geodesists and strengthen the market postition of geophysicits.

3) At the universities studies and courses should be offered unifying the geodetic and geophysical education and taking care of a broader geoscientific knowledge.

4) New approaches of joint geodetic and geophysical activities have to be developed. As an example, a field named applied geodynamics was proposed.

Acknowledgements. I thank Fritz K. Brunner for his valuable suggestions and Heribert Kahmen for his encouragement to present this paper.

References

Altan, O., Fritsch, D., Külür, S., Seker, D., Sester, M.,Volz, S., Toz, G. (1998), Photogrammetry and GIS for the Acquisition, Documentation and Analysis of Earthquake Damages, Symp. on GGSE*), Proceedings, p.40-45.

Granser, H., Exploration Strategy for Hydrocabons, Symp. on GGSE*), Proceedings, p.22-27, 1998.

Kleberger, J. (1998), Methodology of Engineering Geological Exploration, Symp. on GGSE*), Proceedings, p.28-33.

Köhler, M. (1998), Munich-Verona Rail Link - Project Management and Investigation Programme of the Brenner Eisenbahn GmbH (BEG), Symp. on GGSE*), Proceedings, p.34-39.

Mattanovich, E., For joint processing of metric and thematically valuating data, Symp. on GGSE*), Proceedings, p.58-63, 1998.

Otepka, G., Methods and possibilities for data-acquisition, data-visualization and data-management, Symp. on GGSE*), Proceedings, p.52-57, 1998.

Scharler, H., Stolitzka, G., An interactive Tool for the Analysis and Interpretation of Geodetic Deformation Measurements in Tunneling, Symp. on GGSE*), Proceedings, p.46-51, 1998.

Schneider, T.R., Developement of the Geological Investigations in a Geotechnical Project, Symp. on GGSE*), Proceedings, p.15-21, 1998.

*) Symposium on Geodesy for Geotechnical and Structural Engineering, Eisenstadt, Austria, April 20-22, 1998, IAG-SC4

PART 6

An IAG Structure to Meet Future Challenges

Martine Feissel
Fernando Sansò
Carl C. Tscherning

An analysis of the current IAG structure and some thoughts on an IAG focus

K.-P. Schwarz
Department of Geomatics Engineering, The University of Calgary, Canada.

Abstract. The current structure of the IAG is analyzed and its suitability for supporting a strong research focus of the IAG is examined. The analysis shows that there are some serious drawbacks and unbalances in the current structure which limit its internal functioning. In addition, the research interaction with organizations outside the IAG is often not effective and response times are slow. It appears that a clear focus for IAG research is required and that such a focus would require major changes in the IAG structure. Some suggestions on the required changes are made in the Conclusions and Recommendations.

Keywords. Analysis of IAG structure; IAG research focus; Services; Commissions; Special Study Groups.

1 Background

The papers in this part have been invited to examine the question: 'How well does the current IAG structure support IAG research priorities?' The discussion of this question has been an ongoing theme in the IAG Executive Committee meetings of the period 1995-1999. Although opinions on specifics differ, there was general agreement that a discussion of this questions by a broader audience, such as a General Assembly, would not only be timely but also beneficial to the IAG. When planning the program for the Birmingham General Assembly, a full day was therefore set aside for this discussion.

The day was subdivided into three sessions with the following topics:

- Advantages and Drawbacks of the Current IAG Structure
- The Role of IAG Services in Support of Research in Geodesy and Geodynamics
- Alternatives for Change

The first session had five invited papers of which the first one gave a general overview and analysis of the current structure while the other four examined the structure with specific emphasis on Sections, Commissions, Services, and Special Study Groups. Unfortunately, the paper with the Commission emphasis was cancelled, so that only four of the five papers are printed here. The second session had six invited papers, all of which were presented. Since some of the material had been published earlier and was available in print (Mueller, 1997), it was decided to summarize the highlights of this session in one paper which the chairperson, Martine Feissel, agreed to write. The third session consisted of three papers and a panel discussion. The papers in this session had a double focus. On the one hand, they discussed what IAG research should be; on the other hand, which structure would best support a specific focus. Two of the three presentations are published in this volume. After the presentations, a one-hour panel discussion followed which provided valuable input on the questions raised in the three sessions.

This paper combines two presentations given by the author. The introductory presentation on the current IAG structure and the final presentation on the future IAG focus. It was felt that both aspects should be considered together and this has been expressed by the new title.

2 An analysis of the current IAG structure

2.1 Current IAG research objectives

The main objectives of the IAG as stated in the Statutes (Willis, 1996) are:

- to promote the study of all scientific problems and to encourage research to promote and coordinate international cooperation in this field, and promote geodetic activities in developing countries;
- to provide, on an international basis, for discussion and publication of the results
- of the studies, researches and works indicated in the two paragraphs above.

International Association of Geodesy Symposia, Vol. 121
Schwarz (ed.), Geodesy Beyond 2000 – The Challenges of the First Decade

In terms of research, the definition is rather broad (all scientific problems) and its implementation rather vague (promote, encourage). It reflects the current reality of the IAG which, as a scientific organization with minimal financial support, depends on the quality of its scientific work and on the benefits of international scientific cooperation. In such a structure, the priorities are essentially decided by peer consensus and are minimally influenced by funding decisions of the organization. In this way, it provides a rather open environment for future-oriented research. On the other hand, because of the large diversity of research topics that are treated simultaneously, the research profile of the organization remains rather unfocussed.

2.2 The current administrative structure

The current IAG structure can be subdivided into an administrative and a scientific part. The administrative structure is shown in Figure 1 and consists of the Council, the Executive Committee, and the Bureau.

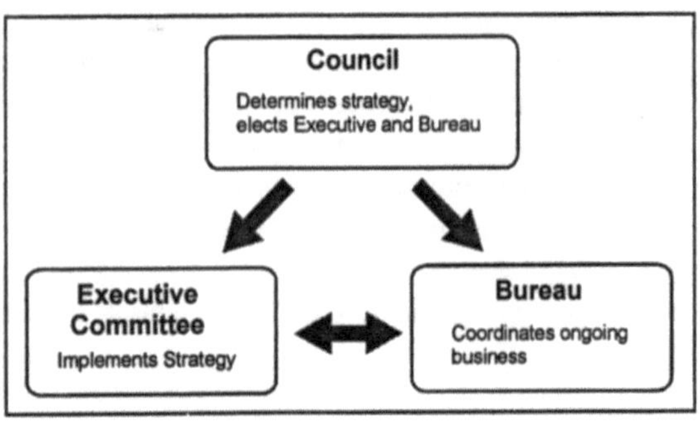

Fig. 1: The administrative structure of the IAG

The Council is made up of national delegates accredited by member countries. It meets once every four years and determines the overall strategy of the Association, as defined in the Statutes (II,7):"Responsibility for the direction of the Association affairs shall be vested in the Council of the Association." One of its major tasks is the election of officers for the Executive Committee and the Bureau. Another is the approval of major changes in the scientific structure of the Association.

The Executive Committee (EC) consists of officers elected by Council and meets, on average, once a year. It implements the strategy adopted by Council by providing coordination between the numerous scientific bodies of the Association and by making major policy decisions, as defined in the Statutes (II,10): "The duties of the Executive Committee shall be to further the scientific objectives of the Sections and other scientific bodies of the Association through effective coordination and through the formulation of general policies to guide the scientific work of the Association."

The Bureau consists of three elected officers – the President, the First Vice-President, and the Secretary General. It meets as required, but at least in conjunction with each EC meeting. The Bureau coordinates the ongoing business, as defined by Council and EC decisions, see Statutes (II,9):"The duties of the Bureau shall be to administer the affairs of the Association in accordance with these Statutes and By-Laws and with the decisions of the Council and the Executive Committee."

In general, the administrative structure has worked well for the IAG, especially after e-mail has become a major means of communication. This has strengthened the link between Bureau and Executive Committee and has provided a means to make decisions between EC meetings in an efficient and democratic way. What may be required in the future is a more active involvement of Council members in the ongoing discussions of the Executive Committee in order to better prepare the quadrennial Council meetings and provide continuity for the decision making process. The latter would require that a majority of national delegates would be on the Council for more than one period. This seems to be the case for the delegates at the two last Council meetings.

2.3 The current scientific structure

The scientific work of the IAG takes place in Sections, Services, Commissions, and Special Study Groups. As indicated in Figure 2, Sections provide a subdivision of geodesy into five major slices which currently are: Positioning and Reference Frames, Advanced Space Techniques, Determination of the Gravity Field, Theory and Methods, and Geodynamics. In a way, sections provide a framework for the objective to study "all scientific problems", but they are not the source of active research. Research takes place in the Commissions and Special Study Groups (SSG) which are the organizational units for long-term and short-term research, respectively. Each Section has therefore a number of Commissions (or Special Commissions) and SSGs which define the research done in a specific section. The Section Steering Committee coordinates interaction between the research units, organizes collaboration across sections, and represents the section in the Executive Committee.

Services are not defined in the Statutes and By-Laws. Traditionally, their major role has been data collection and dissemination. More recently, there has been a stronger emphasis on research and products derived from such data. Each service is assigned to a specific Section.

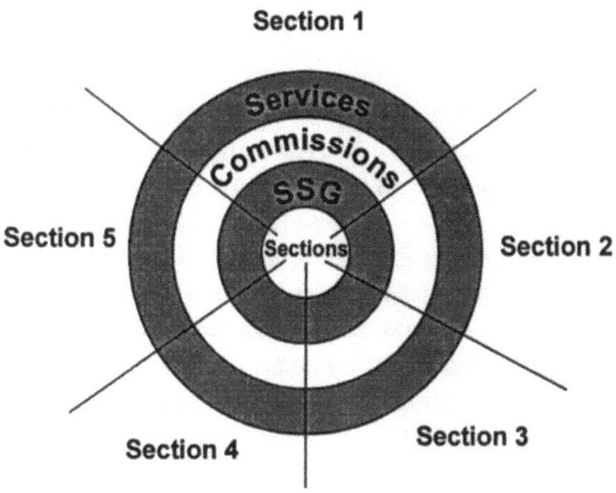

Fig. 2: The scientific structure of the IAG

The idea behind the Section structure is the creation of more managable units which, as a whole, cover all important aspects of the discipline. Its advantage is that research areas that are either small or not fashionable have a chance to find their place. Its drawback is that it concentrates attention on section research and does not encourage a common research focus. Thus, individual scientists are often more attached to a specific section than to the Association as a whole. Geodesy becomes subdivided into compartments which, in the worst case, degenerate into fiefdoms, where turf wars become more important than a common goal. Fundamental change in a discipline is difficult to accommodate in an established section structure, as the emergence of space geodesy has shown. In such cases, the existing structure can become a dead weight for progress.

Commissions and Special Commissions are defined as IAG bodies for long-term research. Commissions are formed where global or regional cooperation is required to achieve long-term goals. Special Commissions are more topical in nature and are formed where the solution of a specific scientific problem requires the cooperation of scientists from different countries. The boundary between the two varieties is somewhat fluid. There is no question that Commissions have very effectively contributed to many areas of geodetic research. There is no question either that many of them have existed well beyond their useful lifetime. The fact that they are defined in terms of long-term research goals makes them often immune to change and very slow to adapt to change in observational or numerical techniques. In terms of structure, there have been a number of cases where a specific Section and a specific Commission were almost exchangeable. Each one could have well existed without the other and nothing essential would have been lost. The fact that most other associations of the IUGG have a structure consisting of long-term and short-term research units only, points to Sections as a superstructure that may not be essential.

Special Study Groups are research units to solve specific scientific problems of limited scope. Their lifetime is usually restricted to one period of four years. They are the units in which highly technical and specialized research is done and where young scientists often make their entrance into the Association. There is little doubt that any scientific organization needs working groups of this type. Their problem, however, is effectiveness. The number of SSGs, which achieve enough synergy to come up with results that can be considered as a group effort, is still very small. Many of them simply exist as clearing houses for pre-publications or as information collectors. The current attempts to replace what is essentially a top-down approach by a bottom-up approach may help to alleviate this problem. Tying such working groups to major scientific projects in the Commissions or Services may be another way out of this dilemma.

IAG Services have developed out of the need for data collection and dissemination. In the pre-IUGG period, they were central activities of the Association and were often fully funded by it, as for instance the International Latitude Service. This was not possible any more after the IUGG had been founded and a smaller amount of money had to be shared by a larger number of associations, see Tardi (1963) for some background. When the Statutes and By-Laws were drafted, Services did not find a place in them. With the increasing need for data and research products, Services have experienced a renaissance during the last decade. One reason is the increasing willingness of countries to share their data because the integrated product they are receiving in return is of considerable value to them. Thus, financing is largely in-kind and can be done with a relatively small overhead. IAG Services have profited from this development. In order to attract a larger customer base, they usually had to move away from their IAG origins. Many of them have developed from IAG Services to FAGS Services and now serve a much larger scientific community. Their link to specific IAG sections is often tenuous and their influence on IAG decisions indirect at best. Their integration into a future IAG structure is important because they often represent the IAG profile in the larger scientific community. To do this in such a way that their range of activities is enhanced rather than restricted, is the challenge to be met.

Finally, important changes in the scientific and organizational structure are essentially tied to the General Assemblies. This means that the response time is typically four years and may be as much as eight years. The organizational dynamics created by this makes it very difficult to respond to new challenges in a flexible manner and to implement decisions as quickly as desirable. Although the IAG Executive has found ways to interpret the Statutes and By-laws in a creative manner in order to respond more quickly, IAG still has the image that it is slow to change. Since the four-year cycle is a result of the IUGG cycle, it will be difficult to change. Flexibility could be greatly increased, however, by further changes in the Statutes and By-Laws.

3 IAG with a focus

3.1 A historical perspective

The impetus for organizing international scientific work in geodesy came from J.J. Baeyer who in 1861, at the age of 66, sent his famous memorandum to the King of Prussia urging the establishment of the Mitteleuropäische Gradmessung. Baeyer had had a long and distinguished career in the army and was an expert in triangulation networks. At the time of his retirement in 1858, triangulation had become a well-established technique, due in no small measure to the efforts of this energetic general. In today's terms, triangulation had become a matter of engineering design and analysis whose main procedures were well established. Baeyer was aware of that and what he proposed in his memorandum was not more of the same, but something quite different. When discussing the idea of a Central European triangulation network, Baeyer(1962) states: „*In this framework, one could compute about 10 meridian arcs at different longitudes and even more parallel arcs at different latitudes; it would also be possible to compare the curvature of the meridians on both sides of the Alps, to study the effect of the alpine ranges on the deflections of the vertical and to determine the curvature of parts of the Adriatic and the Mediterranean, as well as the North Sea and the Baltic; in a word, there is a wide field for scientific investigations which have not been considered in any of the arc measurement campaigns and which, no doubt, will lead to many interesting and important results. If Central Europe is therefore willing to unite and use its resources for the solution of this task, it will call into being an important and magnificent work.*"

What is remarkable in this statement is the recognition that science often begins where engineering ends. In Baeyer's time, measuring a triangulation chain as a framework for topographic mapping was engineering. Using these measurements for the determination of the figure of the Earth was science. Science was the new set of questions that could be formulated based on the existing engineering system. What is equally remarkable, coming from a general of the Prussian Army, is the insight that science needs to cross political and historical boundaries to be effective. Considering that topographic maps and their triangulation base were highly classified objects of military intelligence, this was indeed a breakthrough in an established thinking pattern. It is a sad footnote that 140 years later, data availability remains a problem of scientific research. The outcome of this enterprise was magnificent, as predicted. It led in 1862 to the establishment of the Mitteleuropäische Gradmessung which became the Europäische Gradmessung in 1867, and the Internationale Erdmessung in 1886 and made Europe the centre of geodetic research for many years to come.

The new set of questions that Baeyer proposed to ask had to do with the Earth as a system. He established as guiding principles that the System Earth

- must be considered as consisting of the solid Earth and the oceans,
- could be determined by measurement, of both the geometric and the gravimetric variety,
- required the integration of existing components to provide new insights, and
- needed international cooperation and a scientific organization.

These principles created a focus for geodetic research that could still be felt fifty years later when the First World War interrupted this development. They are at the base of our international organization.

When we ask what has changed in the years since this "magnificent enterprise" started, a few things are obvious:

- We have learned that the System Earth is much more complex, consisting of interacting components, including the different parts of the solid Earth (hydrosphere, cryosphere, biosphere), the atmosphere, and the oceans.
- Measurements have become more diverse, more global, and more accurate, especially through the addition of space techniques which have resulted in global models, expressed in highly stable reference systems.
- All processes affecting the System Earth are time variable at different scales. Due to the enormous increase in global measurement accuracy, many of these effects can already be tracked.
- A new level of integration and international cooperation will be required which goes well beyond the boundaries of traditional geodesy. It requires a stronger interaction with other scientific disciplines as well as the integration of new players in the geodetic field.

We are therefore at or already beyond the threshold of another "magnificent enterprise" that will considerably advance our understanding of the System Earth. If we manage to focus our energies on this goal, another major advance in our discipline will result. If we scatter instead of focus our research, the result will simply be interesting research on selected topics of geodesy.

3.2 Science or engineering?

Before addressing the question what the focus for this research could be, let us briefly return to the question of science and engineering in geodesy. There is no question that the current IAG incorporates both. Thus, focusing IAG research on science could be seen as excluding the engineering aspects from its work. At Helmert's time geodesy was both science and engineering. As a science, geodesy investigated the system Earth with the objective of better understanding its structure. As an engineering discipline, it provided the framework for mapping. Although the objectives were different, the two components were seen as complementing each other. The engineering component provided the data which could be used to answer scientific question. The science component ultimately provided improved models for engineering.

If seen in this way, not much has changed. The driving force to investigate the time-changing processes affecting the system Earth is still our incomplete understanding of this system. Thus, it remains an area of active scientific research. On the other hand, the engineering component of geodesy has expanded considerably beyond the range it covered at Helmert's times. It has become a diversified and important field of engineering and it still provides part of the data by which scientific questions can be answered.

To illustrate this, let us look at a recent example where engineering provided the framework for a new set of questions asked by science. When the NAVSTAR GPS was designed its main purpose was positioning, navigation, and time transfer. The engineers building these satellites had to meet the specs for these applications. A small, dedicated tracking network of ground stations was established that satisfied the needs of the military users. The establishment of the International GPS Service for Geodesy and Geodynamics (IGS) was initially proposed to improve orbit prediction for scientific and government users. Within a short time this scientific service, based on voluntary cooperation, provided not only the capability for very precise orbit determination, but also highly accurate network densification. Thus, the global IERS reference frame could now be accessed everywhere on the globe. This provided a major impetus for geoscience research in general and geodynamics specifically. It was not an initial objective of the DoD, although it might still be considered as an extension of the original concept. However, current research activities on the use of GPS for global atmospheric research clearly go well beyond these concepts and added a whole new dimension to the scientific use of GPS. The pattern here is the same as in 1861. New scientific investigations were spawned by successfully putting an engineering system into place, by organizing international cooperation, and by giving free access to data to as many creative minds as possible.

3.3 A focus for IAG research beyond 2000

If the objective of geodesy as a science is a better understanding of the system Earth, what should be the direction of geodetic research at this point in time? Recently, Rummel (1998) has published a paper in which he proposed an integration of all geodetic data and techniques, conventional as well as space based, into an Global Integrated Geodetic and Geodynamic Observing System (GIGGOS). Underlying this concept was the idea of focusing all current activities in such a way that they become visible as geodetic contribution to international science, for instance in the Earth Observing Science (EOS) program. The diagram in Figure 3 shows the major components of such a program and indicates the interactions that define it as one system. One of its characteristics is that it maintains the link between the science and engineering component of geodesy, so vital in the past.

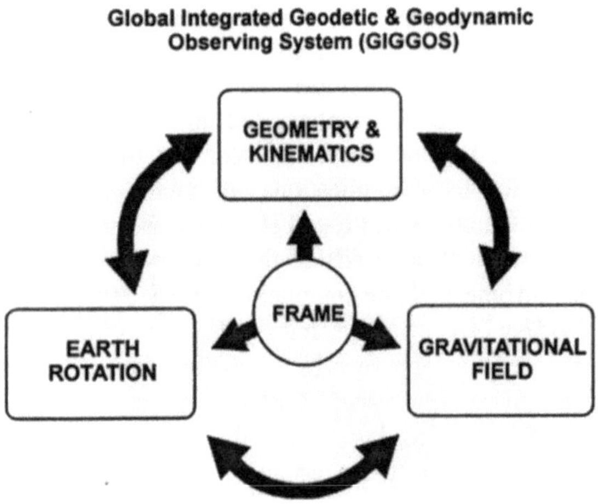

Fig. 3: Concept of GIGGOS

The four components, indicated as Frame, Earth rotation, Geometry and Kinematics, and Gravitational Field, will be briefly discussed. At the centre of this system is a well-defined and reproducible global terrestrial frame which provides the reference for the observing systems and a framework for modeling Earth processes. Its accuracy and stability affects the accuracy with which the other three components can be modelled. The establishment and maintenance of such a reference frame would be done by a combination of space techniques, such as VLBI, SLR, LLR, GPS, DORIS, PRARE.

Closely related to the frame definition is the determination of Earth rotation as the integrated effect of all angular momentum exchange inside the Earth, between land, ice, hydrosphere and atmosphere, and between Sun, Moon, and planets. The measurement systems are the same as for the frame determination, but would be augmented by geodetic astronomy and emerging highly accurate 'super-gyros'.

The geometry of the Earth and its temporal variations would include models for the solid Earth, ice sheets, and the ocean surface and their change in time and space whether secular, periodical or instantaneous. All conventional and space-based point positioning techniques would contribute to this modelling process as well as surface measurement techniques, such as satellite altimetry, satellite interferometric SAR, and remote sensing. Finally, the gravity field of the Earth and its temporal variations would require models for mass balance, fluxes, and circulation patterns which put constraints on the geo-kinematic models. Measurements would be obtained from terrestrial, shipborne, and airborne gravimetry, dedicated gravity satellites, and absolute gravimeters. The largest future contribution to this component is expected from new gravity satellite missions, such as GRACE and GOCE.

The numerous interactions between the components of GIGGOS are discussed in Rummel (1998) where numerous detailed references to specific topics are given. If implemented, such a system

- would integrate much of the current research in geodesy and use the existing engineering infrastructure and services to advantage,
- would accelerate the integration of classical and space measurement techniques,
- could thus become a focal point for international research activities and the IAG
- would more clearly identify the geodetic contribution to Earth system science and show that the interaction of geodesy with other Earth sciences goes well beyond data delivery and is not restricted to solid Earth research,
- would, on the one hand, use the metrology tradition and strengths of geodesy and, on the other hand, open new vistas and challenges for young geodesists.
- could become an identifiable component in the Earth Observing Science (EOS) program, as shown in Figure 4.

3.4 Is a change of the IAG structure necessary?

If IAG decides on a major research focus, such as GIGGOS, then the question arises whether the current IAG structure will give the best possible support to such an enterprise. A number of reasons, why this may not be the case, have already been given in chapter 2. They will be briefly summarized here:

- The current structure splits geodesy in compartments and generates a section focus.
- In some cases Sections and Commissions cover the same area of responsibility and create unnecessary administrative overlap.

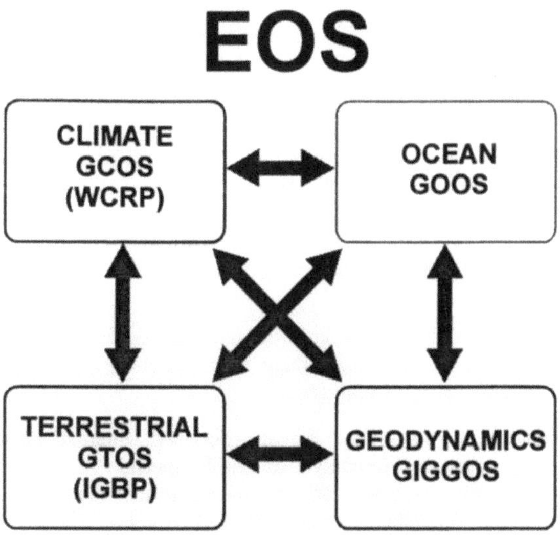

Fig. 4: GIGGOS as an identifiable component of EOS

- Commissions often live well beyond their useful lifetime.
- A large number of Special Study Groups are not working effectively as a team.
- Services are currently not well integrated in the IAG structure and the interaction between the IAG executive and the executive of respective services is minimal.
- The flexibility to respond to changes quickly is rather restricted and must be improved.

It appears that some of these difficulties are due to changes which have taken place since the current structure was introduced. This was done shortly after World War 2 at the meeting of the Permanent Commission in Paris in August 1946. P.Tardi (1946), then General Secretary of the Association, proposed this structure. It was adopted on a tentative basis at that meeting, with the provision that a final decision would be made at the General Assembly in Oslo in 1948. The argument for its adoption was that the number of Commissions was getting too large and that the scientific work could be better organized in this way. The five sections proposed were: Triangulation (including projections); Precise Leveling; Geodetic Astronomy; Gravimetry; Geoid. It should be noted that the Section structure essentially represented the major measurement techniques at that time. Together, they were used to map the surface of the Earth according to Helmert's definition. Thus, the IAG focus was implicitely given as mapping or modeling, as we would say today.

This structure, in a formal sense, has survived until today. However, important modifications of the section responsibilities were made at the General Assemblies in 1971 (Moscow) and 1983 (Hamburg). Today, the five sections cover: Positioning; Advanced Space Technology; Determination of the Gravity Field; General Theory and Methodology; Geodynamics. This means that only one of them emphasizes measurement techniques. The others are topical in nature and subdivide the mapping part in a number of compartments. This has probably contributed to the loss of a common focus and the development of separate Section foci mentioned above.

In addition to this internal change, there have been major external changes. They have to do with the fact that the number of national and international agencies, which are active in global geodesy, have considerably grown and are actively involved in important international projects in which IAG plays only a minor part. Space agencies, such as NASA and ESA, have to be credited for many of the recent advances in global geodesy and sponsor major parts of geodetic research. Strong national agencies, such as GFZ, IGN, and NIMA, have stimulated and supported international projects with strong research components. In the past, other scientific unions, such as the International Astronomical Union (IAU), and other associations of the IUGG, such as IAPSO and IAMAS, were the major scientific partners. Today, meetings organized by other geophysical societies, such as the AGU and the EGS, have increasingly become places where new geodetic results are presented. For GPS research the ION-GPS meetings have taken on a similar function. There are few or no established links between these new stakeholders and the IAG. This means that the forces driving geodesy today are only partially active in IAG.

Thus, when discussing a new IAG structure in support of an IAG research focus, one has also to consider that

- there have been important changes in the topical definition of the sections which may have contributed to a loss of focus for the IAG itself,
- major geodetic research is done today by national and international agencies which have few or no established links with the IAG,
- other scientific societies involved in geodesy interact very little with the IAG.

Do these considerations clearly point towards a major change in the IAG structure? In the author's opinion they do, for internal as well as for external reasons. The specific form such a restructuring will take, depends very much on the priorities that IAG wants to pursue. There seem to be a number of possible options. Such options will not be discussed in detail. Instead, the items listed in Conclusions and Recommendations should be seen as constraints for any of these options.

4 Conclusions and recommendations

Based on the above discussions, the following conclusions have been drawn:

- The current IAG structure falls short of representing the forces that drive geodesy today and has some major weaknesses in supporting an IAG research focus.
- Internally, the section structure subdivides research into compartments which make it difficult to pursue a common research focus.
- Sections, Commissions, and Services have no clear functional distinctions and are very unequally presented in the Executive.
- Externally, IAG is no longer the major international player in geodesy and does not interact sufficiently with other major international players.

It is therefore recommended that IAG

- adopts a mission statement with a clear research focus and a statement of its role vis-à-vis other international players
- seriously considers the abolishment of either sections or commissions
- strengthens the interaction with and the representation of the services, keeping in mind the interest of other stakeholders
- incorporates most of the Special Study Groups into Commissions and Services
- actively plans project-based cooperation with other major international players.

References

Baeyer, J.J. (1861): Über die Größe und Figur der Erde -Eine Denkschrift zur Begründung einer mittel-europäischen Gradmessung. Berlin, 1861.

Mueller, I.I. (ed,1998): Science Services: International Association of Geodesy IAG/FAGS. Special Proc. IAG Scientific Assembly, Rio de Janeiro, Brazil, Sept. 3-9, 1997. Available through the IAG Central Bureau, Copenhagen.

Rummel, R. (1998): Global Integrated Geodetic and Geodynamic Observing System (GIGGOS). Proc. IGGOS Symposium, Munich, Dec. 1998 (in press)

Tardi, P. (1946): Reunion de la Commission Permanente (Paris, 5-8 aout, 1946). Bulletin Geodesique, nouvelle serie, no. 2, pp. 1-39 (specifically page 25)

Tardi, P. (1963): Hundert Jahre Internationale Erdmessung. Zeitschrift fuer Vermessungswesen, Vol. 88, No. 1, pp. 2-10.

Willis, P. (1996): The Geodesist's Handbook. Journal of Geodesy, Vol. 70, no. 12, pp. 839-1036

The pros and cons of having sections in IAG

F. Sansò
D.I.I.A.R., Polytechnic of Milan,
Piazza Leonardo da Vinci 32, 20133 Milan, Italy

Abstract. The present structure of the IAG organization in terms of Sections is analyzed. The deep unhomogeneity of the present Sections in IAG and their significant overlap is illustrated. Pros and cons of organizing IAG into Sections are then examined and short, provisory conclusions are drawn.

Keywords. IAG Sections.

1 The state of the art

According to the by-laws of the IAG, the objectives of the Association are:

a) *"to promote the study of all scientific problems of geodesy and encourage geodetic research;*
b) *to promote and coordinate international cooperation in this field, and promote geodetic activities in developing countries;*
c) *to provide, on an international basis, for discussion and publication of the results of the studies, researches and works indicated in paragraphs a) and b) above".*

Therefore, *"to achieve these objectives, the Association shall comprise a small number of Sections, each of which deals with a distinct part of geodesy. Commissions, Special Commissions and Special Study Groups may be formed as provided in the By-Laws"*, see I, 3) of the Statutes. The definition of Sections emerging from here is that it is a structure which aims at organizing the scientific work of IAG into distinct parts of geodesy. In other words we expect that the Sections structure should correspond to some rational partitioning of geodetic research into subjects.

By stating that the number of Sections should be small, is indicated that subjects should cover wide areas of geodesy so that one could expect a permanent scientific activity of the IAG in the specific field, covered by a Section.

The current Section Structure is:

- Section I Positioning
- Section II Advanced Space Techniques
- Section III Determination of the Gravity Field
- Section IV General Theory and Methodology
- Section V Geodynamics.

This structure is the result of a historical process, which is summarized in Table 1, taken from the very interesting paper by Beutler et al. (1999).

The names of the Sections in 1957 seem to correspond quite well to the above definition and this neat delineation of the borders seems to become increasingly blurred with the new versions. It is also interesting to note that the revisions of 1971 and 1983 were put in place after a period of 14 and 12 years respectively, so that a change at the next General Assembly (2003) after 20 years will be a sign that the actual organization has been able in one way or other to cope with the needs of the Association for a longer period.

The Section is directed by a President which is usually supported by a Secretary and by the Presidents of those Commissions, which are under the Section. All of them are representatives of the Section to the Executive Committee. Furthermore within the Sections there can be Special Commissions and Special Study Groups.

In synthesis, the Commissions deal with long term scientific scopes, but every member country of IAG from all over the world or from a specific region, has the right to nominate its own representative to the Commission; Special Commissions are similar to Commissions in terms of scopes, but without any geographical constraint in membership; Special Study Groups on the other hand should be focussed on specific questions so that, as stated in the by-laws, they should last 1 or at most 2 periods of 4 years.

To analyze the role of the Sections in the actual work of the IAG one has to look carefully into the contents and correspondingly into the organizing structure trying to understand whether this is helping the development of science by international cooperation (pros) or rather hindering it (cons).

International Association of Geodesy Symposia, Vol. 121
Schwarz (ed.), Geodesy Beyond 2000 – The Challenges of the First Decade

Table 1: Historical evolution of Section names in the IAG (from Beutler et al. (1999))

	1957	1971	1983
Section I	Triangulation	Control surveys	Positioning
Section II	Precise levelling	Satellite surveys	Advanced space techniques
Section III	Geodetic astronomy	Gravimetry	Determination of the gravity field
Section IV	Gravimetry	Theory and evaluation	General theory and methodology
Section V	Geoid	Physical interpretation	Geodynamics

2 Contents

Rather than repeating the terms of reference of the IAG By-laws for Sections, I prefer to refer to the description given in the Call for Papers for the General Assembly in Birmingham issued in May '98. It contains the most recent point of view of the Section Presidents on their own subject. We have then:

- Section I – Positioning
  Section I is concerned with the scientific aspects of the measurement and analysis of regional and global geodetic networks as well as satellite, inertial, kinematic and marine positioning. The practical results of this research work is made available through recommendations to National Survey Organisation. Applications of geodesy in engineering is a subject dealt with by Section I. Tremendous advances of GPS surveying have occurred especially in precision and applicability.
  However, there are many issues of accuracy and reliability of GPS surveying (hardware and software) which need to be addressed. Furthermore GPS measurements have shown the potential to be used as remote sensing tool of atmospheric parameters.
- Section II – Advanced Space Technology
  Section II is engaged in new space techniques for geodesy and geodynamics. Its objectives are to anticipate and promote their implementation into geodetic/geodynamic work and, in general, support and coordinate the optimal use of modern space technology for the benefit of geodesy.
- Section III – Determination of the Gravity Field
  The Section is engaged in the determination and modelling of the Earth's gravity field. This includes: absolute and relative terrestrial gravity measurements, gravity networks and control stations, non-tidal gravity variations, determination of the external gravity field and geoid, the different gravity field data types, and reduction and estimation of gravity field quantities.
- Section IV – General Theory and Methodology
  Topics addressed by Section IV: mathematical and physical foundations of geodesy with a particular focus on statistics, numerical and approximation methods, boundary value problems, geometry, relativity, cartography, theory of orbits and dynamics of systems; inversion of altimetric data; wavelets in geodesy; integrated inverse gravity modelling; dynamic isostasy; temporal variations of the gravity field.
- Section V – Geodynamics
  Section V deals with the following topics: reference systems, monitoring and study of time-dependent phenomena, such as polar motion, precession and nutation, length of day, motion of the geocenter, Earth tides, crustal deformations, variations of gravity, sea surface topography including mean sea level, geodetic aspects of international geodynamic projects, such as the Lithosphere project, geophysical interpretation of gravity and related data.

The first comment on the above material is that the number of items of interest to geodesy which is not contained in the above definitions seems to be very small.

The second obvious comment is that the definition of Sections, given in § 1, is certainly not satisfied in that the topical overlap is very large. As seen from Table 2 (which has no pretension of being exhaustive), not only Section IV is, almost by definition, overlapping with the other Sections for all items requiring "a theory", but also the other

Table 2: Diagram of topical overlap

	Space techniques	Gravity field	Theory	Geodynamics
Positioning	Reference frames GPI, SLR,... Kinematic pos. Refraction	Leveling Height datum Inertial navigation systems	All items	Reference frames Crust monitoring and strain estimation Refraction
Space techniques		Orbit analysis Gravity f. model. Altim.: -Geoid -SST	All items	Reference frames Orbit analysis Altimetry Refraction
Gravity field			All items	Gravity inversion Time variability of gravity
Theory				All items

Sections are strongly overlapping, particularly Section II, with the others. A sign of this overlap is that many times in the past SSG's with very similar titles were established in different Sections, sometimes even with almost identical member lists.

Of course, as I can personally testify from 16 years of experience in the Executive Committee, when good work is done in one area, it is often possible to obtain good results, due to the cooperative attitude of the officers of the Sections and of the members of SSG's, but one cannot say that these results were facilitated by the Sections structure.

Another example of this kind has been the creation of the Geoid Commission, which was originally proposed by Section IV, at the time named Theory and Evaluation (the geoid computation was then considered as a case of numerical application of the boundary value problem theory), and finally assigned to Section III which at that time was adopting the name of "Determination of the gravity field", thus widening its peculiar subject.

Finally, one has to observe that in the present structure, Sections seem to play a different role. In fact, while we expect that in Section I positions of points be produced, in Section II satellite missions (or observational campains) and their analysis be the outcome, in Section III the gravity field be determined, in Section IV new theories or methods are produced to treat the subjects of the others, and in Section V the geophysical interpretation of the outcome of the first three are given. In other words the 5 Sections have different roles in the complex of the IAG activities.

3 The organizing structure: pros and cons

3.1 Pros

- The actual Section structure allows its President to coordinate and balance long-term research, activated by Commissions and Special Commissions, and short term research, made by SSG's. This is important to avoid duplication and waste of efforts and at the same time it can promote activities in fields where there is a need of a fast response by IAG to new scientific and technological developments.
- The Sections structure has worked in the past and probably could work in the future, too, on condition that it is implemented much closer to the original definition, significantly decreasing the overlap and following a scheme in which all Sections enter in a homogeneous position into IAG activities.
  Moreover all Scientific Associations have some structure similar to Sections.
- There may be a good point in going through the position of Commission President or Secretary for one period and before becoming President of a Section; in this way one has the possibility of acquiring experience of how the Executive works as a collective guidance of IAG, having time to mature in the process.

3.2 Cons

- The President of a Section is too far from the members who really do the work in the Section, namely members of SSG's or of Commissions or of Special Commissions; i.e. the Section is very often excessively structured.

President of Section

Secretary
Chair of SSG
Member

President of Commission
President of Subcommission
Member

Fig.3: Section structure

For instance, in a Section there are typically 4 different "levels", as we can observe in Figure 3. So when the President is supposed to be mature and good it might not be able to pass its own experience to the young people since their "bureaucratic" distance is too large.

- Since everything happens in periods of 4 years in IAG, a young person entering into IAG as a member of something, has to wait at least 12 years before he/she can contribute significantly to drive the Association. This gives a very poor incentive to young people and produces dissatisfaction.
- The "overstructure" of the Section produces a significant drawback with services; in fact if the Service is well in touch with the Section, the Commission which is in between becomes weaker and faints, if the Service is strongly connected and integrated into a Commission, then it is the Section which looses control.

4 Three brief conclusions

Considering the problem of Sections it seems to me that three brief conclusions could be drawn:

1) If the Structure of Sections will be kept it has to be reorganized in such a way that all Sections enter in a comparable position into IAG. Moreover, although it's clear that science cannot be neatly cut into pieces, overlap of subjects should be significantly decreased and procedures should be established to deal with overlaps.
2) If Sections have to survive, the internal structure has to be significantly simplified by eliminating at least one of 4 levels of Figure 3; moreover, Services should not be part of the Section structure.
3) If Sections are eliminated one suggestion would be that only one level of permanent structures be left for the scientific research work in IAG, for instance the Commissions; permanent here means that no a priori bound is put to its life period. Temporary structures, for instance SSG's could then be created by one or more Commissions, to which they should report with a well defined and specific project plan.
   In this way an orthogonal type of organization, rather than purely vertical, would be created.

References

Geodesist's Handbook (1996). In: Journal of Geodesy, vol. 70, no. 12.

IAG Executive Committee (1998). IUGG 1999 Symposium – IAG Symposia. Call for papers for IUGG General Assembly, Birmingham, UK.

Beutler G., Drawes H., Rummel R. (1999). Reflections on a new structure for IAG beyond 2000. Conclusions from the IAG Section II Symposium in Munich. Presented at G6 – IUGG General Assembly, Birmingham, UK.

IAG services in the current framework of the International Association of Geodesy (IAG)

G. Beutler
Astronomical Institute, University of Berne, Sidlerstrasse 5, CH-3012-Bern, Switzerland

Abstract. It is generally acknowledged that *services* play an essential role in geodesy and, consequently, in the International Association of Geodesy (IAG). Services are not a recent invention within the IAG. They accompany the Association since its creation.

Services in general (not only within IAG) are (should be) created if a well defined user community can be identified needing well defined products. Recent examples within the IAG are the creation of the International Earth Rotation Service (IERS) in 1987 and the creation of the International GPS Service (IGS) in 1991 (officially approved by IAG in 1994).

The IGS and the IERS are probably the best known IAG services with an impact reaching far beyond geodetic applications. There are many more services within IAG, however.

In our analysis we first characterize a scientific service, then we comment the current situation of IAG(-related) services; as a case study we briefly review the activities of the IERS and the IGS. We conclude with some thoughts concerning the future role of the IAG services.

Keywords. International Association of Geodesy, services.

1 Characterization

Let us try to characterize a scientific service with a few keywords:

- The word *Service* implies that there is a well defined *set of products* and a sizeable *user community*.
- Services should be based on a long-term need. Nobody should create a service for just a few years. Decades are the basic time unit.
- *Accuracy, reliability, robustness,* and *timeliness* are the essential characteristics for the *operational aspects* of the service.
- In a scientific services scientists are the most demanding part of the user community. Science is thus an important aspect within the IAG services.
- In a scientific service there must be a clear distinction between *operational* and *research-oriented* aspects in order not to disturb the generation of a well defined set of products.

What makes a scientific service successful? This question is delicate to answer. Let us give a few conditions that have to be met:

- Meeting the user community's requirements has to be the first priority.
- The interactions between the most demanding part of the user community, in the case of the IAG Services the scientific community, and the service must be regular and intensive.
- *Long time series* are of crucial importance in the geosciences. This is why
  - archiving the service's products (raw data also must be considered data in this context),
  - providing easy access to the products

  must be of primary concern to the services. This may even be the most important aspect of a particular service.
- If the community is broad, the products have to be easily understood and used.
- *Redundancy* is an important keyword for the long lasting success of a service. Redundancy is required on the observational side

International Association of Geodesy Symposia, Vol. 121
Schwarz (ed.), Geodesy Beyond 2000 – The Challenges of the First Decade

(to mitigate instrument and operator failures) ***but also*** on the analysis side.

- Competition between different analysis groups is an essential driver for the quality of the service's product (see also discussion in section 3).

2 Status Quo

The structure, tasks, composition and work of the IAG services is documented every four years in the *Geodesist's Handbook* (latest version issued in 1996, new version to be issued in the year 2000) and in the *Travaux de l'Association Internationale de Géodésie.*

It became clear in recent years that the IAG services play an increasingly important role in the IAG framework. This is why Prof. Ivan Mueller was asked to organize a special session at the IAG Scientific Assembly in Rio de Janeiro in July 1997 giving all IAG Services the opportunity to describe their achievement, in particular their products and their "customers". This session is very well documented in Mueller (1998). The booklet gives a concise overview of the current situation.

The IAG Services are very different in nature: They may consist of single institutions, as in the case of the BIPM (Bureau International des Poids et Mesures) or the PSMSL (Permanent Service for Mean Sea Level), or of a loose collaboration of a great manifold of organizations, as in the case of the IGS, the IERS, and the other Space Geodetic Services.

The IAG Services have an impact on a broad user community: Atomic time and frequency, as generated and disseminated by the BIPM, is of greatest importance not only in the geosciences but also in physics and in data transmission. The terrestrial and celestial reference frames, as realized and maintained by the IERS are fundamental in all geosciences, in astronomy, but also for setting up new navigation satellite systems (we think, e.g., of the plans for the European GALILEO system). The technique specific services IGS, ILRS, and IVS are not only of vital importance as suppliers for the IERS but they have greatest impact on navigation, atmosphere sciences, geodynamics, etc. The BGI (Bureau Gravimetrique International) and the IGeS (International Geoid Service) have a major impact on space missions and on industry.

Table 1: The current IAG services

Service/Section	Short Title
IGS/II	Intl. GPS Service
IVS/II	Intl. VLBI Service
ILRS/II	Intl. Laser Ranging Service
BGI/III	Intl. Gravimetric Bureau
IGeS/III	Intl. Geoid Service
IERS/V	Intl. Earth Rot. Service
BIPM/V	Intl. Bureau of Weights and Measures
ICET/V	Intl. Centre for Earth Tides
PSMSL/V	Permanent Service for Mean Sea Level
IBS/—	IAG Bibliographic Service
IIS/—	IAG Information Service

Table 1 lists the currently existing IAG services. The table is taken from Beutler et al. (1999) where the structural aspects of the services are discussed.

Table 1 illustrates that the bandwidth of IAG services is broad indeed, covering pure documentation (e.g., the IIS and the IBS) and services dealing with almost the entire range of geodesy and geodynamics (like *IERS, IGS, IGeS, BGI, and ICET*). Services, like, e.g., the PSMSL are truly interdisciplinary in nature.

In Table 1, the services are arranged according to the IAG Sections. It is immediately clear that this principle of ordering is not a very logical one. The technique-oriented services associated to Section II are in practice very closely related to the IERS in Section V.

3 Two case studies: The IERS and the IGS

Modern services go far beyond the *product/customer*-driven concept. Let us try to explain briefly the principles underlying the *IERS (International Earth Rotation Service)* and the *IGS (International GPS Service)* in order to understand their success.

Both services have clear and relatively simple mission statements. They take a very broad a view when trying to accomplish their mission. The IERS mission is to establish and maintain the ICRF (International Celestial Reference Frame), the ITRF (International Terrestrial Reference Frame), and to determine the best possible time series of transformation parameters between the two frames. These transformation parameters are called Earth Rotation/Orientation Parameters (ERPs). The frames and the transformation parameters are meant to have state-of-the-art accuracies. This demand implies that the observation techniques may change in order to always use the most promising observation techniques. In 1987 the IERS started with VLBI (Very Long Baseline Interferometry), SLR (Satellite Laser Ranging), and LLR (Lunar Laser Ranging) as primary observation techniques. Later on GPS (Global Positioning System) and DORIS observations were accepted as official IERS observation techniques. It is of vital importance for the IERS that different observation techniques are used.

The IERS generates combined products from the technique specific series of coordinates, transformation parameters, etc. This very ambitious task can only be addressed through an intensive collaboration of all relevant scientific organizations. Regular workshops organized by the IERS, technical and annual reports, the maintenance of a set of IERS analysis conventions document the work performed within the service and the progress made.

Through the creation of the IERS Bureau of Atmospheric Angular Momentum (AAM) and the recently created Global Geodynamic Fluids Center the IERS documents that it wants to take into account all geodynamical aspects related to Earth rotation. It is this holistic approach which guarantees the success of the service and the value of the IERS products.

The mission statement of the IGS is simple, as well. As opposed to the IERS the IGS is based on a single technique, namely Global Positioning System (GPS). Archiving and making available raw observations from its global network, generating and making available products like satellite ephemerides, station coordinates, etc., are primary duties of the IGS.

It is again the holistic approach centered around the primary mission which guarantees the value of the service: After having accomplished its primary duties (making available raw observations and highly accurate orbits) the IGS soon started exploiting all scientific aspects inherent in the GPS observations. Contributions to the IERS through high resolution ERP series and through sets of station coordinates and velocities, but also ionosphere mapping, GPS meteorology, time and frequency transfer, and spaceborne applications of the GPS are but a few keywords indicating the broad field of science and application covered today by the IGS.

The reliability and robustness of IGS products is based on a global tacking network sponsored by many international agencies (a fair degree of redundancy proved to be very useful in the past when instrument or data transfer failures related to single contributors occurred), on a fair number of IGS Analysis Centers (a friendly competition between these centers was and is extremely beneficial for the quality of IGS products), and last but not least on the generation of official IGS products based on the individual contributions of the IGS Analysis Centers.

The IGS makes a clear distinction between service-related aspects (documented by regular reports in the electronic IGS report series) and scientific aspects dealt with in working groups or pilot projects.

Annual reports, technical reports, proceedings of workshops document the work of the service and the progress made on the operational and scientific side.

Let us try to summarize the essential characteristics of services of the kind of the IERS and the IGS:

- Clear mission statements and a broad and holistic approach to accomplish the mission are mandatory.
- *Reliability, robustness, timeliness, proper interfacing with the user community, optimization of the performance* are essential for the service-oriented activities within the service.
- Research-oriented activities *within* the service primarily address the following areas:
  - Improvement of the *classical products* of the service, optimization of the production of these products,
  - exploitation of the full potential of the *raw material (input)* the service is dealing with, and, if appropriate
  - establishment of *new sets* of products.

- *Originality, openness for scientific collaboration* with other IAG entities (currently other IAG services, commissions, study groups, etc.) proved to be essential in this context.

Opportunities for collaboration in research were sought by both IAG entities. Let us mention the establishment of the AAM Bureau and the creation of the Global Geodynamic Fluids Center as excellent examples for such research-oriented activities within the IERS. The establishment of the IGS/BIPM Working Group on Time and Frequency Transfer using GPS and the organization of the *IGEX-98*, the first global GLONASS observation and analysis campaign, by CSTG, IGS, ION, and IERS are excellent examples for IGS-related activities with a heavy involvement of external collaboration.

The success of the IGS was, as a matter of fact, so convincing, that two new services, the *ILRS (International Laser Ranging Service)* and the *IVS (International VLBI Service for Geodesy and Astrometry)* emerged recently from the corresponding CSTG Subcommissions in Section II.

4 Potential role of the services in the new IAG structure

There are many relations and interactions between the IAG services. The three services of Section II are, e.g., closely cooperating with the IERS in Section V. Also, there are often bilateral relationships between services and/or commissions in the context of projects or working groups. The project of the IGS and the BIPM (Bureau International des Poids et Mesures), the IGS/BIPM Pilot Project on time and frequency transfer, or the organization of the GLONASS tracking and analysis experiment IGEX-98 by CSTG, IGS, IERS and ION (Institute of Navigation) were already mentioned above.

There are, at times, topics of common interest to a number of services. Let us mention the ongoing activity of creating the International Space Geodetic Network (ISGN) by the services IGS, ILRS, and IVS in Section II, the IERS and (possibly) the BIPM and PSMSL in Section V.

The examples show that attractive geodetic activities are taking place in the services without having a serious influence on the conventional IAG structure because the IAG Services are rather independent and not well embedded in the "normal" IAG structure (Sections, Commissions, Executive). The chairpersons of the services are *not* assigned by either the IAG Sections or the IAG Executive Committee. The IAG thus has very limited influence on the Services.

On the other hand the Services are represented neither in the IAG Executive Committee nor in the IAG Sections' Steering Committees. The services thus have a very limited influence on the higher IAG levels.

What can be done to improve the situation? It seems clear to us that the relationship between the upper IAG levels and the services has to be improved. This means in particular that (some of) the services have to be represented in the IAG Executive. It is completely impossible, on the other hand, to ask for service representatives for each of the services in the IAG Executive. It seems therefore necessary to form groups of services which are logically related and to assign 1-2 representatives to each of the groups.

In Table 2, we formed three such groups. The first group is geometry and time related, the second one gravity and ocean related, whereas the third one is purely administrative in nature.

Let us also mention Commission X (Networks), Commission VIII (CSTG), Commission VII (RCM), and Special Commission 8 (Sea Level&Ice Sheet) are linked to the first group, Commission III, Commission XII (Gravity and Geoid) and CV (Tides) are more linked to the second group.

We do not pursue these relationships any further here. This must be one of the topics of the IAG Retreat early in 2000, see (Beutler et al., 1999). It is, however, good to keep in mind that such relations exist and that there are options to give the Services a more pronounced role in the new IAG structure.

5 Summary and wishes

It seems clear that the services must play an important role in the IAG structure being developed right now. It is also clear, on the other hand, that IAG cannot uniquely rely on services. Many of geodesy-relevant research-oriented activities must be dealt with in other parts of the IAG structure.

Let us conclude this review of the role of the IAG services within the current and the future IAG structure with two wishes:

Table 2: Logical groups of the current IAG services

Service/Section	Short Title
IERS/V	Intl. Earth Rot. Service
IGS/II	Intl. GPS Service
ILRS/II	Intl. Laser Ranging Service
IVS/II	Intl. VLBI Service
BIPM/V	Intl. Bureau of Weights and Measures
PSMSL/V	Permanent Service for Mean Sea Level
BGI/III	Intl. Gravimetric Bureau
ICET/V	Intl. Centre for Earth Tides
IGeS/III	Intl. Geoid Service
IBS/—	IAG Bibliographic Service
IIS/—	IAG Information Service

- **Wish, addressed to the Service:** Maintain or develop a high level of "open mindedness" regarding the scientific collaboration with other parts of IAG, with sister unions, IAU, etc., and try to incorporate young scientists into the services' work a soon as feasible.

- **Wish, addressed to the IAG:** Perform the review and restructuring process of the IAG in the spirit alive in (at least some of) the IAG services, important elements of which are respect for everybody's contribution to IAG, respect for minorities, taking the other (geo-)sciences and their requirements serious.

References

Beutler, G., H. Drewes, R. Rummel (1999). Reflections on a New Structure for IAG Beyond 2000 — Conclusions From the IAG Section II Symposium in Munich. In *this Volume*.

Geodesist's Handbook (1996). In *Journal of Geodesy*, Vol. 70, No. 12, pp. 839-1036.

Mueller, I.I. (1998). Science Services: International Association of Geodesy (IAG) / Federation of Astronomical & Geophysical Services (FAGS). In *Special Volume, IAG Scientific Assembly, Rio de Janeiro, Brazil, Sep 3-9, 1997*, 62 pages, available through IAG Central Bureau c/o Department of Geophysics, Copenhagen, Denmark.

Willis, P. (ed.) (1995). Travaux de l'Association Internationale de Géodésie, Tome 30, Rapports Généraux et Rapports Techniques étblis á l'Occasion de la Vingtiéme Assemblée Générale á Boulder, USA, Juillet 1995. Secrétariat de l'Association, 140, rue de Grenelle, 75700 Paris, France.

The role of IAG Special Study Groups

Detlef Wolf
GeoForschungsZentrum Potsdam, Division 1: Kinematics and Dynamics of the Earth, Telegrafenberg, D-14473 Potsdam, Germany

Abstract. The role of Special Study Groups (SSGs) in the IAG is briefly discussed. The assessment is largely based on the author's experience gained while chairing SSG 4.176 during the IAG period 1995–1999. Specific points of discussion are (1) the relation between SSGs and Special Commissions, (2) the appropriate procedure of establishing new SSGs and (3) the relations between SSGs covering similar topics.

Keywords. Executive Committee, Section, Special Commission, Special Study Group, gravity field.

1 Introduction

At present, the IAG structure consists of five major Sections. Each Section is further subdivided into Commissions (Cs), Special Commissions (SCs), and Special Study Groups (SSGs). These types of structural elements are the scientific working groups, and their functions and objectives are defined in the statutes and by-laws of the IAG. According to them, Cs are established to cover problems that are of broad and long-term interest to the international community of geodesists. SCs, on the other hand, are smaller and cover more specialized problems of medium- to long-term interest. Finally, SSGs are formed to study topical scientific questions and are usually terminated after one four-year IAG period.

In the following, the role of SSGs in the IAG is briefly reviewed and some conclusions are drawn from this. The discussion is essentially based on personal experiences of the author as chairman of SSG 4.176: *Models of Temporal Variations of the Gravity Field* during the IAG period 1995–1999 and as past member of two other SSGs and is therefore necessarily *subjective* and *selective*. An objective and comprehensive discussion of the role SSGs have played in the IAG over the decades appears to be difficult and would require the input of the long-term IAG officers that have been actively involved in the work of SSGs over several four-year IAG periods.

The discussion starts with a comparison between SSGs and SCs with emphasis on the question whether the distinction between their functions and objectives is sufficiently clear (Sec. 2). After this, the problem of the appropriate procedures of establishing new SSGs and selecting its membership is considered (Sec. 3). This is followed by a discussion of the relations between SSGs with overlapping topics (Sec. 4). The article concludes with some recommendations for the future (Sec. 5).

2 Special Study Groups and Special Commissions

According to the 1996 edition of the Geodesist's Handbook, 'Special Study Groups may be formed to study *specific scientific problems of limited scope* which require close cooperation between specialists from different countries' (italics by the author). On the other hand, 'Special Commissions may be formed to study *scientific problems of a long term character* which require close cooperation between specialists from different countries' (italics by the author). The formal difference between the definitions is obviously quite small. In practice, it means that SSGs should cover topics which are of immediate interest to only a small fraction of the geodetic community and for which definite results can be achieved over a short-term period, so that SSGs are normally terminated after a single IAG period. Typically, SSGs have not more than 20 members. In contrast to this, SCs should be concerned with problems that are of interest to a larger group of geodesists over several IAG periods. In view of the broader scope, SCs are usually somewhat larger than SSGs and may be comprised of up to 30 members.

SSG 4.176: *Models of Temporal Variations of*

International Association of Geodesy Symposia, Vol. 121
Schwarz (ed.), Geodesy Beyond 2000 – The Challenges of the First Decade

the Gravity Field chaired by the author during the IAG period 1995–1999 was associated with Section IV: *General Theory and Methodology*. To this Section also belonged SSGs 4.168–4.171 and SC 1: *Mathematical and Physical Foundations of Geodesy*, which is clearly of interest to a large portion of the geodetic community over a longer period of time. For practical reasons, SC 1 is subdivided into five subcommissions with the names (1) statistics, (2) numerical and approximation methods, (3) boundary value problems, (4) geometry, cartography, relativity and (5) theory of orbits and dynamics of systems. These research fields are necessarily more specialized and, thus, the formal distinction between subcommissions and SSGs becomes subtle.

This potential problem is exemplified by the 'history' of SSG 4.176, which was established in 1995 in response to the increased interest in temporal variations of the gravity field resulting from the markedly improved accuracy of surface and satellite gravity measurements that had taken place. Initially, attempts were made to create a sixth subcommission of SC 1 for this purpose. This was clearly in recognition of the fact that the development of theories for studying gravity variations is a truly *long-term* commitment, which can be traced back continuously to the famous studies by Thomson (1864), Darwin (1879) and Love (1911). In contrast to this was the recognition of the *topical* character of upcoming new surface techniques (*e.g.* Lambert *et al.*, 1995, Crossley *et al.*, 1999) and satellite missions (*e.g.* Reigber *et al.*, 1997, Tapley, 1997) to measure gravity with unprecedented accuracy, which had renewed interest in the theoretical foundations. Finally, the Executive Committee decided in November 1995 to emphasize this topical aspect and to establish SSG 4.176 for the IAG period 1995–1999. In view of the long-term character of fundamental theoretical work, it is however planned to continue the work of SSG 4.176 in extended form in the newly created SSG 4.186: *Dynamic Theories of Deformation and Gravity Fields* during the IAG period 1999–2003.

3 Establishment of SSGs

The establishment of new SSGs and the nomination of their chairpersons has traditionally been done by the IAG Executive Committee in response to proposals by the presidents of the individual Sections. This procedure acknowledges the fact that the Section presidents and the Executive Committee are in a position to oversee the general and topical developments in the individual fields of geodesy and, thus, to create new SSGs in order to focus activities in a suitable way. In addition, their professional relations allow them to select appropriate chairpersons who, in turn, are responsible for inviting individual members.

There are however at least two disadvantages associated with this procedure. One is that, with each Section of the IAG represented in the Executive Committee, the establishment of new SSGs is closely related to this sectional structure. In practice, this means that there may be a tendency of associating the importance of an individual Section with the number of SSGs belonging to it. A consequence of this problem is the question whether SSGs should belong to Sections at all, which is one of the issues of the current debate on reforming the IAG structure. The other disadvantage is that the traditional selection procedure tends to nominate chairpersons who – although experienced IAG officers – have too many responsibilities to act efficiently as leaders of SSGs.

An alternative to this well-established procedure is the proposal of topics for SSGs and of potential chairpersons by individuals working in the IAG. The advantage of this more liberal procedure is linked with the experience that activities based on personal initiatives of dedicated persons have a good chance to be successful and produce valuable results. For the upcoming IAG period 1999–2003, the IAG essentially adopted the 'free-market approach' and solicited proposals of individuals, which have subsequently been screened by the Executive Committee.

An important point to be considered by the chairperson of an SSG is the selection of its members. For convenience, these members may be categorized as *senior scientists*, as *junior scientists* or as *students*. Apart from purely scientific aspects, the role of senior scientists in SSGs is that they may assist junior scientists and students when further pursuing their scientific studies and professional careers. A more practical aspect is that, usually, only senior scientists have sufficient funds to attend meetings of their SSG or to provide funds for their students. On the other hand, there is again the danger of collecting a large number of well-respected senior scientists with too many obligations to participate actively in the work of their SSG.

The asset of having junior scientists as members of SSGs is clearly their high motivation and fresh look at the problems studied. An important point is also that SSG meetings may provide the rare or even first opportunity for them to meet peers from foreign countries. A problem with SSG meetings can however be that junior scientists lack the funds for attending such meetings, which may take place far away from their home countries.

The participation of students in the work of SSGs is more difficult and usually requires the affirmative action of their supervisors. Apart from that, the same comments apply as for junior scientists. Quite promising appears to be the idea of gaining some of the senior scientists as associate members and involving junior scientists and students from their group in the active work of the SSG.

4 Relation between SSGs

Usually, several of the SSGs established have some overlap. This may be because they study related topics or because they study the same topic from a different angle. Although such a situation may look undesirable at first sight, it has the potential to lead to fruitful cooperation.

The situation may be illustrated by taking again as an example SSG 4.176, which coexisted with SSG 4.171: *Dynamic Isostasy* and SSG 5.174: *Geophysical Interpretation of Temporal Variations of the Geopotential* during the IAG period 1995–1999. With some simplifications, the relation between SSGs 4.171 and 4.176 can be characterized by the statement that both were concerned with the study of perturbations of the hydrostatic equilibrium state in the earth's interior and employed dynamic theories to quantify these perturbations. The difference is related to the types of forcing considered and the observables used to study the perturbations: Thus, SSG 4.171 investigated isostatic disequilibrium as caused by temporally variable *external* and *internal loading* and as directly measured by uplift or subsidence of the earth's surface. In contrast to this, SSG 4.176 considered also processes other than dynamic isostasy and, thus, *various types of forcing*. As far as observations are concerned SSG 4.176 concentrated on the gravity signatures produced by these dynamic processes.

From a strictly theoretical point of view, the separation between SSGs 4.171 and 4.176 was obviously artificial and suggested that their work be merged into a new subcommission of SC 1 or into a new SSG of broader scope. (The second alternative has been adopted by the establishment of SSG 4.189.) On the other hand, the coexistence of two SSGs with related topics is closer to the concept of a topical 'working group' and has therefore also particular advantages. Rather natural in such a situation would be the organization of a joint meeting, which in the case of SSGs 4.171 and 4.176 proved to be difficult to achieve.

The relation between SSGs 4.176 and 5.174 was different. Slightly simplified, both investigated a range of processes using the same theories and the same observable, the only difference being that SSG 4.176 focussed on the *theoretical* aspects, whereas SSG 5.174 was more concerned with the *observational* aspects. This distinctions was clearly related to the sectional structure of the IAG: Whereas Section IV is responsible for *General Theory and Methodology*, the field of research of Section V is *Geodynamics*, which places more emphasis on the observational phenomena. In practice, a large portion of the researchers active in a particular field is not exclusively concerned either with theory or with observations but synthesizes both aspects when modelling the processes underlying particular observational phenomena. From this point of view, the suggestion of the IAG Executive Committee to avoid double membership in SSGs 4.176 and 5.174 does not seem to be quite appropriate. Although it is clearly desirable to organize a joint meeting in such situations, this could not be realized for SSGs 4.176 and 5.174.

5 Recommendations

The question how SSGs will operate effectively and produce useful results during their existence of four years appears to be closely linked with the procedure followed when establishing new SSGs and selecting their chairpersons and, to a lesser degree, also to the question of whether SSGs should be associated with a particular section or not. Some other aspects of sufficient generality to be applicable to any 'group of people headed by a leader' are regarded as obvious and have therefore not been discussed here. This leaves us with four points to be considered before and during the operation of SSGs:

(1) The establishment of new SSGs and the selection of their chairpersons should take into account *recommendations of individuals* working in the IAG. This 'grassroots approach' requires the development of an efficient procedure of soliciting suggestions before an IUGG General Assembly followed by a critical review of these suggestions by the Executive Committee at the Assembly.

(2) The roles of the chairpersons appear to be crucial for the success of SSGs. Therefore, the selection of dedicated scientists with a *research-oriented* and *non-bureaucratic* approach must be given prime importance.

(3) Although the association of SSGs to particular Sections is not a drawback *per se*, there may be a tendency towards 'territorial thinking' in this case. Since the sectional structure of the IAG is currently under review, no recommendation regarding this point is given here.

(4) With Union and Inter-Association symposia having been major programme points during IUGG General Assemblies for a long time, *joint meetings* of SSGs should likewise become normal. Such activities may take place across the sectional IAG structure or, more directly, in a new IAG structure without sections.

Acknowledgments. Discussions with E.W. Grafarend, P. Holota and K.-P. Schwarz have been of assistance during the preparation of this article.

References

Crossley, D., Hinderer, J., Casula, G., Francis, O., Hsu, H.-T., Imanishi, Y., Jentzsch, G., Kääriäinen, J., Merriam, J., Meurers, B., Neumeyer, J., Richter, B., Shibuya, K., Sato, T., van Dam, T. (1999). Network of superconducting gravimeters benefits a number of disciplines. EOS, 80: 121–126.

Darwin, G.H. (1879). On the bodily tides of viscous and semi-elastic spheroids, and on the ocean tides upon a yielding nucleus. Phil. Trans. R. Soc. London, Part 1, 170: 1–35.

Lambert, A., Courtier, N., Liard, J.O. (1995). Combined absolute and superconducting gravimetry: needs and results. Cah. Cent. Europ. Geodyn. Seismol., 11: 97–107.

Love, A.E.H. (1911). Some Problems of Geodynamics. Cambridge University Press, Cambridge.

Reigber, C., Lühr, H., Kang, Z., Schwintzer, P. (1997). The CHAMP mission and its role in observing temporal variations of the geopotential fields. EOS, 78(46), Suppl., F163.

Tapley, B.D. (1997). The gravity recovery and climate experiment (GRACE). EOS, 78(46), Suppl. F163.

Thomson, W. (1864). Dynamical problems regarding elastic spheroidal shells and spheroids of incompressible liquid. Phil. Trans. R. Soc. London, 153: 583–616.

Scientific services in support of research in geodesy and geodynamics

M. Feissel
Paris Observatory/CNRS UMR8630 and IGN/LAREG
8 Avenue Blaise Pascal, 77455 Marne la Vallée Cedex 2, France

Abstract. The international scientific services related to IAG became key players for research in geodynamics. Their growing role is the result of the increasing interaction between research- and service-oriented activities as well as the permanent international discussion that they maintain. They are a major factor of visibility for IAG. A special session was organized during the IUGG General Assembly in Birmigham (July 1999) to highlight the role of the services for the wide scientific community.

Keywords. Geodynamics, services.

Presentation of the session

Research in Geodesy and Geodynamics makes use of a variety of data that are collected in large scale observational programs managed by institutions or groups of institutions. These data may be used at various levels of elaboration by the researchers, ranging from practically out of the observational device, e.g. GPS Rinex data sets, to highly compact models, e.g. 4D terrestrial reference frames or ocean loading tables. Geodesy, as well as Geophysics and Astronomy, has a long tradition of organizing some of these activities under international services based on voluntary contributions of national institutions. By this way, the operational and research groups benefit from the work and expertise of other groups that deal with validations, intercomparisons and combinations of data in a service mode. The services sponsored by IUGG, IAU or URSI are monitored by FAGS.

In the last decade, the spreading of space techniques in all fields of geodesy and the development of powerfull and easy-to-use communication means has allowed a spectacular transformation of this traditional activity into a priviledged field for scientific discussion, in a spirit of friendly but alert competition. The life of the services is marked by intense electronic discussion as well as quite frequent workshops focused on technical and scientific issues of the service work. The services also organize schools, they distribute tutorials and software (models, analysis protocols).

Following the transformation of the old ILS, IPMS and BIH into the decentralized IERS in 1988, the space geodetic techniques developped multi-centres, multi-products services devoted to the use and valorization of worldwide networks. The first one was the IGS for GPS in 1994, recently followed by the ILRS for SLR in 1998, the IVS for VLBI in 1999 and, in the near future, the IDS for DORIS. With the development of international programs for observation and analysis that rely on geodetic techniques, the services are solicited to contribute with their technical expertise, scientific knowledge and organisational skills in various fields of geosciences. There are many examples of such cases: IERS for the monitoring of global geophysical fluids, IGS for troposphere and ionosphere monitoring, ILRS for multi-purpose orbitography, ICET for the GGP network of cryogenic gravimeters, PSMSL and IGS for intergovernmental sea level monitoring projects, etc. In parallel, the global products of the services are of daily use in the wide scientific community; to take only one example, crustal deformation studies in all countries make a heavy use of the IGS orbits of the GPS satellites and of the IERS/ITRF terrestrial references.

The wide recognition of these services is a well deserved reward to the development of new approaches to their work. Keeping the basic constraints of precision, accuracy, reliability and open availability of results, the new generation of services include in their normal activity scientific research in the domains connected to the responsibilities of the service, upstream with basic geodesy and methodology, downstream with geophysical interpretation of their own results. Being under the permanent pressure of scientific competition, the groups in charge permanently increase the quality of the products to keep track of the state of the art. The quality of their products stays very close to that of research results. Conversely, a large part of the progress in geodetic knowledge takes place in or around the services.

International Association of Geodesy Symposia, Vol. 121
Schwarz (ed.), Geodesy Beyond 2000 – The Challenges of the First Decade

IAG provides the international label to those services that are relevant to geodesy and geodynamics. It is the scientific body to which they report on their geodetic activity. For these reasons, it is important that their tie with the IAG be defined in a context that provides the appropriate scientific interaction with IAG, as well as clear IAG visibility for the users. Following a move started with a special session organized at the IAG Scientific Assembly in Rio de Janeiro in 1997 (Mueller, 1998), another special session in this symposium featured presentations of the role and achievements of a number of services, BGI, IGeS, ICET, IERS, IGS and PSMSL, illustrating in a convincing fashion how scientific research is supported through these international services.

Reference

Mueller, I.I. (ed.), 1998: *Science services, IAG/FAGS*. Ohio State University and IAG Central Bureau.

World Wide Web access to services

BGI	http://bgi.cnes.fr:8110
ICET	http://www.oma.be/KSB-ORB/ICET
IDS	http://lareg.ensg.ign.fr/DORIS
IERS	http://hpiers.obspm.fr
IGS	http:// igscb.jpl.nasa.gov
IGeS	http:// ipmtf14.topo.polimi.it/~iges
ILRS	http://ilrs. gsfc.nasa.gov
IVS	http://ivscc. gsfc.nasa.gov
PSMSL	http:// www.pol.ac.uk/psmsl.info.html

Acronyms

BGI	Bureau Gravimétrique International
BIH	Bureau International de l'Heure
DORIS	Détermination d'Orbite et Radiopositionnement Intégré par Satellite
FAGS	Federation of Astronomical and Geophysical data analysis Services
GGP	Global Geodynamics Project
GPS	Global Positioning System
IAG	International Association of Geodesy
IAU	International Astronomical Union
ICET	International Center for Earth Tides
IDS	International DORIS Service
IERS	International Earth Rotation Service
IGS	International GPS Service
IGeS	International Geoid Service
ILRS	International Laser Ranging Service
ILS	International Latitude Service
IPMS	International Polar Motion Service
URSI	Union Radio Scientifique Internationale
ITRF	International Terrestrial Reference Frame
IUGG	International Union of Geodesy and Geophysics
IVS	International VLBI Service for geodesy and astrometry
PSMSL	Permanent Service for Mean Sea Level
SLR	Satellite Laser Ranging
VLBI	Very Long Baseline radio Interferometry

Reflections on a new structure for IAG Beyond 2000 – conclusions from the IAG Section II Symposium in Munich

G. Beutler
Astronomical Institute, University of Berne, Sidlerstrasse 5, CH-3012-Bern, Switzerland

H. Drewes
Deutsches Geodätisches Forschungsinstitut, Marstallplatz 8, D-80539 Munich, Germany

R. Rummel
Institute for Astronomical and Physical Geodesy, Technical University, Arcisstrasse 21, D-80333 Munich, Germany

Abstract. The current structure and possible future modifications of the IAG were in the focus of the Section II meeting of IAG in October 1998 in Munich.

The main conclusion was, that the IAG should invoke a thorough review process within the next IAG "legislation period" from 1999-2003. This process should include at least one special retreat of the IAG Executive Committee with a well selected list of guests. It must involve all IAG sections and all other relevant IAG entities (like IAG services).

Section II is convinced that a thorough review process is necessary because

- geodesy underwent a dramatic development since the creation of the current structure,
- space geodesy became increasingly important in the same time period: today, it plays a dominant role in all sections,
- the current section structure does not seem to reflect the present-day situation in an adequate way.
- the IAG services (like IERS, IGS, etc.) play an increasingly important role also for research in geodesy and geodynamics but are not well (if at all) integrated in the current structure.

Section II is convinced that these facts fully justify a thorough reorganization. Section II is also convinced, however, that the entire spectrum of IAG must participate in this important process in order to converge to a satisfactory solution.

Keywords. International Association of Geodesy, section, structure.

1 Current IAG objectives and structure

If someone wants to change or modify the objectives and/or the structure of an international organization, she or he is well advised to seriously inspect its current objectives and structure. In the case of the *IAG*, the *International Association of Geodesy*, this is best done by critically reading the latest version of the Geodesist's Handbook (1996). The Geodesist's Handbook is issued every four years following the *IUGG (International Union of Geodesy and Geophysics)* General Assemblies. The Association's work is documented in the *Travaux de l'Association de Géodésie*. The *Travaux* are issued every four years as well, immediately after the IUGG General Assemblies. They describe the work performed during the preceding four years period. The latest version of the *Travaux* available when preparing this manuscript describes the work of IAG in the time period 1991-1995 Willis, (1995). In addition to these official sources of information we refer Mueller (1996) for most valuable background information.

According to Torge (1996) *"geodesy is the science of determining the size and the figure of the Earth, and its external gravity field. ... geodesy therefore is part of the geosciences, providing significant boundary conditions for modeling the Earth's body and its dynamics, including the oceans and the atmosphere. On the other hand geodesy has strong relations to surveying and cartography, to navigation and engineering"*.

International Association of Geodesy Symposia, Vol. 121
Schwarz (ed.), Geodesy Beyond 2000 – The Challenges of the First Decade

We fully agree with this characterization. It shows the strengths, but also the weaknesses of geodesy: On the positive side we note that interdisciplinarity is "so to speak" inherent in geodesy (we like to underline such facts, e.g., when writing proposals to funding agencies). The same aspect also points to a potential weakness: geodesy may be viewed as an "auxiliary science" (very much like mathematics from the point of view of a physicist) by other scientists.

The IAG is an Association of IUGG, the International Union of Geodesy and Geophysics (note that geodesy is explicitely mentioned in the name of the union); the IUGG in turn is member of *ICSU*, the *International Council of Scientific Unions*.

According to the Geodesist's Handbook (1996, p. 855) the objectives and the structure of the IAG are as follows:

a) *to promote the study of all scientific problems in geodesy and encourage geodetic research;*

b) *to promote and coordinate international cooperation in this field, and promote geodetic activities in developing countries;*

c) *to provide, on an international basis, for discussion and publication of the results of the studies, research works indicated in paragraphs a) and b) above.*

To achieve these objectives, the Association shall comprise a number of Sections, each of which deals with a distinct part of geodesy.

Commissions, Special Commissions and Special Study Groups may be formed as provided in the By-Laws.

It is interesting to note that *IAG Services* are not even mentioned in this breakdown of the IAG structure.

Let us conclude this brief review of the current IAG objectives and structure by mentioning that on the administrative side we have the

- *General Assembly*, consisting of the Delegates of the Member Countries,
- the *Council of the Association* consisting of the Delegates designated for meetings of the Council and formally accredited by the Adhering Body of Member Countries,
- *the IAG Bureau* consisting of the *President*, the *First Vice-President*, and the *Secretary General*, all of whom are elected by the Council, and
- *the IAG Executive Committee* consisting of the *Bureau*, the *immediate past president*, the *Second Vice President*, and the *Presidents of Sections*.
- Honorary Presidents, honorary General Secretaries, Presidents of Commissions, Secretaries of Sections, and the Chief Editor of the Journal of Geodesy may attend any meeting of the Executive Committee with voice, but without vote.

Table 1: The IAG sections

Sec	Year	Title
I	1957	Triangulation
	1971	Control Surveys
	1983	Positioning
II	1957	Precise Leveling
	1971	Satellite Surveys
	1983	Advanced Space Technology
III	1957	Geodetic Astronomy
	1971	Gravimetry
	1983	Determination of the Gravity Field
IV	1957	Gravimetry
	1971	Theory and Evaluation
	1983	General Theory and Methodolgy
V	1957	Geoid
	1971	Physical Interpretation
	1983	Geodynamics

Let us quickly review the main elements of the current IAG structure, namely *Sections*, *Commissions* and *Services*. Table 1 lists the five sections of IAG and the (remarkable!) development of their names in time, Table 2 the Commissions and Special Commissions, and Table 3 the IAG Services.

Table 1 shows that the *five section structure* goes back to 1957, the IUGG General Assembly in Toronto, which took place in the Geophysical Year and in the year of launch of *Sputnik I*. It is obvious, however, that the section definition was quite different from what it is today.

In 1960, at the IUGG General Assembly in Helsinki, the Commission on Organization, which was later on renamed to *Cassinis Committee*, was created. This committee reviews the IAG structure prior to every other general assembly and comes up with structural changes for approval by the IAG Council (see below) at the General Assemblies. In 1971, at the General Assembly in Moscow, the section definition was considerably changed. The section on *geodetic astronomy* disappeared and Section IV on *theory and evaluation* was established.

The latest major review of the section structure took place in 1983 at the General Assembly in Hamburg. According to Mueller (1996) a fine tuning took place to accomodate the fact that space techniques cut across the entire spectrum of the IAG organization.

Table 2 lists the current Commissions and Special Commissions. At present we count five Commissions and five Special Commissions. The numbers are no longer logical today. When Commissions or Special Commissions were dissolved, the other ones kept their numbers. Until 1983 the Commission and Special Commission presidents were not represented in the IAG Executive Committee. Since the Vienna General Assembly in 1991 they are members with voice but without vote. This latter change of the IAG statutes was motivated by the observation that a fair portion of the actual work of the IAG was performed in these IAG units.

The same statement might be made with regard to the IAG Services which are listed in Table 3. Some people even believe that some of the services are "IAG flag ships" today. When the present IAG structure was established, many of these services did not even exist. The IAG Executive Committee recognized the increasing importance of the services. This is why Prof. Ivan Mueller was asked to organize a special session at the IAG Scientific Assembly in Rio de Janeiro in 1997 giving the IAG Services the opportunity to describe their achievements, in particular their products and "customers". This effort is documented in Mueller (1998).

What makes the distinction between a service and a commission? The difference resides in the keywords *products* and *user community*. Each service makes available products, e.g., time series of Earth rotation and Earth orientation parameters to a broad user community. The user community may be purely scientific (e.g., the geosciences community) or much broader (e.g., the entire surveyor plus the navigation community in addition to the geosciences community, as in the case of the IGS). For additional information concerning the IAG Services we refer to Beutler (1999).

Table 3 illustrates that the bandwidth of IAG services is broad indeed, covering pure documentation (e.g., the IIS and the IBS) and services

Table 2: The current IAG commissions and special commissions (marked SC)

Commission/Sec.	Title
X/I	Global and Regional Geodetic Networks
SC4/I	Applications of Geodesy to Engineering
VIII/II	Intl. Coordination of Space Techniques for Geodesy and Geodynamics
SC6/II	Wegener Project
SC7/II	Gravity Field Determination by Satellite Gravity Gradiometry
III/III	Intl. Gravity Commission
XII/III	Intl. Geoid Commission
SC1/IV	Mathematical and Physical Foundations of Geodesy
V/V	Earth Tides
VII/V	Recent Crustal Movements
SC3/V	Fundamental Constants
SC8/V	Seal Level and Ice Sheet Variations

dealing with almost the entire range of geodesy and geodynamics (like *IERS, IGS, IGeS, BGI, and ICET*). Other services, like, e.g., the PSMSL are truly interdisciplinary in nature.

This is not the place to discuss the IAG Services in detail. We refer to Mueller (1998) for a description of the "classical" IAG Services and to the *Travaux* covering the time period 1995-1999 for the newly created services ILRS and IVS. Additional information is available in Beutler (1999).

Let us point out that (at least some of) the services in Table 3 are essential elements of the IAG work, but not of the IAG structure. Currently, the services are considered as elements of the sections which is why they are "only" described in the sessions sections of the Geodesist's Handbook (1996). Three services are formally assigned to Section II, two to Section III, and four to Section V. Two services are not attributed to a specific Section. They probably should be viewed as associated with the IAG Bureau.

The services are not well embedded in the sections. In many cases the services' activities might be associated to several sections (the IGS and the

Table 3: The current IAG services

Service/Section	Short Title
IGS/II	Intl. GPS Service
IVS/II	Intl. VLBI Service
ILRS/II	Intl. Laser Ranging Service
BGI/III	Intl. Gravimetric Bureau
IGeS/III	Intl. Geoid Service
IERS/V	Intl. Earth Rot. Service
BIPM/V	Intl. Bureau of Weights and Measures
ICET/V	Intl. Centre for Earth Tides
PSMSL/V	Permanent Service for Mean Sea Level
IBS/—	IAG Bibliographic Service
IIS/—	IAG Information Service

IERS, e.g., are closely related to Sections I, II, and V). Usually, there is no representation of the services in the executive bodies of the IAG sections. The only link between the conventional IAG structure and a service consists of one or two IAG representatives in the services' governing or directing boards. Services are thus highly independent (the positive way of putting it) but they have a very limited influence on the "life" within IAG. In view of the actual impact of the services in the "real world" we conclude that this is a major defect of the present IAG structure.

In Beutler et al. (1999) we discuss some of the relationships between the services. These considerations indicate that coordinating bodies above the services would, at least in some cases, make sense. In view of the rather random distribution of the services over the sections it is clear that the sections cannot play this role. In some cases IAG commissions may take over this coordinating role.

2 Motivations for a change

From the previous discussion of the current IAG structure and from events which took place in the 1995-1999 period we conclude that a thorough review of the IAG organization and structure must take place in the 1999-2003 period. Let us first summarize the events:

- Prominent IAG officers, e.g., the President of IAG (Prof. K.P. Schwarz) and the President of Section II (Prof. Reiner Rummel) of the 1995-1999 period are convinced that a major change is mandatory.

- The IGGOS-Symposium in October 1998 in Munich revealed that this attitude is in essence shared by the individuals, commissions, and services in Section II. (IGGOS stands for *Integrated Geodetic and Geodynamic Observing System* and will be addressed in Section 4.

- The conclusions of the symposium were presented as a proposal to the IAG Executive Committee at its meeting of March 22-23, 1999 to invoke a thorough review of the IAG Structure in the next four years period and to implement it in 2003. The proposal was endorsed by the IAG Executive Committee.

- One entire day of the G6-Symposium in Birmingham is devoted to the new IAG structure.

Let us summarize the arguments in favor of a thorough review of the IAG structure:

- Much of the work done within the Association is actually performed by the Services. The link between the IAG Services and Sections is weak, in some cases nonexistent.

- There is only a "one-way" link between the IAG Executive Committee and the Services (through so-called IAG representatives in the services' Boards).

- Similar statements may hold for some of the (special) commissions.

- According to Torge (1996) "geodesy is part of the geosciences, providing significant boundary conditions for modeling the Earth's body and its dynamics".

- We believe that only geodesy is capable of providing the terrestrial and celestial reference frames (and the connection between them) for all geosciences and for (fundamental) astronomy.

- Despite these facts, the *IAG* is not involved in any of the major "Geo-Programs", like, e.g.,

- *WCRP* (World Climate Research Program),
- *IGBP* (International Geosphere Biosphere Program), and
- *GOOS* (Global Ocean Observing System).

- One might get the impression that the other geosciences are considering geodesy (and thus IAG) as an "auxiliary science" (like mathematics) or as a tool like a PC (Personal Computer).

- In the next decade there will be a number of space missions of profound interest to geodesy and to IAG. CHAMP, GRACE, GOCE, and JASON may serve as examples. A new IAG structure must guarantee that IAG actually will play a major role in the scientific exploitation and the dissemination of results and information related to these missions.

- Even since the latest major reorganization of IAG in 1983, the development of space geodesy was (at minimum) remarkable. It seems strange that space geodesy should be limited to section II. A new structure must acknowledge that space geodesy resp. its tools are essential in all branches of geodesy.

- There is not much interaction between the IAG sections. This is probably due to the fact that no attempt was made to build a structure around a central theme.

- *Reference Systems for the Geosciences, their Realization and Use* might be a central theme for the new IAG structure.

3 The age of space geodesy

In *space geodesy* we study aspects of geodesy and geodetic astronomy by using natural or artificial celestial bodies as observed objects or as observing platforms. Space geodesy is thus defined through the observation techniques, the space geodetic techniques, or methods.

Space geodesy evolved rapidly in the second half of the twentieth century. The space age was initiated by the launch of the first artificial satellite, Sputnik I, on October 4 of the International Geophysical Year 1957. It became possible to deploy and use artificial satellites either to study figure and shape of the Earth from space or to observe them as targets from the surface of the Earth. The use of artificial Earth satellites for geodetic purposes is also referred to as satellite geodesy.

The second essential development in our context consists of the Very Long Baseline Interferometry (VLBI) technique as a new tool to realize an extraordinarily accurate and stable inertial reference system and to monitor Earth rotation using quasars. VLBI is at present the only "non satellite geodetic" technique meeting modern accuracy requirements of geodesy, geodynamics, and fundamental astronomy. In view of our general definition above VLBI clearly is a space geodetic technique.

Today, space geodetic techniques are the primary tools to study size, figure, deformation and gravity field of the Earth, and the Earth's motion as a finite body in the inertial reference system. Space geodetic techniques thus are the fundamental tools for geodesy, geodetic astronomy, and geodynamics.

Space geodetic observations contain information concerning the position (and motion) of the observed object and the observer. Thus they also contain information concerning the transformation between the terrestrial and the inertial systems. The Earth orientation parameters, i.e., polar motion, UT1, precession and nutation define this transformation.

From these definitions we see that space geodesy covers an extremely broad spectrum. It includes virtually all reference frame aspects and the determination of the gravity field.

4 Guidelines for the development of an alternative structure

We propose to adopt the following general principles for the review process:

- The "old" structure shall remain in place as long as the process to develop a new structure is not concluded.

- The restructuring process will be considered as terminated when the IAG Council has formally adopted the new structure.

- The process of defining a new structure must include the entire spectrum of geodesy.

- This process must in particular result in new *Statutes and By-Laws*.

- The new structure must have a *central theme*. This might be formulated as a mission statement.

4.1 A possible central theme

There are several ways how the central theme of the future IAG might be defined. Our discussion follows the thoughts contained in Beutler et al. (1998) and Rummel (1998) which were presented at the Munich 1998 IGGOS Symposium. It should be viewed as an attempt to have *reference systems and reference frames (including gravity)* as the central theme of the proposed review of the IAG.

We propose that a *Global Integrated Geodetic and Geodynamic Observing System (GIGGOS)* shall be considered to be geodesy's resp. IAG's contribution to large international science programs like, e.g., the *WCRP* (World Climate Research Program), the *IGBP* (International Geosphere Biosphere Program). The *IAG* contribution shall provide the geodetic component to the Earth system research.

In the currently active programs the emphasis obviously lies on climate change and environment. Such programs leave to a large extent open how solid Earth and ice processes as well as their interaction with ocean and atmosphere are to be quantified and modeled, and that geodesy is in a position to fill this gap by offering a very concrete and central element, namely the *GIGGOS*, the central objectives of which are:

- To provide a well defined and reproducible *global terrestrial frame*,
- the integral effect on *Earth rotation* of all angular momentum exchange inside the Earth, between land, ice, hydrosphere and atmosphere, and with Sun, Moon, and planets,
- the *geometrical shape* of the Earth's surface (solid Earth, ice, and oceans), globally or regionally, and its *temporal variations*, whether they are horizontal or vertical, secular, periodical or episodic, and
- by adding the *Earth's gravity field* — stationary and time variable — mass balance, fluxes, and circulation.

The above four elements of *GIGGOS* are briefly referred to as

(1) *frame*,

(2) *geokinematics*,

(3) *Earth rotation*, and

(4) *gravity field*.

The aimed at precision of the observing system must be of the order of about 10^{-9}, provide a high spatial and temporal resolution, consistency and stability over decades. (This was by the way the topic of the entire first day of symposium G6 in Birmingham).

The *GIGGOS* shall play a strategic role for the *IAG* in two respects:

- *GIGGOS* should be seen as the contribution of geodesy to the large scale international programs mentioned above. This contribution should be recognized by all associations within *IUGG* or *ICSU*.
- *GIGGOS* could serve as a common and very challenging focal point for "practically all" current research activities inside IAG.

Many, if not most, of the elements of the *GIGGOS* already exist and are dealt with in a very satisfactory way by *IAG* entities, in particular by the *IERS*, *IGS*, *CSTG*. We mention the celestial and terrestrial reference systems, the time series of Earth rotation parameters connecting the two systems, we also mention the development of the *ISGN* which should be able by design to take over most of the observational part of *GIGGOS*.

Regional activities are very well addressed as well within *IAG* – we mention the *EUREF Subcommission*, the *SIRGAS Project*, and the *WEGENER Special Commission*.

It is our vision to view all these activities as the central theme of the new IAG structure and organization.

If we would only consider the first three of the four components of *GIGGOS*, we could safely state that *GIGGOS* is almost in place. The gravity field is an exception: our knowledge (as seen from the point of view of satellite geodesy) goes back essentially to optical and *SLR* tracking of geodetic satellites (like *Lageos 1,2, Starlette, etc.*) and to satellite altimetry. The aimed at accuracy (of 10^{-9}), with a very high time and space resolution, has not yet been achieved.

The new gravity-oriented satellite missions will dramatically improve our knowledge of the gravity field, however. This aspect will have to be addressed in the development of space geodesy.

4.2 Elements of the restructuring process

Let us point out that the process of restructuring IAG is not yet in an advanced stage. Right now, the authors are aware of four proposals, namely

- a structure proposed by the IAG president, K.-P. Schwarz at the Munich meeting, where the structure is centered around a central project (global observing system),
- a structure proposed by Martine Feissel at the Munich meeting, which in essence maintains the section structure but gives much more weight to services, projects, and research,
- a proposal to merge sections 1,2 and 3 thus yielding three new sections (measurement methods, modeling, geodynamics)
- to form sections according to the key words frames, geokinematics, Earth rotation, gravity field (as outlined above)

These proposals are not yet formulated in a way that they could be compared in a meaningful way. This is why we are convinced that more time is required for the restructuring process.

Let us, however, try to nail down the essential elements of the process:

- The process must be fundamental. We have to ask, e.g., whether the section structure still is appropriate.
- We have to give the proper weight to the IAG units (to services in particular) performing much of the work the Association stands for.
- We have to make sure that IAG is recognized as the international organization providing the reference frames for the other geosciences.
- We have to make sure that the upcoming gravity missions are playing an important role within the new IAG structure.

5 The team and the plan

We propose to proceed as follows:

- The new IAG structure based on the principles outlined above shall be developed in the 1999-2003 period.
- The process must be initiated at the Birmingham IAG Executive Committee meeting, where
  - a Steering Committee of about six IAG officers and
  - a Chairperson shall be designated by the IAG Executive Committee,
  - the Chairperson shall initiate the review process based on the outcome of the Birmingham G6 symposium.
  - the Steering Committee will review this summary. Afterwards this summary will be made available as background information to the authors of position papers.
- In summer 1999 the IAG Retreat to review the IAG Structure has to be prepared by the steering committee.
- A *list of 20-30 invitees*, a *draft agenda*, and a *list of authors of position papers* shall be proposed to the IAG Executive Committee by the Steering Committee end of November 1999.
- The *IAG Retreat* shall take place in fall 1999 early in the year 2000.
- A report about the IAG Retreat is prepared and presented to the IAG Executive Committee mid 2000.
- The IAG Executive Committee discusses, modifies, approves the recommendations and action items emerging from the retreat.
- The new IAG structure is defined and the new By-Laws are written.
- The resulting documents are published (summarized) in the Journal of Geodesy around the end of the year 2000.
- At the 2001 IAG Scientific Assembly one Session is devoted to presenting the new structure to the plenum.
- At the same assembly an extraordinary Council Meeting is organized to allow for a discussion and a formal acceptance (or refusal) of the new IAG Structure and By-laws.

6 Epilogue

The plan for the review process outlined above was intensively discussed at the IAG Symposium G6 in Birmingham. The plan and the timeline was approved by both, the IAG Executive Committee and the IAG Council. Both of these IAG Bodies approved the plan and the timeline specified.

The *IAG Review Committee* now has the demanding task to fill these plans with life. The committee (in alphabetic order) consists of:

- Georges Balmino
- Gerhard Beutler (chair)
- Fritz Brunner
- Jean Dickey
- Martine Feissel
- Rene Forsberg
- Reiner Rummel
- Fernando Sanso
- Klaus-Peter Schwarz

The committee has started to work. The *IAG Retreat* is scheduled to take place from February 14-16, 2000 at Jet Propulsion Laboratory (JPL) in Pasadena.

Acknowledgements The authors are indebted to Martine Feissel, Ivan Mueller, Klaus-Peter Schwarz, Christian Tscherning, Wolfgang Torge, and Fernando Sanso for critically reading and commenting this manuscript.

References

Beutler, G., H. Drewes, R. Rummel, Ch. Reigber (1999). Space Techniques and their Coordination within IAG at Present and in Future. In *proceedings of 1998 IGGOS Symposium*, in press.

Beutler, G. (1999). IAG Services in the Current Framework of the International Association of Geodesy (IAG). In *this Volume.*

Geodesist's Handbook (1996). In *Journal of Geodesy*, Vol. 70, No. 12, pp. 839-1036.

Mueller I.I. (1996). The Commission: a Historical Note, in *CSTG Bulletin*, No. 12, pp. 7-20, Deutsches Geodätisches Forschungsinstitut, Munich.

Mueller, I.I. (1998). Science Services: International Association of Geodesy (IAG) / Federation of Astronomical & Geophysical Services (FAGS). In *Special Volume, IAG Scientific Assembly, Rio de Janeiro, Brazil, Sep 3-9, 1997*, 62 pages, available through IAG Central Bureau c/o Department of Geophysics, Copenhagen, Denmark.

Rummel, R.(1998). Global Integrated Geodetic and Geodynamic Observing System (GIGGOS). In *Proceedings of 1998 IGGOS Symposium*, (in press).

Torge, W.(1996). The International Association of Geodesy (IAG) — More than 130 Years of International Cooperation. In *the Geodesist's Handbook, Journal of Geodesy*, Vol. 70, No. 12, pp. 840-845.

Willis, P. (ed.) (1995). Travaux de l'Association Internationale de Géodésie, Tome 30, Rapports Généraux et Rapports Techniques átblis à l'Occasion de la Vingtiéme Assemblée Générale à Boulder, USA, Juillet 1995. Secrétariat de l'Association, 140, rue de Grenelle, 75700 Paris, France.

AUTHOR INDEX

KEYWORD INDEX

MIX
Papier aus verantwortungsvollen Quellen
Paper from responsible sources
FSC® C105338

If you have any concerns about our products,
you can contact us on
ProductSafety@springernature.com

In case Publisher is established outside the EU,
the EU authorized representative is:
Springer Nature Customer Service Center GmbH
Europaplatz 3, 69115 Heidelberg, Germany

Printed by Libri Plureos GmbH
in Hamburg, Germany